"The theme of the casebook is highly relevant. The focus of the book is on enhancing our understanding on the applicability of marketing science (marketing for social good context; like social marketing, sustainability marketing). It uncovers an important and highly relevant subject in a comprehensive manner. This is especially important in view of i) Hyper connected world, and ii) recent development all over the globe triggered by COVID-19. In such a pandemic oriented scenario, socially responsible consumer behaviors assume significance in the marketing science domain.

"Marketing is undergoing tremendous transformations. Marketing is basically an umbrella concept and can be utilized to sensitize consumers about socially responsible behavior in view of current pressure on sustainability. Various contributions of researchers and practitioners buttress this point very strongly!

"There seems to be lot of fuzziness and lack of clarity regarding - social marketing, sustainability marketing, green marketing, ethical marketing and similar other concepts in the domain area. This book has made a laudable attempt to present concepts in a lucid manner as-a-single-resource book. The marketing science presented in this book is dictated by the societal needs (see for example – Wroclaw in Poland or Sydney in Australia) rather than individual needs. Corporates are engaging their efforts in this direction.

The book Is organized into 4 sections. It presents theoretical frameworks and models. Various case studies drawn from diverse applications such as hoteling, retailing, milk, handloom etc.an in diverse settings gives a compressive picture. Case studies from both developed and developing economies present an interesting landscape. It brings out the power of social media in a constructive and emphatic manner.

"The case book is intended to provide a good resource for researchers, students, faculty members and corporates, and thus relevant. I compliment the book editors for bringing out this compendium.

"I hope that the book will sensitize the readers about real-life cases in socially responsible marketing and how consumers are guided to various initiatives on socially responsible behavior. I am confident that the book will go a long way in Imparting Social and Sustainability marketing competencies."

Professor (Dr.) S G Deshmukh
Professor at Indian Institute of Technology Delhi, New Delhi India
sgdeshmukh2003@gmail.com, deshmukh@mech.iitd.ac.in

"This book is a welcome addition to the ongoing dialogue and discourse on sustainability. The book is nicely aligned to the United Nations Sustainable Development Goals (SDGs). A great collection of book chapters and case studies makes a great case for the importance of social and sustainability marketing in the present times. This book provides a much-needed focus on responsible consumers and responsible consumption. To achieve the UN SDGs, it is not enough to have responsible companies, we need a dialogue on how to motivate responsible consumption and create responsible consumers.

"In addition, the book is unique with its focus on both developed and developing economies. In doing so the contribution of the book is invaluable and I have no doubt in my mind that it will serve both the academic and the practitioner communities."

Dr Sanjit K. Roy
Associate Professor of Marketing at UWA Business School
The University of Western Australia, Australia
sanjit.roy@uwa.edu.au

"This book covers through an excellent collection of essays a wide range of topics on Social and Sustainable Marketing. While numerous papers and some books on the subject have appeared in recent years, to the best of my knowledge, this book is the first that intends to cover theory and practical applications in such a wide range and depth on social and sustainable marketing. The book not only reflects an extensive variety of issues but also a wide geographic diversity of its contributors from four continents, including Europe, America, Asia, and Oceania.

"The reasons for the wide influence and diffusion of sustainability among academics, practitioners, and society in general, are not difficult to find. First, since the definition of sustainable development was proposed, more than three decades ago, as the ability to generate wealth to meet the needs of present generations without compromising the ability to generate wealth for future generations, these concepts have gone expanding and penetrating society. Second, increasing social pressure on organizations has placed marketing in a unique position to play a proactive role, not only developing sustainable products but also influencing and shaping social and sustainable consumption in the marketplace.

"In addition, the confluence that this book shows between academic research, practical applications, through the description of cases, and new technologies provide the necessary synergy for rapid acceptance in both the public and private sectors."

Dr. Rubén Huertas Garcia
Serra-Húnter Associate Professor at Business Department
Faculty of Economics and Business, Universitat de Barcelona, Barcelona
rhuertas@ub.edu

"The book on Social and Sustainability Marketing: A Casebook for Reaching Your Socially Responsible Consumers through Marketing Science recognizes the importance and trends of social and sustainability marketing. It is comprehensive, comprising original research, case studies, and pedagogical discussions and best practices in relation to social and sustainability marketing. Specifically, it provides readers with up-to-date knowledge of social and sustainable marketing from multiple perspectives as well as educators with case studies and materials that could incorporate into teaching and learning activities for further developing learners' competencies on the subject. The discussions on pedagogical approaches and best practices for imparting social and sustainability marketing competencies are insightful and useful to educators."

Dr. Zach Lee
Assistant Professor in Marketing at Durham University Business School,
Durham University England
zach.lee@durham.ac.uk

"This text is an important intervention in the conversation around social and ecological sustainability which draws on both micromarketing and macromarketing scholarship to help the reader understand the challenges with illustrations from insightful cases both from emerging and developed economies. Such a balance of Northern and Southern perspectives has been lacking to date.

"The editors should be congratulated for this compilation which should be essential reading for the discerning student of sustainable consumption & production"

Professor Pierre McDonagh
Associate Editor, Journal of Macromarketing (USA),
Professor of Critical Marketing & Society, University of Bath, UK

Social and Sustainability Marketing

Social and Sustainability Marketing

A Casebook for Reaching Your Socially Responsible Consumers through Marketing Science

Edited by
**Jishnu Bhattacharyya
Manoj Kumar Dash
Chandana R. Hewege
M.S. Balaji
Weng Marc Lim**

A PRODUCTIVITY PRESS BOOK

First Edition published 2022
by Routledge
600 Broken Sound Parkway #300, Boca Raton FL, 33487

and by Routledge
2 Park Square, Milton Park, Abingdon, Oxon, OX14 4RN

Routledge is an imprint of the Taylor & Francis Group, an informa business

© 2022 selection and editorial matter, Jishnu Bhattacharyya; individual chapters, the contributors

The right of Jishnu Bhattacharyya to be identified as the author of the editorial material, and of the authors for their individual chapters, has been asserted in accordance with sections 77 and 78 of the Copyright, Designs and Patents Act 1988.

All rights reserved. No part of this book may be reprinted or reproduced or utilised in any form or by any electronic, mechanical, or other means, now known or hereafter invented, including photocopying and recording, or in any information storage or retrieval system, without permission in writing from the publishers.

Trademark notice: Product or corporate names may be trademarks or registered trademarks, and are used only for identification and explanation without intent to infringe.

Library of Congress Cataloging-in-Publication Data
A catalog record for this title has been requested

ISBN: 978-0-367-55363-0 (hbk)
ISBN: 978-1-032-03602-1 (pbk)
ISBN: 978-1-003-18818-6 (ebk)

Typeset in Minion
by KnowledgeWorks Global Ltd.

eResources are available at www.routledge.com/9780367553630

We dedicate this book to our family, who supported and encouraged us in our journey.

— **Editors**

Contents

Foreword xv
Acknowledgments xvii
About the Editors xix
Contributors... xxiii

SECTION I An Overview of Social and Sustainability Marketing

Chapter 1 Empowering Marketing Organizations to Create and Reach Socially Responsible Consumers for Greater Sustainability 3

Weng Marc Lim

SECTION II Advances in Knowledge of Social and Sustainable Marketing: Understanding Sustainability Marketing

Chapter 2 The Sustainability Marketing Framework: A Tool for Teaching and Learning about Sustainability Marketing 13

Al Rosenbloom

Chapter 3 Cascades: What Is It and How Did It Reach Sustainability in a Highly Competitive Sector? 53

Myriam Ertz

Chapter 4 Sustainability Marketing in Contending for the Position of the European Capital of Culture (ECoC) 85

Ezeifekwuaba Tochukwu Benedict

vii

Chapter 5 Cool Branding for Indian Sustainable Fashion Brands 115

Jasmeet Kaur and Gursimranjit Singh

Chapter 6 Personal Experience of Sustainability Practices and Commitment toward Corporate Sustainability Initiatives: Reflections of Sri Lankan Marketing Professionals............................ 143

Kadka Ranjith and W.D.C. Jayawickrama

SECTION III Advances in Knowledge of Social and Sustainable Marketing: Sustainable Consumption and Consumer Behavior

Chapter 7 Evolving Prosumer Identity in Sustainable Consumption: Deconstructing Consumer Identity... 175

Chamila Roshani Perera and Chandana R. Hewege

Chapter 8 Sustainable Practices and Responsible Consumption by the Hotel Industry: The Consumers' Perspective 199

Srishti Agarwal and Neeti Kasliwal

Chapter 9 How Does Sustainability Affect Consumer Satisfaction in Retailing?... 225

Antonio Marín-García, Irene Gil-Saura, and María-Eugenia Ruiz-Molina

Chapter 10 Bridging the Intention–Behaviour Gap in Second-Hand Clothing... 249

Elin Pedersén, Amanda Persson and Adele Berndt

SECTION IV Advances in Knowledge of Social and Sustainable Marketing: Understanding Sharing Economy and Marketing

Chapter 11 Sustainability through Sharing Farm Equipment: A Research Agenda 279

Priyanka Sharma

SECTION V Advances in Knowledge of Social and Sustainable Marketing: Understanding Social Marketing

Chapter 12 Wroclaw: Transforming a City towards a Circular Economy-Zero Waste Social Marketing Campaign in Poland 311

Dorota Bednarska-Olejniczak and Jaroslaw Olejniczak

Chapter 13 Influencing Sustainable Food-Related Behaviour Changes: A Case Study in Sydney, Australia 345

Diana Bogueva, Dora Marinova and Talia Raphaely

SECTION VI Advances in Knowledge of Social and Sustainable Marketing: The Power of Online Consumer Reviews

Chapter 14 Enforcing Brands to Be More Sustainable: The Power of Online Consumer Reviews 387

Feyza Ağlargöz

Chapter 15 Leveraging Social Media to Create Socially Responsible Consumers 415

Bikramjit Rishi and Neha Reddy Kuthuru

SECTION VII Advances in Knowledge of Social and Sustainable Marketing: Addressing Global Crises

Chapter 16 Management of Shocking Global Crises: The Use of Public Marketing 4.0 within a Social Responsibility and Sustainability Approach 435

Manuel Antonio Fernández-Villacañas Marín and Ignacio Fernández-Villacañas Marcos

SECTION VIII Advances in Knowledge of Social and Sustainable Marketing: Understanding the Benefits of Sustainability Reporting Practices by Social Enterprises

Chapter 17 Assessing Sustainable Outcomes of Reporting Practices by Social Enterprises 477

Judith M. Herbst

SECTION IX Advances in Knowledge of Social and Sustainable Marketing: Safeguarding against Unsocial and Irresponsible Customers

Chapter 18 Unsocial and Irresponsible Behaviour: What Happens When Customers Lie? 509

M. Mercedes Galan-Ladero and Julie Robson

SECTION X Pedagogical Directions and Best Practices: Imparting Social and Sustainability Marketing Competencies

Chapter 19 Case Method as an Effective Pedagogical Tool: Some Insights for Better Learning Outcomes for Social and Sustainability Marketing Educators 537

Chandana R. Hewege

SECTION XI Selected Case Studies to Reflect on Practice and Use as Learning Tools: Case Studies from Emerging Economies

Chapter 20 From Skin Whitening to Skin Brightening and, Now, Skin Glowing: How L'Oréal Sustains Its Skincare Line from Colourism and Genderism to Racism and Classism 547

Huey Fen Cheong and Surinderpal Kaur

Chapter 21 Fashion Accessory Brand Development via Upcycling of Throwaway Clothes: The Case of Chapputz 563

Selcen Ozturkcan

Chapter 22 Sustainable Marketing in China: The Case of Monmilk 575

Ruizhi Yuan and Yanyan Chen

Chapter 23 Nurpu: A Dream towards a Sustainable Handloom Weaving Society 597

Sathyanarayanan Ramachandran and S. A. Senthil Kumar

Chapter 24 Social and Sustainability Marketing: Secure Meters: The Dharohar Case 613

Kirti Mishra and Shivani Singhal

SECTION XII Selected Case Studies to Reflect on Practice and Use as Learning Tools: Case Studies from Emerging Economies (Complex and/or Long)

Chapter 25 Saheli: The Zero-Side-Effect Pill—Marketing of Oral Contraceptives in the Context of Sexual Education to Create Socially Responsible Consumers 621

Neharika Binani, Anshika Singh and Palakh Jain

Chapter 26 Much Needed 'Pad Man' for Indian Females to Be Dignified: A Case Study on Period Poverty 647

Sneha Rajput and Pooja Jain

SECTION XIII Selected Case Studies to Reflect on Practice and Use as Learning Tools: Sustainability Marketing in the NFL

Chapter 27 Sustainability Marketing in the National Football League (NFL): The Case of the Philadelphia Eagles 721

Jairo León-Quismondo

Chapter 28 Leave It: Upskilling a Dog Owning Community......... 739

Jessica A. Harris, Sharyn Rundle-Thiele, Bo Pang, Patricia David and Tori Seydel

Chapter 29 Coexisting: The Role of Communications in Improving Attitudes towards Wildlife 755

Bo Pang, Patricia David, Tori Seydel, Sharyn Rundle-Thiele and Cathryn Dexter

Chapter 30 Closing the Confidence Gap in STEM: A Social Marketing Approach to Increase Female Retention... 765

Carina Roemer, Bo Pang, Patricia David, Jeawon Kim, James Durl and Sharyn Rundle-Thiele

Chapter 31 GlobalGiving and Performance Metrics..................... 777

Sivakumar Alur

Chapter 32 Co-Creating and Marketing Sustainable Cities: Urban Travel Mode Choice and Quality of Living in the Case of Vienna... 785

Tim Breitbarth, David M. Herold and Andrea Insch

SECTION XIV Selected Case Studies to Reflect on Practice and Use as Learning Tools: Case Studies from Developed Economies (Complex and/or Long)

Chapter 33 Positioning a Company in the Chemical Industry as a Sustainability Driver 795

Filipa Lanhas Oliveira and João F. Proença

Chapter 34 Social and Sustainability Marketing and the Sharing Economy in the Coffee Shop Culture 839

Madhavi Venkatesan

SECTION XV Selected Case Studies to Reflect on Practice and Use as Learning Tools: Global Case Studies

Chapter 35 Hilton Faces Greenwashing Challenge 865

Jinsuh Tark and Won-Yong Oh

Chapter 36 MGM's Dilemma in Responsible Gaming Program 869

Jinsuh Tark and Won-Yong Oh

Index 873

Foreword

This edited book is a timely response by authors, considering the heightened attention that social and sustainability marketing has received from researchers, educators, and practitioners alike. The COVID-19 pandemic and the related social and economic repercussions that the world has experienced undoubtedly made people more aware of the salience of sustainability and social well-being. This book surely adds value in this context. The four main sections of the book offer important theoretical and pedagogical directions for enthusiasts of social and sustainability marketing.

Original research papers included in the book provide readers with some directions to (1) understand the concept of sustainability marketing from different perspectives, (2) explore sustainable consumption and consumer behavior, (3) understand sharing economy and sustainability, (4) explore social marketing, (5) use online media for social and sustainability marketing and (6) understand sustainability reporting practices and unsocial and irresponsible consumer practices. Diversity of areas covered is one of the key strengths of this book.

In addition to the research insights provided through original research papers, this book presents 7 real-life case studies with varying length and from diverse industries and country contexts. First, these cases aptly illustrate how social and sustainability marketing is applied in real-life contexts. This helps educators, learners, and practitioners to understand how the concepts are operating in actual industry settings. Second, the learning guide appended to each case study by the authors offers sound pedagogical tools to marketing educators. The pedagogical directions offered in the Pedagogical Directions and Best Practices sections of this book are certain to provide a framework to analyze these cases and understand the issues underpinning the case scenarios. This undoubtedly facilitates deep learning.

This edited book is compiled in a balanced way to expose a reader to the theory, practice, and pedagogy of social and sustainability marketing. I congratulate the editors for their job well done! I am hopeful that this book would add immense value to advance knowledge in social and sustainability marketing.

Professor Lester W. Johnson
Professor of Marketing, Swinburne Business School
Faculty of Business and Law, Swinburne University of Technology
Melbourne, Australia

Acknowledgments

Our brilliant supposition is to put on record our best admirations, the most thoughtful feeling of appreciation to all the contributing authors of this edited volume for their support, cooperation, and valuable contribution, which were to a countless degree helpful in developing this book. We might similarly want to take this chance to thank all the Professors who took time from their busy schedule and reviewed our edited text and gave valuable inputs.

We want to thank all the external peer reviewers for their opportune direction and constructive feedback to improve the right track. They helped us review a vast pool of manuscripts and select the most suitable contribution to this book successfully. We thank the following individuals for serving as external peer reviewers of manuscripts submitted for consideration for publication in this edited volume. We much appreciate their assistance with the manuscript review.

Diana Bogueva, Curtin University, Australia
Jeet Gaur, Swinburne University of Technology, Australia
Shweta Jha, Indian Institute of Management, Indore, India
Yangyang Jiang, University of Nottingham Ningbo, China
Lokesh Kumar Kethireddi, Indian Institutes of Management, Lucknow, India
Soumyadeep Kundu, Indian Institutes of Management, Kozhikode, India
Zach Lee, Durham University, England
Pankaj Kumar Medhi, Bennett University, India
Subhasree Mukherjee, Indian Institutes of Management, Ranchi, India
Bo Pang, Griffith University, Australia
Chamila Perera, Swinburne University of Technology, Australia
Shivam Sakshi, University of Debrecen, Hungary; Indian Institutes of Management, Bangalore, India
Priyanka Sharma, Indian Institutes of Management, Lucknow, India
Madhavi Venkatesan, Northeastern University Boston, Massachusetts
Sandeep Yadav, Indian Institutes of Management, Kozhikode, India

About the Editors

Jishnu Bhattacharyya
Nottingham University Business School (NUBS China), University of Nottingham, Ningbo, China

Jishnu Bhattacharyya received his engineering education from the Maulana Abul Kalam Azad University of Technology in West Bengal, India, where he earned a Bachelor of Technology in Electrical and Electronics Engineering. He holds a Master of Business Administration from the Indian Institute of Information Technology and Management, Gwalior, India. Prior to receiving his PhD, he worked in a scientific research position at the Indian Institute of Technology (IIT), Delhi, India. Before moving to the University of Nottingham's PhD in Marketing, he participated in research coursework at the Indian Institute of Management Kozhikode, India. He has also served as a visiting consultant (research) for a social science research project. In the past, he was engaged in short-term research projects at IIT Indore, India, and IIT Delhi, India, in the capacity of visiting students. With an interdisciplinary background, he is actively engaged in research and interested in marketing science. He enjoys asking practically motivated and theoretically inspired questions along several interrelated research streams, including but not limited to sustainability communication, socially responsible consumption, social marketing, and disruptive technologies in marketing, particularly artificial intelligence. He often takes inspiration from the natural sciences in marketing research. He is passionate about the value of integrating research into teaching and bringing positive social change through the art and science of marketing. He has won awards for his academic and research excellence, including the best paper award. He serves as a reviewer for many leading marketing and business management journals. He has co-edited two books, co-authored a book, and published journal articles and case studies.

About the Editors

Manoj Kumar Dash
Atal Bihari Vajpayee-Indian Institute of Information Technology and Management Gwalior, Madhya Pradesh, India; Visiting faculty (Marketing), Indian Institute of Management Indore, India

Manoj Kumar Dash has been research active with a regular journal and book publications. A few of his recent publications are in journals like the *International Journal of Production Research, Online Information Review, Benchmarking: An International Journal, Telemetric and Informatics*, and *Neural Computing and Application*, etc. He is currently writing a book with Palgrave Macmillan. He has to his credit several best paper awards like the Best Research paper Award in Marketing in International Conference of Arts and Science organized by Harvard University (USA), Boston. Besides being a regular faculty at the Indian Institute of Information Technology and Management, Gwalior, India, he is also a visiting faculty in IIM Indore. He has also served as adjunct faculty at Lancaster University, UK, and visiting professor at SIOM Nashik. He had conducted several faculty development programs on multivariate analysis, econometrics, research methodology, multi-criteria optimization, multivariate analysis in marketing, etc. He delivered lectures as a resource person and keynote speaker in multiple programs organized by reputed institutes in India. He is specialized in marketing analytics. He was involved as Chair Member in International Conference at Harvard University, Boston, USA. He has supervised several PhDs who are serving in academic positions across the globe.

Chandana R. Hewege
Swinburne Business School, Swinburne University of Technology, Melbourne, Australia

Chandana R. Hewege holds a PhD in Management from the Monash University, a Master of Business Administration Degree and a BSc. (Bus Admn) Honors Degree from the University of Sri J'pura, Sri Lanka. Chandana's academic and professional experiences are predominantly centered on teaching, research and industry consultancy. He has published about 45 referred research papers and referred conference papers. Chandana's academic research encompasses a wide spectrum of research areas, such as corporate social responsibility and sustainability, management controls of transitional economies, marketing, logistics, international business and research methodology. He has published research papers in Journal of consumer Behaviour, Journal of Retailing and Consumer Services, Australasian Marketing Journal, Marketing Intelligence and Planning, Journal of Consumer Marketing, International Journal of Consumer Studies, International Journal of Work Organisation and Emotion, Equality, Diversity and Inclusion, Journal of Marketing and Logistics, Journal of Sustainability in Higher Education, Social Responsibility Journal, Accounting, Auditing, Accountability Journal, Journal of International Business Education, Critical Perspectives on International Business and Yong Consumers.

Chandana is accredited as a teacher in higher education by the Staff Education and Development Association, United Kingdom (SEDA, UK). He possesses about 20 years of teaching experience undergraduate and postgraduate, and local as well as overseas. Chandana has won several awards for his academic and research excellence, including the best critical literature review award from Monash University; Emerald Literati Award-Social Responsibility Journal 2017; the Teaching Excellence Award by the Faculty of Business & Law of Swinburne University of Technology: the best reviewer 2007 award by Contemporary Management Research Journal; and the best student Gold Medal award from the Department of Business Administration-University of Sri Jayewardenepura. Chandana served the Australian Business Deans Council (ABDC) journal ranking panel in 2013 for marketing, tourism, logistics and commercial services research areas.

Sathyaprakash Balaji Makam (M.S. Balaji)
Nottingham University Business School (NUBS China), University of Nottingham, Ningbo, China

M.S. Balaji is Associate Professor in Marketing at Nottingham University Business School China (NUBS). Prior to joining NUBS in September 2015, he held a Lecturer (research) position at Taylor's Business School, Taylor's University, Malaysia, and Assistant Professor position at IBS Hyderabad, IFHE University, India. He was a visiting research scholar at Whitman School of Management, Syracuse University, USA between August 2007 and July 2008. Prior to his academic career, he worked in the healthcare industry for more than 3 years in India. Dr. Balaji has published his research in leading marketing and business management journals including *Journal of Business Research, European Journal of Marketing, Information and Management, International Journal of Hospitality Management, Journal of Services Marketing, Service Industries Journal*, and others. He received the AIMS-IMT Outstanding Young Management Researcher Award in 2012 for his research contributions. He serves as a reviewer for many leading marketing and business management journals. He servings as an Associate Editor of *Asia Pacific Journal of Business Administration* and on the Editorial Advisory Board of *International Journal of Contemporary Hospitality Management*.

Weng Marc Lim
School of Business, Swinburne University of Technology, Sarawak, Malaysia; Swinburne Business School, Swinburne University of Technology, Australia; Editor in Chief of International Journal of Quality and Innovation

Weng Marc Lim is Adjunct Professor of Swinburne Business School at Swinburne University of Technology, Australia, and Professor and the Head of School of Business at Swinburne University of Technology, Sarawak. Both campuses offer AACSB-accredited business degrees. He holds a doctorate in business and economics from Monash University and post-doctorate certificates in leadership and pedagogy from Cornell University and Harvard University. His research interests include business, consumer, and government research. He has published in numerous A*/A-ranked journals such as *European Journal of Marketing, Industrial Marketing Management, Journal of Business Research, Journal of Business and Industrial Marketing, Journal of Retailing and Consumer Services, Journal of Strategic Marketing, Marketing Intelligence and Planning*, and *Marketing Theory*, among others. He has also presented his work and led high-level policy discussions at the *United Nations Educational, Scientific and Cultural Organization*, and the *World Economic Forum*.

Contributors

Srishti Agarwal is a PhD, MBA, and UGC NET qualified Assistant Professor at MITWPU's School of Management (PG), Pune, with seven years of academic and research experience. She has over ten research papers published in refereed journals and edited books with Emerald, Pearson, Elsevier, and IGI Global. She also chaired a session in "Marketing and Consumer Behavior" at the 12th Annual ISDSI conference at SPJIMR. She is a peer reviewer for prestigious journals such as the International Journal of Hospitality Management.

Feyza Ağlargöz is Assistant Professor of Marketing in the Department of Business Administration at Anadolu University, Turkey. She received her PhD degree from Anadolu, and has been a guest post-doctoral researcher at the School of Economics and Management, Lund University. She is currently working on degrowth, videography, and voluntary simplicity. Her research interests are arts marketing, consumer culture, social marketing, and sustainable consumption.

Sivakumar Alur has more than 25 years of experience in business school teaching, research, consulting, and executive education. He did his post-doctoral research at Technology University Delft, The Netherlands. He has authored two prize winning cases "Primacy: Global Design from India" and "Universal Print Systems Limited: Exploring Operations Strategy Options" listed in Ivey cases. Several of his cases are listed in leading case clearing houses like Case Centre. He is a seminar and workshop facilitator of AACSB International – a leading global accreditation body. His research has appeared in journals like *Journal of Nonprofit and Public Sector Marketing*.

Dorota Bednarska-Olejniczak (PhD in economics) is an assistant professor at the Finance Department at the Wroclaw University of Economics and Business and member of the Senate of the University. She is an expert in the field of public finances, as well as a lecturer certified trainer, and consultant. She is an author of many scientific publications in the field of local government finances, as well as expert opinions and trainings for central government and local government administration. She is a member of the International Institute of Public Finance.

Ezeifekwuaba Tochukwu Benedict is a writer with experience in Marketing and Business Administration, Finance, and a variety of other business and marketing topics and subtopics. He has written a number of books.

Adele Berndt is an Associate Professor at Jönköping International Business School (JIBS), Sweden, and an affiliated researcher at Gordon Institute of Business Science (GIBS) at the University of Pretoria. Her research focuses on the interaction of consumers and branding and she has authored articles appearing in various marketing journals.

Neharika Binani, an entrepreneur by passion, is the co-founder of a local art merchandise brand in India called arika.in. It aims to provide consumers with the ability to procure contemporary artwork at an affordable price. She is currently working on two projects with Creatnet Services as their business development associate and Himalaya Opticals as a marketing associate.

Diana Bogueva is an Adjunct Postdoctoral Research Fellow at the Curtin University Sustainability Policy (CUSP) Institute. Her research interests are in the areas of social marketing, consumer behavior, and food innovation. She is the Australian coordinator of the Global Harmonization Initiative.

Tim Breitbarth is particularly interested in corporate responsibility, environmental sustainability, organizational change and stakeholder communication – often linked to the global sport sector. Among others, he is a former scientific chair of the European Association for Sport Management and co-founder of the German CSR Communication Congress. He holds social science and business degrees from University of Göttingen and University of Otago, and has held positions in Germany, New Zealand, United States, United Kingdom, and Australia throughout both his academic and industry careers.

Yanyan Chen is Assistant Professor in marketing in the Department of Marketing & International Business, Toulouse Business School, France. Her research interests mainly focus on the impact of emotional appeals to consumers, as well as consumers' social and environmental values during consumption.

Huey Fen Cheong is Senior Lecturer at the Faculty of Languages and Linguistics, University of Malaya, Malaysia. Her work has been focusing on marketing communication, particularly gender representation in marketing male grooming and female beauty products. She is interested to see how male grooming industries overcome gender norms or stereotypes to promote male grooming, commonly known as metrosexuality (may refer to the metrosexual practice, phenomenon, or identity). Her research is not just empirical, but also philosophical. One of her works looked at the interdisciplinary differences and potential collaboration between linguistics and marketing in semiotics.

Patricia David is a Research Fellow at Social Marketing @ Griffith, Griffith University, Queensland, Australia. Her research interests are motivated by understanding what drives behavioral change. Patricia's PhD advanced the Social Marketing field by taking the first steps toward the development of a Theory of Behavior Change. Her work is award winning. Her behavior change research has won a commendation award in the ESMC conference in 2018 and the Doctoral Colloquium Contribution to Theory & Knowledge Award in the ANZMAC conference in 2016. Patricia has led teams in research projects, working with both quantitative and qualitative approaches. She has previously worked in marketing management positions, and her current work focuses on the design, implementation, and evaluation of campaigns and social marketing programs across a broad range of social issues.

Cathryn Dexter is an ecologist and conservation biologist with a background in road ecology research, with a particular emphasis on urban koala movement. Cathryn has over 12 years' experience as a researcher and project manager and has worked on major linear infrastructure projects for clients that included environmental consultancies and both State and Local government. Her work has been published in refereed journal articles, conference proceedings, and numerous research reports. Cathryn's current role is with the Redland City Council as their Koala Conservation Project Officer, where her work involves the development of strategic planning to retain Redland's koala population through a strong focus on community engagement. Cathryn is a founding member and current committee member for the Australasian Network for Ecology and Transportation conferences, and is also a PhD candidate with Griffith University Environmental Futures Research Institute, Applied Road Ecology Group in Queensland.

James Durl is a PhD candidate and Research Assistant with Social Marketing @ Griffith, Griffith University, Queensland, Australia. Holding a Bachelor of Business with Honors, James brings the learnings of numerous experiences working across academic and industry projects. These experiences have provided him with robust skills in systematic literature reviews, co-designing with end-users to improve program models and create new services, managing high quality reports for clients with quick turnarounds, and providing education of illicit substances to Australia's adolescent population. With a growing list of publications and awards for his work, noteworthy projects that James has worked on include the Blurred Minds alcohol education program (see www.blurredminds.com.au/students) and the TeamUp program. Through his short career, James can boast industry partnerships with the Queensland Catholic Education Commission and VicHealth

Myriam Ertz is Assistant Professor of Marketing at Université du Québec à Chicoutimi (UQAC), and head of LaboNFC (laboratory of research on new forms of consumption). Her research interests cover sustainability-related themes, the collaborative economy and new technologies. She has published on these topics in many top-tier journals (e.g., *Resources, Conservation & Recycling, Journal of Cleaner Production, Business Strategy & the Environment, Industrial Marketing Management, Technological Forecasting & Social Change, Journal of Environmental Management, Journal of Business Research*), prestigious conference proceedings (e.g., *American Marketing Association*), and book chapters. She is also editor of a book about the reconfiguration of commercial exchanges and author of a book on responsible marketing.

Ignacio Fernández-Villacañas Marcos is Consultant in Technology at M&M Planning and Project Management and holds a Bachelor's degree in Aerospace Engineering from the Technical University of Madrid, Spain, Major in Aerospace Sciences and Technologies, and master's degree in Big Data and Business Analytics from the University of Alcalá, Spain. Now he is the student of the Thermal Power Master's in Science at Cranfield University, UK.

Manuel Antonio Fernández-Villacañas Marín is Consultant in Logistics & Management at M&M Planning and Project Management, Associate professor at Technical University of Madrid, and Colonel at Spanish Air Force (R). Degree in Statistics and Operational Research (University of Granada, 1987); a Bachelor degree in Economics and Business Sciences (University of Murcia, 1990); a Masters degree in Security and Defence (University Complutense of Madrid, 2006); a Masters degree in Business Studies (IE University, 2011); a Masters degree in Public Management (IE University, 2011); a PhD in Economic and Business Sciences (University of Granada, 1995); and a PhD in Administration and Logistics of S&D Systems (Rey Juan Carlos University, 2012). He also obtained his PhD in Economic and Business Sciences, and in Administration and Logistics of S&D Systems. He has been a professor at the Academy and the Air Warfare Center of the Spanish Air Force, professor at the Universities of Granada, Murcia, CEU San Pablo, Polytechnic of Cartagena, Rey Juan Carlos, Alcalá de Henares, ENAE Business School, Metropolitan University (Ecuador), University of Monterrey (Mexico), University of Wales (UK), Autonomous University of Yucatán (Mexico), etc. He is the author of five technical books and co-author in six others, five teaching material publications, and many research articles, monographs, and presentations at international conferences in the field of logistics, management, and technology.

M. Mercedes Galan-Ladero is Associate Professor at the University of Extremadura (Spain), teaching marketing courses in the Faculty of Economics and Business Sciences in the Badajoz campus. Her main research focuses on Cause-Related Marketing, Social Marketing, Non-Profit Marketing, and Social Responsibility. She also works in Services Marketing, Consumer Behavior, International Marketing, and Ethnocentrism. She received her European PhD (Hons) by University of Extremadura (Spain) in 2011. She also has a Degree on Business Administration (University of Extremadura, Spain), Master in European Communities and Human Rights (Pontifical University of Salamanca,

Spain), and Master in Marketing Management and Market Research (UNED, Spain).

Irene Gil-Saura (PhD, University of Valencia) is Professor of Marketing and Market Research at the University of Valencia (Spain), where she teaches classes on commercial distribution and service marketing in undergraduate, master's, and doctorate degrees. Her research interests are in line with the analysis of consumer behavior and the management of service companies, with special emphasis on tourism companies and commercial distribution, and on variables related to technologies, perceived value, and capital branding and sustainability. She has published her work in international magazines such as *Journal of Retail & Leisure Property*, *Journal of Services Marketing*, *Tourism Review*, *The Service Industries Journal*, the *International Review of Retail, Distribution and Consumer Researcher*, and *Journal of Vacation Marketing*.

Jessica A. Harris is a social marketer who offers extensive experience in changing behaviour on several social issues. Her current work focuses on the design, implementation and evaluation of campaigns and community behaviour change programs across a broad range of social issues that focus on health and well-being and environmental contexts. She uses both quantitative and qualitative research to co-create, build, and engage with stakeholders to increase the sustainability of programs.

Judith M. Herbst is Visiting Fellow in the School of Management at QUT. Her research interests are aligned with how organizations integrate change through establishing systems to co-create shared value with stakeholders along the value chain. Previously, she worked as a researcher for the Australian Centre for Philanthropy and Nonprofit Studies to complete "Giving and Volunteering" reports for the Australian Government Department of Social Services while completing her doctorate in the School of Accountancy.

David M. Herold's research deals with the topic of sustainable logistics with an environmental focus as well as practical research in entrepreneurial innovation. Prior to his academic career, he worked for more than ten years in the logistics industry for a Fortune 500 company in various professional and management positions. He holds business degrees from leading universities in Germany, Austria and Australia.

Andrea Insch's research expertise is inter-disciplinary, connecting marketing, urban studies and tourism. Designed to contribute to the debate on the role and impacts of place branding for cities, nations and their stakeholders, her main research activity is focused on several subthemes – the interplay of cities branding and urban development, resident engagement with city branding and value derived from country-of-origin images. Andrea is on the Editorial Advisory Board of the *Journal of Management History* and is the Book Review Editor for the journal *Place Branding & Public Diplomacy* and the Regional Editor (Australia and New Zealand) for the journal.

Palakh Jain is Assistant Professor in the School of Management at Bennett University. A Fellow of IIM Ahmedabad in Economics and alumni of Delhi School of Economics, University of Delhi, Palakh was awarded Junior Research Fellowship by the UGC in 2005 and associated as a Consultant with ICRIER leading a project on Indo-Pak FDI. She has also co-authored a book titled *Outward FDI-Why, Where and How? The Indian Experience*.

Pooja Jain (PhD, Pursuing) is presently working as an Assistant Professor in the Department of Commerce, Prestige Institute of Management, Gwalior. She has more than eight years of experience in teaching and research. She has published papers in reputed Scopus listed national and international journals. She has presented more than 18 research papers in various national and international conferences and attended various in-house and outside workshops. Her area of interest is financial management, behavioral finance, accounting, and business strategy.

W. D. C. Jayawickrama is Lecturer in Marketing Management at a state university in Sri Lanka. His research interests are in the area of Social Marketing, Transformative Consumer Research, Consumer Ethics and Rights, and Pedagogies in Higher Education. His work has been published in reputed journals and he continues to work on several collaborative projects with parties overseas.

Neeti Kasliwal is a PhD, MBA, and UGC NET qualified Visiting Professor at Indian Institute of Rural Management (IIRM), Jaipur; Ex Dean In-Charge at IIHMR University; Associate Professor at Banasthali Vidyapith; and Director of Mother and Angels Pre-School with 17 years of administrative and teaching experience. She has been a PhD Research Supervisor and four students have been awarded PhDs under her supervision. She is a Guest Speaker at Centre for Management Studies, Rajasthan State Institute of Public Administration, Government of Rajasthan and at Poornima University.

Jasmeet Kaur is a Research Scholar at the Mittal School of Business, Lovely Professional University, Phagwara, Punjab, India. She has completed her MBA in Retail Management from Amity University, ADDOE, Noida, India. With Post Graduation Diploma in Fashion Retail Management from National Institute of Fashion Technology (NIFT), Delhi, India, and a graduate in Fashion Technology, her professional experience includes full-time and part-time Lecturer positions at YWCA, Raffles Millennial International, Delhi and New Delhi YMCA. Her area of research includes fashion marketing, sustainable fashion, and slow fashion.

Surinderpal Kaur is Dean at the Faculty of Languages and Linguistics, University of Malaya, Kuala Lampur, Malaysia. Her research interests focus mainly on media discourses revolving around critical discourse studies, multimodal studies, language and gender studies, and terrorism studies.

Jeawon Kim is a PhD Candidate, Researcher at Social Marketing @ Griffith, Griffith University, Queensland, Australia, and Project Manager for Waste Not Want Not program. Holding Bachelor of Business (1st class Honors) and Psychological Science degrees, her research interest is in tackling food waste from a social marketing perspective. She has undertaken extensive formative research on household food waste behavior and applied community-based social marketing to develop, design, and implement a households-targeted food waste reduction campaign called Waste Not Want Not in early 2017 for the Redland City Council. Jeawon values consumer research to deliver a positive behavioral change. She has led teams of field experiment for the Department of Health and co-design study for Johnson and Johnson. Her work involves conducting systematic literature reviews, mixed-methods studies including co-design sessions and online surveys, delivery of consumer-oriented campaigns across a wide range of social issues and reporting results for industry partners.

S. A. Senthil Kumar currently heads the Department of Management-Karaikal Campus of Pondicherry University. He has 18 years of academia experience and is a specialist in healthcare management, insurance management, services marketing, research methodology and operations research. Dr. Senthil is a Post-Doctoral Fellow from University of Massachusetts, Dartmouth, USA. He is a winner of the Distinguished Educator Scholar Award from the National Foundation for Entrepreneurship Development and a winner of Best Teacher Award at Pondicherry University for six consecutive years. He has published more than 40 papers in peer-reviewed journals and has also authored the book *Emerging Paradigms in Insurance Industry and Management*.

Neha Reddy Kuthuru is a student pursuing her bachelor's in Management Studies at the School of Management and Entrepreneurship at Shiv Nadar University, a recognized Institute of Eminence. Neha specializes in marketing management and finance with her research interests aligned with marketing & strategy, consumer behavior, digital marketing & social media marketing. Her previous experience includes working in the non-profit organization sector and the marketing industry.

Jairo León-Quismondo (PhD in Physical Activity and Sport Sciences) is Professor of Sport Management at Universidad Europea de Madrid, Spain, and collaborates with other universities in their sport management courses. He works together with the Spanish National Sports Council (Minister of Culture and Sport), the Spanish government agency responsible for the sport. His research experience includes a research stay at Southern Connecticut State University, New Haven, Connecticut, USA. His main research interests are based on organizational management and quality service, specifically in fitness centers, sporting events, and other sport organizations. His research has been presented in national and international conferences such as the European Association of Sport Management (EASM) Conference and Ibero-American Conference on Sport Economics, where he received the 1st Researcher Award.

xxxii • Contributors

Antonio Marín-García (PhD, University of Valencia) has a degree in Market Research and Techniques from the Faculty of Economics of the University of Valencia (Spain), where he teaches commercial distribution and marketing classes in different undergraduate degrees. His areas of research interest are focused on the impact of innovation and sustainability in commercial distribution companies, as well as the analysis and implementation of new teaching methodologies applied to university centers. He has published his work in international and national journals, such as *Journal of Product & Brand Management, Cuadernos de Gestión/ Management Lettres*, and *UCJC Business & Society Review*. In addition, he is the author of numerous book chapters and the author of books related to educational innovation and knowledge transfer.

Dora Marinova is Professor of Sustainability at the Curtin University Sustainability Policy (CUSP) Institute, Perth, Australia. Her research interests cover innovation models, including the evolving global green system of innovation and the emerging area of sustainometrics. She has over 400 referred publications and has successfully supervised 66 PhD students.

Kirti Mishra is Assistant Professor at the Indian Institute of Management, Udaipur, India, in the Organizational Behavior and Human Resource Management Area. An early career researcher who has previously held appointments at Monash University, Swinburne University and the University of Queensland; her area of research predominantly focuses on corporate sustainability, organizational responses to climate change, and strategizing for grand challenges. Kirti's research work has been presented at multiple European Group of Organization Studies and Academy of Management Conferences and published in *Australasian Journal of Environment Management* and *South Asian Journal of Human Resources Management*, among others.

Won-Yong Oh is Lee Professor of Strategy and Associate Professor at the Lee Business School, University of Nevada, Las Vegas (UNLV), in the USA. His research primarily focuses on corporate governance, corporate social responsibility, and strategic leadership. He has published more than 60 peer-reviewed journal articles, book chapters, and case studies. Previously, he has taught at the University of Kansas in the USA and the University of Calgary in Canada.

Jaroslaw Olejniczak holds a PhD in economics and is an Assistant Professor in the Finance Department at the Wroclaw University of Economics and Business (Poland). He is an expert in the field of public finances. Researcher, lecturer, certified trainer and consultant. He is the author of over 90 articles and conference publications and co-author in several monographs. In his scientific work, he mainly deals with local government finance issues (tax policy, equalization schemes), local and regional sustainable development, and citizen participation in local government activities. He has authored expert opinions and trainings for central government and local government administration. He is a member of the Global Participatory Budgeting Research Board, the International Institute of Public Finance and the Senate of the Wroclaw University of Economics and Business.

Filipa Lanhas Oliveira, BA, University of Minho, Portugal, is a postgraduate student at the Faculty of Economics, University of Porto. She has recently done a curricular internship at Henkel Headquarters on Global Market and Customer Activation for the Business Unit of Henkel Adhesives, Düsseldorf, Germany.

Bo Pang is a Research Fellow in the Social Marketing @ Griffith Centre at Griffith University. Bo holds a PhD in Social Marketing and offers extensive experience in conducting research and delivering programs in the field of social marketing, all of which are published in leading scholarly journals. Bo's work involves implementing theoretical constructs into empirical community interventions and he offers experience delivering changes benefitting the community across a diverse range of projects delivered for local, state, and national governments, including reducing koala death by 40%, increasing donation to Australia charities, and increasing female participation in STEM outcomes. Bo has also worked with a wide array of profit and not-for-profit organizations. His work has been published in over 60 refereed journal articles, conference papers, and research reports.

Elin Pedersén has a master's degree in Business Economics, specializing in International Marketing, from Jönköping International Business School (JIBS), Sweden. With seven years of experience in retail and sales, she has an interest in consumer behavior and was motivated to study marketing in order to understand customer needs from a sustainability perspective.

Chamila Roshani Perera holds a PhD in Marketing from the University of Melbourne, Australia, and a Master of Commerce Degree from University of Kobe, Japan. She published several research papers in ABDC ranked journals and won several best paper awards. Her research interests are in consumer culture, climate change-related behavior, sustainability, environmentalism, minimalistic lifestyles, corporate social responsibility, research methodological issues, and pedagogy.

Amanda Persson has a Master of Science degree in Business and Economics, business administration. After her graduation, she started as a payroll assistant at Axfood.

João F. Proença, PhD, is Full Professor at the Faculty of Economics of the University of Porto, Portugal, and Researcher at the Advance-CSG, ISEG, University of Lisbon, Portugal. He has published extensively in several scientific journals and books, and he was the Rector of the European University, Laureate International Universities, Portugal, and the Dean of the Faculty of Economics, University of Porto. He has a strong international teaching and research experience and has being Visiting Professor and Researcher in several places around the world.

Selcen Ozturkcan continues her academic career as Associate Professor of Business Administration (since 08/2018) with a permanent position at the School of Business and Economics of Linnaeus University, Sweden, and as a Network Professor (since 03/2020) at the Sabancı Business School, Turkey. Previously, she served as Full Professor of New Media (2017–2018) at the Bahcesehir University, Turkey, Associate Professor of Marketing (2011–2016), and Assistant Professor of Marketing (2010–2011) at the Istanbul Bilgi University, Turkey. Her research, which appeared as journal articles, books, book chapters, and case studies, is accessible at http://www.selcenozturkcan.com.

Sneha Rajput, PhD, is presently working as Assistant Professor at Prestige Institute of Management Gwalior, India. She earned her doctorate from Jiwaji University in the year 2015 and has twelve years of experience in teaching and research. She has published more than 50 research papers in reputed, ABDC, national and international journals, and conference proceedings and has authored/co-authored six edited books. She is a content writer for E-Pathshala (an online learning website by MHRD) and resource person for PhD, with coursework for management and tourism specialization. She is a member of the Board of Studies in Management at the Jiwaji University, Gwalior.

Sathyanarayanan Ramachandran has 20 years of industry and academic experience in Organizations like BIM, CII, CBS and in the European Union's ECCP project. Before joining IFMRGSB, Sathya helped the Confederation of Indian Industry as the Head of Education Initiatives in setting up an Education Practice and laid the conceptual foundation for a PPP-mode Incubator in Chennai. Sathya has extensive experience in PG-level teaching in marketing courses, corporate training and consulting assignments. Sahtya is a gold medalist in marketing, a UGC NET-JRF holder, a winnner in London Business School's Case Development Initiative in 2009, and the of the Dewang Mehta Award for Best Teacher in Marketing.

Kadka Ranjith is a fresh graduate who recently obtained his Bachelors from the University of Sri Jayewardenepura, Nugegoda, Sri Lanka. He is currently engaged in preparing children for local government examinations, and is both passionate and engaged in exploring solutions for crucial societal concerns.

Talia Raphaely (PhD) is a University Associate at Curtin University, Perth, Australia, and a sustainability practitioner. Her areas of passion relate to individual empowerment, animal welfare and food.

Bikramjit Rishi is Associate Professor in Marketing Management at the Institute of Management Technology, Ghaziabad, India. His research interests are in the areas of social media marketing, consumer behavior, and retailing. His research has been published in the *Australasian Marketing Journal (AMJ)*, *Journal of Brand Management (JBM)*, *International Journal of Business Innovation and Research*, *International Journal of Indian Culture and Business Management*, *Singapore Management Review*, and others. He has co-edited *Contemporary Issues in Social Media Marketing* (Routledge) and adapted another social media marketing title (Sage). He has designed and delivered many training programs for Hindustan Coca

Cola Beverages Limited (HCCBPL), Maruti Suzuki India Limited (MSIL), Apollo Tyres Limited, APL Apollo Tubes Limited, Jubilant Foods and RITES Limited, and so on.

Julie Robson is Associate Professor of Marketing at Bournemouth University, Poole, UK. Her research focuses on trust and corporate branding, primarily in the financial services sector and particularly general insurance. Her interest in financial services started with her PhD on building societies. A career in the insurance industry then followed in senior management roles (marketing, new product development and research). She has published in internationally recognized journals (including the *Journal of Business Research* and *Marketing Theory*) and has guest edited several journal special issues on financial services marketing and insurance.

Carina Roemer is a Research Fellow at Social Marketing @ Griffith, Griffith University, Queensland, Australia. Carina offers experience in behavioral and social science research, analyses, and tailored reporting in short timeframes to client satisfaction. Currently, driven by the complexities of behavior change and to understand what drives systems change, Carina employs a systems lens to her research. Carina has published in peer reviewed journals, presented her work at national and international conferences and written up research reports. Partnering with industry, Carina is currently working with the Australian Government, Queensland Agriculture and Fisheries, Fisheries Research Development, and Research and Currie.

Al Rosenbloom is Professor Emeritus and was the first John and Jeanne Rowe Distinguished Professor at Dominican University, River Forest, Illinois. His research interests include case writing, the application of the case method in management education, global branding, marketing in countries with emerging and subsistence markets, and the challenge of integrating the topic of poverty into the management education. Al co-leads the Anti-Poverty Working Group, Principles of Responsible Management Education (PRME) and participates broadly within PRME. He was a Fulbright Scholar in Nepal and Bulgaria and was twice honored with the Teaching Excellence Award from Brennan School of Business students.

María-Eugenia Ruiz-Molina (PhD, University of Valencia) is Associate Professor at the University of Marketing and Market Research at the University of Valencia (Spain), where she teaches classes on commercial distribution and international marketing in undergraduate and master's degrees. Her research interests focus on the impact of innovation, technology, and sustainable practices on consumer behavior and the management of service companies. The findings of her research have been published in international journals such as *International Journal of Retail & Distribution Management*; *Management of Environmental Quality: An International Journal*; *Journal of Hospitality, Leisure, Sports and Tourism Education*; and *Journal of Consumer Marketing*.

Sharyn Rundle-Thiele is a social marketer and behavioral scientist. She is the Founding Director of Social Marketing @ Griffith, Griffith University, Queensland, Australia, which is the largest university based group of social marketers in the world. She has led projects that have changed behaviors for tens of thousands of people in areas including health, the environment and for complex social issues. Sharyn has led programs that have increased healthy eating, changed adolescent attitudes to drinking alcohol, reduced food waste, increased dogs' abilities to avoid koalas, and many more. She has published more than 175 books, book chapters and journal papers. Awards and appointments, including The Philip Kotler Social Marketing Distinguished Service Award and the Australian New Zealand Marketing Academy Fellow, acknowledge her innovative and high-quality practice, science and leadership

Tori Seydel is a PhD candidate and research assistant with Social Marketing @ Griffith, Griffith University, Queensland, Australia. Tori has experience in working across industry projects and within academia. Her early career has focused on how to apply theory to practice and her PhD is exploring social messaging as a catalyst for social change. Projects that Tori has been involved with include Leave It, which has reduced koala deaths by 40%, and O-it delivered in partnership with Australian charities to increase the quality of donated goods.

Priyanka Sharma is a faculty member in the Marketing area at IIM Lucknow, India. She received her PhD in Marketing from the Department of Industrial and Management Engineering, IIT Kanpur, India, and holds a B-Tech in Chemical Engineering from IIT (BHU) Varanasi and PGDM(GM) from XLRI, Jamshedpur. She has rich industry experience in varied sales and marketing roles in prestigious organizations such as Infosys, Oracle, and UST Global. Her research interests are B2B marketing, branding, emerging markets, sharing economy, and sustainable marketing. She has published her works in journals of repute and presented in various national/international conferences.

Anshika Singh is a marketing graduate currently preparing to pursue her MBA in the upcoming academic year while working as a digital media analyst in Aceus Media Pvt Ltd. With her love for art, Anshika is also developing her freelance art project into her art brand soon.

Gursimranjit Singh is Assistant Professor in the Department of Marketing at the Mittal School of Business, Lovely Professional University, Phagwara, Punjab, India. He received his Doctoral in Management from the I.K. Gujral Punjab Technical University, Kapurthala, Jalandhar, India. He has completed his MBA in Marketing and Human Resource Management from the Chandigarh Business School of Administration, Landran, Mohali, Punjab, MA in Public Administration from the DAV College, Chandigarh, and Post-graduation Diploma in Mass Communication from the Panjab University, Chandigarh. His professional experience includes positions at ICICI Securities, CKD Institute of Management & Technology and Senior Research Fellow (UGC-MANF). His areas of research include social media marketing, consumer metrics, brand metrics, and artificial intelligence.

Shivani Singhal is the Head of Dharohar, a non-profit creating meaningful volunteering opportunities for businesses in Udaipur, India. Starting as a classicist, she moved on to teaching before working on developing nonformal learning spaces for children and adults. She currently manages a team of 23 running volunteering for 1500 people a year working with 5000 children. She also works in an advisory role to other non-profit and educational organizations in the UK and India.

Jinsuh Tark holds a BS in hotel administration from the University of Nevada, Las Vegas (UNLV), USA, and is a graduate student in the school of William F. Harrah College of Hospitality at UNLV. His areas of interest include corporate social responsibility and sustainable development in the hospitality industry.

Madhavi Venkatesan is an academic economist and environmental activist. She is presently a faculty member in the Department of Economics at Northeastern University in Boston, Massachusetts. Madhavi earned a Doctorate, Master's and Bachelor's degree in economics from Vanderbilt University, Nashville, Tennessee, and additional Master credentials in sustainability from both Harvard University, Boston, Massachusetts, and Vermont Law School, Royalton, Vermont. In her academic work, she has written and spoken extensively on the relationship between economics and sustainability and has contributed to the literature on the relationship among culture, sustainability, and economics. Her activism incorporates economic literacy and highlights the significance of life cycle assessment. In 2016, she founded the nonprofit Sustainable Practices and is the organization's executive director.

Ruizhi Yuan is Assistant Professor in Marketing at Nottingham University Business School, China. Her research interests focus on green consumption and new emerging economies. Specifically, she focuses on how marketing innovations, such as green marketing and the shared economy, influence consumer behavior and companies' branding strategies. She was awarded a role with the National Natural Science Foundation of China (NSFC) as principal investigator based on the research topic of how collaborative tourism impacts customer engagement.

Section I

An Overview of Social and Sustainability Marketing

This section provides an overview of the knowledge domain and particularly emphasizes the importance of the study area.

1

Empowering Marketing Organizations to Create and Reach Socially Responsible Consumers for Greater Sustainability

Weng Marc Lim
Swinburne Business School, Swinburne University of Technology, Australia
School of Business, Swinburne University of Technology, Malaysia

CONTENTS

Learning Objectives 3
Consumerism..4
Sustainability..4
Social and Sustainability Marketing..6
Organization of the Book..6
Funding Declaration 7
Acknowledgment..7
Credit Author Statement 7
Recommended Readings..7

LEARNING OBJECTIVES

After reading the chapter, the reader should be able to:

- Gain an overview of social and sustainability marketing
- Gain an overview of the organization of the book

CONSUMERISM

The idea of consumerism has been around for many decades, if not centuries. At its core, consumerism is about the consumption of products (e.g., goods, services), which is deemed to be good and desirable when it satisfies consumer needs, and to a reasonable extent, consumer wants. Yet, the evolution of consumer needs, wants, and demands has progressed significantly over the years, with consumer activism increasingly influencing the way in which products are produced and delivered so as to protect and promote consumer interest.

In the past, consumption is often associated with economic (e.g., affordability, economic growth) and eudemonic (e.g., happiness, life satisfaction) reasons. However, the narrative of consumption has changed tremendously, as consumers today do not only demand for products that account for economic and eudemonic well-being but also that consider environmental and social well-being.

Though the society at large today may still be placing economic and eudemonic well-being as top priorities, given the ongoing issues with consumption, or more specifically, overconsumption, which may, to a certain extent, be attributed to hedonistic consumerism (or highly wasteful and discriminatory pattern of consumption—e.g., consumption for instant gratification and material prosperity), it is safe and reasonable to assume that consumption that contributes to environmental and social well-being will be pursued over consumption that does not, especially when economic and eudemonic priorities are satisfied. This change may be attributed to the idea of sustainability, which seeks to preserve the capacity for the biosphere and human civilization to coexist.

SUSTAINABILITY

The idea of sustainability gained prominence as a result of the United Nation's Brundtland Report on "Our Common Future" in 1987, which provided the "official" definition of sustainable development—i.e., the ability of the present generation to live and meet their own needs without compromising the ability of future generations to meet their own needs. Yet, like consumerism, sustainability has been subjected to the debate for change over the years.

In the past, sustainability is often associated with the triple bottom line of profit, planet, and people—or its other popularly known variation: economic, environment, and social well-being. The United Nations, which is the intergovernmental organization for harmonizing the actions of nations, has spearheaded efforts to curate and promote the sustainability agenda, from the Millennium Development Goals (MDGs) for 2000–2015 to the Sustainability Development Goals (SDGs) for 2015–2030. Interestingly, we have witnessed an increase in sustainability-oriented goals, from 8 MDGs to 17 SDGs, which represent a clear indication on the continued need to search for new ideas and ways to tackle the growing problems impeding the sustainability agenda. More recently, the narrative of sustainability has been mooted for change by another world-leading international organization—i.e., the World Economic Forum—as a result of the coronavirus disease 2019 (COVID-19).

The World Economic Forum, which is the international organization for public–private cooperation, has conceived the Great Reset initiative to improve the state of the world as we enter a unique window of opportunity to shape the world's recovery post COVID-19. In line with this initiative, the Forum's Platform for Shaping the Future of the New Economy and Society has developed the "Dashboard for a New Economy," which calls for dialogues and actions to focus on agenda items pertaining to prosperity (i.e., financial resilience—e.g., income and wealth equality, social mobility), planet (i.e., natural resilience—e.g., energy mix and intensity), people (i.e., social resilience—e.g., human capital, public health), and institutions (e.g., public and private organizations). Specifically, the change in focus from profit to prosperity is highly desirable, as this change will improve our perception and evaluation of financial resilience with greater inclusivity. More importantly, this call by the World Economic Forum is in line with the intended contribution of this book, which seeks to curate a research-informed collection of ideas that focus on how marketing organizations, as institutions, could contribute toward the new sustainability agenda that focuses on prosperity, planet, and people.

SOCIAL AND SUSTAINABILITY MARKETING

Marketing organizations are at the forefront of sustainability marketing, as they are the producers and suppliers of products, thereby placing them in a unique position to not only drive but also shape consumption in the marketplace. Indeed, we acknowledge and recognize the central role that marketing organizations play in the sustainability agenda, which is the main reason why we were inspired to commission this book.

Through this book, we seek to bring to light the ways in which marketing organizations could play the role of creators and facilitators of social agents of change. We believe that consumers can be socially responsible—as marketers, through their marketing activities (e.g., product, promotion), can create and facilitate opportunities for consumers to become socially responsible in their consumption, wherein a consumer is deemed to be socially responsible when his or her consumption contributes positively to the sustainability agenda. Therefore, the curation of ideas in our book is in line with the idea of social marketing, as we endeavor to demonstrate the application of marketing principles to drive behavior for social good, wherein the social good that we hope to champion and promote herein is the sustainability agenda.

ORGANIZATION OF THE BOOK

The rest of the book is organized into three major sections. First, the book delivers a collection of chapters consisting of the latest insights from original research on social and sustainability marketing. Through these chapters, readers can expect to learn about the trajectory of sustainability marketing and its associated concepts, frameworks, and tools (e.g., sharing economy, social media, sustainable consumption, sustainability reporting). Second, the book offers a discussion on pedagogical directions and best practices for imparting social and sustainability marketing competencies. The discussion herein sheds light on the case method, which we have also adopted, and how this method can contribute to better learning outcomes for marketing education and practice. Third, the book presents a series of case studies of various length that marketing educators could employ to teach social and sustainability marketing and that marketing practitioners could explore to guide and inform their own marketing strategies to create and reach socially responsible consumers. Through these case studies, marketing educators and practitioners can expect to learn about social and sustainability marketing that have transpired in both developing and developed economies. In summary, we hope that marketing academics, educators, and practitioners will find value in the chapters contributed by esteemed scholars, and that our book will effectively empower future social and sustainability marketing endeavors.

FUNDING DECLARATION

No funding received.

ACKNOWLEDGMENT

None.

CREDIT AUTHOR STATEMENT

The chapter was written in its entirety by Weng Marc Lim.

RECOMMENDED READINGS

Balaji, M. S., Jiang, Y., & Jha, S. (2019). Green hotel adoption: A personal choice or social pressure? *International Journal of Contemporary Hospitality Management*, 31(8), 3287–3305.

Balaji, M. S., & Roy, S. K. (2017). Value co-creation with Internet of things technology in the retail industry. *Journal of Marketing Management*, 33(1–2), 7–31.

Bhattacharyya, J., & Dash, M. K. (2018). A know your student analysis: A case study on the students of a higher education institute in India. *International Journal of Higher Education and Sustainability*, 2(1), 30–44.

Bhattacharyya, J., Dash, M., Kundu, S., Sakshi, S., Bhattacharyya, K., & Kakkar, K. B. (2021). No virus on me: The Indian ways of managing the COVID-19 pandemic, marine to mountain. *Asian Journal of Management Cases*.

Bhattacharyya, J., Krishna, M. B., & Premi, P. (2020). Amul Dairy (GCMMF): Expanding in the USA, leveraging the e-commerce advantage. *International Journal of Management and Enterprise Development*, 19(2), 149–163.

Goh, S. K., & Balaji, M. S. (2016). Linking green skepticism to green purchase behavior. *Journal of Cleaner Production*, 131, 629–638.

Kumar, A., & Dash, M. K. (2017). Sustainability and future generation infrastructure on digital platform: A study of Generation Y. In *Business Infrastructure for Sustainability in Developing Economies* (pp. 124–142). Hershey, PA: IGI Global.

Lee, Z. W., Chan, T. K., Balaji, M. S., & Chong, A. Y. L. (2018). Why people participate in the sharing economy: An empirical investigation of Uber. *Internet Research*, 28(3), 829–850.

Lim, W. M. (2016a). A blueprint for sustainability marketing: Defining its conceptual boundaries for progress. *Marketing Theory*, 16(2), 232–249.

Lim, W. M. (2016b). Creativity and sustainability in hospitality and tourism. *Tourism Management Perspectives*, 18, 161–167.

Lim, W. M. (2017). Inside the sustainable consumption theoretical toolbox: Critical concepts for sustainability, consumption, and marketing. *Journal of Business Research*, 78, 69–80.

Lim, W. M. (2019). Spectator sports and its role in the social marketing of national unity: Insights from a multiracial country. *Journal of Leisure Research*, 50(3), 260–284.

Lim, W. M. (2020). The sharing economy: A marketing perspective. *Australasian Marketing Journal*, 28(3), 4–13.

Lim, W. M. (2021a). A marketing mix typology for integrated care: The 10 Ps. *Journal of Strategic Marketing*.

Lim, W. M. (2021b). Conditional recipes for predicting impacts and prescribing solutions for externalities: The case of COVID-19 and tourism. *Tourism Recreation Research*.

Lim, W. M., & Heitmann, S. (2017). Sustainability and ethics in rural business and tourism in the developing world. In A. Oraide & P. Robinson (Eds.), *Rural Tourism and Enterprise: Management, Marketing and Sustainability* (pp. 133–144). Oxfordshire, UK: CABI International.

Lim, W. M., Ng, W. K., Chin, J. H., & Boo, A. W. X. (2014). Understanding young consumer perceptions on credit card usage: Implications for responsible consumption. *Contemporary Management Research*, 10(4), 287–302.

Lim, W. M., & Ting, D. H. (2011a). Cyberactivism: Empowering advocacies for public policy. *World Journal of Management*, 3(2), 201–217.

Lim, W. M., & Ting, D. H. (2011b). Green marketing: Issues, developments and avenues for future research. *International Journal of Global Environmental Issues*, 11(2), 139–156.

Lim, W. M., Ting, D. H., Bonaventure, V. S., Sendiawan, A. P., & Tanusina, P. P. (2013). What happens when consumers realise about green washing? A qualitative investigation. *International Journal of Global Environmental Issues*, 13(1), 14–24.

Lim, W. M., Ting, D. H., Mah, J. J., & Wong, C. Y. (2013). Managing alcohol consumption: A responsible consumption approach. *Journal of International Management Studies*, 13(2), 75–82.

Lim, W. M., Ting, D. H., Ng, W. K., Chin, J. H., & Boo, W. X. A. (2013). Why green products remain unfavorable despite being labelled environmentally friendly? *Contemporary Management Research*, 9(1), 35–46.

Lim, W. M., Ting, D. H., Sim, J. C. K., Hew, K. M., & Cheong, K. K. W. (2013). Understanding the influence of green marketing strategies on consumer perception and decision-making. *Review of Business Research*, 13(2), 21–28.

Lim, W. M., Ting, D. H., Wong, W. Y., & Khoo, P. T. (2012). Apparel acquisition: Why more is less? *Management & Marketing*, 7(3), 437–448.

Lim, W. M., & Weissmann, M. A. (2021). Toward a theory of behavioral control. *Journal of Strategic Marketing*.

Lim, W. M., Yong, J. L. S., & Suryadi, K. (2014). Consumers' perceived value and willingness to purchase organic food. *Journal of Global Marketing*, 27(5), 298–307.

Patra, A. K., & Dash, M. K. (2017). Increasing corporate-community consistency by entrenching micro-plan in CSR policies with reference to OPGC Limited. In V. S. Gajavelli, K. Chaturvedi, & A. N. Singh (Eds.), *Recent Trends in Sustainability and Management Strategy* (pp. 57–68). New Delhi, India: Allied Publishers.

Perera, C. R., & Hewege, C. R. (2016). Integrating sustainability education into international marketing curricula. *International Journal of Sustainability in Higher Education*, 17(1), 123–148.

Perera, C. R., & Hewege, C. (2018a). Climate change risk perceptions among green conscious young consumers: Implications for green commodity marketing. *Journal of Consumer Marketing*, 35(7), 754–766.

Perera, C. R., & Hewege, C. R. (2018b). Religiosity and environmentally concerned consumer behaviour: 'Becoming one with God (nature)' through surrendering environmental identities. *International Journal of Consumer Studies*, 42(6), 627–638.

Perera, C. R., Hewege, C. R., & Mai, C. V. (2020). Theorising the emerging green prosumer culture and profiling green prosumers in the green commodities market. *Journal of Consumer Behaviour*, 19(4), 295–313.

Perera, C. R., Johnson, L. W., & Hewege, C. R. (2018). A review of organic food consumption from a sustainability perspective and future research directions. *International Journal of Management and Sustainability*, 7(4), 204–214.

Perera, L. C. R., & Hewege, C. R. (2013). Climate change risk perceptions and environmentally conscious behaviour among young environmentalists in Australia. *Young Consumers*, 14(2), 139–154.

Yadav, R., Balaji, M. S., & Jebarajakirthy, C. (2019). How psychological and contextual factors contribute to travelers' propensity to choose green hotels? *International Journal of Hospitality Management*, 77, 385–395.

Section II

Advances in Knowledge of Social and Sustainable Marketing: Understanding Sustainability Marketing

This section presents selected original research to acquaint the readers with an up-to-date understanding of social and sustainable marketing by providing diverse perspectives. It includes chosen models and theoretical frameworks that can be used as lenses to see through and critique current knowledge and practices.

2

The Sustainability Marketing Framework: A Tool for Teaching and Learning about Sustainability Marketing

Al Rosenbloom
Dominican University, River Forest, IL, USA

CONTENTS

Learning Objectives 14
Themes and Tools 14
Theoretical Background 14
A Prelude to Sustainability Marketing: A Very Short Review
   of Marketing Fundamentals 16
Sustainability 18
Why Sustainability Is a Marketing Issue... 20
   CEOs Embrace Sustainability 20
   Changing Consumer Attitudes and Values 21
   The Evolving Society–Business Relationship 22
What Is Sustainability Marketing? 23
Sustainability Marketing in the Classroom 26
   Case Method Teaching 28
      Grupo Familia Mini-Case... 29
      Case Use and Analysis... 34
   Active Learning 40
   Critical Thinking..41
   Macromarketing Implications of SDG 12 43
   Transformative Insights into Consumer Behavior 44
Conclusion 46
Funding Declaration 47

Acknowledgments 47
Credit Author Statement 47
Notes 48
References.. 48

LEARNING OBJECTIVES

After reading the chapter, the reader should be able to:

- Summarize the evolution in marketing thought that has led to sustainability marketing
- Articulate the differences between traditional marketing management and sustainability marketing
- Present a comprehensive definition of sustainability marketing for use in the marketing classroom
- Apply the Sustainability Marketing Framework to case analyses, active learning exercise, and critical thinking assignments
- Understand how assignments that link the Sustainability Marketing Framework with course content complement each other

THEMES AND TOOLS

- Application of macro-environmental change (PESTLE and SWOT)
- Importance of segmentation and targeting
- Consumer behavior drives marketing strategy
- Importance of triple bottom line, sustainable development, and stakeholders

THEORETICAL BACKGROUND

This chapter's theoretical background is the principles and perspectives of marketing. One of the chapter's goals is to position sustainability marketing as an extension of traditional marketing practices, but of course with some new additional elements as required by sustainability marketing as developed in the chapter: The triple bottom line (profit-planet-people), stakeholder theory, and sustainable development.

The chapter's primary perspectives are the "fundamentals" of marketing that are universal: Putting the customer at the center of the entire organization; building marketing programs backward from customer wants and needs; using the principles of segmenting and targeting to concentrate firm resources on consumer groups that will be most responsive to the firm's offerings; developing products/services that create value though meaningful differentiation; adapting the marketing mix to specific target markets; and recognizing that marketing is a total organizational response to finding and keeping customers.

Yet sustainability marketing requires two additional theoretical frames: The triple bottom line and stakeholder theory. The triple bottom line (TBL) summarizes the requirement that companies consider profit, people, and planet – and the tradeoffs involved when trying to optimize the benefits for all three – in all company decisions. Stakeholder theory expresses the more inclusive idea that a number of groups beyond investors and customers can influence and are, in turn, influenced by company decisions. Therefore, stakeholders much be considered in all strategic decisions of a firm.

This focus on marketing fundamentals was done intentionally, so that students can more easily see that sustainability marketing uses universal principles of marketing – principles that students may already be familiar with. Only at chapter's end are the challenges raised by sustainability marketing to the standard marketing model outlined. Since the chapter's goal is focused otherwise, these challenges are only mentioned briefly.

Marketing has always been at the center of every organization. Peter Drucker (Drucker, 1954) has famously stated that there are only two core functions for every business: innovation and marketing. Drucker goes on to say that the purpose of every business is to get and keep a customer – the very essence of marketing. As "marketing" continues to evolve, social marketing applies marketing principles to a broad range of organizations that extend beyond entities engaging in commerce. For example, health care institutions, college/universities, nonprofits, governmental, and social service agencies all use social marketing to help them achieve their missions. Peter Drucker also brought his insights to social marketing when he said, "The 'non-profit' institution neither supplies goods or services …. Its product is a changed human being … a cured patient, a child that learns, a young man or woman grown into a self-respecting adult; a changed human life altogether" (Drucker, 1990/2011, p. x). Social marketing seeks to change human behavior for the betterment of society (Saunders, Barrington, & Sridharan, 2015).

The application of marketing principles to benefit society is also found in one of marketing's newest fields of study: sustainability marketing. Sustainability marketing gives tangible expression to the increased demands from stakeholders that businesses should think beyond the confines of their spreadsheets and become active partners in addressing some of the world's biggest global challenges (e.g., poverty, climate change, inequality, sustainable development, sustainable consumption) (Brammer, Branicki, Linnenluecke, & Smith, 2019).

This chapter's focus is on sustainability marketing. The chapter's first half discusses the importance of sustainability in marketing. In so doing, it answers the question why sustainability marketing is an important issue for contemporary marketing, it defines what sustainability marketing is, and it untangles three words often used in discussions about sustainability marketing: "sustainable," "sustainability," and "sustainable development." The chapter's second half focuses squarely on marketing pedagogy by presenting a framework for teaching and learning about sustainability marketing suitable for use in the marketing classroom. This Sustainability Marketing Framework has three unique elements: (1) It explicitly incorporates the Sustainable Development Goals (SDGs) as a gateway for marketing educators to teach about – and for students, in turn, to learn about – sustainability marketing; (2) it acknowledges that sustainability marketing shares some common features with social marketing, since the overriding goal of both is to improve the well-being of individuals and society; and (3) it integrates macro- and micro-level marketing perspectives to provide a comprehensive view of sustainability marketing. Because sustainability marketing is a relatively new topic, the chapter's second half provides examples of how marketing educators can apply the Sustainability Marketing Framework in case analyses, in active learning exercises, and in developing students' critical thinking skills.

A PRELUDE TO SUSTAINABILITY MARKETING: A VERY SHORT REVIEW OF MARKETING FUNDAMENTALS

Historically, marketing has focused primarily on consumers engaging in commercial transactions. The consumer orientation places the in-depth understanding of consumer needs and wants at marketing's core (Kotler

& Armstrong, 2014; Lee & Edwards, 2013; Ramaswamy & Namakumari, 2018). Through marketing science, whose focus is always to better understand consumers in all their complexities, come ideas for competitive products and/services, which are, in turn, redirected back to consumers through a tailored marketing mix. Levitt (1960) reinforces this point by noting that

> an industry is a customer-satisfying process, not a goods-producing process.... An industry begins with the customer and his or her needs .... Given the customer's needs, the industry develops backwards, first concerning itself with the physical delivery of customer satisfactions. Then it moves back further to creating the things by which these satisfactions are in part achieved .... Finally, the industry moves back still further to finding the raw materials necessary for making its products. (p. 55)

Products and services thus become the mechanisms for developing long-term, valuable, valued, and profitable relationships with consumers. Marketing's goal is thus to create a customer and by extension to create the business itself. As Levitt's (1960) quote indicates, a business, including its supply chain, develops backward from the customer. Without a customer (the sole domain of marketing), there *is* no business.

In 1969, Philip Kotler and Sydney J. Levy broadened the application of private sector marketing to nonprofit organizations (Kotler & Levy, 1969). Their thinking was as follows: (1) All organizations have marketing problems; (2) marketing principles (such as voluntary exchange, consumer sovereignty, segmenting, targeting, and positioning; the marketing mix) are universal and therefore apply to every organization, including nonprofits; and (3) when disciplined market analysis (driven by marketing research) is used, nonprofit organizations can achieve their organizational goals through marketing just as businesses do. In so doing, Kotler and Levy (1969) also broaden the concept of "customer": A customer is now any individual who is central to achieving the organization's mission and with whom the organization wants to establish a noncoercive relationship. "Customers" can be students, patients, donors, government officials, volunteers, and clients.

Philip Kotler, along with Gerald Zaltman, soon applied that same logic (that marketing has universal application) to social problems (Kotler & Zaltman, 1971). Social marketing took the tools of marketing and marketing's relentless focus on understanding behavior, to tackle both individual behavioral problems (e.g., alcoholism, drug abuse, unwanted pregnancy, unsafe sexual practices), and important social issues (e.g., human rights abuse, racial inequality, littering, pollution) by promoting voluntary behavioral change to increase individual and society's well-being (Andreasen, 1994). Social marketing can be thought of as blending marketing with social science (Truss, Marshall & Blair-Stevens, 2010).

Businesses, nonprofits, and social service agencies are all open systems. They both influence and are, in turned, influenced by the external environment. Marketers in every organization, therefore, have obligations to constantly monitor and measure changes in the macroenvironment that might impinge on their consumers/customers. This is why marketing starts with external analyses such as PESTLE, PEST, and SWOT. This outside-in perspective, as the quote from Levitt (1960) above indicates, is a hallmark of all market-responsive and market-oriented organizations. These organizations do not fear change. In fact, they see change as an ally because they know that change can, with the right organizational leadership and right organizational culture, brings opportunity – and marketing leverages that opportunity. One of those new opportunities is sustainability marketing.

SUSTAINABILITY

Three words that have different meanings but appear repeatedly when discussing sustainability are: "sustainability," "sustainable," and "sustainable development." The main thrust of anything being sustainable is intuitively clear: It means to continue, to be durable, or to have the ability to maintain itself. When used as an adjective in English, "sustainable" connotes longevity, something continuing on forever (Lélé, 1991). This is the meaning inherent in "sustainable competitive advantage." However, a bit of linguistic history adds depth to contrasting "sustainable" with "sustainability."

Sustainability was first used in relation to forestry, where the idea was never to harvest more trees than the forest could naturally self-generate (Kuhlman & Farrington, 2010). "To harvest" is the key word. Harvesting is a human intervention, and if done too aggressively, the stock of trees available for succeeding generations would be diminished. "Sustainability" thus became the word that captured "[t]he concern with preserving natural resources for the future" (Kuhlman & Farrington, 2010, p. 3437). This is a perennial human development and global economy issue, since there is always a need to balance the demands for food, clothing, shelter, energy, etc., with the earth's ecosystem's ability to supply them now and in subsequent years. As summarized by Solow (1992): "'Sustainability' is … an injunction to preserve productive capacity for the indefinite future" (p. 163). Since productive capacity is "the ability of an ecosystem to produce the raw materials necessary for economic activities" (Sustainable Development Indicator Group, 1996), "sustainability" tightly links the economy with the environment.

"Sustainability" is often associated with "sustainable development." Development is concerned with the enhancement of well-being and finds its animating interest in reducing economic inequality within a country, thereby improving the overall quality of life of individuals. In 1972, the Club of Rome published, *The Limits of Growth*. This report "fundamentally confronted the unchallenged paradigm of continuous material growth and the pursuit of endless economic expansion" (https://clubofrome.org/about-us/). The report predicted that many of the natural resources that underpinned economic growth would be exhausted in one or two generations.

The UN World Commission on Environment and Development (WCED), better known as the Brundtland Commission, extended this thinking by asking: "How can the aspirations of the world's nations for a better life be reconciled with limited natural resources and the dangers of environmental degradation?" (Kuhlman & Farrington, 2010, p. 3438). The Brundtland Commission's answer was the first and most popular definition of sustainable development. Sustainable development is "development that meets the needs of the present without compromising the ability of future generations to meet their own needs" (WCED, 1987, p. 16). By defining sustainable development as they did, the Commission asserted the tight connection between the "environment" and "development": "[The] 'environment' is where we all live; and 'development' is what we all do in attempting to improve our lot within that abode. The two are inseparable" (WCED, 1987, p. 6).

WHY SUSTAINABILITY IS A MARKETING ISSUE

The origins of *why* sustainability is a marketing issue (as opposed to *what* sustainability marketing is) can be found in the convergence of (1) the significant change in CEO thinking about the importance of sustainable development as a viable, long-term corporate strategy; (2) changes in consumers' values wanting more eco-friendly/sustainable products and services, as they lead more sustainable lifestyles; and (3) an evolution in thinking about the responsibilities businesses have for creating a more just and prosperous world for all individuals.

CEOs Embrace Sustainability

CEOs now recognize that sustainability must be an integral part of their company's strategic plans. Research on CEOs and corporate strategy formulation has found that a majority of CEOs believe that finding sustainable solutions to current and future business challenges will transform their industries within the next five years (Hayward et al. 2013). CEOs also state that implementing sustainability strategies is now a competitive imperative (Kiron, Kruschwitz, Haanaes, & Velken, 2012).

Furthermore, in a 2013 study of global CEOs on sustainability, consumers were identified as *the* most important stakeholder for sustainability becoming a total enterprise strategy by 64% of survey respondents ($n = 1,000$). "Business leaders strongly believe that their sustainability performance and reputation are key factors in shaping consumer and customer demand: CEOs report that both the reputation and brand of their company on sustainability (81%) and the sustainability performance of products and services (77%) are important in the purchasing decisions of their consumers and customers" (Lacy, Gupta, & Hayward, 2019, pp. 510–511). Corporate branding, the ability to access new market segments, new product development, the positioning of products around sustainable attributes, and strategic market entry into new, international markets are all marketing's responsibility.

In the transition to becoming sustainable, company marketers play both strategic and tactical roles. As the strategic level, the Chief Marketing Officer (CMO) is the person best positioned to bring the voice of the customer into corporate-level strategy discussions. The CMO has an important voice in co-creating, along with other C-Suite executives, the firm's strategic sustainability efforts. Kenneth Weed, CMO, Unilever, said succinctly, "CFOs how where the dollars have gone; CMOs should show where the dollars are going to come from. Serving tomorrow's consumer will future proof the company" (quoted in Yosie, Simmons, & Ashken, 2016, p. 6). As the above research indicates, "future proofing" the firm revolves around sustainability. Kumar (2004) affirms this perspective by noting that CMOs must "participate in those conversations that shape the firm's destiny" (p. 8), and that best way to do this is to champion initiatives that are strategic, cross-functional, and bottom-line focused. Sustainable marketing, as described below, checks all those boxes.

Changing Consumer Attitudes and Values

At the tactical level, product line and brand managers also have responsibilities for monitoring and understanding changing consumer attitudes toward sustainability. Marketing managers typically use marketing science and the insights that data provide to guide their marketing efforts. Research confirms that segments of consumers are increasingly purchasing sustainable products and living more sustainable lifestyles (Ottman, 2011). As consumers lead, marketers follow.

One of the first efforts by marketers to understand consumers and their changing relationships with the natural world was "ecological marketing" (Kumar, Rahman, & Kazmi, 2013; Peattie, 2011). Ecological marketing was a response to the environmental damage done by the Exxon Valdez oil spill and the Chernobyl nuclear reactor accident. While ecological marketing "moved marketing thought and practice beyond an abstract economic and social worldview to embrace the physical realities of the tangible world" (Belz & Peattie, 2009, p. 28), it only had limited effect.

More enduring is the rise of the green consumer (Ottman, 2011). Prompted by evidence of increased environmental harm caused by businesses, green consumers were motivated by green consumption values, that is "the tendency to explore the value of environmental protection through one's purchases and consumption behaviors" (Haws, Winterich, & Naylor, 2014, p. 337). Initial green marketing efforts revolved around package redesign to reduce waste, new product development that stressed the product's eco-friendliness and pro-environmental advertising. Overall consumer response was tepid (Grant, 2008), in part, because many of these initial marketing efforts were seen as "greenwashing" by consumers.

Greenwashing is the artful effort to mislead consumers to believe that either the company itself (firm-level greenwashing) or the company's products and/or services (product-level greenwashing) are more beneficial for the environment than they really are. Because marketing communications are instrumental in promoting green product benefits and in positioning the company as environmentally responsible, marketers are at the center of greenwashing charges.

Green marketing and greenwashing responses provide the transition to sustainability marketing. On the one hand, the rise of green consumers indicates that there are market segments concerned about the natural environment, who wanted products and services responsive to these concerns. On the other hand, greenwashing raised important questions about marketing ethics and whether "marketing" could legitimately be a force for good or whether commercial marketing was, as many critics claimed, manipulative and focused primarily on short-term gains for shareholders. The growing demand that commercial enterprises contribute positively to society dovetails nicely with the evolution in thinking about corporate social responsibility (CSR).

The Evolving Society–Business Relationship

The relationship between society and business is complex and is constantly evolving. CSR summarizes the contemporary view that businesses "are integrated within, rather than detached from, the rest of society" (Lindgreen et al., 2019, p. xxix). Implicit in CSR are the ideas that every business has obligations to behave ethically (i.e., to act responsibly); that this moral obligation to behave ethically comes from an implicit social contract between business and society that gives business "the license to operate" (Donaldson & Dunfee, 1999; Shocker & Sethi, 1974); and those ethical obligations extend beyond shareholders and even beyond customers. The concept of stakeholders (Lindgreen et al., 2019) codifies the more inclusive idea that a number of groups beyond investors and customers can influence and are, in turn, influenced by company decisions. As Van Tulder, Van Tilburg, Francken, and Da Rosa. (2014) state, "Stakeholders are the eyes and ears of society" (p. xxii). While company survival is linked inevitably to profitability, the metrics by which all responsible companies are measured now include social and environmental dimensions. The TBL (Elkington, 1998) summarizes the requirement that companies consider profit, people, and planet – and the tradeoffs involved when trying to optimize the benefits for all three – in all company decisions.

As firms have matured into a more comprehensive understanding of their relationship with society, so, too, has marketing. Within marketing, macromarketing is the field that looks at marketing's relationship to society and marketing's responsibility to serve the common good (Murphy & Sherry, 2014). As its name suggests, macromarketing transcends the managerial, firm-level focus of traditional, marketing management, which is oriented toward delivering sustainable competitive advantage. Instead, macromarketing addresses marketing efforts at the aggregate level of society. Macromarketing "asks how marketing should be carried out to meet the goals of society and to optimize social benefits" (Belz & Peattie, 2009, p. 26). Sustainability marketing responds to the challenge of using marketing frameworks and tools for the benefit of society, and it is to this discussion the chapter turns next.

WHAT IS SUSTAINABILITY MARKETING?

There have been many efforts to define sustainability marketing. Some definitions simply added "sustainability" to the marketing mix (Fuller, 1999); some tweak the American Marketing Association's definition of marketing as a value creating process so that "human and natural capital are preserved and enhanced" (Martin & Schouten, 2012, p. 18). The textbook author Phil Kotler (Kotler & Armstrong, 2014) bases his definition of sustainability marketing on the Brundtland Commission's formulation of sustainable development of not disadvantaging the needs of future generations based on current consumer needs.

Belz and Peattie (2009) make two contributes to define sustainability marketing. First, they distinguish between "sustainable marketing" (Emery, 2012; Martin & Schouten, 2012; Richardson, 2020) and "sustainability marketing." Although this is a subtle difference, Belz and Peattie (2009) state that sustainable marketing could be thought of as marketing that is just "durable or long-lasting" (p. 31). "Sustainability marketing," in contrast, refers specifically to the sustainable development agenda. (Refer to the above section on distinguishing between "sustainable," "sustainability," and "sustainable development").

Belz and Peattie's (2009) second contribution is the "eagle's eye" perspective they take on describing sustainability marketing itself. Sustainability marketing is characterized as:

- Beginning with the analysis of social and environmental problems in the macroenvironment
- Adding the consideration of social and environmental criteria to traditional marketing management responsibilities at the company level
- Having a long-term orientation that strengthens marketing's goal of developing productive and profitable relationships with customers and specific target markets
- Using marketing research and marketing science to probe deeply consumer decision-making around purchase, use, disposal of, and even the delay in purchasing products/services
- Finding market opportunity at the intersection of socioecological problems and consumer wants
- Supporting sustainability innovations across the entire marketing mix
- Aligning marketing activities to support the company's active involvement in public and political processes to change institutions (so that sustainability is supported in society)
- Emphasizing the TBL, that "simple heuristic that both managers and business students can use as a prompt to remember the interrelated social, environmental, and economic dimensions fundamental to sustainability" (Collins & Kearins, 2010, p. 500)

More succinctly, Belz and Peattie (2009) defined sustainability marketing as "building and maintaining sustainable relationships with customers, the social and the natural environment" (p. 31).

Several aspects of sustainability marketing are worth highlighting. Foremost, sustainability marketing is an expression of social responsibility. As such, sustainability marketing decisions are *always* made within the context of the triple bottom line and with a stakeholder perspective. Both of these contrast with more traditional and managerial marketing approaches. Whereas traditional marketing management places profitability and long-term competitive advantage as the sole determiner of marketing strategy, sustainability marketing additionally considers the more complex calculus of being profitable while also maximizing long-term environmental and societal value. Also, traditional and managerial marketing considers customers as the most important stakeholder, since customers drive profitability. Sustainability marketing, in contrast, makes marketing decisions through the lens of thoughtful considerations of diverse stakeholders, which must include society. Making marketing decisions within a stakeholder framework complicates decision-making because, while the ideal is to create a "win-win" situation for all stakeholders, more often than not, stakeholder considerations involve tradeoffs and comprise (Hahn, Pinkse, Preuss, & Figge, 2015). Not every stakeholder gets everything they want, including the firm and its marketers.

Sustainability marketing is also explicit in its support for sustainable development (Belz & Peattie, 2009). A tangible expression of sustainable development is the Sustainable Development Goals (SDGs). By extension, then, sustainability marketing supports the SDGs.

Agreed to by 193 members of the United Nations in 2015, the SDGs consist of 17 goals (see Figure 2.1) to which 169 specific targets have been attached. These 17 goals provide a global "to-do list" of improving the quality of life for every individual on earth. In other words, the SDGs express a

FIGURE 2.1
The sustainable development goals.

future in which every human being has opportunities to flourish, prosper, and live a life of self-determination. The SDGs seek to remove the barriers that limit full human development. A number of goals, for example, aim to meet basic needs, such as ending extreme poverty and hunger, ensuring universal access to health care, a quality education, as well as clean water and sanitation, and empowering people through the elimination of gender inequality and workplace discrimination. Another set of goals cluster around environmental concerns: climate change, affordable and clean energy, and maintaining the biodiversity of animals in the ocean and on land. To achieve this more equitable world, individuals need to feel free, safe, and included in their everyday lives. There must, therefore, be a commitment by all to develop just, peaceful, and inclusive societies.

From a business/entrepreneurship perspective, the SDGs can be thought of as a matrix of opportunities. Hoek (2018) identifies business opportunities related to SDGs as "the trillion dollar shift," and according to the *Better Business, Better World* report (Business & Sustainable Development Commission, 2019), working on the SDGs would open up $12 trillion in market opportunities for commercial businesses and create up to 389 million new jobs in four sectors: food and agriculture, ($2.3 trillion), cities ($3.7 trillion), energy and materials ($4.3 trillion), and health and well-being ($1.8 trillion). This meshes completely with marketing's outside-in perspective, and its goal of finding, evaluating, and leveraging marketplace opportunity (Lee & Edwards, 2013; Ramaswamy & Namakumari, 2018). "Achieving the Global Goals would create a world that is comprehensively sustainable: socially fair; environmentally secure; economically prosperous; inclusive; and more predictable" (Business & Sustainable Development Commission, 2019, p. 11).

The question now becomes: Is there a way to integrate the broad scope of sustainability marketing to facilitate learning in the marketing classroom? The answer to this question follows.

SUSTAINABILITY MARKETING IN THE CLASSROOM

The perspective taken in this section is that of the marketing educator, who wants to introduce sustainability marketing to students and to use various learning strategies to do so.

Sustainability Marketing Framework • 27

FIGURE 2.2
The sustainability marketing framework.

Figure 2.2 can serve as a framework for doing both. Figure 2.2 is simple enough to be drawn on a white board, distributed as a handout, or included in a PowerPoint presentation. The framework is hierarchically structured. Sustainability marketing's superordinate goal is to create a better world by using marketing methods and perspectives to support sustainable development (Belz & Peattie, 2009). The framework highlights TBL (the three Ps of profit, people, and planet), a stakeholder perspective, and the Sustainable Development Goals (SDGs) as unique sustainability marketing characteristics. Figure 2.2 also includes macro/societal-level and firm-level marketing orientations.

Class focus might determine the order in which topics of Figure 2.2 are developed. In principles, introductory, or marketing management classes, sustainability marketing can be seen as an emerging application of universal marketing principles (see A Very Short Review of Marketing Fundamentals above) – but with three significant additions: TBL, stakeholders, and the SDGs. In these courses, a bottom-up development of ideas might work best. A lecturer would discuss how Sustainability Marketing supports Sustainable Development which contributes to A Better World (see Figure 2.2 concepts). Alternatively, in nonprofit/social marketing, social entrepreneurship, or even macromarketing (Shapiro et al. 2020) courses, a top-down approach is best. All these courses share a common premise: That marketing can be a force for good and can become an agent of world benefit (see https://weatherhead.case.edu/centers/fowler/). Hence, a class lecture can move from macro/societal issues to more specific discussions of individual firms, social enterprises, and social agencies: A Better World is the overriding purpose of Sustainable Development which is supported by firm-level Sustainability Marketing. As before, sustainability marketing's unique characteristics need discussion: TBL, stakeholders, and the SDGs.

It's imperative to note the reflexive and deductive nature of this chapter. It is written *specifically* to support Figure 2.2. Each of the framework's components has a previous chapter section devoted to it. Marketing educators can take what is written as starting points for their own lectures, or they can assign this chapter as a "pre-read" for further class discussion. Figure 2.2 is also useful for applied classroom learning activities, such as case analyses, marketing strategy evaluations, and macromarketing applications as follows.

Case Method Teaching

Cases and case method teaching not only have a long tradition in marketing education (Brennan, 2009; Crittenden, Crittenden, & Hawes, 1999) but also have a privileged place in the marketing classroom (Pitt, Crittenden, Plangger, & Halvorson, 2012). Case analysis sharpens students' real world analytic problem-solving skills in ways that other pedagogical techniques do not. In decision-oriented cases, students are asked "to figuratively step into the position of a particular decision maker" and to make a decision, solve a problem, analyze a process, or confront a situation (Leenders, Erskine & Mauffette-Leenders, 2001, p. 2). Descriptive cases, in contrast, lead the student through decisions that the case's focal organization has made already. These cases become examples of either effective or ineffective decision-making. In turn, these cases illustrate what to do, or become a cautionary tale of what not to do. Cases have pedagogical power because they sharpen student's critical thinking skills (Klebba & Hamilton, 2007).

Because every case has its own specific-learning objectives, it is impossible to predict the specific sustainability marketing issues of any one case. It is presumed that most sustainability marketing cases deal with universal marketing management issues: For example, strategies for targeting socially conscious consumers, innovations in the marketing mix, creativity in social media engagement and response tracking, new product development that position products/services as sustainable, the application of big data and data analytics in the service of sustainability, etc. These marketing management tasks are included in Figure 2.2's umbrella term, "Sustainability Marketing."

Cases can prompt robust discussion. Marketing educators can use Figure 2.2 to expand any case discussion beyond the analysis of marketing management issues. The instructor can ask simply: "Have all of the components in Figure 2.2 been explicitly considered by the company profiled in this case?" Discussion prompts could be:

- Has the company considered carefully each aspect of the TBL? Did the company consider society (people) in its decision, or did it focus primarily on planet and profit? What is gained, from a marketing perspective, when all three TBL elements are considered versus considering just one or two of them?
- Who are the stakeholders beyond customers for this case? Did the CEO, CMO, and/or the product line/brand manager consider these stakeholders fully as they crafted their marketing strategy? What marketing issues might arise, such as changes to brand equity and company reputation, when not all stakeholders are considered?
- Which SDG(s) does/do this sustainability marketing case address? Given sustainability marketing's express commitment to support sustainable development, how do the actions of the case firm contribute to one or more of the SDGs?
- Overall, how successful is this sustainability marketing effort in making a positive change for society long term?

The following mini-case on Grupo Familia is an applied example for linking case discussion with Figure 2.2. A more detailed discussion of Grupo Familia is found in Vélez-Zapata, Cortés, and Rosenbloom (2020).

Grupo Familia Mini-Case

Grupo Familia is a Colombian-based, Multi-Latina that positions itself as a well-being company focusing on personal care, hygiene, and grooming solutions for individuals and families throughout their entire life cycle.

Founded in 1958 in Medellín (Colombia) by John Gómez Restrepo and Mario Uribe, Grupo Familia began as an importer of Scott Paper's (the USA) Waldorf brand toilet paper into Colombia. In 1970, Grupo Familia opened offices outside Medellín (first in Bogotá and then in Cali and Barranquilla), and in 1982, it expanded into Peru and Chile, thereby beginning its international expansion. In 1997, Grupo Familia formed a joint-venture with the Swedish multinational Svenska Cellulosa Aktiebolaget (SCA), which continues still. Currently, Grupo Familia markets products in 22 countries in Latin America and the Caribbean. It has production facilities in Colombia, Ecuador, Argentina, and the Dominican Republic.

Starting with a single product, toilet paper, Grupo Familia's commitment to innovation and new product development can be seen in the company's current, diversified product line (see Table 2.1), all of which support Grupo Familia's mission to being a relevant brand across the entire family life cycle. For example, Pequeñín is an infant care brand offering diapers, shampoos, and creams for newborn and young children; Nosotras is a line of feminine hygiene products; the Familia® brand includes products for home use by the entire family, such as toilet paper, napkins, towels, and antibacterial wipes; and TENA is a line of men's and women's incontinence products. Because pets are often considered as family members, Grupo Familia developed Petys, a comprehensive pet care brand. In its

TABLE 2.1

Grupo Familia Brands and Products

Brand	Product Line
Familia®	Toilet paper, napkins, kitchen towels, tissues, air fresheners, wet cloths, Microfiber kitchen towels, antibacterial gels
Pequeñín®	Diapers, baby wet cloths, baby protective creams, baby shampoos
Petys®	Pet wet cloths, odor eliminators, pet shampoo, flea repellent spray, dog absorbent mats, cat liter
Nosotras®	Sanitary towels and napkins, daily protectors/pants liners, tampons, wet cloths, intimate and body soap, women V zone after shave care products line
Pomys®	Wet wipe makeup removers for normal, dry, oily, and mixed skin types
Familia Institucional®	Work place hygiene and grooming products: toilet paper, hand towels, soaps, antibacterial gels, napkins, cleaners, wet cloths, semi-disposable cloths, tissues, air fresheners
TENA	Broad range of incontinence products for men and women

home market of Colombia, Grupo Familia is the market leader in hygiene products for home and personal care. It is also ranked as one of the top 10 most remembered brands in the country.

Grupo Familia's commitment to both being a socially responsible company and serving low-income consumers can be traced back to John Gómez Restrepo. Gómez Restrepo grew up very poor and worked early-on to support his single mother and five brothers. He said, "I am going to conquer poverty and reach my dreams." As a serial entrepreneur, over time, he developed many managerial insights, one of which was "In this company, the dignity of the person is respected above all."

As a company whose core products revolve around paper (napkins, toilet paper, kitchen towels feminine hygiene products), Grupo Familia is strongly dependent on forests and water as raw material inputs in its production processes. Yet the production processes also involve bleaches like chlorine dioxide for extracting fiber, which can pollute the environment. Grupo Familia understood its dependence on and its responsibility for not harming the natural environment across its entire supply chain. Grupo Familia created "Papel Planeta," a social responsibility campaign that sought to educate children about the correct separation of waste and to make them aware of climate change and importance of water protection. Grupo Familia created a virtual platform for "Papel Planeta" that includes a set of educational videos. Grupo Familia also incorporates the Forest Stewardship Council (FSC) logo on its products.[1]

The company's 2014 Sustainability Report described sustainability as an important element that connected all strategies and structures across the entire company. It made a commitment to its customers, its shareholders, its employees, government regulators, community residents, and the media, which regularly cover it to reduce environment impacts and to implement strategies that have positive effects on communities. The current CEO, Andres Felipe Gómez-Salazar, stated, "We have declared in *Familia* that sustainability has a first name: There is financial sustainability, environmental sustainability and around the communities in which we operate, market sustainability and sustainability of people." Table 2.2 presents some of Grupo Familia's sustainability commitments.

Grupo Familia's sustainability commitments are supported further by the company's approach to innovation. Through a uniquely developed Innovation Model, Grupo Familia has reframed innovation as a core competency that constantly generates new solutions for products, services, and

TABLE 2.2

Grupo Familia Triple Bottom Line Metrics

Sustainability Dimensions	2014	2015	2016	2017
Economic				
Added value created (USD millions)	629	704	766	774
Environmental				
Waste recovery	79%	80%	95%	93%
Reduced water consumption	40%	40%	34%	35%
Reduced electric energy consumption	3,6%	7%	8%	7,3%
Reduced on biological oxygen demand	42%	55%	46%	—
Reduced on direct emissions of CO_2	39%	36%	38%	33%
Social				
Benefit of management social programs (health, education, and living place)	—	1,069	1,382	1,588

Source: Vélez-Zapata, Cortés, and Rosenbloom (2020).

processes. Recent product innovations, for example, include the following: (1) For the Familia brand, a double-leaf family towel was launched made of 100% recycled material; (2) in the Pequeñín line, baby pants was redeveloped to offer a more superior fit with greater absorption; (3) Petys developed a new flea repellent shampoo and spray for dogs made with 100% all-natural ingredients; (4) Pomys Basic, a line of wet wipe makeup removers priced more affordably yet delivering high quality value-in-use, was developed for low-income women; and (5) Nosotras reformulated its 100% herbal intimate soap to provide greater pH balance, gentleness, and a more balanced fragrance. The 2017 Grupo Familia Annual Report summarized the strategic importance of innovation when it stated that within Grupo Familia, innovation drives profitable growth, ensuring sustainability, competitiveness, and leadership in the market (Grupo Familia Annual Report, 2017).

Grupo Familia's commitment to marketing sustainability is also found in the following three examples:

1. **Use of post-consumer content and integration of recyclers in the supply chain.** Susan Irwin, Grupo Familia's Director of Sustainability, has said, "Because Grupo Familia is fundamentally a paper company, [we have] an important responsibility for committing to

recycling processes. That's why we use around 70% of post-consumer paper content in our production processes as a raw material." Post-consumer content is defined as "[m]aterial from products that were used by consumers or businesses and were collected for recycling instead of being discarded as waste" (Paper Recycling-Coalition, n.d.). To obtain the 300 tons/day of post-consumer paper content needed for its products, Grupo Familia works with 1,382 recycling workers. These recyclers, or "recicladores," walk city streets, often with their pet dog, collecting paper thrown into garbage cans, which they then sort, bundle the clean paper together, and bring to a cooperative.

A cooperative is an organization that coordinates the collection, quality of material, weighs the paper, and pays a reasonable rate per kilo. Recyclers must be cooperative members to use their services. In addition to paying a fair price per kilo of paper, cooperatives provide social services to their members, some of which Grupo Familia funds. Grupo Familia only buys post-consumer paper content from cooperatives. Grupo Familia's commitment to improving the quality of life for recyclers is highlighted in a short film on Grupo Familia's Facebook page, entitled, "Heroes of the Planet." The firm profiles Rodrigo de Jesús Gonzalez, a recycler, who talks about his pride in being a recycler and how Grupo Familia gives him dignity and respect in his work.[2]

2. **Understanding and respecting low-income consumers**. Grupo Familia marketers use marketing research to gain strategic consumer insight into one of their target markets: The low-income consumer. Low-income consumers live in barrios, which are unplanned, densely populated, violent, and high crime neighborhoods located in the foothills of cities. Whereas middle class consumers shop in traditional supermarkets, discount warehouses, and convenience store chains, thereby making bulk distribution of products possible, low-income consumers shop in small, independent grocery stores, at kiosks, and in mom-and-pop stores that hold smaller quantities of product due to their significantly smaller size. This requires firms to develop intensive channel distribution systems that are able to deliver small quantities of products, to many outlets, several times a week.

Grupo Familia's research confirms that low-income consumers are family oriented, are present-day focused yet optimistic about the

future, and buy products that provide great consumer value. Low-income consumers frequently avoid "cheap" products and instead buy more expensive leading brands because of the brand's reputation and reliability. Grupo Familia keeps the same packaging, the same product quality, and the same suggested retail price for all its products targeting the low-income segment as it does up higher-income segments. In addition, Grupo Familia contracts with community/barrio entrepreneurs who are micro-transport agents to act as distributing agents, who rebundle larger pallets of products into smaller units and deliver Grupo Familia's products to barrio merchants. Grupo Familia's respect for the low-income consumer strengthens the brand consumer–firm relationship, which leads to strong long-term segment loyalty.
3. **Community empowerment**. The location of a manufacturing plant is a strategic decision. Grupo Familia intentionally located one of its three Colombian plants in Guachené, a city, and a region that has high drug and alcohol use. Poverty is extensive. Children receive a low quality of education. More than half of the city's residents work in seasonal agricultural work harvesting sugarcane. Unemployment is high with many jobs in the informal sector, so individuals work without social security coverage. In addition to providing employment in its plant for city and regional residents, Grupo Familia created two community projects to improve the quality of life in Guachené: (1) Misión + Hogar, the goal of which is to improve the housing and living condition of city residents. By 2017, Misión + Hogar had built 332 new homes that gave families better hygiene and health conditions. (2) Grupo Familia helped rebuild primary and secondary schools in the area, thereby improving the learning environment for those children.

Case Use and Analysis

Learning objectives for the Grupo Familia case are as follows: (1) To illustrate the integration of strategic as well as tactical sustainability marketing perspectives; (2) to highlight the CEO's role in creating a corporate culture that supports sustainability marketing; (3) to analyze the relationship between sustainable development and sustainable marketing; and (4) to apply the Sustainability Marketing Framework (Figure 2.2).

Experience indicates that this case can be used either as a written assignment for assessment purposes or as an in-class discussion learning activity. When used as a paper assignment, students should be asked to evaluate Grupo Familia's sustainability marketing efforts. This unstructured writing prompt casts a wide net, intentionally. Student papers are evaluated along two criteria: (1) How well the paper integrates the Sustainability Marketing Framework (Figure 2.2) into its discussion of sustainability marketing and (2) the completeness of case information and data used within the paper to support points and arguments made. A model answer is graphically summarized in Figure 2.3, and the in-class discussion that follows suggests the paper's analytic content.

The prompt used for the written assignment (evaluate Grupo Familia's sustainability marketing efforts) also can be used as a single, pre-assigned case discussion question. For faculty wanting to assign more specific questions before class discussion, these questions are possibilities: (1) Who is Grupo Familia's target market and how effective is Grupo Familia's marketing mix in creating value for them? (2) What sustainability marketing efforts stand out in the case? (3) Does the CEO have a role in sustainability marketing? If so, what is it? If not, why not? (4) How does Grupo Familia demonstrate a commitment to sustainable development? Experience with in-class case discussions indicates that students learn the most when they are asked to reflect on the prediscussion assignment questions rather than simply answer them in class. The following classroom teaching plan proceeds from that premise.

The teaching plan. Class discussion can begin with the following question: *How well is Grupo Familia doing in terms of sustainability marketing?* Students will typically begin by noting some of the tactical, traditional elements of sustainability marketing. Students frequently cite the communication aspects of marketing. These students note the FSC logo placed on the Familia brand and "Papel Planeta" videos. Some students will observe that Grupo Familia promotes these products through social media. These students have watched the "Heroes of the Planet" video and have spent time scrolling through the products promoted on Facebook. With some prompting to consider other marketing mix variables (product, price, and place), students point to three new products that illustrate green marketing: the double-leaf Family Towel made with 100% recycled material, Petys flea repellent shampoo and spray made with 100% all-natural ingredients, and Nosotras's reformulated 100% herbal intimate soap. Students

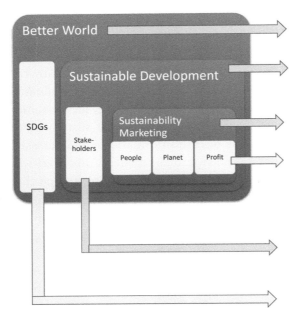

FIGURE 2.3
Sustainability Marketing Framework applied to Grupo Familia mini-case.

have more difficulty articulating distribution and pricing as a sustainability marketing mix elements.

At this point, instructors can ask: *Who is the target market for Grupo Familia's sustainability marketing efforts?* This leads not only to an analysis of low-income consumers but also to a reminder that all successful marketing, including sustainability marketing, begins with an in-depth understanding of the target market. Instructor points made in this discussion include: (1) That although pricing is uniform across brands, low-income consumers are willing to pay a higher price for Grupo Familia's products because of the value received from them; (2) that distribution into barrios involves local microentrepreneurs. This is clever because Grupo Familia leverages the creditability of local entrepreneurs who are trusted by barrio residents, while simultaneously protecting its delivery personnel from the violence and theft in the barrios themselves; and (3) that Grupo Familia uses market research to understand the buying behavior of this segment. At this point, discussion leaders can refer to Figure 2.2: The Sustainability Marketing Framework. Faculty can either draw the "Sustainability

Marketing" component on a white board or capture the essence of the discussion so far (see Figure 2.3), or they can simply refer to this element in Figure 2.2 as a way of summarizing the discussion points made so far.

Discussion can next turn to one of the distinctive features of sustainability marketing: The TBL. Instructors can ask: *How well does Group Familia consider each of the three Ps in the Triple Bottom Line?* Here students typically note first Grupo Familia's recycling efforts and its commitment to using 70% post-consumer content (Planet). Often this reflects student thinking that sustainability marketing is synonymous with green marketing. It is not, as the remaining discussion illustrates. An astute student will observe that Grupo Familia is exceeding its goal of using 70% recycled paper already with its new line of family towels made completely of recycled content. Other students will offer Table 2.2 as a strong indicator of Grupo Familia's TBL commitment. The most astute students will quote Andres Felipe's statement that begins, "We have declared in *Familia* that sustainability has a first name…," as an example of TBL thinking. They are correct. Should this occur, faculty may ask class to remember this insight, as it will be discussed more fully later on. Discussion leaders should note that "profit" in the TBL framework is always easiest to determine for public firms (Annual Reports, investor meetings, etc.) and often so is "planet" through sustainability reports. *But what about "people"?*

Grupo Familia's commitment to "people" is seen in its respect for recyclers (recicladores) and its Misión + Hogar program. Faculty can make several observations about Grupo Familia's relationship with recyclers: (1) Recyclers are essential to achieving the sustainability goal of using 70% post-consumer content in product manufacturing. Grupo Familia is driven here by an economic incentive not altruism; (2) Grupo Familia integrates recyclers into the entire supply chain, thereby linking an informal market activity (collecting paper waste) with a formal market activity (product manufacturing); and (3) the sincerity behind Grupo Familia's respect for recyclers is expressed not only in its purchase of post-consumer paper content from cooperatives for a fair price but also in its Facebook film, "Heroes of the Planet." This is an excellent moment to show this short film in class. This film honors recyclers and should be seen as an affirmative example of corporate social responsibility, a societal driver behind sustainability marketing.

Before moving onto a discussion of Misión + Hogar, faculty might pause and ask what seems to be a tangential question: *What's the rationale for creating the Petys brand? How does it fit into Grupo Familia's brand portfolio?* The answer is: Grupo Familia defines "family" in a broad, creative way. The case states that pets are often considered as family members. Discussion leaders are encouraged, at this point, to refer to a line in the case often overlooked by students: That recyclers often walk the streets with their pet dog. This is an opportunity to observe the very tight marketing link between brands and target market: Recyclers are included in Grupo Familia's low-income target market; yet Grupo Familia strategically binds its target market to its products across its *entire* brand portfolio and life cycle by having product lines for both two-legged and four-legged "family" members.

As an entry into the Misión + Hogar discussion, faculty can say: *Companies can place their manufacturing plants in a wide variety of locations. Why did Grupo Familia choose one of the poorest regions in the country, Cauca, for its production plant?* Based on the previous recycler discussion, students quickly develop the following points: Poverty in the district, more specifically in Guachené, embodies Grupo Familia's target market of low-income consumers. Additionally, cyclical unemployment and the city's large number of poor individuals gives Grupo Familia a ready and relatively low-wage labor force. Providing local residents with steady employment means families are able to buy Grupo Familia products. Also, choosing to build a manufacturing plant in Guachené becomes another tangible expression of responsible corporate citizenship: the quality of life improves for local residents through steady employment. Misión + Hogar further positions Grupo Familia as a good corporate citizen as the program's overall aim is to enhance the quality of life for community residents, some of whom many not be Grupo Familia employees. At this point, instructors can add the terms "people," "planet," and "profit" to the white board drawing and summarize these new discussion points, as given in Figure 2.3. Instructors can also draw or refer to the box labeled Sustainable Development with the word "stakeholders" identified. Discussion has noted several stakeholder groups: investors, community, and employees (see Figure 2.3).

Class discussion can now explore more fully sustainable development from the strategic marketing and enterprise strategy perspective. Experience indicates that enterprise aspects of marketing are often overlooked because marketing classes tend to focus on tactical considerations of marketing management (unless, of course, strategic marketing is the class topic). Faculty may begin by asking, *How important is the CEO in sustainability marketing*? For Grupo Familia, the answer is "very important." Instructors can make two key points here: (1) Grupo Familia has elevated sustainability to an enterprise level strategy. This is most evident in Andres Felipe Gómez-Salazar's statement that "We have declared in *Familia* that sustainability has a first name: There is financial sustainability, environmental sustainability and around the communities in which we operate, market sustainability and sustainability of people." Students should be asked to fully consider this statement's importance. Experience suggests that pausing the discussion for student reflection yields insightful responses. In essence, sustainable development has become the "strategic DNA" of Grupo Familia; it permeates and unifies everything that Grupo Familia does. (2) Andres Felipe's role as CEO affirms sustainable development as the key enterprise strategy at Grupo Familia, thereby creating a corporate culture that is perfectly aligned around sustainable development. Because sustainable development is concerned with balancing current needs against future ones, students can infer that Grupo Familia's commitment to using recycled content in products along with its community engagement efforts acknowledge that the actions it takes today will have a net positive effect on future generations.

Also worth noting is how Grupo Familia (by inference also under the leadership of a CEO) reframed innovation to be an enterprise strategy, which, in turn, supports Grupo Familia's entire brand portfolio with new products and new product lines (think Petys here). The CEO's role as champion for sustainable develop is essential. Strategic marketing and the Chief Marketing Officer's role fit seamlessly within Grupo Familia's corporate culture. Faculty can add these points to the Sustainable Development component on the white board, as shown in Figure 2.3, or verbally state the connection to Figure 2.2.

Faculty can conclude this case discussion with the Framework's one element yet to be discussed: The SDGs. This discussion is most efficiently handled by simply asking students to refer to Figure 2.1 and ask: *Which SDGs do Grupo Familia's sustainability marketing activities address*? Students readily state SDG #1: No Poverty, and then move on to propose SDG #4: Quality Education (Misión + Hogar), SDG #8: Decent Work (Production plant in Guachené and incorporation of recyclers in the supply chain), SDG #10: Reducing Inequalities (the impact of SDG #8), and SDG #12: Responsible Consumption and Production (TBL, the goal of having 70% recycled paper content in products, and centrality of innovation throughout the firm). Some student may infer that SDG #5: Gender Equality is also involved, as Grupo Familia presents itself as a fair, equitable, and morally just firm. These SDGs can be added to white board figure of the Framework, or stated as illustrations of sustainability marketing's commitment to the SDGs. Figure 2.3 summarizes the entire class discussion and concludes this teaching plan for the Grupo Familia case.

Active Learning

Active learning is a teaching approach that asks students to "to think, talk, listen, read, write and reflect about course content through problem-solving exercises" (Laverie, 2006, p. 60). As its name implies, active learning up-ends the one-way, often passive approach to developing student knowledge through lecture (Auster & Wylie, 2006). What follows are two active learning assignments that stress the linkage between the SDGs and sustainable development (see Figure 2.2).

Assignment 1. One perspective on the SDGs is to view all 17 goals as the complete spectrum of biological, environmental, and commercial life on earth. As such, the SDGs cover a wide range of development sectors including food, health, education, housing, jobs, climate, energy, water, and sanitation. A wide range of businesses (from SMEs, to corporations, to entrepreneurial start-ups, to social enterprises/hybrid organizations) already exist to serve consumers covered by each of the 17 SDGs.

This assignment asks students (1) to find an existing business whose products/services meet the needs of consumers that support either one or several SDGs and then (2) to evaluate that business's sustainability marketing efforts through the analytic lens of Figure 2.2. Students can write a paper, present a PowerPoint/Prezi, or make an oral presentation (in the classroom or online). The whiteboard diagram for the Grupo Familia mini-case (see Figure 2.3) is an example of what students may produce for their specific business. Although students are adept at doing online research, marketing educators could direct students to the *Business for 2030* website (http://www.businessfor2030.org/), where students can find well-known multinational enterprises already working to support the SDGs. Marketing faculty wanting to prompt or introduce students to South Asian companies supporting the SDGs can suggest students begin by looking at *Indian*

Solutions for the World to Achieve the SDGs (https://smartnet.niua.org/sites/default/files/resources/60786.cs641sdgsreportweb.pdf). The essential insight to be gained from this assignment is that, while many businesses might be well-known to students already, typically, they are not analyzed in relation to sustainable development. One of sustainability marketing's contributions is thus to widen the lens through which marketing strategy is evaluated.

Assignment 2. This active learning assignment asks students to find a product category of interest and to think through the impact that that product category has on the SDGs. A natural student inclination is to select familiar technology products, such as laptops, smartphones, etc. However, this assignment works best with simple products, for example, bottled water (Pallant, Choate, & Haywood, 2020). Bottled water is a low cost, readily available, very well-known, simple product that is often used in marketing textbooks to illustrate basic marketing principles (e.g., consumer purchase motivation, product differentiation, target marketing, positioning, branding, etc.). Bottled water consumption also has important environmental implications not only through the manufacture of its plastic packaging but also in the bottle's disposal as waste. This assignment's end product is a table similar to Table 2.3, which uses bottled water as the analyzed product. Students should not only link their chosen product with selected SDGs but also should raise marketing issues relevant to each SDG chosen. Through this active learning assignment, students come to realize that sustainability marketers have obligations to understand the broad impacts their product has on the SDGs.

Critical Thinking

Critical thinking skills are often said to be at the heart of all education (Kuhn, 1999) because critical thinking skills weave together logic, reasoning, and creativity in the deliberative process of analysis so that individuals can make better decisions and judgments. "Critical thinking allows students to reach beyond a single perspective, to challenge assumptions, and to better analyze a wide range of challenges and problems" (Roy & Macchiette, 2005, p. 265). Discussing sustainable consumption from a macromarketing perspective not only completes this chapter's analysis of each of components shown in Figure 2.2 but also presents students with a grand challenge (George, Howard-Grenville, Joshi, & Tihanyi, 2016) that requires critical thinking skills to resolve.

TABLE 2.3

Spectrum of Sustainability Marketing Issues Raised by Bottled Water in Relation to Selected SDGs

SDG	Sustainability Marketing Issue
SDG #3: Good Health and Well-Being	Understanding consumer perceptions about the safety of bottled water in relation to tap water.
SDG #6: Clean Water and Sanitation	Exploration of what constitutes "clean" water. What should those standards be? How does this perception vary by country (developed versus still developing)?
SDG #9: Industry Innovation and Infrastructure	How might industrial processes incentive reusable water containers? What innovations can occur in packaging to reduce waste? Can completely recyclable bottled water containers be developed?
SDG #10: Reduced Inequalities	How can bottled water marketed to subsistence and base of the pyramid consumers be priced so as to reduce the percentage of the household budget devoted to just bottle water?
SDG #12: Sustainable Consumption and Production	Bottled water companies have been known to greenwash the sustainability of their product. How can this be combatted?
SDG #14: Life below Water	Emptied water bottles end up in oceans and rivers, often harming and killing marine life. How can this be avoided?
SDG #15: Life on Land	Production process for plastic bottles emits toxins and consumes significant amounts of energy. Are there more sustainable ways of manufacturing?

Source: Adapted from Pallant, Choate & Haywood (2020).

Sustainable consumption grew out of the realization that while consumption wants may be infinite (to use marketing-oriented language), the earth's ability to respond to those infinite wants is not (Kotler, 2011). Sustainable consumption is sometimes termed "voluntary simplicity," "sufficient consumption," "frugality", "downshifting," "mindful," "slow," and "ethical and responsible consumption." Whatever the term used, each concept suggests that collective consumption behaviors must change if society wants to slow down or ideally eliminate the destructive effects that aggregate consumption decisions have on the earth's ecosystem. Immediately, several points are worth noting: (1) Sustainable consumption is not a choice between consuming and not consuming. To consume is to live (Borgmann, 2000); (2) sustainable consumption concerns purposefully living within limits. Because of this, sustainable consumption is tightly bound to sustainable development, since the fundamental premise of sustainable development is not to disadvantage future generations to meet their own consumption needs because of decisions individuals and society make today; and (3) sustainable consumption from a macromarketing perspective is about marketing systems and larger social issues, such as how "marketing" can help create a fair, just society in which all individuals can prosper.

Because macromarketing seeks to "improve marketing strategies and policies that affect social welfare" (Fisk, 1981, p. 3), marketing faculty are strongly encouraged to keep analyses and discussions focused squarely on society and creating social good. Classroom experience suggests that it is easy for discussions to slide into debate about what individual firms are doing rather than looking at the larger, social issues that macromarketing mandates. The gravitation pull of individual firms and consumers is very strong in marketing (Shapiro et al. 2020). To help students grapple with the complex issues of macromarketing's relationship with sustainable consumption, the following two critical thinking exercises are proposed.

Macromarketing Implications of SDG 12

SDG 12 provides entrée into the macromarketing issue of sustainable consumption. SDG 12, which is "to ensure sustainable consumption and production," includes eight specific targets that are related to "production efficiency, in relation to use of natural resources (12.2), food production and supply related losses (12.3), management of chemicals and wastes (12.4), sustainable corporate practices and reporting (12.6) and sustainable public procurement (12.7)" (Gasper, Shah & Tankha, 2019, p. 85). Any of these specific subgoals can become a critical thinking assignment that asks students to understand and thoroughly evaluate "how well markets, marketing, and marketing systems" (Shapiro et al. 2020) ultimately serve society.

One student assignment, for example, can be to tackle food waste (SDG 12.3). Most individuals have had personal and sometimes even organizational experience with this grand challenge. Gustavsson et al. (2011) report that, globally, approximately 1.3 billion tons of food grown for human consumption per year is either wasted or lost. Macromarketing poses these questions in relation to food waste and food loss:

- What role do distribution systems play in the amount of food lost to supply chain inefficiencies? How can distribution systems

be redesigned to increase efficiencies and reduce the amount of lost food?
- Has traditional marketing created consumer perceptions that only "perfect food" is consumable? How can social marketing and consumer attitudinal change strategies be designed to increase consumer acceptance of imperfect or "ugly food" (Calvo-Porral, Medín, & Losada-López, 2017)?
- Using a hierarchical model of food waste, such as figure 2.4 in Papargyropoulou et al. (2014),[3] how can marketing systems be redesigned to reduce food waste?

Any of the above questions can prompt sustained, critical thinking about "causes," "solutions," and marketing's role in both creating and then resolving them. Student teams could work on any question and present their findings in formal papers, team presentations, and video analyses.

Transformative Insights into Consumer Behavior

A macromarketer's perspective on consumption explores the deeper, more fundamental drivers of consumption behavior and the effect that collective behavior has on society's well-being (Mish & Miller, 2014). Macromarketers explore questions such as: What role does consumption play in our collective lives? Why do some society's value material possessions more than others? Does consuming more and having more "things" lead to greater societal happiness and enhanced well-being? As Kilbourne, McDonagh and Prothero (1997) note, "The prevailing belief within industrial societies is that the sure and only road to happiness is through consumption" (p. 4). Jackson (2005) synthesizes these questions by asking simply:

> Why do we consume? What do we expect to gain from material goods? How successful are we in meeting those expectations? What constrains our choices? And what drives our expectations in the first place? (p. 20)

Marketing faculty can take any of the above questions about the ideology of consumption (Mick, Broniarczyk & Haidt, 2004) and turn them into a critical thinking assignment. The assignment asks students to write a thoughtful, detailed, reflective paper that connects their chosen question with sustainable consumption.

Similarly, students could be asked to use the following quote as a paper prompt that similarly explores sustainable consumption: "What is required is that consumers shift both the quality and the quantity of consumption. We must consume better and less if sustainability is to become a reality" (Kilbourne & Mittelstaedt, 2012, p. 296). To make this assignment more concrete and tangible, students could be asked to consider the fast fashion industry an industry with which they might have had personal experience, and consider what needs to happen from a macromarketing perspective (Cline, 2013), so that consumers not only consume less and better in the service of sustainable consumption.

Some scholars consider sustainable consumption to be *the* defining issue for sustainability marketing (McDonagh, Dobscha, & Prothero, 2012; Pereira Heath & Chatzidakis, 2012). They acknowledge that firm-level sustainability marketing is an excellent starting point. As Richardson (2020) observes, "Every time a consumer makes a decision, it has the potential to contribute to a more or less sustainable pattern of consumption" (p. 43). Admittedly, many firms are making significant advances to deliver more sustainable customer solutions to their target markets. These companies are often profiled online and frequently become the focal company in more rigorous academic case study.

Yet as Belz and Peattie (2009) state, these companies "however well intentioned, may struggle to make substantive progress toward more sustainable consumption and production" (p. 280). Belz and Peattie's (2009) statement merges seamlessly with scholars who are increasingly asking marketing researchers (Gossen, Ziesemer, & Schrader, 2019; Kemper, Hall, & Ballantine, 2019) and marketing educators (Heath & McKechnie, 2019; Kemper, Ballantine, & Hall, 2020) to reconsider the assumptions embedded within traditional marketing practice in light of sustainability's grand societal challenges. These scholars would, in fact, point to this chapter's beginning section, A Very Short Review of Marketing Fundamentals, as precisely the starting point to reexamine the implicit beliefs within marketing as a discipline (Kemper, Hall, & Ballantine, 2019). Critical discussion asks educators and students to challenge this perspective in relation to sustainability marketing (Kemper, Ballantine, & Hall, 2020). Essentially, has "marketing" fostered such a deep-seated consumer-oriented vision of life that nothing less than a transformation in thinking about the purposes of "marketing" itself is needed?

> We need to be concerned more with why we operate an economy, before focusing narrowly on how to be more efficient, because economic growth is not the purpose of life. As in so many aspects of life, the dominance of quantity thinking over quality thinking is unhealthy. It is self-evidently pointless—stupid even—to consume more to generate wealth in an attempt to fix the problems caused by consuming more!
>
> (Varey, 2010, p. 114)

Sustainability marketing has an important role to play in prompting that discussion.

CONCLUSION

The center piece of this chapter is the development of an integrated Sustainability Marketing Framework for use in the marketing classroom (see Figure 2.2). As an emerging topic within the marketing discipline, the chapter summarized three influences that support the need for sustainability marketing: (1) CEOs now understand that integrated, corporate-level strategies intentionally built around sustainability are essential for firm survival; (2) changes in consumer beliefs and values increasingly demand sustainable products and services thereby becoming markets of opportunity for market-responsive firms; and (3) there has been an evolution in corporate social responsibility that requires all firms to be stewards of the planet's natural, economic, and human capital. The chapter then presented a comprehensive definition of sustainability marketing (Belz & Peattie, 2009). Based on this definition, the chapter presented the Sustainability Marketing Framework.

The Sustainability Marketing Framework highlights sustainable marketing's three unique features that distinguish it from traditional marketing management: sustainable development as expressed by the SDGs, a stakeholder orientation, and TBL measures. The Sustainability Marketing Framework seeks to integrate macromarketing into its model and is designed to be quickly hand-drawn in class or easily copied into faculty and/or student presentations and papers. The Framework also acknowledges that many of its elements overlap with social marketing because both perspectives use marketing strategies for the benefit of society. Three applications follow and these applications are structured around common classroom learning activities: Case analysis, active learning exercises, and critical thinking assignments. The chapter concludes with a reflection on sustainable consumption as *the* sustainable marketing challenge and how this challenge has the potential for researchers and educators to rethink the implicit assumption of marketing.

Overall, the chapter attempts to support marketing educators, who can use the chapter to introduce or extend sustainability marketing ideas in their classrooms (Bhattacharyya, Dash, Hewege, Makam, & Lim, 2021). The chapter also has a student-learning focus with assignment suggestions variously designed to highlight different aspects of sustainability marketing by applying different pedagogical approaches to learning. The assignments concentrate on "real-world contexts for critically examining sustainability related to marketing and consumption" (Heath & McKechnie, 2019, p. 112). The chapter's final statement, though, comes from Paul Polman, former CEO, Unilever, who crystalized the sustainable development challenge of which sustainable marketing is a part, when he said, "Business cannot succeed in societies that fail."

FUNDING DECLARATION

Not applicable.

ACKNOWLEDGMENTS

The author is appreciative of feedback from anonymous reviewers for their insights in refining this chapter.

CREDIT AUTHOR STATEMENT

None.

NOTES

1. Video examples of Grupo Familia's "Papel Planeta" along with its use of the FSC certification can be found at: https://www.youtube.com/watch?v=QP4W-8jNiHw; https://marketingtoolkit.fsc.org/media/2195; https://www.youtube.com/watch?v=GhXwtKukodw&feature=emb_logo.
2. The video can be viewed at: https://www.facebook.com/FamiliaColombia/videos/h%C3%A9roes-del-planeta/961303557653240/).
3. A freely available version of Papargyropoulou et al.'s (2014) paper that includes figure 4 can be found at: http://eprints.whiterose.ac.uk/79194/1/accepted%2520manuscript.pdf.

REFERENCES

Andreasen, A. R. (1994). Social marketing: Its definition and domain. *Journal of Public Policy & Marketing*, 13(1), 108–114.

Auster, E. R., & Wylie, K. K. (2006). Creating active learning in the classroom: A systematic approach. *Journal of Management Education*, 30(2), 333–353.

Belz, F. M., & Peattie, K. (2009). *Sustainability marketing*. Chichester, UK: Wiley & Sons.

Bhattacharyya, J., Dash, M. J., Hewege, C., Makam, S. B., & Lim, W. M. (2021). *Social and sustainability marketing: A casebook for reaching your socially responsible consumers through marketing science*. Abingdon/Delhi: Routledge.

Borgmann, A. (2000). The moral complexion of consumption. *Journal of Consumer Research*, 26(4), 418–422.

Brammer, S., Branicki, L., Linnenluecke, M., & Smith, T. (2019). Grand challenges in management research: Attributes, achievements, and advancement. *Australian Journal of Management*, 44(4), 517–533.

Brennan, R. (2009). Using case studies in university-level marketing education. *Marketing Intelligence & Planning*, 27(4), 467–473.

Business & Sustainable Development Commission. (2019). *Better business, better world*. Retrievable from: http://report.businesscommission.org/uploads/BetterBizBetterWorld_170215_012417.pdf

Calvo-Porral, C., Medín, A. F., & Losada-López, C. (2017). Can marketing help in tackling food waste? Proposals in developed countries. *Journal of Food Products Marketing*, 23(1), 42–60.

Cline, E. L. (2013). *Overdressed: The shockingly high cost of cheap fashion*. New York: Portfolio.

Collins, E. M., & Kearins, K. (2010). Delivering on sustainability's global and local orientation. *Academy of Management Learning & Education*, 9(3), 499–506.

Crittenden, V. L., Crittenden, W. F., & Hawes, J. M. (1999). The facilitation and use of student teams in the case analysis process. *Marketing Education Review*, 9(3), 15–23.

Donaldson, T., & Dunfee, T. W. (1999). *Ties that bind: A social contract approach to business ethics*. Boston, MA: Harvard Business School.

Drucker, P. (1954). *The practice of management*. New York, NY: Harper & Row Publishers.

Drucker, P. (1990/2011). *Managing the non-profit organization: Principles and practices*. Abington, UK: Routledge.

Elkington, J. (1998). Partnerships from cannibals with forks: The triple bottom line of 21st-century business. *Environmental Quality Management*, 8(1), 37–51.

Emery, B. (2012). *Sustainable marketing*. Harlow, UK: Pearson Education.

Fisk, G. (1981). An invitation to participate in affairs of the Journal of Macromarketing. *Journal of Macromarketing*, 1(1), 3–6.

Fuller D. (1999). *Sustainable marketing: Managerial–ecological issues*. Thousand Oaks, CA: SAGE.

Gasper, D., Shah, A., & Tankha, S. (2019). The framing of sustainable consumption and production in SDG 12. *Global Policy*, 10, 83–95.

George, G., Howard-Grenville, J., Joshi, A., & Tihanyi, L. (2016). Understanding and tackling societal grand challenges through management research. *Academy of Management Journal*, 59(6), 1880–1895.

Gossen, M., Ziesemer, F., & Schrader, U. (2019). Why and how commercial marketing should promote sufficient consumption: a systematic literature review. *Journal of Macromarketing*, 39(3), 252–269.

Grant, J. (2008). Green marketing. *Strategic Direction*, 24(6), 25–27.

Grupo Familia Annual Report. (2017). Informe de Gestión, Retrievable from https://www.grupofamilia.com.co/es/noticias/PublishingImages/Lists/EntradasDeBlog/AllPosts/GESTION%20GRUPO%20FAMILIA%202017.pdf

Gustavsson, J., Cederberg, C., Sonesson, U., van Otterdijk, R., & Meybeck, A. (2011). *Global food losses and food waste: Extent, causes and prevention*. Rome: FAO.

Hahn, T., Pinkse, J., Preuss, L., & Figge, F. (2015). Tensions in corporate sustainability: Towards an integrative framework. *Journal of Business Ethics*, 127(2), 297–316.

Haws, K. L., Winterich, K. P., & Naylor, R. W. (2014). Seeing the world through GREEN-tinted glasses: Green consumption values and responses to environmentally friendly products. *Journal of Consumer Psychology*, 24(3), 336–354.

Hayward, R., Lee, J., Keeble, J., McNamara, R., Hall, C., Cruse, S., Gupta, P., & Robinson, E. (2013). The UN Global Compact-Accenture CEO study on sustainability 2013. *UN Global Compact Reports*, 5(3), 1–60.

Heath, T., & McKechnie, S. (2019). Sustainability in marketing. In K. Amaeshi, J. Muthuri, & C. Ogbebcie (Eds.), *Incorporating Sustainability in Management Education: An Interdisciplinary Approach* (pp. 105–131). Cham: Palgrave Macmillan.

Hoek, M. (2018). *The trillion dollar shift*. Abingdon, UK: Routledge.

Jackson, T. (2005). Live better by consuming less? Is there a "double dividend" in sustainable consumption? *Journal of Industrial Ecology*, 9(1–2), 19–36.

Kemper, J. A., Ballantine, P. W., & Hall, C. M. (2020). The role that marketing academics play in advancing sustainability education and research. *Journal of Cleaner Production*, 248, 119229.

Kemper, J. A., Hall, C. M., & Ballantine, P. W. (2019). Marketing and sustainability: Business as usual or changing worldviews? *Sustainability*, http://dx.doi.org/10.3390/su11030780.

Kilbourne, W., McDonagh, P., & Prothero, A. (1997). Sustainable consumption and the quality of life: A macromarketing challenge to the dominant social paradigm. *Journal of Macromarketing*, 17(1), 4–24.

Kilbourne, W., & Mittelstaedt, J. (2012). From profligacy to sustainability: Can we get there from here? In D. Mick, S. Pettigrew, C. Pechmann, & J. Ozanne (Eds.), *Transformative Consumer Research for Personal and Collective Well-being* (pp. 283–300). Abingdon, UK: Routledge.

Kiron, D., Kruschwitz, N., Haanaes, K., & Velken, I. (2012). Sustainability nears a tipping point. *MIT Sloan Management Review*, 53(2), 69–74.

Klebba, J. M., & Hamilton, J. G. (2007). Structured case analysis: Developing critical thinking skills in a marketing case course. *Journal of Marketing Education*, 29(2), 132–139.

Kotler, K. & Armstrong, G. (2014). *Principles of marketing* (15th ed.). Upper Saddle River, NJ: Prentice Hall Pearson.

Kotler, P. (2011). Reinventing marketing to manage the environmental imperative. *Journal of Marketing*, 75(4), 132–135.

Kotler, P., & Levy, S. J. (1969). Broadening the concept of marketing. *Journal of Marketing*, 33(1), 10–15.

Kotler, P., & Zaltman, G. (1971). Social marketing: An approach to planned social change. *Journal of Marketing*, 35(3), 3–12.

Kuhlman, T., & Farrington, J. (2010). What is sustainability? *Sustainability*, 2(11), 3436–3448.

Kuhn, D. (1999). A developmental model of critical thinking. *Educational Researcher*, 28(2), 16–46.

Kumar, N. (2004). *Marketing as strategy: Understanding the CEO's agenda for driving growth and innovation*. Cambridge, MA: Harvard Business Press.

Kumar, V., Rahman, Z., & Kazmi, A. A. (2013). Sustainability marketing strategy: An analysis of recent literature. *Global Business Review*, 14(4), 601–625.

Lacy, P., Gupta, P., & Hayward, R. (2019). From incrementalism to transformation: Reflections on corporate sustainability from the UN Global Compact-Accenture CEO study. In G.L. Lennssen, & N.C. Smith (Eds), *Managing Sustainable Business* (pp. 505–518). Dordrecht: Springer.

Laverie, D. A. (2006). In-class active cooperative learning: A way to build knowledge and skills in marketing courses. *Marketing Education Review*, 16(2), 59–76.

Lee, A., & Edwards, M. G. (2013). *Marketing strategy: A life cycle approach*. Cambridge: Cambridge University Press.

Leenders, M. R., Erskine, J. A., & Mauffette-Leenders, L. A. (2001). *Writing cases*. University of Western Ontario: Richard Ivey School of Business.

Lélé, S. M. (1991). Sustainable development: A critical review. *World Development*, 19(6), 607–621.

Levitt, T. (1960). Marketing myopia. *Harvard Business Review*, 38(4), 45–56.

Lindgreen, A., Maon, F., Vanhamme, J., Florencio, B. P., Vallaster, C., & Strong, C. (Eds.). (2019). *Engaging with stakeholders: A relational perspective on responsible business*. Abington, UK: Routledge.

Martin, D., & Schouten, J. (2012). *Sustainable marketing*. Upper Saddle River, NJ: Prentice Hall/Pearson.

McDonagh, P., Dobscha, S., & Prothero, A. (2012). Sustainable consumption and production. In D. Mick, S. Pettigrew, C. Pechmann, & J. Ozanne (Eds.), *Transformative Consumer Research for Personal and Collective Well-being* (pp. 267–281). Abingdon, UK: Routledge.

Mick, D. G., Broniarczyk, S. M., & Haidt, J. (2004). Choose, choose, choose, choose, choose, choose, choose: Emerging and prospective research on the deleterious effects of living in consumer hyperchoice. *Journal of Business Ethics*, 52(2), 207–211.

Mish, J., & Miller, A. (2014). Marketing's contributions to a sustainable society. In P. E. Murphy & J. F. Sherry (Eds.), *Marketing and the common good: Essays from Notre Dame on Societal impact*. Abington, UK: Routledge.

Murphy, P. E., & Sherry Jr, J. F. (Eds.). (2014). *Marketing and the common good: Essays from Notre Dame on societal impact*. Abington, UK: Routledge.

Ottman, J. (2011). *The new rules of green marketing: Strategies, tools, and inspiration for sustainable branding*. San Francisco, CA: Berrett-Koehler Publishers.

Pallant, E., Choate, B., & Haywood, B. (2020). How do you teach undergraduate university students to contribute to UN SDGs 2030? In L. Filho, W. Salvia, A. L. Pretorius, R. W. Brandli, L. L. Manolas, E. Alves (Eds.), *Universities as Living Labs for Sustainable Development* (pp. 69–85). Cham: Springer.

Papargyropoulou, E., Lozano, R., Steinberger, J. K., Wright, N., & Bin Ujang, Z. (2014). The food waste hierarchy as a framework for the management of food surplus and food waste. *Journal of Cleaner Production*, 76, 106–115.

Paper Recycling Coalition. (n.d.). Paper recycling terms. Retrievable from: https://www.paperrecyclingcoalition.com/faqs/paper-recycling-terminology/

Peattie, K. (2011). Towards sustainability: Achieving marketing transformation-a retrospective comment. *Social Business*, 1(1), 85–104.

Pereira Heath, M. T., & Chatzidakis, A. (2012). 'Blame it on marketing': Consumers' views on unsustainable consumption. *International Journal of Consumer Studies*, 36(6), 656–667.

Pitt, L., Crittenden, V. L., Plangger, K., & Halvorson, W. (2012). Case teaching in the age of technological sophistication. *Journal of the Academy of Business Education*, 13, 77–94.

Ramaswamy, V. S., & Namakumari, S. (2018). *Marketing management: Global perspective, Indian context* (6th ed.). Delhi, India: SAGE.

Richardson, N. (2020). *Sustainable marketing planning*. Abingdon, UK: Routledge.

Roy, A., & Macchiette, B. (2005). Debating the issues: A tool for augmenting critical thinking skills of marketing students. *Journal of Marketing Education*, 27(3), 264–276.

Saunders, S. G., Barrington, D. I., & Sridharan, S. (2015). Redefining social marketing: Beyond behaviour change. *Journal of Social Marketing*, 5(2), 160–168.

Shapiro, S., Beninger, S., Domegan, C., Reppel, A., Stanton, J., & Watson, F. (2020). Macromarketing pedagogy: Empowering students to achieve a sustainable world. *Journal of Macromarketing*, DOI:0276146720949637.

Shocker, A. D., & Sethi, S. P. (1974). An approach to incorporating social preferences in developing corporate action strategies. In S. P. Sethi (Ed.), *The Unstable ground: Corporate social policy in a dynamic society*. Los Angeles: Melville Publishing.

Solow, R. M. (1992). An almost practical step toward sustainability. *Invited lecture on the occasion of the fortieth anniversary of resources for the future*, Washington, DC.

Sustainable Development Indicator Group. (1996). *Productive capacity definition*. Retrievable from: https://www.hq.nasa.gov/iwgsdi/Ecological_Capacity.html#:~:text=1.6%20Productive%20Capacity%3A%20The%20ability,the%20surface%20of%20the%20Ecosystem

Truss, A., Marshall, R., & Blair-Stevens, C. (2010). A history of social marketing. In J. French, C. Blair-Stevens, D. McVey, & R. Merritt (Eds.), *Social marketing and public health: theory and practice* (pp. 20–28). Oxford: Oxford University Press.

Van Tulder, R., Van Tilburg, R., Francken, M., & Da Rosa, A. (2014). *Managing the transition to a sustainable enterprise: Lessons from frontrunner companies*. Abington, UK: Routledge.

Varey, R. J. (2010). Marketing means and ends for a sustainable society: A welfare agenda for transformative change. *Journal of Macromarketing*, 30(2), 112–126.

Vélez-Zapata, C., Cortés, J.A., & Rosenbloom, A. (-2020). What does it mean to be a Latin family? Achievements on sustainability of the Multi-Latina Grupo Familia. In P. Flynn, M. Gudić, & T. K. Tan (Eds.), *Global champions of sustainable development* (pp. 9–19). Abington, UK: Routledge.

WCED. (1987). *Our common future*. Oxford: Oxford University Press.

Yosie, T. F., Simmons, P. J., & Ashken, S. (2016). *Sustainability and the modern CMO*. Retrievable from: http://www.corporateecoforum.com/wp-content/uploads/2017/01/Sustainability-and-the-CMO_FINAL.pdf

3

Cascades: What Is It and How Did It Reach Sustainability in a Highly Competitive Sector?

Myriam Ertz
LaboNFC, Department of Economics and Administrative Sciences, Université du Québec à Chicoutimi, Canada

CONTENTS

Learning Objectives	54
Themes and Tools Used	54
Theoretical Background	54
Introduction	56
Development and Positioning of the Company	59
Factors to Successful Performance in Sustainability	62
Family Culture and Values	64
A Will Supported by Real Planning with Stakeholders	68
Reuse Philosophy	72
Resource Optimization	73
Summary	74
Discussion	74
Conclusion	79
Lessons Learned	79
Discussion Questions	80
Project/Activity-Based Assignment/Exercise	81
Funding Declaration	81
Acknowledgment	81
Credit Author Statement	81
References	81

LEARNING OBJECTIVES

After reading the chapter, the reader should be able to:

- Explore the notion of sustainable development in the case of a for-profit organization
- Identify how the principles of sustainable development may be implemented in practice to spur sustainability both in production and consumption
- Examine the key success factors for a successful implementation of sustainable development
- Assess the importance of culture and values, stakeholder management, reuse philosophy and resource efficiency to foster corporate growth within the sustainable development ethos
- Determine the challenges and perspectives faced by a company committed to sustainable development over the long run

THEMES AND TOOLS USED

Content analysis of secondary data.

THEORETICAL BACKGROUND

This case study adopts the sustainable development framework which was originally outlined in the Brundtland Report (1987). The concept of sustainability revolves around the idea of a "triple bottom line" (TBL)—the social, environmental and economic components of sustainable practices (Elkington, 1997, 1998a, 1998b). This theoretical model originally aimed at measuring corporate performance, taking into consideration not only the traditional financial bottom line but also less quantifiable indicators that measure social and environmental impact categories, i.e., the social bottom line and the environmental bottom line.

The first dimension of economic sustainability refers to business practices that sustain long-term economic and financial growth while avoiding negative impacts on the social, environmental and cultural aspects to the communities in which the business operates (UMW, 2015). The second dimension of environmental sustainability should be considered from a maintenance of natural capital viewpoint. To Daly (1990), for renewable resources, it refers to the rate of harvest that does not exceed the rate of regeneration (sustainable yield). For pollution, it refers to the rates of waste production that do not surpass the assimilation capacity of the environment; and for nonrenewable resources, it refers to the depletion of resources that is being compensated by the development of renewable substitutes for that resources.

The last dimension of social sustainability refers to the identification and management of business impacts on stakeholders in the supply chain and of the organization at large. It deals with the recognition of the significance of the organization's relationship with individuals, communities and society (ADEC, 2020).

Recently, some frameworks also included a fourth cultural dimension (ADEC, 2020), or alternatively a governance dimension. However, in this study, the focus will be put on the original triple bottom line framework.

This framework presents two general advantages for methodological accuracy in assessing sustainability practices. First, the TBL framework highlights the relationships among the three main elements of sustainability (Elkington, 1997, 1998a). Visually, the three bottom lines can be mapped as three circles, which overlap in a Venn diagram. Ideally, a company should seek to operate at the intersection of this diagram, where all the three bottom lines are satisfied.

Second, each dimension (i.e., circle) can be measured by specific and measurable reference points (Slaper and Hall, 2011). In fact, the three dimensions of sustainability can be measured using indicators that may vary across countries or industries but sharing the commonality of quantitative reporting and assessment that is critical for managerial decision-making. This has been proven very useful in a managerial context, where managers are increasingly under pressure for providing measures of the impact and success of their programs and activities.

Interest in the triple bottom line has, therefore, increased across for-profit, nonprofit and government sectors (Rezapouraghdam et al. 2018). Researchers and managers have adopted the TBL sustainability framework to evaluate their sustainability performance (Ghannadpour et al. 2020). Although the purpose of this case study is not to measure specific sustainability dimensions of Cascades, these two features show that the sustainable development framework is a well-known and well-implemented one, across the globe and disciplines. Therefore, it constitutes a useful framework for this research as well. In sum, the TBL framework is well-suited for the purpose of this research since it emphasizes both the tripartite and measurable aspects of the sustainability concept.

Companies are increasingly under pressure to deliver not only financially but also environmentally and socially. This can become puzzling and confusing for managers. Yet, this case study shows that reconciling the diverse pieces of the sustainability puzzle, albeit challenging, is feasible. Better, it may also ensure sustainable competitive advantage on the market, even in an international business environment. This study case focuses on Cascades, a Quebec Canadian company that has thrived on implementing its motto that everything must and can be recycled while people must be taken care of. The case presents the four pillars that enabled Cascades to implement the principles of sustainable development across its three encompassing dimensions for business growth.

INTRODUCTION

This case study aims at showing how a company was able to reconcile successfully the three dimensions of sustainable development.

More specifically, the study presents the four key success factors which enabled that success, namely culture and values, stakeholder management, reuse philosophy and resource efficiency. Furthermore, the study also presents challenges and perspectives faced by the company to uphold its sustainability standards in the future. Recommendations for business practitioners are also provided.

Cascades is a Quebec paper manufacturer founded in 1964 by Bernard Lemaire and his brothers Laurent and Antonio. The three brothers revived an old disused mill of the Dominion Paper Cie in Kingsey Falls, in the province of Quebec, Canada. This is the birth of Papier Cascades Inc. and the company headquarters have remained located at Kinsey Falls, Quebec, since then. After almost six decades in business, Cascades works still predominantly in the fields of manufacturing, processing and marketing of packaging products and tissue paper mainly composed of recycled fibers. To be more specific, Cascades' activities are grouped into four groups (Cascades, 2020a):

1. *Cascades Groupe Carton Plat Europe* manufactures coated flat cardboard products made from virgin and recycled fibers such as folding boxes and microcannulated packaging.
2. *Cascades Containerboard Packaging* designs containerboard and flat cardboard products, corrugated cardboard packaging as well as folding boxes. Nearly 80% of the products of the containerboard group consist of recycled fibers.
3. *Cascades Specialty Products Group* concentrates its activities in four sectors: consumer product packaging, industrial packaging as well as recycling and recovery. Among other things, the company designs honeycomb cardboard and polystyrene trays for the food packaging industry.
4. *Cascades Tissue Group* manufactures and transforms tissue paper: facial tissue, paper towels, paper towels, toilet paper, napkins and paper rags. The company designs products for the retail and professional markets under the Cascades Fluff™, Cascades Tuff™ and Cascades PRO™ brands.

The activities of these four groups enable both the professional and the consumer market to produce and consume more responsibly. As such, Cascades is a well-known reference for companies or individuals seeking to develop more sustainable production or consumption practices.

Currently, the company has units in more than 90 countries, employs about 12,000 employees, and offers more than 500 products and services (Cascades, 2020a).

The story of Cascades is inspiring due to the pioneer nature of the business as geared toward sustainability, at a time when those ideas did not really exist. Before even founding Cascades in 1964, the Lemaire family "saw life in green", "Sustainable development was not yet in the customs of the time that these Quebecers shared a deep conviction: reuse, recover and recycle are gestures to be valued for the good of the environment and society" (Cascades, 2020a). Table 3.1 presents a timeline of this development of Cascades into sustainability.

As can be seen in the timeline, the Cascades company follows a form of triple bottom line (people, planet, profit) by seeking to ensure not only profits but also human well-being (people) and environment preservation (planet) (Slaper and Hall, 2011). For example, its specialization in manufacturing products out of recycled fibers is tangential to the environmental

TABLE 3.1

Development of Cascades toward Sustainability

Year	Key Developments
1964	Foundation of Papier Cascades Inc.
1967	The Lemaire family develops a profit-sharing program with employees.
1969	Cascades' fifth anniversary. The significant economic contribution to the communities surrounding the company is underscored in Kingsey Falls.
1971	Cascades creates *Forma-Pak*, a molded pulp factory made from 100% recycled fibers.
1974	After a decade in business, Cascades has 165 employees and generates sales of $7.2 million.
1976	Cascades moved to Cabano by setting up a containerboard factory there.
1977	Cascades specializes in the production of toilet paper, paper towels and industrial paper, all made from recycled fibers.
1983	Cascades is listed on the Montreal Stock Exchange and penetrates the United States.
1985	Creation of a Research and Development Center and acquisition of Cascades' first factory in Europe, at La Rochette, France.
1986	In North America, employee participation in the company is growing: more than half of them are shareholders of Cascades.
1995	Cascades enters the energy market by acquiring *Boralex*, an electricity producing company to develop and operate renewable energy.
1997	Cascades merges its activities with *Domtar* in the containerboard industry, giving birth to *Norampac Inc*.
2006	Cascades becomes the full owner of *Norampac*.
2008	Cascades merges with the Italian *Reno De Medici, SpA* for its cardboard manufacturing activities.

Source: Cascades (2020a). Une vision avant-gardiste. Available at: https://www.cascades.com/fr/developpement-durable/planete/entreprise-circulaire (accessed on 15-07-2020).

dimension (i.e., planet), while implementing a profit-sharing program with the employees shows a deep commitment to the social dimension (i.e., people). The economic dimension (i.e., profits) remains evidently important since the company has continuously sought to grow and expand its business either by organic growth, mergers or acquisitions. The company calls this the "planet, prosperity, partners triad". It comes therefore as no surprise that, to Corporate Knights, the company ranks among the 100 most responsible companies in the world, and ranks 49th. It is also the only company in the containers and packaging category appearing in the ranking, where there are only 12 Canadian companies (Cascades, 2020d).

This chapter focuses on the evolution of a company that has made a deep commitment to sustainable development from its inception. At a time when managers and business developers are increasingly keen on furthering their corporate social responsibility activities, a complete permeation of those principles within companies remains challenging. Moreover, it remains unclear what drivers do most particularly contribute to successful balance of the three dimensions of the triple mission: people, planet and profit. This chapter uses Cascades as a case study in order to provide crucial insights into these areas of inquiry.

The main body first delineates the development and positioning of the company within a sustainable development framework. That detailed account of the history and evolution of Cascades shows how a company may be successful while genuinely following a path toward sustainable development and remaining true to its core values. That understanding provides the basis for the discussion of the key success factors for successful sustainable performance. A final section outlines the challenges and perspectives faced by Cascades for its future development.

DEVELOPMENT AND POSITIONING OF THE COMPANY

In order to understand the importance, the problematic and the dynamics of sustainable development at Cascades, it is important to know its evolution. The following description of this evolution is largely based upon Bernatchez and Cossette's (2007) detailed report about Cascades. Cascades is built around people determined to build a business as prosperous as it is respectful of the human and natural resources that are sources of its success (Bernatchez and Cossette, 2007). Among them, the vision of the three Lemaire brothers helped transform the small family business of recycling, founded in 1957, into a multinational packaging and paper company. So, fraternity has always been a fundamental component of the organizational culture of Cascades.

In the year 1964, the Lemaires officially started the manufacturing of products made from recycled fiber when they buy, in Kingsey Falls (Quebec), an abandoned mill from the *Dominion Paper Cie*. This transaction gives birth to *Papier Cascades Inc*. In 1971, expansion begins for Cascades with the creation in Kingsey Falls of Cascades *Forma-Pak*, its first molded pulp mill made from 100% recycled fibers. This event kicks off dispatch of the development of a real paper complex in the heart of the small town of Estrie: from 1972 to 1977 were born successively *Kingsey Falls Paper* (multilayer cardboard), *Industries Cascades* (tissue papers), *Cascades Plastics* and *Cascades Conversion*. Paper crates Cascades in the region of Cabano, Quebec, come next.

At the beginning of the 1980s, the company is listed on the Montreal Stock Exchange. Later that year, Cascades entered in the United States with *Cascades Industries Inc.*, in Rockingham, North Carolina. In Quebec, Cascades launches into the manufacture of Kraft paper in East Angus and acquires, the following year, a boxboard factory in Jonquière, in the region of Saguenay-Lac-Saint-Jean, Quebec.

In 1985, the company took root on the European continent when it acquires the plant in boxboard from La Rochette, France. This business development will materialize in 1986 with the founding of *Cascades S.A.* and the acquisition of a second boxboard manufacturing plant in Blendecques, France. Between the time period from 1987 to 1989, other factories in Duffel, Belgium, and Djupafors, Sweden, as well as in Blendecques, France, will be added to the fold of Cascades. In North America, employee participation in the company is also growing: more than half of them are Cascades' shareholders.

At the beginning of the 1990s, *Cascades Énergie* emerged. This subsidiary then oversees a whole new natural gas cogeneration plant to generate the energy necessary for the proper functioning of the Kingsey Falls paper complex. This project will lead Cascades to invest more in the promising sector of energy with the acquisition, in 1995, of *Boralex*. In 2007, *Boralex* shares are 34% owned by Cascades. In August 2017, the Cascades company, at the origin of the creation of *Boralex*, sold all of its shares to the *Caisse de dépôt et placement du Québec (CDPQ)*, which thus acquired 17.3% of the company. In December 2017, the British magazine *A Word About Wind* ranked Patrick Lemaire, the President and CEO of *Boralex*, 58th in its annual ranking of the most influential players in the wind industry (PRnewswire, 2017). Since 2019, the company has participated in the "Green Electricity of Controlled Origin" (EVOC) offer, guaranteeing a renewable and French origin, marketed by *Plüm Énergie* (Environnement Magazine, 2019).

Cascades Énergie is the prelude to a series of acquisitions, resulting in the activities of Cascades being greatly diversified. The purchase of

Rolland and Paperboard Industries Corporation in 1992, *Perkins Papers* in 1995 and *Provincial Papers* in 1997, and the creation of *Norampac* with *Domtar*, all of these business expansions restructure Cascades in five distinct groups working in the fields of specialized packaging, containerboard, tissue papers, boxboards and fine papers. In 1997, the growth of Cascades in Europe expands further with the acquisition of a boxboard factory in Arnsberg, Germany. In 1998, a conversion unit was started in Wednesbury, England.

With the transition to the year 2000, the 2001 acquisition of factories in Pennsylvania and in Wisconsin happened. Two years later, the group moved to Alberta, New York, Arizona, Oregon and Tennessee. In 2004, the boxboard sector acquired *Dopaco, Inc.*, an important producer of packaging products for the fast food industry.

Appreciation of the Canadian dollar, higher energy and fiber prices, strong competition: the 2000s also presented challenges for Cascades, which can be seen in the obligation to implement a major rationalization plan. The process resulted in the closure of plants in Red Rock, Montreal, Thunder Bay, Buffalo, Pickering and Boissy-le-Châtel. In 2005, Cascades sold its distribution assets of fine papers (*Cascades Resources*) and tissue papers (*Wood Wyant*).

Recently, the company has earned several prizes that recognize its deep commitment to sustainable development. The *Kingsey Falls Public Market*, an initiative of Cascades, was recognized among the winning projects of the *2020 Novae Awards*, which recognizes impact initiatives that respond to social or environmental issues. The public market initiated by Cascades in 2015 with various local stakeholders promotes access to greater dietary diversity and short-circuit distribution (Cascades, 2020b). Cascades also received a *DUX Grand Prix,* Products category, for its most recent innovation in food packaging: a 100% recycled and 100% recyclable cardboard tray (Cascades, 2020b). Cascades is also part of the list of the 50 best corporate decisions for the planet published by *Earth Day Canada* and *Earth Day Initiative*, in collaboration with the media *Corporate Knights*. This recognition underlines Cascades' role as a pioneer in recycling (Cascades, 2020b). Last but not least, the *Cascades Tissue Group – Wisconsin* plant received the *2020 Excellence in Energy Efficiency Award* from the *FOCUS ON ENERGY®* organization. This award recognizes the efforts of the company since it has carried out more than ten projects for sound energy management over the past three years (Cascades, 2020b).

Moreover, the company ranks 16th in the ranking of the 50 best Canadian corporate citizens according to *Corporate Knights* magazine. Worldwide, Cascades is among the 100 most responsible companies according to *Corporate Knights* and ranks 49th. Cascades was also awarded the *Good Design Awards*, awarded annually by the Chicago Museum of Design and Architecture Athenaeum (Cascades, 2020d). Finally, Cascades was named one of the 2020 top 300 employers in Canada according to *Forbes* magazine (Cascades, 2020d).

FACTORS TO SUCCESSFUL PERFORMANCE IN SUSTAINABILITY

From the above, it emerges that sustainable development at Cascades fits logically within an innovative and strategic framework of business development. Companies often struggle to maintain a balance between the three poles of sustainability (people, planet, profit) not to mention governance with ESG (environment, society, governance) standards. Yet, this balance seems to have been found by Cascades. An external observer might argue that Cascades was able to unfold a strategy centered around sustainable development because it evolves in a niche market or a "safe haven" that is free from competitive pressure. Yet, quite the opposite is true. The pulp and paper industry has been under significant external constraints such as the rise in the cost of energy, the emergence of Chinese and Brazilian producers, the rise in the raw material of forest fiber and so on. In that sense, Cascades' road toward success seems all the more unsettling. Therefore, one might ask: Despite all the hurdles on the path toward sustainability, how did Cascades succeed into fostering sustainable development as a core mission and value in all of its business activities? The answer may lie in a series of four key success factors that unlocked Cascades' sustainable potential (Lévesque, 2019). By so doing, the company was able to infuse sustainability into its products, processes, systems, procedures and services.

The following subsections discuss into greater depth these four key success factors which have contributed to Cascades' success in implementing sustainable development in an industry itself grappling with sustainability issues. These four key success factors are also summarized in Figure 3.1.

Cascades • 63

1. Family culture and values

An open management style and worker benefits.

2. Real planning with stakeholders

Internal levers to establish and maintain the corporate culture.
External following of sustainable standards and dialogue with key stakeholders.

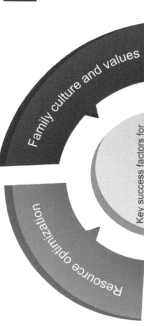

4. Resource optimization

Energy and resource efficiency management.

3. Reuse philosophy

Sustainable reuse, recycling and energy performance strategy.

FIGURE 3.1
Key success factors for implementing sustainable development at Cascades.

Family Culture and Values

As underscored in 2012 by Ken Peattie and Frank-Martin Belz in their best-selling book *Sustainability Marketing* (2nd ed.), nothing is as powerful as a strong sustainability-oriented culture. That type of culture typically values long-term orientation, human-scale exchanges and stakeholder perspectives. Culture is a vast concept but reflects a series of key underlying mental schemas such as morals, beliefs, norms, values, management philosophy or management style. Aktouf (1988) made a thorough ethnographic account of these abstract higher-order layers at Cascades during the company's aggressive expansion phase in the 1980s. However, strikingly, the first observation that he made was a very open management style, at all levels, and a freedom—even an incentive—to speak fully and generally. What is striking then is the small number of hierarchical levels: roughly four levels, from the simple employee to the vice presidents of the head office. Here are two elements of primary importance combining a whole set of practices which, for the classic manager, would seem a managerial heresy: short-circuiting of hierarchical layers and fair worker compensation.

With regard to the short-circuiting of the hierarchy:

> The short-circuiting of the hierarchy is common and perfectly accepted (a worker can go to headquarters and ask state a problem directly to the president or to the vice-president).
>
> The titles and designations of the positions are vague and considered by all as unimportant (one can simply be "in charge of" or "responsible for").
>
> Positions (including high-level positions) are not the subject of a description or job description. We just explain in broad outline what everyone will have to do.
>
> The doors of offices, wings, buildings … are all strictly open to whoever wants them. The general rule is to carefully avoid having to refuse an interview to anyone in the company, for whatever reason (the president, Bernard Lemaire, is in the habit of saying that if the employee considers that what he has to say is important, that is enough to listen to him).
>
> No information is considered confidential vis-à-vis employees. Apart from a few technical details, all the information is available, concerning production, profits, sales, etc. and most of it is posted (to the nearest month) in all workshops.
>
> In the factories, there are only workers, factory managers and their deputy. Outside office hours, including weekends, only the workers are alone.

The central machine attendant in the process is then "responsible", without being invested with official powers.

(as reported in Aktouf, 1988)

Turning now to the compensation offered to workers:

The wages paid are among the highest in the sector (for comparison, the Montreal paper converting company, of the same type as one of the Cascades factories in Kingsey Falls, pays exactly half as much). The average annual wages of the worker in 1986 is approximately $30,000.

A substantial portion of the profit before amortization and taxes (which can go, for the worker, up to 7% of his annual salary) is redistributed to all employees.

In 1982, on the occasion of the company's first public issue of shares, five shares were offered as gifts to employees for each year of service.

All employees can use, for personal purposes and without charge, any tool, installation or vehicle of the company, including weekends, they just need to inform.

Any employee who builds their house will receive, on request, free of charge, the building materials produced by the company.

All employees can acquire shares of the company which are financed two years without interest by the employer.

Everyone has the right to make mistakes and is encouraged to "try their hand", whatever the position or level.

Everyone is encouraged to make their point, to make suggestions. Everyone is listened to.

There is no separation of tasks or positions. Everyone can, if they wish—and are encouraged to do so—learn something else, replace, swap … it is enough that there is an agreement between the employees concerned.

The layoff of anyone is considered by senior management to be "a serious thing to avoid at all costs". We will first exhaust all possibilities, including the transfer from one factory to another.

There is a general meeting of each of the workshops at least once a year, including where there is no union, to discuss—directly with management and a representative of senior management (usually one of the Lemaire brothers) —working conditions, salary increases, grievances …

(as reported in Aktouf, 1988)

At a time (in the 1980s) when the principles of corporate downsizing, offshoring, outsourcing and reengineering started to gain traction in the industry, to spur profit maximization (Déry, 2010), Cascades chose a different path. An open management style and worker benefits that are quite comparable if not superior to the best in the industry and even elsewhere, became the trademark of the company and facilitated the realization of the other key factors of sustainability (and success).

In fact, Cascades knew a spectacular success at the start of the 1980s, in a sector in great difficulty, that of pulp and paper. To reach that success, the training, permanent self-training and self-qualification are for many dependent on the "philosophy management" specific to the company (Aktouf, 2006). In what the people of the Cascades Company themselves call "the Cascades philosophy", there are a few major constants in the mode of management and manager-directed relations, which have always been present there (Aktouf, 2006):

1. *Proximity, respect, trust...* are the terms that come up most systematically in the mouths of all the people working in (e.g., employees), or with (e.g., suppliers) the company, to qualify their way of relationships and of life in and in relation to Cascades, whether they are senior managers or core employees.
2. *Privacy, availability, humility...* are the words used to characterize the behavior of executives and senior managers toward any employee.
3. *Friendly, familial, fraternal, friendly, generous behavior...* above all "business" behavior is considered by all to be the "trademark" of the Cascades style.
4. Appropriation vis-à-vis the work situation, in terms of concern for a job well done, corporate image, performance (shared) results and so on, is a dominant feature of the organizational climate. Each and every one feels personally concerned with what is being done or not done, and what will be done.
5. Visibility and transparency-franchise-consultation in everything related to achievements, accounts, projects and decisions are considered as immovable assets.

In sum, this is "shared power" at Cascades. It means that the place of real decision-making organization, remains the basic employee and his team among peers, the "hierarchy" (reduced moreover to 3 or 4 levels maximum everywhere) being there to first "work as everyone" and then to answer the operators' questions and needs to "do their job well" (Aktouf, 2006). All the members of the hierarchy, at all levels and at any time (e.g., at home, during night shifts, on public holidays), are likely to be "challenged" by any employee who feels the need.

Cascades' management method, largely given as "shared management", is therefore, in a nutshell, a management method which presents itself as genuinely "participatory" (Aktouf, 2006). That is to say, as a "humanistic approach to management", participation does not stop only at the operational aspects of the management of production but extends widely to the sharing of reflection on future strategies and projects, on the ins and outs of decisions taken or to be taken, on the corrective measures to be made along the way to the achievements in progress, on the rates and destinations of the profits made, plant by plant (Aktouf, 2006). All information of all kinds (e.g., technical, accounting, financial, strategic) is shared in real time, with whoever wants it, when they want it. Additionally, who wants it, can be "transmitter" of information and be guaranteed of the appropriate listening, even at the highest levels (see also Aktouf, 2006, for a detailed review). Mario Plourde, current CEO of Cascades since 2016, who manages 11,000 employees worldwide and 4,500 in Quebec wraps up the philosophy eloquently as follows (Laperrière, 2016):

> We continue to do what we have always done. We do integration days when we have new executives. We continue our way of communicating with employees, with formal and informal meetings, our open-door policy. We don't like to hide behind a door. We want supervisors to be accessible, to be on the floor, for our employees to be near them. As a company, we keep the commitments we had at the beginning. We are a company that was born from sustainable development and we are staying the course. We are socially engaged people. We support hospitals, schools, athletes ... We haven't changed the Cascades formula, because it's a good formula that has served us well for 50 years.

All this makes this company among the most stable in terms of jobs, the most prosperous in terms of expansion and results, in its sector (despite that sector being reputed to be cyclical and having had difficult times for over a decade).

However, while these are important principles that imbue the corporate culture, this does not mean that the company never had to make difficult choices. But even in hard times, the persistence of Cascades to hold to its key values and principles tells a lot about the truthfulness of their approach. In 2009, *Norampac*, a division of Cascades decided that the production capacities of the company had outgrown demand, which fell. In fact, the economic crisis had hurt the factory, which produces corrugated cardboard boxes used, among other things, for shipping (Drolet, 2009). Cascades' customers produced fewer products in Quebec or had fewer requests so that the company ended up with less demand. It was then necessary to reduce production to be more competitive and the older facilities, such as the one in Quebec City, were the first to be closed (Drolet, 2009). Remarkably, the company still tried to relocate or transfer its employees to other factories.

Despite the goodwill, though, many factories faced similar difficulties and could not absorb the worker surplus. The *Norampac* division of Cascades, alone had, therefore, no other choice then to layoff almost a thousand employees over the year 2009 (Drolet, 2009). It is worth mentioning that in contrast to some multinationals laying off employees after record-high profits, Cascades never did so. For example, the 2018 results shattered records both in terms of sales and operating profit before amortization, yet, no employee was laid off to maximize those records even further in 2019 (Noël, 2019). In fact, the only cases of layoffs occurred in such extreme situations as the wake of the global financial crisis, which jeopardized the very survival of the whole company itself.

A Will Supported by Real Planning with Stakeholders

The will to sustainability is notably enshrined in Cascades' sustainable development approach as follows (Cascades, 2020a):

> Cascades values the opinion of its stakeholders in order to clearly identify the core priorities of its sustainable development approach.

Cascades works therefore like a family not only internally with its employees but also externally by considering itself as a member of a broader family-like network consisting of stakeholders and partners. However, this focus does not manifest itself into a vacuum or in unapplied corporate memos. Rather, it is carefully channeled through careful and rigorous planning.

Internally, Cascades implements levers to establish and maintain the corporate culture. "The values of Cascades are lived. There is a real concern for keeping the local environment. It is not an industrial area and there is a lot of greenery" (Genium360, 2019). Consideration for the environment before the rise of environmental consciousness was therefore avant-garde at the time. Other key elements support that integration of the corporate culture: employee engagement, their knowledge of business objectives, the concrete applications of action plans, or the recognition program (Genium360, 2019).

Externally, following the ISO 26000 standard, the company engaged in a more thorough dialogue with its stakeholders, which allowed the company to grasp the effect of its operations on both the environment and society, by taking notice of the individuals and organizations that are involved or impacted (Cascades, 2020c). In 2009, the company performed a materiality analysis, which consisted in the consulting of key stakeholders including employees, product consumers, shareholders, corporate clients, nonprofit organizations, suppliers, students and others (Cascades, 2013a), as shown in Figure 3.2.

In partnership with an independent external firm, they contacted stakeholders via an online survey, consultation workshops and interviews (GRI, 2016). The results of the analysis of the primary data thus obtained, pinpointed priority sustainable development issues for both Cascades and its stakeholders. The materiality analysis supported Cascades in identifying its most "material issues" and determining what should be reported and focused on. Figure 3.3 shows an example of such materiality analysis. The horizontal axis, ranging from left to right, means an increase in the level of materiality for Cascades, so the more to the right, the more important. Conversely, the vertical axis, ranging from bottom to top, means an increase in the level of materiality for the stakeholders identified in Figure 3.3.

In the 2010–2012 Sustainable Development Plan of Cascades, nine strategic objectives were eventually selected and reported as important to focus on for the 2010–2012 period, including (Cascades, 2013a):

- *Energy*: Reduce the quantity of energy purchased to make products (environment)
- *Waste*: Increase the recovery of waste materials (environment)
- *Water*: Reduce the amount of waste water (environment)
- *Sustainable procurement*: Source materials from responsible suppliers (economy)
- *Innovation*: Develop and market new products (economy)
- *Financial performance*: Optimize the return on capital employed (economy)

70 • *Social and Sustainability Marketing*

FIGURE 3.2
The stakeholders of Cascades. (From Cascades (2020c). Sustainable development approach. Available at: https://blogue.genium360.ca/article/professionnel/etude-de-cas-4-facteurs-cles-de-succes-de-cascades-pour-une-performance-energetique-durable/ [accessed on 14-07-2020]).

- *Health and safety*: Reduce occupational injuries and illnesses (social)
- *Employee mobilization*: Increase the level of employee commitment (social)
- *Community involvement*: Increase contributions in the communities where facilities are located (social)

The same sustainable development targets were then found again and reused to report on Cascades' sustainable performance over the 2013–2015 period. In the 2016–2020 Sustainable development plan, the company added

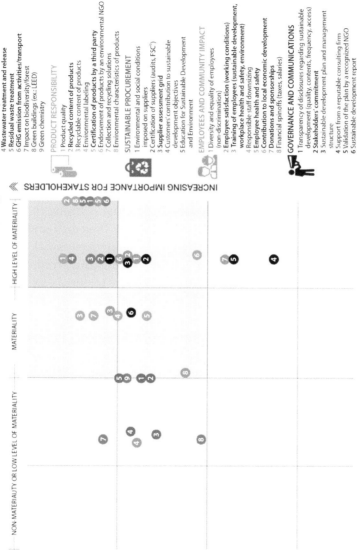

FIGURE 3.3

The results of the materiality analysis of Cascades. (From Cascades (2020c). Sustainable development approach. Available at: https://blogue.genium360.ca/article/professionnel/etude-de-cas-4-facteurs-cles-de-succes-de-cascades-pour-une-performance-energetique-durable/ [accessed on 14-07-2020]).

a tenth objective, *greenhouse gas emissions* (Cascades, 2016). The company should unveil its new sustainable development plan by soon, for the future.

Reuse Philosophy

According to Batellier et al. (2013), Cascades has grown by buying low-cost assets in financial difficulty. Factory rehabilitation followed the principles of "recycling" and "reuse" of assets. For example, in 1985, Cascades entered La Rochette, France when acquiring two boxboard plants, one in 1985 and another one in 1986. Other mills located in Sweden, Belgium were acquired in the 1980s and so was the case for *Perkins Papers, Rolland and Paperboard Industries Corporation, Provincial Papers* in Canada and other companies in Germany, the United States and the United Kingdom, in the 1990s.

On a more environmental ground, despite the current trend toward the minimization of packaging, Cascades—being an important packaging producer—considers that growth perspectives remain strong in that sector (Noël, 2019). According to Mario Plourde, the President and CEO of Cascades in 2019, the company tries to reduce its environmental footprint by using the maximum amount of recycled materials and promoting eco-design. For example, in 2019, 82% of Cascades' raw materials are recycled and the company tries to increase this percentage (Noël, 2019).

The integration of social and environmental responsibility within the processes also contributes significantly to sustainable energy performance. In 2013, Cascades marks the 15th edition of *Energy Efficiency Week*, an initiative of the Ministry of Natural Resources which takes place from November 24 to 30, by reiterating its objective of reducing energy consumption and by encouraging other companies to do more (Cascades, 2013b). It is in this perspective that Cascades announced, through its 2013–2015 Sustainable Development Plan, that it was continuing its efforts by aiming to reduce by 3%, by the end of 2015, the quantity of energy purchased to make its products (Cascades, 2013a).

In an attempt to improve the energy efficiency of its various manufacturing units, Cascades innovated 15 years ago, by setting up an energy intervention group (*Cascades GIE Inc.*). Since its inception, it has successfully completed hundreds of projects that enabled estimated savings of more than 7,500,000 gigajoules and cumulative energy savings of more than $60 million (Cascades, 2013a). By the end of 2015, Cascades had in place energy efficiency measures and projects totaling nearly 600,000 gigajoules, which equals the consumption of 5,900 Canadian households. Among these initiatives are the reduction in the use of live steam, recovery by installing heat exchangers, and daily and real-time monitoring of plant consumption (Cascades, 2013a).

Cascades' energy management practices have earned the company recognition for the "Hydro-Québec Ecolectric" network, which brings together large companies whose leadership and energy performance are exceptional. With this expertise developed over the years, Cascades now shares this success with its external partners by offering them its energy skills.

Currently, with Industry 4.0, there is more data today and leveraging this data will add new opportunities to develop energy efficiency projects (Genium360, 2019). The company's research and innovation center as well as the engineering consultancy in engineering and project management of Cascades allow the development of new projects for sustainable energy performance at Cascades, but also in Quebec.

Resource Optimization

Energy efficiency management is one of Cascades' common approaches to analyze the main sources of energy consumption (Genium360, 2019). The operational and investment angles are considered for each project, meaning that the company seeks how it can better operate, recover factories, reuse old machines, modernize factories and transform old assets. The recovery of waste is also a priority. For example, for a heat loss recovery project, there will be an exhaustive review of the processes in order to assess the best recovery potential. The goal is to recover that waste to heat the building or heat a process. For example, a steam generator can recover 15–20% of the energy available in the evacuation of a tissue paper machine (Genium360, 2019).

The attempt to increase resource productivity aligns squarely with the objective of the circular economy. According to the Ellen MacArthur Foundation & McKinsey Center for Business and Environment's (2015) online definition, "a circular economy aims to redefine growth, focusing on positive society-wide benefits. It entails gradually decoupling economic activity from the consumption of finite resources, and designing waste out of the system". As such, the circular economy model is akin to various schools of thought including natural capitalism (Hawken et al. 2013), blue economy systems (Pauli, 2010), the performance economy

(Stahel, 2010), the cradle-to-cradle design philosophy (McDonough and Braungart, 2010), industrial ecology (Lifset and Graedel, 2002) and biomimicry (Benyus, 1997), to name but a few.

Cascades implements environmental practices that contribute to a circular economy and related philosophies through greater resource optimization. In fact, the company is known for offering products made from recycled fibers that carry certification labels such as *processed chlorine free (PCF), Green Seal, Green-e®* or *EcoLogo*. Cascades products are also made with less water than the Canadian industry average, while they are biodegradable and compostable. Overall, these resource optimization processes and tactics align well with the engineering ethos and business objectives of efficiency. Most importantly, they also "enable [the company] to manufacture a higher quality product with a smaller ecological footprint" (GE, 2016, p. 2). Therefore, Cascades has reduced its carbon footprint by 50% over the past 25 years by investing heavily in energy efficiency, renewable energies and employee expertise (Roy, 2018). The transport electrification pilot project should further this trend. During the official announcement, Mario Plourde, CEO of Cascades, challenged other Quebecer companies (also humorously called "Québec Inc.") to set up similar programs.

SUMMARY

Albeit motivated by sustainable principles, Cascades also faced a couple of challenges, particularly during its expansion phase as well as during times of crises. The company was able to retain its sustainable ethos by sticking to a few core principles. Albeit this is not always an easy task to do, the management team was able to inspire this vision to its staff and partners. An overall four-key factor framework also emerged from the analysis of Cascades' operations to explain that success.

DISCUSSION

Cascades is known to the general public for its tissue products (toilet paper, paper towels, napkins). It is the first producer in Canada and the fifth in North America. However, the overall offering is much broader and includes different types of packaging, ranging from food containers to cardboard boxes to industrial packaging (David Suzuki Foundation, 2017). In Canada, Cascades is the leader in most of these markets and also occupies a strong position in the US market. Its expertise in several areas has led the company to offer services in energy management, engineering as well as external research and development.

As a manufacturer of recycled fiber products, and whose major products are mostly made of recycled fibers, there are a series of key challenges that the company is currently facing to uphold its sustainable positioning.

First, the availability of raw material is a key issue. The company has long been the only one in Canada to make paper and cardboard from 100% recycled fibers, but other companies have joined that market. Among them, *DS Smith* provides corrugated packaging for consumer and industrial goods; *International Paper* is a packaging and paper company; *Smurfit Kappa Group* is a company producing corrugated and paper-based packaging and *WestRock* manufactures and sells paper and packaging solutions for both the corrugated and consumer markets. All of them are involved in recycling activities (Craft, 2020). This increase in competition has necessarily resulted in greater demand for recycled fibers. Certain categories of paper are also in decline, such as printing and writing papers including newsprint and office paper, which affect some of the company's revenues, especially those of tissue papers.

Second, Cascades is also preoccupied by the quality of the raw material. This issue has faded somewhat thanks to the equipment that the sorting centers have acquired (David Suzuki Foundation, 2017). However, the issue of quality will continue to be important. Cascades needs to work at the source to help recover more and of better quality. To that end, the company even suggested a few tips for consumers to improve the quality of the recycled material. According to Marie-Ève Chapdelaine:

> Citizens can take very simple actions, for example, separating recovered materials. Many packages include paper and plastic. When placing them in the bin, it is preferable to separate them, this facilitates the work of the employees of the sorting centers. The recycling bin should include only containers, packaging and printed matter. If in doubt, it is recommended to call your municipality, which will provide you with information.
>
> (David Suzuki Foundation, 2017)

Third, we are entering an era where the minimization of packaging is put forward. However, using sustainable development as a competitive advantage (Belz and Peattie, 2012), the Cascades company is convinced that it can still grow in this sector of activity. The company is notably working to reduce its ecological footprint by using as much recycled material as possible and by promoting eco-design (Noël, 2019). In 2019, 82% of Cascades' raw materials were recycled and it appears that the company seeks constantly to increase that percentage (Noël, 2019).

Finally, the digital transformation, also known as "Industry 4.0", is also a strong trend. With Industry 4.0, there is more data today, but the valuation of this data should allow the addition of new opportunities to develop energy efficiency projects (Genium360, 2019). The company's research and innovation center as well as Cascades' engineering and project management consulting component allow the development of new projects for sustainable energy performance at Cascades, but also in Quebec (Genium360, 2019). The need to transform enables a better achievement of sustainable objectives. This shift to technology (especially Industry 4.0) to better reach key sustainable objectives (especially the circular economy) has been largely emphasized in the literature as the next big trend that is formed by two megatrends, just like two giants who meet (Gu et al. 2019). For example, Narayan and Tidström (2020) advanced the tokenizing of coopetition using blockchain technology for transitioning to the circular economy.

In recent years, Cascades has implemented a program to optimize its business processes in order to strengthen the customer approach, increase efficiency and improve profitability (CTEQ, 2018). This program affected the company's supply chain, finances and human resources. This has resulted in the integration of several technologies, including the implementation of an integrated management software package and communication platforms such as *IP telephony* and the *Office 365* suite (CTEQ, 2018).

To continue its ascent in Industry 4.0, Cascades pays particular attention to respect its employees and their environment. The digital strategy is being deployed (CTEQ, 2018)

- by involving all departments and employees in the process
- by staying abreast of market developments
- by quickly and efficiently integrating new features
- by developing an innovation strategy
- by intensifying a unique expertise in sustainable innovation

Meanwhile, in accordance with its culture and values of putting people first, Cascades has managed technology to empower employees instead of replacing employees by machines.

When the first new technologies were incorporated into factories, employees would comment and some reluctance was felt (CTEQ, 2018). They had a lot of questions and were worried about the changes. Cascades therefore chose to involve them in the search for solutions, and this approach was successful.

Since then, the company has engaged its employees in the transformation process from the start of the project (CTEQ, 2018). This helps target the most damaging problems and engage them in the development of possible and viable solutions. This approach ensures the success of the transformation, because the worker engaged in the process is convinced of the solution and participates in its realization.

When the first robots were introduced to the company, the employees were enthusiastic (CTEQ, 2018). For them, it was a project that came to fruition thanks to their commitment and their recommendations. They felt responsible for choosing the new technological solution.

In recent years, and in accordance with the open management principles at Cascades (Aktouf, 1988, 2006), a training center has been set up at Cascades to train super users. Employees who wish to become experts in new technologies are trained. They can then support their colleagues and train them in turn, as needed. Skills are thus developed within staff while strengthening team spirit (CTEQ, 2018).

Cascades has also implemented a succession plan in each plant to identify the potential of employees to access certain positions. A training plan is then developed with supervisors and human resources (CTEQ, 2018). It is also repeated in each division for the integration of new employees.

Change and innovation are part of Cascades' DNA. Its way of integrating new technologies and new management practices into its operations ensures that it continues to be a pioneer in its field (CTEQ, 2018). With Cascades, we might therefore expect to see, over the coming years, new waves of innovations to foster sustainable development in both oceans of the professional and retail markets.

There are a number of key takeaways from these case studies for managers who wish to reach socially responsible consumers. First, to be considered by such consumers, organizations need to integrate sustainable principles in the foundations of their organization. However, consumers are also sensitive to sincere and genuine attempts toward sustainability. Therefore, as shown in the case of Cascades, to reap consumers' trust and willingness to pay, sustainability is not a one-shot activity but results from a long and convoluted process that evolves over longer periods of time. One key challenge to maintain socially consumers' trust and interest resides in the maintenance of sustainability principles, even in periods of crises or paradigmatic shift were old principles tend to appear outdated or unfitting with the new business realities. This is the time when businesses need to make an introspection into their core values and navigate through difficult times. One example in the case pertains to the layoff of workers in the aftermath of the subprime crisis in 2007–2008. This contradicted the company's social sustainability but proved essential to the very survival of the company through the economic sustainability. From the previous, it follows secondly that a small number of core principles might be easier to follow and to adjust than a large number of them. Parsimony is therefore preferable to exhaustiveness. This is also important from a communication perspective, where fewer key messages will be easier to disseminate to consumers instead of many messages. Third, idiosyncratic realities (different places, different industries, different partners, etc.) of companies and diverse consumer profiles will create a whole range of diversified opportunities for sustainable success factors. But even within a specific industry, not every company will thrive on a similar set of factors. The four factors that have been studied in this case are unique to Cascades and may not work as effectively in another company due to different business realities. This flexibility also needs to be considered. Fourth, successful sustainable development in the past is only partially indicative of success in the future. It might be that principles and values on which the company thrived in the past need not to be abandoned but reinvented in the future. For example, in an attempt to avoid layoffs (a core social principle of the company), Cascades resisted by reallocating workers to other plants or adapting their tasks. Besides, sustainable development goes beyond cosmetic actions and superficial communications. It is all about a very authentic motivation by managers to provide a positive contribution to the market, communities and the world. This is important in the eyes of consumers who value sincerity and authenticity. Finally, while consumers remain considered as consumers, sustainable-oriented companies treat them more as partners and key stakeholders who cocreate value with the company. Cascades walks the talk by showing that it values its consumers not only as buyers and payers but also as insightful partners through its stakeholder management program. Consumers need therefore to be heard not only as consumers but also more generally as individuals with meaningful opinions. This respect that Cascades shows internally to its employees is also demonstrated externally to its consumers to further the companies' credibility and be in a better position to market confidently to consumers.

CONCLUSION

This chapter focuses on the case of the Quebecer Cascades company to illustrate how an organization might successfully embrace sustainable development for corporate growth. After retracing the timeline of the company development, the case emphasized the four key success factors that have enabled Cascades to achieve balance between the three dimensions of sustainable development: the economic, social and environmental dimensions. More specifically, the case discusses how a strong family culture and values, real planning with key stakeholders, reuse philosophy and resource efficiency principles were unfolded by the company to keep a balanced approach to sustainability. A glimpse into the challenges and perspectives faced by the company is also provided to offer insightful avenues for the future development of the company. The elements presented in this chapter may not constitute a recipe for immediate and guaranteed success. However, they surely provide a point of reference, some food for thought and a blueprint to help managers, decision-makers and policy-makers to navigate through the increasingly complex challenges of sustainable development.

LESSONS LEARNED

- To implement sustainable development successfully, organizations need to weave sustainable principles in the very core of their business model, mission, vision and strategies.
- Being sustainable is easy in the short run or for small periods of time. The most difficult aspect in being sustainable is actually to remain sustainable over the medium and the long run, especially in times of crises or business paradigm shifts.
- A company needs to identify only a few key factors that fit with the sustainability ethos, that fit well with its business model, that

could constitute a basis for competitive advantage, while being more or less easy to uphold even in unconventional times and disruptions.
- The case presented four key success factors that are specific to the company under study. Yet, different companies will have different factors of success. Companies should identify those factors that fit most adequately with their markets, business model, objectives and capacities.
- While remaining true to a few key sustainability principles elaborated in the past is a good approach to sustainable development, this should not preclude the company to remain oriented toward the future and anticipate how those principles might be challenged or possibly adapted to continue to thrive.
- The different factors and approaches presented in this case study are not a recipe for success in sustainable development. A managers' genuine approach to consistently striving to uphold a few key sustainable principles is more appropriate and will likely result in more successful outcomes.

DISCUSSION QUESTIONS

1. Sustainable development is a major constraint on businesses as it deprives them from their entrepreneurial freedom and imposes too strict requirements on managers. Discuss this statement.
2. A stakeholder management approach might enable a company to better consider its key partners and communities impacted. What could be the pros and cons of such an approach?
3. To which sustainable dimension does industrial relations and layoffs refer to? Do you think that "not laying off at all cost" is a reasonable sustainable principle? Justify your answer.
4. If you were hired as a strategy consultant by Cascades, what would you recommend to the company in order to face the rising challenges of the coming years, especially regarding automation and digitization trends?
5. Sustainable development is a road to success for companies. Discuss this statement.

PROJECT/ACTIVITY-BASED ASSIGNMENT/EXERCISE

Choose a company that you think was able to implement successfully the principles of sustainable development in the three dimensions of sustainability (3BL) and used that as a competitive edge. Try to identify the key factors pertaining to sustainable development which enabled the company to be successful. How do they compare to Cascades' success factors? What do you conclude?

FUNDING DECLARATION

This case study has been conducted in the framework of a broader project at LaboNFC that is supported by a grant from the Fonds de Recherche du Québec – Société et Culture (FRQSC) (Quebecer Fund of Research in Society and Culture) (Grant no. 2020-NP-267004).

ACKNOWLEDGMENT

The author wishes to thank all members of the LaboNFC for their support in enabling her to conduct this research.

CREDIT AUTHOR STATEMENT

Myriam Ertz was responsible for conceptualization, writing—original draft preparation, writing—reviewing, and editing, project administration, funding acquisition, reviewing and editing.

REFERENCES

ADEC (2020). What is social sustainability? Available at: https://www.esg.adec-innovations.com/about-us/faqs/what-is-social-sustainability/#:~:text=Social%20sustainability%20is%20a%20proactive,with%20people%2C%20communities%20and%20society (accessed on 30-09-2020).

Aktouf, O. (1988). La communauté de vision au sein de l'entreprise: exemples et contre-exemples. *La Culture des Organisations, Québec, IQRC, Questions de culture*, 14, 71–98.

Aktouf, O. (2006). 4.16. Pour une approche humaniste du management: le cas de l'entreprise Cascades. Les 4 temps du management.

Batellier, P., Raufflet, E. B., & Hébert, L. (2013). Cascades tissue group: Sustainable growth? *International Journal of Case Studies in Management*, 11(4).

Belz, F. M., & Peattie, K. (2009). *Sustainability marketing*, 2nd ed. Hoboken, NJ: Wiley & Sons.

Benyus, J. M. (1997). *Biomimicry: Innovation inspired by nature*. New York, NY: Harper Perennial.

Bernatchez, J. C., & Cossette, C. (2007). Le plan de relève chez Cascades. *Revue internationale sur le travail et la société*, 5(3), 74–84.

Brundtland Report (1987). Report of the world commission on environment and development: Our common future. Available at: https://sustainabledevelopment.un.org/content/documents/5987our-common-future.pdf (accessed on 30-09-2020).

Cascades (2013a). Sustainable development plan: SDP 13-15. Available at: https://www.cascades.com/sites/default/files/developpement-durable/2013-2015-Sustainable-Development-Plan-Cascades_EN.pdf (accessed on 14-07-2020).

Cascades (2013b). Cascade accroît son leadership en matière d'efficacité énergétique. Available at: https://www.cascades.com/fr/nouvelles/cascades-accroit-leadership-matiere-defficacite-energetique (accessed on 15-07-2020).

Cascades (2016). Sustainable development plan – 3 pillars – 10 priorities. Available at: https://www.cascades.com/sites/default/files/developpement-durable/2016-2020-Sustainable-Development-Plan-Cascades_EN.pdf (accessed on 14-07-2020).

Cascades (2020a). Une vision avant-gardiste. Available at: https://www.cascades.com/fr/developpement-durable/planete/entreprise-circulaire (accessed on 15-07-2020).

Cascades (2020b). Nos bons coups. Available at: https://www.cascades.com/fr/propos/bons-coups (accessed on 28-07-2020).

Cascades (2020c). Sustainable development approach. Available at: https://blogue.genium360.ca/article/professionnel/etude-de-cas-4-facteurs-cles-de-succes-de-cascades-pour-une-performance-energetique-durable/ (accessed on 14-07-2020).

Cascades (2020d). Reconnaissances. Available at: https://www.cascades.com/fr (accessed on 14-07-2020).

Craft (2020). Cascades competitors. Available at: https://craft.co/cascades/competitors (accessed on 29-07-2020).

CTEQ (2018). Industrie 4.0: Portrait de Cascades. Le Blogue du CTEQ, May 28, 2018. Available at: https://www.bloguecteq.com/single-post/2018/05/28/industrie-40-portrait-de-cascades

Daly, H. E. (1990). Toward some operational principles of sustainable development. *Ecological Economics*, 2(1), 1–6.

David Suzuki Foundation (2017). À la rencontre de Cascades: quand nos rebuts deviennent des trésors. Fondation David Suzuki, June 15, 2017. Available at: https://fr.davidsuzuki.org/blogues/a-rencontre-de-cascades-nos-rebuts-deviennent-tresors/ (accessed on 29-07-2020).

Déry, R. (2010). *Les perspectives de management*. Montreal: Éditions JFD.

Drolet, A. (2009). Le couperet tombe sur l'usine de Cascades. Le Soleil, April 3, 2009. Available at: https://www.lesoleil.com/affaires/le-couperet-tombe-sur-lusine-de-cascades-b5038bd52bfa9b7f4afc1c4a1638aff2 (accessed on 15-07-2020).

Elkington, J. (1997). *Cannibals with forks: The triple bottom line of 21st century business*. Conscientious Commerce Series. Oxford: Capstone.

Elkington, J. (1998a). Accounting for the triple bottom line. *Measuring Business Excellence*, 2(3), 18–22.

Elkington, J. (1998b). Partnerships from cannibals with forks: The triple bottom line of 21st-century business. *Environmental Quality Management*, 8(1), 37–51.

Ellen MacArthur Foundation & McKinsey Center for Business and Environment (2015). *Growth within: A circular economy vision for a competitive Europe*. Cowes: Ellen MacArthur Foundation.

Environnement Magazine (2019). Sept producteurs d'énergie renouvelable lancent une offre d'électricité verte d'origine contrôlée. Environnement Magazine, October 21, 2019. Available at: https://www.environnement-magazine.fr/energie/article/2019/10/21/126405/sept-producteurs-energie-renouvelable-lancent-une-offre-electricite-verte-origine-controlee (accessed on 28-07-2020).

GE (2016). Cascades tissue group achieves reliable and predictable manufacturing performance. Available at: https://www.ge.com/digital/sites/default/files/download_assets/Cascades-Tissue-Group-Achieves-Reliable-and-Predictable-Manufacturing-Performance-customer-study.pdf (accessed on 26-07-2020).

Genium360 (2019). Quatre facteurs de succès chez Cascades pour une performance énergétique durable. Genium360, May 12, 2019. Available at: https://blogue.genium360.ca/article/professionnel/etude-de-cas-4-facteurs-cles-de-succes-de-cascades-pour-une-performance-energetique-durable/ (accessed on 14-07-2020).

Ghannadpour, S. F., Hoseini, A. R., Bagherpour, M., & Ahmadi, E. (2020). Appraising the triple bottom line utility of sustainable project portfolio selection using a novel multi-criteria house of portfolio. *Environment, Development and Sustainability*, 1–42. https://doi.org/10.1007/s10668-020-00724-y.

GRI (2016). Defining what matters: Do companies and investors agree on what is material? Available at: https://www.globalreporting.org/resourcelibrary/GRI-DefiningMateriality2016.pdf (accessed on 14-07-2020).

Gu, F., Guo, J., Hall, P., & Gu, X. (2019). An integrated architecture for implementing extended producer responsibility in the context of Industry 4.0. *International Journal of Production Research*, 57(5), 1458–1477.

Hawken, P., Lovins, A. B., & Lovins, L. H. (2013). *Natural capitalism: The next industrial revolution*. New York, NY: Routledge.

Laperrière, E. (2016). Questions pour un patron: Transition réussie chez Cascades. La Presse, May 2, 2016. Available at: https://www.lapresse.ca/affaires/economie/emploi/201604/29/01-4976348-questions-pour-un-patron-transition-reussie-chez-cascades.php (accessed on 29-07-2020).

Lévesque (2019). Les facteurs clés de succès. 5e édition de la journée-conférence des Rencontre de Génie (RDG) Genium360, June 11, 2019.

Lifset, R., & Graedel, T. E. (2002). Industrial ecology: Goals and definitions. In R. U. Ayres and L. W. Aryes (Eds.) *A handbook of industrial ecology* (pp. 3–15). Northampton, MA: Edward Elgar Publishing, Inc.

McDonough, W., & Braungart, M. (2010). *Cradle to cradle: Remaking the way we make things*. New York, NY: North Point Press.

Narayan, R., & Tidström, A. (2020). Tokenizing coopetition in a blockchain for a transition to circular economy. *Journal of Cleaner Production*, 263, 121437.

Noël, M. (2019). Cascades vers une année record. La Tribune, May 9, 2019. Available at: https://www.latribune.ca/affaires/cascades-vers-une-annee-record-f8b3a3d-dc8ea8689dbf393c18ca54746 (accessed on 15-07-2020).

Pauli, G. A. (2010). *The blue economy: 10 years, 100 innovations, 100 million jobs.* Brookline, MA: Paradigm Publications.

PRNewswire (2017). Patrick Lemaire classé dans les personnalités les plus influentes selon le rapport « Top 100 Power People ». Cision PRNewswire, November 15, 2017. Available at: https://www.prnewswire.com/news-releases/patrick-lemaire-classe-dans-les-personnalites-les-plus-influentes-selon-le-rapport--top-100-power-people--657679683.html (accessed on 28-07-2020).

Rezapouraghdam, H., Alipour, H., & Arasli, H. (2018). Workplace spirituality and organization sustainability: A theoretical perspective on hospitality employees' sustainable behavior. *Environment, Development and Sustainability,* 21(4), 1583–1601.

Roy, G. (2018). Cascades lance un défi au Québec Inc. Unpointcinq, January 17, 2018. Available at: https://unpointcinq.ca/economie/cascades-lance-defi-quebec-inc/ (accessed on 29-07-2020).

Slaper, T. F., & Hall, T. J. (2011). The triple bottom line: What is it and how does it work. *Indiana Business Review,* 86(1), 4–8.

Stahel, W. (2010). *The performance economy.* Berlin: Springer.

UMW (2015). Economic sustainability. Available at: https://sustainability.umw.edu/areas-of-sustainability/economic-sustainability/#:~:text=Economic%20sustainability%20refers%20to%20practices,sector%20through%20the%20UMW%20foundation (accessed on 30-09-2020).

4

Sustainability Marketing in Contending for the Position of the European Capital of Culture (ECoC)

Ezeifekwuaba Tochukwu Benedict
University of Lagos, Nigeria, SAU, USA

CONTENTS

Learning Objectives 86
Background and Purpose 86
Methodology Approach and Design 87
Outcome/Result 87
Conclusion 88
Introduction 88
Marketing Cohesion of Sustainability and Culture 92
The European Capital of Culture (ECoC) Initiative 94
European Capital of Culture (ECoC) Analysis 96
    Progressive Evaluation and Sustainability Requirements 96
    Evaluation of Progress and the Marketing Requirements 97
Procedures and Methods 99
Research Methodology 100
Results 104
Discussion 108
Conclusion 109
Funding Declaration 110
Acknowledgment 110
Author Statement 111
References 111

86 • *Social and Sustainability Marketing*

LEARNING OBJECTIVES

After reading the chapter, the reader should be able to:

- Understand what it means by the European Capital of Culture (ECoC)
- Evaluate and assess the level of sustainability in regard to competing for the position of the ECoC
- Look at the program and initiatives of ECoC
- Look at competition and marketing including various nations or cities that are bidding for the prestigious position "ECoC", applying the triple bottom line (TBL) technique which is executed by applying the content analysis technique including the DEX methodology in evaluating the cities

BACKGROUND AND PURPOSE

This research paper uses the European Union (EU) because it designated the European Capital of Culture (ECoC) and it is a position for a yearly period in which it organizes varieties of cultural festivals and events with a strong and vibrant Pan-European perspective and notion. The evaluation or the extent of the sustainability in regard to the bidding the position of ECoC is stated in this research paper as the EU sponsored the initiative of the ECoC, that is, a major and effective pillar of every cultural project. It is strongly attested that the ECoC initiative significantly maximizes economic and social benefits, particularly if the scenarios are embedded as part of a durable term culture-focused development initiative of surrounding region or city. The integration of the Cultural Manifesto Programs and Initiatives develops and ensures the links and connections between domains (e.g., social sciences, planning, territorial planning, education, culture, tourism, etc.) as well as assists to mold and ensure a sustainable connections with the social and economic sectors. Individuals act on the values, principles, and beliefs they perceive and view to be widespread in their tradition or culture. Therefore, marketing tends to play a not insignificant responsibility in the position competition while there is a necessity in regard to ECoC project integration into major strategic initiative of the region, including the possible marketing.

I concur that the necessary and appropriate integration and linkage of the cultural policy into the social system affects as well as transforms the cultural beliefs and values, shifting and migrating them toward sustainability and sustainable behavior. I evaluated the interrelation between sustainability and culture; therefore, defining and emphasizing on the role and responsibility of culture for sustainability. Although few were emphasized on the possible tools and the approaches that may offer and provide assistance in the scenario and situation of how to attain and ensure sustainability in regard to culture.

METHODOLOGY APPROACH AND DESIGN

This research paper focused on comparative evaluation of usage of different and various regions. The triple bottom line (TBL) technique is executed by applying the content evaluation (analysis) technique as a tool. The classical version of the content evaluation is applied while aggregating the major terms (sustainability, marketing, ecology, sustainability marketing), including analyzing and evaluating the in-depth application of sustainability, including sustainability marketing.

The outcome/result of content evaluation is also applied for the explanation and elaboration of qualitative vast attributes technique applying the DEX technique.

OUTCOME/RESULT

In analyzing and evaluating the bidding documents of the ECoC, I (a) argue and emphasize that the marketing of ECoC requires to be transformed into "sustainability marketing" which is defined or described by various and numerous authors and (b) define and emphasize on the marketing plan importance and essence emphasized as a complete and a well-detailed action.

CONCLUSION

The ECoC commission needs to put into consideration the essence of the culture for sustainable development variously and to evaluate and access the marketing plan and ideology of the applicants under the sustainability conceptual technique.

INTRODUCTION

The ECoC is a region that is prepared by the EU for a yearly calendar as it coordinates and also organizes multiple cultural events with a strong and vibrant Pan-European perspective and dimension.

Preparing for the European capital can be an opportunity and privilege for the city to accumulate considerable economic, social, and cultural purposes and benefits, and it can assist to promote urban renegration and raise its profile and visibility on an international level as well as change the image of the city.

In 1985, Mercouri Melina, the Minister of Culture for Greeece including her French Counterpart Lang Jack, came up with the notion and the idea of designing and preparing an annual capital of culture so as to almagamate the Europeans nearer and emphasizing on the diversity and richness of the European Culture as well as raising the awareness or the consciousness of their common values and history. It is strongly and majorly believed and attested that the ECoC actually maximized the economy including social benefits particularly when the scenarios are embedded or included as an aspect of the durable culture-focused development initiative surrounding region or city.

The program, plan, and initiative of the "City of Culture" continued through the 1980s and 1990s. Most regions were selected behind closed doors by the respective governments. There was as unofficial and an informal iota of nations usually changed or corrected by trade-off and agreements. In 1990, the United Kingdom championed the idea, ideology, or notion of an open and transparent competition. It was won by Glasgow, whose technique (approach) involving with citizens and connecting the initiatives or programs to their existing (current) regeneration of ideas and policies transformed the program's direction as this was so vital and essential.

The EU Commission manages and handles the title, and each year the EU Council of Ministers formally and offfficially designates the ECoC. Fourty cities and more have been bestowed and designated so far. The ECoC program and initiatives were launched and sponsored in the summer of 1985 with Athens being the first and primary holder. During the German Presidency of 1999, the ECoC program was modified to ECoC.

A new and current framework ensures that it is possible and attainable for cities in candidate nations including the potential and major candidates for the EU member of the European Free Trade Association (EFTA) member nations to hold the title every year as of 2021. This will be chosen through an open and transparent competiton entailing that cities from different nations may compete with each other.

The ECoC was meant to be in the United Kingdom in 2023. Although as a result of its decision to leave and depart from the EU in 2016, the cities in the United Kingdom would not be eligible and capable to hold the title after 2019. The European Commission of Scotland attested that this would be the case and scenario on November 23, 2017, solely a week before the United Kingom was due to declare the city to be put forward. The candidate and respective cities were Leeds, Dundee, Nottingham, Milton Keynes, and a Joint Bid from the Northern Irish Cities Derry, Belfast, and Strabane. This triggered anger among the United Kingdom candidates city's bidding teams as a result of the amount they had already spent in getting ready for their bids. The objectives and rules including the processes for the program transformed in 2014, 2009, and 2004, each time unique and significant. The trend has been for a vast array of activities, maximized forward planning, more transparency including independent and open selection and monitoring processes.

Culture actually affects and influences sustainability. According to Dessein et al. (2015), Sakkers and Immler (2014), Sazonova (2014), Scammon (2012), Maraña (2010), Powell and Fithian (2009), Nurse (2006) and Hawkes (2001), culture actually affects and influences sustainable development. Also, culture can be handled as a significant technique for the word sustainable development (Opuku, 2015) willing to connect various aspects of policy and initiative (Dessein et al., 2015). For this purpose, the EU drafted and prepared the program and the initiative of the ECoC also, plan or initiative, which is actually the most effective project of every cultural project (Maronić-Lamza et al., 2011). It is mainly attested that ECoC initiative tremendously increases economy including social opportunities particularly if the scenarios are included as an aspect of a durable principle culture-focused development tactics and plan of the surrounding region or city.

The cultural exercises and activities of the initiative integration must ensure the connections and links between various domains (such as education, culture, social services, tourism, territorial planning, etc.) and they also assist to strengthen the sustainable connections and partnerships with the social and economic sectors. In regard to Dessein et al. (2015, p. 44), "Culture is a significant element in the learning and the adaptation of the modern practices."

Therefore, marketing seems to play a not unimportant role and responsibility in the competition of the position while there is a demand for involving in integrating of ECoC initiatives into the primary and basic strategic plan and initiative of the city including its necessary marketing. As stated by Van Melik and Van Aalst (2012), municipalities seek for the chances and privileges to execute bigger events. Regardless of the fact that marketing is explained and emphasized as a necessary action for the ECoC project and initiative, the marketing costs also must be properly and appropriately planned in the budget as well as confirmed and attested by the awarding commission. Maronić-Lamza et al. (2011) emphasized and stated major difference in budgeting of marketing of the various candidate regions and also the tools and means of advertising (promotion). In this manner, various city governments have competed in the remarks of their responsibilities as well as ensured a public value for the general citizens. The necessary marketing promotes the communication of the ECoC not solely during the designated period but in preparation and readiness stage. Success in marketing entails the performance of the necessary and various audiences.

If culture is seen to be essential toward sustainable development, Why would marketing not be a way of disseminating or sharing the notion and the perspective of sustainability within a specific or a precise society and also transforming (shifting) that society toward the paradigm of sustainability? The function and the responsibility in changing or alternating societal and individual values, behaviors and beliefs have been examined by Wróblewski (2016), Jones et al. (2014), Senkus (2013), Helmig and Thaler (2013), and many others. The strategy of marketing the ECoC, therefore, must also entail sustainability or social aspects (complex interconnections between economic, ecological/environmental, and social dimensions). The criteria of sustainability marketing are still not complete as they are not found on the present reports and applications and are, therefore, the major concentration of this research paper.

It was emphasized the different types of applied marketing techniques for the implementation and the management of such cumbersome scenarios, the ECoC needs to consider the sustainability marketing as it was evaluated by Lim (2016), Rakic and Rakic (2015), Nkamnebe (2011), Belz and Peattie (2010), Karstens and Belz (2005), and many others. The requirement and the demand for sustainability marketing needs to be included to the condition by various ECoC commission including many others. The requirement and demand for sustainability marketing must be included to the condition of various ECoC commission and assessed while selecting the winner. The marketing techniques that will promote any region's sustainability need to be planned mandatorily in readiness stage including exhibiting in bidding the forms of the candidates' regions.

In regard to the complexity theory, sustainability can be seen as a complex and a cumbersome system (Swilling and Peter, 2014). Sustainability marketing as a process entails environmental, economical, ethical, social as well as technological aspects (Lim, 2016) confirming its cumbersomeness also. This permits and ensures one to argue and state that marketing in regard to ECoC must be focused on and accessed by applying the TBL technique. This technique is very essential and is emphasized in detail by different and various authors, beginning with Elkington John in his book of 1997, titled "The Triple Bottom Line (TBL) of the Twenty First Century Business: Cannibals with Forks."

The incidence and situation of Lithuania have been selected to provide for the context for evaluation of challenge. The major implication of tourism including culture plays a significant responsibility, both for the labor market development and the economic development and growth; cultural tourism not solely assists to safeguard cultural heritage but it also unveils the modern prospects for the collaboration and the integration among various sectors, promotes innovations as well as ensures an attractive or a luring tourism products; it is the motive behind cultural tourism classified as major reasons of the Lithuania's development. Also, significant measures and techniques dedicated to the safeguard of the cultural heritages were financed by European structural function from the year frame of 2007–2013. During the present EU structural financing period from 2014 to 2020, extra measures and techniques were worked out leading to the promotion of the nation's cultural tourism and also the creation and the provision of the cultural tourism routes at foreign level. Digital marketing is majorly financed concerns, although the economic importance of Lithuania culture was not identified by the government till 2007. Vilnius, the capital of Lithuania, was the primary Lithuania city which was awarded and bestowed the ECoC till 2009. An evaluation (Nechita 2015, p. 105) emphasizes that no Lithuania region has struggled led for the position of ECoC during the period from 2013 to 2019.

MARKETING COHESION OF SUSTAINABILITY AND CULTURE

In regard to Opuku (2015), culture can be handled as a major yardstick in the terms of sustainable development as it is the amalgamation or joint willing to connect individual's consciousness toward the buildup as well as the natural environments. Culture (most especially in evolutionary culture) empowers individual with the capability and the willingness to comprehend the common globe and its challenges afresh. Culture is a sphere where persons including various meanings and depictions are ensured with sustainable development as a major principle (Sazonova, 2014). As it was concurred that sustainable development demands for systematic manner of reasoning, holistic behavior, and attitude; culture therefore becomes a better and a significant tool toward the integration of modern modes and values of life, a novel pathway toward the development of economics. Also, culture can be signified on the basis of the modern paradigm of sustainability and sustainable development.

Dessein and Soini (2016) emphasized on the roles and responsibilities of culture in regard to sustainability as well as providing the three framing depictions of sustainability and culture. According to Dessein and Soini (2016), the emphasis was on three traditional and former sustainability foundations and pillars (ecological, economic, and social) by applying three various light circles as the dark depicts the culture. The first and major mode sees culture as the fourth pillar of sustainability. The second mode depicts culture acting as an intermediary function and responsibility to reach ecological, social, and economic sustainability, while the third mode sees culture as an important foundation in attaining the general aims and objectives of sustainability. Therefore, culture is "a part and an aspect of a regular evolving process geared towards transformation" (Dessein and Soini, 2016, p. 9). Also, migrating from the first to the second and lastly to the third mode, it is a necessity for cumbersome to move toward sustainability. The ideal outcome of this change would result to (i) modern modes of meta-governance or self-governance, (ii) wholly new or modern initiatives or policies that accommodate intrinsically all sustainability principle and techniques, and (iii) the reshaped or shaped depiction of nature, seeing it as a constituent of nature. While emphasizing on the direction of change or transformation toward sustainability as well as defining and evaluating the relationship between sustainability and nature in eight dimensions, Dessein and Soini (2016) neither intend to provide a solution to the inquiry of how to ensure the transformation or to provide, ensure, and create the necessary tools to do so.

Eroglu and Picak (2011) created a remainder that culture is seen as an array of shared and common practices, beliefs, value rules of coexistence, and models of conduct including an expected and anticipated behavior. Sustainability is also similar to behaviors, values, and beliefs (Edwards, 2009; Epstein, 2008; Giltrow, 2015; Morse and bell, 2003; Sazonova, 2014; Smith and Senge, 2008). Picak and Eroghi (2011, p. 146) emphasized "irrational and an in-depth embedded shared values shapes and transforms technical and social systems as well as the political institutions and initiatives, all of which simultaneously reinforce and reflect the beliefs and values." While sharing and discriminating the above perspective, we further emphasize that the effect and the implication of political actions and decisions are of paramount importance and significance toward sustainable development. Regional and local cultural policy promotes societal migrations toward sustainable behaviors, sustainable collection actions, and the involvement of citizens or can lead to the transformation and to change to a sustainable society. Cultural location is very appropriate for this discussion including diversity as well as the creative techniques and approaches. It enables and permits for a modern points of view toward development and prohibits sustainability from a "lifeless frozen dogma and doctrine" (Sazonova, 2014, p. 7).

Concurring to Wroblewski (2016), Jones et al. (2014), Senkus (2013), and Helmig and Thaler (2013), we emphasize that the social marketing is an essential technique for changing or modifying the behavior and values of the region. While in the current situation, sustainability marketing transpired in the beginning of the twenty-first century, we emphasize that this form of marketing will be an essential instrument for the modification or conversion to culture policy framework sustainability. Social marketing included in the sustainability marketing, which depicts a social perspective, ecological/environmental/green environmental marketing (Belz and Peattie, 2010; Karstens and Belz, 2005; Nkamnebe, 2011; Lim, 2016; Rakic and Rakic, 2015) depicts the environmental perspective, while the business marketing depicts economic sphere.

Presently, sustainability marketing is something not unknown and strange to the professionals in marketing. As in other aspects of research, marketing has evolved while responding to cases of sustainability which have occurred. The process has been evaluated, researched, and also defined by Lim (2016), Rakic and Rakic (2015), Nkamnebe (2010), Belz and Peattie (2010), and Karstens and Belz (2005). Sustainability marketing sustainability entails the systematic (innovative or new) reasoning of the marketing managers including the long-term orientation (Belz and Peattie, 2010). In regard to Rakic and Rakic (2015), marketing sustainability is aimed at an entire setting of the safeguard including the social goals of the entire community. It entails the involvement of local and national government, population and organization including the appropriate capital (financial, human, infrastructural, etc.). An array of the attributes is connected to the sustainability marketing as well as environmental, economic, ethical, social, and the technological perspectives (Lim, 2016), including the transformation capacity of the marketing exercises and activities (Belz and Peattie, 2010).

The concepts above require employing and applying in practice by national, regional and local governments while emphasizing on the sustainability plans and objectives. We emphasized on EU Interdisciplinary Program of ECoC that integrates all three conceptual dimensions and aspects (which are marketing, sustainability, and culture), which could play a tremendous and an essential role in promoting the notion of sustainability without more important power.

THE EUROPEAN CAPITAL OF CULTURE (ECoC) INITIATIVE

The political notion of ECoC majorly evolved since the beginning of 1985. It is attested that the ECoC majorly maximizes economic or social benefits when the scenarios are embedded as an aspect of a long and a durable period culture-focused development initiative of surrounding region or city. The ECoC initiative was prepared to encourage integration of Europe and identity. The launching cities have ensured multiple objectives including visions varying from enhancement of urbanization and material infrastructure on the improvement of cultural life to the fight against poverty by maximizing employment and luring so many tourists.

The initiative and program integration into the durable term city initiative is promoted, triggered, and propelled and has sustainable effect on local economic, social including cultural development (Turşie, 2015). Thus, majority of the regions also emphasize and highlight on the long-term culture, economic and social effect while bidding the position of the ECoC (Nechita, 2015). The initiative or plan integration into durable term strategies, also, could promote and encourage polycentric spatial development entailing peripheral (former urban or rural) regions around the bidding city as well as making them (and therefore the whole area) more luring and lucrative for the business investments, modern tourists, and inhabitants (Nemeth, 2010).

The responsible, actual, and real introduction of the initiative in general development tactics could, according to Richards (2000, p. 12), act and work as guard against "festivalization" development that hinders "cultural sustainability." The plan including mandatory and responsible ways of marketing could be seen as a better means for integrating of cumbersome sustainability phases into regional growth or development.

Actually, the value chain of cultural festivals including its events or more commercialization (e.g., consumer city) results to an optimistic economic outcomes and outlook (for employment and income) and also attains restriction opposition (Van Melik, & Van Aalst, 2012). Steiner et al. (2015), Draghici et al. (2015), Nechita (2015), and Aubert et al. (2015) emphasize various negative (pessimistic) aspects and challenges such as the overestimation of the possible opportunities and benefits and the underestimation of costs; challenges in measuring the social and economic impact, fewer comprehensive changes, an absence of funds to back up the modern cultural facilities both following the period of the ECoC and also from a durable term idea, the absence and lack of a durable term sustainability projects; the pessimistic implication on the regional masses welfare, etc.

As declared and emphasized by the commission in charge of such situations (Commission Staff working document of 2012), "The European Union (EU) has a legal and a moral responsibility to carry out action to safeguard and promote cultural diversity" particularly since the United Nations Educational, Scientific and Cultural Organization (UNESCO) summit including promotion, safeguard including the protection of diversity of cultural exhibitions came into light in 2007. Every prepared cultural scenarios need to depict the contemporary life, emphasizing and depicting the special cultural heritage and culture of the region. Every year, two regions of the participating nations are selected to contest for this position. O'Callaghan (2012), Maronić-Lamza et al. (2011), Boland (2010), and Herrero et al. (2006), the awarded title to the city raises the necessities and the prerequisites for raising or maximizing the international portfolio of a city, executing plans for arts/events and cultural activities; revitalizing communications in the city, long-term cultural development, attracting of foreign and domestic tourists, the expansion and the growth of the local cultural audience, enhancing the sense of belongings including self-confidence, providing a new cultural facilities, and ensuring a festival atmosphere, the repositioning, rebranding, and the regeneration of the city and lastly, the expansion of local cultural audience and urban redesign. Also, the title may assist with preparing "a modern development path" if thriving with social, identity, and economic conflict (Aubert et al., 2015, p. 27) and also the transformation of the "local population from the final consumers of the city and region into the creators of the city" (Aubert et al., 2015, p. 28) and a change to technique – focused regeneration that is encouraged and triggered by entrepreneurs including teenage (youthful) organizers (Džupka and Hudec, 2016).

These glaring opportunities (both social and economic) emphasize such fierce and a severe battle for the position in which the youthful and teenage democracies of post-Soviet Nations are willing and interested to compete. Thus, Estonia, Poland, Slovenia, Romania, Hungary, Croatia, Bulgaria, and Lithuania (an aggregate of youthful and teenage democracy nations from 28 candidates generally in the period of 2020–2033) are at the ECoC Candidate Nations List (Maronić-Lamza et al., 2011). Mercouri Melina Prize of 1.5 million euro bestowed by the commission intends to be attracting and a luring incentive for these nations.

EUROPEAN CAPITAL OF CULTURE (ECoC) ANALYSIS

Progressive Evaluation and Sustainability Requirements

An evaluation and assessment of the ECoC documents (reports, studies, and application forms) emphasizes that the sustainability requirements had not been added on the major preparation and documentation. Report of Palmer (Palmer/Rae Associate, 2004) emphasizes that local plans are not sustainable than those luring or drawing much audiences, and majority of the payments from the reporting or documenting period from 1995 to 2004 were not over time sustainable. Respondents and interviewees stressed that regardless of the severe extents of activity or investment, they hardly seem to have been connected by durable sustainable strategy for the city. Thus, the advice springing from the report entails recommendations to distinguish and classify between the longer-term and the shorter-term impacts, to create the sustainable initiatives as well as to create for sustainable initiatives and programs.

As a result, the creative Europe plan and initiative that has substituted the above-emphasized program presently already pays a unique need to encouraging inclusive, sustainable, and a smart growth while encouraging the cultural diversity of Europe. Sustainability can be seen as criteria for the current and modern selection prerequisite. The application from the candidates' regions or cities must be included in a durable term strategy toward cultural development as well as to add the initiatives for building and ensuring sustainable alliances with social or economic sectors (Commission Staff Working Document, 2012).

The guide for regions getting ready to bid the position of ECoC during 2020–2033 brings more attention to the proof that effective ECoC has applied the title for the general development producing and ensuring sustainable, social, cultural and economic effect. They embedded the programs, activities and functions into the city's or region's general tactics, ensuring the connections between culture, education, social services, tourism, territorial planning, etc. (European Commission, 2014). The major candidates, thus, are demanded in exhibiting their city including their cultural tactics in their contest books and having a strong or broad political backup or support including a sustainable dedication from national, local and regional officials (European Commission, 2014, p. 11).

Evaluation of Progress and the Marketing Requirements

The evaluation and assessment of the main and major ECoC programs and initiative documents (reports, studies and the application forms) indicates that despite the proof that there was a stringent and severe detail for ECoC initiative marketing, "Majority of the Host Cities made huge investments in Marketing" (Palmer/Rae Associates, 2004, p. 128). The study of the research paper by Cox and Garcia (2013) emphasized that the marketing plans presently seem to be complex and majorly handled as a concern as it is backed up by tremendous aspect of the aggregate budget.

The Report of Palmer (Palmer/Rae Associates, 2004) emphasizes cultural festivities and scenarios are majorly not interrelated and not interconnected and thus separately encouraged as isolated or unwanted scenarios. This fragmentation of cultural program resulted to absence of general comprehension of the initiative in numerous ECoC as it is the motive why some respondents stressed on the essence of delegating more funds to the general cumbersome marketing, feeling and seeing it (also with communication) as a priority in the ECoC. Although no term of regulation of marketing is emphasized by the Commission Staff Working Document in 2012 (i.e., the Document of the Commission Staff Working Document, 2012).

Few notable inscriptions regarding marketing transpire in the common or major guide for regions getting ready to bid the position of the European Commission (2014, p. 22). Communication including marketing is seen as a major and main function that tends to be handled as maximizing online exercise. The responsibility of informing the general public pertaining the ECoC as a detailed ideology and principle of the Union is basically and majorly bestowed to marketing. Although it is followed by the sole necessity observed – majorly in which "The Communication and the Marketing of the European Capital of Culture (ECoC) provides the due prominence to the European Capital of Culture as an European Union Action" (European Commission, 2014, p. 23). The working connections between tourism and marketing sector of the region including ECoC team are needed and essential. Although not a single or a sole word can be discovered and seen on sustainability marketing.

As emphasized in this research paper, marketing must consist of the means that aim the locals and visitors. While describing and emphasizing on the outreach (European Commission, 2014, p. 12), the guide stresses sustainability as criteria toward the establishment of modern or sustainable benefits and opportunities for every citizen to attend and involve in the cultural activities. Also, the general strategy for the development of audience with connections to school participations and education is mandatory for applicants. This research paper emphasizes that proper and effective conditions and situations for sustainability marketing could rarely be discovered. All which undoubtedly required entail to ensure a marketing orientation sustainability, which is rampant and important for the professionals of sustainability marketing but actually and normally uncommon to those who are in charge of executing and bidding the ECoC initiatives or plans.

PROCEDURES AND METHODS

In the ECoC applications selections, for research purpose, the two essential cities for the position of the ECoC, 2022 (Klaipeda and Kaunas) were selected in regard to the following criteria:

- Both cities have the experience and knowledge in the sound organization and the coordination of big and major yearly foreign events (such as Klaipeda Jazz Festival, Kaunas Jazz Festival, and Kansas Days in Kansas or the Klaipeda Sea Festival).
- Sustainable public finance/funding of candidate cities toward ECoC title.
- Sustainable growth and development strategy and initiative of Lithuania were revitalized in 2009 entailing that the officials of both cities must apply the necessary attention of the initiative and plan execution situations.
- The past post-Soviet industrial regions are in transition entailing some of that developmental structures are vanishing and are making way for the current and the present ones.

Vilnius, the capital of Lithuania, was bestowed the position in 2009. In July 2015, six of the Lithuanian regions (Kaunas, Anykščiai, Plunge, Roskiskis, Jonava, and Klaipeda) forwarded the applications for the major and primary stage. Lastly, two regions – Kaunas (the second biggest region) and Klaipeda (the third biggest region) – bided the position. Kaunas was selected to be appropriate to collect the award and succumb to the prestige and position of ECoC in 2022.

The national political backup for ECoC was provided through carrying out decision and idea at national level or stage at the statistics of financial contribution up to 50% of the initiative budget (not surpassing 10 million euros excluding capital investment). Politicians from municipalities of Kaunas district including the Kaunas region concurred on the program budget of 16 million euros in aggregate (4 million euros and 12 million euros, respectively). In regard to the application to the Klaipeda City, the three regions intended sharing nearly 16 million euros (the three regions of Palanga, Klaipeda including Neringa disseminating accordingly 700,000, 15 million and 243,000 euros). Although the above statistics indicates the same contribution and allotment from various municipal budgets, aggregate statistics for general execution of the initiative, although is not optimally shared.

Klaipeda intended to draw more financing from sponsors. From the experience of the previous title and position winners, this aspect of the budget normally varies from 10% to 30% of aggregate budget (Milton Keynes Council, 2015).

RESEARCH METHODOLOGY

This research paper focused and concentrated on comparative evaluation of application in various regions. Both candidate regions' budgets are compared, intending to emphasize on the political behaviors of the ECoC) including its marketing (See Figure 4.1).

Classification of culture is not emphasized in the research paper as we concur with Dessein and Soini (2016) and Sazonova (2014) that consider and see culture as the "space," including the foundation toward attaining the general objectives of sustainability. Also, the categorization of politics is exempted because the program of the ECoC is political in history and origin. The political part is necessary in regard to the regulations of the program.

In executing comparative research, TBL technique is applied including methodological basis. TBL is basically applied in weighing and determining output in regard to environment, social including economic determinants (Golden et al., 2015, p. 73). Research in sustainability marketing including sustainability (which are systematic in histories and origins) can actually proceed while applying the TBL technique, while the former is centered on the interrelation and connection between the three major sustainability measures and dimensions: economic, environmental/ecological, and social also known as the "3P," that is, profit, planet and people (Table 4.1).

Sustainability Marketing for ECoC • 101

FIGURE 4.1
The Lithuanian candidate cities for the European Capital of Culture, 2022. (From https://www.researchgate.net/figure/European-Union-EU-capital-cities-that-were-selected-for-the-analyses-Google-Maps_fig1_335399207).

TBL technique is executed by applying the content analysis techniques as requirement. The classical aspect of content evaluation is applied while determining major terms (sustainability marketing, marketing, ecology, and sustainability), including evaluating the context of application of sustainability marketing with sustainability. The necessary research structures are applied.

TABLE 4.1

Dimensions for the Evaluation

Sustainability Dimension	Classification	Sub-Classification
Economic	Profit	Employers, employees, hospitality, visitors, tourism, economics, industry environment, industry
Ecological (Environmental)	Planet	Landscape, natural environment, nature, ecosystem, ecology, etc.
Social	People	Children, grandparents, citizens, residents, community, women, youth, society, grandchildren, etc.

Statements with the terms sustainability including sustainable are compared or counted exempting the template phases.

The terms in regard to the sustainability parts concentrated on "3P" classifications are evaluated, intending in comparing or expressing the optimal among every of the three aspects of sustainability in both applications including the classification of sustainability also. Specific and precise sub-divisions have been selected to show under each of the Ps.

Criteria	Kaunas Project	Klaipeda Project
Leading City	Kaunas	Klaipeda
Including Areas	Kaunas District Municipality	Klaipeda District Municipality
Administrative Areas	Kaunas City Municipality	Klaipeda City Municipality
Population	Nearly 400,000	Approximately 190,000
Budget Total	28 million euros	32 million euros

The calculation and aggregate of the sub-classifications including the classifications are executed by applying the ATLAS.ti Software application (the application exhibits or calculates the amount of every sole word on both application; every words (subclassification) in regard to all *Ps* are selected and subjective by the authors). The outcomes of the aggregation and computation are compared or exhibited.

The information of the content evaluation is applied for the explanation of the qualitative vast characteristics technique applying DEX technique. The technique of sustainability marketing is centered or focused on the theoretical evaluations of the topic as well as in regard to recommendation of the DEX technique as proposed and suggested by Golob et al. (2015) and Bohanec (2015). In regard to this purpose, a main and major hierarchical technique of sustainable marketing (as a useful requirement for decision- and policymakers in awarding and delegating protocols or procedures) is well explained and elaborated.

The technique is centered on the qualitative variables when selecting the main and major attributes and it is also checked the analysis of marketing aspects regarding the evaluated applications. The three-grade weigh is used for evaluating the availability of the necessary sub-characteristics including its similarity (congruent in meaning). As a result of the limited and restricted inscriptions (the largest or biggest statistics of the statements in sole attribute does not exceed 10. We decided to select a range or vary from 0 to 10 or from 10 to 0 (as shown in Table 4.2).

TABLE 4.2

Three Grade Weigh or Scale toward the
Evaluation of the Necessary Sub-Characteristics

Score	Value
7–10	Maximum
4–6	Medium
0–3	Minimum

The technique is prepared by applying the DEXi software program. The outcomes are compared while exhibited on the assessment table, which is arranged and applying the DEXi modeling.

However, the TBL methodology permits organizations/firms to rearrange and reset their exercises (activities) in a proper and appropriate way; there are a few challenges toward determining the effect they have on social environment as well as nature (Hall and Slaper, 2011). The TBL methodology has various limitations or constraints (Jones and Sridhar, 2013):

- It is more useful and important for business organizations.
- The three separate and different accounts cannot be easily added and aggregated. It is challenging to determine the people including the planet in the exact words as the yields.
- It lacks the capability and the willingness to sum up the results across the three principles of the TBL.
- Environmental and social outcome is essential and unique in all forms of situation and it is challenging to quantify.

Analysis and the evaluation of the DEX-I program and with the DEX methodology also resulted to some limitations or constraints. The modeling of the aggregated and basic attributes and features is limited or restricted to three as "too much descendants reason and ensure the combinatorial blow up on the size or weigh of the similar utility functions" (Bohanec, 2015, p. 14); thus, the hierarchical technique of sustainability marketing is focused on majority of essential features (in the subjective comprehension or performance), limited, constraint, and restricted attribute in the recommendations and the scale menu such as "apply the least number of values <…> two to four" restricted and constrained the sustainability marketing model development also.

RESULTS

While evaluating and assessing the applications and the usage by the candidates, it seems that the candidates have various goals and objectives while acting as ECoC including executing the associated and connected cultural initiatives and programs. Kaunas handles the capability and the willingness to create and ensure a unifying portfolio and in becoming contemporary capital as major objectives while Klaipeda on other hand, identifies itself as a province intending to return normalcy to the region through the cultural orients. Both candidates majorly emphasize on the cultural needs and the social dimension in the society.

Various models of initiation of the ECoC have been copied and imitated by the candidates (respective) cities yearly (Herrero et al., 2006). Former encounters of other candidate regions (e.g., Cardiff) have indicated the cause and motive to involve in the bidding is actually linked with marketing including an depth marketing aspect and area of the city (Griffiths, 2006). Kaunas Program and notion to go for ECoC position, however, began from a category of different independent representatives. The group used to the Kaunas City Council as well as collected an unknown assistance from the initial. In regard to the City of Klaipeda, the initiative and program for demanding or ensuring for the ECoC position has also been added in "City Strategic Action Plan and Initiative for 2013–2020." This motive and reason toward effective involvement have been applied in the scenario of Liverpool, where the local initiatives and plan were connected to ECoC competition (Griffiths, 2006). The consistent and coherent orientation and idea of the city or region of Klaipeda toward this could be confirmed and attested by the proof and reason Klaipeda won the position of Lithuanian Capital of Culture in 2017.

Regardless of various histories of the plan, the cities ensured the explanation of the initiatives addition in their longer and wider (durable) term and strategies. Thus, we emphasize that the near cohesion between the sustainability approaches including the ECoC initiatives and programs would transpire if any of the latter were properly planned. An in-depth evaluation of the various documents, thus, somewhat exposed the weak or deficient integration of sustainability including the marketing techniques and principles in the initiatives (Table 4.3).

Sustainability Marketing for ECoC • 105

TABLE 4.3

Assessment of Klaipeda and Kaunas Bid Information

Classification	Klaipeda	Kaunas
Marketing		
"Marketing" word applied	20	40
Marketing means	Every groups (40 positions)	Every groups (25 positions)
Marketing and promotion budget		
Sustainability		
"Ecology" word applied	10	5
"Sustainability" word applied	7	9

Source: Broadcasting and Printed Materials; New Media and Modern Technologies, Special Events, Trade, other Programme emphasized by Maronić-Lamza et al. (2011).

The outcomes of evaluation which are depicted in Table 4.3 allow us to know the behavior of candidate regions toward marketing varies totally. Even if both the regions planned the different ways including the techniques of marketing reflecting and depicting all of the seven groups, evaluation of marketing spectrum emphasized that Klaipeda has described and has also showcased a more detailed marketing strategy. These discoveries bring about the question: What affected and influenced the various preference and choices of both regions?

The reason is that Klaipeda, as the third region of Lithuania, unarguably requires to investigate and explore toward a better and unique competitive advantage. Nevertheless, the impact or effect of the ideology of ruling parties could be experienced also (as a result of the durable term political governance of liberal party, citizens are valued, and handled as consumers).

Comparative evaluation exposes major inscriptions to be same (i.e., logos on transport vehicles, outdoor stands, computer games, virtual platforms, bilingual websites, etc.). However, both cities offered some various marketing means (e.g., Klaipeda: Meeting Points, SMS, Cultural Passport, Post Stamps, Postcards, Exposition of National Costume. Kaunas: Word of Mouth, A hedonometer for Analysis and Evaluation, Flying Balloons and Marketing through Sport). It becomes obvious and glaring while analyzing and evaluating the various proposals that the marketing strategy and plan of Klaipeda's entails more traditional ways than Kaunas that majorly went for the digital willingness or capabilities, therefore; attaining the prerequisites of the guide (European Commission, 2014).

Excess or severe types in regard to marketing ways and manners would be majorly important toward sustainability marketing (Figure 4.2). If solely such prerequisites were added at the initiative guide and sustainability as target were majorly considered in the regions initiatives. The outcomes of content analysis allow us in emphasizing that neither one nor other transpire or transpire *de jure*. Also, pertaining the formerly emphasized context, we view that more of the applicants also intended to modify idea or notion (i.e., brand) of the region as with Liverpool (Boland, 2010) including to market the region as the Lithuania's home of culture that would maximize and raise the nation's international fierce competitiveness. Content evaluation exposes emphasis to be on individual (i.e., social aspect) in both applications. The classification entails almost and nearly 600 words (e.g., youth, residents, citizens, community, society, etc.). Also, there is a close and a near to zero mention and emphasis of the planet (i.e., the ecological or the environmental part) and the classification of sustainability. The scale emphasizes and states the division or the classification of sustainability. The scale or weight emphasizes that the division and the classification among the divisions appear very similar or the same in both applications with the major and main concentration on people. Despite various and different questions dedicated or in regard to sustainability (Application Template Questions No. 5, 7, 17, 20, and 47), classification of sustainability is depicted very poorly. The same is also factual of the planet.

Concurring with the definition of sustainability as well as in regard to the recommendations and suggestions of the DEX technique (as proposed

FIGURE 4.2
Model and technique of the sustainability marketing.

by Bohanec, 2015), we develop or create a main and a major hierarchical technique, model, and tool of sustainable marketing, which could act as a significant technique for policymakers in selecting the winner and bidder of the ECoC.

Sustainability marketing entails a major characteristic in technique or it depicts major output. The minimal attribute depicts the inputs of actions, while the aggregate ones (intermediate attributes) entail and depict the intermediate outcomes/results. In fact, the technique or the model could be more developed and created entailing various policymakers, including their groups that have various and usually some conflicting aims and objectives (see, e.g., Bohanec, 2015).

Also, the model has been checked by assessing on the marketing aspects and parts of evaluated application. The outcomes of the evaluation (aggregated by applying the DEXi software program or tool) are shown and depicted in Table 4.4.

In regard to the outcomes of the analysis, it is proved that the marketing aspect of Kaunas application entails more to sustainability than corresponding aspect of Klaipeda application. Regardless of the effective stance of social marketing, the other two optimally essential aspects of marketing section (i.e., business marketing and environmental marketing) are optimally and majorly poor and weakly expressed and emphasized on the application.

TABLE 4.4

The Sustainability Marketing Scenarios in the Klaipeda's and Kaunas Application (Marketing Sections)

Kaunas	Option	Klaipeda
Maximum	Collaborative Business	Maximum
Minimum	Responsible Business	Minimum
Minimum	Social Business	Minimum
Medium	Business Marketing	Minimum
Minimum	Eco Practices	Minimum
Minimum	Nature Safeguard (Protection)	Minimum
Minimum	Environmental Marketing	Minimum
Medium	Behavior	Minimum
Medium	Values, Principles and Beliefs	Minimum
Medium	Social Marketing	Minimum
Medium	Sustainability Marketing	Minimum

108 • Social and Sustainability Marketing

The outcome of the paper, thus, results to the finalization that sustainability marketing is not properly used in both applications. The outcomes of general research, thus, result to finalization that sustainability marketing is not properly applied in the application.

DISCUSSION

Regardless of the fact and the reality that authors delegated and tasked various models of relations between sustainability and culture, it has already been confirmed that culture plays a tremendous and a significant responsibility and function and could be essential regarding to sustainable development. The assessment of ECoC reports or regulations exposed or emphasized that the program is still quite weak and deficient in the integration of the sustainability aspects generally.

In regard to Dessein and Soini (2016, p. 1), "it is necessary, vital or essential to majorly integrate and allow culture in the Discussion of Sustainability as attaining Sustainability objectives basically focuses on the human actions, accounts and behaviors that are also culturally embedded"; therefore, we emphasize that ECoC initiative and program must be shifted and migrated from the mode 1 culture in sustainability to mode 3 culture as sustainability (which is depicted in Figure 4.2). The actions or decisions of valuable authorities or officials are major significance to the modification.

Therefore, we emphasized on importance of assessing or evaluating on the aims or policies of the ECoC document while integrating increased connection to general and major sustainable development. In admitting and recognizing the power and the authority of marketing, the ECoC must clearly interrelate sustainability and culture, providing a consistent and a coherent sustainability marketing (marketing strategy) grounded or in regard to sustainable reasoning. The use of professionals including the necessary decision-making support systems in sustainable development can result to attainment of this initiative or objective while addition of joint and associated culture – sustainability marketing entails that the assessment as well as the evaluation scenario in knowing the winners or the highest bidders of the ECoC position could be stressing and highlighting further.

CONCLUSION

It has already been confirmed and attested that sustainability as a common and a major worldwide perspective must be completely and fully integrated in societal beliefs and on individuals' behaviors, principles, and values. Culture seen as an array of shared values, beliefs, and anticipated behaviors could then be a major technique in the empowering of individual with the new and modern comprehension of the common and major globe and its challenges or with the courage and the zeal to concentrate on the sustainability.

The appropriate and the necessary cultural policy integration or inculcation in the general and major development of social system may change, transform, or impact on cultural beliefs, principles including values thus moving or migrating them toward sustainable behavior including sustainability. The ideal outcome of this change would reshape, transform, as well as shape the true and actual depiction of the natural environment, seeing and viewing it as constituent of culture and would result to entirely modern, current, and trending programs by accommodating and ensuring for sustainability ideologies, initiatives, or principles, which is also very crucial, important, vital, and so necessary. The shift as well as the various changes could also lead to the modern means of self-governance as well as meta-governance.

Political actions including decisions are majorly essential for sustainable development and for sustainability. Local and regional strategies must entail policies, ideas, and initiatives that engage and involve citizens into an individual and a collective sustainable actions and results to a sustainable behavior as well as even transform and change into sustainable environment inhabiting within sustainable self-governance.

Enhancing sustainability in territory is so necessary while serving as ECoC. ECoC initiative plays an essential function or responsibility while changing to present way of development. Marketing the initiative and program is described and emphasized as comprehensive and well-detailed action.

It is a common and major discovery that the cities or regions in conditions of fierce competition and demanding for changes in the socio-spatial connection and relations (rural exodus, de-industrialization, etc.) have demanded for various ways of attracting capital (tourism, investment) to restore their "local economy." Thus, it is not fortuitous that there is an intense and a severe competition among the cities of the same nation to host a mega and a big event such as the institution of the ECoC. Major attributes or features of almost all the proposal by the candidate and various cities have been on "Urban Generation" (leisure areas, modern cultural facilities, thematic parks, revival of areas, enhancement in infrastructure, transport, and the improvement on the environment) actions that are supported by the EU though "Culture Initiative and Program." The mega and different events have been an opportunity and a privilege for cities to boost their cultural sectors, introduce infrastructural improvements, create employment, and attract investors. At times, social changes have increased as a result of the compulsory and mandatory displacement of the populations (usually unemployed or poor citizens) from areas undergoing the urban regeneration.

Marketing needs to be oriented and focused to the whole community, its social goals, as well as safeguard of the cities must entail every sustainability aspects and must be properly encouraged by applying the different ways of marketing.

Therefore, sustainability marketing is of major significance from the implementing and the executing side and also the initiating aspects of the initiatives. Sustainable marketing, not just sustainability solely, needs to be seen as the necessities and requirements toward the selection criteria while competing toward the ECoC position and future ex-post evaluation of impacts of ECoC. Decision-making support systems also need to assist both professionals and politicians to enhance the ECoC programs that have impact or effect on transformation or the rapid change to sustainability.

FUNDING DECLARATION

This paper was wholly funded and supported by the author and it will be deemed necessary if funding is provided for this paper for the investigation carried out on this paper and for more future study.

ACKNOWLEDGMENT

I want to thank my professors, colleagues, and editors including my parents and siblings who made it possible for me to complete this project and most especially, I would like to thank the Almighty God for his Divine assistance in seeing me through in completing this project.

AUTHOR STATEMENT

This paper is vital for everyone; readers should be able to understand what it means by the European Capital of Culture (ECoC). I emphasized on the evaluation on the sustainability level in regard to the competing and bidding of the position of the ECoC, I viewed the programs and initiatives of the ECoC and the competition, marketing, and various cities that are bidding for this prestigious position "The ECoC" and applying the triple bottom line technique which is executed by using the content analysis technique and the DEX methodology in evaluating the cities.

REFERENCES

Aubert, A., Raffay, Z. & Marton, G. (2015). Implications of the European Capital of Culture (ECoC) title of Pécs on the city's tourism. Timisiensis Geographica, 24(1), 31–42.

Belz, M.F. & Peattie, K. (2010). Sustainability marketing – An innovative and new conception of marketing. Marketing Review St. Gallen, 27(5), 8–15, https://doi.org/10.1007/s11621-010-0085-7

Bohanec, M. (2015). DEXi: Program for Multi – Attribute and the Characteristic Decision Making. User's Manual. Version 5.00, Institute "Stefan Jožef", Ljubljana, Slovenia, Available from: http://kt.ijs.si/MarkoBohanec/pub/DEXiManual500.pdf

Boland, P. (2010). 'You must be having a laugh – Capital of Culture!' Challenging and countering the official and the formal rhetoric of Liverpool as the 2008 European cultural Capital. Cultural & Social Geography, 11(7), 627–645, http://dx.doi.org/10.108 0/14649365.2010.508562

Commission Staff Working Document. (2012). European Capitals of Culture (ECoC) post 2019 Accompanying the Proposal Document for the European Parliament Decision as well as of the Council establishing and setting up a Union action for the European Capitals of Culture (ECoC) for the years 2020 to 2033. Brussels, 20.7.2012 SWD 226 final, Available from: http://eur-lex.europa.eu/legal-content/EN/TXT/PDF/?uri=CELEX:52012SC0226&from=EN

Cox, T. & Garcia, B. (2013). European Capitals of Culture (ECoC): Long-Term effects and Success Strategies study. European Union (EU) Directorate general for internal policies and regulations policy department b: structural and cohesion including the structural policies education and culture.

Dessein, J. & Soini, K. (2016). Sustainability-culture relation: Towards a conceptual technique and framework. Sustainability, 8(2), 167, http://dx.doi.org/10.3390/su8020167

Dessein, J., Soini, K., Horlings, L. & Fairclough, G. (Eds.) (2015). Culture for, as and in Sustainable Development, Conclusions and also the Recommendations from the COST Action IS1007 Exploring Cultural Sustainability, University of Jyväskylä, Finland.

Draghici, C. et al. (2015). The role of European Capital of Culture (ECoC) status and the identity in structuring and in modifying economic profile of Sibiu, Romania. Procedia Finance and Economics, 26, 785–791, http://dx.doi.org/10.1016/S2212-5671(15)00870-9

Džupka, P. & Hudec, O. (2016). Culture-led regeneration through the young generation: Košice as the European Capital of Culture (ECoC). European Regional and Urban Studies, 23(3), 531–538, https://doi. org/10.1177/0969776414528724

ECoC. (2016). The Selection Panel's report, Pre-Selection Period and Stage, Vilnius, Available from: https://ec.europa.eu/programmes/creative-europe/sites/creative-europe/files/files/ecoc-2022-lithuania-preselection.pdf

Edwards, G.M. (2009). An integrative meta theory and principle for sustainability and organizational learning in turbulent and challenging times. The Learning Organization, 16(3), 189–207, https://doi.org/10.1108/09696470910949926

Epstein, J.M. (2008). Making Sustainability Function and Work: The Appropriate Practices in Measuring and Managing Corporate Environmental, Social and Economic Impacts, Greenleaf Publishing.

Eroglu, O., & Piçak, M. (2011). Entrepreneurship, national culture and Turkey. International Journal of Business and Social Science, 2(16).

European Commission. (2014). European Capitals of Culture (ECoC), 2020–2033. Guide for cities getting ready to bid.

Giltrow, M. (2015). Exploring the Relationship of Welfare with Sustainable Attitudes, Values and Behavior, In L.W. Filho (ed.), Transformative Approaches and Techniques to Sustainable Development at Tertiary Institutions, Springer International Publishing.

Golden, S.D., McLeroy, K.R., Green, L.W., Earp, J.A.L., & Lieberman, L.D. (2015). Upending the social ecological model to guide health promotion efforts toward policy and environmental change. Health Education & Behavior, 42(1_suppl), 8S–14S, https://doi.org/10.1177/1090198115575098

Griffiths, R. (2006). City/Culture discussions: Evidence and proof from the competition to choose the European Capital of Culture (ECoC) 2008. European Planning Studies, 14(4), 415–430, https://doi.org/10.1080/09654310500421048

Hall, J.T. & Slaper, F.T. (2011). The Triple Bottom Line (TBL): How Does It Work and What Is It? Available from: http://www.ibrc.indiana.edu/ibr/2011/spring/article2.html

Hawkes, J. (2001). The Fourth Pillar of Sustainability: Culture's Important Role and Responsibility in Public Planning, Cultural Development Network & Common Ground Publishing (Vic), Melbourne.

Helmig, B. & Thaler, J. (2013). Theoretical and conceptual framework of the social marketing effectiveness: Drawing and painting the big picture on its effectiveness and functioning. Journal of Nonprofit & Public-Sector Marketing, 25, 211–236, https://doi.org/10.108 0/10495142.2013.819708

Herrero, C.L., Sanz, Á.J., Bedate, A., Devesa, M. & Del Barrio, M.J. (2006). The economic effect of cultural events: A case study of Salamanca 2002, European Capital of Culture (ECoC). European Regional and Urban Studies, 13(1), 41–57, https://doi.org/10.1177/0969776406058946

Jones, G. & Sridhar, K. (2013). The three essential criticisms of the Triple Bottom Line (TBL) technique and approach: An empirical research study to connect sustainability reports in industries and firms headquartered in the Asia-Pacific axis and Triple Bottom Line (TBL) pitfalls. Asian Journal of Business Ethics, 2(1), 91–111.

Karstens, B. & Belz, M.F. (2005). The Instrumental and the Strategic Sustainability Marketing in the Western European Food Processing Firm/Industry: Hypothesis and Conceptual Framework. Proceedings of the Corporate Responsibility Research Conference, Euromed Management School Marseille France, 4, Available from: http://www.crrconference.org/Previous_conferences/downloads/belz.pdf

Lim, W.M. (2016). A blueprint towards sustainability marketing. Marketing Concept and Theory, 16(2), 232–249, https://doi.org/10.1177/1470593115609796

Maraña, M. (2010). Development and Culture, Prospects and Evolution, Bilbao, Spain, UNESCO Etxea (UNESCO Etxea Working Papers, No. 1).

Maronić-Lamza, M., Mavrin, I. & Glavaš, J. (2011). Marketing aspects of the European Capital of Culture (ECoC) initiative and programme. Interdisciplinary Management Research, 7, 130–140.

Milton Keynes Council (2015). Bidding and Negotiating for the Cultural Status. Feasibility Study, Available from: https://www. milton-keynes.gov.uk/assets/attach/28775/MK%20Capital%20of%20Culture%20Report.pdf

Morse, S. & Bell, S. (2003). Measuring and Determining Sustainability: Learning and Also Adapting by Doing, Earth Scan Publications Ltd.

Nechita, F. (2015). Bidding and negotiating for the European Capital of Culture (ECoC): Common weaknesses and strengths at the pre-selection period and stage. Bulletin of the Transylvania, University of Brasov, Series VII: Social Sciences, Law, 8(1).

Nemeth, A. (2010). Mega-events including its sustainability and its major effect on spatial development: The European Capital of Culture (ECoC). The International Journal of Interdisciplinary Social Sciences, 5, 265–278.

Nkamnebe, D.A. (2011). Sustainability marketing in the Emerging Markets: Challenges, imperatives and also the agenda setting. International Journal of Emerging Markets, 6(3), 217–232, https://doi.org/10.1108/17468801111144058

Nurse, K. (2006). Culture as the fourth pillar towards sustainable development. Small States: Basic Statistics and Economic Review, 1, 28–40.

O'Callaghan, C. (2012). Creative tensions and urban anxieties in the European Capital of Culture (ECoC) 2005: 'It could not just be about Cork, like'. International Journal of Cultural Policy, 18(2), 185–204, https://doi.org/10.1080/10 286632.2011.567331

Opuku, A. (2015). The Role and Responsibility of Culture in a Sustainable Built Environment, In A. Chiarini (Ed.), Sustainable Operations Management, Springer International Publishing, 37–52.

Palmer/Rae Associates (2004). European Cities and Capitals of Culture, Brussels: Palmer/Rae.

Powell, A. & Fithian, C. (2009). Cultural Aspects of the Sustainable Development, Available from: http://webmail.seedengr.com/Cultural%20Aspects%20of%20Sustainable%20Development.pdf

Rakic, M. & Rakic, B. (2015). Holistic and Comprehensive Management of Sustainability Marketing in the Process of Sustainable Development, Management and Environmental Engineering Journal, 14(4), 887–900, Available from: http://omicron.ch.tuiasi.ro/EEMJ/

Richards, G. (2000). The European Cultural Capital Scenario and Event: Strategic Weapon and Tool in the Cultural Arms Race. Journal of Cultural Policy, 6(2), 159–181, https://doi.org/10.1080/10286630009358119

Sakkers, H. & Immler, L.N. (2014). (Re)Programming Europe: European Capitals of Culture (ECoC): Re-reasoning the role and responsibility of culture. Journal of European Studies, 44(1), 3–29, https://doi.org/10.1177/0047244113515567

Sazonova, L. (2014). The Cultural Aspects and Areas of Sustainable Development: Glimpses and Glances of the Ladies' Market. Friedrich-Stiftung-Ebert, Office Bulgaria.

Scammon, D. (2012). Culture and Sustainability: How Do They Function Together? LCC 480 Senior Seminars, 30 April 2012, Available from: http://www.academia.edu/1817961/Sustainability_and_Culture_How_do_they_work_together [cited from A. Chiarini (ed.), Sustainable Operations Management. Springer International Publishing].

Senkus, P. (2013). Marketing 3.0: The Challenge for Public, Private and the Non-profit Sectors and Theoretical Technique and Approach. Rural Development. The sixth International Scientific Conference, Proceedings, 6, 328–335.

Smith, B. & Senge, P. (2008). The Necessary and Important Revolution: How Organizations and Individuals Are Working Together and as One in Other to Create a Sustainable Globe, United States of America (USA).

Steiner, L., Hotz, S. & Frey, B. (2015). Life satisfaction and European Capitals of Culture (ECoC). Urban Studies, 52(2), 374–394, http://dx.doi.org/10.1177/0042098014524609

Swilling, M. & Peter, C. (2014). Connecting Sustainability and Complexity Theories: Implications for Modeling Sustainability Transitions. Sustainability, 6(3), 1594–1622, http://dx.doi.org/10.3390/su6031594

Turşie, C. (2015). Re-Inventing the Centre-periphery Relation and Connection by the European Capitals of Culture (ECoC), Case-studies: Marseille-Provence 2013 and Pecs 2010, Euro limes, 19.

Van Melik, R. & Van Aalst, I. (2012). City festivals and urban development: Does place actually matter? European Urban and Regional Studies, 19(2), 195–206, https://doi.org/10.1177/0969776411428746

Wróblewski, Ł. (2016). Creating and ensuring an image of a region–Euro region Beskydy and also the Euro region Cieszyn Silesia instances. Zarzadzanie I Ekonomia i, 8(1), 91–100, https://doi.org/10.1515/emj-2016-0010

5

Cool Branding for Indian Sustainable Fashion Brands

Jasmeet Kaur and Gursimranjit Singh
Mittal School of Business, Lovely Professional University, Phagwara, Punjab, India

CONTENTS

Learning Objectives 115
Theme 116
Theoretical Background 116
Is Sustainability Cool? 117
Introduction 118
Objectives 121
Literature Review 121
Sustainable Fashion 128
Methodology 133
Marketing Fashion Sustainable Brands Using Cool Branding 134
Discussion 136
Conclusion 137
Lessons Learned 138
Managerial Implications 138
Credit Author Statement 139
References 139

LEARNING OBJECTIVES

After reading the chapter, the reader should be able to:

- Understand characteristics independently contributing to overall brand coolness

- Explore how brand coolness varies across cultures and how it contributes to brand's success
- Understand how brand coolness relates and affects brand personality
- Identify potential of cool branding marketing for sustainable brands

THEME

Cool branding and sustainability

THEORETICAL BACKGROUND

Many researchers have studied cool branding, but scarce work has been found on the effect of cool branding strategies on sustainable fashion. Thus, this is an attempt to understand through literature review how brand coolness relates to enhancing marketing for sustainable fashion brands. The study provides a valuable insight into cool concept as a multidimensional construct and attempts to explore the potential of cool branding marketing for sustainable brands.

One of the largest industries worldwide, the fashion industry, is prospering and providing great opportunities to new industrialized countries, is facing challenges to curb with the environmental issues causing severe problem. The fashion and apparel industry is considered as an ever-growing market, but facing sustainability issues. The problems in the fashion industry lie in the entire production chain.

Sustainability has given fashion and apparel brands various directions. Sustainability emerged as "Mega Trend" (Mittelstaedt et al., 2014); buzz words like eco, organic, environmentally friendly or green are interchangeably used in marketing. Researcher states that sustainable fashion is perceived as an alternative to fast fashion. The aim is to recreate a system with less human impact on the environment. But it is also stated that fashion consumers do not compromise the aesthetics of their fashion clothing to be environmentally conscious (Barnes et al., 2006).

Solomon and Rabolt (2004) argue that fashion and clothing belong to products that are self-expressive, thus emotions related to consumption are important, especially after any purchase event. Thus when consumption-related emotions are concerned, the purchase situation is considered to be a strong positive experience for the consumer. He also states that the nature of such experiences is short term and has least connection to experience of deep satisfaction. Brands try to communicate the benefit of sustainable apparel to the consumers. Most manufacturers and businesses are aware of issues related to quality, design and price, and adequate steps are being taken in this regard.

Cool branding on the other hand has been studied and proven to help marketers build brand image. Known to be rarely outdated and a motivator to many publications (Budzanowski, 2017; Warren et al., 2019), the word "cool" has positive implication, which is shifted to cool brand. Cool brands are claimed to set own trends, and consumers continuously demand for latest ideas and are open to experiences. Cool brands are claimed as authentic when their actions reflect about what the brand really is.

Literature states that marketing holds high value of cool factor and has inspired managers and consumers (Gladwell, 1997; Olson et al. 2005; Warren & Campbell, 2014). Cool brands are considered contemporary, but also reflect the feeling of revolution, rebel and environmentally responsible. Fashion sustainable brands also emotionally involve consumers with their products to motivate purchase.

The aim of the chapter is to review and understand different characteristics independently contributing to overall brand coolness and explore how brand coolness varies across cultures and how it contributes to brand's success. This chapter attempts to understand the concept of cool branding to be used as a marketing strategy to enhance brand building of sustainable fashion brands and how cool branding can influence purchase intention for sustainable products.

IS SUSTAINABILITY COOL?

Sustainability in the fashion and textile industry is motivating many stakeholders to contribute their part by offering sustainable fashion products. Literature states that the slow shift in consumer purchase behavior gives marketers and researchers an opportunity to explore. Marketers are looking for ways to build branding for sustainable fashion brand in order to address the alarming situation created by the industry.

On the other hand, marketers are interested in "cool", as past researchers suggest that it has the ability to add extraordinariness to ordinary things. This study tries to find out positive effects of coolness and also examines the environments where the cool factor can spark excitement and desirability for sustainable brands. Extensive literature study gives insight on how cool branding can impact marketing and consumption of sustainable fashion clothing.

Thus, this chapter attempts to find an approach toward understanding sustainability in fashion by finding the opportunities and ways to relate it to the concept of cool, to help introduce more clarity in both concepts. This is where this chapter is grounded.

INTRODUCTION

The fashion and apparel industry, a product of modern age, holds an important position globally. Before the mid-19th century, the demand of clothing was fulfilled by hand-sewn garments made in home production or by seamstresses and tailors. But by the beginning of the mid-20th century, with rise of factory system of production and introduction of new technology like sewing machine, the beginning of mass production of clothing in standard sizes sold at fixed prices was seen.

Growth, Trends and Forecast (2020–2025) report on the global textile industry states that US$920 billion is estimated for 2018, and Compound Annual Growth Rate (CAGR) is estimated approximately 4.4% by 2024 to reach US$1,230 billion. There are four levels at which the industry works globally. The first level is the production of raw material, i.e., fiber. The second level is conversion of raw material to actual product and production of fashion goods by export houses, designers and other manufacturers. The third level consists of sales of fashion products through retail outlets and fourth level is promotion. All these levels have interdependent sectors, operating to fulfill consumer demand and earn profit for business.

The United States leads in production and exporting for raw cotton and China leads in producing and exporting both of raw materials and garments. European Union consists of Germany, France, Spain, Italy and Portugal as main regions with maximum contribution to the textile industry. India is the third largest textile manufacturing industry globally. Indian Brand Equity Foundation in its report claims that Indian textile and apparel exports was US$38.70 billion in 2019 and was S$22.95 billion in FY20 (in account till November 2019) and is expected to grow by US$82.00 billion by 2021. The contribution of the industry is approximately 7% of the industry output (in value terms) of India in 2018–2019, the second largest employer, contributes 2% to the GDP, and approximately 15% contribution to the export earnings of India (as accounted in 2018–2019).

The textile industry is considered among the oldest industry of Indian economy, which contributes for various market segments, within India and across the globe, offering a variety of products. The Textile industry plays an important role in the India's identity and rich cultural heritage. The first textile mill was established in Calcutta, it was this time when the modern textile industry got its recognition and became advent in the early 19th century. Today, it is an important contributor in Indian economy, as India is the largest producer of jute and cotton fibers and marks itself in the global market as the second largest producer for silk.

Ease of accessibility and supply of raw materials, low cost of manufacturing than other competing countries, low labor cost, etc. are some major factors that work in favor of the development of industry and its increasing contribution in Indian economy. The Indian Textile Industry report by IBEF (2020a, 2020b) states that with increase in economic growth, disposable income has also seen rise in demand and consumption at domestic market. This evidently explains that the future is promising.

From cultivation of fibers, excessive use of pesticides, exploiting of workforce, animal cruelty, with issues related to the fast fashion, dumping of clothes in landfill, etc., nature and environment are exploited. Other problems are high-water consumption and effluents being discharged in water bodies, air emission by factories, workspace safety, etc., thus causing damage to environment (Donnell et al., 2000).

Any textile product harms environment through three major channels: Production issues, maintenance issues and disposal issues. Production issues may include toxicity of chemicals released during production and processing, renewability of raw materials, etc. Harmful chemicals used for dry cleaning or laundering purpose are considered as maintenance issues.

A product being recyclable or biodegradable is disposal issues (Karthik & Gopalakrishnan, 2014).

As apparel is known to express current fashion trends, consumers' personalities and tastes, and is also considered a form of building a perceived identity. Researchers state that there are three needs of fashion consumers that include physical, emotional and psychological needs (Niinimaki, 2010; Solomon, 2004).

Many diverse discipline literature studies on what cool means or signifies (Belk et al. 2010; Nancarrow et al. 2002; Pountain & Robins, 2000). Few researchers have worked on understating coolness origin (Belk et al. 2010; Donnell et al. 2000; Nancarrow et al. 2002; Pountain & Robins, 2000), others have studied coolness association with traits and usage (Dar-Nimrod et al. 2012; Rahman et al. 2009), some have worked to identify the logical precedes or antecedents (Warren & Campbell, 2014).

Ferguson (2011) states that setting a definition for cool is a difficult task and claims that cool factor is transferred to user through product and also from celebrity to brand and later from brand to consumer (Ferguson, 2011). Researchers have highlighted few important features of coolness (Anik et al. 2017). Coolness is suggested as subjective, i.e. brands stay cool or uncool till the time consumers perceive them as cool (Gurrieri 2009; Pountain & Robins, 2000). Studies state that consumers connect cool product with hedonic value (Im et al. 2015), excellence (Mohiuddin et al. 2016) and usefulness (Runyan et al. 2013; Sundar et al. 2014). Coolness is related to goodness or positive valence (Dar-Nimrod et al. 2012; Mohiuddin et al. 2016). This feature directly relates to finding the potential of cool branding marketing for sustainable brands.

The study is unique as cool and sustainability will work as a trigger for the right type of brands to promote products to right type of consumers and its combination is said to add additional value to industry (Sachkova, 2018).

This study reviews the key concepts and characteristics related to "cool" and explores benefits of cool as a concept for sustainable fashion industry. This chapter aims to identify potential of cool branding marketing in sustainable fashion context. The main argument of the review lies in understanding the complex nature of cool as concept, and sustainability and fashion as a system which are studied closely and are intertwined to focus on how cool branding can enhance the marketing and consumption of sustainable fashion clothing.

OBJECTIVES

1. To understand characteristics independently contributing to overall brand coolness
2. To explore how brand coolness varies across cultures and how it contributes to brand's success
3. To understand how brand coolness relates and affects brand personality
4. To identify potential of cool branding marketing for sustainable brands

LITERATURE REVIEW

Brand management gaining interest in the occurrence of cool brands and building "cool" factor among consumers seems more like a strategy. A brand being cool is considered a differentiating factor (Gurrieri, 2009; Nancarrow et al. 2002).

A cool brand is considered unique and is changing the response in consideration to what masses adopt (Rahman et al. 2009). Cool is considered progressive in terms that it supports latest trends (Gurrieri, 2009). Thus, consumers expect brand to be stylish (Rahman et al. 2009), possess high quality of design (Nancarrow et al. 2002), to be socially responsible and also one that is a product of unique creativity.

To understand consumer behavior toward cool brand, many researchers have claimed to identify characteristics of cool brand. Researchers also state that as cool is considered as an experience to be lived, brands are observed "cool" as an attitude or personality type (Nancarrow et al. 2002). A cool brand is said to be inspiring and special, influencing consumer's mindset, and is said to have connected experience between customer and company and build personal relationship (Southgate, 2003).

Researchers state that commercial institutions like advertising, retailing and media focus on the creation of cool-branded objects that help to create the origins, transformations and consumption of cool in turn. But the main question that arises is what is "cool as a concept" and what leaves products with cool factor a commodity of desire. For understanding the "beginning of cool", slavery and discrimination of African American men needs to be understood. Cool is considered to be taken to American with slavery, remains unidentifiable even today (Gurrieri, 2009; Nancarrow et al. 2002). The advent of this word was in "Birth of Cool" album by Miles Davis. Researchers say that the spread of cool spirit was through Jazz society of the 1950s (Gurrieri, 2009; Nancarrow et al. 2002; Pountain & Robins, 2000). In the 1960s, cool was connected with hippies, 1970 with Punk explosion and 1990 with hip hop culture (Pountain & Robins, 2000). With time, there has been multiple meanings attached with the word cool. As consumers have become more materialistic and brand usage defines identity, "cool" has been linked with consumerism from past two decades (Rahman et al. 2009).

Word "cool" and ideas that closely relate to it can be found from past two decades. From 1950s Jazz representing black musician with "cool mask"(Belk et al. 2010) to convey resistance to exploitation and discrimination (Nancarrow et al. 2002; Pountain & Robins, 2000). This was all to show disconnection from culture they lived in and white audience. James Dean and Miles Davis were considered icons representing coolness for certain style of music and attitude (Pountain & Robins, 2000). Coolness represented multi-facet conceptual element, the journey of "cool" continued and showed its impact. Currently, this word is used when something different or unique (Belk et al. 2010) is encountered, something special or else trendy, desirable, up to date, etc. (Runyan et al. 2013).

Table 5.1 shows characteristics of brand coolness (as defined by different authors) and relevant citations from prior research.

TABLE 5.1

Brand Coolness Characteristics and Relevant Citation from Past Researchers

Research Field Defining Coolness	**Exemplary Literature**
A social construct and subjective	Warren and Campbell, 2014
Way of stating that something is desirable or good	Pountain and Robins, 2000
Perception related to impression to get recognition and validation by peer audience	Belk et al. 2010
Dynamic	Belk et al. 2010; Rahman et al. 2009
Being authentic	Southgate, 2003
Original, possessing aesthetic appeal, creative and innovative	Bird and Tapp, 2008
Eye-catching, fashionable and entertaining	Rahman et al. 2009

Table 5.1 explains how different authors define characteristics of brand coolness differently. Warren and Campbell (2014) defines "cool" to be a social construct and subjective in nature. Rahman et al. (2009) defines cool as eye-catching, fashionable and entertaining. Cool is also considered to be authentic (Southgate, 2003) and dynamic (Belk et al. 2010; Rahman et al., 2009). Coolness is also stated as autonomous (Pountain & Robins, 2000) and desired (Dar-Nimrod et al. 2012; Mohiuddin et al. 2016). High status and offering lower prices are characteristics of desirability, being dominant and uniqueness, signals autonomy making brand cool. Literature does not claim which characteristic of being autonomy and desirability makes brand cool or not cool. Studies also suggest that coolness is said to be in connection to hedonism, youth, emotional concealment and excitement (Bird & Tapp, 2008; Nancarrow et al. 2002; Pountain & Robins, 2000), but no theory discusses factors distinguishes cool from uncool brand.

Anthropologist, psychologist and marketing researchers have approached "cool" as a design attribute (Sundar et al. 2014), personality trait (Dar-Nimrod et al. 2012, a cultural phenomenon, an attitude (Pountain & Robins, 2000), etc. Different parts of history state different meaning of cool. It signifies that cool can be related to irony, mystique, authentic heritage, confidence and irreverence. One part also states cool to be rebellion (Belk et al. 2010; Ferguson, 2011).

Research by Warren et al. (2019) suggests different terms for defining characteristics of brand coolness: Rebellious, original, extraordinary/useful, authentic, aesthetically appealing, subcultural, popular, energetic, high status, iconic.

The component characteristics of brand coolness by Warren et al. (2019) has definition and meanings suggested by different authors (Figure 5.1). Extraordinary/useful is said to be a positive quality which helps define a brand apart from its competitors or brands that offer higher functional value with its product (Belk et al. 2010; Dar-Nimrod et al. 2012; Im et al. 2015; Mohiuddin et al. 2016; Runyan et al. 2013; Sundar et al. 2014).

High status is associated with sophistication, prestige, social class and esteem (Belk et al. 2010; Nancarrow et al. 2002; Warren, 2010). Rebellious is defined as a tendency to combat conventions and social norms and oppose, fight and subvert (Nancarrow et al. 2002; Pountain & Robins, 2000; Warren & Campbell, 2014).

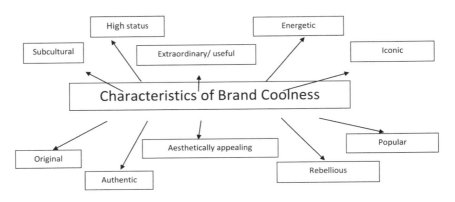

FIGURE 5.1
Component characteristics of brand coolness (Warren et al., 2019).

Other components for cool branding consist of being original, authentic, visually appealing and subculture. Original is being defined as a tendency to do things that have not been done before and to be different and creative (Mohiuddin et al. 2016; Runyan et al. 2013; Sundar et al. 2014; Warren & Campbell, 2014) and being authentic is to behave in ways that are true to their roots or are consistent (Nancarrow et al. 2002; Sriramachandramurthy & Hodis, 2010). Few other definitions to define characteristics of brand coolness are being aesthetically appealing, which are defined to possess visually pleasing appearance and being attractive (Dar-Nimrod et al. 2012; Runyan et al. 2013; Sundar et al. 2014). Subculture has association with people who operate independently from mainstream society and are known as autonomous group (Belk et al. 2010; Runyan et al. 2013; Sundar et al. 2014).

Among features of coolness highlighted by researchers, autonomy is the one that distinguishes a cool product or brand being desirable. Thus, autonomy is willingly following its own path as per expectation and other's desires (Warren & Campbell, 2014). Autonomy is not a subject to directly being observed, but is perceived on the basis of fighting conventions and norms (being rebellious) (Pountain & Robins, 2000), moving beyond convention and norms (being original) (Mohiuddin et al. 2016; Sundar et al. 2014; Warren et al. 2019) and experience pressure and continuously adopting shifting trends (being authentic) (Nancarrow et al. 2002; Sriramachandramurthy & Hodis, 2010).

Being dynamic is considered as another feature. Studies state that brands perceived as cool today may or may not be perceived as cool tomorrow

(O'Donnell & Wardlow, 2000; Pountain & Robins, 2000). Not only has this but characteristics of coolness has also changed over time, across different consumers also. Initially within the specific subculture only, brands become cool, and then later get adopted by broader audience (Belk et al. 2010; Gladwell, 1997).

Directly or indirectly, many diverse perspectives have been studied on cool (Rahman et al. 2009). Since 1960s, idea of being cool has been used in marketing. Association with hippie culture directly connected it with ecology movement. The hard work of the commercial marketers to use the factor of being cool during 1960s led to mediate the use of this term globally (Belk et al. 2010). This further evolved through focused marketing on youth markets in the 1980s and cool was associated with cosmopolitanism and consumerism. Driving aspirations, attractive and fun are some characteristics that 1960s cool-focused marketing associated with. The 1980s marketing connected with cosmopolitanism and consumerism and mainly targeted youth market.

But a major question arises during the review that how a brand is made cool? This question holds theoretical and practical importance, still answer remains unclear. Although few researchers investigated different factors making a brand cool, like personality traits related to cool brands (Dar-Nimrod et al. 2012; Warren et al. 2018) and cool technologies. Few studies also focused on how novelty (Im et al. 2015) and autonomy (Warren et al. 2019; Warren & Campbell, 2014) have influenced a brand for being a cool brand.

A challenge faced by most businesses while establishing and marketing of brands is to consider cultural differences which influence how individual perceives a brand. Brands are perceived differently across culture. As culture influences, it influences a person's practices, norms and values; this becomes an important factor as to how consumer perceives a brand across culture. Thus, needs, wants and usage pattern of product of a consumer from different culture vary.

An interesting fact found out during the literature review was, "Cool" as a term is used to describe brands which are known cool by specific subculture, despite being yet adopted by masses (Warren, 2010) and brands that general population considered to be cool. Warren's (2010) theory states brands to be niche and mass cool. Taking in consideration brands like Harley Davidson and Nike, a question arises is that how cool imagery can affect brand image. For example, mountain dew perceived as an iconic brand, how apple stay ahead in competition being world's first trillion dollar brand as per Forbes 2015. Moreover, Olson et al. (2005) claim that brand possessing cool factor has competitive advantage over other brands and can charge higher price.

Being authentic, original, rebellious, etc., brands initially get cool among small subculture and referred to as niche cool like Steady Hands, Mitsky, INSIDE, etc. Over time, with more fame and acceptability, niche brands when adopted by bigger audience, are then referred to as mass cool like Beyoncé, Nike, Grand Theft Auto, etc. (Gladwell, 1997; Warren & Campbell, 2014). Certain characteristics affect the cool brand to change over time and also change from niche cool to mass cool brand.

Warren et al. (2019) state that mass cool brands are perceived less authentic, subcultured, rebellious, extraordinary and rebellious when compared to niche cool brands. The end consequences also change when brands shift popularity from being niche cool to become mass cool. As mass cool brands are more popular, consumers intend to share more word of mouth for mass cool brands in comparison to niche cool brands. Likewise, mass cool brands can attract higher price as they are found more popular in market.

Cool brands have distinguishing characteristics from uncool brands. With multiple exposure to different marketing techniques, consumers form own perception about brand coolness (Warren et al., 2019). The life cycle of coolness also states that brands that are perceived mass cool like Nike seek high price, are more popular and earn more profit than any other niche cool brand. Brands need to take care of not losing their cool characteristic like desirability, autonomy, etc. Literature lacks justification about the characteristics that differentiate between cool and uncool brands, nor emphasize on how different attributes change when brand shifts from Niche cool (among small subculture) to mass cool (within broader population) (Warren, 2010).

Brand managers need to know the actual position of the brand whether it is uncool, niche cool or mass cool. Brand which is currently uncool needs to become niche cool by engaging in strategies for product, promotion, pricing and distribution that makes it look original, rebellious, authentic, etc. and becoming famous in particular subculture first. For example, social media platform Instagram first got popular among photography enthusiasts. After this, brand can boost popularity to become mass cool. In lieu of not losing cool factor, brands need to maintain connection with the subculture (like Nike is still considered for top athletes) and autonomy is maintained (like Apple positioned products in place of Microsoft). Using certain promotional strategies that link brand with admired subculture, brands unlike Harley Davidson perceived more subculture. Brands can also boost status high and become iconic by packaging (like the Coca Cola contour bottle), high prices, memorable ad styles (one like the witty, artistic campaigns of Absolut vodka, spokesperson, retail cobranding, etc.) (Warren et al., 2019).

Researchers state that brands that are considered cool are considered desirable; it takes extra for a product to be perceived cool than just being positive (Pountain & Robins, 2000). Some studies clarify cool brands to be considered subcultural, popular, extraordinary, aesthetically appealing, original, energetic, rebellious, authentic, iconic and high status. With the subculture outlaw bikers, Harley Davidson became cool with impressions like iconic image and rebellious.

Similarly, Nike brand is considered cool as its shoes have quality, are desirable, look good and signal energy. Apple also is symbolized with autonomy with originality and authenticity. Brands can be perceived to be more original and energetic; this encourages a brand to be innovative continuously and keep oneself ahead in competition. Brands like Google and Samsung Electronics are example of the former. By reminding history and core value of the brand, it stays authentic unlike brand Patagonia (Warren et al., 2019).

Few studies claim that brand coolness is subjective and dynamic in nature (Belk et al., 2010; Gladwell, 1997; Southgate, 2003). Initially, brands are considered cool within small subculture with characteristics of being rebellious, authentic, original, aesthetical pleasing and exceptional. At this stage, brands are referred to as niche cool, remain famous among small group of consumers and remain unfamiliar with bigger and broader population. With time, they cross over and are accepted by wider audience therein, then become mass cool. Some brands like Steady Hands, INSIDE, Mitsky are examples of niche cool and Nike, Grand Theft Auto, Beyoncé are among mass cool and perceive to be as less autonomous and more iconic and popular (Warren et al., 2019).

All studies are in focus with cool branding, so cool branding has been put brand forward in competition and help in gaining fame. The aim of the chapter is to find potential for sustainable fashion brands to understand how cool branding can help marketers to reach socially responsible consumers. This, in turn, will increase purchase behavior toward sustainable fashion brands. An important step for marketers is to understand what does the brand stands in consumer's mind. Best communication strategies are, thus, made as per the taste, desires and taste of the target consumers.

Pountain and Robins (2000) suggest that narcissism, ironic detachment and hedonism are the three personality traits associated with cool persona. The study was conducted in "cool persona" perspective keeping the seriousness aside. With the concept of iconic detachment, it is related to history of cool, dated back from blues and Jazz artist (Pountain & Robins, 2000).

Brand personality is suggested as "a set of human characteristics associated to a brand" (Aaker, 1997), which helps marketers to follow with most effective marketing strategies by understanding emotional and symbolic meaning perceived by consumers related to the brand. Brand current situation is revealed by knowing "the perceptions about a brand as reflected by the brand associations held in consumer memory" (Keller, 2003, 2013). This needs to be studied as it helps to give insight for basics for further marketing planning. Brand personality helps to know how brands represent themselves in different regions and countries, as thus helps to position the brand best fitting the consumer culture. Thus, dimensions which relate to brand personality help to understand consumer's feelings and attitudes toward the brand, which in return help to communicate brand coolness to the target audience.

SUSTAINABLE FASHION

Brundtland report defines sustainability as:

> …. fundamentally a process of changes in which exploitation of resources, rules for investments, developments in technology and institutional changes all are in a correlated balance and enforce the presence and future possibilities in responding the needs and hopes of the human beings.

Still this definition remains official definition all over the world (Tosti, 2012). Since the 1980s and 1990s, the question of sustainability prevails.

Sustainable fashion means to design styles depending on positive future demand; thus, it is a trend in itself. A part of "ethical fashion" trend, sustainable fashion is a global movement focusing on three areas of economic, environmental and social issues. Sustainability also focuses on utilizing the resources of the company to the best possible way. Minimizing use of energy, chemical or reducing amount of waste or treating the waste effluents sustainability strives toward optimization of production process also, by focusing on rationalizing packaging material, stock and transportation to reduce the storage, space, weight, etc.

"Sustainable fashion is about a strong and nurturing relationship between consumer and producer" (Fletcher, 2008). A spectrum is created showing shift of brands and consumers from grey to green which is an attempt to bring clarity to get involved in best practices (Figure 5.2). The aim is to bring awareness and communicate right practices around sustainability in fashion.

Not to be considered as trend, sustainable fashion is planning different styles as positive future demand. Sustainable fashion is a part of larger trend known as "ethical fashion": A global movement involving environmental, economic and social issues. What is important is the sustainability movement today, as it touches different elements of consumer's everyday lives. To make sustainable fashion not just style, quality and cost, many other factors also are taken into consideration.

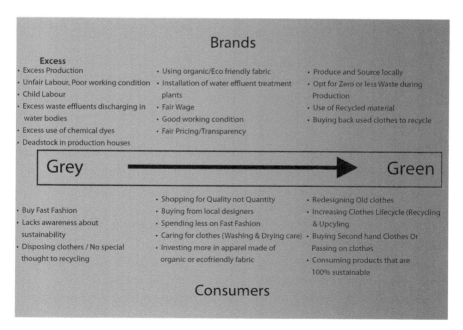

FIGURE 5.2
A shift from gray to green.

"The State of Fashion 2019" report by McKinsey, one of the upstreaming issues of supply chain is transparency and consumer rising concern about fair labor, environmental problem and sustainable resourcing issues, and consumers show interest in brands that are doing good. Thus, in survey 66% were found willing to invest more for goods that are sustainable.

Tosti (2012), in her article, "The Brundtland Report to the Global Organic Textile Standard" states that it is the need of an hour that businesses take serious responsibility for next generation and doing it in the sustainable way makes it realistic and works in favor from economic point of view. Damgaard Nielsen (2012) discusses that transparency, environmental laundering and third-party organizational support like practices help sustainable companies to brand themselves in the right way. He also stresses that small enterprises have advantage in implementing certain sustainable practices as they have close proximity to supplier and vendor.

Ellen MacArthur Foundation, in its report, "A New Textiles Economy: Redesigning Fashion's Future" (2017) states that both individual and multistake holders are getting involved into different practices to drive change and add to sustainability initiatives. Major brands have come forward to bring change, whereas individual brands also have come up with products supporting sustainability. Collaborative efforts would bring a change (Morlet et al., 2017).

Thinking in sustainable and conscious fashion, India has some famous brands catering to the need of the consumer. Doodlage is one, which offers sustainable products made out of creatively using leftover, rejected and waste fabric from production lines, which are shredded and converted in textured fabric to be used for bags and accessories. Another brand is 11.11/eleven eleven, which uses locally originated, natural fiber and handspun Khadi fabric to offer garments, sustaining the local strength. No Nasties, an organic, fair trade and vegan clothing brand offers a range for men, women and kids clothing unlike this sustainable Indian Fashion brand that has a long list like Ka-Sha India, Do You Speak Green, Grassroot, The Pot Plant, Bhusattva, etc.

Conventional businesses are trying to find new opportunities for generating income and sustain in the competitive environment, thus are aware about sustainable fashion, to increase business and get advantage over competitors (Tosti, 2012). The new drive toward sustainability is helping brands to achieve new dimensions for business. The marketing of eco-fashion in lieu of promoting sustainable consumption is continuously being developed by fashion companies and enterprises (Barnes et al., 2006; Fletcher, 2008; Wong et al. 2011).

But sustainable fashion industry faces hard time to get due acceptance from consumers. In terms of commitment toward ethical consumption, consumption of food products is favorable as it is perceived to have direct health benefits (Barnes et al., 2006; Wong et al., 2012). On the other hand, as sustainable apparel consumption doesn't directly relate to wellbeing, this product segment lacks consumer's commitment (Wong et al., 2012).

Studies also confirm that fashion consumers do not compromise with the aesthetics of their fashion clothing to be environmentally conscious (Barnes et al., 2006). Consumers involve in unsustainable consumption like impulse buying, overconsumption, short use of product and early disposable. All this is because of changing fashion trends, affordable price of product, low quality, etc. Investing in short-life span products leads to hike in notable waste. Nearly, 70% of textile and clothing reaches in landfill at the end of the cycle (Fletcher, 2008).

An extensive literature review shows findings of few researchers in understanding consumer behavior toward sustainable fashion brands (Table 5.2).

TABLE 5.2

Review of Literature Related to Sustainable Fashion Consumption Decisions

Reference	Findings
Beard (2008)	Consumers prefer clothes that are made ethically and should be stylish and fashionable and should fulfill ethical need of consumers.
Niinimäki (2010)	Consumer doesn't emphasis on sustainability while making a purchase. Dominant factors while purchasing fashion clothing are price and style. Sustainable consumption is contradicted by the desire to renew appearance as per changing fashion.
Ochoa (2010)	"Being expensive" and "not stylish" is the highest contributing factor for not buying sustainable clothing.
Nakano (2007)	Only 10% more can be paid by customer while purchasing sustainable clothing.
Mintel (2009)	Environmental aspect as value added to product is not considered by the consumers. There is no desire to pay premium price for green product.
Barnes et al. (2006)	No real opportunity lies with consumers to choose ethical clothing as all clothing types are typically produced in Asian countries. Prices for ethical clothing cannot be compared, and are considered unfashionable and unattractive in appearance.

In modern society, consumers claim to be environmentally conscious and show interest in wearing clothes that are made in a responsible way. Although possessing positive attitude toward environment protection does not get applied to fashion consumption. There lies an attitude-behavior gap between the consumer values and apparel purchase, which creates a misalignment in actual branding (Niinimäki, 2010; Solomon & Rabolt, 2004). Many researchers prove that decision-making for purchasing ethical consumption is a process with high complication.

Motivation of purchasing eco-fashion is said to be related to product attributes and retail stores that attract consumers to buy sustainable fashion (Beard, 2008; Niinimäki, 2010). Product, design, quality and price are said to be product-related attributes and store design, environment and shopping convenience are said to be store-related attributes, which directly influence sustainable fashion consumption (Niinimäki, 2010).

Researchers state that consumers of fashion clothing do possess positive attitude toward environmental problem, but this attitude rarely helps during consumption decision related to eco-fashion (Niinimäki, 2010; Solomon & Rabolt, 2004). It is also claimed that consumption decision related to sustainable fashion is complicated (Niinimäki, 2010). Fashion consumers invest less in sustainable fashion despite showing positive attitude toward environment protection (Barnes et al., 2006; Niinimäki, 2010).

This creates major consumption problem of sustainable fashion, where studies have claimed that consumers do possess positive attitude toward sustainable fashion. Fashion consumers purchase behavior for eco-fashion disappoints fashion companies. Despite this attitude-behavior gap, it urges researchers to investigate factors affecting the consumption of eco-fashion by fashion consumers (Niinimäki, 2010; Solomon & Rabolt, 2004).

Lack of transparency in global supply chain and awareness about sustainable fashion comes in the way of purchase decision and strengthens the attitude-behavior gap. Another barrier is that brands lack in distinguishing themselves from other in lieu of "greenization" and make their product competitive from other competitors and conventional options (Ahluwalia & Miller, 2014). All this affects business economically and makes survival of businesses difficult (Thomas, 2014).

Other interesting facts highlight that world's largest young population is from India. Half of the total population of India is below 25 years, which is considered positive in context to consumption of textile and apparel. Report by Deloitte India states that millennial is the group of population of age between 18 and 35 years, redefining India's consumption behavior. Culture, economy, technology, social lifestyle, economy and politics are influencing factors of generation. Millennial is considered to be culturally diverse, digitally sound, well-traveled, high exposure, working with new technology jobs, open to new experiences, ideas and concepts. Driving different consumer segments in direction of rapid growth, the largest demographic group of India and Global possess high level of disposable income and are digitally connected.

Most of the income of the highest wage earners of the country is spent on necessary or essential commodities, fashion possesses 21.4% share in the incremental income spent. A study says that millennial purchase intention is driven by ecological consciousness and social consciousness. They demand high need of variety and thus social consciousness has strong positive effect on the willingness to pay more. Millennial purchase intention is first driven by product's brand name and uniqueness than being sustainable. This doesn't justify that they don't care about environment, but that is considered the last thing. Millennial will make 50% of workforce and is expected to reach 75%. Exposures to advertisement and digital world have made millennial generation skeptical about authenticity. Before making a purchase they look for authenticity, transferring their thoughts and opinions through social media to general public.

METHODOLOGY

After the topic selection, a framework was developed which was used as a basis for study.

The research aim is achieved by referring the literature for concept of cool and sustainable fashion. In-depth literature review has been conducted in order to get insights about the perception of cool for sustainable fashion.

Published research, primary research reports and articles, conceptual articles, journals and commercial reports were studied regarding cool concept and sustainable fashion industry in India. The research aim is achieved by referring to the literature for better understanding of the complex nature of cool as a concept and sustainability and fashion as a system is studied closely and is intertwined. A diagram explaining brands and consumers shift from gray to green will help marketers and consumers to form better understanding about ways of sustainability and their practices. Literature revolves around exploring the potential of cool branding as marketing strategy for sustainable fashion brands.

MARKETING FASHION SUSTAINABLE BRANDS USING COOL BRANDING

While building an identity for a cool brand, factors taken in account are being unconventional (Gurrieri, 2009), possessing an ideal of good actions in relation to aesthetic sense (Southgate, 2003) and uniqueness (Rahman et al. 2009) has to be taken into account. Cool imagery has been used for the promotion of alcohol and the tobacco industry previously, and also for products like fashion accessories, electronics, music and sports accessories (Hoek & Jones, 2011).

Many researchers state that associating "cool" with brand imaginary positively impacts consumer behavior and attitude (Im et al. 2015; Sriramachandramurthy, 2009). For instance, Timberland boots and flannel shirts were designed as per functionality of outdoor comfort and warmth, with less emphasis on fashion statement. But timberland boots were adopted by young urban customers as "Timbs", fashion boots (Bloom, 2004). Similarly, Flannel shirt of grunge music movement gained popularity as "official" shirt (Miller, 1993). Thus, consumers were inspired by desire to look cool and not by functionality of the product.

Bird and Tapp (2008) study the influence of cool on consumer behavior and suggest potential to induce prosocial behavior. They also claim that "cool" does influence teenage and young adults, and suggests to embed cool in design of social marketing programs. Cool is considered to be most attractive and relevant to young people, as it is more attracted to thrill seeking and risk-taking activity.

During the literature review, it is found that few studies address relationship between "marketing" and "cool" as a concept. Some studies state a strong relationship, by claiming that first-mover advantage shall always help the marketers to stay ahead in competition, and by spotting a cool trend before others to get insight to market. But action, opinions and thoughts that indicate "cool" differ from group with time (Gladwell, 1997;

O'Donnell et al., 2000). Young consumers form opinions if something is cool or not, before making purchase, based on hedonic or utilitarian emotion sign. Thus, their purchase behavior is highly motivated by a cool factor. Very few studies claim meaning of cool in sustainable fashion content and explain how cool and sustainability can be worked out together for bringing a change in the fashion industry. Sachkova (2018) states that there is a potential of connecting cool and sustainable fashion to bring change in the fashion industry. His studies provide understanding the relation of cool and sustainable fashion and its unappealing image. The study also explains how to make sustainable brand cool.

Cool and sustainability will work as a trigger for the right type of brands to promote products to the right type of consumers. In previous studies, consumers have shown positive feeling and perception toward consuming products that are more cool and ethical and thus established that sustainability is perceived as a cool factor (Sachkova, 2018). Cool has been considered a "Game Changer" for the fashion industry in context to sustainable fashion.

A misinterpretation in regard to the true meaning of sustainable fashion confuses consumers, media and brands (Lewis, 2008). Literature claims that favorable change in regard to sustainable fashion is in the hands of consumers (Beard, 2008). Although studies reveal that fashion consumers claim that information and availability of eco-fashion are less known to them (Barnes et al., 2006).

Findings by McNeill and Moore (2015) state that people find themselves disconnected while sustainable fashion conversations, as it is perceived that "conscious" or "sacrifice" consumers are being addressed (McNeill & Moore, 2015). Promotion strategies' focus is completely on material and production. This needs to be reconsidered and correct message should be sent by sustainable brands, as it should be as per the target audience and brand value.

The combination of cool and sustainability is said to add additional value to industry (Sachkova, 2018). Sustainable fashion is evaluated to poses full potential, but is it perceived by consumers as undesirable and misunderstood. While cool is considered as a value system in itself; thus, in lieu of introduce positive change in favor of environment and brands, it is best to connect the two concepts (McNeill & Moore, 2015).

Presenting sustainable fashion in subtle way as cool is the best way to reach millennial as target audience as it is studied that they care about doing good. Studies also state that millennial look out for brands that are offering sustainable fashion products (Sachkova, 2018). Extensive literature review also states that adolescents and teenagers are the main target for cool focus marketing.

This target group develops self-identities and values, driven by psychological and physiological experienced changes based on experiences during adolescence, where childhood values by families are replaced. Feelings of autonomy, self-expression and irreverence are considered as the motivational factors in teenagers as a difference between the actual self-defining and new ideal self-teenagers can be identified. This target group seeks to attain attractiveness and wishes to gain popularity among peer group (O'Donnell et al., 2000).

Consumers are motivated by cool, as cool can influence on consumers attitude and behaviors. Existing literature suggests that not only teenager aspire to adopt cool but adults also in their late 20s and even 30s get influenced by cool (Gaskins, 2003). Cool trend has often seen among consumer segment that has tendency to take risk and may be considered in social minorities (Southgate, 2003). The importance of cool in context to social marketing seems promising, specifically relevant among young people and risk-takers. Therefore, a potential is seen, if cool in design and delivery is successfully implemented by social marketing programs.

DISCUSSION

The finding from an extensive literature review confirms that cool branding can be used to enhance marketing for sustainable fashion brands. Adopting sustainable clothing at large is required as present clothing consumption behavior has adverse impact on the environment. On the one hand, where consumers are aware of need of sustainability, it has been found through review that attitude–behavior gap restricts the shift from fast fashion to slow fashion that brings sustainability. Factors like style and price are considered main factors for affecting the purchase decision of the consumers.

Brand coolness impacts several variables and characteristics independently contributing to overall brand coolness and how coolness varies across cultures. Finally, the review shows connection between different constructs (e.g., brand personality) and how it can contribute to brand's success. The literature review focuses on how cool branding can enhance the sustainable marketing and consumption of fashion clothing.

This study has shown opportunities by understanding sustainability in fashion relating with cool as a concept. Thus, there seems a potential of connecting cool and sustainable fashion to bring change in the fashion industry. Though future research will need to further explore factors that moderate or interact with the different characteristics of brand coolness. Relationship of variables like brand personality and brand loyalty can be studied in terms of brand coolness and their impact on purchase decision for sustainable fashion brands.

CONCLUSION

A great opportunity has been created for fashion brands and designers to position themselves as leaders in emerging market of sustainable fashion and also contribute to environmental protection.

Indian fashion brands that offer sustainable products are accepting this transformation and gradually changing themselves. On the other hand, cool as construct has always being considered unique. As consumers continuously demand for new ideas and are open for new experiences, cool brands are considered to set their own trends. Thus, there seems a potential that sustainable fashion products with "cool" brand imaginary have a positive impact on consumer attitude and behavior.

The value of the review lies in the confirmation of the context that cool currently means good. A relation of sustainable fashion and cool has been found and stated, and a potential has been found to connect cool and sustainable fashion to bring change in the fashion industry. Therefore, this review can help marketers and professionals understand issues around getting right message for their products and develop best communication for the brand. The link between cool and sustainable fashion is stated with sustainable capitalism, but idea of consuming is not rejected but is introduced in more creative and new way.

It can be argued that this review all together talks about a new perspective, value and importance of sustainable fashion as this will boost reimagining of big issues prevailing in industry and consumer practice as well (Sachkova, 2018). This chapter illustrates the contemporary world complex nature, where cool as a concept, fashion as a system and sustainability, all hold complexities among themselves but are intertwined together with wide sociopolitical backdrop. The main contribution of this work is to give clarity on how ethical consumption purchase decision can be made by consumer by getting influenced by coolness of brand and its products. In total, this review will help marketers to reach socially responsible consumers with a new perspective.

LESSONS LEARNED

1. A possibility is found in connecting cool and sustainable fashion to bring change in the fashion industry.
2. Cool branding can enhance the marketing and consumption of sustainable fashion clothing.
3. Opportunities have been found in studying characteristics related to cool branding to impact willingness to pay.
4. Variables and factors related to brand coolness have been found (e.g., brand personality, brand loyalty, willingness to pay, etc.) for future studies to understand influence on purchase decision of sustainable fashion brands.

MANAGERIAL IMPLICATIONS

The extensive literature review offers implications to help marketers to reach socially responsible consumers. First, the study is of importance to marketers of sustainable fashion brand who want their brand to be cool. Past literature and strong evidence included in this research claim that cool is studied in different ways as it gives distinction in subtle ways. Brands can influence consumers by incorporating cool symbolism in their identity (Duggal and Verma, 2019). This study expands the understanding of cool in the context of an emerging market of sustainable fashion.

Second, studying constructs like brand personality and cross-culture influence, perception of coolness claims to have impact on consumers' purchase decision for sustainable fashion brand. Present review does not only provide strong evidence that cool has been appropriated by marketers for the value it has for customers but more importantly states that cool has marketing value and sustainable fashion brands are need of the hour; thus, both concepts need to be used together. The scope of the study lies with giving insight to managers of sustainable brands that have long sought to figure out how to give their brands a different outlook from conventional products (Anik et al. 2017; Nancarrow et al. 2002). Our result suggestions will be used for implementing different marketing and communication programs that are designed to increase or maintain a brand's consumer base.

We contribute to the literature that cool and sustainable fashion carries the potential of introducing change to the fashion industry if presented in a balanced way. This review has brought a new perspective and value. Marketing sustainable fashion could encourage the rethinking of bigger issues in the fashion industry and in consumer practices (Sachkova, 2018).

CREDIT AUTHOR STATEMENT

We as authors, Jasmeet Kaur (Corresponding author) and Dr. Gursimranjit Singh, of this manuscript confirm that manuscript was completed by following the ethical review process. We also confirm that the research meets the ethical guidelines, including adherence to the legal requirements. We strongly believe the contribution of this study warrants its publication in the book titled, "Social and Sustainability Marketing: A Casebook for Reaching Your Socially Responsible Consumers through Marketing Science" with Taylor & Francis. Finally, we confirm that this manuscript has been submitted solely to you and is not published, in the press, or submitted elsewhere.

REFERENCES

Aaker, J. L. (1997). Dimensions of brand personality. Journal of Marketing Research, 34(3), 347–356.

Ahluwalia, P., & Miller, T. (2014). Greenwashing social identity. Social Identities, 20(1), 1–4. DOI: 10.1080/13504630.2013.878983

Anik, L., Miles, J., & Hauser, R. (2017). A General Theory of Coolness. Darden Case No. UVA-M-0953, Available at SSRN: https://ssrn.com/abstract=3027026

Barnes, L., Lea-Greenwood, G., & Joergens, C. (2006). Ethical fashion: Myth or future trend? Journal of Fashion Marketing and Management: An International Journal.

Beard, N. D. (2008). The branding of ethical fashion and the consumer: A luxury niche or mass-market reality? Fashion Theory, 12(4), 447–467.

Belk, R. W., Tian, K., & Paavola, H. (2010). Consuming cool: Behind the unemotional mask. In Russel W. Belk (Ed.), Research in Consumers Behavior (Vol. 12, pp. 183–208). Bingley, UK: Emerald.

Bird, S., & Tapp, A. (2008). Social marketing and the meaning of cool. Social Marketing Quarterly, 14(1), 18–28.

Bloom, J. (2004). High life, Timberland give lessons on cultivating 'cool'. Advertising Age, 75(1), 14–14.

Budzanowski, A. (2017). Why Coolness Should Matter to Marketing and When Consumers Desire a Cool Brand: An Examination of the Impact and Limit to the Perception of Brand Coolness (Doctoral dissertation, Universität St. Gallen).

Chan, T. Y., Wong, C. W. Y. (2012). The consumption side of sustainable fashion supply chain: Understanding fashion consumer eco-fashion consumption decision. Journal of Fashion Marketing and Management: An International Journal, 16(2), 193–215.

Damgaard Nielsen, S. (2012). Implementation of Supply Chain Sustainability in the Fashion Industry (IIIEE Master Thesis).

Dar-Nimrod, Ilan, Hansen, I. G., Proulx, T., & Lehman, D. R. (2012). Coolness: An Empirical Investigation. Journal of Individual Differences, 33(3), 175–85.

Duggal, E., & Verma, H. V. (2019). Cool perspectives, Indian cool and branding. South Asian Journal of Business Studies.

Fletcher, K. (2008). Sustainable fashion and clothing. Design Journeys. London: Basım Routledge.

Gaskins, Q. (2003). The Science of Cool: The Art of Understanding Fusionism. *ANA Magazine, ANA*. Retrievable from: http://www.warc.com.simsrad.net.ocs.mq.edu.au

Gladwell, M. (1997, March 17). The Coolhunt. The New Yorker (pp. 78–87).

Gurrieri, L. (2009, November). Cool Brands: A Discursive Identity Approach. In ANZMAC 2009: Sustainable Management and Marketing Conference Proceedings.

Hoek, J., & Jones, S. C. (2011). Regulation, public health and social marketing: A behaviour change trinity. Journal of Social Marketing, 1(1), 32–44.

IBEF. (2020a, June). *Growth of Textile Industry – Infographi [Web log post]*. Retrievable from: https://www.ibef.org/industry/textiles/infographic

IBEF. (2020b, June). *Textile Industry & Market Growth in India [Web log post]*. Retrievable from: https://www.ibef.org/industry/textiles.aspx#:~:text=India's%20textiles%20industry%20contributed%20seven,45%20million%20people%20in%20FY19.&text=The%20domestic%20textiles%20and%20apparel,US%24%20100%20billion%20in%20FY19

Im, S., Bhat, S., & Lee, Y. (2015). Consumer perceptions of product creativity, coolness, value and attitude. Journal of Business Research, 68(1), 166–172.

Karthik, T., & Gopalakrishnan, D. (2014). Environmental analysis of textile value chain: An overview. In Roadmap to Sustainable Textiles and Clothing (pp. 153–188). Singapore: Springer.

Keller, K. L. (2003). Brand synthesis: The multidimensionality of brand knowledge. Journal of Consumer Research, 29(4), 595–600.

Keller, K. L. (2013). Building, Measuring, and Managing Brand Equity.
Lewis, V. D. (2008). Developing Strategies for a Typology of Sustainable Fashion Design. In Hethorn, J. & Ulasewicz, C. (Eds.), Sustainable Fashion: Why Now? (pp. 233–263). New York: Fairchild Books, Inc.
McNeill, L., & Moore, R. (2015). Sustainable fashion consumption and the fast fashion conundrum: Fashionable consumers and attitudes to sustainability in clothing choice. International Journal of Consumer Studies, 39(3), 212–222.
Miller, C. (1993). Marketers who aim for the middle will reach the latest chic market. Marketing News, 27(7), 1–10.
Mintel, O. (2009). Ethical and Green Retailing-UK. Chicago, IL: Mintel International Group Limited.
Mittelstaedt, J. D., Shultz, C. J., Kilbourne, W. E., & Peterson, M. (2014). Sustainability as megatrend: Two schools of macro marketing thought. Journal of Macromarketing, 34(3), 253–264.
Mohiuddin, K. G. B., Gordon, R., Magee, C., & Lee, J. K. (2016). A conceptual framework of cool for social marketing. Journal of Social Marketing.
Morlet, A., Opsomer, R., Herrmann, S., Balmond, L., Gillet, C., & Fuchs, L. (2017). A new textiles economy: redesigning fashion's future. Cowes, UK: Ellen MacArthur Foundation.
Nakano, Y. (2007). Perceptions towards clothes with recycled content and environmental awareness: The development of end markets. In Ecotextiles (pp. 3–14). Cambridge, UK: Woodhead Publishing.
Nancarrow, C., Nancarrow, P., & Page, J. (2002). An analysis of the concept of cool and its marketing implications. Journal of Consumer Behaviour: An International Research Review, 1(4), 311–322.
Niinimäki, K. (2010). Eco-clothing, consumer identity and ideology. Sustainable Development, 18(3), 150–162.
O'Donnell, K.A., & Wardlow, D. L. (2000). A theory of the origins of coolness. Advances in Consumer Research, 27, 13–18.
Ochoa, L. M. C. (2010). Will 'eco-fashion' take off? A survey of potential customers of organic cotton clothes in London. AD-minister, (16), 118–131.
Olson, E. M., Czaplewski, A. J., & Slater, S. F. (2005). Stay cool. Marketing Management, 14(5), 14–17.
Pountain, D., & Robins, D. (2000). Cool rules: Anatomy of an attitude (1st ed.). London: Reaktion Books.
Rahman, K., Harjani, A., & Thoomban, A. (2009). Meaning of "Cool" in the Eye of Beholder: Evidence from UAE, American University in Dubai (SBA Working Paper 09-001).
Runyan, R. C., Noh, M., & Mosier, J. (2013). What is cool? Operationalizing the construct in an Apparel Context. Journal of Fashion Marketing and Management: An International Journal.
Sachkova, L. (2018) How to make sustainable fashion cool. EA: Retrievable from: http://gfc-conference.eu/wp-content/uploads/2018/12/SACHKOVA_How-to-make-Sustainable-Fashion-cool.pdf
Solomon, M. R., & Rabolt, N. J. (2004). Consumer behavior: In fashion. Upper Saddle River, NJ: Prentice Hall.
Southgate, N. (2003). Cool hunting, account planning and the ancient cool of Aristotle. Marketing Intelligence & Planning, 21(7), 453–461.

Sriramachandramurthy, R. (2009), What's cool? Examining brand coolness and its consequences. Carbondale, IL: Southern Illinois University.

Sriramachandramurthy, R., & Hodis, M. (2010). Why Is Apple Cool? An Examination of Brand Coolness and Its Marketing Consequences. In Proceedings from the American Marketing Association Summer Academic Conference (pp. 147–48). Chicago: American Marketing Association.

Sundar, S. S., Tamul, D. J., & Wu, M. (2014). Capturing "cool": Measures for assessing coolness of technological products. International Journal of Human-Computer Studies, 72(2), 169–180.

Thomas, E. M. (2014). Branding in Sustainable Apparel Companies: A Study on the Branding Strategies Adopted by Small-Business Apparel Companies that Practice Sustainability in Canada (Doctoral Dissertation, Ryerson University Toronto).

Tosti, E. (2012). From the Brundtland Report to the Global Organic Textile Standard. Nordic Textile Journal, 1, 100–106.

Warren, C. (2010). What makes things cool and why marketers should care. Boulder, CO: University of Colorado.

Warren, C., Batra, R., Loureiro, S. M. C., & Bagozzi, R. P. (2019). Brand coolness. Journal of Marketing, 83(5), 36–56.

Warren, C., & Campbell, M. C. (2014). What makes things cool? How autonomy influences perceived coolness. Journal of Consumer Research, 41(2), 543–563.

Warren, C., Pezzuti, T., & Koley, S. (2018). Is being emotionally inexpressive cool? Journal of Consumer Psychology, 28(4), 560–577.

Wong, C. W., Lai, K. H., & Cheng, T. C. E. (2011). Value of information integration to supply chain management: roles of internal and external contingencies. Journal of Management Information Systems, 28(3), 161–200.

Wong, C. W., Lai, K. H., Shang, K. C., Lu, C. S., & Leung, T. K. P. (2012). Green operations and the moderating role of environmental management capability of suppliers on manufacturing firm performance. International Journal of Production Economics, 140(1), 283–294.

Wright, J. (n.d.). *How Millennials are Shaping the Fashion World [Web log post].* Retrievable from: https://www.fashionatingworld.com/~fashion3/blog/364-how-millennials-are-shaping-the-fashion-world

6

Personal Experience of Sustainability Practices and Commitment toward Corporate Sustainability Initiatives: Reflections of Sri Lankan Marketing Professionals

Kadka Ranjith and W.D.C. Jayawickrama
Department of Marketing Management,
University of Sri Jayewardenepura, Nugegoda, Sri Lanka

CONTENTS

Learning Objectives	144
Tools Used in the Research	144
Themes	144
Introduction	145
Literature Review	147
The Concept of Sustainable Development	147
Today's Political Reality of Sustainable Development	147
The Role of Marketing in Sustainability Initiatives Led by Businesses	149
Methodology	150
Completion Techniques Used in the Research	151
Associative Techniques Used in the Research	151
Choice or Ordering Techniques Used in the Research	152
Expressive Techniques Used in the Research	153
Construction Techniques Used in the Research	154
Participants Experience and Knowledge in Marketing and Sustainability	158
Profiles of Participants	158
Major Problems	159

Findings and Discussion 159
Sustainability Practices of Professional Marketers 160
    Social Sustainability 160
        Based on the Life Story of Quarrel 160
        Based on the Life Story of Snape 161
    Economic Sustainability 165
        Based on the Life Story of Nonnis 165
Conclusion 167
Practical Implications 169
Acknowledgments 170
Research Problem 170
References 170

LEARNING OBJECTIVES

After reading the chapter, the reader should be able to:

- Identify the personal sustainability practices reflected in marketing professionals through their commitment toward sustainability practices
- Comprehend the applicability of projective techniques in conducting qualitative research
- Determine the suitable areas to incorporate developing a curriculum for sustainability education

TOOLS USED IN THE RESEARCH

- Thematic analysis
- Projective techniques

THEMES

- Personal sustainability practices
- Corporate sustainability initiatives

INTRODUCTION

Although the concept of sustainability has received wide attention among both scholars and practitioners alike over the last few decades, several questions in both theory and practice still remain unanswered. To this end, in addition to these scholarly attempts, the United Nations (1992, 1998 and 2015) have made noteworthy contributions in bringing both academics and practitioners together to address the pertinent issues concerning sustainability. These initiatives have highlighted topics such as the increment of the carbon concentration in the sky, global warming, continuous extinction of animal and plant species as well as the continuous reduction of forest density (Araç & Madran, 2014).

Among the actors in a social system, the corporate sector is identified as a key party that can address issues pertaining to sustainability. Due to its long-term engagement in manufacturing, consumption and disposal of waste, it is perceived as a main contributor toward a number of sustainability-related issues highlighted above (Pachauri et al., 2014). However, although managers appear to be hesitant to consider and implement sustainability practices at their organizations, it is vital for them to understand the "risk of failing to do so" (Wade, 2019). This denotes the need for progressive strategic initiatives by the corporate sector toward sustainability, which would in turn ensure their long-term success (Cleene & Wood, 2004).

An organization may consist of multiple management functional areas that are involved in creating and executing both strategic and operational decisions. Among them, the Production, Operations, Human Resource Management and Marketing Management are considered to be commonly practiced functional areas for strategy implementation. The rising demand for sustainability has initiated a debate over the functional area that is most responsible for executing sustainability initiatives. In an interview with Davey (2010), Robert Nuttall, who helped devise and implement the brand communication strategy for Marks & Spencer's "Plan A Campaign", explained that sustainability can be applied in relation to several functions in an organization, such as Marketing, Corporate Communications, Corporate Sustainability and Finance. Accordingly, the responsibility of implementing sustainability initiatives in an organization seems fragmented.

However, in an interview with Costa (2010) for "Marketing Week", Keith Weed, Unilever's Chief Marketing and Communications Officer, pointed out that sustainability practices are best executed through the marketing function since it facilitates sustainable production and consumption efforts. In support of this argument, Jones et al. (2008) points out that marketers have a significant role to play toward sustainability by changing consumer behavior by way of influencing perceptions, attitudes, beliefs and their perspectives of seeing a business.

Further, marketers are often in direct contact with all the stakeholders of an organization (Merrilees et al., 2005), and identify their concerns in relation to the decision-making of the organization (Clulow, 2005). As marketers continuously engage with the market (consumers) and other stakeholders, they gain an in-depth understanding of sustainability (Andreasen, 2002). Therefore, it can be argued that maketers are in a strong position to decide and implement sustainabilty initiatives at their respective organizations.

However, the level of engagement and commitment of marketers toward sustainability initiatives depends on several factors. To this end, the process of socialization undergone during ones' formative years (Bugental & Grusec, 2007) and the formal education on sustainability that marketers have received seem to have a significant impact over their level of sustainability orientation (Lander, 2017; Mannion, 2019). According to the literature on socialization, it can be argued that if a person has learned the value of environment (an integral part of the sustainability), they are more likely to be committed and supportive of such initiatives later in their life (Bandura, 2002).

In light of the above argument, it is likely that professional marketers' support and commitment toward sustainability can be secured if such initiatives resemble with what they have learned through socialization and/or formal education. However, previous studies have not allocated sufficient focus over this phenomenon. Therefore, the aim of this particular study was to explore, in the Sri Lankan context, what types of personally learned and/or experienced sustainability initiatives are manifested through the corporate sustainability initiatives developed and executed by professional marketers. This exploration is particularly important as marketers have been identified as the key players toward the successful implementation of sustainability in organizations. Further, the study also aimed to provide policy implications toward formal education on sustainability.

LITERATURE REVIEW

The Concept of Sustainable Development

The term sustainable development was first introduced by the Brundtland Commission (1987), being an initiator told sustainability to be addressed through sustainable development and it defined the term as "development that meets the needs of the present without compromising the ability of future generations to meet their own needs". Today "sustainable development" is one of the most used terms and it evaluates economic and social development and also business activity (Crane & Matten, 2004). Sustainable development is a three-sector model of economy, environment and society (Giddings et al. 2002). The sustainable development is described as a model of interconnected rings that symbolize the three sectors (Giddings et al. 2002). As a result, a significant amount of research has been done in the areas of social, sustainability, innovations and sustainable consumption, and companies are being recognized as a force with potential to change the behavior of consumer in a way to address the sustainable development (Bocken, 2017). Figure 6.1 explains the interconnected ring model of sustainable development.

Today's Political Reality of Sustainable Development

The reality of today's life is that economy dominates the environment and the society, the large global companies dominate the government. People are concerned about large companies as a source of solution to sustainable problems (Korten, 1996); the reality of today is that the economy dominates both society and environment. Giddings et al (2002) explain that international forums and organizations are heavily influenced by the large corporations. Since the government and large companies have embraced the sustainable development as the separation of sustainable development

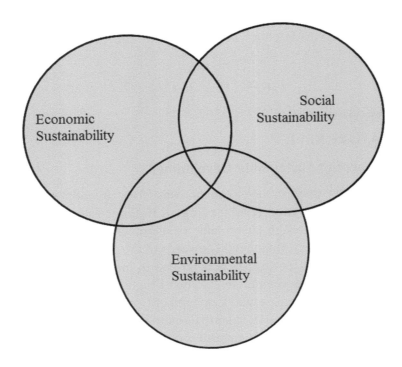

FIGURE 6.1
Interconnected ring model of sustainable development.

into three sectors, and it can be used to justify the concentration on one part rather than the whole. Sustainable development is presented aiming to bring social, economic and environment in a balanced way, reconciling conflicts. In the normal scenario, the model shows symmetrical-sized rings in a symmetrical interconnection but if they are seen separate as the model describes, different perspectives often do prioritize one another. For instance, most of the governments' focus is on economic development (Giddings et al. 2002).

The primary attention of political reality is economy. Society and environment are exploited, both nature and human are considered sinks where problems can be dumped (Daly, 1992). Giddings et al (2002) further emphasize the fact that businesses are referring to the capitalist economies, where the goods and services are exchanged at the operation of the market. Furthermore, organizations do not give equal consideration to the multitude actions that provision people and satisfy their needs that take place outside the market such as subsistence activities, helping friends, raising children, household labor and social relationships. According to the concept of sustainability, economy is a subset of society, where human needs are met through production of commodities and many others are met by the activities happened in an economy (Langley, 2002).

The Role of Marketing in Sustainability Initiatives Led by Businesses

Marketing should be an engine of growth and profitability of the business (Morgan, 2012). Virtually, all persons and organizations engage in marketing activity at various times. However, they do not engage with equal skills at all times (Kotler, 2011). The adoption of marketing plan has been successfully developed as a vehicle for corporate cultural change toward a customer-focused outlook (Giles, 1989). Wind (2005) argues that marketing is the interface between the organization and the environment It can provide new opportunities for the creation of value and growth, and also provides opportunities by identifying opportunities to serve the unmet needs of both current and new customers for the company's current and new products and services. Marketing strategy becomes focused on creating, delivering, sustaining and continuously enhancing values delivered for customers, and marketing can expand the company's focus from customers to all stakeholders, e.g. focus on employees. The argument brought forward is influential in making sense that marketing holds the responsibility of sustainability initiatives. Kotler (2011) mentioned that marketing scholars are cognizant of the imperative for a new and probably radical reformulation of its fundamental philosophy, its operational premises and the heuristics that are used to make marketing decisions. Supporting them Kemper et al. (2018) argued that the transition of marketing requires the understanding of the world views and its ability to transition and how change may be brought about by institutions and individuals. In that case, the understanding of marketing professionals (who actually execute the marketing philosophies in practice) about sustainability remains unexplored.

METHODOLOGY

Even though there is a growing body of literature in the field of sustainability marketing, specific attention has not been given to the area of personal practices reflected through corporate infinitives.

Hence, with limited research conducted, specifically in the area of transforming sustainability education with special references to learning and living of marketers, resulted in the selection of qualitative research. Qualitative research uses a naturalistic approach that seeks to understand a phenomenon in context-specific setting, such as real-world setting where we don't attempt to manipulate the phenomena of interest (Quinn, 2002). A cross-sectional study design is used since the purpose of the study is descriptive, often in the form of a survey. Usually there is no hypothesis as such, but the aim is to describe a population or a subgroup within the population with respect to an outcome and a set of risk factors. Narrative inquiries were collected from the sample since narrative inquiry is defined as a subset of qualitative research designs in which stories are used to describe human actions. Those stories may be oral or written, may be accessed formally through interview or informally through naturally occurring conversations and they may be stories that cover an entire life or specific aspect of life (CHASE, 2005). Narrative inquiry methodology itself was not sufficient to get more rich data within a short period of time; hence, it required multiple points of contact with the participants and ideally different forms of data collecting techniques at the same time, which led us to adopt multiple modal research design that combined projective techniques and interviews. We developed a research guide comprising multiple projective techniques in the form of an electronic booklet that provided stimulus in projecting a comprehensive narration by the respondents. The stories were voice recorded at the same time with the written consent of the respondents. A projective technique is a data collection method that allows the participants to project their subjective or deep-seated beliefs into a stimulus of an object, product or revealed through the expression of imagery (Boddy, 2005). We had the opinion that with a limited time available to spend with a particular respondent, it was very effective in selecting responses from the respondents regarding the research questions within a very short period of time. In the same time since it was designed in electronic form, we could lend a handheld device to the respondents to project their story with ease. The stimuli used in projective techniques were continuously adjusted to make a better disturbance-free environment in the minds of the respondents.

Completion Techniques Used in the Research

The research focused on extracting soft memories associated with the respondent's childhood, with the aim of exploring the personal learning behavior prior to being employed. The projective techniques used in that case were "completion technique". The interviewees are asked to complete a given judgment, story, argument or a conversation (Burns & Lennon, 1993). In the research, respondents were presented with an incomplete stimulus at the start of the interviews; they were given the beginning of a sentence and were asked to complete it or to complete thought and speech bubbles in cartoon drawings. Completion techniques generated fewer complexes and elaborated data. This was very effective in getting the attention of the respondent's right into the focus area, with a noninvolvement of the researcher. Respondents tried to match their experiences to the given stimuli and began to narrate their stories. Among the completion techniques used in the research, a sample is given in Figure 6.2.

Few of the responses from interviewees for the completion techniques are as follows:

> I was born in 1971, my father was an engineer, and my mother was a house wife …
>
> (Nonnis)

> I was born into a village, normal middle class family; we had seven children in our family ….
>
> (Snape)

Associative Techniques Used in the Research

Researchers had to use "associative techniques" to get the understanding of the respondents' social backgrounds. The pictures representing various social classes were used. The respondents were asked to express the

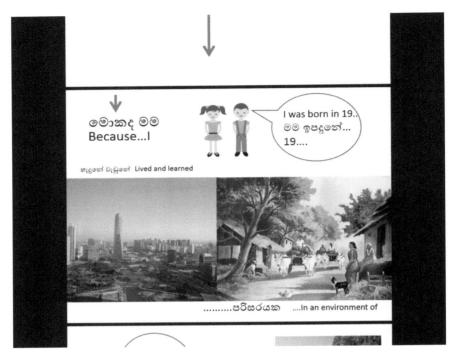

FIGURE 6.2
An example for completion techniques used in the research.

first thing that comes into the mind for the stimuli. Word associative was the most frequently used associative technique and is especially used for identifying the respondent's inner ideas (Gordon, 1988). In the research, it is used in the circumstances where the subject can verbalize a response, since the response is required immediately. Among the various associative techniques used in the research, a sample is given in Figure 6.3.

Choice or Ordering Techniques Used in the Research

In the research, the respondents were requested to express their experiences related with the objects mentioned in life story-stimulated research guide because that was important in exploring the point of contact where they actually had the chance to practice sustainability in their childhood. They picked one by one or one or few which they have associated with and continued expressing experiences. Asking respondents to select one

FIGURE 6.3
Associative techniques used in the research.

form of a list of alternative, or arrange materials such as pictures or statements, into some order, or group them into categories according to other similarities and dissimilarities, is recognized as the choice and ordering technique (Mostyn, 1978). Few of choice and ordering techniques were used to enrich the findings, out of the few techniques used in the research a sample is given in Figure 6.4.

Expressive Techniques Used in the Research

We continued on using almost all the projective techniques since we felt that it would increase the depth of the findings. We had to use a technique that would bring the mindset of interviewees attached to childhood to present day scenario without disturbing the built-up environment. There the use of "expressive technique" helped the researchers in aiding the respondents to narrate their story up to date softly without hindering the mindset that has been developed so far. Expressive technique helps respondents to incorporate stimuli into novel production such as

FIGURE 6.4
Choice and ordering techniques used in the research.

role play (Lannon, 1983). Out of the few expressive techniques that asked the respondents to narrate their own life story in relation to the pathway arranged through stimulus, a sample is given in Figure 6.5.

Construction Techniques Used in the Research

At the end, the respondents were asked to draw an illustration encouraging their imagination and creativity, on the topic of "Draw an image that describes your understanding of the concept of sustainability" as per the construction technique. It summarized the understanding of the respondents about sustainability. The respondents drew illustration or wrote something being related to the topic. Researchers identified it as an effective way to conclude the discussion. Construction techniques are described as asking respondents to describe a picture by drawing a picture by their own (Matthews, 1996). The construction technique used in the research is given in Figure 6.6.

Personal Experience of Sustainability Practices • 155

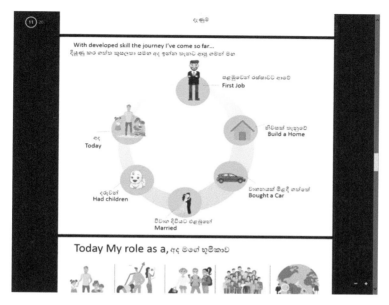

FIGURE 6.5
Expressive techniques used in the research.

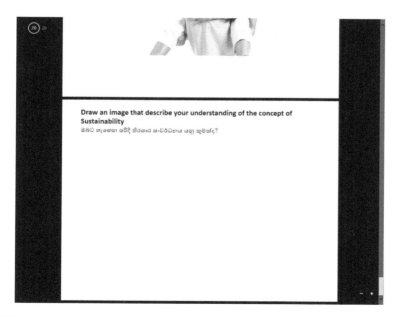

FIGURE 6.6
Construction techniques used in the research.

FIGURE 6.7
Clarence's illustration.

Responding to the construction technique, interviewees illustrated their view about the sustainability. Few of the illustrations drawn by the interviewees are given in Figures 6.7–6.9.

The use of several different methods of data collection enabled more information to gather, presented and interpreted from different viewpoints. Further, the different methods used enabled the discovery of new information and can assist in checking the reliability of the information presented by the participants that increased the strength of the research findings by enabling a triangulation of data and allowing a depth and richness of the findings of the research.

The data were collected over a period of two months up until sufficient amount of data were collected to answer the research questions. The participants were first selected from our personal contacts and then through snowball sampling to represent vivid marketing personals within the age group between 24 and 65 years of age in the research sample. Since snowballing sampling did not help in contacting participants with rich data,

Personal Experience of Sustainability Practices • *157*

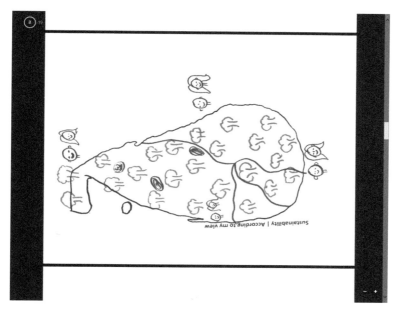

FIGURE 6.8
Hilton's illustration.

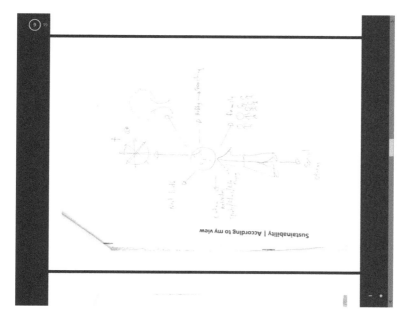

FIGURE 6.9
Isaac's illustration.

FIGURE 6.10
Interviewee residence photo.

we had to use purposeful sampling technique. The research sample was specially focused on marketing professionals.

Participants Experience and Knowledge in Marketing and Sustainability

Nineteen marketing professionals were interviewed for the research that represents different sectors, different age categories and are with different levels of educational background. We had an initial discussion with the respondents to understand their personal sustainability. The marketing professionals who demonstrated a dominant level of sustainability in their personal lives were only selected for interviews. Interviews were conducted at interviewees' personal residence. The expressed narrations were triangulated through personal artifacts and personal culture.

The pictures from an interviewee's residence by the researcher during interview are presented in Figure 6.10.

Profiles of Participants

The sample profile of the participants given in Table 6.1 explains the age, profession and level of education. It is important to note that pseudonyms were used instead of interviewees' real names with an aim to protect the identity of the interviewees participating in the research.

TABLE 6.1

Details of the Interviewees

Pseudonym	Age	Profession	Job area	Education
Aragorg	27	Marketing head	Digital marketing	Degree
Britney	50	Co-partner	Constructions	Diploma
Clarence	26	Entrepreneur	Education	Degree
Douglas	60	CEO	Health care	Degree
Engithan	48	Senior insurer	Insurance	Degree
Farhan	36	Branch manager	Household appliances	Diploma
George	35	Branch manager	Sales and Marketing	Diploma
Hilton	32	Entrepreneur	Taxation	Diploma
Issac	40	Entrepreneur	FMCG products	Diploma
James	42	Entrepreneur	Finance	Post graduate Reading
Knight	26	Assistant Marketer	FMCG	Degree
Lionel	26	Merchandiser	Apparels	Degree
Monis	37	Quality controller	Boats manufacturing	Diploma
Nonnis	46	CEO	Educational	Post graduate
Oracle	28	Trade marketer	Telecommunication	Degree
Putin	34	CEO	Apparels	Degree
Quarrel	32	Sales manager	Telecommunication	Degree
Remus	35	Branch manager	Customer care	Degree
Snape	65	Ambassador	Politics	Post graduate

Major Problems

Researchers found problematic in maintaining a smooth flow of discussion due to the busy schedules followed by marketing professionals participated in the research. Despite the challenges, researchers were able to collect sufficient narrations by meeting interviewees at their respective personal residences.

FINDINGS AND DISCUSSION

The findings in Table 6.2 show what initiatives are strongly supported by marketing professionals as a result of their personal experience and learning. The findings are arranged under the three main pillars of sustainability, that is, social, economic and environment.

TABLE 6.2

Quarrel's Social Sustainability Practices

Learned	Practiced	Applies
From Father: To live in harmony with people.	Baseball player, Facebook page admin, Union member in university	**Sales manager**: Makes customers his close friends. **A part-time FB page admin**: Where he connects with a huge fan base even during his leisure time.

SUSTAINABILITY PRACTICES OF PROFESSIONAL MARKETERS

Social Sustainability

Sustainable development is often explained as being divided into three sectors including economy, environment and society (Chamberlain, 2020). The three sectors are presented in interconnected rings, highlighting the aim that sustainable development must be brought forward in a balanced way. According to the triple bottom line perspective, social sustainability means business profit in human capital. It includes the position of a firm in a particular local society. Social bottom line increases with the density of fairness and benefactor level of labor practices and with the amount of corporate community involvement. It can also be measured from the impact a business makes on the local economy (Chamberlain, 2020).

The society consists of multiple human actions and interactions that build up human life. Without a society, humans would not survive. Even the evolution is based on human interactions (Clulow, 2005). Being marketing is at the forefront of recognizing the key role of all stakeholders in the decision-making process (Clulow, 2005), and the social-interacted initiatives are highly demonstrated from marketers; according to the stories narrated by interviewees, it is highlighted that character is dominant in their personal lives too where they have been experiencing it from their culture and been trained with the influence from the society.

Based on the Life Story of Quarrel

Table 6.2 explains the way how Quarrel was exposed to social sustainability at his childhood, practiced them at his school and currently applies in his profession.

The marketers' stories revealed the fact that their attention has been attracted toward each and every social sustainability practice that they are practicing currently. Quarrel recalls how his father was interacting with the community.

> I have seen the way my father enjoys with his friend after a very tired full day of cultivating they played 'Ombi' they drank even alcohol just enjoy a good tie with friends.

Marketers expressed the idea that living in harmony with the people is an essential thing for the well-being of the society. Irrespective of the religion and the ethnic group, people should live in harmony. Marketers pointed out that we should respect each other's ideas and viewpoints. Quarrel explained the way how he personally executes social sustainability in his daily job as a sales manager.

> I meet different people Buddhists, Muslims, Sinhala, a political party lover b political party love, but I deal with all of them. Because all are important, right…!

Quarrel further explains that he enjoys himself in maintaining Facebook pages as a part-time involvement, since he can check-in with more friends and people, through which he derives a great deal of self-satisfaction.

> I've maintaining few Facebook pages.

Based on the Life Story of Snape

Table 6.3 explains the way how Snape was exposed to social sustainability at his childhood, practiced them at his school and currently applies in his profession.

TABLE 6.3

Snape's Sustainability Practices

Learned	Practiced	Applies
Childhood society: To live being cooperative with people.	Scouting and during his childhood living and education to date.	**As an ambassador**: Tries to ensure the cooperation among countries.

162 • *Social and Sustainability Marketing*

Snape also shared a similar life story as Quarrel's. Snape recalled how cooperative the society was back in his childhood.

> The society was so interactive back then, they helped each other, escorted girls to home in the evening....

Marketers believe that there should be cooperation among all the people and they should take care of each other and make sure that trustworthiness is established among people. Snape explained the importance of having cooperation among the countries and the communities.

> Big countries do not think about the poor countries. Countries with the capacity must bring policies to establish the international cooperation among the people not to destruct the harmony.

The stories recalled by respondents shared similar facts. Most of the marketers have had an attention toward the way how their parents and society acted cooperatively with the community during their childhood. They have been inspired by those attentions. Not only that, they have practiced them with the help of the opportunities to engage in extracurricular activities during school and university life. Those opportunities have led them to realize the importance of having people around them and raise their awareness of the art of living with others. It has ensured their social sustainability. Figure 6.11 explains the way how marketers' social sustainability practices experienced in the childhood bring to the commitment toward corporate sustainability initiatives.

Table 6.4 points out the personal social sustainability practices resulting from the experiences shared in upbringing by each of the interviewees that took part in the research.

FIGURE 6.11
The process of social sustainability practices reflected through corporate practices of the respondents.

TABLE 6.4
Social Sustainability Practices from Upbringing by Interviewees

	Community Exposure			
Marketer	Childhood Community	Personal Exposures to the Community	Activities They Engaged with	In Corporate Initiatives
Aragorg	Suburb	Have so many experiences in streets with community.	Member of AIISEC	Practicing stress-free working environment in the office for employees.
Britney	Suburb	Admired by father's lifestyle.	Leader of school Debate team	Employees are considered very important and rely on them.
Clarence	Suburb	Admired by parents' lifestyle.	Basketball player, Member of Gavel	—
Douglas	Rural/Farmers	A typical village boy spent an enjoyable life with his friends.	Scout member, professional dancer	—
Engithan	Rural	A close relationship with community involvements along with his parents including cultivating and *Pal Rakeema*.	Head Prefect at school	Establishing close relations with the clients he meets through business course.
Farhan	Rural	A typical village boy spent an active involvement in social activities.	Cricket player (SL – A division)	Believes in the fact that customer is the king.
George	Rural	A typical village boy spent an enjoyable life with his friends.	School Prefect	Believes in the fact that customer is the king.
Hilton	Suburb	Suburb children with lot of friends.	Project Organizer – School Prefect	Personal involvement in community services.
Issac	Rural	Admired by his grandfather: A Tea-estate Owner.	Scout, Cricket, Member of a Zoological Club	Ensures community is very well served by his business and personal involvement in community services.

(*Continued*)

TABLE 6.4 (Continued)

Social Sustainability Practices from Upbringing by Interviewees

	Community Exposure			
Marketer	Childhood Community	Personal Exposures to the Community	Activities They Engaged with	In Corporate Initiatives
James	Suburb	Admired by his mother's struggle to live.	School Debate Team, Athletics	—
Knight	Suburb	Rose in a child care.	Member of Gavel	—
Lionel	Suburb	Admired by his father's struggle to live.	Art & Drama club, Physical Fitness club	—
Oracle	Suburb	Experiences with friends collected by playing cricket.	Cricket player	Establishing strong relationships with different people meeting everyday.
Putin	Suburb	An aggressive suburb child spent a lot of time on streets, on bikes and with different social groups.	Athletic	Establishing strong relationships with different people meets every day.
Quarrel	Rural	Comes from a rural family lived with cooperation with the community.	Baseball player, Facebook page admin, Union member in university	Converts clients into friends.
Remus	Suburb	An aggressive suburb child spent a lot of time on streets.	All most all the extracurricular activities (Prefect, Cricket, Athletics, House Captain, Scouts, etc.)	Establishing strong relationships with different people meeting everyday.
Snape	Rural	Comes from a rural middle-class family lived with cooperation with the community.	Scouting	Always tried to ensure the best person receiving the best responsibility to attend during his diplomatic career.

Accordingly, it is clear that marketers have been practicing social sustainability throughout their life to date. Social bottom line reflects the idea that if the business is not maintaining positive relationship with the stakeholders, the pool of stakeholders will shrink eventually. In that case, employees and customers are considered prominent stakeholders. Marketers being the guiding force in a business, their social sustainability nurtured through personal social sustainability competence as mentioned plays a vital role.

Economic Sustainability

The triple bottom line perspective explains that economic sustainability is not traditional corporate capital but business impact on the economy. Further, the business that strengthens the economy will continue to succeed due to its contributing factor on overall economic health and its support toward the network and community (Chamberlain, 2020). The research provides evidences that marketers are driven by a strong sense of economic sustainability whey they practice it by increasing the cash flow of the business and managing marketing campaign budgets. Marketers are not directly involved with financial functions of a business, but marketers' attitude toward economic sustainability of a business is considered vital. They have realized the importance of money prior to their employment during their childhood with exposure they had toward their parents' lifestyle or behavioral patterns. They use their acknowledgment about the importance of economic factor to the business development, but in the meantime they care about their personal economic sustainability as well.

Based on the Life Story of Nonnis

Table 6.5 explains the way how Nonnis was exposed to economical sustainability at his childhood, practiced them at young age and currently applies in his profession.

There was a common view with regard to the economic sustainability among all the professionals who took part in the research. They had been exposed to the struggle engaged in by their parents to earn money. The exposure they had with their parents with regard to economic sustainability in the family has led them to realize the importance of

TABLE 6.5

Nonnis's Economical Sustainability Practices

Learned	Practice	Applies
Childhood experience: Their parents were poor and struggled a lot due to that. Since childhood, he knew that having money in hand can make a difference. "My parents were poor; we had relatives who were talking English when we were struggling" (Nonnis).	Started earning money during university. Believes he has upgraded into certain social status because of money he earns. "I found my first Job at a Japanese company" (Nonnis).	**Tries to increase the ROI of the business.** "In business of course you need to make profits to the shareholders" (Nonnis).

money. It has forced them to be employed just after finishing education, and still they identify money as an important thing for their personal lives as well.

> Father worked so hard for the well-being of my family and then I realized that as the eldest son in the family I also have a responsibility to work toward the well-being of my family which is why now I am contributing toward the well-being of my family.
>
> (Lionel)

In corporate initiatives, marketers are obliged either to increase the cash flow or manage the funds allocated in marketing campaigns.

> Money is very important.
>
> (Aragorg)

> We are here to make money for the long run.
>
> (Oracle)

Financing is not a prominent responsibility in the job description of a marketer, but it is one of their duties as a member of a family. For better understanding, it is summarized in Figure 6.12.

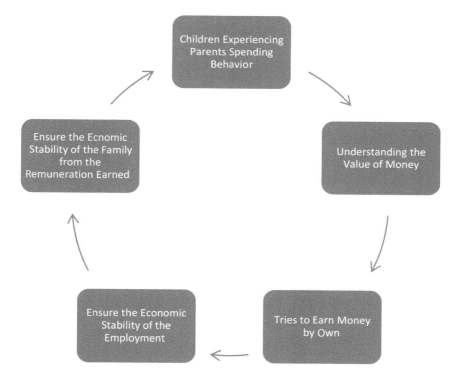

FIGURE 6.12
How personal economic sustainability practices reflected through corporate practices of the respondents.

CONCLUSION

The aim of this study was to explore the personal sustainability practices reflected through corporate sustainability initiatives of marketing professionals in the corporate sector. The chief motivation behind this exploration was to comprehend how the learning process undergone during one's formative years, socialization within family and formal education were instrumental in their commitment to perform tasks pertaining to their careers later in life.

As highlighted in the literature review, the notion of sustainability has broadened over time, both conceptually and practically. To this end, the corporate (business) sector has been identified as a main contributor for many sustainability-related issues, and thus a key player in resolving the same. Within corporate organizations, the marketing function and marketers as the "actors" of the function appear to play a significant role in implementing sustainability initiatives of the organization. However, there is a lack of common agreement in the literature among diverse corporate sustainability initiatives.

Having reviewed narratives of professional marketers in Sri Lanka, the current study highlighted several insights to address this gap in the literature. Despite the common classifications of sustainability initiatives – environmental, social and economic (United Nations, 1987) – these organizations appear to practice more "social" sustainability practices. This was mainly evident through their stakeholder-related sustainability initiatives as highlighted above. Such genuine commitments toward social sustainability initiatives appear to have come from their learning of values related to "fostering relationships" with others during their upbringing. Such behavior may be partly explained by the collectivistic nature of the Sri Lankan culture (Freeman, 1997). For example, some suggest that those in collectivistic societies tend to have stable relationships with their connecting parties (e.g., with family, friends and people considered important in life) (Triandis et al. 1988). Further, "managing stakeholders" is an integral part of marketers in the current competitive business world. Thus, such strong commitment to social sustainability can be justified through their professional point of view as well.

Moreover, despite the dearth of "environment"-related social initiatives implemented in organizations, even the ones that are being implemented do not seem to capture the attention of marketers. Further, the connection between childhood learning experiences do not seem to be reflected in their commitment toward environmental sustainability initiatives of their organizations. This could be partly due to the lack of attention paid toward environmental-related areas when developing curriculum for schools and the higher education sector. Although the current education system in the country has made progression toward addressing this issue, there appears to be more room for development.

As presented under the findings above, economic sustainability initiatives were not common among the corporate sustainability practices of the respective organizations despite the fact that they have some personal practices related to their economic practices. One key reason for the lack of practice and commitment toward economic-related initiatives could be the macro nature of the conceptualization of economic suitability (Purvis et al. 2019). Therefore, converting such broader concept into individual or firm-level practice seems to be challenging.

PRACTICAL IMPLICATIONS

The current study has revealed several practical implications for marketing professionals. First, having understood that marketers' childhood experience and learning is strongly linked to their commitment toward corporate sustainability practices, the selection process of marketing professionals must strive to go beyond the assessment of technical competencies and revelation of such exposure will support productive implementation of sustainability initiatives in the organization can be ensured.

Second, marketers with their understanding and commitment toward sustainability initiatives are learned through both childhood experiences (learning and socialization) and practice at work, are expected to be in a stronger position to better align their corporate sustainability initiatives to secure the attention of target markets and other stakeholders alike. Moreover, in addition to supporting self-learned initiatives, marketers are also in a position to identify new (and more meaningful) sustainability initiatives from their consumers' point of view. This way, the scope of sustainability practices would be enhanced within the organization.

Third, given the association between personal exposure through upbringing, and the commitment to sustainability practices in the organization, strong emphasis can be placed in incorporating the same into the formal education system through the curricula at both school and higher education levels. Further, having observed that environmental initiatives in particular are not strong pursuits of marketers, the important and possible sustainability initiatives toward protecting environment can be introduced through formal education. Accordingly, the study recommends that more practical modules should be incorporated when developing new sustainability education curriculum.

Further, this area can be merged into the training and development activities of marketing professionals. As evidenced in the study, the participants who experienced and fostered social relationships in their personal life were found to possess skills in executing their social sustainability initiatives in their respective firms. Similarly, if the areas that seem to have weak support from the marketers are addressed through their respective training programs, firms can secure a strong commitment of marketers toward successful implementation of their sustainability initiatives.

Moreover, the research possesses an academic value since it answers the integral issue of "how to transform sustainability education into marketing practices?" The study explains that it is transferred through practice to which a person is exposed. This denotes that the depth of sustainability initiatives executed/supported by marketers depends on the extent to which he/she was exposed in practice during his/her past.

ACKNOWLEDGMENTS

We sincerely thank the University of Sri Jayewardenepura, Gangodawila, Nugegoda, for providing academic acceptance for this research. We thank the reviewers of the upcoming publication with Taylor and Francis "Social and Sustainability Marketing: A Casebook for Reaching Your Socially Responsible Consumer through Consumer Marketing Science" for comments that greatly improved the manuscript. We would also like to express our gratitude to all marketing professionals for sharing their life and professional experiences as narrations.

> I always run into these Ph.D.'s. They write and write and write about sustainable development. Then these guys ask me, 'But, how do you do it?' They are scared to death to do anything.
>
> (Jamie Lerner)

RESEARCH PROBLEM

What are the experienced and learned sustainability practices reflected through the commitments toward corporate sustainability initiatives of marketing professional in Sri Lankan corporate sector?

REFERENCES

Andreasen, A. (2002). Marketing social marketing in the social change marketplace. *Journal of Public Policy and Marketing*, 21(1), 3–13.

Araç, S. K., & Madran, C. (2014). Business school as an initiator of the transformation to sustainability: A content analysis for business schools in PRME. *Social Business*, 4(2), 137–152.

Bandura, A. (2002). Social cognitive theory in cultural context. *Applied Psychology*, 51(2), 269–290.

Bocken, N. (2017). Business-led sustainable consumption initiatives: Impacts and lessons learned. *Journal of Management Development*, 36(1), 81–96.

Boddy, C. R. (2005). What do business research students think of the potential for projective techniques in business research? Quite a bit actually. The Academy of Marketing Conference. Dublin: Academy of Marketing.

Bugental, D. B., & Grusec, J. E. (2007). Socialization processes. *Handbook of Child Psychology* (Vol. 3).

Burns, L., & Lennon, S. (1993). Social perception: Methods for measuring our perception of others. *International Textile and Apparel Association Special Publication*, 5, 153–159.

Chamberlain, A. (2020). *Sustainability management system: The triple bottom line*. Retrievable from: https://www.era-environmental.com/blog/sustainability-management-triple-bottom-line

Chase, S. (2005). Narrative inquiry: Multiple lenses, approaches, voices. The Sage Handbook of Qualitative Research. London: SAGE Publications.

Cleene, D. S., & Wood, C. (2004). Sustainability banking in Africa. *African Institute of Corporate Citizenship, Johannesburg* (pp. 18–20).

Clulow, V. (2005). Future dilemmas for marketers: Can stakeholder analysis add value? *European Journal of Marketing*, 39(9/10), 978–997.

Crane, A., & Matten, D. (2004). *Business Ethics: A European Perspective: Managing Corporate Citizenship and Sustainability in the Age of Globalization* (p. 224). Oxford: Oxford University Press.

Daly, H. (1992). *Steady-State Economics* (2nd ed.). London: Earthscan.

Freeman, M. (1997). Demogrpahic correlates of individualism and collectivism: A study of social values in Srilanka. *Journal of Cross - Cultural Psychology*, 28(3), 321–341.

Giddings, B., Hopwood, B., & O'brien, G. (2002). Environment, economy and society: Fitting them together into sustainable development. *Sustainable Development*, 10(4), 187–196.

Giles, W. (1989). Marketing planning for maximum growth. *Marketing Intelligence & Planning*, 7, 1–98.

Gordon, W. (1988). *Qualitative Market Research: A Practitioner's and Buyer's Guide*. London: Routledge.

Greenfield, W. (2004). In the name of corporate social responsibility. *Business Horizons*, 47(1), 19–28.

Jones, P., Hill, C., & Comfort, D. (2008). Marketing and sustainability. *Marketing Intelligence and Planning*, 2(26), 123–130.

Kemper, J. A., Ballantine, P. W., & Hall, C. M. (2018). Global warming and sustainability: Understanding the beliefs of marketing faculty. *Journal of Public Affairs*, 18(4), e1664.

Korten, D. C. (1996). The failures of Bretton Woods. *The Case Against the Global Economy* (pp. 20–30).

Kotler, P. (2011). Reinventing marketing to manage the environmental imperative. *Journal of Marketing*, 75(4), 132–135.

Lander, L. (2017). Education for sustainability: A wisdom model. In Handbook of Theory and Practice of Sustainable Development in Higher Education (pp. 47–58). Cham: Springer.

Langley, P. M. M. (2002). Economy, sustainability and sites of transformative space. *New Political Economy*, 7(1), 49–66.

Lannon, J. (1983). Humanistic advertising: a holistic cultural perspective. *International Journal of Advertising*, 195–213.

Mannion, G. (2019). Resembling environmental and sustainability education: Orientations from new materialism. *Environmental Education Research*, 26, 1–20.

Matthews, B. (1996). Drawing scientists. *Gender and Education*, 8, 231–244.

Merrilees, B., Getz, D., & O'Brien, D. (2005). Marketing stakeholder analysis: Branding the Brisbane Goodwill Games. *European Journal of Marketing*, 39(9/10), 1060–1077.

Morgan, N. A. (2012). Marketing and business performance. *Journal of the Academy of Marketing Science*, 40(1), 102–119.

Mostyn, B. J. (1978). *A Handbook of Motivational & Attitude Research Techniques*. Bradford, UK: MCB Publications.

Pachauri, R. K., Allen, M. R., Barros, V. R., Broome, J., Cramer, W., Christ, R., ... & Dubash, N. K. (2014). *Climate change 2014: Synthesis report. Contribution of Working Groups I, II and III to the fifth assessment report of the Intergovernmental Panel on Climate Change* (p. 151). IPCC.

Purvis, B., Mao, Y., & Robinson, D. (2019). Three pillars of sustainability: In search of conceptual origins. *Sustainability Science*, 14(3), 681–695.

Quinn, P. M. (2002). *Qualitative Research and Evaluation Methods*. Thousand Oaks, CA: SAGE.

Triandis, H. C., Bontempo, R., Villareal, M. J., Asai, M., & Lucca, N. (1988). Individualism and collectivism: Cross-cultural perspectives on self-ingroup relationships. *Journal of Personality and Social Psychology*, 54(2), 323.

United Nations (1987). *Report of the World Commission on Environment and Development*. Brundtland.

United Nations (1992). *Report of the United Nations Conference on Environment and Development*. Rio de Janeiro: United Nations Publications.

United Nations (1998). Kyoto Protocol to the United Nations Framework.

United Nations (2015). *The Millennium Development Goals Report*. New York: The United Nations.

Wade, B. (2019, April 4). *Why companies need to embrace sustainability as a strategic imperative rather than an operational choice*. Retrievable from: https://www.entrepreneur.com/article/331743

Wind, Y. (2005). Marketing as an engine of business growth: A cross-functional perspective. *Journal of Business Research*, 58(7), 863–873.

Section III

Advances in Knowledge of Social and Sustainable Marketing: Sustainable Consumption and Consumer Behavior

7

Evolving Prosumer Identity in Sustainable Consumption: Deconstructing Consumer Identity

*Chamila Roshani Perera
and Chandana R. Hewege*
Department of Management & Marketing, Faculty of Business & Law, Swinburne University of Technology, Hawthorn, Australia

CONTENTS

Learning Objectives 175
Introduction 176
   An Overview of the Notion of Consumption 178
      Self-Identity and Consumption 180
   The Construct of Prosumption 181
      Sustainable Prosumer Identities 184
   Sustainability Conscious Prosumer Practices 185
   Challenges and Opportunities of Sustainable Prosumption 190
Implications for Sustainable Marketing Practitioners 191
Conclusion 193
Funding Declaration 193
Credit Author Statement 193
References 194

LEARNING OBJECTIVES

After reading the chapter, the reader should be able to:

- To describe the nature of the phenomenon of prosumption in the context of sustainable consumption practices

- To identify everyday challenges faced by prosumers
- To classify different sub-cultural groups of prosumers
- To assess the market opportunities that prosumption brings into the marketplace

INTRODUCTION

Self-identity can be seen as a collection of beliefs or a perception of oneself often in relation to a social context. Since self-identity is subject to *time and space*, it cannot be seen as a reflexive phenomenon at all (Giddens, 1991). Giddens (1991) also strassess that "individuals are in a process of continuously and reflexively building a coherent and rewarding sense of identity" in light of the challenges they face from their environment, people and situations. From a consumer perspective, how a consumer projects self-identity is an outcome of his or her perception of the environment and how the consumer deals with the environment (Connolly & Prothero, 2008). To this end, it can, therefore, be argued that environmental problems such as climate change and major health crisis such as COVID-19 pandemic continue to deconstruct the established consumer identities.

In regard to COVID-19 context, it is inevitable that nowadays consumers have a higher sense of uncertainty about and fear of their health and well-being. *Consumption* is an act which is performed by consumers with a view to surviving or satisfying their basic human needs. In its essence, undoubtedly, consumption provides consumers with a sense of comfort and well-being. Consumption practices can largely contribute toward symbolising what consumers want to express and share with other consumers as well. Having to comply with newly imposed restrictions on what constitutes everyday life and routines (e.g. shopping, socialising) or seeing the collapse of major pillars of their outer environment (e.g. challenges faced by national governments when controlling the adverse effects of COVID-19 spread) apparently deconstruct the existing consumer identities and the way in which the consumer shares experiences with other consumers.

Given the unprecedented challenges in the modern marketplace, some consumers could fail to project their long-established end-user identity. Being a typical consumer who shops and buys commodities for personal consumption is not an identity an individual can play in an unprecedented time like this (e.g. COVID-19 pandemic). It is natural to feel vulnerable and not to entertain the comfort of consumption as usual. Therefore, some other consumers may even start to project new consumer identities. Some consumers could become conscious about health and well-being than ever before, whereas some other consumers could become more sustainability conscious. The latter group could become more conscious about the production processes, country of origin or having organic or locally produced certifications when they buy commodities from the marketplace.

The identity of prosumers or *post-consumers* (Firat & Venkatesh, 1995) who are labelled as people consuming goods and services that are self-produced rather than purchasing them from the market (Toffler, 1980) can be considered as an interesting phenomenon to explore in the modern marketplace. For example, many industry reports (e.g. Develocity, 2020; Galus Australis, 2020) show that consumers around the world have been increasingly undertaking several do-it-yourself (DIY) projects during the coronavirus shutdowns. As revealed by a current national survey by RICHGRO (2020), as a result of COVID-19-related restrictions and other concerns, many Australians have turned to be gardeners and developed an increased linking toward home-grown food. Of those surveyed, majority are now apparently growing more fruits and vegetables than they used to grow during this period last year.

In Australia, it is evident that there is an increase in the number of prosumers making intelligent choices supported by smart technology, particularly in relation to energy usage (Mouat, 2016). According to an Australian study (McLean & Roggema, 2019) producing solar power at homes and offices, and selling the excess to the grid, water recycling, processing waste materials and saving money by communal consumption are common prosumption practices. Although they enhance the well-being of the communities, there seems to be number of challenges and opportunities this phenomenon could bring into the structure of the modern marketplace while pushing the boundaries of the conventional structure of the modern marketplace.

Based on this background, the proposed chapter aims to explore the phenomenon of prosumer identity from the lens of sustainable consumption with a view to highlighting the challenges and opportunities it brings into the consumer life and the structure of the modern marketplace. The chapter is organised into five main sections. Followed by the first section, the chapter introduction, the second section provides an overview of the notion of consumption. The way in which consumers construct and express identities through consumption is discussed in a subsection of the second section. Third, the chapter reviews the construct of prosumption including a subsection which details sustainable prosumer identities. Fourth, the chapter details the challenges and opportunities it brings into sustainable marketplace. Finally, the chapter concludes with some implications for marketers.

An Overview of the Notion of Consumption

The notion of consumption can be explained from various perspectives. Some studies explain the notion based on rational assumptions of consumption that view consumption as a utility maximisation. When viewed from the classical economic theories, consumers always make decisions that provide them with the highest amount of personal utility. These decisions provide them with the greatest comfort, benefit or satisfaction given the choices available. This school of thought, therefore, investigates consumption from rational, economic perspectives and attempts to explain the relationship between the antecedents of consumption (e.g. values, attitudes and beliefs) and consumer behaviour intentions.

Some other studies explain the notion of consumption based on the critiques of mass consumerism that views consumption as an unfavourable phenomenon (e.g. materialism, excessive consumption). It is argued that a radical transformation of values, behaviour and institutions leading to a paradigm of reduced consumption is needed (Lorenzoni, Nicholson-Cole, & Whitmarsh, 2007) as far as building a sustainable society is concerned. The studies view materialism as "an excessive regard for worldly possessions", "a personal trait or a value" that make humans "accumulate possessions to seeking happiness" (Belk, 2001; Belk, 1984; Kasser, Ryan, Couchman, & Sheldon, 2004; Richins & Dawson, 1992). Many argue that this form of consumption is a negative phenomenon (Burroughs & Rindfleisch, 2002; Kasser & Ryan, 1993; Mick, 1996). On the contrary, some studies view it as a commonly shared thinking framework of a culture (Arnould & Thompson, 2005; Holt, 1998; Mick, 1996). The value system constitutes a collective psyche through which consumers interpret the social world and is commonly referred to as the dominant social paradigm (Kilbourne, Beckmann, & Thelen, 2002; Pirages & Ehrlich, 1973). Prothero, McDonagh and Dobscha (2010) argue that recent sustainable movements shatter the fundamental assumptions of this social paradigm making a society with greater sustainability concerns. There is, however, no sufficient empirical evidence or scholarly conversations around how this shift of the dominant social paradigm occurs in the modern marketplace. It is expected that this chapter could initiate this timely conversation.

Studies viewing consumption as a shared practice of socially constructed interpretations of consumer products are also not rare in consumption literature. Those studies assume that consumer products largely contribute to symbolising what consumers want to express and share with other social members. This is referred to as an exchange of symbolic meanings through consumption (Douglas & Isherwood, 1996; Holt, 1995; Prothero & Fitchett, 2000). In particular, according to Douglas and Isherwood (1996), consumer products function as *live information systems* and their meanings are derived from the social context. Further, the act of consumption does not always involve a commodity (a trading item in a market). Consumers also use other objects when engaging in consumption, for example, home-grown vegetables or other DIY projects. Therefore, it can be stated that in the act of consumption, commodities as well as consumer-objects can function as *live information systems* that consumers exchange.

In response to previously mentioned critiques of mass consumerism that view consumption as an unfavourable phenomenon, Andreou (2010) argues that materialism is not an unfavourable phenomenon especially in the context of sustainable consumption since happiness derived from materialistic consumption thrives in an emotional level, not necessarily in a material level. For example, at times for some consumers, sustainable consumption is associated with a *premium price* which is reported to have a significant association with "an emotional benefit of high social status" (Griskevicius, Tybur, & Van den Bergh, 2010).

It is, therefore, intriguing to explore sustainability consumption through the lens of symbolic meaning. More specifically, it is required to explore how consumption is (1) formed by meanings that are socially constructed through social interactions and (2) adapted through interpretations of these meanings (Belk, 1988; Blumer, 1969; Holt, 1997). Adopting this perspective has been highlighted in the literature as much needed emerging trend in new sustainability research (Papaoikonomou, Ryan, & Ginieis, 2011).

Consumer culture theorists such as Douglas and Isherwood (1996) also suggest sharing symbolic meanings of products as an alternative perspective that can be used to review materialistic orientations and subsequent criticisms of environmentally destructive behaviour of consumers.

Accordingly, consumption needs to be viewed, favourably, for example, as an inducement to work or make human connections. As such, products need to be viewed as valuable instruments of a society.

It can be seen that some researchers argue that consumer markets consist of culturally mediated needs that are always linked to meanings and cultural images of products (e.g. Bourdieu, 1984; O'Shaughnessy & O'Shaughnessy, 2002; Solomon, 1983). Further, as noted earlier, the act of consumption involves commodities and consumer objects. Accordingly, it can be stated that consumption can be related to adding culturally mediated symbolic meanings to functional benefits of commodities and consumer objects, satisfying the social needs of consumers. These meanings are also referred to as symbolic properties of commodities and consumer objects, for example, self-identity expression through consumer products (Belk, 1988; Dolfsma, 2004; Piacentini & Mailer, 2004). This aspect of consumption is discussed in the forthcoming section.

Self-Identity and Consumption

Generally, the understanding or appreciation of one's psychological and biological characteristics as a particular individual, especially in relation to social context is referred to as self-identity. The notion is sometimes referred to as one's self-concept, self-construction, self-perspective, and self-expression or self-structure. Self-identity is defined as "who one was, is, and may become articulated via an array of personal and social identities" (Oyserman & Destin, 2010). According to the notion of "looking-glass self", in what way we see ourselves comes from our perception of how others see us (Cooley, 1902).

Previous research spaning several disciplines debates about projecting a "sense of self" through consumer objects (Belk, Sherry Jr, & Wallendorf, 1988; McCracken, 1986; Oyserman, 2009). Belk (1988) pioneered the idea of "extended self" that means likelihood of consumers to buy, utilise and throw away consumer products "reflecting who they are and who they want to be". Extended self is described as "an aggregation of objects, mainly human beings, that aid in the production of an individual physically and an individual considers them psychologically, and their potentiality by which one can have a sense of their own identity" (Siddiqui & Turley, 2006).

Over the decades, it has been largely established that consumers tend to project their identities through sustainable consumption (Autio, Heiskanen,

& Heinonen, 2009; Black, 2010; Connolly & Prothero, 2003, 2008; Horton, 2003; Huttunen & Autio, 2010; Soron, 2010). Previous literature on sustainable consumption discusses about expressions of identity namely, "anarchist", "hero", "antihero" and "environmental hero" (Autio et al., 2009) as well as "vegetarians", "vegans", "recyclers", "green voters", "environmentally conscious consumers" or "ethical citizens" (Cherrier, 2006).

Sustainability consumption generates a feeling of "social recognition and connection", distinguishing sustainability conscious consumers from mainstream consumers who are environmentally unfriendly (Black & Cherrier, 2010; Cherrier, 2006, 2007; Griskevicius, Tybur, & Van den Bergh, 2010; Perera, Auger, & Klein, 2016). This can, therefore, be considered an area that self-identity is constructed through sustainable consumption.

From a sociological perspective, self-identity is defined as "the self as reflexively understood by an individual in terms of her or his biography" (Giddens, 1991, p. 53). As mentioned earlier, Giddens (1991) stresses that self-identity should not be viewed as a reflexive phenomenon, as it is subject to "time and space". This is because individuals are going through a development of endlessly and instinctively constructing and rewarding sense of identity based on environmental problems. To further illustrate, how a consumer views self-identity is affected by the manner in which the consumer perceives and deals with other global environmental issues (Connolly & Prothero, 2008, p. 121).

Regarding sustainability consumption, it is revealed that consumers undergo a role conflict between the role of customer and the role of responsible citizen. This conflict is caused by their effort to strike a balance between personal well-being and the environmental well-being (Barr, 2003; Cuthill, 2002; Slocum, 2004). Therefore, exploring how consumers express their identities through sustainable consumption is intriguing. As mentioned in the introduction, this chapter aims to examine the phenomenon of prosumer identity in the area of sustainable consumption. Therefore, the next section reviews the construct of the prosumption first before reviewing the construct of sustainable prosumer identities in detail.

The Construct of Prosumption

Toffler (1980) defines prosumers as "consumers who consume self-produced goods and services rather than buying them from the market". These special consumer groups have been identified by various names such as "values co-creators" (Prahalad & Ramaswamy, 2004; Vargo & Lusch, 2008), "market partners" (Prahalad & Ramaswamy, 2000) and "post-consumers" (Fırat & Dholakia, 2006). Generally speaking, these consumers tend to be engaged in designing or customising products for their own consumption. The concept of prosumption may take several forms ranging from consumption to sharing, to creation, to fully pledged prosumption.

According to previous research, the phenomenon of prosumption commonly appears in the areas of energy and digital resources. Prosumption is also found to operate in the form of "symbolic meaning sharing behaviour" among young consumers who engage in social well-being developments (Lucyna & Hanna, 2016). Overall, in prosumer literature, individuals who apparently perform an active position in creating value by extending the limits of the mainstream market system are called as "value co-creators" (Ramaswamy & Ozcan, 2014, p. 3). As mentioned earlier, since the phenomenon of prosumption could be viewed as a spectrum (i.e. active prosumption to passive prosumption), it is important to borrow a theoretical classification to explore the phenomenon. To this end, we borrow Sayfang's competing subculture classification which is detailed below.

Seyfang (2004) identifies four competing subcultures of consumption along with different attitudes toward sustainable consumption: (1) "individualists" (active eco-friendly commodity purchasers), (2) "fatalists" (passive eco-friendly commodity purchasers), (3) "hierarchists" (responsible consumers) and (4) "egalitarians" (who consume less). Some scholars argue whether the phenomenon of prosumption is a novel format of capitalistic market mechanism (Büscher & Igoe, 2013; Ritzer, 2015; Ritzer & Jurgenson, 2010). Therefore, Seyfang's (2004) categorisation detailed above could be helpful when exploring the concept of prosumption. For example, there can be (1) active prosumers, (2) passive prosumers, (3) responsible consumers and (4) minimal consumers depending on how actively they engage in self-production or coproduction.

Although some aspects of prosumer practices could manifest contrary to current mainstream market mechanism, these prosumer practices are found to be facilitated by the same mainstream market mechanism (Perera et al., 2016; Prothero, McDonagh, & Dobscha, 2010). Distancing oneself from the "eco-friendly" commodity market, a prosumer with sustainable motives tend to engage in self-production. Nevertheless, the sustainable prosumer has to buy essential components from the mainstream marketplace. As such, we argue that prosumer culture and related practices may coexist along with the current market mechanisms or manifest as an independently operating subculture of consumption.

According to Kotler (1986, 2010), prosumption can be viewed as ushering "a new marketing era where several overlapping systems co-exist to create consumer satisfaction". Kotler's (1986, 2010) view is reinforced by some studies that explain prosumption as a "complete collapse of consumption into production" (Humphreys, 2014; Zwick & Denegri Knott, 2009). This view is, however, challenged by some other scholars (e.g. Cova & Dalli, 2009). To illustrate further, by coining the notion of "working consumers", Cova and Dalli (2009) examine the downsides of "consumer production and collaboration". This is referred to as "double exploitation". It is argued that the "working consumers" contribute significantly when they engage in coproduction. Through this coproduction, they stand to receive individual and social rewards in the form of personal gratification, purchase satisfaction and social recognition. Ritzer and Jurgenson (2010) also agree that this amounts to be a high level of prosumer exploitation.

Presenting a counter argument to the above idea, however, some scholars advocate for a more promising scenario where collaboration among producers and consumers can generate knowledge and skills that are advantageous to both parties (Prahalad & Ramaswamy, 2004; Vargo & Lusch, 2008). Further, there are drawbacks as well as opportunities associated with the promising scenario where, "producers and consumers directly communicate to share, swap, trade, or rent" (Albinsson & Perera, 2018, p. 4). This amounts to be an outgrowth of prosumption into a more collaborative form of a consumer society in which consumers play multiple roles of producers, or users, or both. In a way, the traditional ownership structure is clearly disrupted. Moreover, according to Perren and Grauerholz (2015), the collaborative consumption underpinned by strong sense of community also uplifts prosumer confidence to reinforce political and personal ideologies that they cherish. Based on a review of prosumer literature spanning 30 years, Cova and Cova (2012) argue that prosumption is "an extension of postmodern marketing scholarship in the 1980s and 1990s". Cova and Cova (2012, p. 149) view prosumers as *"new consumers* who are agents of their own destinies".

In summary, as evident from prosumption literature, it appears that the consumers' role has changed from a conventional connotation that typically considered them as passive consumers who are at the end of value creation process. Nevertheless, sustainable prosumption activities on a wider scale still remain as largely an unexplored domain in the broad prosumption discourse. As such, more studies on sustainable prosumers from various angles are essential. The following section of the chapter reviews prosumption literature from a sustainable prosumer identity viewpoint.

Sustainable Prosumer Identities

Sustainability conscious prosumer could project an identity of a consumer who transforms into a producer, "an *actor in co-construction*" (Hansen & Hauge, 2017; Kessous et al. 2016), "a producer for self-use" and who is to a lesser extent influenced by manufacturers with lower or no environmental credentials. Therefore, an environmentally aware prosumer can be seen as a member of a community of solar power electricity or have a permaculture garden.

Lehner (2019) argues that prosumption in its essence could end up in a general rejection of overconsumption as a way of seeking pleasure and defining identity. We, however, argue that prosumption can be instrumental in forming a new identity which is facilitated by anticonsumption or self-production. This phenomenon can, therefore, provide special challenges and opportunities to the marketers in the modern marketplace. For example, as detailed earlier, prosumers use self-produced commodities instead of buying them from the mainstream market (Toffler, 1980). Sustainability conscious prosumers, therefore, are likely to look for important information on producers' environmental credibility before deciding to buy (Perera et al., 2016). They could, therefore, project the identity of an informed consumer or empowered consumer. It can be said that sustainability conscious prosumer is not necessarily likely to perceive prosumption as double exploitation as argued by some previous studies. Rather, the prosumer could engage in "empowered consumption" signalling their dislike toward trade practices that harm the well-being of the environment (Cherrier, 2010; Perera et al., 2016).

Sustainability conscious prosumer literature shows that consumers tend to move from an inactive stance toward a more active position and take deliberate measures to mitigate harmful effects of mainstream market system on environmental well-being (Kessous et al. 2016; Perera et al. 2016). Further, in line with Toffler's (1980) explanation, lifestyles of some sustainability conscious prosumers apparently resemble the behaviours demonstrated by people in an ancient agrarian society, where they consciously "prosumed" to with a view to protectiong the environmental well-being.

It can also be seen that some of prosumer identity developments can manifest against the prevailing market practices (e.g. resisting purchase). However, some other prosumer identity developments can be supported by the market practices (Perera et al. 2016; Prothero et al. 2010). For example, to uncouple from the "eco-friendly commodity discourse" (eco-labelled products sold in the market), a sustainability conscious prosumer may refuse to purchase from the marketplace, rather create whatever he or she wants to consume. Nevertheless, inevitably, resources that are essential for such creations have to be purchased from the market. Therefore, it can be argued that prosumer identity development may exceed any form of cohabitation with the prevailing market practices and can also manifest as an "independently emerged prosumption identity". The next section pays a special attention to sustainability conscious prosumer practices to explore how prosumers engage in them with or without the support from the existing marketplace.

Sustainability Conscious Prosumer Practices

Sustainability conscious prosumer practices can be described as any form of self-production an individual engage in with a view to minimising the adverse effects of consumption on environment, for instance, growing vegetables for personal consumption. In their extreme forms, sustainability conscious prosumer practices can partly emerge as alternate procurements (e.g. dumpster diving) as opposed to eco-friendly commodity purchases. Although these types of alternate procurement practices are obviously shaped and facilitated by the prevailing market practices, these practices also symbolise counter actions of emerging sustainability conscious prosumers. These counter actions are aimed at mitigating adverse effects of environmentally unfriendly practices such as food waste and "greenwash" caused by the mainstream market.

Some of the procurement practices of sustainability conscious prosumer may not necessarily involve production for self-use all the time, rather they can be passive prosumers or responsible consumers. However, unlike mainstream consumers, some sustainability conscious prosumers can refrain even from making purchases of eco-friendly commodities because they are mindful about the adverse environmental effects of mass production of eco-friendly commodities, unethical business practices, waste and landfilling. Overall, it can be argued that sustainability conscious prosumers play a role which is different from that of a mainstream consumer's. In its earlier form, this differntiation can be seen as an early stage of becoming semi-prosumers.

Sustainability conscious prosumption can also be practiced signifying behaviours such as less consumption or even controlling eco-friendly commodity purchases from the market. Sustainability conscious prosumer may like to participate in exchange activities (monetary or nonmonetary) with people in their close networks. These could take a form of "collective identity-building narratives" of prosumers who tend to buy fewer products. Their purchase behaviours are enabled and reinforced through collective actions. It can be seen that while the semi-prosumer practices are carried out mainly as individual actions, the behaviours of prosumption can reinforce semi-prosumer practices through collective actions.

There can be very creative prosumers as well. They endeavour to entirely distance themselves from the existing market system. Unlike collective prosumers described above, the creative prosumer behaviour is mainly formed around individual level practices. To illustrate further, a devoted prosumer could be interested in finding his or her daily fruits and vegetables from their gardens, making whatever they want instead of buying them from the conventional market places. Moreover, themes such as "doing things by yourself", exemplify the prosumers' motive of participating in value creations and enriching their prosumer identity expressions.

We, therefore, argue that the phenomenon of prosumption should be viewed as an evolutionary process (Ritzer, 2015; Xie, Bagozzi, & Troye, 2008) or "governmentality" development where consumers gradually take part in the value co-creation and become creative consumers or prosumers (Cova & Cova, 2012). Thus, it can be argued that sustainability conscious prosumption is an evolving subculture of three consumer groups (semi-prosumers, collective prosumers and creative prosumers). On the one hand, these consumer groups need some assistance, at least partially, from the prevailing market practices. On the other hand, they can be significantly enthused by confronting some environmentally unfriendly practices in the prevailing market practices.

In the modern marketplace, a diverse range of sustainability conscious prosumption practices takes place. As detailed earlier, certain types of these presumption behaviours (e.g. semi-prosumption practices) can take place in alignment with the current exchange systems. To support this, Büscher and Igoe (2013) argue that in this new form of prosumer capitalism, the power sustainability conscious prosumers are supposed to exercise is eventually restricted. This demarcates the limitations of the current discussion of prosumption as reflected in mainstream consumer research. Nevertheless, the subculture of sustainability conscious prosumption could clearly evolve within the existing eco-friendly commodity market.

Collective prosumption behaviour encompasses sustainability conscious prosumption actions that are intentionally performed through unconventional buying methods. However, these behaviours are frequently backed by the prevailing market practices. Sustainability conscious prosumption practices such as dumpster diving may not lead to buying of eco-friendly products. Those who engage in "dumpster diving" do not necessarily buy eco-friendly products, although "dumpster diving" may be induced by sustainability concerns, needing to avoid landfilling, prudence and enjoyment. Engaging in those behaviours, sustainability conscious prosumers aim to reduce the unfavourable effects on the environment caused by the prevailing mainstream markets. This subculture, therefore, operates as an environmental movement challenging the environmentally unfriendly, materialistic consumption. Contrary to the typical "future-for-others" enclosing of sustainability conscious prosumption, these practices can be enclosed as "present-for-us" (Büscher & Igoe, 2013, p. 15).

Albinsson et al. (2009) discuss about voluntary behaviour of product disposal by consumers. They also explain that voluntary disposal decisions and methods of disposal are determined by (1) personal characteristics, (2) community characteristics and (3) item characteristics. Clothes swapping, an alternate procurement practice, involves free exchange of apparels by one person with another person. Being sustainability conscious prosumers, some individuals engaging in swapping clothes are mindful about negative environmental consequences of buying new clothes, whereas others are interested in seeking happiness that they could derive by participating in those events. Interestingly, designing a new useable item using those used clothes appears to be the most prevalent practice among prosumers. According to the study, those swapping activities revel alternative consumption practices that are quicker, flexible and economical. Further, these processes give rise to a sustainable exchange system by creating a "sustainable and responsible lifestyle" (Albinsson and Perera, 2018).

As detailed earlier, certain prosumer behaviours can take a form of collective actions demonstrated by collective prosumers. Their prosumption practices are carried out mostly collectively, and empowered through several collective actions. Accordingly, sustainability conscious prosumption activities could be categorised as public prosumption practices and private prosumer practices (Barnett, 2010; Finn & Darmody, 2017). Firat and Dholakia (2006) argue that these activities move consumers from "passive member of consumer culture" to a "cultural constructor" along with a twist in their positioning from "consumer satisfaction" to "consumer empowerment". Further, this transformation of consumer positioning also leads to distancing from to the capitalist market system (Izberk-Bilgin, 2010; Papaoikonomou, Ryan, & Ginieis, 2011).

Under the influence of capitalist market system, prosumers are behaving as "temporary employees" who only see better alternatives once collective-shared exchange systems created among them (Humphreys & Grayson, 2008). To this end, for example, collective prosumption can be considered as an attempt to create a new system of exchanges which resonates with some other scholarly conversations such as peer-to-peer exchanges (Perren & Grauerholz, 2015, p. 143), or energy prosumer community (Ramaswamy & Ozcan, 2014, p. 223). The new trend of *shared exchange systems* is fundamental here; however, stronger empirical findings need to be gathered through future research (Miller, 2016).

It can be argued that *shared exchange systems* are essential for prosumers to evolve and operate (Perren & Grauerholz, 2015). Such exchanges cannot yet exist independent of the mainstream market mechanisms. Therefore, current prosumers' demand for alternate methods of consumption could still be influenced by the mainstream market mechanisms and could be made part of the mainstream market behaviour (Ritzer, 2017). While some argue that the firms can offer a certain level of freedom and creativity to prosumers to achieve effective "performance of the co-produced product, service or experience" (Seran & Izvercian, 2014, p. 1973). Previous research shows that *shared exchange systems* can generate "sustainable economic benefits" (Dellaert, 2019) and "environmental benefits" (Eckhardt et al., 2019). This area of research needs more attention from future researchers.

Albinsson and Perera (2012) investigate the association among online and offline communal events and mainstream market practices. It is found that the "event organisers", labelled as "consumer-citizens", educate the consumers and promote responsible consumption through creating a sustainability ideology to be reflected in day-to-day consumer practices. For those who participate in these events, social belongingness is the prime drive behind those participations, not any other functional benefit.

It is evident that conventional views of exchange and reciprocity are likely to be challenged by these alternative consumption behaviours. Whether such behaviours could be sustained over time rest mostly on two factors: (1) local infrastructure availability and (2) asymmetric nature in value configurations among individuals. However, consumers who are inspired by socially constructed meanings of their practices are likely to present an attractive potential market segment for marketers.

Creative prosumers are motivated by value creation activities. They do not tend to take part in mainstream market activities. Research on consumers largely cover purchasing behaviour, especially focusing on the disticntive roles of consumers and producers. "What consumers actually do when they consume" has not been adequately investigated (Firat & Venkatesh, 1995). As such, consumers are rarely labelled as creators and are often identified as "passive responders" (Shaw, Newholm, & Dickinson, 2006) or individuals engaging in "destruction" (Fien, Neil, & Bentley, 2008, p. 53). It can be argued that creative prosumption activities could be a symbolic act of distancing from those unfavourable notions. They begin to move away from being passive buyers of commodities.

It has been debated in the literature if it is possible to promote sustainability conscious prosumption within the current mainstream market system (e.g. Carrier, 2010; Prothero & Fitchett, 2000). To elaborate further, Black and Cherrier (2010) argue that apart from sustainability concerns, various personal interpretations and experiences tend to trigger anti-consumption (rejection of consumption). Toward this end, particularly, creative prosumption appears as a novel standpoint for examining sustainability conscious prosumption practices. Shedding more light on this argument, Ritzer (2017) states that while "prosumers-as-consumers" may look for alternative commodities, "consumers-as producers" may make their own.

In this chapter, we attempt to argue that the emerging sustainability conscious prosumption culture needs analytical frameworks separate from the ones that are applicable to wider prosumer discourse. Toward this end, this chapter argues that sustainability conscious prosumption culture is taking its shape as a dominant subset distinct from broader prosumer set. These subsets of consumer groups could provide unique sets of challenges and opportunities in the modern marketplace as elaborated in the next section.

Challenges and Opportunities of Sustainable Prosumption

As detailed in the introduction section, one can explain the notion of consumption from diverse angles. From symbolic and cultural meaning perspective, consumption can be seen as an inducement to work, "an aspect of a social need of relating to others" (Douglas & Isherwood, 1996). Therefore, consumer practices such as sustainability conscious prosumer practices could be seen as important instruments in nourishing social connections. As detailed earlier, environmental concerns, enthusiasm, challenge, prudence, team work and makings are the meanings common to the three groups of prosumers highlighted in this chapter. These groups of consumers are certainly growing in line with the latest technologies (e.g. Industry 4.0) (Serafin, 2012, p. 130).

As revealed in a recent review, the latest technologies (e.g. Industry 4.0) play a major role in "sustainable distribution and reverse logistics" by minimising carbon emissions and empowering consumers with necessary information to become sustainability conscious prosumers. They are prepared to pay extra given the eco-friendly products' functional qualities or higher environmental credibility (Dangelico & Vocalelli, 2017, p. 1272). Moreover, the sustainability conscious consumer trends and environmentally conscious undertaking are expected to increase further (Natural Marketing Institute, 2020). In line with this trend, we highlight several implications as detailed below.

Previous studies (e.g. Ritzer and Jurgenson, 2010) point out that the mainstream market mechanisms cannot affect prosumers like mainstream producers and consumers. Prosumers tend to resist the mainstream market mechanisms by creating different exchange systems or markets (Ritzer & Jurgenson, 2010). As such, it is of paramount importance that a clear profile of this particular consumer group is developed. The three prosumer groups, semi-prosumers, collective prosumers and creative prosumers, could be made attractive for sustainable marketers if firms develop tailor-made strategies.

Semi-prosumer practices are shaped through the prevailing market mechanisms, although these prosumers do not necessarily buy eco-friendly commodities. These consumers appear to connect with the prevailing market mechanism to secure an alternative procurement. It is interesting to note that they can be triggered by prudence, enjoyment and challenges. Based on this, we can recommend marketers of eco-friendly products to highlight discourses of frugality, fun and adventure as unique selling propositions (USPs). To illustrate further, eco-friendly products having an optimistic plea of reducing excessive consumption signifying frugality can be attractive to this consumer groups. Also, the marketing messages could target individuals or groups. If these prosumers' preferences were properly addressed, established brands having eco-friendly brand extensions could win this segment.

Collective prosumers may not share the fundamental assumption of the mainstream market operations. Conventional marketing strategies may not be appealing to them, instead they could make them competitors in the mainstream market, especially in the form of anti-brand communities (Cova & Cova, 2012). The unique discourses ideal for this consumer group are "collective identity projects" or "excitements" of their community projects. Based on this, marketers of eco-friendly commodities may adopt strategies that should highlight these discourses as USPs. More specifically, eco-friendly commodities signifying higher "collective identity expressions" can be appealing to this consumer group.

It is said that creative prosumer practices are those practices that mostly deviate from the practices seen in existing market mechanisms. The discourses that are relevant for this consumer group are "autonomy" and "value creation". These prosumers may prefer to participate in sustainability conscious prosumption on their own. This prosumer behaviour is placed in contradiction of the "eco-friendly" commodity market as well; therefore, marketing strategies of eco-friendly commodities may not be applicable these consumer groups. As such, using creative strategies to get the consumer group engaged in value co-creation of eco-friendly products can be recommended.

IMPLICATIONS FOR SUSTAINABLE MARKETING PRACTITIONERS

Following practical implications can be highlighted as key sustainable marketing tools:

1. Sustainability conscious prosumption culture is emerging as a dominant, unique subset of broader prosumer discourse. This segment requires unique marketing approaches that are different from the ones applicable to wider prosumer discourse.

2. Marketers need to pay attention to specific profiling of this segment.
3. Going beyond products and services paradigm, modern-day consumption encompasses attributes such as motivation to work and satisfying a social need of relating to others.
4. Sustainability conscious consumer practices tend to act as mediators in satisfying social needs of consumers. Marketers need to pay attention to these practices.

TABLE 7.1

Key Attributes of Prosumer Segments and Suggested Marketing Approach

Sustainability Conscious Prosumer Segment	Attributes	Suggested Strategies
Semi-prosumers	• Prosumer practices are formed through current market mechanisms. • Not necessarily purchase eco-friendly products. • Engage with the mainstream market to affect alternative procurement. • Influenced by frugality, fun and adventure.	• When marketing eco-friendly features, unique selling propositions developed along the themes of frugality, fun and adventure should be applied. • Optimistic themes. • Signifying eco-friendly extensions.
Collective prosumers	• Act against the existing market mechanisms. • Very challenging to win these consumers. • If not properly engaged, they could form anti-brand communities.	• Marketing communications need to focus on collective identity projects and excitement of prosumption. • Commodities highlighting collective identity expression could be appealing.
Creative prosumers	• Disengaged from the existing market mechanisms. • Autonomy and value creation are important. • Engage in individual prosumer practices. • Tend to oppose mainstream eco-friendly commodity market by engaging in creating value on their own.	• Use strategies to engage them in value creation aspects of eco-friendly products.

5. Prosumers engaging in sustainable consumption are receptive to environmental concerns, excitement, adventure, frugality, social connections and creations. These attributes need to be considered when developing products and marketing messages.
6. Existing market system is not yet fully geared to accurately respond to the needs of growing sustainability conscious prosumer segment. Innovative marketers could exploit this niche and take first mover advantage.
7. There is growing evidence that prosumers tend to resist the existing market mechanism by harnessing the power of Internet (especially Web 2.0 technology). Marketers need to be proactive in formulating appropriate strategies to cater to this emerging segment.

This chapter provides marketers with three sustainability conscious prosumer segment profiles to help formulate their strategies (Table 7.1).

CONCLUSION

This chapter explored sustainability conscious prosumer identity and identified three distinctive categories of this special consumer group: semi-prosumers, collective prosumers and creative prosumers. Profiles of the three prosumer segments were described highlighting key challenges and opportunities each of these segments creates in sustainability market. It was argued that all the three green prosumer segments could be converted to lucrative market segments for sustainability-concerned marketers provided tailor-made USPs are developed for each segment.

FUNDING DECLARATION

No funding support.

CREDIT AUTHOR STATEMENT

Both authors contributed equally.

REFERENCES

Albinsson, P. A., & Perera, B. Y. (2012). Alternative marketplaces in the 21st century: Building community through sharing events. *Journal of Consumer Behaviour*, 11(4), 303–315.

Albinsson, P. A., & Perera, B. Y. (2018). *The rise of the sharing economy: Exploring the challenges and opportunities of collaborative consumption*. Santa Barbara, CA: ABC-CLIO.

Andreou, C. (2010). A shallow route to environmentally friendly happiness: Why evidence that we are shallow materialists need not be bad news for the environment(alist). *Ethics Place and Environment*, 13(1), 1–10.

Arnould, E. J., & Thompson, C. J. (2005). Consumer culture theory (CCT): Twenty years of research. *Journal of Consumer Research*, 31(4), 868–882.

Autio, M., Heiskanen, E., & Heinonen, V. (2009). Narratives of 'green' consumers-The antihero, The environmental hero and the anarchist. *Journal of Consumer Behaviour*, 8(1), 40–53.

Barnett, C. (2010). The politics of behaviour change. *Environment and Planning A*, 42(8), 1881–1886.

Barr, S. (2003). Strategies for sustainability: Citizens and responsible environmental behaviour. *Area*, 35(3), 227–240.

Belk, R. (2001). Materialism and you. *Journal of Research for Consumers*, 1(1), 291–297.

Belk, R. W. (1984). Three scales to measure constructs related to materialism: Reliability, validity, and relationships to measures of happiness. In P. Thomas Kinner (Ed.), *Advances in consumer research* (Vol. 11, pp. 291–297). Provo, UT: Association for Consumer Research.

Belk, R. W. (1988). Possessions and the extended self. *Journal of Consumer Research*, 15(2), 139–168.

Belk, R. W., Sherry Jr, J. F., & Wallendorf, M. (1988). A naturalistic inquiry into buyer and seller behavior at a swap meet. *The Journal of Consumer Research*, 14(4), 449–470.

Black, I. (2010). Sustainability through anti-consumption. *Journal of Consumer Behaviour*, 9(6), 403–411.

Black, I. R., & Cherrier, H. (2010). Anti-consumption as part of living a sustainable lifestyle: Daily practices, contextual motivations and subjective values. *Journal of Consumer Behaviour*, 9(6), 437–453.

Blumer, H. (1969). *Principles of sociology*. New York, NY: Barnes & Noble.

Bourdieu, P. (1984). *Distinction: A social critique of the judgement of taste*. Cambridge: Harvard University Press.

Burroughs, J. E., & Rindfleisch, A. (2002). Materialism and well-being: A conflicting values perspective. *Journal of Consumer Research*, 29(3), 348–370.

Büscher, B., & Igoe, J. (2013). 'Prosuming' conservation? Web 2.0, nature and the intensification of value-producing labour in late capitalism. *Journal of Consumer Culture*, 13(3), 283–305.

Cherrier, H. (2006). Consumer identity and moral obligations in non-plastic bag consumption: A dialectical perspective. *International Journal of Consumer Studies*, 30(5), 515–523.

Cherrier, H. (2007). Ethical consumption practices: Co-production of self-expression and social recognition. *Journal of Consumer Behaviour*, 6(5), 321–335.

Cherrier, H. (2010). Custodian behavior: A material expression of anti-consumerism. *Consumption Markets & Culture, 13*(3), 259–272.
Connolly, J., & Prothero, A. (2003). Sustainable consumption: Consumption, consumers and the commodity discourse. *Consumption, Markets and Culture, 6*(4), 275–291.
Connolly, J., & Prothero, A. (2008). Green consumption: Life-politics, risk and contradictions. *Journal of Consumer Culture, 8*(1), 117–145.
Cooley, C. H. (1902). *Human nature and social order.* New York, NY: Scribner's.
Cova, B., & Cova, V. (2012). On the road to prosumption: Marketing discourse and the development of consumer competencies. *Consumption Markets & Culture, 15*(2), 149–168.
Cova, B., & Dalli, D. (2009). Working consumers: The next step in marketing theory? *Marketing Theory, 9*(3), 315–339.
Cuthill, M. (2002). Exploratory research: Citizen participation, local government and sustainable development in Australia. *Sustainable Development, 10*(2), 79–89.
Dangelico, R. M., & Vocalelli, D. (2017). "Green Marketing": An analysis of definitions, strategy steps, and tools through a systematic review of the literature. *Journal of Cleaner Production, 165,* 1263–1279.
Dellaert, B. G. C. (2019). The consumer production journey: Marketing to consumers as co-producers in the sharing economy. *Journal of the Academy of Marketing Science, 47*(2), 238–254.
Dolfsma, W. (2004). Consuming symbolic goods: Identity & commitment-introduction. *Review of Social Economy, 62*(3), 275–277.
Douglas, M., & Isherwood, B. (1996). *The world of goods.* New York, NY: Routledge.
Eckhardt, G. M., Houston, M. B., Jiang, B., Lamberton, C., Rindfleisch, A., & Zervas, G. (2019). Marketing in the sharing economy. *Journal of Marketing, 83*(5), 5–27.
Fien, J., Neil, C., & Bentley, M. (2008). Youth can lead the way to sustainable consumption. *Journal of Education for Sustainable Development, 2*(1), 51–60.
Finn, M., & Darmody, M. (2017). What predicts international higher education students' satisfaction with their study in Ireland? *Journal of Further and Higher Education, 41*(4), 545–555.
Firat, A. F., & Dholakia, N. (2006). Theoretical and philosophical implications of postmodern debates: Some challenges to modern marketing. *Marketing Theory, 6*(2), 123–162.
Firat, A. F., & Venkatesh, A. (1995). Liberatory postmodernism and the reenchantment of consumption. *Journal of Consumer Research, 22*(3), 239–267.
Galus Australis. (2020). *Global DIY tools market (Impact of COVID-19) growth, overview with detailed analysis 2020–2025.* Retrievable from: https://galusaustralis.com/
Giddens, A. (1991). *Modernity and self-identity: Self and society in the late modern age.* Stanford, CA: Stanford University Press.
Griskevicius, V., Tybur, J. M., & Van den Bergh, B. (2010). Going green to be seen: Status, reputation, and conspicuous conservation. *Journal of Personality and Social Psychology, 98*(3), 392–404.
Hansen, M., & Hauge, B. (2017). Prosumers and smart grid technologies in Denmark: Developing user competences in smart grid households. *Energy Efficiency, 10*(5), 1215–1234.
Holt, D. B. (1995). How consumers consume: A typology of consumption practices. *The Journal of Consumer Research, 22,* 1–16.

Holt, D. B. (1997). Poststructuralist lifestyle analysis: Conceptualizing the social patterning of consumption in postmodernity. *Journal of Consumer Research*, 23(4), 326–350.

Holt, D. B. (1998). Does cultural capital structure American consumption? *Journal of Consumer Research*, 25(1), 1–25.

Horton, D. (2003). Green distinctions: The performance of identity among environmental activists. *The Sociological Review*, 51(2), 63–77.

Humphreys, A. (2014). How is sustainability structured? The discursive life of environmentalism. *Journal of Macromarketing*, 34(3), 265–281.

Humphreys, A., & Grayson, K. (2008). The intersecting roles of consumer and producer: A critical perspective on co-production, co-creation and prosumption. *Sociology Compass*, 2(3), 963–980.

Huttunen, K., & Autio, M. (2010). Consumer ethoses in Finnish consumer life stories–agrarianism, economism and green consumerism. *International Journal of Consumer Studies*, 34(2), 146–152.

Izberk-Bilgin, E. (2010). An interdisciplinary review of resistance to consumption, some marketing interpretations, and future research suggestions. *Consumption, Markets and Culture*, 13(3), 299–323.

Kasser, T., & Ryan, R. M. (1993). A dark side of the American dream: Correlates of financial success as a central life aspiration. *Journal of Personality and Social Psychology*, 65(2), 410–422.

Kasser, T., Ryan, R. M., Couchman, C. E., & Sheldon, K. M. (2004). Materialistic values: Their causes and consequences. In T. Kasser & A. D. Kanner (Eds.), *Psychology and consumer culture: The struggle for a good life in a materialistic world* (pp. 11–28). Worcester, MA: American Psychological Association.

Kessous, R., Davidson, E., Meirovitz, M., Sergienko, R., & Sheiner, E. (2016). The risk of female malignancies after fertility treatments: A cohort study with 25-year follow-up. *Journal of Cancer Research and Clinical Oncology*, 142(1), 287–293.

Kilbourne, W. E., Beckmann, S. C., & Thelen, E. (2002). The role of the dominant social paradigm in environmental attitudes: A multinational examination. *Journal of Business Research*, 55(3), 193–204.

Kotler, P. (1986). The prosumer movement: A new challenge for marketers. *Advances in Consumer Research*, 13(1), 510–513.

Kotler, P. (2010). The prosumer movement. In B. Blättel-Mink & K.-U. Hellmann (Eds.), *Prosumer Revisited* (pp. 51–60). Berlin, Germany: Springer.

Lehner, M. (2019). Prosumption for sustainable consumption and its implications for sustainable consumption governance. In O. Mont (Ed.), *A research agenda for sustainable consumption governance* (Chapter 7, pp. 105–120). Cheltenham, UK: Edward Elgar Publishing.

Lorenzoni, I., Nicholson-Cole, S., & Whitmarsh, L. (2007). Barriers perceived to engaging with climate change among the UK public and their policy implications. *Global Environmental Change*, 17(3-4), 445–459.

Lucyna, W., & Hanna, H. (2016). Prosumption use in creation of cause related marketing programs through crowdsourcing. *Procedia Economics and Finance*, 39, 212–218.

McCracken, G. (1986). Culture and consumption: A theoretical account of the structure and movement of the cultural meaning of consumer goods. *Journal of Consumer Research*, 13(1), 71.

McLean, L., & Roggema, R. (2019). Planning for a prosumer future: The case of Central Park, Sydney. *Urban Planning*, 4(1), 172–186.

Mick, D. G. (1996). Are studies of dark side variables confounded by socially desirable responding? The case of materialism. *The Journal of Consumer Research*, 23(2), 106–119.

Miller, T. (2016). Cybertarian flexibility—When prosumers join the cognitariat, all that is scholarship melts into air. In Curtin, M. & Sanson K. (Eds.), *Precarious creativity: Global media, local labor* (pp. 19–32). Oakland, CA: University of California Press. Retrieved March 18, 2021, from http://www.jstor.org/stable/10.1525/j.ctt1ffjn40.6

Mouat, S. (2016). A new paradigm for utilities: The rise of the "prosumer". *RE New Economy*, 24.

Natural Marketing Institute. (2020). *2020 State of Sustainability in America – 18th Annual Consumer Insights & Trends Report*. Retrievable from: https://www.marketresearch.com/Natural-Marketing-Institute-v1549/

O'Shaughnessy, J., & O'Shaughnessy, N. J. (2002). Marketing, the consumer society and hedonism. *European Journal of Marketing*, 36(5/6), 524–547.

Oyserman, D. (2009). Identity-based motivation: Implications for action-readiness, procedural-readiness, and consumer behavior. *Journal of Consumer Psychology*, 19(3), 250–260.

Oyserman, D., & Destin, M. (2010). Identity-based motivation: Implications for intervention. *The Counseling Psychologist*, 38(7), 1001–1043.

Papaoikonomou, E., Ryan, G., & Ginieis, M. (2011). Towards a holistic approach of the attitude behaviour gap in ethical consumer behaviours: Empirical evidence from Spain. *International Advances in Economic Research*, 17(1), 77–88.

Perera, C., Auger, P., & Klein, J. (2016). Green consumption practices among young environmentalists: A practice theory perspective. *Journal of Business Ethics*, 152(3), 843–864.

Perren, R., & Grauerholz, L. (2015). Collaborative consumption. *International Encyclopedia of the Social & Behavioral Sciences*, 4, 139–144.

Piacentini, M., & Mailer, G. (2004). Symbolic consumption in teenagers clothing choices. *Journal of Consumer Behaviour*, 3(3), 251–262.

Pirages, D., & Ehrlich, P. R. (1973). *Ark II; social response to environmental imperatives*. United States: WH Freeman.

Prahalad, C. K., & Ramaswamy, V. (2000). Co-opting customer competence. *Harvard Business Review*, 78(1), 79–90.

Prahalad, C. K., & Ramaswamy, V. (2004). Co-creation experiences: The next practice in value creation. *Journal of Interactive Marketing*, 18(3), 5–14.

Prothero, A., & Fitchett, J. A. (2000). Greening capitalism: Opportunities for a green commodity. *Journal of Macromarketing*, 20(1), 46.

Prothero, A., McDonagh, P., & Dobscha, S. (2010). Is green the new black? Reflections on a green commodity discourse. *Journal of Macromarketing*, 30(2), 147–159.

Ramaswamy, V., & Ozcan, K. (2014). *The co-creation paradigm*. Palo Alto: Stanford University Press.

RICHGRO. (2020). *Going the gardening distance – results from Richgro's gardening during COVID-19 survey*, https://www.richgro.com.au/going-the-gardening-distance-results-from-richgros-gardening-during-covid-19-survey/. Retrievable from: https://www.richgro.com.au/going-the-gardening-distance-results-from-richgros-gardening-during-covid-19-survey/

Richins, M. L., & Dawson, S. (1992). A consumer values orientation for materialism and its measurement: Scale development and validation. *Journal of Consumer Research*, 19(3), 303.

Ritzer, G. (2015). Prosumer capitalism. *The Sociological Quarterly*, 56(3), 413–445.

Ritzer, G. (2017). Can there really be 'true' alternatives within the food and drink markets? If so, can they survive as alternative forms? *Journal of Marketing Management*, 33(7/8), 652–661.

Ritzer, G., & Jurgenson, N. (2010). Production, consumption, prosumption: The nature of capitalism in the age of the digital 'prosumer'. *Journal of Consumer Culture*, 10(1), 13–36.

Serafin, D. (2012). Defining prosumption for marketing: understanding the nature of prosumption after the emergence of Internet-based social media. *International Journal of Management and Economics*, 36, 124–141.

Seran, S., & Izvercian, M. (2014). Prosumer engagement in innovation strategies. *Management Decision*, 52(10), 1968–1980.

Seyfang, G. (2004). Consuming values and contested cultures: a critical analysis of the UK strategy for sustainable consumption and production. *Review of Social Economy*, 62(3), 323–338.

Shaw, D., Newholm, T., & Dickinson, R. (2006). Consumption as voting: an exploration of consumer empowerment. *European Journal of Marketing*, 40(9/10), 1049–1067.

Siddiqui, S., & Turley, D. (2006). Extending the self in a virtual world. In *NA - Advances in consumer research 33, Connie Peachnmann and Linda Price* (pp. 647–648). Duluth, MN: Association for Consumer Research.

Slocum, R. (2004). Consumer citizens and the cities for climate protection campaign. *Environmental and Planning A*, 36, 763–782.

Solomon, M. R. (1983). The role of products as social stimuli: A symbolic interactionism perspective. *The Journal of Consumer Research*, 10(3), 319–329.

Soron, D. (2010). Sustainability, self-identity and the sociology of consumption. *Sustainable Development*, 18(3), 172–181.

Toffler, A. (1980). *The third wave*. New York, NY: William Morrow.

Vargo, S. L., & Lusch, R. F. (2008). Service-dominant logic: continuing the evolution. *Journal of the Academy of marketing Science*, 36(1), 1–10.

Xie, C., Bagozzi, R. P., & Troye, S. V. (2008). Trying to prosume: toward a theory of consumers as co-creators of value. *Journal of the Academy of marketing Science*, 36(1), 109–122.

Zwick, D., & Denegri Knott, J. (2009). Manufacturing customers: The database as new means of production. *Journal of Consumer Culture*, 9(2), 221–247.

8

Sustainable Practices and Responsible Consumption by the Hotel Industry: The Consumers' Perspective

Srishti Agarwal
MIT World Peace University, School of Management (PG), Pune, Maharashtra, India

Neeti Kasliwal
IIRM-Institute of Rural Management, Jaipur, Rajasthan, India

CONTENTS

Learning Objectives	200
Introduction	200
Responsible Consumption	202
Responsible Tourism	203
Sustainability in Tourism and Hospitality Industry	204
Eco-Certification	206
Sustainable Practices at the Hotel Industry	210
Responsible Consumer Behavior	213
Consumer Buying Behavior: Theoretical Background	213
Sustainability and Theory of Planned Behavior	215
Consumer Attitude and Willingness to Pay toward Green Attributes of the Hotel Industry	217
Implication	218
Conclusion	219
References	220

LEARNING OBJECTIVES

After reading the chapter, the student should be able to:

- Understand the sustainability issues in the tourism and hospitality industry
- Define sustainability marketing, responsible consumption, and sustainable practices in the hotel industry
- Examine the perception and attitude of responsible consumers and factors that compelled hotels to shift toward sustainable practices
- Understand the theoretical background of consumer buying decision
- Explore responsible marketing with respect to a hotel's social responsibility and its competitive advantage

INTRODUCTION

Natural ecosystems are under stress and declining faster across most of the countries. This collapse in the ecosystem includes habitat loss, degradation, industrial growth, overpopulation, climate change, overgrazing, and ocean acidification. According to the World Economic Forum Report 2020, "the loss of over 85% of wetlands occurred due to agricultural and industrial expansion, the exploitation of plants and animals through harvesting, the harmful substances from industrial, mining, and agricultural activities; and by oil spills and toxic dumping. Since 1980, marine plastic pollution alone has increased tenfold, biodiversity loss has increased by 40% globally over the same period and change in climate aggravates nature loss." The rapid degradation of our life support system brought the concept of sustainability in all the disciplines.

The process toward achieving sustainability is called sustainable development (Sidiropoulos, 2014), which normally includes three dimensions: "economic," "social," and "environmental." The most commonly used definition of sustainable development by "Brundtland Commission" (1987) and in the report of the "World Commission on Environment and Development" (1987), i.e., sustainable development is "a process to meet the needs of the present without compromising the ability of future generations to meet their own needs." Around the 1970s, ecological marketing came into the picture (Peattie, 2001; Van Dam & Apeldoorn, 1996), which had a major concern with industrial hazardous substances that harm the environment and purely concentrated on environmental issues such as pollution, ozone depletion, oil spills, and synthetic pesticides (Peattie, 2001; Sheth & Parvatiyar, 1995).

After that, "green marketing" came into the picture late in the 1980s with the modern markets and competitive advantages (Peattie, 2001; Van Dam & Apeldoorn, 1996). With rising demand from the green consumers, green marketing is seen to be occupied with product quality but also criticized for neglecting consumer behavior and for overstated ecological claims (Gordon, Carrigan, & Hastings, 2011). Due to this criticism, the lack of credibility, lower performance of green products, and consumer skeptical mind (Crane, 2000; Peattie & Crane, 2005) were imperative parameters that brought in an attitude-behavior gap featured significantly in green buying (Peattie, 2001). Besides, the broaden the scope of ecological and green marketing, sustainability marketing has been offered that redefined the scale of marketing (Gordon et al. 2011; Van Dam & Apeldoorn, 1996). It brought a major change in the approach of people the way they live, produce, market, and consumes (Peattie, 2001).

Therefore, the agenda of sustainability that emerged in marketing over the last 30 years brought sustainability marketing into the limelight. Fuller (1999) offered "sustainable marketing as the process of planning, execution, and controlling the development, pricing, promotion, and distribution of products in a manner that satisfies the three criteria: (1) Meeting customer needs (2) Attainment of organizational goals and (3) the process is compatible with eco-systems." This concept is not only restricted to products and services which save the environment but also spread globally and to all the sectors which have created responsible consumers and their consumption.

The tourism industry has taken a huge step to bring developments globally in terms of socially as well as economic growth. With the technological advancements, latest business model, reasonable travel, ease of visa facilities, and growing spending power of middle-class people in emerging economies, arrivals of international tourist raised to 5% in 2018 to reach "1.4 billion" mark (UNWTO, 2019, p. 2).

Despite developments, the tourism industry has been blamed several times in degrading the environment. Tourism marketing has normally been seen as an oppressed and fueling hedonistic consumerism. To bring this into the limelight, sustainability marketing intervened for a good purpose. Marketing skills, techniques, and latest technologies have been used to understand market needs, making consumers aware, designing more green products and services, and identifying more influential methods of communication to bring behavioral change. The need for responsible consumers, sustainable consumption, and consumer green acceptance behavior has become prominent. This chapter is to offer reflections on sustainable practices executed by the hotel industry and to examine the notion of "responsible consumption," consumers' attitude, and their behavior toward green practices of hotels.

RESPONSIBLE CONSUMPTION

There are several international agencies like the "United Nations World Tourism Organization" (UNWTO), "United Nations Environment Program" (UNEP), "United Nations Educational, Scientific and Cultural Organization" (UNESCO), the "European Union" (EU), and the "World Development Bank," with the help of these organizations, tourism's relationship with the nature, environment has improved. The hotel industry has put continuous efforts to reduce costs through sustainable practices like controlling water and energy usage which gave the hotel industry additional benefits to sustain in the market. Also, consumers have become aware and interested in those hotels that minimize their consumption of resource and implement eco-friendly practices in saving the environment.

Eventually, the hospitality industry is becoming the sole leading example of operating a sustainable business and proved that for long-term success, sustainability is the key. In the current time, sustainability is the highly important issues for the hoteliers, because of rising costs, sensitive demand, and the burden of being responsible toward economic, social, and environment.

Hotels are not the last resort, rather tourists stay at a hotel because of good experience and satisfaction from lodging and own requirements. This brings hoteliers a new perspective to satisfy their customers by providing a good experience. The hoteliers need to come out with innovative sustainable practices that can give better experience to their customers without compromising their comforts. Therefore, hoteliers are responsible for sustainable business practices because their survival is dependent on customer satisfaction and the environment.

The sustainable practices provide several benefits to a hotel. Bader (2005) has stated that the implementation of sustainable practices makes huge monetary variation such as:

- Use of cost-saving measures
- Added revenue through lower costs
- Viable financial stability and ability to attract lenders
- Easy to get funds due to lower long-term risk and increased asset value
- Long-term sustenance and be profitable

In addition, positive publicity, word of mouth created by the marketing team, public image, goodwill, and awards can increase demand or create other markets for the hotels. The upcoming market can be the one that requires sustainable practices in a verbal manner.

Consumers have played a pivotal role in the development and growth of sustainable consumption (Tran, 2017). Responsible consumption behavior is a result of direct and intended consumer choices and behavior toward their purchases (Miller, 2003). Consumption is closely associated with the issues of "responsible citizenship," hence not a liberty of choice anymore. The interest of tourists toward the environment has subsequently increased over the past few decades and that has mediated through consumer behavior (Razali, Shahril, Rahim, & Samengon, 2019). Studies have shown that those consumers who are concerned about the environment are more likely to evaluate their impact of purchases on the environment (Risqiani, 2017). This has led to companies to enter into a new business paradigm, "Think Green leading to green marketing." The chief element of green marketing is not to instigate the consumer to consume fewer but to consume healthier and responsibly.

RESPONSIBLE TOURISM

The growing trend of sustainability shows how different sectors and people worldwide, getting responsible for the environment. Responsible tourism (RT) is not about initiating sustainable practices to save the environment but also to take responsibility to make tourism a better destination. According to the Government of South Africa Department of Environmental Affairs and Tourism (1996), "responsible tourism is defined as tourism that boosts responsibility to the environment through its sustainable use; responsibility towards involving local communities in the tourism industry; responsibility for the safety and security of visitors, responsible government, employees, employers, unions, and local communities."

Goodwin (2013) held that RT analyzes the concern that arises from the effects of tourism activities and also makes sure to curb the socioeconomic and environmental problems. RTs are the activities that are done by a group of people; it cannot be done in the isolation. To bring positive change to the tourism sector, it is prominent that companies, travelers, governments, and all stakeholders should come together and work toward a bright future of the tourism sector. Booyens (2010) also restated that significant attempts are needed by the governments, the private sectors, professional bodies, and other stakeholders to make RT principles into guidelines, strategies, and regulations. Under the projects of tourism development agencies, the sustainability aspects of environment and poverty alleviation issues should also be included. With the collaboration of private sectors, government can plan, improve, manage, and promote new and varied products of tourism through the involvement of local communities (Nazmfar, Eshghei, Alavi, & Pourmoradian, 2019).

Sustainability in Tourism and Hospitality Industry

Based on the Brundtland Report (1987) and Elkington's "triple bottom line," the concept of sustainable development has three pillars, i.e., "social responsibility," "environment responsibility," and "economic responsibility." These pillars are better known as people, profit, and planet whereas in the hospitality sector, sustainability follows the same principle as respect for the environment, respect for cultures, and responsible economic growth. The environmental corporate social responsibility (CSR) is one of the prominent facets of hotels' responsible behavior and due to this prominence; CSR has become integral ideology for the new business (Marín & Lindgreen, 2017).

The International Hotel Environment Initiative (the IHEI, renamed as International Tourism Partnership, ITP, in 2004) launched is first environmental management manual for hotels in 1993. A subsequent partnership between the "IHEI," the "American Hotel and Restaurant Association (AH&RA)," "The International Hotel Association (IHA)" and the "United Nations environment programs division of technology, industry, and economics (UNEP DTIE)" was formed and a set of standardization tools for environmental management, practical propositions, and checklists for hotels were developed (Pantelidis, 2014, p. 249). ITP has been working with hotel companies worldwide to encourage sustainability.

Three dimensions of sustainability include the "environmental," "economic," and "socio-cultural" aspects of tourism development and to assure the long-term sustainability, the balance among these dimensions should be accomplished (see Figure 8.1 for the sustainable tourism). With regards to sustainability, the tourism sector will have to be consistent in controlling and unveiling the preventive measures to save the environment. Despite this, consumer decision also matters when it comes to experience. It is pivotal for the tourism industry to spread awareness, promote sustainable practices, and responsible consumption to the tourists to ensure the best experience in this service sector.

The tourism industry cannot be left out of the sector in terms of sustainability. That is why various organizations have been formed to assist the tourism sector to take a decision in bringing sustainability in their operations. An organization like the World Tourism Organization (WTO) assists the tourism industry, national, and local government to integrate

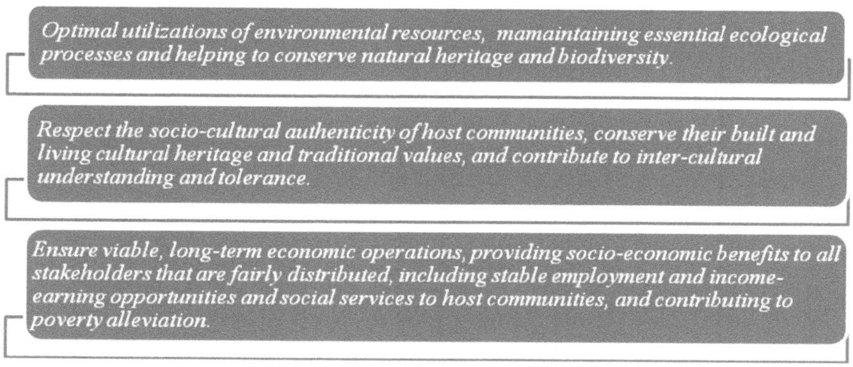

Optimal utilizations of environmental resources, mamaintaining essential ecological processes and helping to conserve natural heritage and biodiversity.

Respect the socio-cultural authenticity of host communities, conserve their built and living cultural heritage and traditional values, and contribute to inter-cultural understanding and tolerance.

Ensure viable, long-term economic operations, providing socio-economic benefits to all stakeholders that are fairly distributed, including stable employment and income-earning opportunities and social services to host communities, and contributing to poverty alleviation.

FIGURE 8.1
Sustainable tourism. (From UNEP & UNWTO, 2005).

206 • Social and Sustainability Marketing

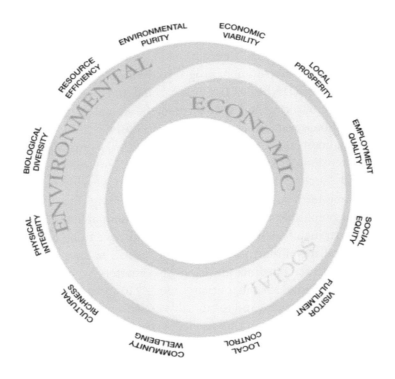

FIGURE 8.2
Aims and pillars of sustainability. (From UNEP/WTO, 2005).

sustainability in their day to day operations by centering on providing technical and advisory support services on policies, guidelines on development, and management practices. Similarly, UNEP commenced program where, through the propagation of technical expertise and business linkages, they can integrate sustainability in decision-making and into purchasing choices of consumers to catalyze sustainability in the tourism industry. The aims for an agenda for sustainability are shown in Figures 8.2 and 8.3.

Eco-Certification

The eco-certification has become one of the important parameters in choosing a hotel for the consumers. The decision of booking a hotel nowadays depends on the eco-friendly practices executed by the hotels.

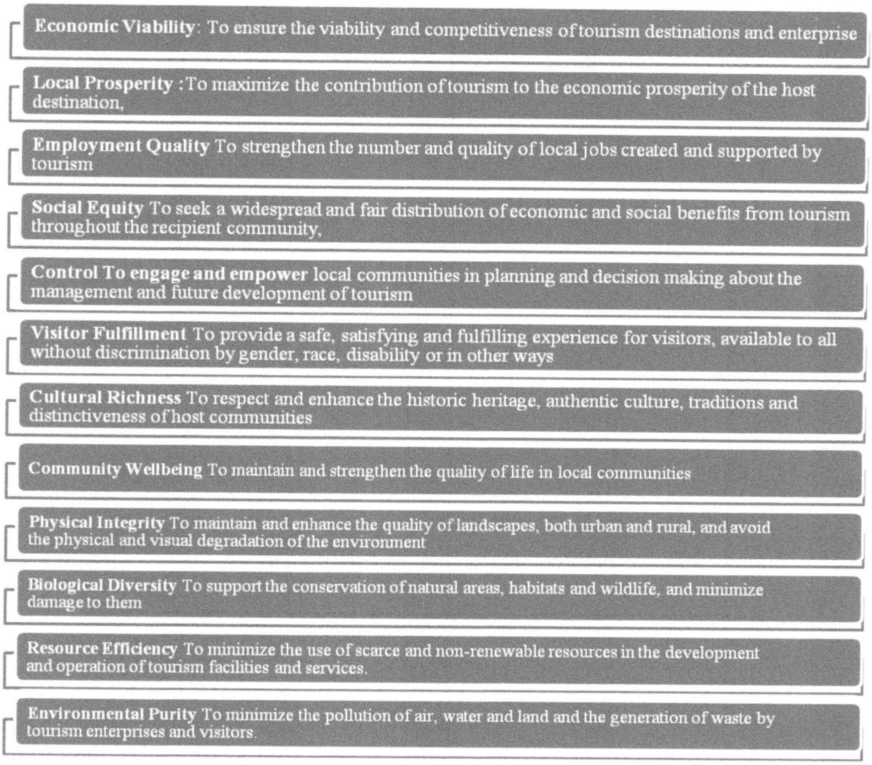

- **Economic Viability**: To ensure the viability and competitiveness of tourism destinations and enterprise
- **Local Prosperity**: To maximize the contribution of tourism to the economic prosperity of the host destination,
- **Employment Quality** To strengthen the number and quality of local jobs created and supported by tourism
- **Social Equity** To seek a widespread and fair distribution of economic and social benefits from tourism throughout the recipient community,
- **Control** To engage and empower local communities in planning and decision making about the management and future development of tourism
- **Visitor Fulfillment** To provide a safe, satisfying and fulfilling experience for visitors, available to all without discrimination by gender, race, disability or in other ways
- **Cultural Richness** To respect and enhance the historic heritage, authentic culture, traditions and distinctiveness of host communities
- **Community Wellbeing** To maintain and strengthen the quality of life in local communities
- **Physical Integrity** To maintain and enhance the quality of landscapes, both urban and rural, and avoid the physical and visual degradation of the environment
- **Biological Diversity** To support the conservation of natural areas, habitats and wildlife, and minimize damage to them
- **Resource Efficiency** To minimize the use of scarce and non-renewable resources in the development and operation of tourism facilities and services.
- **Environmental Purity** To minimize the pollution of air, water and land and the generation of waste by tourism enterprises and visitors.

FIGURE 8.3
Aims of sustainable tourism. (From UNEP & WTO, 2005).

Certifications are being provided to those hotels that are maintaining sustainability in their operations. But getting certified hotels may be a high-cost or no-cost affair. Most certifications require third-party auditors to visit properties, hence cost money. The "Global Sustainable Tourism Council" (GSTC) has developed prerequisites that provide a benchmark for certification schemes. According to O'Neill (2016), the yardstick for eco-certifications are constituted based on energy, water, waste, community involvement, preservation of culture and ecology, sustainable sourcing, accountability, manpower practices, architectural designs and layouts, and some execute all these domains. Table 8.1 shows the different international and regional eco-labels which consist of environmental certifications certified to the hotel companies. These eco-certified hotels

TABLE 8.1

International Local or Regional Eco-Labels

S.No.	Name of Green Certification	Description	Logo
1.	Green Globe	Certifying accommodations in Africa, Middle East, Asia, Europe, and America.	
2.	Green Key	The GSTC recognizes volunteer-based eco-label which operates in over 50 countries and over 3000 companies have been certified.	
3.	Audubon International	Its mission is to educate through high-quality environmental information and assist the sustainable management of land, water, wildlife, and other natural resources.	
4.	Green Seal	Green Seal is the first nonprofit environmental certification program commenced in 1989. They offer Green Seal Standards, Green Seal Certification, and institutional greening programs.	
5.	Earth Check	Earth Check certifies accommodations on their energy consumption, waste management, and green product in the development of local communities.	
6.	Leadership in Energy and Environmental Design	LEED focuses on: Building Design and Construction, Interior Design and Construction, Building Operations and Maintenance, and Neighborhood Development, Homes	

(Continued)

TABLE 8.1 (Continued)

International Local or Regional Eco-Labels

S.No.	Name of Green Certification	Description	Logo
7.	Green Tourism	It is mainly active in the UK, this standard exists since 1998 and awards hotels, restaurants and attractions in other parts of the world.	
8.	Energy Star	ENERGY STAR is a voluntary United States program created by the Environmental Protection Agency (EPA) in 1999. The certification aims to create environmental benefits and financial value through exceptional energy efficiency.	
9.	Travel Life	It is an international certification scheme operating in 50 countries. Lodgings are analyzed on all keystone of sustainability.	
10.	Biosphere Tourism	It is an international certification scheme that certifies sustainable hotels, destinations, campsites, and tour operators, among others.	

Local or Regional Green Labels in Tourism

11.	European Ecolabel For Tourism	This is the European Union voluntary-based labeling scheme. Accommodations awarded with this standard have low energy impact, low waste production and promote environmental practices in their lodgings.	

(*Continued*)

TABLE 8.1 (Continued)
International Local or Regional Eco-Labels

S.No.	Name of Green Certification	Description	Logo
12.	Bio-Hotels	This German hotel association label guarantees hotels serve 100% organic food, beverages, and use only clean cosmetics.	
13	Fair Trade Tourism	It operates in African countries to support accommodations but also activities in the tourism industry.	
14.	Eco-Tourism Australia	It is a certification scheme for accommodations and activities focused on nature-based tourism. It focuses on nature tourism, ecotourism and advanced ecotourism.	

Source: www.holiable.com and www.cloudbeds.com.

that implement green practices can motivate customers to change their perception toward the commitment shown by green hotels (Martínez, Herrero & Gómez, 2018).

Sustainable Practices at the Hotel Industry

The world is at the greater risk today in terms of the degraded environment due to ozone depletion, climatic change, pollution, overpopulation, urbanization, and the industrial revolution. Due to the exploited environment, not only the hotel industry but every sector needs to bring a sustainable revolution to save the environment. The hotel industry historically has been a culprit to have harmful consequences on the environment due to overconsumption of energy and water, hazardous waste creation, and use of nonrenewable resources. According to estimations, an "average hotel expels between 160 kg and 200 kg of CO_2 per square meter of room floor area per year, water consumption between 170 and 440 liters per guest per night in the average five-star hotel and on an average, 1 kg of waste per guest per night (Han, Hsu, & Sheu, 2010)."

With the negative impact on the environment, government pressure and actors in tourism are compelling hoteliers to push green consumption in hotels. Over time, hotels also then realized to enter into sustainability because to be in good books of society, to bring distinction among the competitors, to bring in new customer base those have an inclination toward the green products and services, and to gain other internal and external benefits. Now, it is apparent that various hotels are in the verge of becoming "green hotel" or "eco-friendly hotel" to get a unique position in the growing hotel market (Chan & Wong, 2006; Manaktola & Jauhari, 2007).

"A green hotel is an environmentally friendly lodging property that executes and follows ecologically sound programs/practices like water and energy savings, reduction of solid waste, and cost-saving to help protect our planet" (GHA, 2008). In this competitive era, the green hotel can execute innovative marketing strategies where the implementation of eco-friendly practices that have variably positioned may also attract customers who are seeking green operations (Manaktola & Jauhari, 2007). To become a sustainable industry and to achieve a competitive edge, green hotels differentiate themselves from other nongreen hotels.

Green hotels try to fulfill consumers' needs for eco-friendly hotels which ultimately brings different benefits in the operations like minimizing energy, water consumption, building company image, complying with government regulations, and community attention (APAT, 2002; Enz & Siguaw, 1999; GHA, 2008; Penny, 2007). Thus, in the prevailing competitive lodging industry, the green hotel business is considered to be a flourishing niche sector (Manaktola & Jauhari, 2007). But this is a question, that why hotels are going green? Is it only because of meeting customers' demand or for saving the environment? These are not the only reasons which compel hotels to bring sustainability in their operations. Other reasons can be economic benefits, to maintain good relations with the investors, to strengthen employee and organization commitment, facing public scrutiny, and for a social good book (Gan, 2006; Juholin, 2004).

Alternatively, many conventional hotels are damaging environment due to overconsumption of energy, water, and soil whereas green hotels are executing sustainability as per the environmental guidelines and

TABLE 8.2

Practices Implemented by Environment-Friendly Hotels

Fitting energy saving devices (e.g., dimmer/timed switches, sensors, energy-efficient light bulbs)
Using low-flow showerheads or sink aerators
Installation of dual-flush toilets
Installation of a solar hot water system
Providing energy-efficient appliances
Using eco-friendly cleaning products
Reusing linen and towels
Composting food leftovers and garden waste
Sorting of waste in guestrooms, offices, and kitchens
Educating guests on environmental-friendly practices
Using natural cleaning alternatives (e.g., lemon juice, vinegar, salt)
Improving insulation
Installation of water-saving devices (e.g., flow regulators, waterless urinals)
Using the economy wash cycle
Environmental policy and conveying policy to customers
Buying ethical and environment-friendly products
Establishing a wildlife area in the garden
Membership of environmental bodies/charities

Source: Bohdanowicz (2006), Mensah (2006), and Tzschentke, Kirk, and Lynch (2008).

standards. Some of the green practices like conservation of energy and water, trained and educated employees, use of energy-efficient bulbs, reuse of towel and linen, recycle bins placed in the lobby and rooms, organic food, etc. (Millar & Baloglu, 2011; Verma & Chandra, 2016) are common practices which are being followed by the hotels. Table 8.2 shows the widely used green practices followed by the hotels.

These sustainable practices are usually being followed by hotels nowadays. But there are exceptions also where hotels pretend to be a green hotel. Hotels claim about sustainable practices and try to win consumer trust. This practice is called "greenwashing." "The term greenwash describes efforts by corporations to portray them as environmentally responsible in order to mask environmental wrongdoings" (Whellam & MacDonald, 2007). A practice of greenwash by the hotels ultimately leads to the competitive disadvantage and ruin the trust and reputation of the stakeholders once the indifference in their activities is identified by the people.

RESPONSIBLE CONSUMER BEHAVIOR

In a shift from traditional marketing to responsible marketing, consumers have played a big role in responsible consumption. A growing number of green consumers who have supported sustainable practices have influenced companies to take green initiatives in their operations. Now, companies are moving into a new dimension of business where not only profit but saving the planet is one of the major concerns. For this, consumers are equally showing their ethical behavior for the protection of the environment. Consumer attitude and their behavior toward the products and services have changed over the few decades. The emergence of green consumers has shown how responsible a person can be toward the environment. The green consumers are those who show concern for the environment and put efforts to take action toward saving the environment.

These consumers are regarded as conscious citizens who are inclined to pay additional for ecological practices. The behavior of consumers is changing due to awareness toward saving the environment, which influences consumers in their buying decisions of green services. "Consumer behavior (CB) involves certain decisions, activities, ideas, or experiences that satisfy consumer needs and wants" (Solomon, 1996). "It is concerned with all activities directly involved in obtaining, consuming and disposing of products and services, including the decision processes that precede and follow these actions" (Engel, Blackwell, & Miniard, 1995). A consumer is a person who obtains a product or service for private use. But consuming products or services is not a subject of liberty of choice any longer. Consumption is directly associated with the subject of responsible citizenship.

Consumer Buying Behavior: Theoretical Background

The changing lifestyle of consumers has shown a rise in the demand for green products and services. Consumers are nowadays more informed and knowledgeable about the market. They know what is to be consumed and why it has to be consumed. The era of green consumerism has created a new segment of "green consumers" who are more conscious toward the environment and would like to buy eco-friendly products and services.

Today, identifying customer decision criteria and facilitate the consumption mindset has become the priority of the marketers.

For this changing world, businesses are accepting the demand for sustainability which also leads to achieving competitive advantage and led to achieve strategic benefits also. A focus of business over sustainable has benefits such as classifying newly products and markets, capitalizing burgeoning technologies, promoting creativity and innovativeness, driving organizational effectiveness, and retention and motivation of manpower (Hopkins et al. 2009). There are factors which have an impact in consumer decision toward the purchase of green products and services. This led to the evolution of diverse theories that assert the factors of decision-making. Whereas, many of the models adequately exhibited how consumers' values, attitudes, beliefs, and perceptions about the environment happened into the intention to buy or genuinely buying the eco-friendly products.

The cognitive-affective-behavior (CAB) model is the traditional theory that illustrates people's viewpoint, consciousness, emotional responses, discernment, opinion, or behavior of people about environment-related issues (Milfont & Duckitt, 2010).

- The "cognitive component" of CAB model consists of a person understanding and expertise or notions about a particular topic.
- The "affective component" pertains to the people's sentimental response or emotions about particular or whole attributes.
- The "behavior or conative component" comprises people's desire to behave or the genuine behavior to an object.

With reference to the CAB model, decision begins with cognitions, i.e., knowledge or belief part, followed with the affect that focuses on emotional responses and leads to the behavior, i.e., intention to act. The theoretical consider at ions make it possible to describe the aspects of consumer behavior by looking a shift in the attitude, intention, and actual behavior of a consumer. As more the consumer aware of the environmental problems, the more he/she has a favorable feeling toward the behavior for sustainable consumption. It is also suggested that the personal values, attitudes and awareness, and perceptions toward hotel's image, quality, and satisfaction have positive impact on consumers' behavioral intentions toward green hotels (Gao, Mattila, & Lee, 2016). Therefore, attitude of individuals affects their opinion (the cognitive function) and emotions or feelings (the affective function), and thus affect behavior like buying behavior (Pickett-Baker & Ozaki, 2008).

Sustainability and Theory of Planned Behavior

According to the "Theory of Planned Behavior (TPB)," behavioral intention is a prominent indicator of actual behavior. Attitude, subjective norm, and behavioral control are the three major factors which influence the behavioral intention: Attitude is impacted by "behavioral belief" and "outcome evaluation," the subjective norm is affected by "normative belief" and "motivation" and perceived behavioral control (PBC) is affected by "control belief" and "control strength." When the above factors are put collectively, they create an exhaustive theoretical model which anticipates human behavior. The TPB is a prolongation of the theory of reasoned action (TRA) (Ajzen & Fishbein, 1980; Fishbein & Ajzen, 1975).

TRA is intended to anticipate the human behavior and consumer decision-making process where most decisions or behaviors of an individual are deduced from the vividness of intentional attempt for the particular judgments or behaviors. This theory talks about the decision-making processes where an individual takes decision on the basis of rationality, motivation, and reach to a logical selection among several choices (Fishbein & Ajzen, 1975), whereas the TPB not only predicts voluntary control but also nonvoluntary control in elucidating the behavior of an individual. A key component of the TRA and TPB is an individual intention, which forecasts the particular behavior accurately (Ajzen & Fishbein, 1980; Fishbein & Ajzen, 1975).

According to Ajzen and Fishbein (1980), attitude toward the behavior are concerned toward the positive or negative assessment of the results of individual action. This attitude is a blend of "behavioral beliefs" and "outcome evaluations." "Behavioral beliefs refer to one's perceived probability of an expected outcome's occurrence by engaging in a particular behavior," and outcome evaluations imply the evaluation of the attainable results of an explicit behavior (Ajzen & Fishbein, 1980). The second indicator of intention is the "subjective norm" which consists the social factor and considered to be a function of "normative beliefs" and "motivation to comply." "Normative beliefs are the perceived behavioral expectations of one's prominent referents like family, peer group, relatives, neighbors, and motivation to comply constitutes a desire of an individual to adapt the opinions or judgments of his/her predominant referents with respect to a behavior (Ajzen and Fishbein, 1980)." The third determinant is PBC. This component, which is not admitted in the TRA, represents the perception of an individual of the ease or challenges in playing a particular behavior (Ajzen & Fishbein, 1980).

"PBC" is based on the function of control beliefs and perceived power. "Control beliefs are the perceived availability or unavailability of resources and possibilities that alleviate the functioning of a specific behavior, and the perceived power of each control component pertains to individual appraisals of the importance of the resources and possibilities in attaining behavioral outcomes" (Ajzen & Madden, 1986; Chang, 1998). The existence of PBC with control beliefs in TPB enhances the predictive power for intention/behavior which is not under TRA (Ajzen, 1991; Lam & Hsu, 2004; Lee & Back, 2007). Figure 8.4 represents the TPB model and the TRA model depicted in the dashed square.

Therefore, this theory gives perceptivity that how consumer behavior can be forecasted on the basis of their awareness of the environment, attitude, perception, and behavioral intention. It is seen that consumers having high consciousness toward environment are more inclined to carry out green consumption. There is a favorable relationship between attitude and behavior toward sustainable environment, also a positive attitude toward environment has an influence on desire to purchase eco-friendly products and services (Kotchen & Reiling, 2000; Mostafa, 2007).

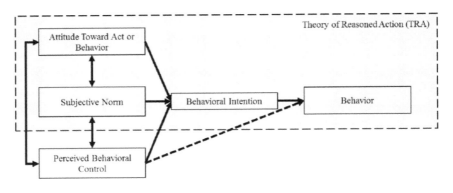

FIGURE 8.4
Theory of planned behavior and theory of reasoned action. (From Ajzen, 1991).

Consumer Attitude and Willingness to Pay toward Green Attributes of the Hotel Industry

Considering the significance of the environment, consumers are getting aware and literate about the green practices followed by green hotels. Due to the changing lifestyle of consumers and an increasing number of green consumers, hotel sectors are bringing more green services to attract a niche segment of consumers. Hoteliers are putting their efforts to save the environment and also trying to sustain in the market. According to the 1992 United Nations Conference on Environment and Development, "hotels can reduce their environmental impacts by installing eco-friendly technology such as solar panels, low flow showerheads, recycling bins, etc., and by this approach can bring in the attention of customers" (Kang, Stein, Heo, & Lee, 2012). Consumers have become sensitive toward the environment and they expect businesses to execute sustainable practices.

The studies conducted by the "International Hotels Environment Initiative (IHEI) reveal that 90% of hotel guests prefer to stay in a hotel that cares for the environment" (Mensah, 2004). Earlier hotels were going green only because of government regulations, saving money by cutting power and energy, but because of the growing number of consumers and their demand toward green products and services, green management is now directly in link with quality product and service, morale of an employee (Enz & Siguaw, 1999), "customer satisfaction" (Manaktola & Jauhari, 2007), "willingness to pay (WTP) a premium for green products" (Laroche, Bergeron, & Barbaro-Forleo, 2001), and "corporate image" (Mensah, 2004; Penny, 2007).

A plethora of studies has been conducted to understand the direct link of consumer attitude toward green attributes of the hotel industry with consumer green purchase behavior. It has been seen that consumers who are eco-literate and conscious about the environment, are more likely to perceive and prefer green attributes of the hotel industry in a favorable manner. Sudhagar (2015) has found "environmentally friendly food, natural guest amenities placed in the room, eco-harvest placed in the room as a give-away, water glass and the flask in the room filled only on the request, fresh fruits and newspaper delivered only on request, reusable cloth bags and the linen reuse options" as mostly preferred green attributes by the customers in a hotel.

Therefore, the above preferred green attributes show that consumers are aware and hence they prefer to stay in green hotels where these green practices are being executed by the hoteliers. The demand for green hotels has increased and consumers have shown their WTP premium for the green services. The changing taste and lifestyle of consumers compelled them to show concern toward the environment.

Tourist usually travels for the sake of comfort, convenience, amenities, desirability, relaxation, a sense of freedom, and "no stress" (Anable & Gatersleben, 2005). Laroche et al. (2001) investigated a favorable and strong relationship between attitudes toward ecological problems and consumers' intention to pay more for green products. Dolnicar (2010) found that values and moral responsibility toward the environment was a strong indicator of eco-friendly tourists and also tourists are becoming constantly aware of green tourism behavior and environmental behavior at home.

Ham and Han (2012) showed that purchase-related loyalty, willingness to visit, ready to pay more price, and giving suggestions, were magnified by consumers who are positive toward environmental values. Manaktola and Jauhari (2007) determined that customers show preference and support a green lodging property because they are informed about the eco-friendly practices of a hotel. Further, consumer attitudes are significant indicators of environmentally conscious consumer behavior (Roberts, 1996). His findings disclosed that people who are conscious for the environmental issues are more likely to participate in eco-friendly activities or actions and people who know that their eco-friendly activities can cause favorable alterations are more likely to show green consumer behavior.

Similarly, a study revealed that customer's eco-friendly attitudes favorably influence their uttered intention (Han, Hsu, & Sheu, 2010). The intention to buy green products and services is the intentional purchase with an idea of least damage to the natural environment. Therefore, tourist who have favorable attitude toward environment will select green hotels. It is critical for the hoteliers to make the guests aware about the significance of being responsible toward consumption of eco-friendly practices and their impact on the environment.

IMPLICATION

This chapter provides different aspects of sustainability within the hospitality industry, attracting, and generating interest to academicians, marketers, policymakers, and practitioners interested in the hospitality industry and more broadly within the industry and organization community. The authors suggested three primary sets of perspectives, namely, defining sustainability practices within the hotel industry, responsible consumption and behavior of consumers toward the green hotel, and commitment to continuing the sustainable practices in the hotel sectors. This chapter would help marketers and hotel industry to understand the consumers' mindset and their preferences toward the green practices of the hotel industry. The implementation of sustainability standards will help hoteliers to achieve economic benefits. Hence, green practices followed by a hotel will enhance market visibility and create value proposition for the hotels that will ultimately attract green consumers. The consumers are showing their responsibility toward the environment but to improve their behavior toward the environment, consumer needs to be acquainted with the benefits of the environmental practices. It is good to encourage consumers to visit green hotels so that they will be familiar with the mechanisms of green certified hotels.

This chapter has focused on the responsible consumption and sustainability practices by the hotel industry where the role of government and other policymakers plays prominent role in making eco-friendly tourist places. There are various governmental bodies, policies, and strategies of sustainable tourism that help in making iconic tourist sites to visit which leads to economic growth. The government and policymakers should collaborate with the NGOs and private companies to work together on RT and consumption, building trust among consumers, sustainability issues in tourism and hospitality sector, building infrastructure, and by promoting benefits of environmental practices. This will create an impact on people and will show their WTP for the green products and services.

The future research of the current study can include understanding employee's behavior toward the green services, consumer experiences in green and nongreen hotels, management initiatives to explore consumer demand, and changing taste of consumers, millennials' spending pattern for the green products and services and also to understand their perspective toward saving environment.

CONCLUSION

The new world of green consumers has shown increased awareness and action toward saving the environment. The awareness of green practices and responsible consumption by the consumers has compelled the hotel industry to think about sustainable practices that have a lesser impact on environmental health. The term "green" has developed a trendy tale in the tourism sector over the past years. The concepts of sustainable marketing and responsible consumption have come into the limelight where hotels have brought green products and services to attract niche segment of consumers called "green consumers."

In order to retain the green consumers and to achieve competitive advantage, hotelier needs to expand their environmental practices. This chapter has focused on the green practices followed by the hotel industry and throws light on consumer's responsible consumption and their behavior toward green practices in the hotel industry. This chapter has highlighted eco labeled certified body that regulates the sustainability issues of the hotels. The hoteliers are using sustainable marketing strategies to cope up with the changing lifestyle of the consumers and to sustain in the competitive environment. The increase in the awareness of consumers has led to their favorable perception about green practices, positive preferences of green services while choosing a hotel, more willing to pay a premium, and positive behavior of consumers toward green practices of hotels.

It is concluded that consumers having a positive attitude toward the environment will tend to choose green hotels. The finding shares evidence on the motivations, consumer behavior, and hurdles that hospitality sector encounters, and on successes in changing consumer lifestyle, and pursuing sustainability goals to enhance economic growth. It explores the growing interest of responsible consumers toward the green service of the hotel industry. It is recommended to the hoteliers to communicate positive messages about the green certified hotels which will support consumers to take eco-friendly decision.

REFERENCES

Ajzen, I. (1991). The theory of planned behavior. *Organizational Behavior and Human Decision Processes*, 50(2), 179–211.

Ajzen, I., & Fishbein, M. (1980). *Understanding attitude and predicting social behavior.* Englewood Cliffs, NJ: Prentice-Hall.

Ajzen, I., & Madden, T. (1986). Prediction of goal-directed behavior: attitude, intentions, and perceived behavioral control. *Journal of Experimental Social Psychology*, 22, 453–474.

Anable, J., & Gatersleben, B. (2005) All work and no play? The role of instrumental and affective factors in work and leisure journeys by different travel modes. *Transportation Research Part A: Policy and Practice*, 39, 163–181.

APAT. (2002). Tourists accommodation EU eco-label award scheme – Final report. *Italian National Agency for the Protection of the Environment and for Technical Services*, Rome, Italy.

Bader, E. E. (2005). Sustainable hotel business practices. *Journal of Retail & Leisure Property*, 5(1), 70–77.

Bohdanowicz, P. (2006). Environmental awareness and initiative in the Swedish and Polish hotel industries—Survey results. *International Journal of Hospitality Management*, 25, 662–682.

Booyens, I. (2010). Rethinking township tourism: Towards responsible tourism development in South African townships. *Development Southern Africa*, 27(2), 273–287.

Brundtland, G. H. (1987). *Our common future: The world commission on environment and development*. Oxford: Oxford University Press.

Chan, W., & Wong, K. (2006). Estimation of weight of solid waste: Newspapers in Hong Kong hotels. *Journal of Hospitality & Tourism Research*, 30(2), 231–245.

Chang, M. K. (1998). Predicting unethical behavior: A comparison of the theory of reasoned action and theory of planned behavior. *Journal of Business Ethics*, 17, 1825–1834.

Crane, A. (2000). Facing the backlash: Green marketing and strategic reorientation in the 1990s. *Journal of Strategic Marketing*, 8(3), 277–296.

Dolnicar, S. (2010). Identifying tourists with smaller environmental footprints. *Journal of Sustainable Tourism*, 18(6), 717–734.

Engel, J. F., Blackwell, R. D., & Miniard, R. W. (1995). *Consumer behavior*. Fort Worth, TX: Dryden Press.

Enz, C. A., & Siguaw, J. A. (October 1999). Best hotel environmental practices. *Cornell hotel and restaurant administration quarterly* (pp. 72–77).

Fishbein, M., & Ajzen, I. (1975). *Belief, attitude, intention and behavior*. Don Mills, NY: Addison Wesley.

Fuller, D. A. (1999). *Sustainable marketing: Managerial-ecological issues*. Thousand Oaks, CA: SAGE Publications.

Gan, A. (2006). The impact of public scrutiny on corporate philanthropy. *Journal of Business Ethics*, 69(3), 217–236.

Gao, Y. L., Mattila, A. S., & Lee, S. (2016). A meta-analysis of behavioral intentions for environment-friendly initiatives in hospitality research. *International Journal of Hospitality Management*, 54, 107–115.

GHA. (2008). What are green hotels? *Green Hotel Association (GHA)*. http://www.greenhotels.com/whatare.htm

Goodwin, H. (2013). *What role does certification play in ensuring responsible tourism? – In WTM blog*. http://www.wtmlondon.com/library/What-role-does-certificationplay-in-ensuring-Responsible-Tourism#sthash.azaYgVZj.dpuf

Gordon, R., Carrigan, M., & Hastings, G. (2011). A framework for sustainable marketing. *Marketing Theory*, 11(2), 143–163.

Government of South Africa Department of Environmental Affairs and Tourism. (1996). *White paper: The development and promotion of tourism in South Africa*. http://scnc.ukzn.ac.za/doc/tourism/White_Paper.htm

Ham, S., & Han, H. (2012). Role of perceived fit with hotels' green practices in the formation of customer loyalty: Impact of environmental concerns. *Asia Pacific Journal of Tourism Research*, 18(7), 731–748. doi:10.1080/10941665.2012.695291

Han, H., Hsu, L. T. J., & Sheu, C. (2010). Application of the theory of planned behavior to green hotel choice: Testing the effect of environmental friendly activities. *Tourism Management*, 31(3), 325–334.

Hopkins, M. S., & Roche, C. (2009). What the 'green' consumer wants. *MIT Sloan Management Review*, 50(4), 87.

Juholin, E. (2004). For business or the good of all? A Finnish approach to corporate social responsibility. *Corporate Governance*, 4(3), 20–32.

Kang, K. H., Stein, L., Heo, C., & Lee, S. (2012). Consumers' willingness to pay for green initiatives of the hotel industry. *International Journal of Hospitality Management*, 31(2), 564–572.

Kotchen, M. J., & Reiling, S. D. (2000). Environmental attitudes, motivations, and contingent valuation of nonuse values: A case study involving endangered species. *Ecological Economics*, 32, 93–107.

Lam, T., & Hsu, C. H. C. (2004). Theory of planned behavior: potential travelers from China. *Journal of Hospitality & Tourism Research*, 28(4), 463–482.

Laroche, M., Bergeron, J., & Barbaro-Forleo, G. (2001). Targeting consumers who are willing to pay more for environmentally friendly products. *The Journal of Consumer Marketing*, 18(6), 503–520.

Lee, M. J., & Back, K. (2007). Association members' meeting participation behaviors: Development of meeting participation model. *Journal of Travel & Tourism Marketing*, 22(2), 15–33.

Manaktola, K., & Jauhari, V. (2007). Exploring consumer attitude and behavior towards green practices in the lodging industry in India. *International Journal of Contemporary Hospitality Management*, 19(5), 364–377.

Marín, L., & Lindgreen, A. (2017). Marketing and corporate social responsibility and agenda for future research. *Spanish Journal of Marketing – ESIC*, 21(1), 1–3.

Martínez, P., Herrero, A., & Gómez, R. (2018). Customer responses to environmentally certified hotels: The moderating effect of environmental consciousness on the formation of behavioral intentions. *Journal of Sustainable Tourism*, 26(7), 1160–1177.

Mensah, I. (2004). *Environmental management practices in US hotels*. Retrievable from: https://www.hotel-online.com/News/PR2004_2nd/May04_EnvironmentalPractices.html

Mensah, I. (2006). Environmental management practices among hotels in the greater Accra region. *International Journal of Hospitality Management*, 25, 414–431.

Milfont, T. L., & Duckitt, J. (2010). The environmental attitudes inventory: A valid and reliable measure to assess the structure of environmental attitudes. *Journal of Environmental Psychology*, 30, 80–94.

Millar, M., & Baloglu, S. (2011). Hotel guests' preferences for green guest room attributes. *Cornell Hospitality Quarterly*, 52(3), 302–311.

Miller, G. (2003). Consumerism in sustainable tourism: a survey of UK consumers. *Journal of Sustainable Tourism*, 11, 17–39.

Mostafa, M. M. (2007). Gender differences in Egyptian consumers' green purchase behavior: The effects of environmental knowledge, concern and attitude. *International Journal of Consumer Studies*, 31(3), 220–229.

Nazmfar, H., Eshghei, A., Alavi, S., & Pourmoradian, S. (2019). Analysis of travel and tourism competitiveness index in middle-east countries. *Asia Pacific Journal of Tourism Research*, 24(6), 501–513.

O'Neill, S. (2016, July 25). Know how guide to sustainable hotel certification schemes. *International Tourism Partnership presents Green Hotelier*. Retrieval from: https://www.greenhotelier.org/know-how-guides/know-how-guide-to-sustainable-hotel-certification-schemes/

Pantelidis, I. S. (Ed.). (2014). *The Routledge handbook of hospitality management*. Routledge.

Peattie, K. (2001). Golden goose or wild goose? The hunt for the green consumer. *Business Strategy and the Environment*, 10, 187–199.

Peattie, K. & Crane, A. (2005). Green marketing: Legend, myth, farce or prophesy? *Qualitative Market Research: An International Journal*, 8, 357–370. Milton Park, UK: Routledge.

Penny, W. Y. K. (2007). The use of environmental management as a facilities management tool in the Macao hotel sector. *Facilities*, 25, 286–295.

Pickett-Baker, J., & Ozaki, R. (2008).Pro-environmental products: Marketing influence on consumer purchase decision. *Journal of Consumer Marketing*, 25, 281–293.

Razali, N. A. M., Shahril, A. M., Rahim, M. A., & Samengon, H. (2019). Eco-friendly attitude and response behaviours of Green Hotel guest in Malaysia. *Journal of Tourism, Hospitality and Environment Management*, 4(13), 77–89.

Risqiani, R. (2017). Antecedents of consumer buying behaviour towards on environmentally friendly Products. *Business and Entrepreneurial Review*, 17(2), 145–164.

Roberts, J. A. (1996). Green consumers in the 1990s: Profile and implications for advertising. *Journal of Business Research*, 36, 217–231.

Sheth, J., & Parvatiyar, A. (1995). Ecological imperatives and the role of marketing. *Environmental marketing: Strategies, practice, theory and research*, Polonsky, M. J., & Mintu-Wimsatt, A. T. eds. New York: Haworth Press (pp. 3–20).

Sidiropoulos, E. (2014). Education for sustainability in business education programs: A question of value. *Journal of Cleaner Production*, 85, 472–487.

Solomon, M. R. (1996). *Consumer behavior* (3rd ed.). Engle-wood Cliffs, NJ: Prentice-Hall.

Sudhagar, D.P. (2015). Exploring customer perceptions of eco-sensitive practices in the Indian lodging industry. *African Journal of Hospitality, Tourism and Leisure*, 4(2), 1–10.

Tran, A. H. (2017). Consumers' behavior towards green purchase intention. *Актуальні проблеми економіки*, (2), 151–158.

Tzschentke, N. A., Kirk, D., & Lynch, P. A. (2008). Going green: Decisional factors in small hospitality operations. *International Journal of Hospitality Management*, 27, 126–133.

UNEP and UNWTO (2005). *Making tourism more sustainable – A guide for policy makers* (pp. 11–12). Paris/Madrid: United Nations Environment Programme. http://www.unep.fr/shared/publications/pdf/DTIx0592xPA-TourismPolicyEN.pdf

UNWTO. (2019). *International tourism highlights*. World Tourism Organization. https://www.e-unwto.org/doi/pdf/10.18111/9789284421152

VanDam, Y. K., & Apeldoorn, P. A. C. (1996). Sustainable marketing. *Journal of Macro Marketing*, 16, 45–56.

Verma, V. K., & Chandra, B. (2016), Hotel guest's perception and choice dynamics for Green Hotel attribute: A mix method approach. *Indian Journal of Science and Technology*, 9(5), 1–9.

Whellam, M., & MacDonald, C. (2007). *What is greenwashing and why it is a problem*. Retrievable from: http://www.businessethics.ca/greenwashing/

9

How Does Sustainability Affect Consumer Satisfaction in Retailing?

*Antonio Marín-García, Irene Gil-Saura,
and María-Eugenia Ruiz-Molina*
Universitat de València, Tarongers Campus, Valencia, Spain

CONTENTS

Learning Objectives 225
Themes and Tools Used 226
Introduction 226
Literature Review and Hypotheses 228
Methodology 235
Results 237
Discussion 241
Conclusion 242
Lessons Learned 242
Acknowledgment 243
Credit Author Statement 243
Research Problem 244
References 244

LEARNING OBJECTIVES

After reading the chapter, the reader should be able to:

- Gain knowledge on the nature of sustainability and to identify its dimensions in the retail sector
- Assess the effect of environmental sustainability, social sustainability and economic sustainability on the image of the retailer's store

225

- Acknowledge the effect of environmental sustainability, social sustainability and economic sustainability on the reputation of the retailer's store
- Identify the elements that build satisfaction in retail

THEMES AND TOOLS USED

- Literature review
- Quantitative study, through a structured questionnaire
- Partial least squares regression technique

INTRODUCTION

Marketing literature shows that sustainability is considered a vital factor in the development of competitive advantages for companies (De Brito et al., 2008; Arcese et al., 2015; Gonzalez-Lafaysse & Lapassouse-Madrid, 2016; Ruiz-Real et al., 2018; Bottani et al., 2019; Marín et al., 2019; Marín-García et al., 2020). As a consequence of the report by the Brundtland Commission at the United Nations, where sustainable is presented as a key element in satisfying present needs without compromising future needs (Brundtland, 1987), sustainability is beginning to acquire a special relevance, above all for businesses (Ilbery & Maye, 2005; Chow & Chen, 2012; Moneva & Martín, 2012; Banterle et al., 2013; Lavorata, 2014).

The emerging literature on this issue aims to throw light on the elements accompanying the development of sustainability in retailing (Kumar et al., 2017; Pantano & Timmermans, 2014; Grewal et al., 2017), and to analyse the effects sustainability triggers in other variables. In this way, some studies have raised the need to explore the existence of links between consumers' perceptions of sustainability with variables traditionally related to consumer satisfaction towards the store (Marín et al., 2019). In particular, the image of the store and the store awareness have been pointed out as the most relevant drivers of consumer satisfaction (Chang & Fong, 2010; Gonzalez-Lafaysse & Lapassouse-Madrid, 2016; Ruiz-Real et al., 2018; Bottani et al., 2019; Marín-García et al., 2020; Marín-García et al., 2020).

However, there is still little evidence in the context of sustainability, and even less in retail (Marín et al., 2019). Moreover, to the best of our knowledge, there is no empirical evidence about the extent of the contribution of each dimension of sustainability from a triple bottom line approach (Elkington, 2004) to consumers' perceptions of the store. Shedding light on the contribution of each dimension of sustainability on store awareness and image may enable store managers to implement effective measures regarding sustainable development to generate customer satisfaction. It seems evident, then, that consumer awareness of sustainable practices and an increase in regulation around environmental issues are important elements encouraging organisations to begin to change their business strategies. Hence, some businesses are implementing marketing activities with a view to creating a sustainable image and changing consumers' perceptions of their retail establishments, in doing so to fulfil the needs and desires of consumers in terms of sustainability and development and customer satisfaction with these issues (Marín-García et al., 2020).

The marketing literature points out that the sustainable activities developed by business (manufacturing in recyclable packaging, quality certifications, respecting workers' agreements and manufacturing products that are free of dangerous substances, such as lead, mercury and chromium) are taken into account by consumers when selecting an establishment (Gonzalez-Lafaysse & Lapassouse-Madrid, 2016). However, retailers should not only implement actions towards improving sustainability but also allow and encourage their customers to intervene in the development of these actions. Indeed, it has been highlighted that co-creation is strongly linked with the sustainability knowledge of a company and their capabilities to progress in the implementation of actions towards sustainable development (Arnold, 2017).

Notwithstanding, to the best of our knowledge and despite calls for further research, studies carried out with the objective of shedding light on the nature and scope of sustainability in the retail environment remain scarce. In addition, some authors argue that more evidence is needed in this field of study, motivated by the relevance that such evidence has for strategic management and for academic research (Amran et al., 2014; Arcese et al., 2015; Gonzalez-Lafaysse & Lapassouse-Madrid, 2016; Marcon et al., 2017; Bottani et al., 2019; Marín-García et al., 2020).

Considering the above, the aim of this research is to analyse sustainability and its dimensions' impact on consumer store awareness and perceived image in the commercial distribution sector. In addition, the nature and scope of the concept are identified and its nexus with especially relevant variables – like image, awareness and satisfaction – is delineated. This study will examine the effects that different types of sustainability have on store image and store awareness. In addition, the study aims to analyse the influence that store image and store awareness have on customer satisfaction demonstrated towards the store.

To achieve the above, after the introduction we will identify the theoretical framework underpinning this investigation, focussing attention on the existing relationships between the variables under study through the formulation of a theoretical model. We will then detail the methodology used for the development of further empirical research and finally, we will set forth a series of implications that will enable retail business managers to improve the knowledge of their customers' perceptions of the sustainable practices implemented by these organisations.

LITERATURE REVIEW AND HYPOTHESES

The different perspectives from which sustainability can be analysed indicate that conceptualising this variable is not simple, as it is possible to identify several definitions of this concept (Kamara et al., 2006). Some of the literature examines sustainability through an environmental lens, without taking into account other factors which form part of the term. For example, Callicott and Mumford (1997) define sustainability as satisfying human needs, taking into account the health of ecosystems.

Traditionally, the study of sustainability has only been concerned with an environmental perspective (Moneva & Martín, 2012; Naidoo & Gasparatos, 2018). However, in recent years this concept is being analysed using a wider approach, focussing, in addition to areas relating to the environment, on areas of study linked to social sustainability and economic sustainability (Morioka et al., 2016; Lüdeke-Freund et al., 2017; Marcon et al., 2017; Bottani et al., 2019; Marín et al., 2019). When it comes to considering the concept of sustainability it is, therefore, vital to retain economic, social and environmental factors and to find a balance between these three variables (OCDE, 2001; Chow & Chen, 2012). These factors contribute to the fact that many organisations are putting in place the necessary mechanisms to change and/or modify traditional business models for more sustainable business models (Chang & Fong, 2010; Amran et al., 2014; Morioka et al., 2016; Lüdeke-Freund et al., 2017; Marín-García et al., 2020).

Progressing the study of variables directly related to sustainability and coming to the concept from the viewpoint of the consumer, it has become evident that social, environmental and economic sustainability are the pillars on which the concept is based. In this way, sustainability is explained on the basis of the 'triple bottom line' explained by Elkington (2004). First, Chow and Chen (2012) analyse environmental sustainability from a generic and integral perspective that goes from reactive to proactive approaches. The first approach analyses the factors that could minimise the damage that products and services can cause in the environment. The second approach examines alternative practices that reduce waste emissions. These processes include the use of alternative combustibles to those traditionally used, reducing the impact on natural habitats, animal species and such alike. Second, social sustainability comprises two principal processes: engagement with the stakeholders and engagement with social issues. Engagement with stakeholders implies actions that achieve stronger and longer lasting relationships with the various agents within and external to the organisation (employees, customers, providers, local communities, governments, etc.). These actions can include taking care of the health and safety of interested parties or the creation of value which helps combat existing inequalities. Engagement with social issues implies businesses to demonstrate ethical behaviour in terms of human rights, a concern for social impact, help and/or involvement in social projects that benefit the community and so on. Finally, the authors consider that an evaluation of economic sustainability is underpinned principally by the long-term financial performance of the company and its economic growth. This factor has recently acquired a wider perspective as it is understood that it includes all the actions that allow a business to create greater value, in this way achieving economic and competitive success. This value can be generated through assets and services produced by the organisation, obtaining income, reducing taxes paid by employees, minimising environmental cost, stimulating sales growth, optimising the production process and improving government regulations.

However, it is clear that there is a scarcity of existing information about how sustainable initiatives can be used to attract consumers to commercial establishments, and hence how to improve the competitive advantage of businesses (Oppewal et al., 2006; Arcese et al., 2015). In order to get a more encompassing view of the concept of sustainability and of the factors of which it is formed, Table 9.1 shows some of the findings about the

TABLE 9.1

Findings Regarding Sustainability

Author/s	Finding	Type(s) of Sustainability
Brundtland (1987)	Sustainable development is that centred on meeting present needs without compromising the ability of future generations to meet their own needs.	Economic Environmental Social
McCann-Erickson (2007)	Sustainability can be understood as a collective term for everything that has to do with the world in which we live. It is an economic, social and environmental issue.	Economic Environmental Social
Closs et al. (2011)	Sustainable initiatives range from incremental changes (advertising and containers for products/services) to radical changes (marketing approach, location of facilities, ways of developing and delivering products and services, and treatment of employees, clients, providers, etc.).	Environmental Social
Sheth et al. (2011)	Responsibility implies that company objectives should not be based solely on economic results but should also bear in mind social and environmental impact.	Environmental Social
Chow and Chen (2012)	Sustainable development is defined as the degree to which companies adopt social, economic and environmental development in their operations.	Economic Environmental Social
Boons and Lüdeke-Freund (2013)	Sustainable business models allow social entrepreneurs to create social value and to maximise social benefits; of importance is that business models have the capacity to act as a market mechanism that helps in the creation and development of new markets for innovations with a social purpose.	Social
Lavorata (2014)	Sustainable development, formed by economic, social and environmental factors, allows businesses to achieve competitive advantage.	Economic Environmental Social
Morioka et al. (2016)	The sustainable business model can help managers to better understand how global sustainable development can contribute to the value of the company, including methods of creation, delivery and value capturing.	Economic Environmental Social

(Continued)

TABLE 9.1 (Continued)
Findings Regarding Sustainability

Author/s	Finding	Type(s) of Sustainability
Marcon et al. (2017)	The concept of sustainability consists of a process of change through the diffusion of technical and organisational innovation, bearing in mind both current and future needs.	Economic Environmental
Naidoo and Gasparatos (2018)	Retailers are increasingly obliged to reduce the environmental impacts, both internal and external, of their operations. To achieve this, they should adopt incremental corporate environmental strategies and actions.	Environmental
Ruiz-Real et al. (2018)	Sustainability is a key element for retailers, who have seen that sustainable development could become an important source of competitive advantage.	Economic Environmental

concept of sustainability and the types of these variables as identified in the literature.

The contributions included in Table 9.1 can be considered as a reference to better understand the concept of sustainability. These definitions have some elements in common. The most relevant definition for this study is that a large part of these studies identify environmental sustainability, social sustainability and economic sustainability as the main dimensions of this construct. In line with this triple bottom line approach, the present study conceives sustainability as a construct made up of three dimensions: environmental sustainability, social sustainability and economic sustainability.

Focusing attention on the link between sustainability and awareness in retail companies, despite the novelty of the proposal and the scarcity of studies to back up said link, some previous research confirm its significant and positive relationship. Coca et al. (2013) show that sustainable actions generated by businesses can lead to increased brand awareness on the part of the consumer. Additionally, sustainable practices developed by the organisation can assist in changing consumer perception of the store favourably, as long as those actions are not perceived as merely commercial and publicity actions (Gonzalez-Lafaysse & Lapassouse-Madrid, 2016).

On the other hand, in an empirical study undertaken with Brazilian consumers by Garcia et al. (2019), analysing levels of consumer awareness of the amount of pollution generated by clothes production, it was evidenced that a large portion of the sample was aware of sustainability, but not aware of textile production chains and their potential for pollution. We therefore propose:

> H1a: Consumer perception of retail businesses' economic sustainability has a positive effect on store awareness of the company.
> H1b: Consumer perception of retail businesses' social sustainability has a positive effect on store awareness of the company.
> H1c: Consumer perception of retail businesses' environmental sustainability has a positive store effect on awareness of the company.

To look at the relationship between sustainability and the other variables of special interest with regard to retail businesses, some research analyses sustainability and store image (Chen and Myagmarsuren, 2010; Kumar, 2014; Lavorata, 2014; Gonzalez-Lafaysse & Lapassouse-Madrid, 2016; Marín-García et al., 2020). In particular, Chen and Myagmarsuren (2010) confirm empirically a positive and significant relationship between sustainability and store image, satisfaction, trust and brand value. The study was based on the notion that all these constructs have been thoroughly explored; however, none of them has been studied from the perspective of sustainability.

Likewise, Gonzalez-Lafaysse & Lapassouse-Madrid (2016) argue that it is of key importance that retail businesses develop sustainable actions with the aim of creating a more sustainable store image. In this respect, the authors note the importance of investing in marketing policies employing a sustainable approach, creating positive associations in the minds of consumers.

Continuing this line of study, Lavorata (2014) proposes a conceptual model in which commitment to sustainable development by retailers is a fundamental element. Lavorata (2014) contrasts the relationship of sustainability with other constructs, such as image, loyalty and behavioural intentions, from the perspective of the consumer, highlighting the positive relationships between sustainable actions and store image. These results are relevant for retail businesses as they suggest that practices linked to sustainable development help to build commitment and a favourable image amongst consumers. Similarly, Kumar's (2014) conceptual research proposes that the highlighting of organic products, the promotion of sustainable commercial practices, and the availability and visibility of organic products are sufficiently relevant actions to improve the brand image of retail businesses.

In consideration of the above, we therefore propose the following hypotheses:

> H2a: Consumer perception of the economic sustainability of retail businesses has a positive effect on store image.
> H2b: Consumer perception of the social sustainability of retail businesses has a positive effect on store image.
> H2c: Consumer perception of the environmental sustainability of retail businesses has a positive effect on store image.

Brand awareness represents the potential capacity of the consumer to recognise or remember the name of an establishment (Aaker & Equity, 1991). Looking at this conceptualisation of brand awareness, Keller (1993) asserts that this construct is linked to brand strength in the memory of the consumer, which leads to the consumer identifying the brand in different circumstances and scenarios (Rossiter & Percy, 1987).

Pappu and Quester (2006) show that the level of knowledge about a retailer can contribute to the level of satisfaction or dissatisfaction on the behalf of the consumer towards that retailer. Previous research shows that consumer satisfaction with commercial distribution businesses is influenced in large part by the knowledge and understanding that consumers have of a store (Bilal & Malik, 2014; Das, 2014; Barreda et al., 2015). Store awareness is postulated as a crucial element in the consumer shopping process, as a consequence of its influence on purchase decisions (Alsoud & Abdallah, 2013). In this way, a retailer's capacity to provide a wide variety of products of various brands with different containers, logos, symbols and designs increases the level of knowledge consumers have of a retailer, generating a direct and positive effect on consumer satisfaction (Das, 2014). Similarly, Jinfeng and Zhilong (2009) and Chi et al. (2009) show that store awareness has positive effects on consumer satisfaction.

Since this study is centred on consumer satisfaction, we anticipate a positive relationship between awareness and consumer satisfaction, and therefore propose:

> *H3: Store awareness of the retailers as generated by the consumer has a positive effect on consumer satisfaction.*

One of the most important factors connected to brand value is brand image and associations (Aaker & Equity, 1991; Chen and Myagmarsuren, 2010), which is defined as any element that forms part of our memory and which identifies a specific brand (Aaker & Equity, 1991), crucial in the process of evaluating the consumer within the framework of consumption and a fundamental element of brand personality (Hartman & Spiro, 2005). Keller (1993) explains that associations have a direct effect on brand in terms of consumers assigning it significance in their minds and relating them with the brand.

As well as being a key element in the development of brand value, store image is considered a relevant factor in the formation of customer satisfaction. In this sense, the associations the store has in the perceptions of the client will directly affect consumer satisfaction with the establishment (Bloemer & Kasper, 1995). The global evaluation that the consumer makes about the establishment will influence the development of satisfaction or dissatisfaction. Similarly, some studies maintain that store image can help with customer loyalty towards the establishment via consumer satisfaction (Bloemer & De Ruyter, 1998).

To continue with the links between other important variables in retailing, it is important to analyse the relations between consumer satisfaction and establishment store. As a consequence of environmental awareness acquired by customers and of the strict regulations with regard to the environment becoming established in today's marketplace, some companies are undertaking marketing actions orientated towards sustainability. The main aim of these actions is the creation of an environmentally friendly image, the purpose of which, moreover, is to increase consumer satisfaction, cater to consumer desires and increase competitive advantage over other companies (Chang & Fong, 2010).

Thus, we propose the last hypothesis of this study:

> *H4: The store image of the retailer as generated by the consumer has a positive effect on consumer satisfaction.*

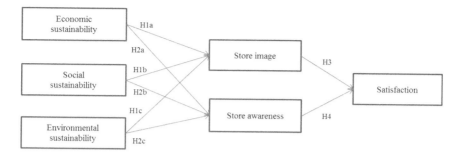

FIGURE 9.1
Proposed model.

Before starting the next section, our structural model is displayed (Figure 9.1).

METHODOLOGY

To test the hypotheses of this study, a quantitative approach is adopted in the context of grocery retailing in Spain. In the Spanish commercial structure, retailing is considered as one of the main economic activities. In addition, it is a competitive sector, which stands out for its dynamism, and which has undergone a profound transformation in the last decade, observed through the remarkable variety of commercial formulas implemented, new complementary uses developed and architectural interventions (Espada, 2017).

In Spain, retail trade has a very important specific weight in the national economy. Commerce in general contributes 13% to total GDP, while more than 5% has its origin in retail activities in particular. Regarding job creation, in 2019 its growth was remarkable, representing 12% of employment, according to the Statistics National Institute (INE, 2020). In this sense, it is important to consider their contribution to the economic development of the country. Moreover, of the total spending of households in 2019, more than 30% went to retailing.

To evaluate the proposed relationships, an empirical study was carried out. In this research, consumers had to answer a set of statements through a structured questionnaire.

The items used to measure the constructs under analysis have been adopted from previous works and previously validated in studies carried out in contexts similar to this research. To measure consumer perception of the sustainability policies implemented by businesses, we have used the items on the scale put forward by Lavorata (2014) and Marín-García et al. (2020), whilst the measurement of awareness was undertaken using items from Yoo et al. (2000) and Netemeyer et al. (2004), Arnett et al. (2003) and Marín-García et al. (2020). To evaluate image, we adapted the scale from Yoo et al. (2000) and Marín-García et al. (2020). Finally, consumer satisfaction was assessed using the scale proposed by Bloemer and Odekerken-Schroder (2002), Dixon et al. (2005) and Zhao & Huddleston, (2012). In all cases, a Likert scale (1 totally disagree; 7 totally agree) was used to measure the items that make up the variables of this research.

To obtain the information necessary to test the hypotheses, previously, a pre-test was carried out to correct any possible incidence of the questionnaire. Through a non-probabilistic quota sampling at the exit point of the establishments, the field work to obtain the data was carried out in 2017 in a region of Spain. Specifically, the hypermarkets, supermarkets and discount stores were the retail formats selected for this study according to their importance in the grocery retail sector. According of the sociodemographic variables, Table 9.2 shows the sample distribution. Of the 510 valid questionnaires obtained, 59.6% are female and 40.4% are male. With regard to median age of the sample, the bracket between 36 and 45 years made up the highest percentage (26.1%).

Moreover, as the data set used in this work is self-reported, the common method bias is a potential problem because of social desirability (Podsakoff et al., 2003). The statistical analysis technique of Harmon's single-factor test is performed to determine the severity of common method bias (Podsakoff et al., 2003). The analysis performed indicates that common method biases are not likely contaminating our results, since the most covariance explained by one factor is 38.70.

The data collected was interpreted through the partial least squares (PLS) regression. This technique is considered by many researchers as optimal for the estimation of structural models (Henseler et al., 2016; Hair et al., 2017). To do this, using the SmartPLS software (version 3.0), we analysed the validity of the scales used to measure the variables under study. Subsequently, we proceeded to examine the relationships raised in this research.

TABLE 9.2

Distribution of Sample: Sociodemographic Variables

	Total		Hypermarket		Supermarket		Discount Store	
	N	%	N	%	N	%	N	%
Gender								
Male	206	40.4	68	40.0	67	39.4	71	41.8
Female	304	59.6	102	60.0	103	60.6	99	58.2
Age								
Between 18 and 25	36	7.1	16	9.4	9	5.3	11	6.5
Between 26 and 35	88	17.3	34	20.0	20	11.8	34	20.0
Between 36 and 45	133	26.1	35	20.6	63	37.1	35	20.6
Between 46 and 55	113	22.2	18	10.6	50	29.4	45	26.5
Between 56 and 65	108	21.2	50	29.4	24	14.1	34	20.0
Over 65	32	6.3	17	10.0	4	2.4	11	6.5
Education								
No schooling	40	7.8	20	11.8	9	5.3	11	6.5
Primary education	84	16.5	41	24.1	22	12.9	21	12.4
Secondary education	148	29.0	41	24.1	46	27.1	61	35.9
University education	238	46.7	68	40.0	93	54.7	77	45.3
Occupation								
Employed	282	55.3	90	52.9	102	60.0	90	52.9
Self employed	63	12.4	16	9.4	24	14.1	23	13.5
Retired/Pensioner	58	11.4	39	22.9	6	3.5	13	7.6
Unemployed	40	7.8	6	3.5	16	9.4	18	10.6
Household tasks	37	7.3	16	9.4	12	7.1	9	5.3
Student	30	5.8	3	1.8	10	5.9	17	10.0

RESULTS

To analyse the psychometric properties of the measurement instrument, we proceeded to examine, for each of the variables of this research, the factor loadings, Cronbach's alpha, the composite reliability (CR) index (Comparative Fit Index) and the average variance extracted (AVE) (Table 9.3). In addition, considering that all the scales have been tested and validated by previous research in similar contexts, we decided to keep

TABLE 9.3

Measuring Instrument of the Structural Model: Reliability and Convergent Validity Convergence

Construct	Item	St. Loading Factor (st. error)	t	Cronbach's α	Composite Reliability	Average Variance Extracted
Economic sustainability	ST1	0.767*** (0.036)	21.436	0.871	0.912	0.723
	ST2	0.896*** (0.011)	78.607			
	ST3	0.872*** (0.014)	62.500			
	ST4	0.860*** (0.017)	49.934			
Social sustainability	ST5	0.739*** (0.030)	24.920	0.883	0.914	0.680
	ST6	0.890*** (0.010)	93.447			
	ST7	0.802*** (0.027)	29.730			
	ST8	0.821*** (0.025)	32.383			
	ST9	0.862*** (0.016)	53.986			
Environmental sustainability	ST10	0.765*** (0.044)	17.354	0.817	0.880	0.711
	ST11	0.849*** (0.031)	27.221			
	ST12	0.910*** (0.017)	52.092			
Awareness	AW1	0.755*** (0.026)	29.213	0.850	0.898	0.689
	AW2	0.853*** (0.013)	66.823			
	AW3	0.879*** (0.015)	58.216			
	AW4	0.829*** (0.018)	45.228			
Image	IM1	0.759*** (0.020)	38.299	0.875	0.901	0.504
	IM2	0.816*** (0.017)	47.757			
	IM3	0.693*** (0.031)	22.144			
	IM4	0.600*** (0.031)	19.141			
	IM5	0.651*** (0.033)	19.457			
	IM6	0.740*** (0.022)	34.070			
	IM7	0.748*** (0.025)	30.217			
	IM8	0.726*** (0.026)	27.695			
	IM9	0.631*** (0.035)	17.875			
Satisfaction	SF1	0.872*** (0.012)	71.690	0.946	0.958	0.822
	SF2	0.934*** (0.006)	165.262			
	SF3	0.920*** (0.009)	101.152			
	SF4	0.930*** (0.006)	146.057			
	SF5	0.875*** (0.010)	84.416			

Note: AVE, average variance extracted; CA, Cronbach's α; CR, composite reliability.
***p < 0.01

all the items that formed the measurement scales of the variables in order to not lose the originality of the scales (Hair et al., 2011).

According to the data given in Table 9.3, we confirm the reliability of all the scales. The results of the study indicate that both Cronbach's coefficient and the CR show values above 0.7 (Nunnally & Bernstein 1994) or 0.8 (Carmines & Zeller 1979). Regarding convergent validity, the loadings of the items were first analysed and it was found that all were statistically significant with values greater than 0.6. The analysis of the reliability of the scales was then carried out using the CR index and the analysis of the AVE, verifying that their values were higher than the thresholds established by the previous literature as critical (i.e., 0.7 and 0.5, respectively) (Fornell & Larcker, 1981). Thus, convergent and discriminant validity of the measurement model is confirmed.

In addition, as observed in Table 9.4, the discriminating validity is confirmed, since for all cases, the square root of the AVE was higher than the estimated correlation between the factors, which appears below the diagonal, according to the criterion of Fornell and Larcker (1981).

Using the SmartPLS software with a full bootstrapping with 5000 subsamples (Ringle et al., 2015), the relationships of the proposed model hypotheses as well as the values for the explained variance, i.e. R2, are examined. Through the blindfolding technique, we obtain the results provided by the predictive relevance criterion based on the Q^2 test. The results obtained are presented in Table 9.5 and Figure 9.2.

The results confirm the majority of the hypotheses put forward in the structural model. With respect to the accepted hypotheses, support was found in the second of them (H2), which has a positive relationship with economic, social and environmental sustainability with store image

TABLE 9.4

Measurement Instrument: Discriminant Validity (Fornell–Larcker Criterion)

	ST - ECO	ST - SOC	ST - ENV	AW	IM	SF
Economic sustainability	*0.850*					
Social sustainability	0.559	*0.824*				
Environmental sustainability	0.503	0.617	*0.843*			
Awareness	0.386	0.289	0.265	*0.830*		
Image	0.409	0.410	0.376	0.666	*0.710*	
Satisfaction	0.451	0.571	0.390	0.643	0.674	*0.907*

Note: Diagonal values in italics are square roots of AVE and values below the diagonal are correlations between variables.

TABLE 9.5

Structural Equation Model Estimation

	Relationship		Hypothesis	Standardised Parameter	t
H1a	Sustainability ECO	→ Awareness	SUPPORTED	0.310***	5.922
H1b	Sustainability SOC	→ Awareness	REJECTED	0.087 ns	1.771
H1c	Sustainability MED	→ Awareness	REJECTED	0.068 ns	1.276
H2a	Sustainability ECO	→ Image	SUPPORTED	0.222***	4.385
H2b	Sustainability SOC	→ Image	SUPPORTED	0.216***	3.686
H2c	Sustainability MED	→ Image	SUPPORTED	0.141**	4.480
H3	Awareness → Satisfaction		SUPPORTED	0.355***	8.939
H4	Image → Satisfaction		SUPPORTED	0.453***	10.090

Note: Store awareness: $R^2 = 0.167$, $Q^2 = 0.104$; Store image: $R^2 = 0.238$, $Q^2 = 0.101$; Satisfaction: $R^2 = 0.539$, $Q^2 = 0.411$.

**p < 0.05; ***p < 0.01; ns, non-statistically significant.

($\beta_{2a} = 0.222$, $p < 0.001$; $\beta_{2b} = 0.216$, $p < 0.001$; $\beta_{2c} = 0.141$, $p < 0.05$). We can, therefore, confirm the importance of all three types of sustainability with store image, as shown in previous studies (Lavorata, 2014; Gonzalez-Lafaysse & Lapassouse-Madrid, 2016). Similarly, it is possible to confirm H1 partially, as the results show the presence of a positive relationship between economic sustainability and store awareness ($\beta_{1a} = 0.310$, $p < 0.001$). As for the hypotheses that postulate the influence of store image and of store awareness as elements preceding consumer satisfaction, positive and statistically significant coefficients allow us to accept the last two stated hypotheses of the structural model ($\beta_3 = 0.355$, $p < 0.001$; $\beta_4 = 0.453$, $p < 0.001$), confirming what was previously stated.

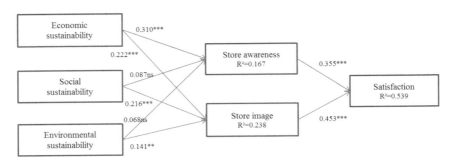

FIGURE 9.2
Structural equation model results.

With regard to the relationships that are not confirmed, there is no support for hypotheses H1b and H1c, which link social and environmental sustainability to store awareness. In both the cases, a coefficient not significantly distinct from zero was obtained ($\beta_{1b} = 0.087$, ns; $\beta_{2c} = 0.068$, ns).

DISCUSSION

Sustainability is a concept that is currently generating special interest due to consumers' greater awareness of the sustainable practices developed by businesses and the increase in regulation around company behaviour (Kumar et al., 2017; Pantano & Timmermans, 2014; Grewal et al., 2017; Pantano et al., 2018; Ruiz-Real et al., 2018; Bottani et al., 2019; Marín et al., 2019). In parallel to the study of this variable, the effects that sustainable actions have on traditional variables within marketing research, such as awareness (Gonzalez-Lafaysse & Lapassouse-Madrid, 2016; Marín-García et al., 2020), image (Kumar, 2014; Lavorata, 2014; Marín-García et al., 2020) and satisfaction (Chang & Fong, 2010), have also been considered.

Accordingly, the findings obtained in the present empirical research deepen the understanding of the existence of disparate consequences and conclude that, within the large-scale grocery sector, store image boosted by sustainable economic, social and environmental practices, has a direct impact on building consumer satisfaction with retail establishments. Thus, sustainability is discernible as a transcendental tool to influence consumer perception of the large-scale grocery sector, as has been shown in previous research (Chang & Fong, 2010). However, it is also important to consider some elements of sustainability, like the economic one, in the formation of consumer satisfaction via store awareness.

In addition, in relation to the links between store image and consumer satisfaction, the results obtained lead us to think that in the retail sector both the image that the customer forms of the store and the consumer knowledge of the store, can positively influence the generation of consumer satisfaction.

Finally, the results of this study show that the development of sustainable practices in Spanish retailing does not have a significant effect on the awareness towards the establishment. This may be a consequence of the positioning of the store chains selected for this research. The brands of the stores chosen for this study are widely known by Spanish consumers. In this sense, the fact of collaborating with associations linked to social problems, or the reduction of materials that can harm the environment, may not have an impact on the ability of consumers to recognise or recall these. In contrast, the development of sustainable practices does influence the image that the consumer perceives on the retail establishment.

CONCLUSION

All in all, the study of sustainability and the impact that this variable may have on society allows us to affirm that it is necessary to continue investigating its effect on other variables and other fields of study. Indeed, sustainability is a broad concept and should be taken into consideration, besides commercial activity, in other areas of study such as teaching. For example, from educational centres, it is important to educate students about sustainable practices. In the future, these students will become consumers and will join the labour market, being necessary that their values include respect for people and also for the environment.

LESSONS LEARNED

The development of this study allows us to state several actions that could be applied by the managers of the retail stores. First, sustainability can be regarded as a key element in the step from conventional business models to sustainable business models. The actions linked to economic, social and environmental sustainability may provide a competitive advantage provided that they are implemented appropriately by managers of retail companies. If these practices are perceived by the consumer, they can have positive effects on consumers' assessments of establishments. The development of products using materials that are less noxious for the environment, an increase in the presence of fair-trade products in stores and participation and collaboration with social institutions are some of the actions that improve consumer perception of businesses. In order to attract consumers, store managers must take into account the opinions of their customers. In other words, by involving the consumer in the decisions of actions related to sustainability, it is likely that the number of customers with values linked to the dimensions of sustainability increases. Through the actions mentioned above, store managers will progress in transforming their businesses towards a more sustainable type of business.

In addition, considering the nature of the retail sector, dynamic and competitive, it is key to analyse the elements that build consumer satisfaction. This study supports the factors on which the building of consumer satisfaction in the retail sector is based. Specifically, awareness is discernible as a key variable in this relationship in that it fosters actions that allow an increase in consumer awareness about the store, the brand and associated products that are beneficial in terms of consumer satisfaction with the store. Similarly, if action is taken around the brand image (e.g. improving the design of the store, developing new promotions or including new products) this effort will impact positively on consumer satisfaction. We consider that organisations should build on practices that improve the image consumers develop of the store. The incorporation of distinctive and inspiring elements such as store logo, the development of promotions, employee attitude or other changes undertaken in elements related to presentation and seduction merchandising, will further increase the level of store awareness by consumers.

ACKNOWLEDGMENT

This chapter is developed within the Research Project ECO2016-76553-R of the Spanish Ministry of Economy, Industry and Competitiveness, National Research Agency.

CREDIT AUTHOR STATEMENT

Conceptualisation, A.M.G. and I.G.-S.; Methodology, A.M.G., M.E.R.-M., and I.G.-S.; Validation, A.M.G., M.E.R.-M., and I.G.-S.; Formal Analysis, A.M.G., M.E.R.-M., and I.G.-S.; Investigation, A.M.G., M.E.R.-M., and I.G.-S.; Data Curation, A.M.G.; Writing – Original Draft Preparation, A.M.G., M.E.R.-M., and I.G.-S.; Writing – Review and Editing, A.M.G.,

M.E.R.-M., and I.G.-S.; Visualisation, A.M.G., M.E.R.-M., and I.G.-S.; Supervision, M.E.R.-M., and I.G.-S.; Project Administration, I.G.-S. All authors have read and agreed to the published version of the manuscript.

For the food sector, innovation and sustainability are considered key strategic factors to overcome the issues of the sector and to achieve worldwide competitiveness goals.

(Arcese et al., 2015)

RESEARCH PROBLEM

How does sustainability influence consumer satisfaction in retailing?

REFERENCES

Aaker, D. A., & Equity, M. B. (1991). Capitalizing on the value of a brand name. *New York*, 28(1), 35–37.

Alsoud, G. F. A., & Abdallah, M. K. A. S. (2013). Customer awareness and satisfaction of Islamic retail products in Kuwait. *Research Journal of Finance and Accounting*, 4(17), 36–52.

Amran, A., Lee, S. P., & Devi, S. S. (2014). The influence of governance structure and strategic corporate social responsibility toward sustainability reporting quality. *Business Strategy and the Environment*, 23(4), 217–235.

Arcese, G., Flammini, S., Lucchetti, M. C., & Martucci, O. (2015). Evidence and experience of open sustainability innovation practices in the food sector. *Sustainability*, 7(7), 8067–8090.

Arnett, D. B., Laverie, D. A., & Meiers, A. (2003). Developing parsimonious retailer equity indexes using partial least squares analysis: a method and applications. *Journal of Retailing*, 79(3), 161–170.

Arnold, M. (2017). Fostering sustainability by linking co-creation and relationship management concepts. *Journal of Cleaner Production*, 140, 179–188.

Banterle, A., Cereda, E., & Fritz, M. (2013). Labelling and sustainability in food supply networks. *British Food Journal*, 115(5), 769–783.

Barreda, A. A., Bilgihan, A., Nusair, K., & Okumus, F. (2015). Generating brand awareness in online social networks. *Computers in Human Behavior*, 50(1), 600–609.

Bilal, A., & Malik, F. M. (2014). Impact of brand equity & brand awareness on customer's satisfaction. *International Journal of Modern Management and Foresight*, 1(10), 287–303.

Bloemer, J., & De Ruyter, K. (1998). On the relationship between store image, store satisfaction and store loyalty. *European Journal of Marketing*, 32(5/6), 499–513.

Bloemer, J. M., & Kasper, H. D. (1995). The complex relationship between consumer satisfaction and brand loyalty. *Journal of Economic Psychology*, 16(2), 311–329.

Bloomer, J., & Odekerken-Schroder, G. (2002). Store satisfaction and store loyalty explained by customer and store related factors. *Journal of Consumer Satisfaction, Dissatisfaction and Complaining Behavior*, 15, 68–80.

Boons, F., & Lüdeke-Freund, F. (2013). Business models for sustainable innovation: State-of-the-art and steps towards a research agenda. *Journal of Cleaner Production*, 45, 9–19.

Bottani, E., Casella, G., & Arabia, S. (2019). Sustainability of retail store processes: An analytic model for economic and environmental evaluation. *International Journal of Simulation and Process Modelling*, 14(2), 105–119.

Brundtland, G. H. (1987). Report of the world commission on environment and development: Our common future. *World Commission on Environment and Development*. New York: Oxford.

Callicott, J. B., & Mumford, K. (1997). Ecological sustainability as a conservation concept: Sustentabilidad ecologica como concepto de conservacion. *Conservation Biology*, 11(1), 32–40.

Carmines, E. G., & Zeller, R. A. (1979). *Reliability and validity assessment*. Thousand Oaks, CA: Sage publications.

Chang, N. J., & Fong, C. M. (2010). Green product quality, green corporate image, green customer satisfaction, and green customer loyalty. *African Journal of Business Management*, 4(13), 2836–2844.

Chen, C. & Myagmarsuren, O. (2010). Exploring relationships between Mongolian destination brand equity, satisfaction and destination loyalty. *Tourism Economics*, 16(4), 981–994.

Chi, H. K., Yeh, H. R., & Yang, Y. T. (2009). The impact of brand awareness on consumer purchase intention: The mediating effect of perceived quality and brand loyalty. *The Journal of International Management Studies*, 4(1), 135–144.

Chow, W. S., & Chen, Y. (2012). Corporate sustainable development: Testing a new scale based on the mainland Chinese context. *Journal of Business Ethics*, 105(4), 519–533.

Closs, D. J., Speier, C., & Meacham, N. (2011). Sustainability to support end-to-end value chains: The role of supply chain management. *Journal of the Academy of Marketing Science*, 39(1), 101–116.

Coca, V. Dobrea, M., & Vasiliu, C. (2013). Towards a sustainable development of retailing in Romania. *Amfiteatru Economic*, 15, 583–602.

Das, G. (2014). Linkages of retailer awareness, retailer association, retailer perceived quality and retailer loyalty with purchase intention: A study of Indian food retail brands. *Journal of Retailing and Consumer Services*, 21(3), 284–292.

De Brito, M. P. Carbone, V., & Blanquart, C. M. (2008). Towards a sustainable fashion retail supply chain in Europe: Organisation and performance. *International Journal of Production Economics*, 114(2), 534–553.

Dixon, J., Bridson, K., Evans, J., & Morrison, M. (2005). An alternative perspective on relationships, loyalty and future store choice. *The International Review of Retail, Distribution and Consumer Research*, 15(4), 351–374.

Elkington, J. (2004). Enter the triple bottom line. In A. Henriques & J. Richardson (Eds.), *The Triple Bottom Line: Does It All Add Up?' Assessing the Sustainability of CSR* (pp. 1–16). London: Earthscan Publications.

Espada, J. R. (2017). Tendencias en la remodelación de mercados minoristas. *Distribución y Consumo*, 2(147), 5–10.

Fornell, C., & Larcker, D. F. (1981). Evaluating structural equation models with unobservable variables and measurement error. *Journal of Marketing Research*, 18(1), 39–50.

Garcia, S., Cordeiro, A., de Alencar Nääs, I., & Neto, P. L. D. O. C. (2019). The sustainability awareness of Brazilian consumers of cotton clothing. *Journal of Cleaner Production*, 215, 1490–1502.

Gonzalez-Lafaysse, L., & Lapassouse-Madrid, C. (2016). Facebook and sustainable development: A case study of a French supermarket chain. *International Journal of Retail & Distribution Management*, 44(5), 560–582.

Grewal, D. Roggeveen, A. L., & Nordfält, J. (2017). The future of retailing. *Journal of Retailing*, 93(1), 1–6.

Hair, J. F. Hult, G. T. M. Ringle, C. M. Sarstedt, M., & Thiele, K. O. (2017). Mirror, mirror on the wall: A comparative evaluation of composite-based structural equation modeling methods. *Journal of the Academy of Marketing Science*, 45(5), 616–632.

Hair, J. F. Ringle, C. M., & Sarstedt, M. (2011). PLS-SEM: Indeed a silver bullet. *Journal of Marketing Theory and Practice*, 19(2), 139–152.

Hartman, K. B., & Spiro, R. L. (2005). Recapturing store image in customer-based store equity: A construct conceptualization. *Journal of Business Research*, 58(8), 1112–1120.

Henseler, J. Hubona, G., & Ray, P. A. (2016). Using PLS path modeling in new technology research: Updated guidelines. *Industrial Management & Data Systems*, 116(1), 2–20.

Ilbery, B., & Maye, D. (2005). Food supply chains and sustainability: evidence from specialist food producers in the Scottish/English borders. *Land Use Policy*, 22(4), 331–344.

INE. (2020). *Encuesta de Población Activa (EPA)*. Madrid: Instituto Nacional de Estadística.

Jinfeng, W., & Zhilong, T. (2009). The impact of selected store image dimensions on retailer equity: Evidence from 10 Chinese hypermarkets. *Journal of Retailing and Consumer Services*, 16(6), 486–494.

Kamara, M., Coff, C., & Wynne, B. (2006). GMO's and sustainability. *European Retail Research*, 26(1), 1–73.

Keller, K. L. (1993). Conceptualizing, measuring, and managing customer-based brand equity. *Journal of Marketing*, 57, 1–22.

Kumar, P. (2014). Greening retail: An Indian experience. *International Journal of Retail & Distribution Management*, 42(7), 613–625.

Kumar, V., Anand, A., & Song, H. (2017). Future of retailer profitability: An organizing framework. *Journal of Retailing*, 93(1), 96–119.

Lavorata, L. (2014). Influence of retailers' commitment to sustainable development on store image, consumer loyalty and consumer boycotts: Proposal for a model using the theory of planned behaviour. *Journal of Retailing and Consumer Services*, 21(6), 1021–1027.

Lüdeke-Freund, F. Freudenreich, B. Schaltegger, S. Saviuc, I., & Stock, M. (2017). Sustainability-oriented business model assessment-A conceptual foundation. In R. Edgeman, E. Carayannis, S. Sindakis (Eds.), *Analytics, Innovation and Excellence-driven Enterprise Sustainability* (pp. 169–206). England: Houndmills.

Marcon, A. de Medeiros, J. F., & Ribeiro, J. L. D. (2017). Innovation and environmentally sustainable economy: Identifying the best practices developed by multinationals in Brazil. *Journal of Cleaner Production*, 160, 83–97.

Marín, G. A., Gil-Saura, I., & Ruiz-Molina, M. (2019). Influencia del formato comercial en la sostenibilidad en el comercio minorista de alimentación en España. *UCJC*

Business and Society Review (formerly known as Universia Business Review), 16(4), 132–173.

Marín-García, A., Gil-Saura, I., & Ruíz-Molina, M.E. (2020). How do innovation and sustainability contribute to generate retail equity? Evidence from Spanish retailing. *Journal of Product & Brand Management*, 29(5), 601–615.

McCann-Erickson. (2007). *Can sustainability sell? Disponible en*. Retrievable from: www.unep.fr/pc/sustain/ reports/advertising/can-sustainability-Sell%20.pdf.

Moneva, A. J., & Martín, V. E. (2012). Universidad y desarrollo sostenible: Análisis de la rendición de cuentas de las universidades públicas desde un enfoque de responsabilidad social. *Revista Iberoamericana de Contabilidad de Gestión*, 10(19), 1–18.

Morioka, S. N. Evans, S., & de Carvalho, M. M. (2016). Sustainable business model innovation: Exploring evidences in sustainability reporting. *Procedia CIRP*, 40, 659–667.

Naidoo, M., & Gasparatos, A. (2018). Corporate environmental sustainability in the retail sector: Drivers, strategies and performance measurement. *Journal of Cleaner Production*, 203, 125–142.

Netemeyer, R. G., Krishnan, B., Pullig, C., Wang, G., Yagci, M., Dean, D., Ricks, J., & Wirth, F. (2004). Developing and validating measures of facets of customer-based brand equity. *Journal of Business Research*, 57(2), 209–224.

Nunnally, J. C., & Bernstein, I. H. (1994). *Psychological theory*. New York, NY: McGraw-Hill.

OCDE. (2001). OECD environmental strategy for the first decade of the 21st century. *Oslo Manual*. Paris: OECD.

Oppewal, H., Alexander, A., & Sullivan, P. (2006). Consumer perceptions of corporate social responsibility in town shopping centres and their influence on shopping evaluations. *Journal of Retailing and Consumer Services*, 13(4), 261–274.

Pantano, E., Priporas, C. V., & Dennis, C. (2018). A new approach to retailing for successful competition in the new smart scenario. *International Journal of Retail & Distribution Management*, 46(3), 264–282.

Pantano, E., & Timmermans, H. (2014). What is smart for retailing? *Procedia Environmental Sciences*, 22, 101–107.

Pappu, R., & Quester, P. (2006). Does customer satisfaction lead to improved brand equity? An empirical examination of two categories of retail brands. *Journal of Product & Brand Management*, 15(1), 4–14.

Podsakoff, P. M., Mackenzie, S. B., Lee, J.-Y., & Podsakoff, N. P. (2003). Common method biases in behavioural research: A critical review of the literature and recommended remedies. *Journal of Applied Psychology*, 88(5), 879.

Ringle, C. M. Wende, S., & Becker, J. (2015). SmartPLS 3. *GmbH*. Bönningstedt: SmartPLS. Retrievable from: Http://www.Smartpls.Com

Rossiter, J. R., & Percy, L. (1987). *Advertising and promotion management*. Nueva York, NY: McGraw-Hill Book Company.

Ruiz-Real, J. L., Uribe-Toril, J., Gázquez-Abad, J. C., & de Pablo Valenciano, J. (2018). Sustainability and retail: Analysis of global research. *Sustainability*, 11(1), 1–18.

Sheth, J. N., Sethia, N. K., & Srinivas, S. (2011). Mindful consumption: A customer-centric approach to sustainability. *Journal of the Academy of Marketing Science*, 39(1), 21–39.

Yoo, B., Donthu, N., & Lee, S. (2000). An examination of selected marketing mix elements and brand equity. *Journal of the Academy of Marketing Science*, 28(2), 195–211.

Zhao, J., & Huddleston, P. (2012). Antecedents of specialty food store loyalty. *The International Review of Retail, Distribution and Consumer Research*, 22(2), 171–187.

10

Bridging the Intention–Behaviour Gap in Second-Hand Clothing

Elin Pedersén, Amanda Persson and Adele Berndt
Jönköping International Business School (JIBS), Jönköping, Sweden

CONTENTS

Learning Objectives 250
Themes and Tools Used 250
Research Questions 250
Theoretical Background 251
Background 251
Introduction 251
Theoretical Background/Literature Review 253
    Bridging the IB Gap 254
        Intentions 254
        Planning 255
        Actions and Action Control 256
        Barriers and Resources: Social Support and Availability 256
        Self-Efficacy 257
        Contrasting Intenders and Actors as Consumer Groups 259
    Methodology 259
    Findings 260
        Intenders 260
        Their Intention and (In)Action 260
        Planning 261
        Barriers and Resources 261
        Task Self-Efficacy 262
        Maintenance Self-Efficacy 262
        Recovery Self-Efficacy 263

Actors 263
    Intention and Action 263
    Planning 264
    Barriers and Resources 265
    Task Self-Efficacy 265
    Maintenance Self-Efficacy 265
    Recovery Self-Efficacy 266
Discussion 267
  Theoretical Contribution 270
  Practical Application 271
Conclusion, Limitations, and Future Research 272
Credit Author Statement 273
References 273

LEARNING OBJECTIVES

After reading the chapter, the reader should be able to:

- Explain the nature of the intention–behaviour (IB) gap
- Indicate the importance of the IB gap in clothes purchasing
- Contrast the two consumer groups, i.e. intenders and actors
- Identify the factors that affect the IB in the two groups of consumers
- Suggest strategies that can be used to encourage the purchase of clothing from second-hand outlets

THEMES AND TOOLS USED

The HAPA model; Content analysis

RESEARCH QUESTIONS

There are two research questions posed for this study.

- How do the factors in the HAPA model impact the purchase of second-hand clothing?
- How do intenders and actors differ with respect to the purchase of second-hand clothing?

THEORETICAL BACKGROUND

The study is based on the health action process approach (HAPA) model suggested by Schwarzer, Lippke, and Luszczynska (2011). This model identifies a number of factors that can be used to bridge the intention-behaviour (IB) gap previously researched in health-related behaviours. In applying this to sustainability behaviours, it is necessary to examine the relevance of these factors and their application specifically in the purchasing of second-hand clothing.

Two groups or clusters can be identified with respect to actual behaviour, namely inactive people (intenders) and active people (actors). The difference between these two groups is linked to the behaviour that is carried out, which suggests an IB gap. Understanding the reasons why they do not carry out the behaviour can help to reduce this gap, potentially encouraging actual behaviour (i.e. turning those who intenders into actors).

BACKGROUND

As the fashion industry is one that contributes to environmental damage, it is necessary to examine alternative ways of consumption. One way is to increase the purchasing of second-hand fashion. While the majority of consumers state they intend to purchase second-hand clothes, this does not always result in action, evidencing an intention–behaviour (IB) gap. The study focuses on this gap and explores the behaviour of Swedish female consumers in order to understand how these different factors affect the IB gap.

INTRODUCTION

Overconsumption, where consumers purchase more products than what they need to satisfy their basic needs, has been identified as contributing to environmental issues (De Graaf, Wann, & Naylor, 2014; Humphery, 2010; Håkansson, 2014). Similarly, production in various industries also contributes to environmental damage with the fashion industry being identified as one of the most environment-damaging contributors (Diddi et al. 2019; Ek Styvén & Mariani, 2020). Clothing is an important product category in which to investigate sustainability issues as clothing is linked to self-identity and self-worth, with consumers continually searching for new products (McNeill & Moore, 2015). Further, as clothing is one of the leading contributors to environmental damage, increasing the reuse of clothing, thereby prolonging the life, can have an impact on the environment (Allwood et al. 2008; Laitala, Klepp, & Boks, 2012).

Consumers have become more environmentally aware, increasingly reusing and reselling fashion clothing (Pookulangara & Shephard, 2013). Second-hand clothing is important in stimulating sustainability as it provides an extended life for clothing. Yet, despite this consumers have been reluctant to implement more sustainable choices (McNeill & Moore, 2015). Improvements have been made in encouraging customers to buy second-hand clothes (Roos, 2019), and this has also been seen in Sweden where the market has grown. Despite this, much remains to be done.

Researchers have focused on the intentions of consumers to act in sustainable ways, yet there is often a gap between the intention and the action (Dholakia & Bagozzi, 2003). For example, in many instances, the intention indicated is not reflected in actual behaviour, reflecting an IB gap (Auger & Devinney, 2007; Carrington, Neville, & Whitwell, 2010). Also known as the green gap (Guyader, Ottosson, & Parment, 2020), the IB gap suggests that individuals intend to carry out certain actions, but for various reasons do not.

The IB gap has been identified in many academic areas including entrepreneurship (Adam & Fayolle, 2016; Schlaegel & Koenig, 2014), health behaviours (Barz et al. 2016; Fueyo-Díaz et al. 2018; Schwarzer et al. 2011; Sniehotta, Scholz, & Schwarzer, 2005), and sustainability behaviours (Carrington et al. 2010; Grimmer & Miles, 2017). It has been suggested that there are two groups or clusters based on their actual behaviour, namely inactive people (intenders) and active people (actors), with the difference being behaviour that is carried out. Understanding the reasons why they do not carry out the behaviour can help to reduce this gap, potentially encouraging actual behaviour (i.e. turning those who intenders into actors). Thus, the purpose of this study is to explore the behaviour of Swedish female consumers and understand how different factors affect the IB gap in the case of second-hand fashion consumption.

This study contributes both theoretically and practically. Theoretically, it enables the understanding of the differences between intention and behaviour, specifically as it relates to the consumption of second-hand fashion (Carrington et al. 2010; Fukukawa 2003; Nguyen, Nguyen, & Hoang, 2019). Further, it also contributes to understanding the classification of two consumer groups and how these groups differ with respect to their behaviour. It also provides theoretical knowledge with regard to how barriers and resources, self-efficacy, and planning affect the IB gap. Practically, this study indicates how marketing managers can tailor marketing strategies for the two consumer groups, encouraging intenders to become actors. Using internal and external factors such as self-efficacy, barriers and resources, and planning to influence consumers' behaviours, marketing managers can target a group of consumers in a way that is most beneficial for them.

The chapter initially presents the nature of the IB gap in the case of second-hand clothing, after which the method and findings of the research are presented. The discussion contrasts the two groups, suggesting strategies for marketers to reach these groups, and the chapter concludes with the limitations and future research associated with the study.

THEORETICAL BACKGROUND/LITERATURE REVIEW

Diverse terms are used to describe consumer behaviour with respect to the purchase of sustainable products. Green consumption refers to the awareness of the environmental impact and focuses on reducing consumption (Dahl & Waehning-Orga, 2015). In contrast, socially conscious consumption reflects an awareness of the social effects of consumption, resulting in the support of brands (and certifications) such as Fair Trade. Ethical consumption includes the positive choices made by consumers to support organisations providing environmentally friendly products while also avoiding companies that do not meet these criteria (Szmigin, Carrigan, & McEachern, 2009). Sheth, Sethia, and Srinivas (2011) describe mindful consumption as "motivated by caring for self, community and for nature" which impacts both behaviour and attitudes (i.e. the mindset) (p. 27). While sustainability as a term has diverse meanings, sustainable fashion is related to secondary markets where no new production is involved or where an item has been used before (Cervellon, Carey, & Harms, 2012).

Though consumers indicate they are concerned about unethical behaviour, this does not always translate into behaviour or action (Bray, Johns, &

Kilburn, 2011). Described as the IB gap, it is seen in professing the desire to purchase environmentally friendly products, but it is not evident in actual purchasing behaviour. The products account for only 9.6% of grocery sales in Sweden in 2018 (Guyader et al. 2020). This raises the question as to how this gap can be bridged.

Bridging the IB Gap

The IB gap has been investigated in consumer studies in order to understand how it can be bridged. Various models have been proposed, including economic models, norm, and belief models (e.g. Schwartz and Howard's Norm-Activation Model) and mixed models (e.g. the Theory of Planned Behaviour (TPB)) (Dahl & Waehning-Orga, 2015) to understand the IB gap. Specifically, the TPB has served as the basis for developing strategies to bridge the IB gap (Carrington et al. 2010; Sniehotta et al. 2005). In the context of this model, previous research has identified the role of detailed action planning, perceived self-efficacy, and self-regulatory strategies (action control) in bridging this gap (Sniehotta et al. 2005).

Specifically, in changing health-related behaviour, use has been made of the HAPA model (Schwarzer et al. 2011), which has been used to explain changes in physical activity, diet, and obesity (Kreausukon, Gellert, Lippke, & Schwarzer, 2012). Its behavioural focus makes it suitable in this context. This model explores different constructs that can be used to bridge the IB gap, namely barriers and resources, self-efficiency (task, maintenance, and recovery), risk perception, outcome expectancies, and planning (action and coping). These are reflected in Figure 10.1 and the discussion that follows is aligned to the various components and movement from intention to action, as reflected in the model.

Intentions

Intentions are regarded as an important predictor of future behaviour (Sheeran, 2002). They reflect whether the action will be carried out (Sniehotta et al. 2005), and as a result, are also referred to as behavioural intentions (Dholakia & Bagozzi, 2003). In marketing, research tends to focus on the customer intentions as the proxy for actual behaviour (Al-Maghrabi, Dennis, & Vaux Halliday, 2011). They reflect the motivation of an individual, and their willingness to carry out the behaviour (Arvola

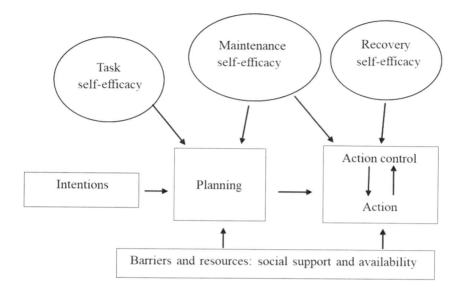

FIGURE 10.1
Bridging the IB gap. (Adapted from Schwarzer et al., 2011).

et al. 2008), being regarded as an antecedent (Ajzen, 1985). The stronger the intention, the greater the likelihood that the behaviour will be carried out as long as there is a strong correlation between intention and behaviour (Ajzen, 1991; Fishbein & Ajzen, 1975). In many instances, intentions are formulated but there may be a period of time between their formulation and subsequent behaviour (Dholakia & Bagozzi, 2003). Various antecedents of intentions have been suggested including moral norm (Arvola et al. 2008), subjective norm and perceived behavioural control (Ajzen, 1991), and perceived usefulness and perceived ease of use (Davis, 1989). It is also important that there should be a strong link between the intentions and the behaviour (Ajzen, 1985) as in some instances, behaviours could be impacted by habit rather than conscious intentions (Sheeran, 2002).

Planning

The importance of planning is that it indicates when, where, and how the desired behaviour should be performed (Kreausukon et al. 2012). In the HAPA model (Schwarzer et al. 2011), planning is divided into action planning and coping planning. Action planning (or implementation intentions) refers to the performance of the behaviour and would indicate when, where, and how the purchase (consumption) of second-hand clothes will occur, for example, at which second-hand stores and which specific types of items would be purchased (Carrington et al. 2010). Coping planning is related to how the consumer will deal with problems that will or could occur. For example, the consumer may plan to visit another second-hand store if their preferred store does not offer suitable clothing, and as part of the planning, they would have an alternative course of action. This could be that they would borrow something from a friend or re-use an existing item of clothing. This would enable them to overcome obstacles but remain on track to perform the behaviour. Action planning is affected by actual behavioural control and the situational context. For example, a consumer develops an implementation intention by saying "When I need to purchase a new pair of jeans I will go to a second-hand store and look for a pair of jeans that looks appealing to me" (Carrington et al. 2010, p. 144), which increases the likelihood of actually realising the intentions and reaching the goal (Gollwitzer & Sheeran, 2006).

Actions and Action Control

Action is the goal of the model as this is where the actual behaviour takes place. In the HAPA model (Schwarzer et al. 2011), the action is divided into two parts: action control and action. Action can be measured subjectively or objectively. In the terms of health behavioural changes, the number of steps can be tracked as an objective method and diary logs as subjective method. In the case of second-hand clothing, the actual purchases can serve as the objective measure. Action control is a concurrent self-regulatory strategy, which means that whilst the person is pursuing the behaviour, he or she is simultaneously evaluating if the current situation is in line with his or her behavioural standards.

Barriers and Resources: Social Support and Availability

In the HAPA model (Schwarzer et al. 2011), barriers and resources have the focus of social support, i.e. support to the person from family and friends. Schwarzer et al. (2011) defined social support as a resource, and the lack thereof could serve as a barrier to engaging in a behaviour. Social support can be emotional, informational, or instrumental. For those classified as intenders, it can enable the adoption of the behaviour, and for those classified as actors, it can enable continuation of the behaviour (Schwarzer et al. 2011). Instead of only looking at social support, factors such as the availability of second-hand stores, availability of style and size, and if friends and family purchase second-hand will also be explored and how it affects the IB gap as previous research indicates it affects purchasing of sustainable products (Tarkiainen & Sundqvist, 2005). A resource for one consumer could be a barrier for another, and through a qualitative study, there is an opportunity to explore what each consumer thinks and feels. According to Diddi et al. (2019), the perceived lack of variety and style was one factor resulting in the consumer not engaging in sustainable clothing consumption, supporting the inclusion of barriers and resources into this study in bridging the IB gap.

Self-Efficacy

Various types of self-efficacy can be identified, including task, maintenance, and recovery self-efficacy, all impacting the IB gap.

- **Task Self-Efficacy**

  Task self-efficacy, also called motivational self-efficacy, is related to the phase of goal setting, affecting both intentions and planning. Task self-efficacy is about whether or not a person believes that he/she can perform the task, even if the task is perceived to be difficult.

  For example, high-task self-efficacy affecting planning could be that the consumer feels like he or she can purchase second-hand clothes even if it will require more effort compared to consuming "regular" (new) clothes. Low-task self-efficacy could be that the consumer does not think that he or she is able to purchase second-hand clothes since it requires more effort and therefore decreasing the chances of consuming second-hand.

  The importance of self-efficacy and how this knowledge affects consumers have been shown previously (Garlin & McGuiggan, 2002). Self-efficacy in general has been proven to have a large impact on the performance of different tasks, no matter if it is task self-efficacy, maintenance self-efficacy, or recovery self-efficacy (Bandura, 1991).

- **Maintenance Self-Efficacy**
  According to the HAPA model (Schwarzer et al. 2011), maintenance self-efficacy is one of the two parts of volitional self-efficacy. It affects planning and action through the person's belief that they will be able to proceed with carrying out the behaviour, even if there is reduced likelihood of success. Maintenance self-efficacy is about how a person perceives his/her own ability to continue implementing the behaviour.

  In this study, maintenance self-efficacy is a construct that affects the planning and action. Here, the customers perceive they can purchase second-hand, even though this may take additional effort. This could include changing the mindset when looking for clothes, or even if the consumer has to visit several stores to find what he or she is looking for, i.e. even if they experience setbacks or failures. In contrast to task self-efficacy, which is more focused on the consumers' perception to start consuming second-hand, maintenance self-efficacy is about the consumer perception of being able to make this a routine when he or she starts consuming second-hand clothes, i.e. maintaining the behaviour into the future.

- **Recovery Self-Efficacy**
  Recovery self-efficacy is the second of the two parts of volitional self-efficacy (Schwarzer et al. 2011). Recovery self-efficacy is about the consumers' perception of being able to proceed to purchase second-hand even if there are times when he or she has a relapse, i.e. purchase new fashion. Even if the consumer would end up in situations where they would rather purchase from a regular store, they would attempt to purchase second-hand in the future. It is about the consumer's belief in themselves to take a setback or failure and keep on going towards the goal of purchasing second-hand. A setback in terms of purchasing second-hand would mean that the consumer encounters some kind of problem, for example, there are no items in the right size, but there is a solution that still enables second-hand purchasing, which is the intended behaviour. A failure in terms of purchasing second-hand would mean that the consumer encounters a problem of some kind that results in a purchase of newly produced fashion, which is not the intended behaviour.

Contrasting Intenders and Actors as Consumer Groups

Individuals can be divided into two groups, namely intenders and actors (Schwarzer et al. 2011). Intenders are inactive consumers who intend to purchase second-hand fashion, but do not act on it and thus do not regard it as part of their regular routine. By contrast, active consumers purchase second-hand fashion, thereby acting on their intentions. Actors purchase second-hand on a regular basis as part of their shopping routine. Actors develop strategies that prepare them for setbacks and adapt to internal and external factors influencing the situation, thus maintaining the behaviour (Schwarzer et al. 2011). Understanding these two consumer groups and their use of the various factors can assist in understanding how to bridge the IB gap.

METHODOLOGY

This chapter employs a qualitative research methodology, which enables the researchers to explore the variables identified in the model. This research strategy was deemed appropriate in order to gain additional knowledge concerning a situation and describing the field of study where little previous research has been conducted, but where there are elements worth discovering (Stebbins, 2001).

The empirical data was collected through in-depth interviews using an interview guide which was developed using the model (Figure 10.1), for example, on the barriers and resources that each participant identified. Interviews were held with Swedish female consumers in order to gain deeper insights in their behaviour (Malhotra, Birks, & Wills, 2012). Two categories of consumers were interviewed including intenders (ten participants) and actors (nine participants), attaining saturation (Denscombe, 2010). Intenders were required to have the intention to engage in sustainable consumption (but do not) while actors also engaged in the behaviour. The average duration of the interviews was 29 minutes. Use was made of a judgement sampling. The initial generation of codes was done separately, where each researcher familiarised with the data gathered individually. This followed by searching and reviewing the individual codes together to define themes and create a mutual understanding. Table 10.1 shows the coding used in the study.

TABLE 10.1

Coding Sheet Used in the Study

Model Factors	Detail Aspects Identified in the Analysis
Intentions	Mindset towards sustainability
	Implementation intentions
Actions (behaviour)	Shopping behaviour
Barriers and resources	Skills and knowledge
	Availability
	Influence of family and friends
Planning	Action planning
	Coping planning
Self-efficiency	Task self-efficacy, i.e. belief in the ability to purchase this way
	Maintenance self-efficacy, i.e. continuing in this action
	Recovery self-efficiency, i.e. recovering from setbacks

Trustworthiness is argued to be evidence in qualitative studies (Guba, 1981) by assessing the credibility, transferability, dependability, and confirmability of the study. The interviews were recorded and transcribed to increase the trustworthiness of the study.

FINDINGS

The findings examine the various factors that can bridge the IB gap, as shown in Figure 10.1, using these to contrast the two consumer groups, namely intenders and actors. Illustrative quotes are also provided to support the perspectives presented in each of the sections.

Intenders

Their Intention and (In)Action

Intenders purchase more new products rather than second-hand fashion. One of the intenders stated that: "No, it's not really a big part of my purchases, I would say max 10% of my clothes are second-hand" (P5). For this consumer group, the main reasons for purchasing second-hand are the lower prices of second-hand rather than environmental considerations. For this group, second-hand fashion was not viewed as a priority as they believed that other types of sustainable consumption were more important, e.g. ecological groceries, and were sceptical of the environmental effects of purchasing second-hand fashion. This is seen in the following quote:

> I want to care about the environment, but I think I'm a little too lazy and a little too ignorant. Besides, I am a little inquiring about how it affects us. Many people use these things with the environment in their marketing just because. So, I'm a little sceptical as to how good it really is. (P1)

Most of these consumers shopped online as it was perceived as easier, more relaxing while also offering better products. As they purchased online, they shopped for fashion alone. They also believed that purchasing second-hand clothes required additional skills. This is summarised in the following quote: "I usually shop online because I find it very nice to sit and look around online, it is relaxing" (P1).

Planning

Intenders mostly do not plan their purchases at all, acting spontaneously and having coping plans for specific items. If specific items are not available in a second-hand store, intenders would either go to a regular store or not purchase the item. They tend to deviate from their initial intentions to purchase second-hand and they choose to purchase newly produced fashion when they encounter problems.

Barriers and Resources

Intenders are surrounded by family and friends with both positive and negative attitudes towards second-hand and sustainable fashion consumption, suggesting both positive and negative influence on purchasing behaviour as seen in the following quotes.

> I would say that my family is a bit anti second-hand actually. Especially my mother I would say, she is very negative about the fact that other people have worn the clothes and so on. (P13)
> My family has always been good at purchasing everything on Blocket.se […] they don't want people to always purchase everything new. (P9)

Many intenders indicated a lack of knowledge of second-hand clothing, claiming that the availability and the range of products could be improved. They indicated that they would be more motivated to purchase if the stores were more similar to regular stores in style, perceiving second-hand stores to be "old fashioned" but that they had previously been able to find suitable products. This is seen in the following quotes:

> If you visit the Red Cross or similar stores, they are not so attractive. (P5)
>
> If you visit a store and your size is never available, you would think "Whatever, I never find anything in here", and it's no reason to look at second-hand. (P13)

Task Self-Efficacy

Intenders are required to give up more time to engage in sustainable fashion consumption as it required additional search while also compromising on their convenience and old routines. Intenders tend to be comfortable with their old routines and think that it is time consuming to purchase second-hand compared to newly produced fashion.

> It's not always available in your size, most often it's just one piece of each item, it's not similar to a regular store where there are several styles and sizes. (P13)

Maintenance Self-Efficacy

Intenders did not believe they could maintain the behaviour of purchasing second-hand fashion regularly, nor were they willing to change their regular routine or spend more time to maintain this behaviour. Intenders would have to put in more effort, go through a more complicated process, and spend more time purchasing second-hand fashion, negatively affecting purchasing behaviour. When asked if they would be able to purchase all their fashion items at second-hand stores, they said they would not, citing laziness, lack of time, and pre-worn items which could not be washed as reasons for not purchasing.

> I would prefer to purchase things that I can wash- I can't was a purse and shoes are something you use until they are shaped after your feet. (P1)

TABLE 10.2

Summary of Intenders

Model Factors	Intenders
Intentions	Intention to purchase second-hand
Actions	Do not purchase second-hand on a regular basis
Sustainability awareness	Aware when exposed to marketing
Mindset	Sceptical of positive effects of second-hand
Knowledge and skills	Believe that it requires skills they do not have
Social support	Surrounded by a few people who purchase second-hand, both positive and negative
Availability	Lack of knowledge and availability could be better
Task self-efficacy	More comfortable with old routines, time-consuming to purchase second-hand
Maintenance self-efficacy	Does not believe they can maintain the behaviour since they are not willing to spend extra time or effort
Recovery self-efficacy	Can purchase second-hand after a purchase of new production
Action planning	Does not plan future purchases
Coping planning	Does not plan how to handle problem
Implementation intentions	Does not plan how to act in different situations

Recovery Self-Efficacy

Intenders stated that they would recover from a setback or failure, and were very certain about their ability to succeed, as seen in the following quote: "I would not have any problems with going back and forth between second-hand and new clothes" (P3). They would not see the setback as a complete failure and change their intentions because of it.

Table 10.2 reflects a summary of the findings of intenders with regard to these factors.

Actors

Intention and Action

Actors purchase second-hand clothing on a regular basis, yet they do not exclude purchasing new products when necessary. Actors indicated that 40%–60% of their clothing was purchased from second-hand sources, seeking to purchase second-hand as often as possible. The following quotes summarise these views:

> I've recently been trying to find second-hand clothes instead of checking in stores. So, maybe 60%. (P14)

Actors indicated that this behaviour is due to their environmental concern and the lower price of these items. Additionally, they appreciated the ability to find unique pieces of clothing.

The participants claimed that they have the intention to purchase fashion that have a long lifespan and try to avoid fast fashion to a greater extent. Some of the participants saw second-hand consumptions as a part of their lifestyle, stating "it feels like I can make a big difference" (P14). In addition to that, actors prioritised second-hand fashion in terms of sustainable consumption and believed that it was very important and that they could make a difference by excluding new fashion items.

> I think more about the environment than economics when I choose to buy second-hand, but I see the opportunity to be able to kill two birds with one stone. (P11)

Actors did their shopping online and offline, preferring online due to availability and ease of search. They believed that second-hand fashion requires both skills and patience which are learnt. "The more second-hand you buy, the more you learn too [...] you probably have to spend a little more time before it becomes a routine." (P18)

Planning

The majority of actors plan their shopping and prepare their second-hand purchases by making mental shopping list, search and monitoring online and offline outlets. The main factor that defined their planning was several online and offline options but the attitude of being flexible with options and thinking outside the box. This is reflected in the following quotes:

> I make a mental list of what I want or need. Then I plan where to look. (P10)
> I think that I might not be too set on how it must be before. You have to be a little flexible when it comes to second-hand. (P14)

Further, actors make detailed plans for possible problems that might occur when shopping for second-hand fashion. In the scenario presented to the participants, some of the actors developed back-up plans with several alternatives of stores, models, and sizes. Others stated that they decided to wait to purchase fashion if something interfered with their original plan or they decided not to purchase fashion at that time. Hence, many planned to continue the search for second-hand fashion in the future instead, also using coping planning when thinking about possible setbacks, as seen in the following quotes:

> Then I look for it in another store. I have a few options. I would probably say that the process of looking for second-hand before going to new production is long, at least six months. (P2)
>
> Then I do not buy it. I would not look for another model, I know what I want. (P2)

Barriers and Resources

Actors were surrounded by family and friends with positive attitudes towards second-hand and sustainable fashion consumption while some believed they were not affected by other people's opinions at all, as reflected in the following comment: "I do not know if I am affected by it, I feel like I am quite unconcerned about the opinions of others" (P7).

Store options were perceived to good online and offline, depending on the location. Participants commented that the quality has improved recently, reflecting a range of styles and options, including items that have never been worn. For actors, second-hand fashion shopping is viewed as a hobby, as reflected in the following quote: "I think the charm of second-hand is that you have to search a little" (P11).

Task Self-Efficacy

Actors believed that it takes more time to engage in second-hand shopping than new fashion, but claim that it did not involve any sacrifice, with it being fun to search. Some actors indicated that they wanted to spend more time on searching for second-hand fashion.

> … I like to spend time on it, I think it's fun. So, I would not say that I "give up something". It takes time, but I want to spend that time, it's an interest for me. (P18)

Maintenance Self-Efficacy

Actors believed they are capable of purchasing second-hand fashion on a regular basis even if they must change their regular routine or spend more time, and they were willing to change their routine to purchase more second-hand clothing. While it was more time-consuming to engage in second-hand shopping, they indicated that they "feel that I have that time" (P7).

While actors purchase second-hand clothing, they also purchase some new items of clothing due to time pressure as well as the desire to purchase some "new" clothes, suggesting there is a limit to what clothing can be purchased as second-hand.

Recovery Self-Efficacy

Actors were very confident that they could continue purchasing second-hand fashion, even if they purchase newly produced fashion sometimes or if what they are looking for was not available at the time. As stated by one actor, "yes, I am certain actually. I don't see that as a problem. I care too much, I think, to quit for setbacks" (P14).

These findings related to these factors are summarised in Table 10.3.

TABLE 10.3

Summary of Actors

Model Factors	Actors
Intentions	Intention to purchase second-hand
Actions	Purchase second-hand on a regular basis
Sustainability awareness	Very aware of sustainability issues
Mindset	View sustainability as a lifestyle
Knowledge and skills	They believe they have the necessary skills to purchase
Social support	Surrounded by family and friends who purchase second-hand, with a positive influence on behaviour
Availability	High levels of knowledge, satisfied with availability
Task self-efficacy	Driven by interest, enjoy spending the extra time
Maintenance self-efficacy	Believe they can maintain the behaviour as they are willing to invest the extra time and effort
Recovery self-efficacy	Can revert to second-hand after the purchase of new items
Action planning	Plan for future purchases
Coping planning	Plan to handle problems
Implementation intentions	Plan how to act in different situations

DISCUSSION

Using the factors identified in Figure 10.1, it is necessary to use the model factors to contrast intenders and actors, thereby contributing to answering the research purpose. These are important as they impact the actions of customers and subsequently marketers in impacting behaviour. The summary of the main findings is found in Table 10.4.

Regarding intentions, intenders were found to have the willingness to engage but did so infrequently or did not regard it as part of their routine (i.e. less than 20% of their clothing purchases), resulting in this classification. Schwarzer et al. (2011) indicated that intenders and actors are "characterized by different psychological states" (p. 163), which was identified in this study. For example, intenders reflected that they were poorly informed on the environmental effects of clothes purchasing and that they did not have sustainable consumption at the forefront of their decision-making. By contrast, actors have the intentions and willingness to purchase second-hand fashion, acting on this intention, which was common to this group. Actor prioritised sustainable consumption and did not view it as an inconvenience and that they had the necessary skills.

Action and coping planning with implementation intentions also differ between the two customer groups. Intenders are spontaneous shoppers, exhibiting little planning. Further, problems associated with purchasing are dealt with when they arise. When faced with problems when trying to purchase second-hand, they always went back to purchasing newly produced fashion. Thus, intenders are non-planners with weak implementation intentions, suggesting a lack of action and coping planning

TABLE 10.4

Comparing Intenders and Actors

	Intender	Actors (Doers)
Intentions	Yes	Yes
Barriers and resources	Weak	Strong
Planning	No	Yes
Task self-efficacy	Low	High
Maintenance self-efficacy	Low	Medium-high
Recovery self-efficacy	High	High
Behaviour	No	Yes

(Schwarzer et al. 2011), which impacts the planning to engage in the behaviour (Carrington et al. 2010; Gollwitzer & Sheeran, 2006). The lack of implementation planning and how to deal with possible problems can contribute to the IB gap and its effect on action.

Actors develop plans for future purchases, planning to avoid setbacks or failures. In the case of second-hand clothes, actors said they had back-up plans if they would encounter a problem. They also indicated they have various solutions rather than purchasing new products. This indicates that actors are both action and coping planners, with strong implementation intentions, affecting the action construct (Schwarzer et al. 2011). Lack of implementation intentions contributes to the IB gap (Carrington et al. 2010) as it is necessary to reach the goal of the intentions (Gollwitzer & Sheeran, 2006).

Social support systems and availability also distinguish intenders and actors. Intender's behaviour is influenced positively but mostly negatively by family and friends, with the lack of social support negatively impacting their purchase behaviour. This finding indicates a weak social support system, thus serving as a barrier, impacting both planning and action, as identified in previous research (Schwarzer et al. 2011). Intenders also described the poor availability of second-hand items with few stores. This lack of availability serves as a barrier, as suggested in previous research (Diddi et al. 2019; Gleim, Smith, Andrews, & Cronin, 2013; Nguyen et al. 2019). Consequently, they were not willing to invest the additional time and effort to purchase second-hand, impacting the benefits received (Ottman, Stafford, & Hartman, 2006; Peattie, 2001). In contrast, family and friends positively influence actor's behaviour with them also having positive attitudes and engaging in the behaviour themselves, reflecting strong social support (Schwarzer et al. 2011), though for some actors, they exerted no influence as actors were unconcerned by others' opinions. The presence of social support positively influences socially responsible consumer behaviour (Ali & Mandurah, 2016), thereby helping actor plan and perform the behaviour as it can be experienced as a resource in overcoming the IB gap. Actors were satisfied with the range of options and the quality offered in second-hand stores, which can help to bridge the IB gap (Nguyen et al. 2019) due to its impact on planning and action (Schwarzer et al. 2011).

Task self-efficacy differences were also identified between the two consumer groups. Intenders were not willing to sacrifice time or old routines or be inconvenienced by a change in behaviour. Their task self-efficacy was based on their previous experience which can be regarded as low. As the level of knowledge, both of the product and the purchasing process, impacts their perceived self-efficacy (Garlin & McGuiggan, 2002), the low task self-efficacy influences planning, specifically action planning, which may contribute to the IB gap (Schwarzer et al. 2011). In contrast, actors are certain that they are able to engage in the behaviour without having to make sacrifices to do so, even willing to spend more time to make these purchases, indicating high-task self-efficacy (Garlin & McGuiggan, 2002). According to Bandura and Wood (1989), those with high levels of self-efficacy are more likely to visualise scenarios that can guide them through potential problems, thereby affecting planning (Schwarzer et al. 2011), bridging the IB gap.

Maintenance self-efficacy also differed between the two groups, with intenders indicating they do not believe in their ability to maintain the desired behaviour regularly as they are unwilling to give up newly produced fashion, associating a change of behaviour with additional effort and time, suggesting low maintenance self-efficacy. As they do not implement the behaviour, they have not practiced the behaviour, leading to lower levels of self-efficacy (Locke, Frederick, Lee, & Bobko, 1984), impacting planning and action (Schwarzer et al. 2011), contributing to the IB gap as suggested in previous research (Bandura & Wood, 1989; Locke et al. 1984; Schwarzer et al. 2011).

The majority of actors believed they would be able to maintain the behaviour of purchasing only second-hand on a regular basis as they indicated they would be willing to change their routine, including investing additional time, to implement this behaviour. Their previous experience purchasing second-hand products would increase their level of maintenance self-efficacy. Further, strong beliefs of a person in their own capability will increase their efforts in planning (Bandura, 1991), and this is relevant to second-hand clothes and dealing with possible problems associated with this action and closing the IB gap due to its affecting both planning and action (Schwarzer et al. 2011).

Recovery self-efficacy strategy differences between the groups were also identified with intenders indicating they will continue with their current behaviour, purchasing the same quantity of second-hand clothing. They were certain they would be able to recover from a failure associated with any attempt but would not purchase second-hand clothing regularly. According to Schwarzer et al. (2011), recovery self-efficacy exists if they believe they will be able to continue engaging in the behaviour even if they engage in the opposite behaviour, in this instance, purchasing new clothes. Intenders said that the amount of newly produced fashion they purchased did not affect their intentions of purchasing second-hand from time to time, suggesting a high level of recovery self-efficacy, affecting action (Schwarzer et al. 2011). While intenders do not engage in purchasing second-hand on regular basis, they would need recovery self-efficacy to continue keeping engaged in the behaviour due to the effect of self-efficacy on performance (Bandura & Wood, 1989). Actors are both certain and willing to continue to purchase second-hand even when they encounter setbacks or failures, suggesting a high level of recovery self-efficacy. This may be a contributing factor to their success in closing the IB gap through its possible effect on action (Schwarzer et al. 2011).

Theoretical Contribution

This study provides theoretical contributions to the field of sustainable fashion consumption in various ways. First, it contributes to the understanding of how the Swedish female consumers behave when they have the intentions to purchase second-hand fashion and the factors specifically applicable in this culture. While Sweden is acknowledged as having high levels of environmental awareness, this awareness is not always evident in the actions in the market. Second, the study enables the classification of two consumer segments and how behavioural characteristics differ for intenders and actors. By focusing on the behaviour of customers in these two groups, it is possible to identify distinct market segments. This enables the development of strategies that can be implemented to encourage intenders to become actors, i.e. implementing sustainable behaviours. Third, this study provides theoretical knowledge with regard to how barriers and resources, self-efficacy, and planning affect the intention–behaviour gap. The factors identified in previous models impacting the IB gap were identified and examined in a new context. This study shows that the HAPA model (Schwarzer et al. 2011), the green consumption behaviour-testing model (Nguyen et al. 2019), and the intention–behaviour mediation and moderation model of the ethically minded consumer (Carrington et al. 2010) can be used to understand these factors in the second-hand clothing consumption, both of which are germane to this context. The identification of the factors and how they differ in the two consumer groups is important as these groups require different strategies, suggesting the targeting of specific behavioural change among intenders. For example, developing specific planning strategies can enable intenders to become actors.

Practical Application

The factors in Table 10.4 provide practical contribution of the study, specifically how marketers can target these two groups. Intenders would benefit from developing stronger implementation intentions, planning for future purchases and possible problems, receiving social support from family and friends, learning more about the availability of second-hand fashion and increased self-efficacy, as suggested by Schwarzer et al. (2011). Intenders benefit more from planning the behaviour and translating their intentions into action rather than getting a message of outcome expectancies. Thus, one can assume that the consumer group of intenders' benefits more from marketing interventions that ease the planning process than getting a marketing message that informs about the effectiveness and positive outcome of the behaviour. For actors, it is more important to maintain the current behaviour and avoid setbacks, i.e. continue purchasing second-hand fashion and avoid purchasing newly produced fashion. Actors benefit from marketing interventions that stabilise and maintain their current behaviour. Based on this, marketing managers need to take the behavioural characteristics of each group into consideration when designing marketing intervention and adapting it for each behaviour.

To bring about behavioural change, it is necessary to select the right tools (McKenzie-Mohr & Schultz, 2014; Mirosa, Liu, & Mirosa, 2018), and in the consumption of second-hand fashion, it is important to design the marketing to ease the planning process for intenders. Sustainable behaviour is usually viewed as more difficult than "regular behaviour" and consumers believe it takes more effort to engage in sustainable consumption (McKenzie-Mohr & Schultz, 2014). For instance, intenders believed that the long search process makes it harder to find second-hand clothing, requiring certain skills. If it is more difficult than purchasing newly produced fashion, it is important to make it more convenient for the consumer. Therefore, convenience is an effective tool that can be used in marketing to ease the process for consumers (Hiller Connell, 2009). By making the search process more convenient, intenders will not experience the same difficulties finding what they are searching for and may be more satisfied with the availability once they learned how to search. This can also make it easier for intenders to develop implementation intentions and plan for future purchases and possible problems. Goal setting as a strategy on its own is usually not effective for behavioural change; however, if a goal is associated with implementation intentions, it is more likely to be obtained (Morwitz, Steckel, & Gupta, 2007). These tools can be used in marketing by asking the consumers when and how they are planning on carrying out the behaviour. For instance, if a consumer shows interest in engaging in a behaviour, marketing managers should ask when they plan to engage in that action (McKenzie-Mohr & Schultz, 2014). Since intenders show an interest in second-hand fashion and have intentions to buy, goal setting can be used to strengthen intender's self-efficacy and implementation intentions. Social norms are a tool that can be used to show consumers that a behaviour is accepted within a larger group (Mirosa et al. 2018). Since intenders experience that there is a stigma of purchasing second-hand fashion based on the opinions from family and friends, it can be important to show that the behaviour is accepted by others. For actors, it is more important to focus the marketing on maintaining the current behaviour and helping the consumer recover from setbacks.

According to McKenzie-Mohr and Schultz (2014), feedback is most effective on consumers who want to achieve the outcome, e.g. reduce emission. When the consumer is provided with information about their consumption and the importance of it, it can work as a motivator to continue to engage in the positive behaviour. Social diffusion can also be suggested. Friends who already engage in the behaviour and have told their social network about it (Mirosa et al. 2018), e.g. consumption of second-hand fashion is carried out and promoted by friends and family and have a positive influence on the consumers. This tool can be utilised by encouraging consumers to display their behaviour to other consumers, e.g. wearing a statement bag indicating that you purchased it second-hand.

CONCLUSION, LIMITATIONS, AND FUTURE RESEARCH

Several limitations can be associated with this study. As with any qualitative study, this study is characterised by a small, judgement sample of consumers with the classification of consumers as intenders or actors by the researchers. Further, all the participants in this study stated that they wore standard sizes and therefore did not experience any difficulties finding the clothes in appropriate sizes, which could have played a role in the intentions, actions, and other factors. Due to the COVID-19 pandemic, all data was collected by telephone, which may have impacted the responses received from the participants.

Numerous future research opportunities can be linked to this research. The study has focused on female consumers due to their behaviour related to this product category but expanding this to men is suggested. Investigating these factors among consumers in other cultures is also recommended. Bridging the IB gap and strategies to bridge this gap is recommended. Testing the model quantitatively both in terms of second-hand clothing as well as other industries is recommended.

This study investigated the IB gap in second-hand consumption by identifying the factors impacting the identification of two consumer groups. Intenders are characterised by sustainable intentions that do not convert into behaviour, with their behaviour characterised by a weak social support system (barrier), poor availability (barrier), low task- and maintenance self-efficacy, high recovery self-efficacy, and no planning. In contrast, the behaviour of actors can be characterised by sustainable intentions and sustainable behaviour as well as a strong social support system, good availability, high task- and recovery self-efficacy, medium to high maintenance self-efficacy, and planning.

CREDIT AUTHOR STATEMENT

Elin Pedersén: Conceptualisation, Methodology, Investigation, Original draft preparation; Writing – Review & Editing; Amanda Persson: Conceptualisation, Methodology, Investigation, Original draft preparation; Writing – Review & Editing; Adele Berndt: Supervision, Original draft preparation; Writing, Reviewing and Editing.

REFERENCES

Adam, A.-F., & Fayolle, A. (2016). Can implementation intention help to bridge the intention–behaviour gap in the entrepreneurial process? An experimental approach. *The International Journal of Entrepreneurship and Innovation*, *17*(2), 80–88.

Ajzen, I. (1985). From intentions to actions: A theory of planned behavior. In J. Kuhl & J. Beckmann (Eds.), *Action Control: From Cognition to Behavior* (pp. 11–39). Heidelberg, Germany: Springer.

Ajzen, I. (1991). The theory of planned behavior. *Organizational Behavior and Human Decision Processes, 50*(2), 179–211.

Ali, I., & Mandurah, S. (2016). The role of personal values and perceived social support in developing socially responsible consumer behavior. *Asian Social Science, 12*(10), 180–189.

Allwood, J., Laursen, S. E., Russell, S., de Rodriguez, C. M., & Bocken, N. (2008). An approach to scenario analysis of the sustainability of an industrial sector applied to clothing and textiles in the UK. *Journal of Cleaner Production, 16*(12), 1234–1246.

Al-Maghrabi, T., Dennis, C., & Vaux Halliday, S. (2011). Antecedents of continuance intentions towards e-shopping: The case of Saudi Arabia. *Journal of Enterprise Information Management, 24*(1), 85–111.

Arvola, A., Vassallo, M., Dean, M., Lampila, P., Saba, A., Lähteenmäki, L., & Shepherd, R. (2008). Predicting intentions to purchase organic food: The role of affective and moral attitudes in the theory of planned behaviour. *Appetite, 50*(2-3), 443–454.

Auger, P., & Devinney, T. M. (2007). Do what consumers say matter? The misalignment of preferences with unconstrained ethical intentions. *Journal of Business Ethics, 76*(4), 361–383.

Bandura, A. (1991). Social cognitive theory of self-regulation. *Organizational Behavior and Human Decision Processes, 50*(2), 248–287.

Bandura, A., & Wood, R. (1989). Effect of perceived controllability and performance standards on self-regulation of complex decision making. *Journal of Personality and Social Psychology, 56*(5), 805–814.

Barz, M., Lange, D., Parschau, L., Lonsdale, C., Knoll, N., & Schwarzer, R. (2016). Self-efficacy, planning, and preparatory behaviours as joint predictors of physical activity: A conditional process analysis. *Psychology & Health, 31*(1), 65–78.

Bray, J., Johns, N., & Kilburn, D. (2011). An exploratory study into the factors impeding ethical consumption. *Journal of Business Ethics, 98*(4), 597–608.

Carrington, M. J., Neville, B. A., & Whitwell, G. J. (2010). Why ethical consumers don't walk their talk: Towards a framework for understanding the gap between the ethical purchase intentions and actual buying behaviour of ethically minded consumers. *Journal of Business Ethics, 97*(1), 139–158.

Cervellon, M.-C., Carey, L., & Harms, T. (2012). Something old, something used: Determinants of women's purchase of vintage fashion vs second-hand fashion. *International Journal of Retail & Distribution Management, 40*(12), 956–974.

Dahl, S., & Waehning-Orga, N. (2015). Ethical consumption. In L. Eagle & S. Dahl (Eds.), *Marketing Ethics & Society*. London: SAGE.

Davis, F. D. (1989). Perceived usefulness, perceived ease of use, and user acceptance of information technology. *MIS Quarterly, 13*(3), 319–340.

De Graaf, J., Wann, D., & Naylor, T. H. (2014). *Affluenza: How overconsumption is killing us–and how to fight back*. San Francisco, USA: Berrett-Koehler Publishers.

Denscombe, M. (2010). *Good research guide: For small-scale social research projects* (4th ed.). Maidenhead, UK: Open University Press.

Dholakia, U. M., & Bagozzi, R. P. (2003). As time goes by: How goal and implementation intentions influence enactment of short-fuse behaviors 1. *Journal of Applied Social Psychology, 33*(5), 889–922.

Diddi, S., Yan, R.-N., Bloodhart, B., Bajtelsmit, V., & McShane, K. (2019). Exploring young adult consumers' sustainable clothing consumption intention-behavior gap: A behavioral reasoning theory perspective. *Sustainable Production and Consumption, 18*, 200–209.

Ek Styvén, M., & Mariani, M. M. (2020). Understanding the intention to buy secondhand clothing on sharing economy platforms: The influence of sustainability, distance from the consumption system, and economic motivations. *Psychology & Marketing, 37*(5), 724–739.

Fishbein, M. E., & Ajzen, I. (1975). *Belief, attitude, intention and behavior: An introduction to theory and research*. Reading, MA: Addison-Wesley.

Fueyo-Díaz, R., Magallón-Botaya, R., Gascón-Santos, S., Asensio-Martínez, Á., Palacios-Navarro, G., & Sebastián-Domingo, J. J. (2018). Development and validation of a specific self-efficacy scale in adherence to a gluten-free diet. *Frontiers in Psychology, 9*, 1–7.

Fukukawa, K. (2003). A theoretical review of business and consumer ethics research: Normative and descriptive approaches. *The Marketing Review, 3*(4), 381–401.

Garlin, F. V., & McGuiggan, R. L. (2002). *Exploring the sources of self-efficacy in consumer behaviour*. Paper presented at the 2002 Asia-Pacific ACR Conference, Beijing, China.

Gleim, M. R., Smith, J. S., Andrews, D., & Cronin, J. J. (2013). Against the green: A multi-method examination of the barriers to green consumption. *Journal of Retailing, 89*(1), 44–61.

Gollwitzer, P. M., & Sheeran, P. (2006). Implementation intentions and goal achievement: A meta-analysis of effects and processes. *Advances in Experimental Social Psychology, 38*, 69–119.

Grimmer, M., & Miles, M. P. (2017). With the best of intentions: A large sample test of the intention-behaviour gap in pro-environmental consumer behaviour. *International Journal of Consumer Studies, 41*(1), 2–10.

Guba, E. G. (1981). Criteria for assessing the trustworthiness of naturalistic inquiries. *Educational Technology Research and Development, 29*(2), 75–91.

Guyader, H., Ottosson, M., & Parment, A. (2020). *Marketing and sustainability: Why and how sustainability is changing current marketing practices*. Lund, Sweden: Studentlitteratur AB.

Håkansson, A. (2014). What is overconsumption?–A step towards a common understanding. *International Journal of Consumer Studies, 38*(6), 692–700.

Hiller Connell, K. Y. (2009). *Exploration of second-hand apparel acquisition behaviors and barriers*. Paper presented at the ITAA 2009 Proceedings# 66, Bellevue, Washington.

Humphery, K. (2010). *Excess: Anti-consumerism in the West*. Cambridge, UK: Polity.

Kreausukon, P., Gellert, P., Lippke, S., & Schwarzer, R. (2012). Planning and self-efficacy can increase fruit and vegetable consumption: A randomized controlled trial. *Journal of Behavioral Medicine, 35*(4), 443–451.

Laitala, K., Klepp, I. G., & Boks, C. (2012). Changing laundry habits in Norway. *International Journal of Consumer Studies, 36*(2), 228–237.

Locke, E. A., Frederick, E., Lee, C., & Bobko, P. (1984). Effect of self-efficacy, goals, and task strategies on task performance. *Journal of Applied Psychology, 69*(2), 241–274.

Malhotra, N. K., Birks, D. F., & Wills, P. (2012). *Marketing research: An applied approach* (4th ed.). Harlow, UK: Pearson.

McKenzie-Mohr, D., & Schultz, P. W. (2014). Choosing effective behavior change tools. *Social Marketing Quarterly, 20*(1), 35–46.

McNeill, L., & Moore, R. (2015). Sustainable fashion consumption and the fast fashion conundrum: Fashionable consumers and attitudes to sustainability in clothing choice. *International Journal of Consumer Studies, 39*(3), 212–222.

Mirosa, M., Liu, Y., & Mirosa, R. (2018). Consumers' behaviors and attitudes toward doggy bags: Identifying barriers and benefits to promoting behavior change. *Journal of Food Products Marketing, 24*(5), 563–590.

Morwitz, V. G., Steckel, J. H., & Gupta, A. (2007). When do purchase intentions predict sales? *International Journal of Forecasting, 23*(3), 347–364.

Nguyen, H. V., Nguyen, C. H., & Hoang, T. T. B. (2019). Green consumption: Closing the intention-behavior gap. *Sustainable Development, 27*(1), 118–129.

Ottman, J. A., Stafford, E. R., & Hartman, C. L. (2006). Avoiding green marketing myopia: Ways to improve consumer appeal for environmentally preferable products. *Environment: Science and Policy for Sustainable Development, 48*(5), 22–36.

Peattie, K. (2001). Golden goose or wild goose? The hunt for the green consumer. *Business Strategy and the Environment, 10*(4), 187–199.

Pookulangara, S., & Shephard, A. (2013). Slow fashion movement: Understanding consumer perceptions—An exploratory study. *Journal of Retailing and Consumer Services, 20*(2), 200–206.

Roos, J. M. (2019). *Konsumtionsrapporten 2019*. Retrievable from: http://hdl.handle.net/2077/62834

Schlaegel, C., & Koenig, M. (2014). Determinants of entrepreneurial intent: A meta-analytic test and integration of competing models. *Entrepreneurship Theory and Practice, 38*(2), 291–332.

Schwarzer, R., Lippke, S., & Luszczynska, A. (2011). Mechanisms of health behavior change in persons with chronic illness or disability: The Health Action Process Approach (HAPA). *Rehabilitation Psychology, 56*(3), 161–170.

Sheeran, P. (2002). Intention-behavior relations: A conceptual and empirical review. *European Review of Social Psychology: European Review of Social Psychology, 12*(1), 1–36.

Sheth, J. N., Sethia, N. K., & Srinivas, S. (2011). Mindful consumption: A customer-centric approach to sustainability. *Journal of the Academy of Marketing Science, 39*(1), 21–39.

Sniehotta, F. F., Scholz, U., & Schwarzer, R. (2005). Bridging the intention–behaviour gap: Planning, self-efficacy, and action control in the adoption and maintenance of physical exercise. *Psychology & Health, 20*(2), 143–160.

Stebbins, R. A. (2001). *Exploratory research in the social sciences* (Vol. 48). Thousand Oaks, CA: SAGE.

Szmigin, I., Carrigan, M., & McEachern, M. G. (2009). The conscious consumer: Taking a flexible approach to ethical behaviour. *International Journal of Consumer Studies, 33*(2), 224–231.

Tarkiainen, A., & Sundqvist, S. (2005). Subjective norms, attitudes and intentions of Finnish consumers in buying organic food. *British Food Journal, 107*(11), 808–822.

Section IV

Advances in Knowledge of Social and Sustainable Marketing: Understanding Sharing Economy and Marketing

11

Sustainability through Sharing Farm Equipment: A Research Agenda

Priyanka Sharma
Marketing area, Indian Institute of Management (IIM), Lucknow, India

CONTENTS

Learning Objectives 279
Themes and Tools Used 280
Theoretical Background 280
Introduction 282
Literature Review 285
Methodology: In-Depth Interviews 287
Development of Conceptual Framework 290
    What Does Sharing Mean to Farmers? 290
    Proposed Framework for Intention to Adopt Shared Farm
        Equipment Services 292
    Effect of Moderators 297
Discussion 299
Implications and Future Research Directions 300
Acknowledgment 301
References 301

LEARNING OBJECTIVES

After reading the chapter, the reader should be able to:

- Recognize the practice of marketing for the benefit of society
- Explore three key themes: sustainability, village-level marketing, and digital platforms; and their interrelationships in the context of shared farm equipment services.

- Deliberate on customer-perceived value and its effect on product adoption
- Understand key marketing theories such as extended valence framework, transaction cost economics, and the theory of reasoned actions
- Conceptualize how to strengthen an agricultural economy through greater use of technology and collaborative approaches
- Understand how to develop hypotheses based on the literature review and field study

For managers/practitioners

- Consider this model as a basis to improvise their current and future-shared service offerings.

THEMES AND TOOLS USED

- Perceived value/customer value in exchange
- The pooling of resources/collaborative consumption
- Extended valence framework
- Transaction cost economics
- Thematic analysis of interview scripts

THEORETICAL BACKGROUND

This study introduces some of the most fundamental and highly important theories. The first construct in this regard is the value, which is widely used in various fields such as economics, marketing, and finance and is rooted in customer utility, customer psychology, and social psychology. For marketing scholars and practitioners, this can be explained in terms of consumption value, transaction value, perceived value, or service value. According to the most prominent definition given by Zeithaml (1988), a customer's perceptions of what she receives and what she loses indicates the product utility for her. Kim, Chan, and Gupta (2007) extended this concept to develop a value-based adoption model of mobile Internet, which says that customer adoption intention is positively affected by customer

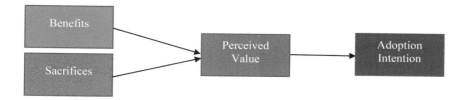

FIGURE 11.1
Value-based adoption model. (From Kim et al., 2007).

perceived value, which in turn is influenced by the customer benefits and customer sacrifices as shown in Figure 11.1.

For example, the adoption intention of a smartphone would depend on the perceived value a college student sees on that smartphone. This value would be estimated based on factors such as price, iOS vs. Android technical capabilities, the functionality provided by a smartphone, how useful is the smartphone for the student to connect with friends, for entertainment, or educational purpose. These benefits or sacrifices, which we refer to as value drivers and barriers, respectively, are important for a marketer to develop the right product and right marketing communications to attract customers. This study is concerned with the perceived value of shared farm equipment services. The customers are farmers and explore their perceptions about the adoption of such services. Other prominent psychology-based consumer decision-making theories are the valence framework (Goodwin, 1996), the theory of reasoned action (Fishbein & Ajzen, 1975), and the extended valence framework (Kim, Ferrin, & Rao, 2009), which combines the earlier two theories. Additionally, this study also rests on the tenets of another fundamental theory, transaction cost economics (TCE), proposed by Williamson (1979). It suggests that people wish to perform transactions in the most economical ways. Earlier research studies have used TCE to explain customer adoption of online products or to manage risks in maintaining supply chain networks (Teo & Yu, 2005; Wever, Wognum, Trienekens, & Omta, 2012), as shown in Figures 11.2 and 11.3.

As a marketer, one can apply TCE to understand transaction uncertainty for a stakeholder view and then design measures to reduce that uncertainty and offer a good product/service to benefit society. For example, Reliance markets fruits and vegetables. So, farmers can get together or form small

282 • Social and Sustainability Marketing

FIGURE 11.2
TCE model for online buying behavior. (From Teo and Yu, 2005).

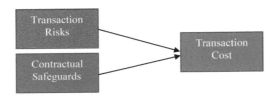

FIGURE 11.3
TCE model for managing risks. (From Wever et al., 2012).

cooperatives to demand a stake in the Reliance retail chain and reduce the uncertainty associated with procurement and information asymmetry on quality and pricing of fruits/vegetables and therefore, ensure revenue maximization for the whole ecosystem. In the present study, we have utilized TCE to assess farmers' understanding of risks and returns in using shared farm equipment service.

To conclude, based on a comprehensive literature review about (1) sharing economy, (2) services marketing, and (3) rural marketing and in-depth interviews of different stakeholders in a village set up, we propose a model for the adoption of shared farm equipment services.

INTRODUCTION

Sustainability is a well-known concept in policy-oriented research to express what public policies should strive to achieve. The basic tenet came from the Brundtland Report (WCED, 1987, p. 23). It is explained in terms of three aspects: social, economic, and environmental. The three dimensions emanate from the triple bottom line concept, given by Elkington (1994). Thus, this is a natural topic of study for scholars, economists, managers, and policymakers because the scarcity of resources, the idea of the well-being of all for an extended period, and mutual coexistence have increasingly gained importance. Thus, sustainability experts are looking for more transformative policies in social and economic spheres. One of the pressing needs is to develop sustainable business models as well as participatory decision-making in politics and society to foster social inclusiveness (De Vries, 2013; Heinrichs, 2013). Mobilization and sharing of existing resources in a profitable manner is one such participatory and integrative strategy.

This leads to the concept of sharing economy, which can build sustainable societies (Heinrichs, 2013; Daunorienė, Drakšaitė, Snieška, & Valodkienė, 2015). It is defined as a peer-to-peer exchange and collaborative usage of goods and services, backed by a community-based online platform. Individuals exchange, redistribute, rent, and share knowledge, goods, services, and skills either themselves or via commercial or social media platforms. There seems to be an increase in collaborative consumption services for vehicles, accommodation, food, and instruments (Hwang & Griffiths, 2017). Similar behavior is noticed for product- or service-based transactions with other business or customers, including recycling of old goods. One such application area is the sharing and access of farm equipment (Sengupta, Narayanamurthy, Moser, & Hota, 2019). However, there is a concern and ongoing debate that some firms use "sharing economy" as a strategy to pretend to make society a better place. Therefore, there is a need that companies provide clear and precise information and product options to potential customers, and governments also need to promote sustainable consumption practices among the public (Mi & Coffman, 2019).

Vice-President of India Shri. Venkaiah Naidu stated that farmers are the backbone of the country, and therefore all efforts much be done to prioritize the development of the rural sector (Business Standard, 2019). In India, a larger percentage of farmers (70%) own less than one hectare of land, and reports suggest that over 12,000 farmers commit suicides per year (TOI, 2017). A primary reason for farmer suicides in India is the operational challenge due to the inflated prices of agricultural inputs (Talule, 2020). One such critical input is agricultural equipment, but procuring machinery such as tractors, threshers, and submersible pumps has become unaffordable to small and marginal farmers due to a surge in acquisition costs. Ultimately, crops get destroyed with low quality and quantity of farm produce, resulting in financial, psychological, and social loss to farmers and their families and increase the price of other edible products (biscuits, flour, bread, fruit juices, jams, etc.) putting pressure on end-customers. Another major sustainability issue is the burning of paddy straw and such agricultural stubble at the end of a crop cycle (The Wire, 2018). This is of utmost concern for all as it increases pollution, breathing problems, and low visibility as nearby areas observe thick fog for several days. Although we have schemes such as the National Agricultural Market (https://www.enam.gov.in/web/) and contract farming to alleviate farmers' concerns and create a bridge between farmers and the market, the situation is still far from being a robust model to sustain in the long run.

Therefore, developing sustainable solutions for our farmers necessitates immediate concern and inquiry from marketing scholars and practitioners. This chapter, while addressing this pertinent topic, also emphasizes "how 21st-century organizations are addressing socially responsible consumers and meeting their need while keeping their business profitable," which is the objective of this book.

We state that a possible solution is to rent out machinery in an economical manner, but this also seems complicated due to farmers' limited access. To empower Indian farmers and strengthen this agrarian economy, the Indian government and several firms have come up with farm equipment rental apps for farmers. For example, the central government (DNA, 2019), some state governments (The Better India, 2016), farming as service startup companies (YourStory, 2018), Mahindra and Mahindra (The Hindu, 2016), and several other firms have either planned or launched Agri mobile applications to help farmers rent tillers, harvesters, weeders, tractors, etc. Digital applications have been credited to address the challenges of rural markets and reduce the service costs for small-scale farmers (Aker, Ghosh, & Burrell, 2016; Baumüller, 2018; Nakasone, Torero, & Minten, 2014; Deichmann, Goyal, & Mishra, 2016). However, the penetration of digital applications is low, as about 16% of the rural segment uses the Internet for digital payments (Mint, 2018). Acceptance of novel technology has always been of concern for academics and practitioners. It involves assessing pertinent factors of adoption, behavioral concerns, and usage patterns of technology by individuals (Park & Del

Pobil, 2013). Several models have tried to explain the behavioral intention to use technology, such as the technology acceptance model (Chuttur, 2009; Luarn & Lin, 2005). Despite an impetus given to sharing economy initiatives in agriculture systems across the globe, we are still not clear on what sharing exactly means from a farmer's perspective and what would motivate them to adopt such shared-services platform (Carlisle et al. 2019; Artz, Colson, & Ginder, 2010; Kabbiri et al. 2018; Kashyap, 2016). To develop and promote the use of a digitized access-based service for farmers, firms must understand their expectations, fears, and concerns and use that information as the basis to create such platforms. This study addresses this gap by qualitatively assessing the antecedents to farmers' intention to adopt such rental service. Based on the extant literature and in-depth interviews of various subjects from a village set up, we derive a set of propositions to be tested by future researchers. Therefore, this framework acts as a starting point for firm managers and policymakers to develop a profitable and sustainable-shared service model to strengthen the Indian economy.

The rest of the chapter is organized in the following manner. We first present the methodology followed by the conceptual framework that explains the various constructs and their linkages. We conclude with the implications and an agenda for future research.

LITERATURE REVIEW

In stark contrast to the extensive literature on consumer marketing or business marketing, few articles have been published combining the trifecta rural economy, sharing economy, and sustainable outcomes. Further scrutiny reveals relatively little attention to the sharing of farm equipment and machine in the context of villages. A few of the seminal studies are listed in Table 11.1.

We found that most of the extant work primarily deals with (1) descriptive work on the peculiarities of rural markets compared to urban markets, (2) application of bottom-of-pyramid marketing concepts to enhance consumer adoption of FMCG goods, (3) discussions of factors that facilitate or hamper micro-lending/financing for rural entrepreneurs, and (4) from operations perspective looking at the efficiency and performance

TABLE 11.1

Seminal Papers on the Related Ideas

Reference	Key Findings
Curtis and Lehner (2019)	Features of the sharing economy to create sustainability: ICT-mediated, nonfinancial drive for ownership, access, nature of goods.
Asian, Hafezalkotob, and John (2019)	Explores the role of sharing economy (SE) to enable small farmers to address their challenges by sharing resources and peer-to-peer processes and mechanisms using SE.
Sung, Kim, and Lee (2018)	Analyses acceptance model from customer and supplier point of view using the network effect, which is an important aspect of the sharing economy.
Pouri and Hilty (2018)	Applies the life cycle or the structural impacts model of information technology to Digital Sharing Economy and explains pros and cons of digital sharing to achieve sustainability at multiple systemic levels.
Martin (2016)	Defines sharing economy from different lens: (1) economic gains; (2) sustainable consumption; (3) a means to create a decentralized, fair, and sustainable economy; (4) unregulated marketplaces; (5) neoliberal paradigm; and (6) innovation paradigm.
Cheng (2016)	Identifies importance area of research: (1) SE business models and their influence in society/economy, (2) types of SE, and (3) sustainable development.
Daunorienė et al. (2015)	Explains how to measure and validate sharing economy business models by employing circles of sustainability.
Başarik and Yildirim (2015)	Discussion on Turkish-shared farm equipment system.
Artz, Colson, and Ginder (2010)	Enablers to participate in sharing consumption: cost savings and access to reliable labor. Trust and communication reduce transaction costs.
Larsén (2010)	Effects of machine-sharing collaborations on farm efficiency in Swedish context.

aspects of such sharing farm equipment systems. Few studies look at the farm equipment sharing platforms from the user or farmer's perspective. Thus, we proceed based on such studies and the research studies done on sustainability, specifically at the village level (e.g., Kumar, 2004; Zhao & Wong, 2002; Sumner, 2005; Marsden, 2003) to understand what is sharing from the customer's perspective in the present context and what are the enablers/inhibitors for the adoption of such shared services.

METHODOLOGY: IN-DEPTH INTERVIEWS

We followed a qualitative methodology to explore and develop insights into the value mindset and expectations of rural stakeholders. The findings of the field study and literature review are synthesized to develop a conceptual model to achieve a sustainable rural economy through the sharing of farm equipment.

In this study, we adopt a qualitative, naturalistic methodology (Miles & Huberman, 1994) to gain insight into farmers' evaluations of "shared resources" and "shared farm equipment services." The field study involved in-depth interviews with 15 people hailing from three different villages in a North Indian state. Due to data collection issues in rural markets, even a small expert sample is accepted in this domain, as seen in recent studies (Troncoso, Castillo, Masera, & Merino, 2007; Mayer, Habersetzer, & Meili, 2016). Since our objective was theory construction, we opted for a purposive sampling (Glaser & Strauss, 1967). We ensured that the sample consisted of farmers, Gram Pradhan (village head), and distributors of fertilizers and farm machinery, which have been involved in the purchase decision-making for such equipment. Of the 15 subjects, 2 are village heads, 3 are distributors, and 10 are farmers. All were males. The distributors were graduates, and the Gram Pradhans did schooling till 12th standard, and the farmers' education ranged from 8th to 12th standard.

All interviews had a standardized format. We resorted to the extant literature to provide support to develop the initial set of questions regarding perceived value in the shared farm equipment services (Edward & Sahadev, 2011), availability of information regarding the machine and system/application delivering those services (Lee, Chan, Balaji, & Chong, 2018), financial bearing and uncertainty in such transactions (Boateng, Kosiba, & Okoe, 2019), and the role of village leaders/influencers and social capital (Xiao, Wang, & Chan-Olmsted, 2018). However, depending on the conversation, some questions had to be explained or related questions asked for more information or relevant examples. In the beginning, each interviewee was informed of the purpose of the academic research project and that we do not belong to any particular company or political party to gain their trust and assured to maintain their anonymity for honest answers. People were requested to consider the scenarios of buying farm equipment (tractor, thresher, rotator, etc.) and asked relevant questions, e.g., what do you understand by shared resources, what kind of resources are generally shared in your village. The questions were modified based on the nature of the subject. Table 11.2 describes the coding scheme used in this study.

In the end, we thanked them profusely for their time. Each interview lasted for about 30-45 minutes. First, each interview script was read twice

TABLE 11.2

Qualitative Study Coding Scheme

	Coding Scheme
	Interview
Opening	• Introduction of primary researcher and respondent, purpose of the study, assurance of confidentiality, and no political party or industrial firm's connection
	• Requested to visit the farm
Demographic Data	• Title and role of participants, their education and family background, information regarding their farming practices (kharif/rabi crop, plotting, number of people involved in farming, etc.)
	• How long have you been in this farming or distribution profession? What is your current farm/crop status? How does a normal day look like?
Tailored Interview Questions (farmers, Gram Pradhan, distributors)	• What is your understanding of sharing of resources? With whom do you share resources in the village? What kind of things are generally shared among people in your village? Why do you share? Do you purchase or rent the farm machinery? What are the important considerations in making the final purchase/rent decision? Have you heard of any companies which help you to rent some farm machine? If no, then suppose a company or government comes up with such a platform then what would you expect from them so that you can rent through it? If suppose a company or government comes up with such a platform or application, then what would you expect from them so that you can rent your tractor or thresher to others? What do you think companies should do to promote small farmers to take machine on rent for a limited time through that application? How aware are the villagers about different farm equipment available in the market? Are all farmers alike in their criteria and machine preference? What are some of the must-haves for any industrial firm to stand out in this shared equipment services model?
Prompts	• Tell me more about that, can you explain that in more detail, can you give me examples, how does that work, sorry didn't get that can you repeat

(Continued)

TABLE 11.2 (Continued)
Qualitative Study Coding Scheme

	Coding Scheme
	Interview

Categories or Themes	
Sharing of Resources at the Village Level	• Human capital • Physical capital • Financial capital
Value Enhancers for Farmers	• Information quality • Platform quality • Contractual governance • Return on investment
Other Factors Affecting Adoption of Shared Machine Application	• Social capital • Influencer interactivity • Social advocacy

Coding Rules
- Include all text necessary to capture the context, but if the interviewee is expanding on some idea or example, this need not be coded unless it is essential to serve the research objectives
- Irrelevant text need not be coded into a theme
- Pay attention to the question of building an understanding of possible codes for the answer
- Create as many codes as possible (e.g., purchase price, price for repair, instalments, machine details, machine's previous use)
- Later combine the related codes into themes
- Attempt to code text into one mutually exclusive theme; however, if necessary, it may have more than one idea (e.g., farmers may consider the price they pay as that something related to both contracts/rent agreements as well as to assess overall return on their investment to acquire that machine)

to familiarize with the respondent's ideas. Then the phrases that were relevant to our study were manually coded with a keyword to indicate the implied meaning. After that, the between-case analysis was done to arrive at the most common themes across all interviews. This coding was repeated after one month and checked for consistency in the interpretation of codes, and there was no significant discrepancy between the two content analysis stages. The intelligence thus gathered offers new insights into the meaning, antecedents, and consequences of the sharing economy, shared equipment, and sustainability at the village level. The following section discusses the culmination of literature review and interview quotes in explaining the proposed conceptual framework.

DEVELOPMENT OF CONCEPTUAL FRAMEWORK

The "sharing economy" concept is used in multiple ways by different researchers (Schor, 2016; Belk, 2014). It is articulated in several forms: collaborative consumption, peer-to-peer economy, and access-based consumption. Shared service offerings reduce transaction costs by directing connecting sellers to buyers for trade or leasing products (Rauch & Schleicher, 2015). Hailing a tractor through mobile phones made headlines in the New York Times in 2016. However, there are several differences between hiring a taxi and hiring farm equipment. The latter requires operational skills. There are issues concerning the size of the farm, widely distributed farms, seasonality, and loss of income for farmers if the equipment does not work as expected (Binswanger & Rosenzweig, 1986). In this study, we state the term "renting" as suitably explaining this sharing economy concept. It creates "value" by offering underutilized goods and services to potential segment on a paid basis for a predefined time, which resolves the ownership demand (Stephany, 2015). Thus, the perceived value is the central construct in our proposed model for creating sustainability through sharing farm equipment. But first, we will discuss what sharing means for farmers

What Does Sharing Mean to Farmers?

To understand the concept of sharing at the village level, we followed purposive sampling (Yin, 2009). The subjects displayed various farming/agricultural characteristics to explore what does sharing mean for the stakeholders in our villages. Sharing was manifested as words such as "helping," "giving money," "giving small tools," "my wife borrowed utensils," "sharing is good," "care," "together work in fields," and "all helped in repairing my house during rains." To them sharing is a way to care and express that they live for mutual gains and strive to uplift themselves as a community. Limited resources and knowledge make people stay connected and make judicious use of available means. People are dependent on each other in general but more so on financially well-off and educated

TABLE 11.3

Sharing of Resources in a Village

Factor	Codes	Quotes from Interviews
Human capital	• Physical labor • Knowledge sharing • Collective work	• We work together in fields and help each in different activities in cultivation and harvesting • Usually we sit together in evening and tell what we hear from our relatives in city or there is some news from our or nearby villages • We discuss political issues and news. We switch on radio and listen as a group • We normally help each other in weddings or if somebody dies to manage • We are a small village and we depend on each other for our general well-being
Physical capital	• Farm equipment • House/room	• I took tractor from another farmer last year • I used the outhouse of Gram Pradhan for my daughter's wedding • we have a common granary to store our harvest • we have made a common building with 3 rooms which we use as small school in daytime and for other activities on festivals etc.
Financial capital	• Financial investment • Risks	• I along with another family combine our money to buy seeds, fertilizers etc., and we share the earning in equal proportions • it is a community at work in my village. Most big activities are usually jointly funded and hence the outcomes are also shared. For example people buying big machines jointly and using them on a rotation basis • sometimes people work on others' farms and in return get a share of produce but if no rains or bad weather they get less as there is total loss to the primary farmer or land owner

ones for guidance. The findings indicate that villagers share and leverage their resources in different forms broadly categorized into three forms of resources: (1) human capital, (2) physical capital, and (3) financial capital, as shown in Table 11.3.

Human capital includes physical labor, knowledge sharing, and collective work. Physical capital includes room/house and farm equipment (tractors, threshers, etc.), and financial capital includes financial investments and shared risks. This categorization gains support from the pooling of resources, collaborative consumption, and trust among the different actors. Resource pooling promotes shared consumption, and it ranges from coproduction to co-consumption as the two extremes (Heinrichs, 2013; Cohen & Munoz, 2016; Belk, 2014). A farmer stated that people having pucca houses (made of bricks) and that are spacious usually allow them to be used by other villagers during their daughter's/sister's wedding as a rest house for the groom's family and related ceremonies in return for money or some farm produce. This saves a lot of search time and costs for the users. Furthermore, the users pay for any repair or maintenance, if needed, on top of the existing set up, which acts as a coproduction scenario. This leads to collaborative consumption or common goods as trading or renting (Botsman, 2013). One of the distributors asserted the importance of trust among villagers. One or two villagers are generally regarded as more knowledgeable because their son/brother is working in some city, and they keep learning from their experiences in the city. Often, they pass on that information to other farmers; for example, he heard of a *Pusa Krishi* mobile app launched by the Indian government (TOI, 2016) from one of the Gram Pradhans during one of the regular evening sessions in his village. Gram Pradhan's son worked in the national capital region and got to know this mobile app during one of the *Krishi Mela* (fair for farmers) organized in that region. One of the main features of the sharing economy rests on trust-based relationships among community members (Botsman, 2013; Miralles, Dentoni, & Pascucci, 2017). Some researchers have mentioned that even without knowing each other, the trust toward the company or corporate brand and its processes promotes access to shared resources (Miralles, Dentoni, & Pascucci, 2017). However, we find that in the present context, farmers know almost everyone in their as well as nearby villages well and they listen and follow only those people they trust. Also, we find that someone must introduce farmers to the new forms of technology, applications, and means to improve their livelihood. This trust comes from their higher power in the village set up, knowledge, or financial standing of these opinion leaders. However, there are specific barriers in resource sharing at the village level that, even now, lead to low farmer participation in using available resources, as discussed in the following section.

Proposed Framework for Intention to Adopt Shared Farm Equipment Services

The basis of sharing economy is to harness the value of slack or underutilized products (Dillahunt & Malone, 2015). Considering the widespread use of digital media and mobile phones, the sharing economy is transforming society and businesses alike. While users have regarded shared services as more economical and convenient, privacy and security risks have always prevented a large-scale use of such services. One of the farmers stated, "I am not so literate and do not operate a phone. But I taught my daughter. She has just completed 12th, and she uses a mobile phone. My son works in city ... for such service I am afraid if my daughter must give all her details and what if it is misused? ... Will this company give all thresher related information clearly and in simple *hindi* (their regional language) so that there are no problems later?" There are fears and risks farmers see in using such applications and so shared equipment service providers must understand the farmer's expectations and perceived value to adopt such services. Understanding motivators and inhibitors of users' intention to adopt such services will allow firms and the government to invest their resources into value driver agents. We hereby present the synthesis of literature review and field interviews as a conceptual model based on the extended valence framework (Kim, Ferrin, & Rao, 2009) and TCE (Williamson, 1989). The former theory suggests that customer participation in technology-based commerce is driven by their perception of risks, gains, and trust that are combined into a perceived value construct in the present framework. Since the provision of shared farm equipment services is based on the information and mobile technologies, we incorporate product information quality and platform quality as the antecedents to perceived value. Similarly, from an exchange perspective, we include contractual governance and return on investment (ROI) affecting the value concept for a farmer, as shown in Figure 11.4.

Applications play a significant role for success in sharing economy markets (McAlone, 2015). Application quality includes both the information quality and the platform robustness in terms of users' assessment of the rental platform to meet their requirements and experience with that platform (Aladwani & Palvia, 2002; Delone & McLean, 2003; Lee et al. 2018). One of the distributors mentioned, "I use mobile app for connecting with my supplier and I have seen many such applications. I could not use these applications on my earlier phone so had to buy a new smartphone to run these applications. This was costly. I can order goods through this app and see the supply status but there are so many tabs, click multiple times and registration took a long time. Sometimes I got very frustrated and thought

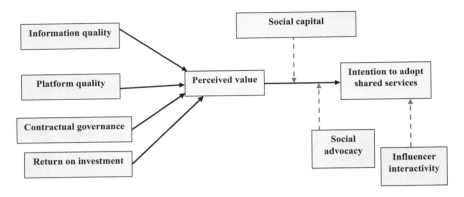

FIGURE 11.4
Proposed theoretical model of shared service adoption.

I were more educated …" A Gram Pradhan mentioned, "… people in this village are very simple and have not been exposed to technology much so applications should be mentioned in their language, and give information on how to use the mobile app, products they rent, if complaints then who to contact and how to make payment. It should work on any mobile phone and should have a button where one can click and listen to the information to benefit all kinds of farmers."

Most farmers stated their concern if the products showed different from the product available for rent, or the user manual or video is not clear that they can learn how to operate the machine on their own. Thus, information quality is a critical feature that indicates the degree to which farmers would perceive that the descriptions, photos, contact details, company details, etc., given on the platform is of value to them (Lin, 2007). Product- and usage-specific details should be complete, accurate, and timely provided (Kuan, Bock, & Vathanophas, 2008). The application should be available in the local language, and some voice features should be enabled. Field study highlighted that farmers were concerned about downloading the app and what to do if something goes wrong in between. A farmer's son who studies in some city college has told him not to install any application as some fake applications and viruses may corrupt the operating system and hack all banking/personal/sensitive information from the mobile. The user group seems to be skeptical regarding the overall security and quality of the platform/application. This, referred to as the platform quality, indicates the extent to which farmers perceive that the processing of mobile apps or the underlying information system is of some quality (Chen, 2010). The value drivers, therefore, include ease of use, reliability, access convenience, and interactive overall user experience for the shared service platform (Lin, 2007; Lee et al. 2018).

Quality of both information and platform provides farmers with correct and timely information, a consistent and easy-to-operate mobile interface, and interactive experience during the rental transactions, which increases the perceived value of such services leading to higher usage (Kuan, Bock, & Vathanophas, 2008). These two dimensions indicate that the service provider understands users' concern, is capable, listens to its customers, and so instills trust on its users (Zhou, Li, & Liu, 2010; Zhou, 2014; Möhlmann, 2015). That is if the shared farm equipment application can give timely and correct information with proper security mechanisms as well as can assist them in getting and using their machine, it has a potential to create a higher perceived value in that shared service platform and the company providing the same. Thus, we hypothesize that:

H1: Users' perceived information quality is positively related to their perceived value of the platform.
H2: Users' perceived platform quality is positively related to their perceived value of the platform.

Another common theme that appeared in the field study was if the product did not work as expected and incurred high financial cost and time on renting through the mobile app. The farmers mentioned that if they know somebody is renting a tractor or rotator, they directly visit, check the machine, or ask their friends who have used it and request the Gram Pradhan or the owner to lend it for some defined time. They find this mode of transaction easy because villages thrive more on personal meetings, group discussions, and *chaupals* (formal gathering of people) compared to cities where people are more independent. This relates to a customer's ROI of financial, mental, behavioral, and psychological resources toward some outcome (Mathwick, Malhotra, & Rigdon, 2001). The farmers may look for the application in terms of financial affordability as they have limited capacity, time involved in searching for the machines that are available to rent in the nearby places, and cognitive contentment that they got the right equipment through a novel technology platform (Mathwick, Malhotra, & Rigdon, 2001; Luarn & Lin, 2005). We conceptualize the farmer's ROI in terms of financial gains and psychological gains, affordability, and cognitive dissonance in using the shared service app (Hamari, Sjöklint, & Ukkonen, 2016; Kalyanaraman & Sundar, 2006; Schmutz, Heinz, Métrailler, & Opwis, 2009). Literature suggests that financial and psychological costs of renting determine the attractiveness of sharing economy model (Lamberton & Rose, 2012). Customers try to maximize the derived utility by trading ownership with affordable and temporary acquisition of services (Hamari, Sjöklint, & Ukkonen, 2016). Therefore, the farmer's ROI can determine their perceived value in such mobile apps. Therefore, we hypothesize that:

> H3: Users' return on investment is positively related to their perceived value of the platform.

The fourth most common theme among the field interviews was the level of formal contracts. One of the Gram Pradhan's who lends his idle plow and tiller mentioned that farmers have questions regarding payment terms, who will repair if something breaks while in use, what if they do not return within the stipulated time because many practices such as harvesting/sowing seeds, etc., are not at times time bound. Depending on rains or other weather conditions, the time for renting can change. So, both the provider and user need some flexibility as well as precisely defined terms and conditions for different scenarios that may arise. On the other hand, when we talked to farmers, they believed that at times on returning the machine, they are asked to pay more amount than was initially decided for some reasons which they may not understand. One of the farmers mentioned that he forgot the actual date to return the machine, and therefore had to pay a high penalty that he was now told at the beginning that left a hole in his pocket. Contracts are formal norms or governance mechanisms that clearly specify the actions and consequences (e.g., penalty) (Macneil, 1978). They help to lower the uncertainty in buyer-seller relationships and increase customer value (Lusch & Brown, 1996; Ghosh & John, 2005). The shared farm equipment service requires the providers and customers to work in a collaborative manner and be acquainted with each other's procedures, ideas, expectations, and approaches to negotiate on the best possible value for in the mutual interest. Both parties explicitly need to elaborate on their needs, thought-process, and understanding about customer value and gains from that transaction (Aarikka-Stenroos & Jaakkola, 2012). By clearly mentioning the rights and duties of either party, formal governance tends to ensure confidence in the transactions and help to resolve any unforeseen issues and conflicts that may arise (Mooi, Kashyap, & van Aken, 2020). This will curtail any opportunistic behavior from any other party. Therefore, we state that:

> H4: Contractual governance is positively related to customer's perceived value of the platform.

Value in the platform will increase user adoption of that platform. Based on the tenets of the prospect theory (Kahneman & Tversky, 1979, Wang & Wang, 2010), we assume that an overall judgment of value affects farmers' intention to adopt shared equipment services. In e-commerce, perceived value is the consumer's cost-benefit analysis when shopping online (Zeithaml, 1988). Farmers would wish to rent tractors, threshers, etc., through those platforms that offer maximum value. The literature suggests that the perceived value influences the purchase intention through mobile apps (Prodanova, Ciunova-Shuleska, & Palamidovska-Sterjadovska, 2019). Thus, we mention:

> H5: Users' perceived value of the shared service platform is positively related to their intention to adopt shared services.

Effect of Moderators

Villages are highly community-based set up, and firms need to understand this to promote user adoption of their offerings. Even though during the in-depth interviews, farmers seemed to be interested in using a rental service platform, almost all focused on the importance of thrust given by knowledgeable people and what they hear from their friends in their community. People are more likely to believe the information to be true when most people believe so (Chaiken, 1987). We define the first moderator of social influence or peer influence as social advocacy (Xiao, Wang, & Chan-Olmsted, 2018). It affects an adopter's assessment of source and information credibility (Giffin, 1967). Reviews, recommendations, and consumer testimonials, used as a branding strategy, indicate the importance and intensity of social influence while marketing goods/services (Wang, 2005). Firms need to identify a few farmers and train them on how to use the platform and let them educate other farmers in their vicinity. They need to empower farmers who are potential users to create higher user adoption of their shared service platform. A successful example is project Shakti by Hindustan Unilever in India (Rangan & Rajan, 2007). Firms also need to identify opinion leaders from among the farmers who can connect and influence the behavior of their fellow members. This involves frequent discussions between that leader/influencer and other villagers leading to our second moderator, influencer interactivity. Just like the comment section in any online platform allows viewers of a video to interact and influence each other and enables hosts to interact with customers directly. Similarly, influencers would mediate between the farmers and platform owners. Interactivity has a positive effect on information credibility (Fogg et al. 2003). Just as for e-commerce, Kim, Bickart, Brunel, and Pai (2012) advised brands to enhance their interactions on blogs to improve the brands' perceived credibility; we propose that in a rural set up, the influencers need to show high interactivity with the farmers to increase their sense of the utility of the shared services application. From the heuristic-systematic model (HSM) perspective, interpersonal interactivity is a heuristic that affects one's judgment of online information credibility (Sundar, 2008). Similarly, the social aspect is an essential criterion for the acceptance of the sharing economy (Ozanne & Ozanne, 2011). Social capital creates a sense of belonging, which gives emotional security and a sense of mutual closeness (Park, Kim, & Son, 2015). Hence, a successfully shared service platform should share a set of values that are enabled through common goals and activities. Farmers should feel motivated that being part of such a network gives the advantages of sharing new information about technology, products, and services besides expanding the possibility of renting those resources (Kang & Na, 2018). Social capital has been regarded as a necessary or sufficient condition to spread the sharing economy concept (Ju, 2016). Firms that can help to foster and leverage the social capital of farmers can nurture trust and values among them toward a common goal, and the users would be compelled to use the shared equipment application in a way that is helpful to all actors. Therefore, we hypothesize that:

H6: Social advocacy moderates the relationship between users' perceived value and their intention to adopt shared services.

> H7: *Influencer interactivity moderates the relationship between users' perceived value and their intention to adopt shared services.*
> H8: *Social capital moderates the relationship between users' perceived value and their intention to adopt shared services.*

DISCUSSION

This book enables students/scholars/practitioners to develop an in-depth understanding of social marketing and sustainability marketing. All the chapters highlight the need to create, communicate, and deliver customer value to socially responsible consumers through different marketing strategies and processes. This chapter, in particular, delves into creating and delivering value to farmers – who are an integral part of our economy. By illustrating a range of actual narratives obtained from the farming community (customers) and what existing literature discusses marketing situations in a shared-services context, this chapter will help students to acquire the theoretical and practical skills they need to make informed marketing decisions as business leaders. It combines the aspects of (1) customer purchase decision-making, (2) perceived value, and (3) sharing economy to provide sustainable shared farm equipment services for the benefit of society. The adoption of technologies and new programs/offerings to build a sustainable farming ecosystem is a challenging and dynamic issue for the industry, government, policymakers, and farmers. Product and service innovation, increased marketing communication, farmer education and training, and vast information availability are geared to balance economic efficiency with environmental and social sustainability – and this chapter compels students and marketing managers to think on those lines. Digital/mobile applications to share farm equipment can lead to a socially acceptable and economically efficient farm sector and enhance environmental performance providing "triple dividends" to sustainability. Big brands in the FMCG sector, such as ITC (https://www.itcportal.com/sustainability/sustainable-agriculture-programme.aspx) are investing in sustainable agriculture programs such as farm mechanization and agri-business center. However, the adoption of such practices in India is still minimal, and marketers need to closely scrutinize the challenges and needs of our farmers before designing solutions. This study presents a conceptual framework as a starting point for them to further their investigation.

IMPLICATIONS AND FUTURE RESEARCH DIRECTIONS

Sharing economy is an emerging phenomenon which has received major attraction from all stakeholders in society. The study mentions how does sharing happens and what it means for farmers. Sharing economy potentially has a strong competence to transform and develop sustainable agrarian economies. This could be a radical change accomplished by technological progress and suitable legal and economic instruments. The proposed model presents new insights into the research of farmers' intention to adopt shared farm equipment services. Drawing on the field study and using extended valence framework and TCE, we identify the antecedents to the perceived value of such a shared services platform. Also, we incorporate social factors to provide a comprehensive understanding of the phenomenon. This study opens doors for further empirical investigation and validation of the technology-enabled sharing economy services in a rural set up.

The providers need to account for more transparency and an assortment of farm equipment, rationalize the cost of operations, and formulate stringent contractual agreements to safeguard the interests of farmers such that they perceive high value in such shared services platform. Economically beneficial and time saving are basic requirements for such an application. But to add value for farming applications, firms need to put more emphasis on the quality of product-related information and the robustness of the platform. Manuals, language compatibility, ease of searching for the relevant machine, registering the application, and downloading the application are some of the activities that should be seamlessly integrated into the offering. Furthermore, considering the close-knit community fabric of the villages, firms that can identify, train, and work with local leaders/influencers would be more successful in enhancing their rate of adoption. Interaction and advocacy of such charismatic individuals, chosen from among those villagers, will enable the farmers and instill a sense of safety, trust, and commitment in their farmers toward hiring farm equipment through the application. Thus, this article proposes a conceptual model and suggests implications for theory and practice to develop a sustainable shared economy model at the village level. Also, it opens a new research agenda for further inquiry, empirical validation, and analysis of different aspects in a highly relevant domain. Further research needs to explore and develop different types of value creating rental models for farm equipment, investigate the role of advertising and promotions to increase farmer adoption of such services, look for any opportunistic behaviors from the supplier side and how to safeguard that, and from an economic perspective understand the efficiency of using shared equipment compared to other ownership modes toward strengthening agricultural sector and hence the regional economy.

ACKNOWLEDGMENT

I wish to thank the Director, Prof. Archana Shukla, and the Deans, Prof. Neeraj Dwivedi and Prof. Sanjay Singh, IIM Lucknow, to provide such an excellent platform to hone my skills in research and teaching and a motivation to conduct a study like this.

Secondly, I profusely thank the entire editorial team and reviewers of the book "Social and Sustainability Marketing: A Casebook for Reaching Your Socially Responsible Consumers through Marketing Science" for their timely support and feedback to present and refine my work. My special thanks go to all the eminent professors: Dr. Chandana R. Hewege, Dr. Weng Marc Lim, Dr. Manoj K. Dash, Dr. MS Balaji (in alphabetical order), and Jishnu Bhattacharyya – the corresponding editor – for a seamless experience while processing the submission.

REFERENCES

Aarikka-Stenroos, L., & Jaakkola, E. (2012). Value co-creation in knowledge intensive business services: A dyadic perspective on the joint problem solving process. *Industrial Marketing Management*, *41*(1), 15–26.

Aker, J. C., Ghosh, I., & Burrell, J. (2016). The promise (and pitfalls) of ICT for agriculture initiatives. *Agricultural Economics*, *47*(S1), 35–48.

Aladwani, A. M., & Palvia, P. C. (2002). Developing and validating an instrument for measuring user-perceived web quality. *Information & Management*, *39*(6), 467–476.

Artz, G. M., Colson, G., & Ginder, R. G. (2010). A return of the threshing ring? A case study of machinery and labor-sharing in Midwestern farms. *Journal of Agricultural and Applied Economics, 42*(1379-2016-113645), 805–819.

Asian, S., Hafezalkotob, A., & John, J. J. (2019). Sharing economy in organic food supply chains: A pathway to sustainable development. *International Journal of Production Economics, 218*, 322–338.

Başarik, A., & Yildirim, S. (2015). A case study of sharing farm machinery in Turkey. *International Journal of Natural & Engineering Sciences, 9*(3), 1–5.

Baumüller, H. (2018). The little we know: An exploratory literature review on the utility of mobile phone-enabled services for smallholder farmers. *Journal of International Development, 30*(1), 134–154.

Belk, R. (2014). You are what you can access: Sharing and collaborative consumption online. *Journal of Business Research, 67*(8), 1595–1600.

Binswanger, H. P., & Rosenzweig, M. R. (1986). Behavioural and material determinants of production relations in agriculture. *The Journal of Development Studies, 22*(3), 503–539.

Boateng, H., Kosiba, J. P. B., & Okoe, A. F. (2019). Determinants of consumers' participation in the sharing economy. *International Journal of Contemporary Hospitality Management, 31*(2), 718–733.

Botsman, R. (2013). *The sharing economy lacks a shared definition.* Retrievable from: http://www.fastcoexist.com/3022028/the-sharing-economylacks-a-shared-definition

Business Standard (2019). *Farmers backbone of country, govt should priortise development of rural India: Vice-President Venkaiah Naidu.* Retrievable from: https://www.business-standard.com/article/news-ani/farmers-backbone-of-country-govt-should-priortise-development-of-rural-india-vice-president-venkaiah-naidu-119092200497_1.html

Carlisle, L., de Wit, M. M., DeLonge, M. S., Calo, A., Getz, C., Ory, J., ... & Iles, A. (2019). Securing the future of US agriculture: The case for investing in new entry sustainable farmers. *Elem Sci Anth, 7*(1), 17.

Chaiken, S. (1987). The heuristic model of persuasion. In M. P. Zanna, J. M. Olson, & C. P. Herman (Eds.), *Ontario Symposium on Personality and Social Psychology. Social Influence: The Ontario symposium* (Vol. 5, pp. 3–39). Hillsdale, NJ: Lawrence Erlbaum Associates, Inc.

Chen, C. W. (2010). Impact of quality antecedents on taxpayer satisfaction with online tax-filing systems—An empirical study. *Information & Management, 47*(5-6), 308–315.

Cheng, M. (2016). Sharing economy: A review and agenda for future research. *International Journal of Hospitality Management, 57*, 60–70.

Chuttur, M. Y. (2009). Overview of the technology acceptance model: Origins, developments and future directions. *Working Papers on Information Systems, 9*(37), 9–37.

Cohen, B., & Munoz, P. (2016). Sharing cities and sustainable consumption and production: Towards an integrated framework. *Journal of Cleaner Production, 134*, 87–97.

Curtis, S. K., & Lehner, M. (2019). Defining the sharing economy for sustainability. *Sustainability, 11*(3), 567.

Daunorienė, A., Drakšaitė, A., Snieška, V., & Valodkienė, G. (2015). Evaluating sustainability of sharing economy business models. *Procedia-Social and Behavioral Sciences, 213*, 836–841.

De Vries, B. J. M. (2013). *Sustainability science*. Cambridge, UK: Cambridge University Press.

Deichmann, U., Goyal, A., & Mishra, D. (2016). Will digital technologies transform agriculture in developing countries? *Agricultural Economics, 1*(47), 21–33.

Delone, W. H., & McLean, E. R. (2003). The DeLone and McLean model of information systems success: A ten-year update. *Journal of Management Information Systems, 19*(4), 9–30.

Dillahunt, T. R., & Malone, A. R. (2015, April). The promise of the sharing economy among disadvantaged communities. In *Proceedings of the 33rd Annual ACM Conference on Human Factors in Computing Systems* (pp. 2285–2294).

DNA (2019), *Govt launches mobile app for farmers to buy, rent agricultural machinery*. Retrievable from: https://www.dnaindia.com/india/report-govt-launches-mobile-app-for-farmers-to-buy-rent-agricultural-machinery-2792106

Edward, M., & Sahadev, S. (2011). Role of switching costs in the service quality, perceived value, customer satisfaction and customer retention linkage. *Asia Pacific Journal of Marketing and Logistics, 23*(3), 327–345.

Elkington, J. (1994). Towards the sustainable corporation: Win-win-win business strategies for sustainable development. *California Management Review, 36*(2), 90–100.

Fishbein, M., & Ajzen, I. (1975). *Belief, attitude, intention, and behavior: An introduction to theory and research*. Boston, MA: Addison-Wesley.

Fogg, B. J., Soohoo, C., Danielson, D. R., Marable, L., Stanford, J., & Tauber, E. R. (2003). How do users evaluate the credibility of web sites?: A study with over 2,500 participants. In *Proceedings of the 2003 Conference on Designing for User Experiences* (pp. 1–15). New York, NY: ACM. doi: 10.1145/997078.997097.

Ghosh, M., & John, G. (2005). Strategic fit in industrial alliances: An empirical test of governance value analysis. *Journal of Marketing Research, 42*(3), 346–357.

Giffin, K. (1967). The contribution of studies of source credibility to a theory of interpersonal trust in the communication process. *Psychological Bulletin, 68*(2), 104–120.

Glaser, B. & Strauss, A. (1967). *The discovery of grounded theory*. Chicago, IL: Aldine Publishing Company.

Goodwin, N. R. (1996). *Economic meanings of trust and responsibility*. Ann Arbor, MI: The University of Michigan Press.

Hamari, J., Sjöklint, M., & Ukkonen, A. (2016). The sharing economy: Why people participate in collaborative consumption. *Journal of the Association for Information Science and Technology, 67*(9), 2047–2059.

Heinrichs, H. (2013). Sharing economy: A potential new pathway to sustainability. *GAIA-Ecological Perspectives for Science and Society, 22*(4), 228–231.

Hwang, J., & Griffiths, M. A. (2017). Share more, drive less: Millennials value perception and behavioral intent in using collaborative consumption services. *Journal of Consumer Marketing, 34*(2), 132–146.

Ju, D. (2016). Understanding about the relationship between sharing economy and social capital: Focusing on the consumption intentions of the sharing accommodation. *International Journal of Tourism Management Science, 31*, 23–40.

Kabbiri, R., Dora, M., Kumar, V., Elepu, G., & Gellynck, X. (2018). Mobile phone adoption in agri-food sector: Are farmers in Sub-Saharan Africa connected? *Technological Forecasting and Social Change, 131*, 253–261.

Kahneman, D., & Tversky, A. (1979). Prospect theory: An analysis of decision under risk. *Econometrica, 47*(2), 263–292.

Kalyanaraman, S., & Sundar, S. S. (2006). The psychological appeal of personalized content in web portals: Does customization affect attitudes and behavior? *Journal of Communication, 56*(1), 110–132.

Kang, S., & Na, Y. K. (2018). The effect of the relationship characteristics and social capital of the sharing economy business on the social network, relationship competitive advantage, and continuance commitment. *Sustainability, 10*(7), 2203.

Kashyap, P. (2016). *Rural marketing*, 3/e. India: Pearson Education India.

Kim, D. J., Ferrin, D. L., & Rao, H. R. (2009). Trust and satisfaction, two stepping stones for successful e-commerce relationships: A longitudinal exploration. *Information Systems Research, 20*(2), 237–257.

Kim, H. W., Chan, H. C., & Gupta, S. (2007). Value-based adoption of mobile internet: An empirical investigation. *Decision Support Systems, 43*(1), 111–126.

Kim, S. J., Bickart, B. A., Brunel, F. F., & Pai, S. (2012). *Can your business have 1 million friends? Understanding and using blogs as one-to-one mass media* (SSRN scholarly paper No. ID 2045346). Rochester, NY: Social Science Research Network.

Kuan, H. H., Bock, G. W., & Vathanophas, V. (2008). Comparing the effects of website quality on customer initial purchase and continued purchase at e-commerce websites. *Behaviour & Information Technology, 27*(1), 3–16.

Kumar, R. (2004). eChoupals: A study on the financial sustainability of village internet centers in rural Madhya Pradesh. *Information Technologies & International Development, 2*(1), 45–73.

Lamberton, C. P., & Rose, R. L. (2012). When is ours better than mine? A framework for understanding and altering participation in commercial sharing systems. *Journal of Marketing, 76*(4), 109–125.

Larsén, K. (2010). Effects of machinery-sharing arrangements on farm efficiency: Evidence from Sweden. *Agricultural Economics, 41*(5), 497–506.

Lee, Z. W., Chan, T. K., Balaji, M. S., & Chong, A. Y. L. (2018). Why people participate in the sharing economy: An empirical investigation of Uber. *Internet Research, 28*(3), 829–850.

Lin, H. F. (2007). The impact of website quality dimensions on customer satisfaction in the B2C e-commerce context. *Total Quality Management and Business Excellence, 18*(4), 363–378.

Luarn, P., & Lin, H. H. (2005). Toward an understanding of the behavioral intention to use mobile banking. *Computers in Human Behavior, 21*(6), 873–891.

Lusch, R. F., & Brown, J. R. (1996). Interdependency, contracting, and relational behavior in marketing channels. *Journal of Marketing, 60*(4), 19–38.

Macneil, I.R. (1978). Contracts: Adjustments of long-term economic relations under classical, neo-classical, and relational contracting law. *Northwestern University Law Review, 72*(6), 854–905.

Marsden, T. (2003). *The condition of rural sustainability*. AA, Netherlands: Uitgeverij Van Gorcum.

Martin, C. J. (2016). The sharing economy: A pathway to sustainability or a nightmarish form of neoliberal capitalism? *Ecological Economics, 121*, 149–159.

Mathwick, C., Malhotra, N., & Rigdon, E. (2001). Experiential value: Conceptualization, measurement and application in the catalog and Internet shopping environment⇔. *Journal of Retailing, 77*(1), 39–56.

Mayer, H., Habersetzer, A., & Meili, R. (2016). Rural–urban linkages and sustainable regional development: The role of entrepreneurs in linking peripheries and centers. *Sustainability*, *8*(8), 745.

McAlone, N. (2015). *There's an army of apps to get almost anything you want on demand–here are the best and worst*. Retrievable from: http://uk.businessinsider.com/sharing-economy-best-and-worst-apps-2015-6?r=USandIR=T

Mi, Z., & Coffman, D. M. (2019). The sharing economy promotes sustainable societies. *Nature Communications*, *10*(1), 1–3.

Miles, M. B., & Huberman, A. M. (1994). *Qualitative data analysis: An expanded sourcebook*. Thousand Oaks, CA: SAGE.

Mint (2018). *Only 16% of rural users access Internet for digital payments: Report*. Retrievable from: https://www.livemint.com/Politics/PhY0kTxoJqq6U9GISIpSaK/Only-16-of-rural-users-access-Internet-for-digital-payments.html

Miralles, I., Dentoni, D., & Pascucci, S. (2017). Understanding the organization of sharing economy in agri-food systems: Evidence from alternative food networks in Valencia. *Agriculture and Human Values*, *34*(4), 833–854.

Möhlmann, M. (2015). Collaborative consumption: Determinants of satisfaction and the likelihood of using a sharing economy option again. *Journal of Consumer Behaviour*, *14*(3), 193–207.

Mooi, E., Kashyap, V., & van Aken, M. (2020). Governance and customer value creation in business solutions. *Journal of Business & Industrial Marketing*, *35*(6), 1089–1098.

Nakasone, E., Torero, M., & Minten, B. (2014). The power of information: The ICT revolution in agricultural development. *Annual Review of Resource Economics*, *6*(1), 533–550.

Ozanne, L. K., & Ozanne, J. L. (2011). A child's right to play: The social construction of civic virtues in toy libraries. *Journal of Public Policy & Marketing*, *30*(2), 264–278.

Park, E., & del Pobil, A. P. (2013). Technology acceptance model for the use of tablet PCs. *Wireless Personal Communications*, *73*(4), 1561–1572.

Park, H. J., Kim, J. O., & Son, Y. H. (2015). The analysis of the longitudinal trend of the sense of community in adolescence and its predictors. *Asian Journal of Education*, *16*(4), 105–127.

Pouri, M. J., & Hilty, L. M. (2018). Conceptualizing the digital sharing economy in the context of sustainability. *Sustainability*, *10*(12), 4453.

Prodanova, J., Ciunova-Shuleska, A., & Palamidovska-Sterjadovska, N. (2019). Enriching m-banking perceived value to achieve reuse intention. *Marketing Intelligence & Planning*, *37*(6), 617–630.

Rangan, V. K., & Rajan, R. (2007). *Unilever in India: Hindustan lever's project Shakti-marketing FMCG to the rural consumer* (pp. 9–505). Boston, MA: Harvard Business School.

Rauch, D., & Schleicher, D. (2015). Like Uber, but for local governmental policy: The future of local regulation of the 'sharing economy'. *George Mason Law & Economics Research Paper*, (15-01).

Schmutz, P., Heinz, S., Métrailler, Y., & Opwis, K. (2009). Cognitive load in eCommerce applications—measurement and effects on user satisfaction. *Advances in Human-Computer Interaction*, https://doi.org/10.1155/2009/121494

Schor, J. (2016). Debating the sharing economy. *Journal of Self-Governance and Management Economics*, *4*(3), 7–22.

Sengupta, T., Narayanamurthy, G., Moser, R., & Hota, P. K. (2019). Sharing app for farm mechanization: Gold Farm's digitized access based solution for financially constrained farmers. *Computers in Industry, 109,* 195–203.

Stephany, A. (2015). *The business of sharing: Making it in the new sharing economy.* Berlin, Germany: Springer.

Sumner, J. (2005). Value wars in the new periphery: Sustainability, rural communities and agriculture. *Agriculture and Human Values, 22*(3), 303–312.

Sundar, S. S. (2008). The MAIN model: A heuristic approach to understanding technology effects on credibility. In M. J. Metzger & A. J. Flanagin (Eds.), *Digital Media, Youth, and Credibility* (pp. 73–100). Cambridge, MA: The MIT Press.

Sung, E., Kim, H., & Lee, D. (2018). Why do people consume and provide sharing economy accommodation?—A sustainability perspective. *Sustainability, 10*(6), 2072.

Talule, D. (2020). Farmer suicides in Maharashtra, 2001–2018: Trends across Marathwada and Vidarbha. *Economic and Political Weekly, 55*(25), 116–125.

Teo, T. S., & Yu, Y. (2005). Online buying behavior: A transaction cost economics perspective. *Omega, 33*(5), 451–465.

The Better India (2016). *Farmers in Karnataka will soon be able to use an app to rent farm equipment on an hour or day basis.* Retrievable from: https://www.thebetterindia.com/63336/farmers-agri-mobile-app-karnataka/

The Hindu (2016). *M&M launches e-commerce startup for farm equipment rental.* Retrievable from: https://www.thehindu.com/news/cities/mumbai/business/mm-launches-ecommerce-startup-for-farm-equipment-rental/article8354742.ece

The New York Times (2016). *How do you hail a tractor in India? All it takes is a few taps on your phone.* Retrievable from: https://www.nytimes.com/2016/10/18/world/what-in-the-world/trringo-app-india.html

The Wire (2018). *What is the solution to Punjab's paddy stubble burning problem?* Retrievable from: https://thewire.in/agriculture/what-is-the-solution-to-punjabs-paddy-stubble-burning-problem

TOI (Time of India) (2016). *Govt launches mobile app 'Pusa Krishi'.* Retrievable from: https://timesofindia.indiatimes.com/centre/Govt-launches-mobile-app-Pusa-Krishi/articleshow/51507384.cms

TOI (Time of India) (2017). *Over 12,000 farmer suicides per year, Centre tells Supreme Court.* Retrievable from: https://timesofindia.indiatimes.com/india/over-12000-farmer-suicides-per-year-centre-tells-supreme-court/articleshow/58486441.cms

Troncoso, K., Castillo, A., Masera, O., & Merino, L. (2007). Social perceptions about a technological innovation for fuelwood cooking: Case study in rural Mexico. *Energy Policy, 35*(5), 2799–2810.

Wang, A. (2005). Integrating and comparing others' opinions. *Journal of Website Promotion, 1*(1), 105–129.

Wang, H. Y., & Wang, S. H. (2010). Predicting mobile hotel reservation adoption: Insight from a perceived value standpoint. *International Journal of Hospitality Management, 29*(4), 598–608.

WCED (World Commission on Environment and Development) (1987). *Our Common Future.* Oxford: Oxford University Press.

Wever, M., Wognum, N., Trienekens, J., & Omta, O. (2012). Managing transaction risks in interdependent supply chains: An extended transaction cost economics perspective. *Journal on Chain and Network Science, 12*(3), 243–260.

Williamson, O. E. (1979). Transaction-cost economics: The governance of contractual relations. *The Journal of Law and Economics, 22*(2), 233–261.

Williamson, O. E. (1989). Transaction cost economics. *Handbook of Industrial Organization, 1,* 135–182.

Xiao, M., Wang, R., & Chan-Olmsted, S. (2018). Factors affecting YouTube influencer marketing credibility: A heuristic-systematic model. *Journal of Media Business Studies, 15*(3), 188–213.

Yin, R. K. (2009). *Case study research: Design and methods.* London, SAGE Publications Ltd.

YourStory (2018). *Agri startups are bringing in IoT, BigData and equipment to farmers and revolutionising the sector.* Retrievable from: https://yourstory.com/2018/02/iot-big-data-equipment-farmers-agri-startups?utm_pageloadtype=scroll

Zeithaml, V. A. (1988). Consumer perceptions of price, quality, and value: A means-end model and synthesis of evidence. *Journal of Marketing, 52*(3), 2–22.

Zhao, S. X., & Wong, K. K. (2002). The sustainability dilemma of China's township and village enterprises: An analysis from spatial and functional perspectives. *Journal of Rural Studies, 18*(3), 257–273.

Zhou, T. (2014). Understanding the determinants of mobile payment continuance usage. *Industrial Management & Data Systems, 114*(6), 936–948.

Zhou, T., Li, H., & Liu, Y. (2010). The effect of flow experience on mobile SNS users' loyalty. *Industrial Management & Data Systems, 110*(6), 930–946.

Section V

Advances in Knowledge of Social and Sustainable Marketing: Understanding Social Marketing

12

Wroclaw: Transforming a City towards a Circular Economy-Zero Waste Social Marketing Campaign in Poland

Dorota Bednarska-Olejniczak and Jaroslaw Olejniczak
Wroclaw University of Economics and Business, Faculty of Business and Management, Faculty of Economics and Finance, Komandorska, Wroclaw, Poland

CONTENTS

Learning Objectives	312
Themes and Tools	312
Social Marketing	312
Circular Economy and Zero Waste Concept	315
Municipal Waste	319
Food Waste	321
Education	322
Wrocław – City's Characteristics in the Context of Transition into CE	323
Social Campaign Entitled "Wrocław does not Waste" as a Tool for Shaping the Attitudes and Behaviours of Residents in the Scope of the Zero Waste Concept	326
Summary	333
Discussion	337
Conclusion	339
Lessons Learned	341
Discussion Questions	341
Exercise in Groups	342
Funding Declaration	342

Acknowledgment 342
Credit Author Statement 342
References 342

LEARNING OBJECTIVES

After reading the chapter, the reader should be able to:

- Demonstrate how to use social marketing techniques to promote ideas that serve the common good and protect common resources
- Explain how target groups in social marketing can be defined
- Indicate what types of goals are there and how they are defined in social campaigns
- Teach how to choose marketing instruments according to the goal and target group

THEMES AND TOOLS

- Social campaign
- Transtheoretical model of change
- Engagement and education
- Change of attitudes
- Deconsumption
- Zero waste
- SWOT

SOCIAL MARKETING

The goal of social marketing is to efficiently impact the behavioural patterns among certain groups of recipients or an entire community. Kotler and Lee (2011, p. 34) indicate for key directions of such changes in behaviours of recipients: (1) their acceptance of new behaviours (i.e. giving away unnecessary clothes to persons in need or preparing meals from leftovers), (2) rejecting potentially non-demanded behaviours (i.e. reckless shopping of too large amounts of food, leading to their wasting), (3) modifying the present behaviour (i.e. increasing the frequency of car-sharing when travelling), (4) giving up the old, undesired behaviour (i.e. throwing non-segregated waste). Two directions tend to be presented in the subject literature every now and then: (5) in which we get people to continue the demanded behaviours and (6) in which we want people to switch behaviour. Change of attitudes and levels of knowledge among the recipients is a substantially important aspect of social marketing; however, the final measure of this success is an actual modification of behaviours of addressees of such actions. One of the activities undertaken in social marketing is the implementation of social campaigns or public information campaign. Weiss and Tschirhart (1994) underline that those campaigns refer to government-sponsored communication efforts usually aimed at shaping the beliefs, attitudes, social norms and behaviour of society or parts of it. This may involve both increasing citizens' involvement in public affairs and shaping attitudes that are beneficial to households and the environment.

An important stage of the planning process of social marketing is to formulate its goals. Lee and Kotler (2011, p. 87) distinguish three goals of social marketing: (1) behaviour objective – as stated previously it is a strive for recipients to accept, reject, modify, abandon, switch or continue something, (2) knowledge objectives – concern provision of information regarding facts recipients ought to be familiar with, as their knowledge and understanding significantly increase the motivation of people to undertake or change activities. This justifies the importance of information and educational actions undertaken in the frames of social marketing, targeted at making the society conscious of the issues and revealing ways to cope with them, (3) belief objectives – concern shaping sensations and attitudes of recipients. Shaping attitudes and behaviours is a long-term process; thus, the undertaken actions and tools used for this purpose should also be long term.

Theoretical bases for reaching the goals of social campaign Wroclaw zero waste may be found in the Transtheoretical Model of Change (TMC), Hierarchy of Effects Model (Attention Interest Desire Action (AIDA)) and Community Reliance Model (CRM). In social marketing, the first of the models is used rather frequently (Cismaru & Wuth, 2019; Levit,

Cismaru, & Zederayko, 2016; Luca & Suggs, 2013). This model elaborated by Prochaska and DiClemente in 1982 (Prochaska & DiClemente, 1982) indicates the necessity to perform segmentation of information recipients not only based on the most frequently met criteria but also based on the level of readiness for changes. This stems from the fact that the process of changing behaviour is of multistage nature, which results in the necessity to adjust the message and tools to individual stages. From the point of view of changing behaviours of an entity stemming from their own will, one may notice that the city, through various activities and actions under the campaign, strives to support the transition of its inhabitants through the subsequent stages of the process. The first stage of this process may be "pre-contemplation" – lack of thinking about changing behaviour, unwillingness to change one's behaviour, perception of change as a nuisance, failure to realize the results of present behaviour and the experience of negative effects in the previous attempts to change. Hence, the key goal of undertaken actions towards persons remaining on this stage is to make them aware of the issue. This leads to the second stage – "contemplation" whereby changes stemming from an awareness of the issue are possible and considered by inhabitants, while they analyze the pros and cons of changes, search for information. While a lack of readiness to change is still visible. Properly realized social marketing actions may lead to the transition of inhabitants towards the stage of "preparation", characterized by readiness for changes, gathering information, preparation of the plan of changes and selection of specific solutions. Yet, another stage is "action", thus undertaking of actions – carrying out changes. Nevertheless, an important issue is to obtain a permanent change, since as a result of the impact of various internal and external factors, the final stage – "maintenance" – may prevail in the long perspective. Such "shifting" between stages concerns the whole process of changes – since as a result of failures of actions it is easy to face discouragement and abandonment of changes.

The second model – AIDA – is visible in a diverse use of social advertising, depending on the degree of consumer engagement (Pelsmacker, Geuens, & Bergh, 2013, p. 74). One may consider this to be supplementation of the previously indicated TMC – as, depending on the stage of TMC, the goal of communication actions will vary.

The third model – CRM, the elements of which may be found here, is focused on community change (Chilenski, Greenberg, & Feinberg, 2007). From the perspective of the city, the success of introducing zero waste economy will be reflected through a permanent change in attitudes and behaviours of the whole community and not individuals forming it. In line with CRM, this process may cover joint actions of inhabitants that lead to gaining by the community of necessary changes, elaborating plans of actions and their implementation as generally demanded.

Social marketing, thus, uses a set of instruments which originate from the classic 4P model by McCarthy (1960), though contrary to the traditional marketing in this case one ought to consider individual tools:

- Product – it is an object of actions, i.e. behaviours and attitudes or changes in attitudes/behaviours of the target market, which are the goal of social marketing.
- Price – costs and other barriers that the target market connects with the proposed product and methods of their minimization.
- Promotion – messages, sources of data, promotional tools and techniques used for passing messages of social marketing.
- Place – places in which the addressees may undertake a decision regarding accepting the expected behaviours; places in which one may carry out actions under social marketing, offer products and additional materials.

CIRCULAR ECONOMY AND ZERO WASTE CONCEPT

One of the most frequently indicated reports as underlying document of Circular Economy (CE) is a 1976 report on the "Potential for Substituting Manpower for Energy", for DG V Labour and Social Affairs, by Walter R. Stahel and Geneviève Reday at the Battelle Institute Geneva (final version published in Stahel & Reday-Mulvey, 1981). This report defined the structure and nature of an "economy in loops" (or CE). Its authors pointed to two different loops with different impacts – the first loop "product-life extension" (reuse of goods, repair of goods, reconditioning of goods, upgrading of goods) and the second one – "reusing molecules" (recycling of goods). One of the most frequently cited definitions is the definition prepared by the Kirchherr-Reike-Hekkert team, which indicates that "a circular economy describes an economic system that is based on business models, which replace the 'end-of-life' concept with reducing, alternatively reusing, recycling and recovering materials in production/distribution and consumption processes, thus operating at the micro level (products, companies, consumers), meso level (eco-industrial parks) and macro level (city, region, nation and beyond), with the aim to accomplish sustainable development, which implies creating environmental quality, economic prosperity and social equity, to the benefit of current and future generations" (Kirchherr, Reike, & Hekkert, 2017, pp. 224–225).

A good example is the concept of "zero waste", i.e. a lifestyle that aims to reduce the amount of waste in households by: reducing consumption, sharing instead of owning, repairing or recycling. Similarly to CE, the concept of zero waste is defined in various manners in theory and practice (Cole et al. 2014; Lehmann, 2011; Phillips, Tudor, Bird, & Bates, 2011). Braungart, McDonough and Bollinger (2007) indicate that: "The zero waste concept encompasses a broad range of strategies including volume minimization, reduced consumption, design for repair and durability and design for recycling and reduced toxicity". On the other hand, the latest definition of zero waste, which was developed by Zero Waste International Alliance (ZWIA), defines it as "the conservation of all resources by means of responsible production, consumption, reuse, and recovery of products, packaging, and materials without burning and with no discharges to land, water, or air that threaten the environment or human health" (ZWIA, 2018). In accordance with the division indicated by this organization, from the viewpoint of a consumer/resident, the following principles are important in the scope of zero waste's implementation (ZWIA, 2018): Rethink (Refuse, e.g. reconsider purchasing needs and look for alternatives to product ownership), Reduce (e.g. to plan consumption and purchase of perishables to minimize discards due to spoilage and non-consumption), Implement Sustainable Purchasing (that supports social and environmental objectives as well as local markets where possible), Reuse (e.g. maintain, repair or refurbish to retain value, usefulness and function), Recycle/Compost (Rot) (e.g. provide incentives to create clean flows of compost and recycling feedstock, support and expand composting as close to the generator as possible, prioritizing home or on site or local composting wherever possible). In practice, it is the extension of the 3R concept within the CE, which is discussed in the literature as 5R (Refuse, Reduce, Reuse, Recycle, Rot). It is necessary to emphasize that this is one of many areas which are important from viewpoint of the city's transition into the zero waste and CE (Figure 12.1).

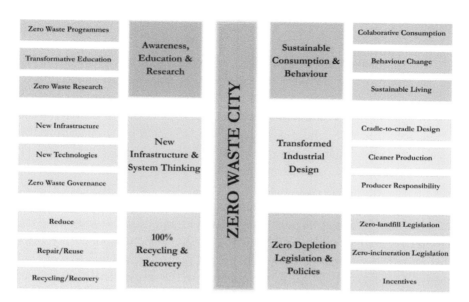

FIGURE 12.1
Drivers for transforming current cities into zero waste cities. (From Zaman & Lehmann, 2013, p. 125).

In recent years, the CE has been one of the priorities of the European Union's economic policy. Its current scope and tasks are presented in the document entitled "A new Circular Economy Action Plan for a cleaner and more competitive Europe" (European Commission, 2020) dated March 11, 2020. This plan includes a series of interrelated initiatives that aim to establish a solid and coherent product policy framework, which will make sustainable products, services and business models the new standard and which will transform the consumption patterns in a manner preventing the generation of waste.

The problem of increasing environmental pollution and the overconsumption of natural resources is known to all of us. However, we are not always willing to take action to reduce these phenomena – especially in our everyday lives. How, then, can we convince people to reduce consumption, rational purchases or the amount of municipal waste they generate? This problem is particularly evident in large cities, where, among other things, the authorities are trying to influence the attitudes and behaviour of residents through social campaigns promoting the idea of zero waste. This chapter will discuss the actions which one of the cities in

Poland is taking to persuade its residents to reduce the amount of various types of rubbish they produce.

Throughout Poland, the authorities of many major cities are beginning to see the problem of waste management, in the context of striving to ensure sustainable development of the city, and undertake actions beyond the obligatory organization of municipal waste collection in the city, in the scope resulting from the Act of September 13, 1996 on maintaining the cleanliness and order in communes, and its amendment (Dz.U. 1996 nr 132 poz. 622, 1996). In this aspect, they undertake activities included in the environmental strategies, as well as other actions in the form of separate programmes, which lead to reduction in the amount of waste by the city's stakeholders, e.g. by providing them with knowledge, influencing the change of their attitudes and behaviours. The abovementioned activities are compliant with the zero waste philosophy, which (although to a limited extent) is systematically gaining popularity in the scope of management of Polish cities. However, it is necessary to emphasize that significantly more often the zero waste occurs as a separate and systemic issue, which is not associated with the implementation of CE by the largest Polish cities, due to their relatively little experience in the area of CE. It should not come as a surprise, considering that it was not until the end of 2019 that the document regulating the Poland's transition into CE was adopted. This document is the "Roadmap for the circular economy", which was adopted by the Council of Ministers on September 10, 2019 (Ministry of Entrepreneurship and Technology, 2012), and which indicates the objective, as well as scope of actions that have to be undertaken in this field. Moreover, it should be noted that this "Roadmap for the circular economy" constitutes one of the elements intended to implement the main objective of much broader (Ministry of Development, 2016). Poland's priorities within CE include four following areas, which are indicated in the "Roadmap for the circular economy":

- Innovativeness, improvement of cooperation between the industry and the science sector, and in result the implementation of innovative solutions in the economy
- Creating the European market for secondary raw materials, on which their flow would be easier
- Ensuring high-quality recycled raw materials, which result from sustainable production and consumption
- Development of the service sector

The "Roadmap for the circular economy" is based on the CE model commonly used in the EU, which has been developed by the Ellen MacArthur Foundation (www.ellenmacarthurfoundation.org/, 2012), and which assumes the existence of two cycles: biological cycle – including renewable resources, and technical cycle – including non-renewable resources (Ministry of Entrepreneurship and Technology, 2012, p. 6). The objectives of this document can be perceived in the long perspective as an indication of long-term measures, which relate to the widest possible segment of socioeconomic life, and in the short-term as a selection of priority areas, the development of which will allow Poland to seize opportunities in this scope, and at the same time will constitute a response to the current and expected threats.

The above mentioned document is divided into five chapters, which are devoted to: sustainable industrial production, sustainable consumption, bio-economy, new business models, and implementation, monitoring and financing of the CE. It is evident that sustainable consumption is a priority area in the process of Polish economy's transformation into the CE, particularly in the scope of three elements directly associated with the zero waste philosophy, which are (description of measures presented in the subsequent part of this study was prepared based on the information from Ministry of Entrepreneurship and Technology, 2012):

a. Municipal waste
b. Food waste
c. Education

Municipal Waste

In 2013, a waste management reform was introduced in Poland that transferred the control of municipal waste to communes, which in practice meant the exclusion of entrepreneurs from competing for an individual customer, and the need to participate in tenders for collecting and managing municipal waste announced by the communes. In 2018, according to Statistics Poland (GUS), Poland generated 12,485,000 tonnes of waste and 4.3% increase in generation was noted compared to the previous year. This means an increase in the amount of generated municipal waste per capita from 311 kg in 2017 to 325 kg in 2018 (GUS, 2019b, p. 134). It should be emphasized that the highest indicator of generated waste per capita – amounting to 394 kg – was noted in 2018 in the Lower Silesia Province, and the city of Wrocław (provincial capital) significantly contributed to that by generating 531 kg of municipal waste per capita. In Poland, 7.1 million tonnes of municipal waste collected in 2018 were designated for recovery (i.e. 57% of generated municipal waste), and out of it:

- 3.3 million tonnes (26%) were allocated for recycling
- 2.8 million tonnes (23%) for thermal transformation with energy recovery
- 1.0 million tonnes (8%) for biological treatment processes (composting or fermentation) (GUS, 2019b, pp. 154–155)

A total of 5.4 million tonnes were allocated for the neutralization processes, out of which 5.2 million tonnes (approximately 42% of generated municipal waste) were designated for landfilling, while the remaining 0.2 million tonnes (approximately 2% of generation) for disposal by thermal transformation without energy recovery. The abovementioned data demonstrates that too much waste is still landfilled and there is a need for better management of it. Reduction in the level of landfilled waste could have a positive impact on the development of the Polish economy, particularly in the context of lowering the demand for primary raw materials, in favour of the greater use of recycled raw materials. At the same time, the improvement of municipal waste management is dictated by the requirements of European law – amendment to the Directive on Packaging and Packaging Waste, which was adopted in 2018, provides for a significant increase in the levels of recycling of packaging waste, along with simultaneous limitation of the possibility of landfilling the municipal waste (to 10% in 2035).

The "Roadmap for the circular economy" indicates three measures planned for the period 2019–2022 in the scope of municipal management:

- Monitoring of effectiveness and performance of current regulations, as well as developing recommendations for adapting and changing the national regulations concerning the municipal waste (2021–2022)
- Preparation of a proposal for regulations concerning the hazardous waste (2019–2021)

- Identification of all municipal waste streams, including post-consumer waste, which has not been recorded so far, but has economic significance and importance in terms of achieving recovery and recycling objectives in waste management (2020–2021)

Food Waste

In accordance with estimates of the European Commission from 2010 (Directorate-General for Environment, 2014), there are over 9 million tonnes of food wasted in Poland every year. Food is wasted at the stage of production, as well as the stage of distribution and consumption. Food losses at the stage of consumption usually result from difficulties in determining the demand, incorrect planning of purchases and meals as well as inadequate storage. Selective collection of food waste and its management constitute a significant element of waste management, particularly when taking into account the current results of research conducted by the Federation of Polish Food Banks in 2019 (Federation of Polish Food Banks, 2019). The abovementioned data demonstrates that mainly households are responsible for wasting food, while the level of consumer waste in recent years has maintained at a similar level – 42% of respondents in 2019 admitted that they sometimes throw out food, and this result was similar to the result obtained in 2018. Most often thrown out products include: bread (23.7%), cold cuts (12.8%) and fresh fruits (12.6%). The main reasons for throwing out food include: spoilage (65.2%), missing the expiration date (42%) and preparing too much food (26.5%). Other reasons also include: too big purchases, purchase of a poor-quality product and lack of ideas for using a given ingredient. It should be also emphasized that, according to the results of this research, Poles are spontaneous consumers – almost every third respondent (27.8%) never or very rarely prepares a shopping list, while as many as 72.7% of respondents spontaneously buy products that they had not planned to buy before. Moreover, Polish consumers avoid buying fruits and vegetables, the appearance of which is not attractive – almost 74% of respondents do not buy small or shapeless (although completely full-value) products at all or do it rarely. It has also been demonstrated that Polish consumers have problems with distinguishing between the terms associated with the expiration dates – as many as 64% of respondents considered the terms "use-by date" and "best-before date" to be identical – or could not point out the difference between them. Therefore, it is evident that the implementation of educational activities is an extremely important factor, without which it is basically impossible to change the attitudes and behaviours of consumers in the scope of reducing food waste.

Measures planned in the "Roadmap for the circular economy" for 2019–2022 provide the following in this scope:

- Implementation of an information campaign in the years 2020–2021, in order to disseminate knowledge among consumers and producers in the scope of counteracting food waste (among others by disseminating the 4P principle – planning of purchases in advance, processing food in order to extend shelf life, storing products in appropriate conditions as well as sharing unnecessary food with those in need).
- Development (in the years 2020–2021) of the concept of distribution mechanisms and appropriate handling of products with brief expiration date.
- Development of a system of incentives and obligations for entrepreneurs, aimed at counteracting food waste – as a result, the enterprises are supposed to become more actively involved in cooperation with charity organizations, which provide food to those in need.
- From 2021, the implementation of periodic statistical research concerning the scale, structure and directions of processes associated with food waste in Poland.

Education

Transformation towards a CE requires appropriate environmental education of consumers as well as producers. Unfortunately, the studies indicate a relatively weak belief of Polish consumers in regard to the real impact on the environment, in which they live. Studies concerning environmental awareness demonstrate that Polish consumers are aware of the risks associated with overexploitation of resources; however, they do not know the practical ways to counteract this phenomenon. Thus, it is necessary to implement education aimed at changing the attitudes and behaviours of consumers, particularly by providing knowledge that raises the awareness of Poles in the scope of environmental protection, as well as by increasing their knowledge concerning the rights in the scope of access to information about the product and the producer. Education in this field should cover all age groups and start already at the stage of basic education, by providing practical knowledge and presenting trends that are currently emerging in the market. It is especially important to educate adults, particularly in the practical aspects of sustainable consumption, whereas social campaigns constitute an appropriate tool to achieve this objective.

Two planned measures are indicated in the "Roadmap for the circular economy" in this scope:

- Development of the concept of governmental information platform regarding the CE in the period 2020–2021. It is supposed to enable the exchange of information between the government administration, business sector and self-government. Ultimately, it should include: guides on CE, information on incentives for entrepreneurs as well as current support programmes and information brochures.
- Implementation of a social campaign on sustainable consumption patterns in the period 2020–2021. Its objective is to popularize (among all social groups) the sustainable consumption patterns concerning (among others): sharing, waste management, storage of food, purchase of functions instead of ownership of products, etc.

Wrocław – City's Characteristics in the Context of Transition into CE

In Central Europe, in the middle of a triangle formed by Berlin, Warsaw and Prague, lies the fourth largest city in Poland – Wrocław. This cultural, scientific, industrial and economic centre, which has over 640,000 residents, is one of the oldest and most beautiful cities in Poland. It is also the capital of the region (Lower Silesian Province [NUT 2]) – the seat of its self-government authorities and regional government administration. It is one of the most important metropolitan centres in Poland, which favours the inflow of new residents and enterprises. There are 25 universities in Wrocław that educate over 110,000 students, which may significantly affect an increase in the consumption of public services, among others collective transport or municipal waste management. Among its residents, about 59.2% people are of working age, 16.6% are people of pre-working age (children and youth), 24.2% people are of post-working age (retirees) – the share of the last two groups has significantly increased in the recent years (GUS, 2019a). The economic situation of residents, as well as the city compared to other cities, should be assessed as favourable. In 2018, the average salary of an employee in Wrocław was approximately 5% higher than the national average, while the city's income per capita placed this city in the second place among the capitals of 16 regions (GUS, 2019a). Wrocław is characterized by a large urban diversity (101 separate urban units) (Wrocław City Council, 2018, p. 26) with various functions (residential, commercial, service, industrial). At the same time, it should be emphasized that 2/3 of Wrocław consist of green and agricultural areas, while residential areas cover only 12%. The urban layout of Wrocław is partly a consequence of over 1000 years of development of the city located on the main communication routes of Europe, crossed by the Odra River and its tributaries, as well as the destruction and reconstruction after World War II – the dense pre-war development of southern part of the city was largely replaced by housing estates of large-panel buildings (usually 4- and 10-story multifamily buildings). At the same time, new housing estates with various buildings are developing dynamically.

Despite the high level of development and implementation of solutions aimed at improving the quality of residents' life, e.g. within the SmartCity Wrocław programme (the city is every year classified in the Top-100 of the IESE Cities in Motion Index 2019; it is also second in Poland after Warsaw (IESE Business School, 2019)), improvement of the public transport network, implementation of public bikes, scooters and electric cars, as well as programmes of heating stoves' replacement with low-emission ones, the city is still experiencing problems associated with environmental protection. Among others, they consist of seasonal high air pollution and the problem with excessive generation of municipal waste by residents (President of the City of Wrocław, 2019a). In recent years, the amount of this waste has significantly decreased; however, there is still over 540 kg of waste per capita generated annually (second place in Poland). In accordance with the regulations, all managers (owners) of inhabited and uninhabited real estate, generating municipal waste in the city of Wrocław, are obliged to pay disposal fees. In the case of residential buildings, such fees depend on the number of residents or the number of m^2 of residential area/per capita, as well as the declaration of using/not using waste sorting. The rates of fees are calculated by the city based on the costs of collection and disposal. The fees are collected on a monthly basis and their amount fluctuates around the average national rates.

In 2019, the city undertook actions aimed at integration of its activities within the city's consistent transformation towards a CE. A representative for CE was appointed, who is supposed to coordinate all activities in this area. Moreover, the analyses in the scope of identification and diagnosis of the city's functioning within CE are carried out. At the same time, the city implements activities of investment nature (construction of infrastructure – composting plants), systemic nature (e.g. implementation plan of low-emission economy) and educational nature in the scope of CE.

Activities undertaken within CE take into account the guidelines of the "Roadmap for the circular economy", as well as the EU guidelines. One of the elements of activities in the scope of sustainable development and CE is the "Wrocław does not waste" campaign. Changing the attitudes and behaviours of residents requires the city to use the right tools. One of the key ones may be a social campaign – an important element of social marketing used in the public sector. The aim of this chapter is to present the possibility of using the social campaign as a tool supporting the implementation of the zero waste idea in the process of transforming the city into a CE. The campaign has a long-term character and focuses on shaping the attitudes and behaviours of the inhabitants in terms of understanding, acceptance and practical implementation of the zero waste idea in everyday life. The objectives of the campaign are (1) change of inhabitants' behaviour in the field of waste (behavioural goals), (2) increase of inhabitants' knowledge about the idea of zero waste (cognitive goal) and (3) shaping convictions, opinions and feelings of inhabitants of Wrocław in the area of zero waste (goal connected with attitudes). Target groups covered by the activities are children and youth, adult residents and officials working in city offices. This chapter will allow readers to understand the role of the social campaign in the implementation of the idea of zero waste as part of the city's transition to a CE and to understand how the implementation of the behavioural, knowledge and belief objectives is to relate to the key elements of the zero waste concept in this type of campaign. The study describes and analyzes the various communication channels and tools used in the campaign to reach its target markets. It shows that the Internet is an important communication channel for the campaign, but equally important are happenings and special events and other public relations tools. It also points to the need to apply differentiated segmentation of the population, among other things, due to the degree of involvement in the process of change according to the TMC theory.

SOCIAL CAMPAIGN ENTITLED "WROCŁAW DOES NOT WASTE" AS A TOOL FOR SHAPING THE ATTITUDES AND BEHAVIOURS OF RESIDENTS IN THE SCOPE OF THE ZERO WASTE CONCEPT

The campaign called "Wrocław does not waste" was commenced in April 2019. This way, as one of the few cities in Poland, Wrocław began long-term and comprehensive activities aimed at promoting among residents and implementing the zero waste concept. Long-term – because this campaign has no fixed timeframe and it is intended as a long-term process. Comprehensive – because its activities cover all residents (children, youth, adults), adapting communication tools and channels to the specificity of each group of recipients. The reason for targeting such a wide group of recipients is the need to influence all residents who, by making numerous consumer choices every day, can through them contribute to the protection of natural resources and to the reduction of negative impact on the environment. This campaign is supposed to teach responsible methods of buying and using products, as well as services, in order to reduce the amount of generated waste, and to make residents aware of their responsibility for the city's environment and ecosystem. Moreover, it is also supposed to provide residents with knowledge regarding the CE, which is not widely known at this moment. The campaign logo (Figure 12.2) plays an essential role in the visual identification of this campaign, which positions its main objective by using the name of the campaign "Wrocław does not waste", and additionally by adding the "Zero waste" slogan, as well as in the graphic layer, by placing symbols associated with recycling, ecology and nature.

The logo is maintained in a colour tone evoking associations with nature – it is dominated by shades of green, white, beige and black. The campaign logo is placed on its website, on the campaign profiles in social media, on promotional materials used during the campaign, in media coverage of the events associated with the campaign and on the doors of partner institutions that participate in this programme. Name of the campaign – "Wrocław does not waste", which is an important element of its identity, allows it to remain in the minds of residents and facilitates positioning. It is accurate, short and easy to remember; thus, it is identifying, as well as informative.

FIGURE 12.2
"Wroclaw does not Waste" campaign logo. (From https://www.wroclaw.pl/srodowisko/wroclaw-nie-marnuje (with permission of Wroclaw City Council)).

One of the first activities undertaken as part of the campaign was to create a website: www.wroclaw.pl/srodowisko/wroclaw-nie-marnuje. In addition to the campaign's name and logo, the website demonstrates the essence of CE, describes activities that help consumers to select products and services, which are better for the environment, while at the same time ensure financial savings and improve the quality of life. They include, among others: the use of renewable energy, reducing the amount of waste by reusing it, sharing, i.e. joint use of objects and sharing services, recycling, i.e. recovery of raw materials, and upcycling, i.e. secondary processing of waste. The abovementioned website also contains a video presenting the idea of this campaign, as well as initiatives and venues that contribute to it. Moreover, the website also includes tabs devoted to the composter sharing programme, the selective municipal waste collection point (PSZOK) and the research-educational programme "ECO-School". In the period from January 1, 2019 to July 15, 2020, the number of views of the campaign website amounted to 11,161, the number of views of the composter sharing programme tab – 1,255, the number of views of the Eco-School programme tab – 346.

An important tool of the campaign, which is also located on its website, is the interactive map of "Wrocław does not waste" (Figure 12.3).

328 • *Social and Sustainability Marketing*

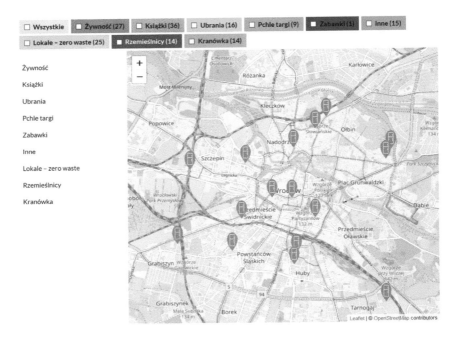

FIGURE 12.3
Screenshot of "Wroclaw does not Waste" campaign interactive map. (From https://www.wroclaw.pl/srodowisko/wroclaw-nie-marnuje [with permission of Wroclaw City Council]).

This virtual map presents the locations of places/institutions that allow residents to donate or exchange unnecessary food products, books, toys, etc. It also makes it easier to find flea markets (markets allowing for resale or exchange of second-hand goods) in the city, as well as venues promoting the zero waste principles, and places where you can drink tap water. Currently, the map helps the residents to find the following categories of places (all the information for this section was obtained from the campaign website: www.wroclaw.pl/srodowisko/wroclaw-nie-marnuje):

- Food (27 locations) – community refrigerators (24-hour shared refrigerators available to all residents, where they can leave unnecessary food products or from which they can take the products that they need), soup kitchens, social cafes.
- Books (35 locations) – bookcrossing locations, i.e. places for free transfer of books by leaving them in a specially designated place.

These may consist of specially marked bookcases, tables, cabinets, from which the residents take books and then pass them on. Among others, such places are located in one of the Wrocław's cinemas, in neighbourhood clubs, shopping malls, botanical garden, cafes, primary school, traditional libraries, offices of the Housing Estate Councils, Staromiejski Park, Main Railway Station, etc.
- Clothes (16 locations) – places where you can bring or from which you can take clothes free of charge. These include the offices of associations, homeless shelters, clubs and foundations, as well as generally accessible hangers placed in public places, where the residents leave, e.g. winter clothing for those in need, who can take it free of charge.
- Flea markets (9 locations) – markets that allow to buy second-hand products or exchange such products for other products, e.g. clothes, toys, books.
- Toys (2 locations).
- Other (14 locations) – places where you can leave certain products for those in need (e.g. for babies, for the homeless, for children from hospice), as well as stores that promote the zero waste concept.
- Zero-waste venues (24 locations) – mainly cafes and restaurants.
- Craftsmen (14 locations) – the map presents locations of the workshops of watchmakers, tailors, shoemakers as well as upholstery and furniture repair shops.
- Tap water (13 locations) – places where the residents can drink tap water for free. Mainly restaurants, bistros, cafes.

As previously mentioned, each venue that joins the campaign and introduces the zero waste principles receives a sticker with the campaign logo, which informs about the partnership with the campaign and which is placed on the door. The map allows to fully use the potential of virtual environment in communication with the recipients – on the one hand, it allows them to quickly find the appropriate category of venue and their precise location, while on the other hand, it allows the recipients to independently report a venue that belongs to a given category and operates in accordance with the zero waste principles.

In the scope of communication with the residents, an important element was the appropriate selection of senders, i.e. determining who is supposed to convey the content of the campaign. It is essential for its success to select people, whose image will be consistent with the campaign's objective and will increase the interest in it, as well as its credibility. This also applies to the selection of campaign partners. As part of the "Wrocław does not waste" campaign, the cooperation was established with the journalist Sylwia Majcher, who is the author of the best-selling books on zero waste and conducts trainings on this subject, as well as with several key partners – Food Bank, Municipal Water and Sewerage Company (MPWiK S.A.), Housing Estate Councils, associations, the Food sharing Wrocław initiative. Together with our partners, a leftover cooking festival "Housing Estates do not waste" was organized, which was combined with the educational activities and the distribution of tap water. From May to September 2019, six Wrocław housing estates were the location of meetings organized in the form of a picnic, which were participated by about 1600 people. During these festival meetings, the residents participated in the workshops and learned from the experts how to prepare wholesome and tasty dishes from the leftovers. Moreover, they learnt how to reduce plastic in everyday life, use rainwater and how the composters work. Moreover, there were also workshops on pickling vegetables and fruits carried out, as well as workshops on sewing your own shopping bags, creative ecological workshops for children, trainings with a trainer for the elderly, which included the organization of weekly exercises in the park, gym, etc. Culinary workshops were led by the ambassadors of this event, consisting of the Wrocław chefs, among others the winner of the sixth Polish edition of the "MasterChef" TV show and the winner of the Polish edition of the "Top Chef" TV show. Reports from all events were shared on the Facebook social network.

The "Wrocław does not waste" campaign used various communication channels and tools, which allowed to reach its target markets. As it was previously demonstrated, the Internet is an important communication channel for the campaign; however, the happenings and special events, as well as other tools of public relations, are equally important. In addition to the abovementioned leftover cooking festival, the campaign also uses the following measures in the scope of public relations (Wrocław City Council, Department of Sustainable Development, 2020):

- Conducting educational activities during the city events (Odra Day, Spring Reading in the Botanical Garden, 37th PKO Wrocław Marathon).
- Publishing an article in the local newspaper entitled "Circular economy, i.e. from a steel straw to sustainable development" (April 15, 2019).

- Lecture during the Women's Voices concert, combined with culinary workshops of cooking from the leftovers (August 2, 2019).
- Using only biodegradable dishes during the culinary festival "Europe on a fork" (May 29, 2019–June 6, 2019) and the 37th PKO Wrocław Marathon (September 15, 2019).
- Organization of ecological workshops and workshops on energy production during the "Made in Wrocław 2019" trade fair. The "Wrocław does not waste" campaign was a part of the Smart City stand. The idea of "giveboxes" as a tool of exchange in the workplace was presented there. Moreover, the visualization of electricity generation and consumption in everyday life was also demonstrated (October 17, 2019).
- As part of the City's Volunteering Days (November 22, 2019), the lectures entitled "#WrocławDoesNotWaste, i.e. the Circular Economy" were organized.
- Organization of the collection of electronic waste among the employees of the Wrocław City Hall (April 16, 2019), during which also the educational activities were carried out – it included the discussions with an environmental educator, presentation of information on waste sorting, disposal of electronic waste, expired medications and bulky waste, familiarization with the concept of Selective Municipal Waste Collection Points (PSZOK). This event resulted in the collection of 177 kg of electronic waste.
- Organization of the first meeting with the residents entitled "An hour of zero waste", which was held at the Tarnogaj Activity Centre (February 29, 2020). It was devoted to not wasting food, using reusable products, household chemicals, cosmetics, and more broadly – introduction of the zero waste and less waste principles into our everyday life.

One of the five key elements of the zero waste concept is composting. This area was also taken into account in the "Wrocław does not waste" campaign – as previously mentioned, during the "Housing Estates do not waste" festival, the residents acquired knowledge regarding the composting. In order to support the implementation of this idea into the daily lives, the campaign includes a programme of providing free composters for composting green waste and biodegradable kitchen waste to the residents owning home gardens, as well as to educational institutions located within the city (Wrocław City Council, 2019). In 2019, 303 composters were made available (300 were issued to residents and three to educational institutions), while in the first half of 2020, as much as 230 composters were already issued (including four pieces to educational institutions). The purpose of this programme is to provide practical education to the residents, as well as to reduce the amount of municipal waste generated in Wrocław and designated for landfilling. The participants of this programme are required to submit reports on the amount of waste subjected to composting during the term of the contract, and after the termination of such contract, the composter becomes their property.

Another measure undertaken as part of this campaign is the placement of special bookcases within three buildings of the City Hall – the so-called "giveboxes", which are used to exchange books, CDs with music and movies. It is planned to place more bookcases in other buildings of the City Hall.

Implementation of the campaign objectives is also backed by the legislative support, which is one of the tools of the social marketing mix. In the case of the "Wrocław does not waste" campaign, such support is provided by the Ordinance of the President of Wrocław dated June 17, 2019 on the activities aimed at eliminating plastic in the Wrocław City Hall and organizational units of the Wrocław Commune, as well as at the events organized by the city (President of the City of Wrocław, 2019b). It was intended to implement actions consisting of eliminating the use of disposable items made of plastic in the City Hall's activity, and replacing them with reusable packaging or packaging made of biodegradable materials. Effects of the abovementioned ordinance include (accordingly to: Wrocław City Council, Department of Sustainable Development, 2020):

- Conclusion of a contract for the lease of 87 cylinder-free water dispensers for the Wrocław City Hall
- Conclusion of a contract for the supply of ecological disposable catering articles for the Wrocław City Hall
- Purchase of glass jugs for the employees of educational institutions, intended for boiled water
- Elimination of balloons during the meetings organized in educational institutions and replacing them with paper decorations
- Purchase of reusable water bottles for field employees
- Introduction of a series of ecological solutions on the premises of municipal companies – among others water dispensers, water bottles for employees, ecological gadgets

In 2019, the educational activities regarding the zero waste concept, addressed to the school youth, were also planned as part of the campaign. Ecological lessons were planned and prepared for the fourth-grade students of selected primary schools. They constitute the complementation of activities that have been conducted since 2017 by the Office for Nature and Climate Protection of the Wrocław City Hall, as part of the programme entitled "Sustainable development in Wrocław – our common cause", addressed to a wider group of recipients: for preschool children, for children in grades 1–3, for students in grades 4–7, as well as high school students. Lessons devoted to the zero waste concept are supposed to instil new consumption habits, draw attention to the problem of waste and develop creative eco-solutions. Moreover, this programme also envisages interactive activities associated with waste sorting, alternatives to plastic and conscious use of resources, as well as trips to venues operating in the trend of CE – farms located on the culinary heritage trail of Lower Silesia (Wroclaw City Council, 2019). Implementation of the educational lessons was planned for the 2019/2020 school year; however, it was temporarily suspended due to the problems associated with the pandemic.

SUMMARY

This chapter shows the manner in which it is possible to apply in practice the TMC in a social campaign which is a tool of social marketing implemented by a city that transforms its economy from linear into circular. First of all, in the analyzed campaign the use by the city of segmentation carried out on the basis of the criterion of readiness of its recipients for changes is visible. We notice here both:

- Segment completely unaware of the existence of the issue of excessive consumption and producing too large volumes of waste by the inhabitants. This segment has yet to become aware of the issue of CE and the principles of zero waste as well as how to apply them in practice in the everyday life and what impact this will have on the city ecosystem.
- Segment aware of the existence of the issue of excessive and improper consumption and, in this respect, searching for information

regarding the possibility of its solving and introducing changes. The inhabitants belonging to this segment already have a certain level of knowledge and continue to gather further information, but they are at a stage of considering the pros and cons of possible changes of their actions. Lack of motivation and readiness to change is characteristic for them (i.e. they notice high bills for non-segregated waste disposal, which raises amble technical problems – i.e. where and how to place several containers for waste segregation in a kitchen. As a result, they search for further information and keep postponing the decision).
- Inhabitants who thank to access to various information, through their understanding and acceptance, are not only motivated to introduce changes in the scope of their consumption and limit the level of produced waste but also are also able to select and plan the implementation of specific solutions (i.e. making their own shopping bags, preparing own multiuse containers for buying dry products, finding places in which it will be possible to exchange one's books for other books etc.).
- Segment of inhabitants who have undertaken specific actions targeted at changing their consumption into more responsible and sustainable one (i.e. through making properly planned shopping, sharing the excess of food with the needy thanks to placing it in social fridges, purchase of composter and its practical use in a household on a daily basis, shopping mainly in "zero waste" shops and boycotting premises the operations of which are harmful to the natural environment, etc.). Within this group, we failed, however, to recognize persons undertaking occasional, from time to time actions, as well as those for which balanced consumption became a new lifestyle that accompanies them in each area of their consumer activity.

Second of all, the analyzed case study shows that the social campaign "Wrocław zero waste" applies various tools in realizing its goals, adjusting them to the specificity of the above-stated segments. In case of the segment of totally problem conscious inhabitants, the basic role is played by information actions realized, i.e. with the assistance of a dedicated online website that explains the essence of CE and the idea of zero waste, educating during municipal events, publications of articles in the local press, lectures during concerts or carrying out lessons on ecology at schools.

On the other hand, the segment conscious of the issue and searching for information regarding the possibilities of its solving requires slightly more advanced information actions – the city website remains useful here, but information placed on it is too general; therefore, social media became more applicable here, informing the inhabitants on an ongoing basis of actions undertaken by the city in the scope of zero waste and giving the possibility of fast interaction with recipients. The inhabitants who consider, for instance, using a composter in the future, may participate in the exchange of information and experiences with other internet users, actively visiting the profiles of the city on Facebook. A similar role is played in case of this segment by actions assuming the form of events and special campaigns organized by the city – i.e. Participation in the festival of cooking from leftovers "Zero waste hoods" increased theoretical knowledge, allowing at the same time for its practical use during events (i.e. in the course of joint cooking) and provided the possibility of full interaction between the organizer (originator) and participants (campaign recipients) and interactions between the participants themselves (exchange of experiences, support, join solving of problems). The practical aspect of engaging inhabitants in actions is of significance from the perspective of copying with their rational and emotional reluctance and strengthening motivation to undertake actions. Furthermore, a crucial function in combating resistance and shaping attitudes and beliefs of inhabitants was performed in case of this segment by persons appearing as campaign ambassadors – reliable and generally known experts in the scope of zero waste, known and liked Wrocław inhabitants – winners of culinary contests appreciating large popularity in Poland. Reliability of campaigns and actions they promoted was enhanced in case of this segment also through the correctly selected campaign partners – among others, generally known and uniquely positively associated Food Banks. Yet another segment, that is, inhabitants motivated to introduce changes, able to plan and prepare for them, required from the city as campaign creator elaboration and use of different instruments, adjusted to the level of engagement of this group. We are dealing here with inhabitants who are determined, convinced and willing to act; hence, in order to facilitate their actions, readymade solutions must be made available. Such support instrument in case of the analyzed campaign is the programme of free access of composters (the city offers a tool and, ultimately, inhabitants who are strongly engaged in modifying their consumption patterns may implement such changes easier), access of bookshelves for book and CD exchanging in City Halls as well as placing on the website of the campaign "Wrocław zero waste" of an interactive map, facilitating finding places of exchange, sharing, access to rainwater or "zero waste" premises for the interested inhabitants. One may at the same time note that these actions would not be effective when reaching the first segment – unaware of the issue related to excessive consumption and supply of waste.

Yet another important issue to be indicated here is organizing of social marketing goals, described in the theoretical part, including social campaign goals, which is its tool. The subject literature shows a division of behaviour objective, knowledge objective and belief objective. Realizing each of these objectives requires the use of proper tools and channels. In case of "Wrocław zero waste" campaign, the behaviour objective is the demanded change in behaviours of inhabitants in the scope of consumption and practical implementation of zero waste ideology into their everyday behaviours, the knowledge objective is an increase of knowledge among inhabitants regarding sustainable consumption, CE and zero waste ideology so that they can understand the idea behind introducing changes, their importance for the city and the environment and the role they perform in this process, while the belief objective constitutes shaping of beliefs, opinions and feelings of the Wrocław inhabitants in the scope of sustainable consumption and the zero waste concept targeted at overcoming the obstacles of rational and emotional nature, strengthen motivation and consolidate introduction of changes in behaviours. Realization of the knowledge objective is ensured in the discussed case study by such tools as: campaign website, lectures during events organized by the city for its inhabitants, articles in the local press, contents placed on campaign profiles on Facebook, ecological lessons organized in the selected schools, lectures during the cooking festival from leftovers "Zero waste hoods". Behavioural objectives are realized by the city through the use of larger scope direct interactions with inhabitants and the possibility of offering them the tools to support the introduction of new behaviours – these include for example:

- Events activating inhabitants and giving them the possibility of actual testing of the newly discovered solutions (i.e. cooking by oneself of a new meal from leftovers, learning how to make preserves in order not to waste food, trying to sew a shopping bag on one's own, etc.).

- Interactive map facilitating fast localization of social fridges, exchange points, flea markets, places of rainwater storage, premises operating on the basis of zero waste principles, etc. by inhabitants.
- Programme of free access to composters.
- Placing in the selected buildings of the City Hall of giveboxes for book exchange.

Realization of belief objective would not be possible, though, without the use of image actions of the campaign, performed by carefully selected ambassadors, At the same time, the city impacts, through its legislative actions, the attitudes of inhabitants towards actions it undertakes – showing the good, worth following actions in its buildings (i.e. container-free waster dispensers, elimination of plastic in public buildings) and during events organized by it (i.e. use of biodegradable paper cups instead of plastic ones during marathons).

DISCUSSION

It is easy to notice that within the analyzed campaign, traditional and frequently used in social campaigns tool of advertisement was abandoned. Let us remember that this very popular, non-personal, paid form of presenting and promoting a specific concept, striving to change attitudes and behaviours into the socially desired ones, can relatively easily and fast reach a broader group of recipients. It may though assume various forms – a TV, cinema, radio, online, external commercial, thus reaching many different groups of recipients at once. Such a solution is indeed tempting from the perspective of the city as an originator, wishing to reach all inhabitants, regardless of their age or preferred media. A benefit of using advertising in "Wrocław zero waste" campaign would thus be the possibility of reaching to the perception of many inhabitants in a relatively short period of time. Would such a solution have any downsides? Apart from limitations in a form of high costs, it is worth to ponder also about the extent to which such a commercial would be effective in realizing the behavioural objectives which require the support of inhabitants in their actions through providing ready solutions. Even though a commercial may successfully reach the inhabitants who are completely unaware of the problems of excessive consumption and city contamination and contribute to the realization of the goals related to education, already at the stage of supporting the recipients in introducing changes in their daily consumption patterns and their consolidation, it seems to be an insufficient tool in the end. Through offering the awareness of the problem, stressing its weight and possibilities of solving it, a commercial does not provide the ready solutions, nor does it facilitate a practical education and adjusting to new consumer habits, providing no possibilities of ongoing explanation of doubts accompanying inclusion of the new principles of consumption into everyday life. Here, events organized by the city are clearly a better choice, offering possibilities of physical experiencing new solutions, undertaking attempts of independent performance of a new dish or preparing preserves, making it easier for the inhabitants to undertake actions through providing composters or elaborating an interactive map. While analyzing this case study, it is also worth looking into the use in the communication actions of above all the internet (website of the campaign, profiles in the social media) while at the same time marginalizing the use of traditional media (press, TV, radio). Ultimately, one may expect a hindered access to the segments of recipients who avail of modern media to a minor extent, staying loyal to the traditional media. This issue concerns especially the elderly, conservative, less-educated or less wealthy.

Yet another interesting issue worth focusing on is the emotional overtones of messages used in the social campaign. The campaign "Wrocław zero waste" is based, above all, on rational messages, indicating the importance of the idea of CE and zero waste for the city and its inhabitants and the practical possibilities of changing consumer behaviours by the inhabitants, city authorities and enterprises functioning in it. In case of social campaigns, we are, however, frequently dealing with the use of messages of emotional nature – through the use of positive emotions (joy, happiness, good humour) or negative emotions (fear, anxiety, disgust, aversion, etc.). Taking into consideration the complexity of attitudes that knowledge, belief and behaviour component comprises, the use of such emotions in messages seems right. Looking at the city specificity as an object under transformation and the complexity of changes that accompany its transition into the CE, bearing in mind implementations of the zero waste concept, it seems justified to place rational elements in the message, which explain the idea behind changes, their course, goal and importance to the recipients – inhabitants and the city. Such an approach allows for a comprehensive reaching the recipients who remain on various stages of readiness to undertake changes and increase chances for their long-term engagement. The brevity of emotions accompanying, for instance, messages revealing negative outcomes of excessive consumption (presenting i.e. stagnant litter and dirt, lack of drinking water, starving children and loads of thrown away food) could constitute a significant limitation in realizing long-term goals of campaigns related to changing behaviours, even though it might at the same time contribute to changing the attitude of recipients. Let's not forget, however, that the measure of success of the social campaign is an actual, long-term change of what people do and not just what they think about something.

Considering the order indicated in the subject literature of carrying out actions under planning of the social campaign, one should pay attention to establishing calculable goals and specifying methods of measurement of their realization. In case of the campaign "Wrocław zero waste", the objectives are of general nature, it is worth pondering about how they could be made more precise. Looking at the applied tools one might suggest, for example, specifying the demanded number of persons participating in each event organized in the framework of the campaign (planned number of participants of the leftover cooking festival "Zero waste hoods" divided into individual events, the planned number of participants of fourth grades covered by the educational programme in the selected Wrocław schools, etc.), the offered number of composters, made available for free to the inhabitants in a given calendar year, the desired number of inhabitants declaring litter segregation, the desired volume of mixed waste falling per one inhabitant each year. One ought to also specify the measurable goals concerning the issue of changing the level of knowledge and changing the attitudes of campaign recipients which might be verified by means of survey researches.

CONCLUSION

The social campaign entitled "Wrocław does not waste" is an example of applying the element of social marketing in the implementation of the city's objectives, which begins a difficult and complex process of transformation towards CE. In order to ensure its success, it is necessary (among others) to effectively influence the behaviours and attitudes of residents, in the scope of their preferred consumption model – it is necessary to awaken in them the need to make conscious and responsible consumer choices, as well as to facilitate the understanding of how they translate into the city's ecosystem, natural environment and heritage for future generations. Knowledge, preferences and attitudes must be reflected in the real behaviours of residents; hence, the social campaign should not only provide knowledge but should also use tools based on interaction, which provide the opportunity to influence the experiences of recipients. In the case of the "Wrocław does not waste" campaign, the measures were applied that allow the city to implement cognitive and behavioural objectives, as well as objectives associated with the feelings and beliefs of residents in regard to the zero waste concept. It is evident that due to the specificity of this problem, the measures within this campaign are long term and are subject to modifications, in response to the emerging problems and needs of residents. At the same time, this campaign is consistent with the measures undertaken within the smart city programme, as well as the city's strategy in the scope of sustainable development.

A SWOT analysis can also be carried out on the basis of the above.

> Strengths – the long-term nature of the social campaign, the budget provided for financing the actions, having many of its own information channels, the ability to easily reach individual segments, through its own network of public institutions, the determination of the authorities to introduce change and innovation, the existence of departments and departments dedicated to the implementation of the campaign.
> Weaknesses – focus on selected communication channels, lack of in-depth research and needs analysis, limited scope for monitoring and measuring campaign effects, the division of competences between the various departments.
> Opportunities – growing interest of some residents in zero waste, access to modern technologies enabling the involvement of residents in the process of information exchange, government activities supporting the transition to the CE, increasing costs of waste collection, increasing environmental awareness of some residents (appearance of NGOs and volunteers), support from the celebrities.

Threats – low-environmental awareness of part of the society, poor motivation to change, large diversity of inhabitants due to their social and economic status and the type of property inhabited, large variety of activities covered by the "zero waste" concept, reluctance of part of the inhabitants to use used things, COVID-19 pandemic, ease in avoiding sanctions for unsegregated rubbish, limited number of NGOs and supporting volunteers.

LESSONS LEARNED

The starting point for an effective social campaign is the correct definition of its objectives. The correct definition of campaign objectives makes it possible to adjust tools to their individual types (knowledge, belief, behavioural). At the same time, it is important to properly segment the campaign audience. The selection of segmentation criteria for the specificity of the campaign is important here. Correct segmentation enables the adjustment of communication tools to the specificity of recipient segments being, in the analyzed case study, at different stages of readiness for change. Another issue is to draw attention to the need to take external factors into account when planning and implementing the campaign. No less important is the recognition of possible limitations in the implementation of campaign objectives.

DISCUSSION QUESTIONS

1. Which of the segments distinguished in the case study and why can you include yourself? What kind of social campaign activities of the city would motivate you to be "zero waste"?
2. What can the city do to increase the number of volunteers giving away their things, food, books?
3. What can the city do to encourage people to use their belongings?
4. How can the effectiveness of a social campaign be measured in each segment?
5. Is it important to reach not only individual segments but also different age groups with the campaign?

EXERCISE IN GROUPS

Imagine that you want to convince the people of your town to recycle the rubbish. Think about how many of you are already sorting waste and why. Why are some people not sorting their rubbish? How can you divide into segments of residents who do not segregate their waste? Discuss in groups how you can influence their attitude. Which media can be used to reach them most effectively? Think about the slogan for a social campaign on recycling. How can you measure the effects of such a campaign?

FUNDING DECLARATION

The project is financed by the Ministry of Science and Higher Education in Poland under the programme "Regional Initiative of Excellence" 2019–2022 project number 015/RID/2018/19 total funding amount 10721 040,00 PLN.

ACKNOWLEDGMENT

The authors would like to thank the chairman and employees of the Sustainable Development Department of the Wrocław City Hall for providing the information necessary to prepare the text of this case study.

CREDIT AUTHOR STATEMENT

Bednarska-Olejniczak: Conceptualization, methodology, validation, formal analysis, investigation, writing original draft, writing review, supervision; Olejniczak: validation, formal analysis, software, investigation, writing original draft, supervision, writing review & editing, project administration.

REFERENCES

Braungart, M., McDonough, W., & Bollinger, A. (2007). Cradle-to-cradle design: Creating healthy emissions – a strategy for eco-effective product and system design. *Journal of Cleaner Production, 15*(13), 1337–1348. https://doi.org/10.1016/j.jclepro.2006.08.003

Chilenski, S. M., Greenberg, M. T., & Feinberg, M. E. (2007). Community Readiness as a Multidimensional Construct. *Journal of Community Psychology, 35*(3), 347–365. https://doi.org/10.1002/jcop.20152

Cismaru, M., & Wuth, A. (2019). Identifying and analyzing social marketing initiatives using a theory-based approach. *Journal of Social Marketing, 9*(4), 357–397. https://doi.org/10.1108/JSOCM-06-2018-0063

Cole, C., Osmani, M., Quddus, M., Wheatley, A., & Kay, K. (2014). Towards a zero waste strategy for an English local authority. *Resources Conservation and Recycling, 89*, 64–75. https://doi.org/10.1016/j.resconrec.2014.05.005

Directorate-General for Environment. (2014). *Preparatory study on food waste across EU 27—European Environment Agency [External Data Spec]*. Retrievable from: https://www.eea.europa.eu/data-and-maps/data/external/preparatory-study-on-food-waste

Dz.U. 1996 nr 132 poz. 622. (1996, September 10). *Ustawa z dnia 13 września 1996 r. O utrzymaniu czystości i porządku w gminach. Dz.U. 1996 nr 132 poz. 622*. Retrievable from: http://isap.sejm.gov.pl/isap.nsf/DocDetails.xsp?id=wdu19961320622

European Commission. (2020, March 11). *A new Circular Economy Action Plan for a cleaner and more competitive Europe, COM(2020) 98 final*. Retrievable from: https://eur-lex.europa.eu/legal-content/EN/TXT/?uri=CELEX:52020DC0098

Federation of Polish Food Banks. (2019). *Raport: Nie marnuj jedzenia 2019*. Retrievable from: https://niemarnuje.bankizywnosci.pl/wp-content/uploads/2019/11/banki-zcc87ywnosci_-raport-nie-marnuj-jedzenia-2019.pdf

GUS. (2019a). BDL GUS. *Bank Danych Lokalnych Główny Urząd Statystyczny*. Retrievable from: www.stat.gov.pl

GUS. (2019b). *Ochrona środowiska 2019*. Retrievable from: https://stat.gov.pl/obszary-tematyczne/srodowisko-energia/srodowisko/ochrona-srodowiska-2019,1,20.html

IESE Business School (2019). *IESE Cities in Motion Index 2019*. Retrievable from: https://media.iese.edu/research/pdfs/ST-0509-E.pdf

Kirchherr, J., Reike, D., & Hekkert, M. (2017). Conceptualizing the circular economy: An analysis of 114 definitions. *Resources, Conservation and Recycling, 127*, 221–232. https://doi.org/10.1016/j.resconrec.2017.09.005

Lee, N. R., & Kotler, P. (2011). *Social marketing: Influencing behaviors for good*. Los Angeles, London, New Delhi, Singapore, Washington, DC: SAGE Publications.

Lehmann, S. (2011). Optimizing urban material flows and waste streams in urban development through principles of zero waste and sustainable consumption. *Sustainability, 3*(1), 155–183. https://doi.org/10.3390/su3010155

Levit, T., Cismaru, M., & Zederayko, A. (2016). Application of the transtheoretical model and social marketing to antidepression campaign websites. *Social Marketing Quarterly, 22*(1), 54–77. https://doi.org/10.1177/1524500415620138

Luca, N. R., & Suggs, L. S. (2013). Theory and model use in social marketing health interventions. *Journal of Health Communication, 18*(1), 20–40. https://doi.org/10.1080/10810730.2012.688243

McCarthy, E. J. (1960). *Basic marketing, a managerial approach*. Homewood, IL: R.D. Irwin.

Ministry of Development. (2016). Draft strategy for responsible development until 2020 (with prospects until 2030), Warszawa.

Ministry of Entrepreneurship and Technology. (2012). *Mapa drogowa GOZ*. Retrievable from: https://www.gov.pl/attachment/72d8cd08-f296-43f5-af28-21ab2fada40e

Peattie, K., & Peattie, S. (2009). Social marketing: A pathway to consumption reduction? *Journal of Business Research*, 62(2), 260–268. https://doi.org/10.1016/j.jbusres.2008.01.033

Pelsmacker, P. D., Geuens, M., & Bergh, J. V. (2013). *Marketing communications: A European perspective*. Harlow, UK: Pearson.

Phillips, P. S., Tudor, T., Bird, H., & Bates, M. (2011). A critical review of a key waste strategy initiative in England: Zero waste places projects 2008-2009. *Resources, Conservation and Recycling*, 55(3), 335–343. https://doi.org/10.1016/j.resconrec.2010.10.006

President of the City of Wrocław. (2019a). *Sprawozdania Prezydenta Wrocławia z realizacji zadań z zakresu gospodarowania odpadami komunalnymi za lata 2014—2018—Utrzymanie czystości i porządku we Wrocławiu—Biuletyn Informacji Publicznej Urzędu Miejskiego Wrocławia*. Retrievable from: https://bip.um.wroc.pl/artykul/374/21544/sprawozdania-prezydenta-wroclawia-z-realizacji-zadan-z-zakresu-gospodarowania-odpadami-komunalnymi-za-lata-2014-2018

President of the City of Wrocław. (2019b, June 17). *Ordinance no. 1156/19 of the President of Wrocław dated 17.06.2019 on the activities aimed at eliminating plastic in the Wrocław City Hall and organisational units of the Wrocław Commune, as well as at the events organised by the city*. Retrievable from: http://uchwaly.um.wroc.pl/uchwala.aspx?numer=1156/19

Prochaska, J. O., & DiClemente, C. C. (1982). Transtheoretical therapy: Toward a more integrative model of change. *Psychotherapy: Theory, Research & Practice*, 19(3), 276–288. https://doi.org/10.1037/h0088437

Stahel, W. R., & Reday-Mulvey, G. (1981). *Jobs for tomorrow: The potential for substituting manpower for energy*. New York, NY: Vantage Press. Retrievable from: https://catalog.hathitrust.org/Record/006208423

Weiss, J. A., & Tschirhart, M. (1994). Public information campaigns as policy instruments. *Journal of Policy Analysis and Management*, 13(1), 82–119. https://doi.org/10.2307/3325092

Wrocław City Council. (2018, January 11). *Uchwała w sprawie uchwalenia Studium uwarunkowań i kierunków zagospodarowania przestrzennego Wrocławia*. Retrievable from: http://uchwaly.um.wroc.pl/uchwala.aspx?numer=L/1177/18

Wrocław City Council. (2019). *Udostępnianie kompostowników—Środowisko Wrocław*. Retrievable from: https://www.wroclaw.pl/srodowisko/program-udostepniania-kompostownikow

Wroclaw City Council. (2019, March 5). *Wrocław nie marnuje*. Retrievable from: https://www.wroclaw.pl/srodowisko/wroclaw-nie-marnuje

Wrocław City Council, Department of Sustainable Development. (2020, July 20). *Reply no. WSR-GO.605.165.2020.JWM from the Department of Sustainable Development to a request for the campaign details: Wroclaw does not waste*.

www.ellenmacarthurfoundation.org/. (2012). *Ellen MacArthur Foundation*. Retrievable from: https://www.ellenmacarthurfoundation.org/

Zaman, A. U., & Lehmann, S. (2013). The zero waste index: A performance measurement tool for waste management systems in a 'zero waste city'. *Journal of Cleaner Production*, 50, 123–132. https://doi.org/10.1016/j.jclepro.2012.11.041

ZWIA. (2018). *Zero waste hierarchy of highest and best use*. Retrievable from: http://zwia.org/zwh/

13

Influencing Sustainable Food-Related Behaviour Changes: A Case Study in Sydney, Australia

Diana Bogueva, Dora Marinova and Talia Raphaely
Curtin University, Bentley, Australia

CONTENTS

Learning Objectives 346
Themes and Tools Used 346
Theoretical Framework: Sustainability Social Marketing 347
Introduction 351
Method 353
Data Collection and Analysis 356
Results from the Start-Up Interviews 357
    Personal Context Factors 358
    Behaviour Influencing Factors 361
        Cultural Factors 361
        Health-Related Factors 361
        Physiological Factors 362
        Environmental Factors 362
        Animal Welfare Factors 363
    Barrier Factors 363
        Personal Attitudes 364
        Social Relationships 364
        The Media and Marketing 365
        Economic Considerations 365
    The Game-Changing Factors 366

The Sustainability Social Marketing Intervention 367
"No Biodiversity Destruction in My Plate!" 368
"Is There a Choice between a Steak and a Burning Planet?" 369
"Red Meat Also Can Cause Cancer and Heart Attack ..." 369
Summary of the Outcomes from the Sustainability
   Social Marketing Intervention 370
Discussion of the Sydney Longitudinal Study Results 376
Conclusion 378
Lessons Learned 379
Discussion Questions 379
Suggested Activities 380
Interesting Facts 380
Key Definitions 380
Authors Whose Work to Follow 381
Acknowledgment 381
Credit Author Statement 381
References 382

LEARNING OBJECTIVES

After reading the chapter, the reader should be able to:

- Understand the impacts of human diets as they relate to people's health and the well-being of the planet's ecological systems
- Absorb the concept and benefits of flexitarianism, including the factors that contribute towards dietary shifts
- Familiarise themselves with sustainability social marketing and its 4S (sustainability, strength, self-confidence and sharing) marketing mix
- Devise interventions that impact on dietary behaviour and food choices
- Understand the pros and cons of longitudinal studies
- Explore the benefits and challenges of qualitative research

THEMES AND TOOLS USED

- Planetary health – tools for reducing the impacts of food and people's dietary choices to achieve better planetary health (union of population and ecological health).

- Flexitarianism – tools for sustainability social marketing to encourage reduction in meat consumption.
- Food – tools that educate people to make better dietary choices and prepare suitable meals.
- Sustainability social marketing – tools to encourage behaviour changes for broader social and environmental benefits.
- Longitudinal study – research design involving repeated observations of the study participants to test the effects of the intervention tools used to influence people's dietary behaviour.

THEORETICAL FRAMEWORK: SUSTAINABILITY SOCIAL MARKETING

Social marketing is a way to influence behaviour towards the common good (Kotler & Lee, 2008; Rundle-Thiele et al. 2013), in other words to create a better world for better people. Its essence is in getting people to change their individual behaviour for the benefit of society as a whole and for the achievement of pro-social outcomes (French & Gordon, 2019). For social marketing to be successful, it needs to develop an appropriate marketing mix. The 4P (product, price, promotion and place) marketing mix established since the 1950s and widely used nowadays is considered a powerful way to influence consumers (Twin, 2020). However, this marketing mix needs to be modified when transitioning towards sustainability. The 4S (sustainability, strength, self-confidence and sharing) has already been put forward as a most-needed alternative (Bogueva, Raphaely, Marinova, & Marinova, 2017). Social marketing goes beyond influencing individual behaviour. It draws attention to an existing social problem, and through an ingenious campaign offers a simple and compelling solution for the target audience to be willing to change their behaviour and contribute to a sustained social transformation (Saunders, Barrington, & Sridharan, 2015). In addition, a systems approach is required if society is facing complex and dynamic social issues (Truong, Saunders, & Dong, 2019; Saunders & Truong 2019) as is the case with climate change and other sustainability challenges, including food production and consumption.

What does it take for someone to stop eating meat? The vast majority of people are not aware that meat is an environmentally unsustainable food choice. Many are not informed that excessive meat consumption is detrimental to human health, leading to the spread of non-communicable diseases, such as colorectal cancer. In fact, the EAT Lancet Commission Report unequivocally states that "[u]nhealthy diets now pose a greater risk to morbidity and mortality than unsafe sex, alcohol, drug and tobacco use combined" (Willett et al. 2019, p. 5). People need to be educated in relation to their diets. Social marketing is a way to help people constructively and encouragingly, not through frightening messages, warning and scaring them. This experiment exploratory longitudinal case study with 30 Australians aimed at pursuing a dietary change without directing the participants what to do, but by using positive social marketing campaign interventions showing them how change can be done. Using real-life lessons, we aimed at encouraging dietary transitions and making something so difficult to shift as is Australians' love for meat into a change that is fun, easy, attractive and accepted to do.

The original Sustainability Social Marketing Model (SSMM) is developed for assisting in wider campaigns (Bogueva et al. 2017), but in this longitudinal study it was applied to a smaller circle of participants and specifically for meat reduction (Figure 13.1). Given the extremely high environmental impacts of red meat in particular (e.g. Eshel, Shepon, Makov, & Milo, 2014) and the World Health Organisation's warnings about overconsumption of these types of foods (WHO, 2015), we specifically targeted the intake of such products by replacing them with plant-based alternatives.

There are six stages in the SSMM aimed at producing effective social marketing campaigns and interventions. The stages are: define, scope, develop, implement, evaluate and finally adapt and improve, which are also iterative as a sequence (see Figure 13.1). In our case, the duration of the process spanned across five years. Furthermore, there are also internal iterations between the stages of define and scope and implement and evaluate which allow for fine-tuning the interventions. The SSMM is tailored for influencing behaviour change. Its first stage is extremely important as it defines the targeted population section for which the social marketing interventions are developed. For example, there are big differences if the targets are young or older people compared to an intervention designed for the general public. The social marketing interventions are then scoped to influence people in a dietary shift, particularly in relation to meat reduction. Designing and developing an intervention includes the creation of specific educational materials, planning of demonstration sessions, information about plant-based alternatives to meat consumption and de-marketing of animal-based foods. All these materials centre on the

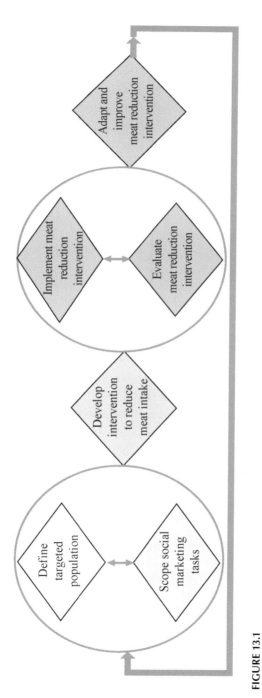

FIGURE 13.1
Sustainability social marketing model (SSMM) for reduction in meat intake.

positives of increased intake of plant-based foods and the negatives of continued high meat intake. They blend health, environmental and ethical issues, ranging from human and planetary well-being to animal suffering (Singer, 2015), overuse of antibiotics and the threat from new emerging zoonotic diseases. Implementing the intervention required active involvement of all participants over the entire five-year duration of the longitudinal study. Evaluating the outcomes was essential to have a good understanding of what works, what progress is made and whether there may be unexpected reactions, for example, related to sensitive issues or previous experiences the participants may have had. It is also very important to adjust and refine the intervention based on people's reactions and their response to change. Adaptation and improvement are essential not only for the particular intervention but also for future social marketing sustainability-related campaigns (Bogueva et al. 2017).

The 4S marketing mix (Bogueva et al. 2017) supporting the SSMM illustrates the role of each S in the case of meat reduction:

- Sustainability – meat production and consumption are contributing to global warming, climate instability, resources depletion and land degradation. Excessive red meat consumption is also associated with health problems linked to non-communicable diseases, such as cancers, diabetes type 2 and obesity.
- Strength – by reducing meat consumption, individual people, policy actors and influencers have the power to stop the destruction caused by livestock, improve human health and reverse negative ecological impacts; the positive impacts of such a change would be immediate.
- Self-confidence – each person has the power to make the choice to change their diet by reducing meat intake; anybody can do it and by taking such actions we can all participate and contribute towards transitioning to a better and more sustainable society; taking pride in doing less harm.
- Sharing – the planet is there to be shared and no one has the right to compromise the global commons on which all life depends.

More information about the methodology and the results from the SSMM longitudinal study is presented below. We are drawing on the experience of 30 participants and using their words to support our findings and recommendations.

People are used to their eating habits and in everyday life, they rarely question or pay attention to their existence. This is very much the case in relation to meat consumption. Western society culturally embraces carnivory and meat meals have a regular presence on the dinner table. Meatless options are considered a deviation from the common practice and often frowned upon. This is in stark contrast with what science says about the need to transition to more plant-based options to maintain planetary health, including the well-being of the bio-physical systems and the human population (Willett et al. 2019). The text to follow presents the results from a 2010–2020 longitudinal study in Sydney, Australia which used sustainability social marketing to influence people's dietary choices towards more plant-based options.

INTRODUCTION

In recent years, meat production and consumption have been identified to be among the biggest threats to sustainability with abundant research evidence suggesting the need for behavioural changes away from livestock-based products (e.g. Greenpeace, 2018; Willett et al. 2019). Choosing more moral, flexitarian options (Raphaely & Marinova, 2014; Berley & Singer, 2014; de Bakker & Dagevos, 2012) and eating less meat can slow climate change, improve biodiversity, reduce water pollution, prevent exhaustion and misuse of arable land as well as diminish humans' negative impacts on planetary health (Willett et al. 2019). Adopting a socially responsible plant-rich diet can help reconcile moral obligations to future generations and other non-human species.

However, such a transition is not a simple task given the meat's special status in human diet. In the West, society seems to have instituted the structure of meals as comprising of one piece of meat and two vegetables (Lupton, 1996; Schösler, de Boer, & Boersema, 2012). There are many factors that shape people's food choices and their behaviour as consumers, including health, psychological, sensory and environmental considerations as well as marketing influence (Marinova & Bogueva, 2019). Eating is also considered a matter of individual choice for adults, particularly when and where there is ample availability of food. This assumption however is increasingly being challenged.

As with the right to clean air and a healthy natural environment, it is time to ask the question whether food should also be part of humanity's responsibility for the planet and its current and future inhabitants. The use of the available arable land on this planet and the conversion of vast areas of native vegetation for agricultural purposes are prime examples of the "tragedy of the commons" (Hardin, 1968; Gusmai, 2018) which continues to be exploited by a selected few to the detriment of the global public good. Australia and its livestock are major contributors to climate change and the deterioration of the natural environment by overexploiting the global commons. Ironically, Hardin's original paper described cattle herders left unregulated whose behaviour could lead to overgrazing the common land. This is exactly what is happening with the livestock sector currently whose selfish behaviour and self-interest are destroying the common resources on which we all depend. The global cattle and sheep herds have been consistently on the rise with Australia being a significant player contributing 3% and 7% of the global production of beef and sheep meat, respectively, in 2017 (MLA, 2019). Australians also continue to be some of the highest consumers of meat in the world, three and five times higher than the global averages in the case of beef and sheep meat, respectively (MLA, 2019).

These statistics are not surprising given the fact that the Australian government is yet to limit the behaviour of the livestock industry which means animal husbandry continues to exploit the country's and global commons. The sector is a large contributor to global warming, soil and water depletion as well as adding towards the spread of non-communicable diseases, such as cancer, diabetes and obesity. It is also the biggest factor for the clearing of Australian native vegetation. In 2019, the EAT Lancet Commission proposed a universal planetary health diet, focussed on human dietary and planetary health and encouraged the intake of a vast variety of plant-based options and low amount of animal-based food products (Willett et al. 2019). This planetary diet calls for reorienting priorities from producing large quantities towards feeding humanity with quality, nutritious and healthy food on the existing agricultural land putting in place strong governance to restrict the expansion of the livestock sector (Willett et al. 2019).

While policy actors in Australia are yet to respond to the scientific evidence and calls by the research community to stop the expansion of the livestock sector, the alternative to prevent further destruction is to influence consumer habits. This is not an easy task as dietary habits are deep-rooted in social norms and determined by individual behaviour. Sustainability and social marketing offer tools which, if deployed properly, can influence consumption habits with lasting implications. Insights can be drawn from anti-smoking and sun protection campaigns which have been very successful in Australia. With the right approach and using sustainability social marketing interventions (Bogueva et al. 2017), individual consumers can develop new attitudes and willingness to change their dietary habits. Various strategies to encourage dietary changes have already been used by different researchers and non-government organisations (NGOs) through marketing campaigns which encourage meatless days (de Boer, Schösler, & Aiking, 2014), "meatless Mondays" (Euromonitor International, 2011; Parker, 2011) or "Green Mondays" in France (Bègue & Treich, 2019), meat portion size reduction, e.g. "less but better" (Laestadius et al. 2016, de Boer et al. 2014) or choosing "prime cuts" (Sutton & Dibb, 2013). Further efforts have emerged in advocating for meat substitutes (Schösler et al. 2012; Bogueva et al., 2019) and calling consumers to become agents of change (de Bakker & Dagevos, 2012).

Asking consumers to eat less animal products may trigger not only resistance to change and in some cases confusion regarding amounts and sources of protein but also challenges the widely perceived link between meat and masculinity, manliness traits of power, virility and hegemony (Fessler & Navarrete, 2003; Ruby & Heine, 2011; Adams, 2010; Bogueva & Marinova, 2018). Finding a solution and promoting pro-environmental, sustainable consumption behaviour are a complex and tough task, especially as it also requires the discrediting of habits and debunking of many myths. A broad societal change can only take place if individual consumers respond by shifting their dietary preferences.

This study portrays the efforts to trigger individual change through a longitudinal study based on sustainability social marketing. We first provide the methodology of the longitudinal qualitative study conducted in Sydney, Australia. This is followed by a discussion of the study's results using quotes from the 30 participants. We also identify the marketing messages (or catchphrases) which can trigger dietary change and can potentially be used in wider campaigns aimed at improving human and ecological health.

METHOD

This longitudinal case study used social marketing intervention to influence change by following up a group of 30 participants over five years. A longitudinal study is particularly useful for evaluating the relationship between predetermined factors and the development of change, when the outcomes need to demonstrate over a prolonged period of time (Caruana, Roman, Hernández-Sánchez, & Solli, 2015). At the outset, we made clear the logic of the intervention (Hill et al. 2016) by explaining its conceptual basis, components and emphasis on action through dietary behaviour changes. The study was qualitative and we used a combination of in-depth and group interviews to collect data to gauge behaviour changes in response to the social marketing intervention. In research, it is common to use control groups to assess whether any change resulted from the intervention. However, in our case we applied a single-group design which is common in longitudinal studies as it allows for before-after comparisons (Paulus et al. 2014). There was also an implicit comparison with the rest of the Australian population for which it is widely known that it maintains very high levels of meat consumption.

Instead of using a one-off intervention, we opted for regular exposure. The frequency of the intervention was decided to be three times over the five-year period to allow enough time for the participants to absorb the social marketing information and develop appropriate response. During the periods in between, we kept in touch with the "participants to assure their commitment to the project and to keep updated records of their contact information" (Hill et al. 2016, p. 812). The actual components of the intervention comprised:

- Component 1: Information session about the environmental drawbacks of meat consumption and the advantages of plant-based food options – conducted in second half of 2015.
- Component 2: Cooking demonstration classes with plant-based alternatives to meat to develop skills and taste experiences – conducted at the start of 2017.
- Component 3: Information session about the health and performance benefits of plant-based foods – conducted in late 2018.

The three components of the intervention were applied approximately one and a half years apart during the study's period, with initial and final interviews held at the start of 2015 and 2020, respectively. Such frequency allowed time for learning and action as well as to maintain interaction with and between the participants in the study. The chosen period between interventions (of approximately one and a half years), the range of activities, interviews and observation measures were adequate to determine the presence of change over time (Ployhart & Vandenberg, 2010). This allowed to assess whether any change developed and if yes, whether there was formation of lasting habits.

The chosen group of participants satisfied the following conditions:

1. To consume meat at least four times per week – this puts the participants in a category aligned with the average Australian population whose red meat intake is 560 g per week plus 84 g per week of processed meat (National Cancer Control Indicators, 2020).
2. To belong to the same circle of friends for cross-monitoring of eating behaviour over the years – as eating is also a social activity, it was important to allow the participants the opportunities to socialise with other members of the research group who were exposed to the same intervention.
3. The participants being happy to be repeatedly involved in the intervention components and interviewed – as this was a longitudinal study, it was important to have access to the participants on a regular basis.

As advised by Hill et al. (2016), we treated the participants not like "research subjects" but as friends who had consented to the study and were also collaborating, sharing group activities and were part of a social network. Human research ethics permit was obtained from Curtin University. Each participant was informed about the nature and frequency of participation expected over the five-year period and they explicitly accepted these terms through a signed consent form prior to the study. The participants were also happy to share basic demographic information to provide additional context for the study. All participation was on a voluntary basis and no rewards were offered. The participants were free to withdraw from the study at any stage without the need to provide any justification for such a decision.

The empirical data collection occurred throughout the five-year period through:

- In-depth one-on-one semi-structured interviews with the 30 participants at the start of the study in early 2015.
- Three small group discussions during each component of the intervention – the participants were divided in 12 groups comprising of

2–3 persons and were interviewed in a group setting; there were in total 36 such group interviews.
- Final in-depth one-on-one semi-structured interviews at the completion of the study in early 2020. Because of unforeseen circumstances, we were not in a position to conduct final interviews with two of the participants – one passed away from a heart attack a couple of months before the interviews were scheduled and the other was hospitalised with a mental health issue. In total, there were 28 individual interviews conducted.

Food is generally a personal choice. The questionnaire used for all individual and group interviews tried to capture personal opinions and behaviour. It contained five sets of questions:

1. Basic demographic information, such as age, gender, education and income
2. Dietary habits, including frequency of meat consumption
3. Food preparation, including cooking skills and time dedicated to cooking
4. Opinion about meat, including pros and cons, and reasons for consuming meat
5. Factors that may impact their meat consumption, including barriers and incentives

It was important to also capture catchwords, expressions and arguments that may sway the average Australian consumer away from high meat intake. Such clues could potentially be used as a dietary game changer in broader social marketing campaigns targeted at a wider population.

DATA COLLECTION AND ANALYSIS

The period covered by this longitudinal study is from 2015 until 2020. All participants were initially contacted individually by telephone to inform them about the study, discuss their potential participation, obtain a participation agreement and arrange the place for their involvement, including intervention and interviews. Using our social networks in Sydney,

Australia, we approached more than 60 people to recruit the 30 participants. Although many expressed interest in the study, the five-year commitment was a factor limiting their availability and desire to participate. The recruited participants found the study topic appealing and proved to be very reliable.

All interviews and group discussions were conducted creating and maintaining a natural conversation flow, passing from topic to topic while following the pre-determined schedule. This gave the participants space and time to describe freely their thoughts, feelings and opinion about food and meat consumption. The interviews and discussions lasted between one and two hours and were conducted at places where the participants felt safe, including their homes.

There were 94 interviews conducted in total, including 30 individual start-up, 28 individual final and 36 group (12 times 3 to coincide with each of the three intervention components) interviews. All interviews were digitally recorded and then transcribed verbatim with any identifying data removed. Handwritten notes were taken during the interviews to instantly record key points made by the research participants. All interviews were analysed by the researchers involved in this study. Participants' answers were transcribed and coded through assigning units of meaning including individual words, phrases, sentences or whole paragraphs. The transcripts and field notes were also entered into NVivo 11 software for additional analysis to complement the researchers' discretion in the development of a thematic framework.

A condition for participation was frequency of meat consumption of no less than four times per week. All participants were then assigned to two groups, namely "heavy" meat eaters with intake of 6–7 times per week and "moderate" meat eaters with intake of 4–5 times per week. The next section presents the results from this longitudinal study.

RESULTS FROM THE START-UP INTERVIEWS

Four core themes and several supporting subthemes emerged from the initial in-depth interviews identified as: (1) personal context factors; (2) behaviour influencing factors; (3) barriers factors and (4) game changing factors. These themes were reviewed to detect key food consumption drivers and formed the thematic framework for the identification of social marketing messages which can influence behaviour change. They are described in turn below using information collected through the interviews, including word-for-word quotes from the participants (presented verbatim in quotation marks or with indentation).

Personal Context Factors

All participants differed vastly according to the personal context related to food choices. They came from different cultural backgrounds with some being born in Australia, Europe, Asia and America, their ages spanned from 18 to 72 at the start of the study (23 to 77, respectively, at its end) and they have had diverse life experiences as represented by education, marital status, income and food preparation skills. Men and women were almost equally represented with 14 (47%) male and 16 (53%) female participants. At the end of the study, the gender breakdown became equal (14 men and 14 women) as both participants with whom we could not conduct the final individual interviews were female.

The difference in the participants' age indicated that different generations were represented and this influenced their food-related decisions and behaviour. While the younger and mid-age generations (18–40 years old) were more open to considering changing their current diet by adopting more plant-based options, the older generations (50–70+ years old) saw meat as a staple food, essential for a balanced nutrition and were not prepared to make a shift. Table 13.1 presents some basic characteristics of the sample. The majority of the participants were married (24 or 80%, including 11 men and 13 women). There was a gender difference with only one male participant being a moderate meat eater (the remaining being heavy meat eaters) while the respective number for the females was 4. Although there were further differences between the participating men and women in their levels of awareness about the negative impacts of meat, they were not as pronounced. The quote below summarises the way many of the participants felt in relation to eating meat:

> If it gets to food, I eat the meat first ... And with the costs rising now, I'm certainly not extravagant with eating, meat is the first thing that I'll buy and put on my table.
>
> (Female, 71 years old)

TABLE 13.1

Characteristics of the Sydney Study Sample

N	Gender	Age	Family Status	Job Status	Household Income per Annum	Education	Cooking Skills and Available Time for Cooking	Dietary Choice	Awareness about Meat's Impacts
14	Male	18–29 years – 2 30–39 years – 3 40–49 years – 4 50–59 years – 3 60–69 years – 1 70–79 years – 1	Single – 3 Married – 11	Full time – 11 Part time – 2 Retired – 1	<$50K – 1 $50K–$99K – 2 $100K–$149K – 8 $150K–200K – 2 >$200K – 1	University degree – 10 Technical and further education (TAFE)/college – 2 High school – 2	Good cooking skills – 7 Not strong cooking skills – 7 Have time for cooking – 5 Lack time for cooking – 9	Moderate meat eaters – 1 Heavy meat eaters – 13	Some awareness – 5 No awareness – 9
16	Female	18–29 years – 2 30–39 years – 4 40–49 years – 3 50–59 years – 3 60–69 years – 2 70–79 years – 2	Single – 3 Married – 13	Full time – 11 Part time – 2 Retired – 3	<$50K – 2 $50K–$99K – 5 $100K–$149K – 7 $150K–200K – 1 >$200K – 1	University degree – 13 Technical and further education (TAFE)/college – 2 High school – 1	Good cooking skills – 8 Not strong cooking skills – 8 Have time for cooking – 8 Lacking time for cooking – 8	Moderate meat eaters – 4 Heavy meat eaters – 12	Some awareness – 7 No awareness – 9

Socio-economic and income factors impact on people's food choices. The participant group is not representative of the general Australian population as it did not include any people who are unemployed. In fact, it was not our intention to seek a general representation as we specifically aimed at people who have high levels of meat consumption and are not economically constrained in their food choices. Such a starting point would exclude economic factors as barriers to making a behavioural dietary change in favour of plant-based foods which are widely available at reasonable prices in Sydney. The majority of the participants (n = 22 or 73%, including 11 males and 11 females) were employed full time, with only a few (n = 4 or 13%, including two males and two females) working part-time and the rest (n = 4 or 13%, including one male and three females) were retired. Consequently, most participants had sufficient income to source food for themselves and their families with only three (or 10%, including one male and two females) having annual household income below A$50,000.

Food preparation is an important consideration when it comes to dietary choices. It relates to being able to cook as well as to have the time to do so. It was interesting to observe a high percentage of people (17 or 57%, including nine men and eight women) not having enough time in their busy schedules to prepare food for themselves. Half of the participants also reported that their cooking skills were not very good. The combination between time unavailability and lack of good cooking skills made meals a lower priority in people's busy weekly schedules. Many of them resorted to quick solutions, such as putting sausages and steaks on the grill or making pasta dishes, such as spaghetti Bolognese, for their dinner menus. Meal preparation had to fit between busy routines of travel to and from work, picking up children and other social activities.

The educational level of the participants also was higher than that of the general Sydney population. This was an important consideration in the study as it potentially relates to improved environmental and health awareness. Only three participants (10%, including two men and one woman) had completed high school; the remainder had higher qualifications, including university degrees (23 or 77%, including 10 men and 13 women) or further technical education (4 or 13%, including two men and two women). These relatively high educational achievements for the group did not translate in a high level of awareness about the negative environmental and health impacts of excessive meat consumption. Only 12 (40%, including five men and nine women) participants had some such awareness obtained mainly through general knowledge and word of mouth. At the start of the study, the majority was not aware of any negative consequences from consuming meat. This also meant that the current Australian Dietary Guidelines, limiting red meat intake to 455 g per week, had not reached them; neither have other social marketing campaigns, such as "2 Fruit and 5 Veg" (Healthy WA, n.d.).

Behaviour Influencing Factors

The analysis of the start-up interviews revealed five groups of factors which influenced people in their decisions to eat meat. They are related to culture, health, physiology, the natural environment and animal welfare.

Cultural Factors

Many made the argument that besides liking its taste, eating meat is part of their "heritage and background". It provides "special cohesion", "it is our Australian culture", "makes people happy". Some admit that they have not given the thought about eating meat not a second of their time as it is such a normal thing to do. Furthermore, meat is "readily available anywhere in Australia". According to the participants, preparing meat-based meals also means you are looking properly after your family and guests.

Some male participants stressed that meat is "part of our manliness and masculine identity". Others linked it to social status and were pride of being able to consume it in Australia:

> Australians eat bacon for breakfast and ham for lunch, and meat for dinner and two veggies (laughing). That's the way we were all eating and continue eating.
>
> (Male, 51 years old)

Health-Related Factors

The participants were asked their opinion about what amount of red meat is appropriate and healthy to eat. Almost all were not aware of the limit imposed in the Australian Dietary Guidelines and the recommendations of the World Health Organisation. The majority referred to amounts of 200 g to 350 g per meal, some even "500 grams a day for an adult" and twice per day, as healthy.

It was interesting to observe that the older participants, in the age bracket of 60–73 years old, were convinced their food choices do no impact on their health. They claimed to be proud of their meat-rich diet which had secured them good health and overall well-being. Some credited meat for helping them to prevent putting on weight: "I maintained my body weight all my life eating meat and I am continuing to do so without any regrets".

There were, however a few, mainly female voices which warned that as people get older meat-based products "are not as good as they were before". Some were "aware of health implications with bowel cancer" and the use of "some bad hormones" and "steroids". Nevertheless, the prevailing attitude was that Australian red meat is "beautiful" and any health-related concerns are "a hoax".

Physiological Factors

The physiological importance of meat was given as an argument for high consumption. Meat "makes you strong, full of energy and stamina" was an often-used justification. The human species being "omnivores" and meat's importance for "a balanced diet" were similarly emphasised by the participants. Several female participants stressed the role of meat as a source of iron, particularly as some have heard "about pregnancy resulting in anaemia".

Another aspect was the psychological and emotional attachment to meat. For example, "I feel distressed if I am not able to eat meat", "the absence of meat makes me moody" and "I experience some sort of anxiety when not eating meat". On the other hand, thinking about meat had a calming effect with the image of "a cow eating grass". These examples indicate a possible addiction to meat and expectations that it should form part of the participants' diet.

Environmental Factors

All participants were asked to describe what amount of meat is environmentally sustainable for Australians to consume. The older participants (60–73 years of age) believed their food choices had no impact on the natural environment. Some blamed "foreign land buyers" who "are buying more of the Australian land and meat products". Others accused mining for using "cattle production areas".

By comparison, the younger participants were more aware of the negative environmental problems caused by cattle with close to half ($n = 13$ or 43%, including five males and eight females) acknowledging the issue. They admitted to be worried "about environmental issues" and "land clearance", however, were quick to justify their behaviour with the love for meat:

> To be honest, probably no amount of red meat is environmentally sustainable …, but we are all still eating it. There are 7 billion people on the planet, and we can't all eat red meat. And if it is good enough for one person to eat red meat then, you know, it's not fair for a first world country to get too indulgent on expensive food, and people living in third world countries are starving to death. So, from the planetary point of view, I don't think is sustainable for anyone to eat red meat. This was my reason to cut some of the meat I ate, but I am in love with the taste of red meat, and I simply can't avoid it.
>
> (Female, 38 years old)

Animal Welfare Factors

Animal welfare-related issues, including cruelty, intensive factory farming techniques, mass use of feedlots and antibiotics, were mainly of concern to the female participants and to only one man in the study group. Many admitted to having "mixed feelings", to being "selfish" and morally "obliged to be vegetarian". Some opted for humanely raised meat and criticised farming practices assuming that "some of those farms are foreign-owned". A male participant was particularly concerned about "the way red meat is produced" and the use in feedlots of corn – an "unnatural ingredient" for cows.

Overall, there was substantial gap between the latest scientific evidence and the opinions of the study participants. Such a gap equally manifested itself in relation to the barriers which prevent the group from switching towards more plant-based options.

Barrier Factors

Many of the barrier factors which inhibit the study participants from consuming less meat are in many ways similar to the factors which encourage Australia's dependence on animal-based proteins. They relate to personal factors, such as beliefs, values and knowledge, and also to the social environment and the influence of the media. Economic considerations were further emphasised, particularly when it comes to changing people's diet.

Personal Attitudes

In the initial start-up interviews, there was a lot of reluctance to change the dietary habits as the participants held strong beliefs about the value of meat. For example, one of the male participants categorically stated: "I know that [it] is the right food for us". Another echoed a similar belief: "I believe this is the way things need to be". Such a firm logic indicated that it would be a strong social marketing contest to challenge these personal attitudes.

Social Relationships

The social perception and reaction of relatives, friends and generally other people to a person's food choice is another major inhibiting factor. How social connections would be impacted by changing someone's diet appeared to have a strong influence on the study participants and their eating behaviour. People's choices often are linked to their social circles, including what is seen as acceptable or strange. Those who do not share similar tastes can feel socially isolated and can even become a subject of mockery and ridicule. Food preferences which exclude meat are often met with surprise and questioned by others. Not everybody is prepared to face such situations. People generally prefer not to be judged and an easy way to avoid this is to blend with the dominant food tastes.

Explanations, such as: "It's a bit of difficult to look after only one person that is having a different diet. It's kind of killing the conversation and the whole fun", are put forward. There was also explanation that many new migrants come from countries where they could not afford meat and "immigrants are raising a lot of the demand for red and other meat".

A major aspect of factors preventing people from eating more plant-based options is that the prevalent social norms encourage high meat consumption. Ironically, vegetarians and vegans in many ways are treated as smokers whose behaviour and presence on the shared table are undesirable. The reason behind this is the lack of awareness about the many benefits and co-benefits from reducing high meat consumption.

One of the participant men admitted that despite his positive attitude to meat, his son was vegetarian. Although couched along the terms of acceptance, his words sounded more like a criticism towards the younger generation:

> My attitude is that meat is good. We all want to have red meat, but for some reasons if someone happens to be a vegetarian, I also respect their

point of view. For us is important to have meat daily, although my son is vegetarian.

(Male, 51 years old)

The Media and Marketing

Barriers to changes in consumption behaviour were associated also with the media, advertising and marketing messages impacting people. In the absence of social marketing, the public space is taken up by private advertising aimed at promoting meat consumption. For example, people trust "media sources for information regarding food and health. I think a good iron and zinc and B12 protein like meat is well advertised and the health benefits message of meat is very clear". The media have also created meat icons, such as "Sam Kekovich ads. They are pure gold and funny". Commercial advertising is successfully manipulating the English language with phrases that encourage meat consumption and stick in people's mind. A participant referred to: "We will not lamb alone on Australia Day".

Sometimes, however, people find it difficult to follow the latest information presented in the media and are getting confused. One male participant explained that he was "aware of the methane issue" but has heard that things "are good now". A female participant explained: "These new modern, clean farms, not anymore dirty and gross are an improvement". It may also be the case that often by presenting both sides of the story the media do not actively pursue the latest scientific evidence.

Economic Considerations

Economic security was a major barrier to the participants. Many see meat as a valuable food the supply of which should be protected and guaranteed for the Australian population. Sentiments along the lines: "we got enough meat as long as we are not giving it to others" and "the whole of Asia is going crazy on red meat, which is going to increase the prices" were commonly expressed.

Some participants saw meat as "still cheaper than most of the fruits and veggies" while others lamented that they could not afford "buying beautiful fillet steak". Overall, people were reluctant to make any changes in their diet as they have currently balanced their budgets and "need to be careful about what type of food we spend it" with meat being a priority for "putting good protein food at the table".

The start-up interviews revealed a very complex landscape which did not match the facts about meat production, its environmental and human health impacts. This was not surprising given that the group of participants was deliberately selected to comprise regular meat-eaters. On the other hand, the majority of the Australian population are regular meat-eaters with some of the world's highest per capita levels of consumption. It still made the task of any social marketing challenging. Before responding to the task by devising a long-term intervention, we also asked what would potentially make the participants change their diets. We refer to these factors as game changing.

The Game-Changing Factors

The participants recognised a number of factors that could change their dietary behaviour and motivate them towards making it more sustainable. They could be organised under three categories, namely environmental concerns, including climate change; health considerations, including cancers; and experience with meals without meat.

In relation to the environment, some of the participants explained "global warming and if there is no food for the animals to eat, we have to eat less" and "[r]esources on our planet are limited and we don't realise that we are cutting trees to have land to grow meat... We have to think and see far away from our plate". However, not many were aware about the scale of the problems and livestock's contribution to environmental deterioration, overuse of natural resources and climate change.

Health was another big game changer, but there was also a lot of disbelief that overconsumption of meat could be harmful. As one of the participants explained:

> I think if the WHO claims happened to be true, especially if meat-eating can cause cancer or other illnesses. It's scary to even think about. I am sure people will change everything, their lifestyle, eating … everything to avoid this.
>
> (Female, 51 years old)

Others thought about allergies and people getting sick as meat "takes a long process to be digested in a body… and it stays in a colon". Again, there were widespread resentment and doubts that such negative health effects could be real given the popularity of meat as human food.

With many of the participants not being very confident in their cooking skills, food preparation also could become a game-changing factor. Even those who have time to cook and enjoy doing it, describe using vegetables "in the simplest possible way steaming or boiling or baking them. I don't know any other way". When plant-based meals are made, they found them to be "plain and tasteless". By comparison, "meat dishes… are tastier and nice" and "never end up in the bin". Many acknowledged that they "don't have much experience with vegetarian dishes". From all start-up interviews, it became clear that knowing how to prepare tasty plant-based meals could also be a game changer.

These game-changing factors, namely environmental concerns, health considerations and food preparation skills became the three components of the intervention applied in this longitudinal study. They are described with further detail below.

THE SUSTAINABILITY SOCIAL MARKETING INTERVENTION

The social marketing intervention used during the longitudinal case study was informed by all the identified themes during the start-up interviews, including the game-changing factors. Each of its three components was developed using the SSMM for the target population, namely the participants in the Sydney study, and lasted one and a half years. This allowed to adjust the scoping of the tasks according to the knowledge and skills of the participants, develop the actions they had to do, implement and evaluate the results from their activities leading towards adaptation and improvement for the next component and ultimately for the next social marketing campaign (see Figure 13.2). The three components were all carried out under a marketing catchphrase representing the particular emphasis at that stage of the intervention, namely: "No biodiversity destruction in my plate!" – for the environmental element; "Is there a choice between a steak and a burning planet?" for the food preparation

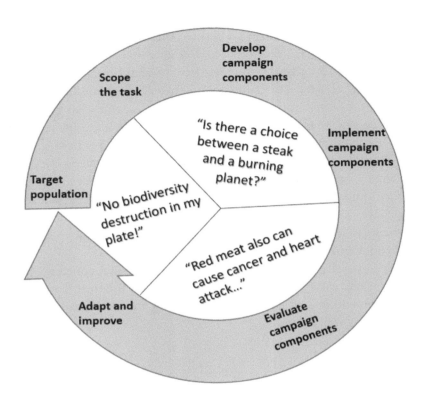

FIGURE 13.2
Sustainability social marketing intervention.

part and "Red meat can also cause cancer and heart attack…" for the health-related module.

"No Biodiversity Destruction in My Plate!"

During the first component held in the second half of 2015 after the start-up interviews, the catchphrase "No biodiversity destruction in my plate!" was used. The participants were exposed to science-based evidence and educational materials related to the environmental impacts of livestock and meat production and consumption, including contributing to climate change and biodiversity loss. These materials included a summary of the findings of the UN Food and Agriculture Organisation's "Livestock's Long Shadow" report (Steinfeld et al. 2006) and Michael Pollan's (Pollan, 2006) book "The Omnivore's Dilemma: A Natural History of Four Meals". The aim was to explain why consuming more plant-based meals would make a positive impact on the natural environment. Participants were given further tasks which included watching four documentaries in their free time, namely: "Cowspiracy: The sustainability secret" (2014), "Meat the truth" (2007), "Before the flood" (2016) and "Food choices" (2016).

Group debriefs were held to discuss the educational materials and films during which the participants shared their thoughts, feelings and perceptions about food and personal experiences. The participants were also observed to develop an understanding about the degree of impact this first social marketing component had on them.

"Is There a Choice between a Steak and a Burning Planet?"

In early 2017, the second sustainability social marketing component used the question "Is there a choice between a steak and burning planet?" which aimed to introduce new experiences in preparing and eating plant-based foods. This involved a series of demonstration and cooking sessions with the participants to teach them how to cook a variety of delicious plant-based dishes and enjoy eating them with family and friends. During the food preparation sessions, conversations were held about advantages of plant-based dishes, preparation time, accessibility and price of ingredients. Group debrief sessions and observations were also held. Interestingly, prompted by this component of the intervention, some of the participants started to recommend documentaries about meat consumption they have watched, including "The end of meat" (2017), "What the health" (2017), and "H.O.P.E.: What you eat matters" (2016 in German and 2018 in English).

"Red Meat Also Can Cause Cancer and Heart Attack ..."

The third social marketing component used "Red meat also can cause cancer and heart attack..." as a phrase to prompt the participants to think about the health consequences from excessive meat consumption. It also aimed at showing that human performance can improve when people are adhering to plant-based diets. In addition to watching "The game changers" (2019) documentary, the participants were given information leaflets summarising the latest scientific evidence about the health benefits of plant-based foods. They were directed to watch the weekly videos produced by Dr Michael Greger at nutritionfacts.org.

Again, group debriefs and discussion sessions were held with observations as what impact this component had on the participants. The aim was not to convert them to become vegan or vegetarians, but to encourage flexitarianism (Raphaely & Marinova, 2014), that is a reduction in meat consumption to healthy levels and incorporate plant-based dishes in their diet on a regular basis.

Final one-on-one interviews were held with all participants at the start of 2020. The outcomes from the longitudinal study are reported below.

SUMMARY OF THE OUTCOMES FROM THE SUSTAINABILITY SOCIAL MARKETING INTERVENTION

The findings reveal that the social marketing intervention played a role in shifting the meat consumption perceptions and attitudes for 14 (or 47%) of the original number of participants. Table 13.2 shows the changes, if any, the participants experienced throughout the five-year period. The social marketing campaigns results exhibited a complex trajectory towards a different degree of meat consumption changes (see Figures 13.3 and 13.4). As explained earlier, with this single-cohort group, we were looking for changes before and after the intervention. Any such changes were happening within a sociocultural environment which strongly encourages meat consumption. Despite the increasing new evidence presented in documentaries and academic publications, the message about decreasing meat consumption had not penetrated the wider Australian community.

At the commencement of the study, there were 25 heavy meat-eaters; this number became 13 at the end of the longitudinal period, including counting the woman who was hospitalised with mental health problems. Hence, the success in reducing meat consumption was around 50%. We did not expect to be able to influence all participants, given the multitude and multifaceted factors that shaped their dietary behaviour. The achieved reduction in meat consumption can be considered reasonable and shows that with social marketing effort we can sway a major proportion of the Australian population.

TABLE 13.2

Summary of Dietary Outcomes in the Sydney Longitudinal Study

			Meat Consumption at the Intervention's Occasions			
Subject No.	Age at the Start of the Study/Gender	Initial In-Depth Interview	Social Marketing Component 1 "No Biodiversity Destruction in My Plate!"	Social Marketing Component 2 "Is There a Choice Between a Steak or a Burning Planet?"	Social Marketing Component 3 "Red Meat Also Can Cause Cancer and Heart Attack …"	Final In-Depth Interview
		Early 2015	Mid 2015–2016	2017–Mid 2018	Mid 2018–2019	Early 2020
1	20 years/Male	Heavy	Heavy	Heavy	Heavy	Heavy
2	23 years/Male	Heavy	Heavy	Heavy	Heavy	Heavy
3	36 years/Male	Heavy	Heavy	Heavy	Heavy	Heavy
4	62 years/Male	Heavy	Heavy	Heavy	Heavy	Heavy
5	70 years/Male	Heavy	Heavy	Heavy	Heavy	Heavy
6	39 years/Male	Heavy	Moderate	Heavy	Heavy	Heavy
7	47 years/Male	Heavy	Moderate	Heavy	Heavy	Heavy
8	50 years/Male	Heavy	Heavy	Moderate	Heavy	Heavy
9	38 years/Male	Heavy	Heavy	Moderate	Moderate	Moderate
10	49 years/Male	Heavy	Heavy	Heavy	Moderate	Moderate
11	53 years/Male	Heavy	Heavy	Moderate	Moderate	Moderate
12	44 years/Male	Heavy	Heavy	Moderate	Vegetarian	Vegetarian
13	51 years/Male	Heavy	Heavy	Moderate	Moderate	Moderate
14	45 years/Male	Moderate	Moderate	Moderate	Moderate	Moderate
15	37 years/Female	Heavy	Heavy	Heavy	Heavy	Heavy
16	60 years/Female	Heavy	Heavy	Heavy	Heavy	Heavy

(Continued)

372 • Social and Sustainability Marketing

TABLE 13.2 (Continued)
Summary of Dietary Outcomes in the Sydney Longitudinal Study

Subject No.	Age at the Start of the Study/Gender	Initial In-Depth Interview	Meat Consumption at the Intervention's Occasions			Final In-Depth Interview
			Social Marketing Component 1 "No Biodiversity Destruction in My Plate!"	Social Marketing Component 2 "Is There a Choice Between a Steak or a Burning Planet?"	Social Marketing Component 3 "Red Meat Also Can Cause Cancer and Heart Attack…"	
		Early 2015	Mid 2015–2016	2017–Mid 2018	Mid 2018–2019	Early 2020
17	71 years/Female	Heavy	Heavy	Heavy	Heavy	Hospitalised
18	72 years/Female	Heavy	Heavy	Heavy	Heavy	Heavy
19	46 years/Female	Heavy	Moderate	Heavy	Moderate	Moderate
20	34 years/Female	Heavy	Moderate	Heavy	Moderate	Moderate
21	50 years/Female	Heavy	Moderate	Heavy	Moderate	Moderate
22	28 years/Female	Heavy	Heavy	Moderate	Moderate	Moderate
23	39 years/Female	Heavy	Heavy	Moderate	Moderate	Moderate
24	44 years/Female	Heavy	Heavy	Moderate	Moderate	Moderate
25	43 years/Female	Heavy	Moderate	Vegetarian	Vegetarian	Vegetarian
26	51 years/Female	Heavy	Moderate	Vegetarian	Vegetarian	Vegetarian
27	38 years/Female	Moderate	Moderate	Moderate	Moderate	Moderate
28	65 years/Female	Moderate	Moderate	Moderate	Moderate	Deceased
29	18 years/Female	Moderate	Moderate	Vegetarian	Vegetarian	Vegetarian
30	52 years/Female	Moderate	Moderate	Vegetarian	Vegetarian	Vegetarian

Influencing Sustainable Food-Related Behaviour • 373

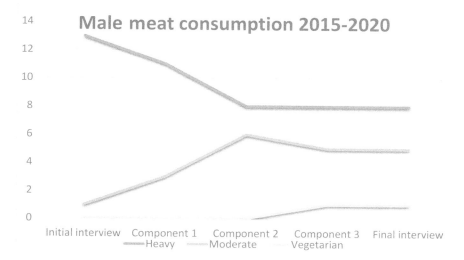

FIGURE 13.3
Men's meat consumption behaviour change trajectory.

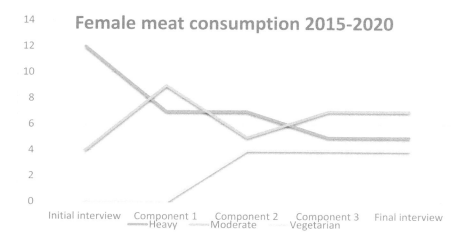

FIGURE 13.4
Women's meat consumption behaviour change trajectory.

Such a change however was much more difficult for the male participants, where only five (or 36%) of them changed their dietary habits, compared to nine (or 56%) among their female counterparts. At the end of the study, there was only one vegetarian man who switched to only a plant-based diet following a serious heart problem, and four vegetarian women. Below is an excerpt from the final interview with the vegetarian man:

> Since I become a vegetarian, I feel pretty good, very energetic. I completely submerged with my new dietary identity and even when my old friends are laughing at me when gathering and ordering a vegetarian meal, I don't care. Being healthy and happy is much worth then pleasing friends. I want to be able to see my kids growing and be part of their and my wife's lives. Now it's hard to imagine I used to eat meat.
>
> (Male, 44 years)

Two female heavy meat-eaters also became vegetarian after the social marketing intervention and at the final interview stated that there was no way back for them. One (43 years old) was shocked by the cruelty associated with livestock production and appalled by its high environmental footprint. She explains:

> One morning two years ago (2018) I woke up and I couldn't put meat in my mouth anymore. I am still very shaken from what I learned, and I am slowly sharing with my daughter, husband, and my friends my experience. I can't believe it how I was totally strayed and it took me a whole 3 years since I started the research study to realise things.
>
> (Female, 43 years)

The second female (51 years old) switched to vegetarianism after suddenly becoming allergic to meat. She explicated how difficult it was for her to not experience any bad reactions when tasting meat, and particularly red meat:

> I believe my body just had enough of meat. It was screaming from inside stop poisoning me and I just decided to stop. I haven't experienced any issues for 2.5 years. Back then I was thinking that this was a reaction provoked from the information I received from the study, but actually, the study helped me to realise that I wasn't meant to eat that much meat. Something just clicked inside me and the next day I embraced vegetarianism. I feel strong and will never go back.
>
> (Female, 51 years)

Two female participants who were moderate meat-eaters similarly became vegetarians. One of them (18 years old) shared that she was thinking of changing her diet prior to the intervention but she had been afraid that excluding meat from her meals would result in her putting on weight. The social marketing intervention was an "eye-opener" for her and she was reassured that plant-based dishes "are protein competitive to meat, something I thought wasn't the case before". The other (52 years old) commented: "The more meat you eat, the more consequences you experience. This is the deal we are all signing in with red meat and other meat."

Around 50% of the participants were not affected by the social marketing intervention and their meat consumption remained unchanged. Some of them reported dissatisfaction with the experience and that they felt "trapped and not engaged" and "sick of people telling me what to do, what to eat and what not to eat". Others continued to remain convinced that there were no health risks associated with high meat consumption: "It's healthy, nutritious, the best source of iron, zinc, you name it …" and "I am too young to think that a disease for elderly people like non-communicable are something like that should bother me".

There is also the group of participants whose dietary choices continued to include meat but shifted towards more plant-based options. Health considerations were a major contributing factor. According to one of these participants:

> If we know that red meat, and meat, is not good for us, why we continue to love eating it? This question stood still for days in my mind. We all need a wakeup call and this study was my awakening. I am a slow acceptor of new things, so I am taking it slowly and steady. So far so good. I am eating less meat.
>
> (Male, 51 years old)

There were also reasons related to the protection of the natural environment and concerns about animal welfare:

> Growing beef and other animals is harming the environment in many ways. Not only because of the land and resources we use to grow these animals, but also we are just looking at beef sizing, how to make it grow quicker and fattier… and to do this we feed the animals with unnatural food adding lots of hormones and antibiotics to avoid making them sick. But all of these are going into our own bodies and into the land, the water and everything.
>
> (Female, 51 years)

One participant also shared that she and others from the group now live much better without "the huge burden being murderer" and "killing so many animals to satisfy our appetite".

Being more confident about making plant-based dishes also played a role in these participants' decision to lower their meat intake. Many commented that they continue to "enjoy some of the recipes we've learned at the demonstration sessions".

DISCUSSION OF THE SYDNEY LONGITUDINAL STUDY RESULTS

To the best of our knowledge, this is the first study to apply a prolonged intervention and follow up of the participants over five years. This makes comparisons with other studies impossible. There are, however, useful lessons that could be learned from the Sydney longitudinal study.

First, it is challenging to maintain the interest and commitment of all participants. Furthermore, the effects of many interventions tend to dissipate in the medium and longer-term, as experienced in studies related to smoking and alcohol consumption (Stead et al. 2006). We also witnessed such an effect with two men and one woman (or 10% of the sample) reducing their meat intake and then reverting to the original levels of consumption of animal proteins. This is a relatively small share of attrition. We also did not observe any change between the last component of the intervention and the final interview (see Figures 13.3 and 13.4) which indicated that the achieved results had stabilised.

Second, as this was a longitudinal study, each participant was observed on many occasions (Cook & Ware, 1983) and under different settings, namely individual and group-based. This allowed us to follow the individual responses throughout the different components of the sustainability social marketing intervention. As we were dealing with a small group of participants, it is difficult to draw generalisations. However, it became clear that women, and particularly younger women, are more likely to embrace such a dietary shift and were more open to change. This is a positive finding as women are also those who have a bigger impact on the dietary preferences of the future generations.

Third, as in real life, it was very difficult in this longitudinal study to control for external influences and dependencies including cultural, religious, societal and family pressure. In total, we saw a drop in meat consumption for five men and nine women (14 participants or around 50%) of the Sydney group. Given that we witnessed a high level of stability, it is highly likely that these people would also influence their friends and relatives making the impact of this study even more significant.

Irrespective as to what the future holds for this particular group of people, we were able to observe that sustainability social marketing worked, including the 4S mix. The messages about sustainability, strength, self-confidence and sharing were commonly used in the information materials and they were also reflected in the words of the participants during the final interviews. Below are three quotes that summarise the feelings of those who made the transition towards more sustainable and healthier food choices:

> I can say I consider myself confident enough to continue being a vegetarian despite the pressure I ... experience from family and friends.
>
> (Female, 52 years old)

> You know, there are not many opportunities this planet can provide us and we have to learn how to keep it going and to learn how to share with others what is best for all of us. Eating less meat and enjoying healthy life is what little we can do to help the planet.
>
> (Female, 18 years old)

> I can call it strength to be a man and not be afraid to be judged by others because of your meatless meal.
>
> (Male, 44 years old)

The longitudinal design of the study permitted the observation and recording of behaviour change over time and with multiple observations (Caruana et al. 2015). Although we could not isolate the participants from other factors affecting their food consumption, the three components of the intervention were uniquely suited to help us understand the dynamics of behavioural change and adjust the messages we were sending to them. Nevertheless, we also experienced some difficulties, particularly in maintaining the size of the original sample of 30 participants over the five-year period. Additionally, another challenge was the high demand on research time for the design and delivery of the components of the intervention and the large number of interviews conducted with the participants individually and in group settings.

This study also has some theoretical and practical implications. From a theoretical point of view, it was testing the SSMM and it confirmed that such an approach is a useful way of dealing with behavioural change in complex and often sensitive circumstances, such as the ones related to personal food choices. The importance of sending the same message through an iterative process was demonstrated in the encouraging outcomes and particularly the fact that the achieved new food behaviour remained stable without relapsing. Furthermore, the study was using the 4S marketing mix in a real-life situation and it showed that it worked on a practical basis. The majority of the participants were able to embrace the sustainability social marketing messages which resonated with their improved knowledge, better awareness and newly developed values. Hence, this study opens up the perspective for larger-scale SSMM interventions which can lead to improved outcomes for the greater public good.

CONCLUSION

Change is a multifaceted construct (Caspi & Roberts, 1999) and personality traits are indisputably consistent across time and age (Roberts, Caspi, & Moffitt, 2001; Fraley & Roberts, 2005). People are frequently making choices reflecting their personal values but also the knowledge they have. The study provided understanding of the complex environment within which people consume food and the factors that conspire against them being able to pursue sustainable behaviours. Social marketing aims at awaking people's behaviour towards the common good by succinctly exposing them to the appropriate and relevant messages and triggering positive changes.

This was manifested in the Sydney exploratory longitudinal study where 30 participants were subjected to a five-year long intervention aimed at reducing the level of their meat consumption. The components of the intervention were oriented towards preventing further environmental deterioration – "No biodiversity destruction in my plate!", improving food preparation skills – "Is there a choice between a steak and a burning planet?", and improving personal health – "Red meat also can cause cancer and heart attack …". The behavioural changes we witnessed affected close to half of the subjects who participated in the study. They were sustained over a prolonged period of time which indicates that these participants have started to see the world differently. Such a shift towards an increased intake of plant-based options was more pronounced amongst women, and particularly younger women.

The findings from this study as well as the marketing messages or the catchphrases can be used as a reflective tool for other investigations. They can also inform policy actors and influencers, particularly when governments are not yet acting in the best interest of their constituents. It is difficult to make generalisations from this exploratory study; however, it leaves us with a positive optimistic view of the world where gradual changes in people's diets can contribute towards eliminating the tragedy not only of the commons but also of all species who inhabit this planet.

LESSONS LEARNED

- It is important to educate people about the impacts of their dietary choices, and sustainability social marketing provides a useful platform to achieve this.
- People make food choices based on their personal values and preferences; however, knowledge and skills in food preparation of vegetarian meals are crucial contributing factors.
- Sustainability marketing catchphrases are a good way to convey important messages in a succinct way that engages with the general audience and can become behaviour triggers.
- Shifting people's diets towards making them more sustainable is a challenging but not impossible task with the first step to accomplish the change is to convince your friends to give it a try.

DISCUSSION QUESTIONS

1. Comment on the success of the sustainability marketing intervention related to reduction in meat consumption.
2. What are the main factors that are likely to impact people's food preferences and why?
3. What is the role of social networks in impacting people's food choices?

4. Can food be seen as masculine or feminine and why?
5. What are the challenges and benefits in longitudinal/qualitative studies?

SUGGESTED ACTIVITIES

1. Develop sustainability social marketing catchphrases to influence people's dietary choices.
2. Analyse your own diet from a planetary health perspective. Do you need to make any changes and why?
3. Watch a sustainability food-related movie, such as "Cowspiracy", "Meat the truth", "Before the flood", "Food choices", "Food Inc.", "What the health", "The game changers", "The end of meat", "H.O.P.E.: what you eat matters", "The need to grow" etc., and discuss what are the main issues covered and what are the main messages that you would remember.
4. Organise a food session in class by bringing dishes to share that are more sustainable than what you would generally eat. Explain what makes your dish more sustainable.
5. Create a menu for a new restaurant; how would you advertise it for a clientele interested in sustainability?

INTERESTING FACTS

- In the case of beef, it takes 38 calories fed to the animal to produce 1 calorie for human consumption.
- The land mass used on this planet for livestock grazing and feed (27%) is bigger than the land occupied by forests (26%).
- According to Google trends, veganism is increasingly gaining popularity.

KEY DEFINITIONS

- 4S marketing mix – a new way of marketing products which emphasises the following qualities: sustainability, strength, self-confidence and sharing.
- Flexitarianism – a diet which aims at reducing the consumption of animal-based foods.

- Longitudinal study – a research study which follows up the same participants over a prolonged period of time.
- Sustainability social marketing – a subclass of social marketing that aims to trigger and encourage behavioural changes for the common good of the human and other species.
- Tragedy of the commons – overexploitation of common resources in the pursuit of personal gains to the detriment of society.

AUTHORS WHOSE WORK TO FOLLOW

In addition to the authors of this chapter, below are some writers whose works are worth reading:

Carol Adams
Hans Dagevos
Jeff French
Ross Gordon
Philip Kotler
Nancy Lee
Michael Pollan
Sharyn Rundele-Thiele
Peter Singer

ACKNOWLEDGMENT

The authors acknowledge the contribution of an Australian Government Research Training Program Scholarship in supporting this research. They are also thankful to the study's participants who persevered with this longitudinal project despite any challenges.

CREDIT AUTHOR STATEMENT

All authors conceptualised the study; DB conducted the interventions and interviews; DB and DM analysed the data and drafted the chapter; all authors made substantial contributions throughout all sections, read and approved the final manuscript for publication.

REFERENCES

Adams, C. J. (2010). *The sexual politics of meat* (20th anniversary ed.). New York, NY: Continuum.

Bègue, L., & Treich, N. (2019). Immediate and 15-week correlates of individual commitment to a "Green Monday" national campaign fostering weekly substitution of meat and fish by other nutrients. *Nutrients, 11*(7): 1694. https://doi.org/ 10.3390/nu11071694

Berley, P., & Singer, Z. (2014). *The flexitarian table: Inspired, flexible meals for vegetarians, meat lovers and everyone in between.* Boston, MA: Houghton Mifflin Harcourt.

Bogueva, D., & Marinova, D. (2018). What is more important – perception of masculinity of personal health and the environment? In D. Bogueva, D. Marinova & T. Raphaely (Eds.), Handbook of research on social marketing and its influence on animal origin food product consumption (pp. 148–162). Hershey, PA: IGI Global.

Bogueva, D., Marinova, D., & Raphaely, T. (Eds.) (2018). *Handbook of research on social marketing and its influence on animal origin food product consumption.* Hershey, PA: IGI Global.

Bogueva, D., Marinova, D., Raphaely, T., & Schmidinger, K. (Eds.) (2019). *Environmental, health and business opportunities in the new meat alternatives market.* Hershey, PA: IGI Global.

Bogueva, D., Raphaely, T., Marinova, D., & Marinova, M. (2017). Sustainability social marketing. In J. Hartz-Karp & D. Marinova (Eds.), *Methods for sustainability research* (pp. 280–291). Cheltenham, UK: Edward Elgar.

Caruana, E. J., Roman, M., Hernández-Sánchez, J., & Solli, P. (2015). Longitudinal studies. *Journal of Thoracic Disease, 7*(11), E537–E540. https://doi.org/10.3978/j.issn.2072-1439.2015.10.63.

Caspi, A., & Roberts, B. W. (1999). Personality change and continuity across the life course. In L. A. Pervin & O. P. John (Eds.), *Handbook of personality theory and research* (pp. 300–326). New York, NY: Guilford Press.

Cook, N. R., & Ware, J. H. (1983). Design and analysis methods for longitudinal research. *Annual Review of Public Health, 4,* 1–23. https://doi.org/10.1146/annurev.pu.04.050183.000245

de Bakker, E., & Dagevos, H. (2012). Reducing meat consumption in today's consumer society: Questioning the citizen-consumer gap. *Journal of Agriculture and Environment Ethics, 25*(6), 877–894. https://doi.org/ 10.1007/s10806-011-9345-z

De Boer, J., Schösler, H., & Aiking, H. (2014). "Meatless days" or "less but better"? Exploring strategies to adapt Western meat consumption to health and sustainability challenges. *Appetite, 76,* 120–128. https://doi.org/10.1016/j.appet.2014.02.002

Eshel, G., Shepon, A., Makov, T., & Milo, R. (2014). Land, irrigation water, greenhouse gas, and reactive nitrogen burdens of meat, eggs, and dairy production in the United States. *Proceedings of the National Academy of Sciences of the United States of America* (PNAS), 111(33), 11996–12001. https://doi.org/10.1073/pnas.1402183111

Euromonitor International. (2011). *The war on meat: How low-meat and no-meat diets are impacting consumer markets.* https://www.euromonitor.com/the-war-on-meat-how-low-meat-and-no-meat-diets-are-impacting-consumer-markets/report

Fessler, D. M. T., & Navarrete, C. D. (2003). Meat is good to taboo: Dietary proscriptions as a product of the interaction of psychological mechanisms and social processes. *Journal of Cognition and Culture, 3*(1), 1–40. https://doi.org/10.1163/156853703321598563

Fraley, R. C., & Roberts, B. W. (2005). Patterns of continuity: A dynamic model for conceptualizing the stability of individual differences in psychological constructs across the life course. *Psychological Review, 112*(1), 60–74. https://doi.org/10.1037/0033-295X.112.1.60

French, J., & Gordon, R. (2019). *Strategic social marketing: For behaviour and social change* (2nd ed.). Thousand Oaks, CA: SAGE.

Greenpeace. (2018). *Less is more: Reducing meat and dairy for a healthier life and planet.* Scientific background of the Greenpeace vision of the meat and dairy system towards 2050. https://storage.googleapis.com/p4-production-content/international/wp-content/uploads/2018/03/6942c0e6-longer-scientific-background.pdf

Gusmai A. (2018). Right to food and "tragedy" of the commons. In A. Isoni, M. Troisi & M. Pierri (Eds.), *Food diversity between rights, duties and autonomies* (pp. 243–263). Cham, Switzerland: Springer. https://doi.org/10.1007/978-3-319-75196-2_15

Hardin, G. (1968). The tragedy of the commons. *Science, 162*(3859), 1243–1248. https://doi.org/10.1126/science.162.3859.1243

Healthy WA. (n.d.). *Go for 2&5 fruit and vegetable campaign.* Department of Health Western Australia. https://healthywa.wa.gov.au/Articles/F_I/Go-for-2-and-5

Hill, K. G., Woodward, D., Woelfel, T., Hawkins, J. D., & Green, S. (2016). Planning for long-term follow-up: Strategies learned from longitudinal studies. *Prevention Science, 17*(7), 806–818. doi:10.1007/s11121-015-0610-7

Kotler, P., & Lee, N. (2008). *Social marketing: Influencing behaviors for good* (3rd ed.). Thousand Oaks, CA: SAGE.

Laestadius, L.I., Neff, R.A., Barry, C.L., & Frattaroli S. (2016). "No meat, less meat, or better meat: Understanding NGO messaging choices on how meat consumption should be altered in light of climate change. *Environmental Communication, 10*(1), 84–103. https://doi.org/10.1080/17524032.2014.981561

Lupton, D. (1996). *Food, the body and the self.* London, UK: SAGE.

Marinova, D., & Bogueva, D. (2019). Planetary health and reduction in meat consumption. *Sustainable Earth, 2*:3, https://doi.org/10.1186/s42055-019-0010-0

Meat & Livestock Australia (MLA). (2019). 2019 *State of the industry report: The Australian red meat and livestock industry.* https://www.mla.com.au/globalassets/mla-corporate/prices–markets/documents/trends–analysis/soti-report/mla-state-of-industry-report-2019.pdf

National Cancer Control Indicators. (2020). *Processed meat and red meat consumption.* https://ncci.canceraustralia.gov.au/prevention/diet/processed-meat-and-red-meat-consumption

Parker, C. L. (2011). Slowing global warming: Benefits for patients and planet. *American Family Physician, 84*(3), 271–278.

Paulus, J. K., Dahabreh, I. J., Balk, E. M., Avendano, E. E., Lau, J., & Ip, S. (2014). Opportunities and challenges in using studies without a control group in comparative effectiveness reviews. *Research Synthesis Methods, 5,* 152–161. https://doi.org/10.1002/jrsm.1101

Ployhart, R., & Vandenberg, R. (2010). Longitudinal research: The theory, design, and analysis of change. *Journal of Management, 36*(1), 94–120. https://doi.org/10.1177/0149206309352110

Pollan, M. (2006). *The omnivore's dilemma: A natural history of four meals.* New York, NY: Penguin Books.

Raphaely T., & Marinova, D. (2014). Flexitarianism: A more moral dietary option. *International Journal of Sustainable Society, 6*(1–2), 189–211. https://doi.org/10.1504/IJSSOC.2014.057846

Roberts, B. W., Caspi, A., & Moffitt, T. (2001). The kids are alright: Growth and stability in personality development from adolescence to adulthood. *Journal of Personality and Social Psychology, 81*(4), 670–683. https://doi.org/10.1037/0022-3514.81.4.670

Ruby, M. B., & Heine, S. J. (2011). Meat, moral and masculinity. *Appetite, 56*(2), 447–450. https://doi.org/10.1016/j.appet.2011.01.018

Rundle-Thiele, S., Kubacki, K., Leo, C., Arli, D., Carins, J., Dietrich, T., Palmer, J., & Szablewska, N. (2013). Social marketing: Current issues and future challenges. In K. Kubacki & S. Rundle-Thiele (Eds.), *Contemporary issues in social marketing* (pp. 41–58). Newcastle Upon Tyne, UK: Cambridge Scholars Publishing.

Saunders, S. G., Barrington, D. J., & Sridharan, S. (2015). Redefining social marketing: Beyond behavioural change. *Journal of Social Marketing, 5*(2), 160–168. https://doi.org/10.1108/JSOCM-03-2014-0021.

Saunders, S. G., & Truong, V. D. (2019). Social marketing interventions: Insights from a system dynamics simulation model. *Journal of Social Marketing, 9*(3), 329–342. https://doi.org/10.1108/JSOCM-05-2018-0054.

Schösler, H., de Boer, J., & Boersema J. J. (2012). Can we cut out the meat of the dish? Constructing consumer-oriented pathways towards meat substitution. *Appetite, 58*(1), 39–47. https://doi.ord/10.1016/j.appet.2011.09.009

Singer, P. (2015) *Animal liberation* (new ed.). London, UK: The Bodley Head.

Stead, M., McDermott, L., Gordon, R., Angus, K., & Hastings, G. (2006). *A review of the effectiveness of social marketing: Alcohol, tobacco and substance misuse interventions*. NSMC Report 3, Institute for Social Marketing (ISM), University of Stirling & The Open University, London. http://www.asrem.org/corsi_aggiornamento_convegni/comunicazone%20per%20la%20slute/Alcol,%20tabacco%20ed%20abuso%20sostanze%20nel%20Marketing%20sociale.PDF

Steinfeld, H., Gerber, P., Wassenaar, T., Castel, V., Rosales, M., & de Haan, C. (2006). *Livestock's long shadow: Environmental issues and options*. Food and Agriculture Organisation of the United Nations (FAO). http://www.fao.org/3/a0701e/a0701e00.htm

Sutton, C., & Dibb, S. (2013). *Prime cuts valuing the meat we eat*. World Wide Fund for Nature (WWF-UK) and the Food Ethics Council, Surrey, UK. http://assets.wwf.org.uk/downloads/prime_cuts_food_report_feb2013.pdf

Truong, V. D., Saunders, S. G., & Dong, X. D. (2019). Systems social marketing: A critical appraisal. *Journal of Social Marketing, 9*(2), 180–203. http://doi.org/10.1108/JSOCM-06-2018-0062.

Twin, A. (2020). *The 4Ps*. Investopedia. https://www.investopedia.com/terms/f/four-ps.asp

Willett W, J., Rockström, J., Loken, B., Springmann, M., Lang, T., Vermeulen, S., Garnett, T., Tilman, D., DeClerck, F., Wood, A., Jonell, M., Clark, M., Gordon, L. J., Fanzo, J., Hawkes, C., Zurayk, R., Rivera, J. A., De Vries, W., Sibanda, L. M., … Murray, C. J. L. (2019). Food in the Anthropocene: The EAT–Lancet Commission on healthy diets from sustainable food systems. *The Lancet, 393*, 447–492. https://doi.org/10.1016/S0140-6736(18)31788-4

World Health Organization (WHO). (2015). *Q&A on the carcinogenicity of the consumption of red meat and processed meat*. http://www.who.int/features/qa/cancer-red-meat/en/

Section VI

Advances in Knowledge of Social and Sustainable Marketing: The Power of Online Consumer Reviews

14

Enforcing Brands to Be More Sustainable: The Power of Online Consumer Reviews

Feyza Ağlargöz
Faculty of Economics and Administrative Sciences,
Anadolu University, Eskişehir, Turkey

CONTENTS

Learning Objectives 387
Themes and Tools Used 388
Voicing for Sustainability 388
Introduction 388
Introducing the Power Shift: From Companies to Consumers 390
From Word of Mouth to Electronic Word of Mouth 394
Online Reviews: The Other Consumers 395
Marketing in the Age of Sustainability 400
Forcing Brands to Be More Sustainable 402
Online Reviews and Sustainability: Illustrative Cases
    and Recent Trends 404
Discussion and Conclusion 406
Additional Content 408
Funding Declaration 408
References 408

LEARNING OBJECTIVES

After reading the chapter, the reader should be able to:

- Discuss the power shift from companies to consumers
- Describe the path from word of mouth to electronic word of mouth

- Explain the role of online reviews in marketing
- Explain how marketing changes in the age of sustainability
- Discuss how consumers force companies to be more sustainable by the power of online reviews

THEMES AND TOOLS USED

Sustainability as being the main theme of the chapter can be analyzed by using SWOT analysis. Online reviews can be analyzed by using Critical Discourse Analysis (CDA).

VOICING FOR SUSTAINABILITY

When did you last raise your voice after a problematic purchase? When did you last write a comment based on your wonderful experience? Moreover, when did you last read other consumers' reviews before making a purchase decision? Have you ever stopped buying your favorite brand because of negative comments? While you are thinking about your answers, consider the sustainability problems that we are facing globally. Have you ever thought that these two concepts can be interrelated with each other? This chapter invites you to an intellectual discussion on these topics based on the rising power of consumers and sustainability awareness. Sometimes seemingly, unrelated topics can be interdependent. Throughout this chapter, we will scrutinize these interdependencies to give you an idea of developing more sustainable customer solutions and how you can integrate consumer reviews into this process.

INTRODUCTION

Brands cannot continue with what they have done so far. The passive consumerism of the past has already changed. The logic of marketing has evolved. This evolution is not only due to past conditions such as wars, welfare expectations, and technological advances but is also due to the empowerment of consumers. Brands, adopting the traditional philosophies of marketing, had been using their supremacy against the consumers who had limited alternatives and limited media to be heard. Conversely, the empowered consumers of today have more to say, more to do, and generally tend to ask for more. Limited and one-way communication with brands has turned into timeless and limitless interaction. In today's developed nations, satisfying the needs and wants of today's consumers is not enough. The same needs and wants can be met by many brands, and thus these are not competitive advantages for brands anymore. Socially responsible brands, corporate social responsibility (CSR) initiatives, greener solutions, cruelty-free products, environmentally friendly technologies, fair trade, and/or sustainability are the new demands of contemporary consumers. Furthermore, the sustainability expectations of the consumers are expanding every day.

When contextualizing all the processes companies and consumers pass through, word of mouth (WOM) communication is one of the oldest structures in human socialization history (Dellarocas, 2003), and it has kept its importance. However, with the rapid growth of the Internet, Web 2.0, and social media, WOM has evolved into online WOM marketing. In Web 2.0, consumers are not just the target audience anymore; they have transformed into active users, in other words, "prosumers". Thanks to Web 2.0, consumers can make comments about and evaluate almost everything, easily share content, create a community, take part in communities, and affect the opinions of others positively or negatively (Özata, 2011). Today's consumers are more informed on, more connected to, and more powerful for brands than ever before. These new powerful consumers are better informed about brands, and they use numerous digital platforms to air and share their ideas about brands with others (Kotler & Armstrong, 2018).

The rise of the digital economy increases the importance of consumer reviews, which are easily shared and spread. Today's more digitalized consumers generally do not make their buying decisions without evaluating these valuable experience outcomes through *online reviews*. Online reviews are one of the uses of consumer-generated content. Consumers who experience products and services share their comments on various customer review websites, social media, or customer ratings and review parts of the companies. Thus, consumers rely increasingly more on online consumer reviews to enable their purchase decisions (Liu, 2017; Parment, 2013; Smith & Anderson, 2016). This tool is not only essential for brands because of its effect on other consumers' decisions, but it is also essential because of its ability to affect brands to be more sustainable. Evidence that supports sustainability has become of vital importance around the world is increasing (Nielsen, 2018). The expectations for being more sustainable became more visible with online consumer reviews. Consumers can now urge brands to be more responsive to worldwide problems, social causes, the environment, and human welfare by using the online world. Therefore, based on previous studies (e.g., Gerdt, Wagner, & Schewe, 2019; Strähle & Gräff, 2017; Ben Abdelaziz, Saeed, & Benleulmi, 2015; El Dehaibi, Goodman, & MacDonald, 2019; White et al. 2019; Anastasiei & Dospinescu, 2019), this chapter's purpose is to search and highlight the traces of consumer power created through online reviews making brands be more sustainable and to offer more sustainable consumer solutions. Following the aforementioned purpose, the chapter will scrutinize how online consumer reviews push brands toward sustainability. The discussion will be initialized by explaining the online transformation in marketing as the power shift from companies to brands and from WOM to electronic word of mouth (eWOM) followed by online consumer reviews. The chapter will be finalized by highlighting the power of online consumer reviews on brands' sustainability efforts by using illustrative global cases.

INTRODUCING THE POWER SHIFT: FROM COMPANIES TO CONSUMERS

The rapid development of modern communication technologies, the high Internet penetration rates in developed and developing countries, and wireless systems and mobile communication have increased the emphasis on technology for both companies and consumers. Technology tends to change current behaviors and add new ones (Amaral, Tiago, & Tiago, 2014, p. 137). Dellarocas (2003, p. 1407) specifies "bidirectionality" as one of the most critical abilities of the Internet compared to earlier methods of mass communication. Via the Internet, communication between companies and consumers became more interactive. Now, it is not only cheaper and easier to reach consumers, but consumers can also disseminate their thoughts, reactions, and experiences with companies and other consumers on a global level. New technologies provide consumers more power and control over their relationships with brands. Today, consumers have a chance to access more information about brands than ever before. Through the Internet and social media, consumers interact with companies and share their product and brand reviews with other consumers. This power and control have even added a new dimension to customer relationship management, named customer-managed relationships (Kotler & Armstrong, 2018).

These developments have changed the power balance between companies and customers in their relationships, and O'Connor (2008, p. 48) explains this situation as one of the main reasons that Time Magazine chose "You" as the Person of the Year in 2006. The rapid growth and impact of user-generated content on the Internet such as blogs, YouTube, and MySpace were the reason for choosing "You" as "Person of the Year" among all the other candidates. Users came together, and user-generated content transformed the Internet. Time Magazine explained the cover as "Yes, you. You control the Information Age. Welcome to your world". The year 2006 can be seen as the first acceptance of a new age. Today's empowered consumers can shop around the world online, search and find information about products and brands, share their views with hundreds and thousands of people, and make or break brands overnight. In other words, consumers search online, live virtually, shop online, post tweets, like and become fans, and explore hundreds of mobile applications (Amaral et al., 2014, p. 137). For example, Nas Daily is a Facebook page with 16 million followers. Nuseir Yassin produces one-minute travel videos on social media. In 2018, in his daily video named "Is Apple worth it?", he criticized Apple products in general (Nas Daily, 2018). The impact of this criticism could be high for Apple.

It is self-evident that the relationships between brands and consumers have started to change with the Internet, but Web 2.0 and social media as a Web 2.0 innovation have revolutionized this relationship. The World Wide Web radically transformed the power imbalance, and thus, consumers can now access rich data to use while making their purchase decisions. Social media enables consumers to create content and spread their words about anything across the world to anyone who wants to listen (Labrecque et al. 2013, p. 257). As the use of Web 2.0 applications has increased, many kinds of online user reviews have emerged (Ye, Law, & Chen, 2011, p. 634).

Hellberg (2014) states the change in the balance of brand-consumer relationships as a power shift through social media. Social media give power to consumers; consumers are taking control of social media, and brands must adapt to this situation. Consumers can force companies to relaunch dead products, and they can even harm the brands. Consumers control the information and content on social media. Social media presents an entirely new environment for consumers to share their evaluations of products or brands, and consequently social media opens the door for WOM communication (Chen, Fay, & Wang, 2011, p. 85).

The Internet has transformed from an atmosphere of information into an atmosphere of influence due to social media sites such as Facebook, Twitter, and YouTube (Hanna, Rohm, & Crittenden, 2011, p. 272). Social media has a high impact on company-customer relationship development and shifts the methods of social communication and interaction (Amaral, et al., 2014, p. 137). Facebook, as the leading global social platform, has 2.603 billion monthly active users; YouTube has 2 billion monthly active accounts; and Instagram has 1.082 billion monthly active users (Statista, 2020). Instagram is the go-to social platform for 18 to 24-year-old users. For marketers looking to connect with millennials, Instagram is the top method (Olapic, 2016). These numbers demonstrate the importance of social media in marketing communication. With this user power, social media is indispensable for brands. Brands can create social media accounts to build long-lasting customer relations. In addition, companies can obtain insights on how consumers talk about their brands and competitors' brands as well. The Internet and especially social media provide opportunities to track consumer conversations about brands. By accessing the product reviews, comparisons on product features, and prices quickly, consumers can harmonize their preferences with the products and reduce their information asymmetry with companies (Labrecque et al. 2013, p. 261).

Labrecque et al. (2013, p. 259) define consumer power in the digital era using four distinct sources of consumer power. These are demand-based, information-based, network-based, and crowd-based power. Demand-based power lies in the cumulative influence of the consumption and purchase behaviors generated by the Internet and social media. Information-based power is composed of power through content consumption and content production. Information-based power is connected with easy access to product or service information through content consumption. In addition, information-based power is the ability to produce user-generated content through content production. This facilitates the empowerment of consumers by supplying a platform for self-expression, increasing individual reach, and improving the potential to influence markets for individuals. Network-based power derives from activities such as the dissemination, completion, and modification of content. Last, crowd-based power aims to achieve benefits for both individuals and the group. For instance, crowd-based power includes crowd creation such as Wikipedia, crowdsourcing such as Amazon Mechanical Turk, and crowd selling such as Etsy.

Furthermore, the same power shapes consumers' daily activities in various ways. In an online environment, search technology systems control the information that consumers can reach. YouTubers, Instagrammers, and other opinion leaders continually affect purchase decisions by means of their advice and testing videos (Labrecque et al. 2013, p. 258). Control over people in social media environments is associated with influence. Labrecque et al. (2013, p. 258) define influence as a function of reach and the degree of relationship and persuasion power of the person in the network, depending on the relevance of the content the person creates online. Consumers have become stronger, but companies also try to find ways to manage relationships in compliance with the rules of the new game.

In short, the World Wide Web has allowed modern, recent, and more effective practices of transmitting, conveying, and acquiring the information. Potential customers can now access the tremendous amount of information provided by the World Wide Web to facilitate the evolution of alternatives during their decision-making process. Instead of the company managing how information is submitted and used, consumers are now in charge (O'Connor, 2008, p. 48). Consumers are no longer solely passive parts of the marketing process. They are now active creators in this exchange relationship. Consumers are taking increasingly more active roles in cocreating marketing content with the brands (Hanna et al. 2011, p. 265). The diffusion of the Word Wide Web as a mainstream consumer media had significant implications for economic life. Recently described core values connected with the development of the Web are peer-to-peer production and sharing data (O'Connor, 2008, p. 47). Web 2.0 and user-generated content are changing the ways people approach information. How consumers search, find, read, collect, share, develop, and consume information are shifting (Ye et al. 2011, p. 635). After highlighting the power shift from brands to consumers, another shift, namely, from WOM to eWOM, will be discussed in the upcoming section.

FROM WORD OF MOUTH TO ELECTRONIC WORD OF MOUTH

WOM communication has been one of the oldest methods of communication in history (Dellarocas, 2003). WOM underlies interpersonal communications, which makes it very effective at influencing consumers' product evaluations and purchase decisions (Cui, Lui, & Guo, 2012, p. 41). Given the lack of direct experience, consumers should try a product or take some ideas from friends, neighbors, relatives, and colleagues (Cakim, 2010, p. 6). When making a purchase decision, consumers may use WOM communication as a vital source of information. However, it can be hard to access WOM communication for many consumers due to limited time and limited personal networks (Pitt, Berthon, Watson, & Zinkhan, 2002, p. 8).

The rapid development and extensive use of the Internet facilitate access to WOM communication. Moreover, the Internet tremendously increased the effect of WOM communication. Consumers can now potentially affect thousands of people within minutes (O'Connor, 2008, p. 48). Therefore, the emergence of the Internet has brought a new form of web communication that facilitates the provision and distribution of information between companies and consumers and among consumers themselves (Park & Nicolau, 2015, p. 67). Reading online reviews brought a new aspect to WOM. In this situation, WOM has become one of the most important information sources about brands, products, and services (Cakim, 2010, p. 6). WOM communication influences the purchasing decisions of the consumer bypassing information. With the growth of online social network sites and customer interactions, traditional WOM turned into eWOM (Amaral et al., 2014, p. 138). Electronic or online WOM is composed of any positive or negative comments of potential, current, or former customers about products, services, and brands on the Internet that are available to other people (Amaral et al., 2014, p. 140). Dellarocas (2003) explains the change from traditional WOM and eWOM as an ancient concept in a modern setting.

Internet forums and online discussions, as the first versions of eWOM, have greater credibility, and are more relevant to consumers and have exceptional abilities to generate empathy among readers compare to company-generated Web content (Bickart & Schindler, 2001, pp. 32–33). Chen and Xie (2008) categorize online customer product reviews as a new type of WOM information because of its increasingly essential part in the purchasing processes of consumers. eWOM can be seen as user-generated content (Amaral et al., 2014). Today, many different websites help consumers to swap information and share ideas with others. New technologies such as smartphones and social media have empowered consumers to create their own content (Olapic, 2016). Content creators gain control power based on the information they share with potential buyers (Labrecque et al. 2013, p. 261).

Comparing traditional WOM and eWOM, eWOM passes traditional WOM due to the tremendous amount of content. For a product or a service, one may find information from multiple sources. The social environment limits conventional WOM, but eWOM can quickly spread and reach a vast amount of customers via the Internet. The scale and scope of eWOM are much more considerable (Lu, Ba, Huang, & Feng, 2013, p. 598). Bickart and Schindler (2001, p. 37) explain the difference between traditional forms of WOM communication and online discussions (Internet forums) that make consumer information searches easier. Typically, WOM communication involves the information shared by a friend, a relative, or a colleague in face-to-face communication. In contrast, online discussions involve personal experience and opinions in a written format. With Internet forums and online discussion platforms, consumers were able to share their experiences, opinions, and knowledge with other consumers on any subject. Currently, online reviews are becoming more apparent with social media platforms such as Instagram and YouTube compared to the past. The online shopping and e-commerce report of the Pew Research Center revealed that 55% of American adults watch online product review videos before making purchasing decisions. Customers who are under 50 are more likely to watch product review videos more than their older counterparts. In terms of gender, men are more willing to watch online reviews videos than women (Smith & Anderson, 2016, p. 12). At this stage, we will now scrutinize online reviews as a form and foundation for eWOM.

ONLINE REVIEWS: THE OTHER CONSUMERS

Today, customers can effortlessly demonstrate their disappointment or satisfaction with a product to a broader audience (Beldad, Avicenna, & De Vries. 2017, p. 163). Increasingly more consumers are now reading product reviews before making purchasing decisions. These product reviews have become a vitally important content of online shopping websites (Liu, Huang, An, & Yu, 2008, p. 443). Online consumer reviews have become a relatively new way of gathering product information and are growing in popularity (Chen & Xie, 2008, p. 478). Online consumer reviews are mainly considered to be a form of eWOM (Park & Nicolau, 2015, p. 689). Chen and Xie (2008, p. 489) view consumer reviews as information created by users for users and as a new element in the marketing communication mix. Compared to the information offered by brands, online reviews are evaluated as more trustworthy and credible (Park & Nicolau, 2015, p. 689). "People trust people, not brands" is a proverb that explains the importance of online reviews very well (Olapic, 2016).

People who do not have previous experience with a brand or product are more likely to seek information before buying it. Especially for the high-involvement products with high prices, perceived risk and feelings of uncertainty are high. These risk perceptions and feelings of uncertainty lead customers to use online reviews as a primary source of information (Beldad et al. 2017, p. 163–164). The role of online reviews is to assist consumers who feel indecisive and less knowledgeable by offering them a reliable, independent, impartial source of peer feedback on products and brands and showing good or bad alternatives (Valant, 2015).

Online consumer reviews have two necessary strengths since they provide well-timed and cost-free information with regard to peer consumers' experiences with products and they are viewed as more reliable and trustworthy than company-served information (Liu, 2017, p. 247). Online reviews provide some advantages to consumers. Consumers may find online reviews for many products and services. For example, the most reviewed categories for products are electronics, computers, appliances, baby products, and home and garden products. Online reviews allow consumers to share information directly with peers and narrow their research by filtering the criteria. Using online reviews is convenient and timely. They enable consumers to make faster and better decisions (Valant, 2015).

Consumers generally prefer websites with reviews. Consumers mostly read 1 to 10 online reviews before making a purchase decision. Younger consumers and new generations read and post online reviews more. Consumers who had negative experiences are more likely to post a review by going to a site and creating an account (Spiegel Research Center, 2017, p. 15). Online consumer reviews are not just influencing online sales but are also influencing offline sales. Online reviews are substantial sources of information for customers and they are also hidden treasures for obtaining customer insights for companies. These comments become valuable assets for businesses (Amaral et al., 2014, p. 140). Today, companies and brands such as hotels, hospitals, online retailers, and even doctors send emails to customers asking them to provide reviews. This new world is rapidly becoming an invaluable marketing instrument for companies (Olapic, 2016). For example, companies also use consumer-generated content on the Internet. Online reviews are so crucial for companies that there are many popular articles about managing online reviews such as strategies to promote positive reviews, turning negative online reviews into positive reviews, and uncovering fake online reviews.

According to the Pew Research Center (2016), some factors are essential for US customers at the first purchase of a product (Smith & Anderson, 2016, p. 3). These important factors that affect first time shopping include WOM and online reviews, getting advice from people they know, reading the online reviews of others, and looking up online reviews during offline shopping.

Survey results (Figures 14.1 and 14.2) demonstrate that US consumers are heading toward common opinions obtained by online reviews and ratings in their purchase decisions. A total of 82% of Americans say they consult

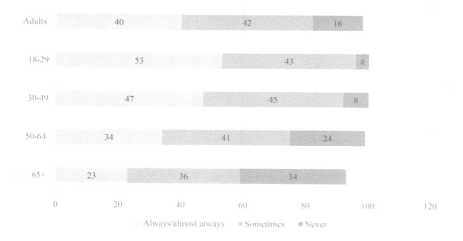

FIGURE 14.1
Age group and online reviews. (From Pew Research Center (2016). Online Shopping and E-Commerce).

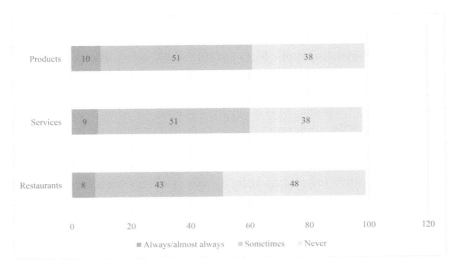

FIGURE 14.2
Posting online reviews. (From Pew Research Center (2016). Online Shopping and E-Commerce).

online reviews and ratings prior to the first purchase of a product. Even though reading online reviews before shopping is common in different demographic groups, it is especially common among consumers under 50. Almost half of the American citizens feel more confident about their purchases due to the help of customer reviews and believe that companies are more accountable to their customers (Smith & Anderson, 2016, pp. 3, 12).

Similarly, according to the Power Reviews (2014) report, nearly all consumers (95%) rely on reviews during the decision-making process. Some people (24%) even check reviews for all their purchases. This ratio is higher among the 18–44 age range consumers, and the ratio is decreasing for older consumers. All these numbers and reports indicate that online reviews have become omnipresent and are an important variable of the purchasing process (Power Reviews, 2014). Gretzel, Yoo, and Purifoy (2007) investigated online travel reviews and stated that more than half of the consumers (57%) read online travel reviews before trips. Seventy-eight percent of the respondents think that other travelers' reviews have a high level of importance in determining where to stay. The percentage of people who post online reviews is lower than the percentage who read online reviews. Younger American shoppers are more likely to write product reviews than older shoppers (Smith & Anderson, 2016, p. 14). Regarding the Deloitte Consumer Review

(2014), The Growing Power of Consumers (2014), the most trusted source of information on products and services is family/friends (60%), customer reviews (60%), and independent product/service experts (43%).

In the tourism sector, the majority of respondents not only read reviews, but they also write reviews. Some travelers do not post online reviews because of time constraints, a lack of interest and confidence in writing, and laziness. On the other hand, motivations to write online reviews include the following: the desire to help others by sharing positive experiences, the desire to reciprocate since other reviews helped them a lot, the desire to support the best travel service providers, the desire to tell others about a great experience, and the desire to help companies be successful. Additionally, there is the desire to warn others of poor services (Gretzel et al., 2007). People are more eager to write online reviews when their experience is not aligned with expectations, when their experience is particularly good or bad, when there is lack of other communication channels for a company or a brand and when they could not solve the problems related with a company or brand (Couzin & Grappone, 2014, pp. 70–71).

Consumers are increasingly trusting online consumer reviews as credible information sources to enable their decision-making processes (Liu, 2017; Shi et al. 2017; Olapic, 2016; Ye et al. 2011; Gretzel & Yoo, 2008). Today's consumers have become increasingly more reliant on eWOM information to make various decisions such as what movie to watch, which book to read, which cosmetic product to buy, which hotel to stay at, or which restaurant to visit for a meal (Lu et al. 2013, p. 596). Customers, primarily those from new generations, use eWOM sources before making their decisions on almost everything. Online reviews and ratings have become the main source of information for potential customers' purchase decisions (Power Reviews, 2014; Liu et al. 2008, p. 443; Goes, Lin, & Au Yeung, 2014). Studies have revealed that online reviews significantly affect consumer choices, product sales, and consumer decision-making processes (Cui et al., 2012; Lu et al. 2013; Duan, Gu, & Whinston, 2008; Chen et al. 2011; Liu, 2017; Spiegel Research Center, 2017; Sparks & Browning, 2011; Beldad et al. 2017). The Power Reviews (2014) report revealed that the top five factors impacting purchase decisions were prices, ratings and reviews, recommendations from friends and family, brands, and finally free shipping and the retailer. It is noteworthy that online reviews and ratings are just behind the price in importance. Now, the argument related to the power shift will be integrated with sustainable consumption.

MARKETING IN THE AGE OF SUSTAINABILITY

The concept of sustainability was first described in a report of the United Nations Environment and Development Commission in 1987. The Brundtland Report, which is this report's more commonly known name, defined sustainable development as "meeting existing needs without compromising the ability of future generations to meet their own needs" (UN, 1987). Starting in those years, governments, lawmakers, and consumers increasingly talked about, thought about, and made sustainability a part of their lives.

Two main facts have driven sustainability toward the focus of consumer consumption. These considerations are the evolution of marketing and consumers within time. As described before, the balance of power between the brands and consumers has reversed over time. Before the 1980s, marketing was more company-centric. Consumers whose needs and wants were not considered played a passive role in the traditional marketing era. Consumers had to buy what was offered to them and chose from limited options. However, consumers have become active and empowered over the brands in modern marketing orientation. With the development of communication opportunities, increasing levels of education, technological advances, consumer awareness, and the diversification of the needs and desires of consumers, the concept of social marketing has started to come to the forefront (Kotler & Armstrong, 2018).

The American Marketing Association (AMA, 2017) states that marketing creates value exchanges for customers, clients, partners, and society at large. While marketing fulfills the goals of individuals and businesses, it also considers the interests of society with the social (societal) marketing approach, which is one of the most recent approaches in marketing. According to this approach, based on the changing consumer needs and expectations, businesses try to establish profitable relationships while satisfying the needs of their customers and seek to consider the societal benefits while achieving these two goals. In this context, they conduct activities that will add value and support social goals by avoiding activities that may harm society (Altunışık, Özdemir & Torlak, 2006; Dibb, Simkin, Pride & Ferrell, 2001; Solomon & Stuart, 2000).

The societal marketing approach includes the concept of marketing since it considers the needs of individual customers, but it goes one-step further in the marketing approach by aiming to improve the welfare of the society in which the business operates. This means that businesses take responsibility to be good citizens without waiting for customers to consider or understand the broader consequences of their consumption behavior. The problem is that the business needs to balance three factors: customer needs, business profits (or other purposes), and the needs of society as a whole (Blythe, 2005). The concept of social marketing advocates that the marketing strategy should provide value to customers by maintaining and improving the well-being of both the customer and society. Societal marketing requires a sustainable marketing approach. Sustainable marketing is a socially and environmentally responsible marketing approach that preserves and improves the ability to meet the needs of future generations while meeting the current needs of customers and businesses (Kotler & Armstrong, 2018). Sustainable marketing means creating a positive change in society in terms of promoting the product, increasing profits, and emotionally affecting the consumer to choose responsible brands (Anastasiei & Dospinescu, 2019, p. 815). We are in an age of global warming, diminishing resources, and widening gaps between the rich and poor. In this era, consumers' well-being should be the main focus of public and consumer relations, and strategies to increase consumer welfare must be enacted in both a responsible and sustainable manner (Balderjahn, Lee, Seegebarth, & Peyer, 2020, p. 478).

Furthermore, it is not enough for the consumers, especially in developed and affluent countries, to satisfy their needs via brands. Empowered consumers know well that there are many alternatives that can fulfill their needs and they prefer to have more socially responsible ones that better protect the long-run welfare compared to the alternatives. This expectation of consumers makes companies create more sustainable solutions. The relationship of modern marketing can be called a customer-managed relationship in which today's empowered and active customers shape their relationships with companies by interacting with companies and other customers (Kotler & Armstrong, 2018). Customer-managed relationships in developed countries with affluent consumers push companies to sustainability. Greener products are one of the examples of this new direction. The desire of consumers for more sustainable products is growing. Even during recessionary periods, consumers are willing to buy sustainable products. Therefore, to be competitive in the following years, the brands' product portfolios should include greener product offerings. The demand of the marketplace is the greatest driver for developing greener products (Iannuzzi, 2012).

FORCING BRANDS TO BE MORE SUSTAINABLE

Customers' desire brands that present paths toward sustainable lifestyles. This expectation is placing considerable pressure on brands to embrace "sustainable" behaviors. The behaviors and applications of the brands are expected to contribute to society's overall well-being and to cause the least possible negative impact on the environment (Davis, 2011). The market composed of all potential and actual consumers and investors is an alternative source of sanctioning pressure by providing the potential to have a significant impact on environmental performance. Consumers who have high environmental concerns might favor products and brands that are also environmentally responsible (Stafford, 2007, p. 84). Consumers who are aware of their rights can turn into active actors of the market to force brands to be sustainable. Empowered consumers, which we discussed, can engender innovation, investment, and competition in the market by driving sustainable production and consumption (Coke-Hamilton, 2020). According to the results of a CSR study, the majorities of consumers buy or boycott products based on values. Consumers are likely to switch their current brands to brands that are associated with a good cause, especially in the case that these brands all have similar other features such as price and quality. Consumers consider the importance of a company's responsible business practices such as being a good employer, operating in a way that protects and benefits society and the environment, and investing in causes in the community and around the globe. Consumers would like companies to address many issues such as poverty, environment, and environmental change (Cone Communications, 2017). Similarly, the Brands & Stands report revealed that consumers believe and expect companies to take stands on and support social issues that fit their businesses (Shelton Grp, 2018).

Al Iannuzzi, vice president for sustainability at Estée Lauder Companies, explains that Estée Lauder is facing an increasing demand for being sustainable from customers, company stakeholders, and shareholders. This increasing demand has turned sustainability into a market driver for innovation and growth. Hence, the companies are taking account of every step in the marketing process starting from obtaining raw materials to delivering products to customers. How and where they obtain raw materials, improving energy use, decreasing waste, providing good working conditions to employees, and being transparent in their communication with customers have become vital (Friedlander, 2020). Gazzalo, Colombo, Pezzetti, and Nicolescu (2017) express that millennials, as the most influential generation, are forcing brands to be more sustainable. Consequently, companies need to find suitable ways to make their products and processes more sustainable by offering value to customers. White, Hardisty, and Habib (2019) express that consumers, particularly millennials, are increasingly searching for brands that adopt a purpose and support sustainability. The results of the research conducted by CGS in November 2018 revealed that sustainability is a criterion that is increasingly considered in purchase decisions. Even some retailers are listening to the demands of consumers such as the demands for ethical supply chains, which were previously generally ignored (Chadha, 2019).

Lately, consumers are pushing companies to enact corporate responsibility. Uber faced pressure from consumers because of the perceived corporate irresponsibility of its CEO. Companies have strived to respond by showing their sensitivity to people, profits, and the planet. Intel has been increasing its use of clean energy. The fashion industry is one of the most discussed consumer goods industries. In addition to the pressure from governments, pressure from consumers for fashion brands to be greener is increasing. For global brands such as H&M and Nike, the pressures to manufacture more sustainable and ethical products are higher, and thus they are increasing their efforts on sustainable sourcing and manufacturing (Fortune, 2019).

According to a Nielsen report (2018), 81% of global respondents strongly agree with the idea that companies should assist in enhancing the environment. A wide range of demographic groups, mostly Generations X, Y, and Z, share this expectation for CSR. Consumers are using their spending power to cause a change toward what they want. Nielsen (2014, 2015) reported that 66% of the global respondents are eager to pay more for products and services served by brands who adopt positive social and environmental impacts or for sustainable brands, which is up from 55% in 2014 and 50% in 2013. Similarly, supporting Nielson's results, a large majority (64%) of Europeans said that companies would face consumer backlashes if they did not take steps to go green (ING International Survey, 2019). According to the 2018 Global Web Index survey, the purchasing decisions of half of the digital consumers are affected by environmental concerns. Brands are focusing on offering more sustainable and environmentally friendly products and services as the purchasing power of millennials and Generation Z has increased (UNIBusiness Editor, 2020). According to consumers, individuals are the most effective entity in solving social and environmental problems. Consumers believe that businesses are not very effective at enacting change. However, they still have faith that brands will be more active in implementing social and environmental change (Cone Communications, 2017) since consumers force companies to be more sustainable and open to change by providing better solutions. In the next part of the chapter, the traces of this change in the expectations of consumers and its effect on brands experiencing consumer power shifts will be exemplified.

ONLINE REVIEWS AND SUSTAINABILITY: ILLUSTRATIVE CASES AND RECENT TRENDS

Social media is one of the most preferred media for sharing and airing the comments, ideas, and reviews of consumers today. Social media platforms are generally the first places for sharing online reviews that come to the minds of consumer. Balaswamy and Palvai (2017, p. 264) state that social media have allowed people to communicate their ideas and opinions on products and services across the world with other people. Social media provide the rapid transfer of information on brands from a consumer to others. Traditional media played a significant role in raising awareness about the concept of sustainable development in the past. However, sustainable development concepts have begun to spread through peer learning due to the emergence of social media. Social media can be beneficial to achieving the goals of sustainable development. Onete, Dina, and Vlad (2013) state that social media can provide a useful space for sustainable business and provide support for the development of sustainable products and services. Social media empowered consumers. The recent phenomenon is called empowerment because social media give customers the opportunity to share their experiences with others and affect their purchasing decisions with online reviews and recommendations (Strähle & Gräff, 2017, p. 237). Empowered consumers seek to have a say in how brands should look and be addressed in the market (Strähle & Grünewald, 2017, p. 114). In 2017, The World Wide Fund for Nature (WWF) organized a campaign for using sustainable palm oil. Irresponsible palm oil production causes water and air pollution in addition to deforestation. An e-mail bombardment from consumers was directed to the companies who did not respond to the WWF. Therefore, companies have learnt that consumers care what kind of palm oil they use (Hui & Leow, 2017).

Ben Abdelaziz et al. (2015, p. 87) found that the eagerness of consumers to seek information about sustainability prompts them to view social media as an important resource for this particular topic. This dependency on social media for sustainability information affects consumers' views of products' sustainability positions. Therefore, it was found that the intention of consumers to buy a particular product/brand was bound by their dependency on social media for sustainability information. They concluded that social media influence consumers' behavior and purchase decisions with respect to sustainable products. Similarly, Balderjahn et al. (2020, p. 456) argue that empowerment improves consumer sovereignty. Consumer empowerment plays an important role in supporting sustainable methods that improve consumer well-being.

Brazytė, Weber, and Schaffner (2017) found that hotel guests perceive sustainability in the tourism sector positively, and they are generally in favor of green practices. The results of the study showed that online consumer reviews that contained clues or traces of sustainability had significantly higher ratings than those without indicators of sustainability. Therefore, hotel managers in the accommodation section should invest in sustainable principles. D'Acunto et al. (2020) revealed that there is increasing attention given to CSR factors, particularly to social and environmental factors. The data were collected from 480,000 reviews across six European cities via longitudinal automated text analysis. However, Gerdt et al. (2019) find that German tourists do not reflect their sustainable travel interests in their online reviews of hotels, even though they have higher interest. Austrian and German tourists' expectations regarding the ecological engagement of hotels' toward sustainability are high. They conclude that even though the content of reviews does not reflect sustainability concerns, sustainability is a factor in the decisions of hotel guests. Leal Londoño and Hernandez-Maskivker (2016) analyze the ecofriendly practices developed by hotels based on the data gathered from customer reviews. According to the results, Copenhagen has a higher rate of sustainable reviews. Recognition as a green or a sustainable city by customers influences consumer choice. In another example, Naidoo, Ramseook-Munhurrun, and Li (2018, p. 49) analyzed online consumer reviews to understand which sustainability dimensions are key components for customers of scuba diving experiences. Recognizing tourists' sustainability perception can encourage dive operators to better understand and focus on increasing their contributions to conservation issues and better managing their social impacts.

El Dehaibi et al. (2019) analyze the social, environmental, and economic sustainability in positive and negative perceptions of the product features of a French press coffee carafe with online reviews. By analyzing online reviews, designers can obtain perceived sustainable features for further design and create products that are consistent with customers' sustainable values. Moreover, the majority of consumers believe that they have a personal responsibility to help address global environmental challenges (ING, 2020). Despite the fact that the literature suggests that consumers are main driving forces of corporate environmentalism, Sandhu, Ozanne, Smallman, and Cullen (2010) find that managers do not think such as that. Similarly, White, Habib, and Hardisty (2019, p. 30) state that sustainability is future-focused by nature, while conversely, consumers are often present-focused. We should acknowledge that there is still room for improvement.

DISCUSSION AND CONCLUSION

It is crystal clear that the Internet has considerably changed customers, businesses, and marketing, and it will undoubtedly affect these aspects more in the upcoming decades. Empowered customers are more curious and investigative. They expect more from companies and search for more information than ever before. Grand challenges transform the relationship between consumers and companies to being more visible. Customers are creating a new world where they tend to trust each other when making decisions. They still use an old but standing tool of WOM but in an online, more effective, and interactive manner.

As can be inferred from this chapter, today, empowered consumers are more sensitive to being environmentally conscious and eager to use their power and voice through their consumption. Similarly, the lack of consumer empowerment was a barrier along with sustainability. Consumerism always has a significant function in the development of society, and consumption can be a driving force to shift society into a more sustainable future (JingJing, Xinze, & Sitch, 2008). The digital era has affected both consumers and companies. Consumers with constant access to the Internet and smartphones now incorporate technology in every part of their live. Companies should give voices to consumers, listen to their words, and learn from them since consumers are aware of their power to reach and influence companies and other consumers. Companies can increase both the welfare of society and goodwill toward their brands by integrating sustainability into their business models (Nielsen, 2015). Listing is an important capability. Companies should be good listeners to what consumers are saying about products or services, good inspirers to make consumers promote products and services, and good cocreators to engage in conversations with active followers (Deloitte, 2014). Companies that do not respond to the sustainability demands of consumers face real threats to their profitability (ING, 2020).

More empirical effort is needed to bridge online reviews and sustainable consumption. Shaw, Newholm, and Dickinson (2006) indicate that consumers adopt consumption as a voting mechanism. They vote on brands with their money. Consumers reflect their ethical or political concerns with their powers to choose, buy, not buy, and boycott. In addition, White, Habib, and Hardisty, (2019, p. 32) state that the positive consumer associations with sustainability should be studied. Although this paper is a theoretical one, it is believed that there are still practical implications that it offers. It is recommended that exploratory research can be utilized to explore the enforcement effect of sustainability messages in online consumer reviews of brands.

There is one world and everyone is welcomed. According to the literature on social and political theories of power and empowerment (Avelino, 2017, p. 507), the type of power relations between consumers and brands have transformed over time. Consumers and brands have different types of powers, regardless of their amounts. They have mutual dependence and coexistence. Both sides need to cooperate and be synergized to maintain sustainability. Last but not the least, as Mihajlovic (2020) states that all the market actors should work together to contribute to the achievement of corporate sustainability, the synergy of all market actors is required to achieve their full power.

Consumers are looking for truth, democracy, openness, and simplicity. In achieving their goals, companies need to give consumers ways to make, to share, and to sell their ideas by allowing them to do so on their terms and in their time (Olins, 2013). Online reviews are a good example of consumers creating impact; and most importantly, these online reviews can be regarded as easy-to-access customer insights. Consumers are pushing companies, companies are serving more sustainable products, and companies are pushing suppliers to meet the sustainability goals of the company (Cohen, 2017). Let the power of consumers heal the world!

ADDITIONAL CONTENT

- The Naked Brand (2013) https://www.imdb.com/title/tt2262281/
  A documentary on the erosion of public trust in big companies. Telling truth, transparency, and trust are the keys for success that make online consumer reviews indispensable part of value creation process in the relationship between customers and companies.
- Good On You https://goodonyou.eco/
  A platform provides brand ratings on sustainability on ethics in fashion industry by analyzing a brand's impact on planet, people, and animals.

FUNDING DECLARATION

The author received no specific grant from any funding agency in the public, commercial, or not-for-profit sectors.

REFERENCES

Altunışık, R., Özdemir, Ş., & Torlak, Ö. (2006). *Modern pazarlama* (4th Edition). İstanbul: Değişim Yayınları.

AMA (2017). *Definitions of marketing*. https://www.ama.org/the-definition-of-marketing-what-is-marketing/

Amaral, F., Tiago, T., & Tiago, F. (2014). User-generated content: Tourists' profiles on TripAdvisor. *International Journal on Strategic Innovative Marketing*, *1*(3), 137–147.

Anastasiei, B., & Dospinescu, N. (2019). Electronic word-of-mouth for online retailers: Predictors of volume and valence. *Sustainability*, *11*(3), 814–832. https://doi.org/10.3390/su11030814

Avelino, F. (2017). Power in sustainability transitions: Analysing power and (dis)empowerment in transformative change towards sustainability. *Environmental Policy and Governance*, 27(6), 505–520.

Balaswamy, B., & Palvai, R. (2017). *Role of social media in promoting sustainable development*. International Conference People Connect: Networking for Sustainable Development. Hosted by St. Claret College, Bengaluru, & IJCRT (pp. 264–271).

Balderjahn, I., Lee, M. S.W., Seegebarth, B., & Peyer, M. (2020). A sustainable pathway to consumer wellbeing. The role of anticonsumption and consumer empowerment. *The Journal of Consumer Affairs*, 54(2), 456–488. https://doi.org/10.1111/joca.12278

Beldad, A., Avicenna, F., & De Vries, S. (2017). The effects of online review message appeal and online review source across two product types on review credibility, product attitude, and purchase intention. In: F. F. Nah & C. Tan (Eds.), HCI in business, government and organizations 4th international conference, HCIBGO 2017, Held as Part of HCI International 2017 Vancouver, BC, Canada, July 9–14, 2017 Proceedings, Part II (pp. 163–173). Springer.

Ben Abdelaziz, S. I., Saeed, M. A., & Benleulmi, A. Z. (2015). Social media effect on sustainable products purchase. In: W. Kersten, T. Blecker, & C. M. Ringle (Eds.), *Sustainability in logistics and supply chain management: New designs and strategies*. Proceedings of the Hamburg International Conference of Logistics (HICL) (Vol. 21; pp. 63–93). Berlin: ePubli GmbH.

Bickart, B., & Schindler, R. M. (2001). Internet forums as influential sources of consumer information. *Journal of Interactive Marketing*, 15(3), 31–40.

Blythe, J. (2005). *Essentials of marketing* (3rd Edition). Upper Saddle River, NJ: FT Prentice Hall.

Brazytė, K., Weber, F., & Schaffner, D. (2017). Sustainability management of hotels: How do customers respond in online reviews? *Journal of Quality Assurance in Hospitality & Tourism*, 18(3), 282–307, https://doi.org/10.1080/1528008X.2016.1230033

Cakim, I. M. (2010). *Implementing word of mouth marketing: Online strategies to identify influencers, craft stories and draw customers*. Hoboken, NJ: John Wiley & Sons, Inc.

Chadha, R. (2019). *Sustainability is driving consumers' purchase decisions. And that's especially true among younger cohorts*. https://www.emarketer.com/content/sustainability-is-driving-consumers-purchase-decisions

Chen, Y., Fay, S., & Wang, Q. (2011). The role of marketing in social media: How online consumer reviews evolve. *Journal of Interactive Marketing*, 25(2), 85–94.

Chen, Y., & Xie, J. (2008). Online consumer review: Word-of-mouth as a new element of marketing communication mix. *Management Science*, 54(3), 477–491.

Cohen, G. (2017). *How health care is pushing sustainability ahead of consumer demand*. https://sustainablebrands.com/read/organizational-change/how-health-care-is-pushing-sustainability-ahead-of-consumer-demand

Coke-Hamilton, P. (2020). *World consumer rights day: Do consumers play a role as influencers of sustainable consumption?* https://unctad.org/en/pages/newsdetails.aspx?OriginalVersionID=2303

Cone Communications (2017). *2017 Cone Communications CSR Study*. https://www.conecomm.com/research-blog/2017-csr-study#download-the-research

Couzin, G., & Grappone, J. (2014). *Five star: Putting online reviews to work for your business*. Indianapolis: Sybex, a Wiley Brand.

Cui, G., Lui, H., & Guo, X. (2012). The effect of online consumer reviews on new product sales. *International Journal of Electronic Commerce*, 17(1), 39–58.

D'Acunto, D., Tuan, A., Dalli, D., Viglia, G., & Okumus, F. (2020). Do consumers care about CSR in their online reviews? An empirical analysis. *International Journal of Hospitality Management, 85*(2020), 1–9.

Davis, M. (2011). *Consumers put firms under pressure to lead the way in sustainable living.* https://www.theguardian.com/sustainable-business/consumer-behaviour-companies-sustainable-living

Dellarocas, C. (2003). The digitization of word of mouth: Promise and challenges of online feedback mechanism. *Management Science, 49*(10), 1407–1424.

Deloitte (2014). *The Deloitte consumer review. The growing power of consumers.* https://www2.deloitte.com/content/dam/Deloitte/uk/Documents/consumer-business/consumer-review-8-the-growing-power-of-consumers.pdf

Dibb, S., Simkin, L., Pride, W. M. & Ferrell, O. C. (2001). *Marketing concept and strategies* (4th Edition). New York, NY: Houghton Mifflin.

Duan, W., Gu, B., & Whinston, A. B. (2008). Do online reviews matter?—An empirical investigation of panel data. *Decision Support Systems, 45*(4) 1007–1016.

El Dehaibi, N., Goodman, N. D., & MacDonald, E. F. (2019). Extracting customer perceptions of product sustainability from online reviews. *Journal of Mechanical Design, 141*(12), 12110. https://doi.org/10.1115/1.4044522

Fortune (2019). *How consumers are pushing the global fashion industry toward greater sustainability.* https://fortune.com/2019/09/06/government-consumer-pressure-global-fashion-sustainability/

Friedlander, B. (2020). Execs: Consumers pushing companies toward sustainability. *Cornell Chronicle.* https://news.cornell.edu/stories/2020/03/execs-consumers-pushing-companies-toward-sustainability

Gazzalo, P., Colombo, G., Pezzetti, R., & Nicolescu, L. (2017). Consumer empowerment in the digital economy: Availing sustainable purchasing decisions. *Sustainability, 9*(5), 693–712.

Gerdt, S., Wagner, E., & Schewe, G. (2019). The relationship between sustainability and customer satisfaction in hospitality: An explorative investigation using eWOM as a data source. *Tourism Management, 74*(2019) 155–172. https://doi.org/10.1016/j.tourman.2019.02.010

Goes, P. B., Lin, M., & Au Yeung, C. (2014). "Popularity effect" in user-generated content: evidence from online product reviews. *Information Systems Research, 25*(2), 222–238.

Gretzel, U., & Yoo, K. (2008). Use and impact of online travel reviews. In: P. O'Connor, W. Hopken & U. Gretzel (Eds.), *Information and communication technologies in tourism* (pp. 35–46). Vienna, Austria: Springer-Verlag Wie.

Gretzel, U., Yoo, K. H., & Purifoy, M. (2007). Online travel review study. Role & impact of online travel reviews. *Laboratory for Intelligent Systems in Tourism.* https://www.tripadvisor.com/pdfs/OnlineTravelReviewReport.pdf

Hanna, R., Rohm, A., & Crittenden, V. L. (2011). We're all connected: The power of the social media Ecosystem. *Business Horizons, 54*(3), 265–273.

Hellberg, S. (2014). *Power is shifting through social media, consumers are taking the driver's seat.* http://www.brandba.se/blog/power-is-shifting-through-social-media-consumers-a-force-to-be-reckoned-with

Hui, K. X., & Leow, A. (2017). *Firms; palm oil use: 7,700 e-mails show people care.* https://www.straitstimes.com/singapore/firms-palm-oil-use-7700-e-mails-show-people-care

Iannuzzi, A. (2012). *Greener products. The making and marketing of sustainable brands.* Boca Raton, FL: CRC Press, Taylor & Francis Group.

ING (2020). *Learning from consumers: How shifting demands are shaping companies' circular economy transition. A circular economy survey.* https://www.ingwb.com/media/3076131/ing-circular-economy-survey-2020-learning-from-consumers.pdf

ING International Survey (2019). *Consumer attitudes towards the circular economy.* https://think.ing.com/uploads/reports/IIS_Circular_Economy_report_FINAL.PDF

JingJing, D., Xinze, L., & Sitch, R. (2008). *Ethical consumers: Strategically moving the restaurant industry towards sustainability.* [Master's thesis, Blekinge Institute of Technology Karlskrona, Sweden]. http://www.diva-portal.org/smash/get/diva2:829490/FULLTEXT01.pdf

Kotler, P., & Armstrong, G. (2018). *Principles of marketing* (17th edition). Harlow: Pearson.

Labrecque, L. I., vor dem Esche, J., Mathwick, C., Novak, T. P., & Hofacker, C. F. (2013). Consumer Power: Evolution in the Digital Age. *Journal of Interactive Marketing, 27*(4), 257–269.

Leal Londoño M. P., & Hernandez-Maskivker G. (2016). Green practices in hotels: The case of the GreenLeaders Program from TripAdvisor. Proceedings of the 7th International Conference on Sustainable Tourism (Vol. 1–13). DOI: 10.2495/ST160011

Liu, F. (2017). Numbers speak where words fail: exploring the effect of online consumer reviews on consumer decision making. In F. F. Nah & C. Tan (Eds.), HCI in Business, Government and Organizations 4th International Conference, HCIBGO 2017 Held as Part of HCI International 2017 Vancouver, BC, Canada, July 9–14, 2017 Proceedings, Part II (pp. 246–263). Cham, Switzerland: Springer.

Liu, Y., Huang, X., An, A., & Yu, X. (2008). *Modeling and predicting the helpfulness of online reviews.* 2008 Eighth IEEE International Conference on Data Mining. DOI: 10.1109/ICDM.2008.94

Lu, X., Ba, S., Huang, L., & Feng, Y. (2013). Promotional marketing or word-of-mouth? Evidence from online restaurant reviews. *Information Systems Research, 24*(3), 596–612.

Mihajlovic, B. (2020). The role of consumers in the achievement of corporate sustainability through the reduction of unfair commercial practices. *Sustainability, 12*(3), 1009–1029.

Naidoo, P., Ramseook-Munhurrun, P., & Li, J. (2018). Scuba diving experience and sustainability: An assessment of online travel reviews. *The Gaze: Journal of Tourism and Hospitality, 9*, 43–52. https://doi.org/10.3126/gaze.v9i0.19720

Nas Daily (2018). *Is Apple worth it?!* https://www.facebook.com/watch/?v=1053347648150720

Nielsen (2014). *Doing well by doing good. Increasingly, consumers care about corporate social responsibility, but does concern convert to consumption?* https://www.nielsen.com/us/en/insights/report/2014/doing-well-by-doing-good/#

Nielsen (2015). *The sustainability imperative. New insights on consumer expectations.* https://www.nielsen.com/wp-content/uploads/sites/3/2019/04/global-sustainability-report-oct-2015.pdf

Nielsen (2018). *Global consumers seek companies that care about environmental issues.* https://www.nielsen.com/us/en/insights/article/2018/global-consumers-seek-companies-that-care-about-environmental-issues/

O'Connor, P. (2008) User-generated content and travel: A case study on Tripadvisor.Com. In: P. O'Connor, W. Höpken, U. Gretzel (Eds.), *Information and communication technologies in tourism* (pp. 47–58). Vienna: Springer.

Olapic (2016). *Consumer trust: Keeping it real.* http://visualcommerce.olapic.com/rs/358-ZXR-813/images/wp-consumer-trust-survey-global-FINAL.pdf

Olins, W. (2013). *The new mainstream consumer.* https://www.prnewswire.com/news-releases/the-new-mainstream-consumer-224013341.html

Onete, C. B., Dina, R., & Vlad, D. E. (2013). Social media in the development of sustainable business. *Amfiteatru Economic Journal, 15*(7), 659–670.

Özata, F. Z. (2011). Tüketiciyi Yönlendiren Güç: Öteki Tüketici. *İnternet Uygulamaları ve Yönetimi Dergisi, 2*(2), 1–25.

Park, S., & Nicolau, J. L. (2015). Asymmetric effects of online consumer reviews. *Annals of Tourism Research, 50*(1), 67–83.

Parment, A. (2013). Generation Y vs. Baby Boomers: Shopping behavior, buyer involvement and implications for retailing. *Journal of Retailing and Consumer Services, 20*(2), 189–199.

Pitt, L. F., Berthon, P. R., Watson, R. T., & Zinkhan, G. M. (2002). The Internet and the birth of real consumer power. *Business Horizons, 45*(4), 7–14.

Power Reviews (2014). *Power of Reviews. How ratings and reviews influence the buying behavior of the modern consumer.* https://www.powerreviews.com/wp-content/uploads/2016/04/PowerofReviews_2016.pdf

Sandhu, S., Ozanne, L. K., Smallman, C., & Cullen, R. (2010). Consumer driven corporate environmentalism: Fact or fiction? *Business Strategy and the Environment, 19*(6), 356–366. https://doi.org/10.1002/bse.686

Shaw, D., Newholm, T., & Dickinson, R. (2006). Consumption as voting: An exploration of consumer empowerment. *European Journal of Marketing, 40*(9/10), 1049–1067. https://doi.org/10.1108/03090560610681005

Shelton Grp (2018). *Brands & Stands. Social purpose is the new black. How can your company stand and deliver?* http://storage.googleapis.com/shelton-group/Pulse%20Reports/Brands%20%26%20Stands%20-%20Final%20Report%202018.pdf

Shi, Y., Zeng, Q., Nah, F. F. Tan, C., Sia, C. L., & Yan, J. (2017). Effect of timing and source of online product recommendations: an eye-tracking study. In: F. F. Nah & C. Tan (Eds.), HCI in Business, Government and Organizations 4th International Conference, HCIBGO 2017 Held as Part of HCI International 2017, Vancouver, BC, Canada, July 9–14, 2017 Proceedings, Part II (pp. 95–104). Cham, Switzerland: Springer.

Smith, A., & Anderson, M. (2016). *Online shopping and e-commerce.* Pew Research Center.

Solomon, M. R., & Stuart, E. W. (2000). *Marketing: Real people and real choices* (2nd Edition). Upper Saddle River, NJ: Prentice Hall.

Sparks, B. A., & Browning, V. (2011). The impact of online reviews on hotel booking intentions and perception of trust. *Tourism Management, 32*(6), 1310–1323.

Spiegel Research Center (2017). *How online reviews influence sales.* https://spiegel.medill.northwestern.edu/_pdf/Spiegel_Online%20Review_eBook_Jun2017_FINAL.pdf

Stafford, S. L. (2007). Can consumers enforce environmental regulations? The role of the market in hazardous waste compliance. *Journal of Regulatory Economics, 31*(1), 83–107. DOI: 10.1007/s11149-006-9006-8

Statista (2020). *Most popular social networks worldwide as of July 2020, ranked by number of active users (in millions).* https://www.statista.com/statistics/272014/global-social-networks-ranked-by-number-of-users/

Strähle, J., & Gräff, C. (2017). The role of social media for a sustainable consumption. In: J. Strähle (Ed.), *Green Fashion Retail* (pp. 225–247). Singapore: Springer.

Strähle, J., & Grünewald, A.-K. (2017). The prosumer concept in fashion retail: Potentials and limitations. In: J. Strähle (Ed.), *Green Fashion Retail* (pp. 95–117). Singapore: Springer.

UNIBusiness Editor (2020). *The consumer's impact on sustainability and the economy.* https://business.uni.edu/news-views/consumers-impact-sustainability-and-economy

United Nations (1987). *Report of the world commission on environment and development.* http://www.un.org/documents/ga/res/42/ares42-187.htm

Valant, J. (2015). *Online consumer reviews. The case of misleading or fake reviews.* European Parliament. https://www.eesc.europa.eu/sites/default/files/resources/docs/online-consumer-reviews---the-case-of-misleading-or-fake-reviews.pdf

White, K., Habib, R., & Hardisty, D. J. (2019). How to SHIFT consumer behaviors to be more sustainable: A literature review and guiding framework. *Journal of Marketing, 83*(3), 22–49. https://doi.org/10.1177/0022242919825649

White, K., Hardisty, D. J., & Habib, R. (2019). *The elusive green consumer.* https://hbr.org/2019/07/the-elusive-green-consumer

Ye, Q., Law, R., Gu, B., & Chen, W. (2011). The influence of user-generated content on traveler behavior an empirical investigation on the effects of e-word-of-mouth to hotel online bookings. *Computers in Human Behavior, 27*(2), 634–639.

15

Leveraging Social Media to Create Socially Responsible Consumers

Bikramjit Rishi
Institute of Management Technology (IMT), Ghaziabad, India

Neha Reddy Kuthuru
School of Management and Entrepreneurship, Shiv Nadar University, Greater Noida, India

CONTENTS

Learning Objectives	415
Introduction	416
Sustainability as a Marketing Strategy	417
Issues Concerning Sustainability Marketing	418
Social Media for Sustainability Marketing	419
Guidelines to Design an Active Social Media Strategy for Sustainable Marketing	421
The Way Forward	426
Shaping Sustainability Behavior	427
Making Sustainability Resonate	427
Data-Driven Impact	427
Leveraging Social Media for Social Change	428
Conclusion	428
References	429

LEARNING OBJECTIVES

After reading the chapter, the reader should be able to:

- Highlight issues concerning sustainability marketing and reaching socially responsible consumers

- Justify the usage of social media as a lucrative tool for sustainability marketing
- Provide guidelines that marketers can follow while designing social media marketing strategies to effectively reach socially responsible consumers
- Establish specific trends that can be anticipated to use social media in sustainability marketing

INTRODUCTION

In 1972, the United Nations Conference on the Human Environment in Stockholm, Sweden, explored the idea of sustainability for the first time. This conference's deliberations reported heavily in the *1987 Brundtland Report, Our Common Future* (World Commission on Environment and Development, 1987). This report defined sustainable development as "development that meets the needs of the present without compromising the ability of future generations to meet their own needs" (p. 8). This definition and the report have paved the way for further research in sustainability and how businesses can become more sustainable and environmentally conscious with their practices.

Savitz and Weber (2006) discussed *the triple bottom line*, which has grown to become extremely popular for establishing the guidelines to achieve sustainability at a firm level. The same report further classified the broad concept of sustainability into the three Es—Environmental, Economic, and Equity. It necessarily implies that businesses focus on ecological and social equity goals without hurting their financial situation. Thus, with emphasis on these three dimensions, organizations can integrate sustainability into their core business strategy to help them achieve their sustainable business goals.

As organizations increasingly make efforts to ensure sustainability in their practices, it becomes crucial for them to communicate these efforts to the key stakeholders involved. For that purpose, the concept of sustainability and sustainable business practices influences marketing principles (Kotler, 2011). It has led to the emergence of a discipline called *sustainability marketing*. Fuller (1999) was one of the first few to research this discipline of marketing and defined sustainability marketing as:

> the process of planning, implementing and controlling the development, pricing, promotion, and distribution of products in a manner that satisfies

the following three criteria: (1) customer needs are met; (2) organizational goals are attained; and (3) the process is compatible with eco-systems.

It primarily aims to promote sustainable products and services that satisfy the consumers' needs and significantly improve the social and environmental performance while emphasizing on creating value for the consumers and achieving the business goals (Lim, 2016).

Sustainability as a Marketing Strategy

Seventy percent of the consumers have expressed that they are likely to spend more on brands that support a cause they care about (Patel, 2017). It is evident that organizations transition toward sustainability ideas and accelerate efforts to communicate these ideas to their target consumers effectively. Sustainability marketing at its heart is to strive and create a balance between promoting the sustainable attributes of the products while taking into consideration the consumers' needs. It emphasizes on two aspects—one is to make the target consumers aware of how a product/service will meet their needs while addressing the ecological, social, and environmental issues, and the second is to facilitate an engaging dialogue between the stakeholders and the organization as a whole (Belz & Peattie, 2012).

Gordon et al. (2011) discuss that sustainability marketing can be achieved by applying three subdisciplines of marketing—green marketing, social marketing, and critical marketing. Through green marketing, efforts are made into marketing/promoting sustainable products/services while incorporating sustainability principles into the marketing processes. Social marketing encompasses the use of marketing principles and procedures to encourage sustainable behavior among various stakeholders and assess the impact that commercial marketing creates on sustainability. Furthermore, critical marketing helps critically analyze the marketing theory and the principals involved such that sustainability continues to remain the vital goal (Gordon et al., 2011).

The linkage between sustainability and marketing is more critical now than ever before, owing to the increasing awareness of consumers' environmental concerns. As more and more consumers are on the lookout for sustainable, ethically sourced, and packaged products, these businesses' responsibility to reach out to these consumers is further heightened.

Kemper and Ballantine (2019) elaborated on the *Auxiliary Sustainability Marketing* that involves integrating sustainability principles throughout the whole marketing mix. This perspective toward sustainability marketing focuses on the environmental, social, and economic dimensions of products and consumption. It is also essential to craft a communication message that delves deeper into the different product attributes and makes it sustainable. While the earlier research on green marketing focused heavily on the products, sustainability marketing is rather holistic and extends its role in areas of distribution, pricing as well (Kemper & Ballantine, 2019).

Issues Concerning Sustainability Marketing

Although about 61% of the consumers express concerns about environmental issues, the actual sustainable purchasing behavior decreases (Lau, 2016). While studies show that consumers care about sustainability, but marketers consider it a difficult message to communicate. *Green Marketing* emerged as a phenomenon that focused on catering to the segment of "green consumers" who understood the ecological impact on the products they were consuming (Peattie, 2001). The effects of this were somewhat counterproductive with exaggerated environmental claims that have eventually led to the poor performance of these environmentally conscious products and high consumer skepticism (Peattie & Crane, 2005).

The application of sustainability in the marketing function suffers from sustainability marketing myopia, which implies that the marketers focus on promoting the product's sustainability features rather than catering to the consumers' needs (Villarino & Font, 2015). Sustainability marketing also results in greenwashing, which refers to a strategic disclosure of positive information about sustainability while ignoring the negative information (Lyon & Maxwell, 2011). It leads to the consumers receiving irrelevant and incomplete information, which does not aid them in making a well-informed purchase decision. Communication about sustainability also lacks a consumer appeal for most consumers (Villarino & Font, 2015). It is possible because sustainability marketing inherently focuses on the product attributes and the intricate details of how the product is sustainable. It fails to effectively communicate how the product/service leads to any value creation for the consumers. Hence, despite the organizations' efforts, sustainability marketing fails to share sustainability and reach the right kind of consumers honestly.

Based on the existing studies, we aim to highlight how social media, as a tool, is useful for resolving issues currently concerning sustainability marketing and helping educate the consumers to become socially responsible consumers. We seek to establish it as a lucrative tool for sustainability marketing. The chapter will highlight social media marketing features that justify using the medium for reaching socially responsible consumers. Consequently, we aim to explain guidelines that can accelerate sustainability marketing efforts to reach socially responsible consumers when implemented. In the end, the chapter anticipates the specific trends concerning sustainability marketing. Finally, the readers will get specific takeaways in terms of guidelines and a framework to incorporate social media into the marketing strategies to reach socially responsible consumers. These guidelines and the proposed framework would help the marketers understand their consumers better and create content that effectively communicates the social causes.

SOCIAL MEDIA FOR SUSTAINABILITY MARKETING

Gilsenan (2019) pointed out that as the demand for sustainability increasing from 49% in 2011 to 57% in 2018, consumers are increasingly becoming inclined toward sustainable products. It presents a unique opportunity for organizations to scale their sustainability initiatives and effectively communicate the idea to their target audience. And failure to do so would lead to the companies losing their customers who are increasingly becoming environmentally conscious (Baldassarre & Campo, 2016).

The digital era has seen the emergence of social media networks that have grown to become a key element in any marketing strategy. Social media marketing has become one of the most accessible ways of reaching consumers online (Dwivedi et al., 2020). As sustainability initiatives' popularity increases on social media, it has become a useful tool in influencing the consumers' views on sustainability and the environment (Valentine, 2019).

The following techniques explain the use of social media for sustainability marketing.

- **Social listening**
  The emergence of social media has led to a rise in the significance of social listening. Stewart and Arnold (2018) defined social listening as:

  > an active process of attending to, observing, interpreting, and responding to a variety of stimuli through mediated, electronic, and social channels.

  The dynamic nature of social listening makes it a feasible and viable tool for understanding the changing consumer behavior concerning sustainability on platforms like Facebook, Instagram, YouTube, etc. to name a few. By studying consumer discussions in the relevant areas, organizations can comprehend the consumers' needs and preferences. The insights gathered through this process can guide the companies in various methods of product development and marketing. Active social listening ensures that companies should produce sustainable and environmentally friendly products by understanding consumers' expectations. It subsequently portrays the value creation for the consumers who would then be inclined to make a purchase decision. Hence, social media provides organizations a means to track the consumer and industry trends regarding sustainability and ensure immediacy in implementing the generated insights into their processes.

- **Locating the target audience**
  It is all the more critical for sustainable brands to understand their target audience to better customize their offerings. Through the power of social media, companies have sophisticated tools at their disposal, which help them immensely in understanding their target audience's demographics. Through the features offered by these social media platforms, organizations can get a deeper understanding of various demographic variables such as gender, age, location, income, etc. to name a few. This analysis would aid them in making well informed and targeted decisions toward their potential consumers.

- **Crafting an engaging dialogue with the consumer**
  One of the unique characteristics of social media is the opportunity to engage in a dialogue with consumers (Rishi & Kuthuru, 2021). Edgecomb (2017) indicates that on social media platforms,

consumers are at the forefront, driving the narrative for the organization. Engaging with the consumers on a human level helps the brands understand their preferences and purchase intention, critical drivers in several promotional and product-related decisions. Since sustainability is a rather complex topic, it is crucial to maintain two-way conservation with the target consumers, and social media provides the most accessible way to do so.
- **Sense of community**
  Building a strong and engaging community is the crux of any successful social media initiative and is a value that enhances the brand personality for sustainable brands. A strong purpose-filled community of consumers on social media platforms would help the companies influence and encourage sustainable lifestyles. By leveraging the power of peer-to-peer recommendations and social influence, sustainable brands can use social media to reach their eco-conscious consumers. These communities are also beneficial in receiving instant feedback about the products/services, thus maintaining the desired customer satisfaction, trust, and loyalty (Laroche et al., 2012; Laroche et al., 2013).

GUIDELINES TO DESIGN AN ACTIVE SOCIAL MEDIA STRATEGY FOR SUSTAINABLE MARKETING

Now we can say that social media can play an important role in sustainability marketing. But it is also important to discuss specific guidelines that the marketers have to keep in mind while developing the organizations' sustainability marketing plan on social media.

- **Understanding the consumer**
  Like any other marketing strategy, sustainability marketing should first aim to gain a fundamental understanding of the target market, their wants, and needs. It would play a critical role in devising other strategies, which aligns with the consumers' expectations. The companies can treat their social media pages to gather insights on sustainable products' changing consumer behavior by paying attention to consumer interactions by understanding the consumer sentiment

toward sustainable outcomes. The immediacy provided on the platforms allows the marketers to incorporate the relevant feedback into product development and marketing processes. The social listening process offers necessary insights to form a foundation for the marketing strategy on social media platforms (Stewart & Arnold, 2018; Rishi & Bandyopadhyay, 2017).

For instance, The Worldwide Fund (WWF) has done an incredible job of connecting to the younger generation of more socially conscious consumers through its #EndangeredEmoji campaign.

The campaign launched in 2015, and it centered around creating awareness about the endangering species worldwide. As part of the campaign, the WWF launched a series of emojis that represented the gradually becoming endangered species. For every emoji tweeted, WWF made a small donation. It is considered an "innovation-based campaign" with fundraising as the central theme of messaging (Mortimer, 2015). The campaign focused on Twitter, prompted viewers to donate for their fundraiser by tweeting the emojis and supporting the campaign by signing up. The campaign was highly successful, with over 559,000 mentions and 59,000 sign-ups (Mortimer, 2015). This campaign serves as an example of how understanding the consumers and designing communication strategies that resonate with them can help reach socially responsible consumers.

- **Highlight the value creation for the consumer**
There is a considerable gap between the consumers who express concerns for the environment and the consumers who attempt to resort to sustainable consumption. In a way, organizations often resort to persuading consumers by guilt-tripping them for the bigger purpose of sustainability. Most communication messages take the tone of emphasizing the greater good such as—"Save the environment" or "Recycle for the next generation" ("The problem with sustainability marketing? Not enough me, me, me", 2015). Companies do not highlight the consumer's value creation, resulting in the intention-action gap in sustainable consumption. Hence, it leaves the consumer wondering what is in it for them.

Luchs et al. (2010) also discuss that when consumers value a particular product's strength, they are less likely to choose a sustainable alternative for the product. One suggested way to deal with these negative associations is to communicate about the positively perceived product attributes like—safety, quality, innovativeness.

Niemtzow (2015) discusses the different ways to communicate the idea of sustainability to the consumers on various social media platforms:

1. **Emphasizing the value from sustainability:** It involves showcasing the tangible value that the consumers can derive from such purchases. It could include the promise of performance, efficacy, value for money, or quality.
2. **Balancing the functional, emotional, and social benefits**: It necessarily implies that any effective communication strategy which in encompasses sustainability should have the elements of operational benefits (such as ease of use, quality, performance), emotional benefits (how it resonates with the intangible ideas and values), and social benefits (which involves the impact created on the environment)

- **Avoid greenwashing and focus on the communication message**
  Consumers often perceive sustainable or green products as complicated, less aesthetically pleasing, and more expensive than their nonsustainable counterparts. Greenwashing, which is the organizations' exaggerated environmental claims, leads to negative consumer attitudes toward the brand (Nyilasy et al., 2014). These negative brand attitudes could also significantly impact brand perception and purchase intention. Organizations resort to greenwashing is by excessively using complicated jargon and buzzwords like biodegradable, eco-conscious, etc. to name a few (Villarino & Font, 2015). The organizations must have a strong background to make such claims regarding sustainability. One suggested way of avoiding greenwashing is by carefully structuring the message that concisely conveys the product's information without emphasizing the irrelevant and intricate details. It is also essential for brands to be transparent in their communication on social media since it is a medium from which consumers gather information to make their purchase decisions. Brands should be open and vocal about their practices and emphasize incorporating sustainability values into their products and business models.

Since how organizations communicate has a significant influence on adopting sustainable behavior, companies need to connect with the audience personally. A study report that consumers are likely to be involved in the action when they derive positive feelings from doing so (White et al., 2019). To achieve this, companies can adopt the storytelling mechanism to communicate their sustainability story in a compelling way that appeals to the target consumers. Hence, when using social media for sustainability marketing, the marketers have to craft a personal sustainability story, connect to the bigger picture, and positively emphasize the future (Schwartz, 2013).

- **Normalizing green behavior**
Studies show that consumers who do not adopt sustainable products do so because of what they consider healthy (Rettie et al., 2011). Consequently, it becomes challenging to alter consumer behavior since they perceive to be healthy, mainstream, and what everybody does. Based on their study, Rettie et al. (2011) explain that sustainability marketing can play a role in normalizing sustainable behavior, thus, driving consumption.

    Rettie et al. (2012) discuss how sustainability marketing can increase sustainable products' adoption by emphasizing their normalcy by using endorsements from celebrities or authority figures. The authors reported that through marketing, organizations could encourage sustainable products by repositioning them as usual. This insight helps companies' social media strategy by highlighting sustainability as expected and what everybody does. They can establish this normalcy by emphasizing on sustainable products/services as the best alternative for conventional products, such as drinking tap water instead of bottled water. Also, to ensure the message of normalcy is communicated to the social media users, it is critical for the campaign to have excellent visibility and the message to be consistent throughout.

- **Leveraging social influence to drive consumption**
In a *Harvard Business Review* article, the authors highlighted the importance of social influence as a nudge mechanism to prompt users about sustainable consumption (White et al., 2019). Research has shown that humans show an inclination toward behavior that constitutes the *social norms*, and hence considered acceptable behavior (Kristofferson & White, 2015). It is in line with the Social Influence Theory, which

suggests that humans will conform to the behavior of those around them, owing to their desire to fit in (Cialdini & Trost, 1998).

By leveraging the power of social influence on social media, organizations can drive pro-sustainable consumption. Demarque et al. (2015) discuss nudge mechanisms to encourage sustainable use. The study found that by telling the fellow online shoppers about green products' purchase, a 65% increase in purchasing at least one sustainable shopping by the other shoppers. Studies have also shown that peer groups like family, friends, and colleagues exerted social influence can result in increased consumption of sustainable products (Salazar et al., 2013). Thus, organizations can use social media to prompt consumers to refer to their peers' products/services. The use of influencers or opinion leaders on social media platforms can also promote sustainable products and increase purchase intention.

We can amplify the effect of social influence on consumers' commitment to purchasing sustainable products by making this behavior public. The public display of this commitment will make consumers more accountable and increase the likelihood of sustainable product consumption (Green & Peloza, 2014). Sustainable companies can use this insight into social media by publicly acknowledging and appreciating the consumers' association with the brand. It would likely lead to a relatively long-term association with the brand and an increase in consumer satisfaction.

Apart from the guidelines mentioned above, researchers have identified specific other nudge mechanisms to drive sustainable consumption. One such example is by involving incentives/rewards in the marketing strategy. We can combine it with the external stimuli (such as save money on the purchase) with intrinsic motives (such as protecting the environment) (Edinger-Schons et al., 2012).

Figure 15.1 summarizes the guidelines to reach socially responsible consumers successfully. By understanding the consumers, organizations can effectively center the marketing and product development process around the consumers' preferences. Normalizing green behavior and highlighting value creation for the consumers enable the marketers to communicate sustainability to the consumers. Organizations are expected to be cognizant and abstain from Greenwashing. Social influence on social media can be leveraged to impact the segment of socially responsible consumers positively.

FIGURE 15.1
Guidelines to reach socially responsible consumers.

THE WAY FORWARD

As the awareness regarding various environmental issues heightens, it is longer an option for companies to ignore them. With consumers also becoming more environmentally conscious and vocal, it has become crucial for organizations to connect with the next generation of socially responsible consumers. About 81% of consumers think that organizations should work toward a sustainable future by improving the environment. Millennials, Gen Z, and Gen X seem to be the most inclined toward sustainability initiatives, but the older generations are catching up as well (Neilsen, 2018). But a powerful paradox remains concerning sustainability—consumers express inclination toward sustainable products and lifestyle while only a few follow through with the purchase. Hence, organizations need to attempt bridging this gap between intention and action.

It could be done effectively by stating the value creation for the consumers clearly and concisely.

Shaping Sustainability Behavior

The key to promoting sustainable behavior is to break the unsustainable habits and encourage sustainable ones (White et al., 2019). The most effective way to shape sustainable behavior is to make it the default option (White et al., 2019). For example, researchers found that by making green electricity the default option in buildings, 94% of the residents continued to use it (Pichert & Katsikopoulos, 2008). This insight could also be implemented in sustainability marketing by promoting sustainability as the default and better option than conventional products. Another means of shaping sustainability behavior is to use feedback mechanisms (White, Habib, & Hardisty, 2019). It involves providing feedback to consumers concerning their behavior. Social media can, thus, be leveraged to engage in a dialogue with consumers continually.

Making Sustainability Resonate

Despite the growing awareness for sustainability, organizations are still finding it challenging to communicate sustainability to the consumers and connect with the new generation of socially responsible consumers. It has become more challenging now than ever to heighten brand relevance and shift the consumer perception toward sustainability. Through the tools discussed in this chapter, the companies can better understand their target audience, needs, and expectations and accordingly curate the communication message. The central idea is to make sustainability and its principles resonate with the consumers in the most organic way possible.

Data-Driven Impact

With data being the backbone of various business functions, it is also essential to make it a critical element of different sustainability initiatives. Organizations can use data analytics to bridge the consumers' intention-action gap by showcasing the quantifiable impact created by various sustainability efforts. Given the amount of greenwashing that companies have undertaken these days, consumers lose confidence in sustainability claims. By communicating clear, concrete, and credible data, companies can better put into perspective how they have made a difference. The data would help consumers trust the organization's claims and make a conscious purchase decision.

Leveraging Social Media for Social Change

The impact social media has on the generation of today is unprecedented and undeniable. The digitalized world has led to social media platforms' growth that has amplified both consumer and organizations' voices. Social media can connect individuals from across the globe and facilitate extensive scale collaborations that can significantly improve outcomes. Hence, organizations can use social media to scale their sustainability initiatives and bring about a positive change. It provides opportunities for companies to reach a broad and diverse set of consumers. Engaging with consumers at a very human level helps the organizations gather insights into consumer perceptions regarding sustainability. By listening to what people are saying on the platform and placing transparency, authenticity, and feedback at the center of communication, organizations can embrace social media's reach to create a sustainable impact.

CONCLUSION

With the guidelines proposed in this chapter, marketers can accelerate their efforts toward communicating their sustainability efforts to all the key stakeholders involved. Reaching the growing segment of socially conscious consumers involves understanding such consumers at a somewhat more in-depth level. Understanding the consumers' motivations and perceptions regarding the organization's social efforts can be fundamental in designing communication strategies to reach the consumers effectively. Through tools such as social listening on social media, organizations can analyze various consumer interactions on the platforms to gain actionable insights that can drive product development and marketing strategies.

An increasing number of socially responsible consumers express their interest in brands that support a social cause. Still, more often than not, they do not follow up that interest with purchases. Hence, organizations should attempt to bring the intention-action gap and communicate the value added to the consumers. With organizations flocking to reach the segment of socially responsible consumers, greenwashing has become a common phenomenon on social media. Through tools like storytelling, marketers can avoid communicating irrelevant details and focus on the consumers' value creation. Normalizing green behavior is another fundamental step to reaching and creating socially responsible consumers. By presenting sustainable options as a better alternative, marketers can sense normalcy and encourage sustainable practices.

Another key characteristic of social media that makes it suitable to reach socially responsible consumers is the power of leveraging social influence on the platform. Using opinion leaders and social media influencers, marketers can use social impact to drive socially responsible consumers' consumption.

Thus, in challenging times, it becomes the need of the hour for the organizations to place sustainability at the core of their businesses and navigate toward becoming a purpose-driven organization. The unification of sustainability initiatives and social media continues to grow stronger. More and more companies incorporate social media into their business strategy that positively impacts their bottom line and society.

REFERENCES

Baldassarre, F., & Campo, R. (2016). Sustainability as a marketing tool: To be or to appear to be? Business Horizons, 59(4), 421–429. doi: 10.1016/j.bushor.2016.03.005

Belz, F. M., & Peattie, K. (2012). *Sustainability marketing: A global perspective.* Chichester: John Wiley and Sons.

Cialdini, R. B., & Trost, M. R. (1998). Social influence: Social norms, conformity, and compliance. In R. B. Cialdini, M. R. Trost, D. T. Gilbert, S. T. Fiske, & G. Lindzey (Eds.), *The Handbook of Social Psychology. Social Influence: Social Norms, Conformity and Compliance.* New York, NY: McGraw Hill (pp. 151–192).

Demarque, C., Charalambides, L., Hilton, D. J., & Waroquier, L. (2015). Nudging sustainable consumption: The use of descriptive norms to promote a minority behavior in a realistic online shopping environment. *Journal of Environmental Psychology, 43,* 166–174.

Dwivedi, Y. K., Ismagilova, E., Hughes, D. L., Carlson, J., Filieri, R., Jacobson, J., ... & Wang, Y. (2020). Setting the future of digital and social media marketing research: Perspectives and research propositions. *International Journal of Information Management* (https://doi.org/10.1016/j.ijinfomgt.2020.102168).

Edgecomb, C. (2017). *Social media marketing: The importance of a two-way conversation*. Impactbnd.com. Retrievable from: https://www.impactbnd.com/blog/social-media-marketing-the-importance-of-a-two-way-conversation

Edinger-Schons, L. M., Sipilä, J., Sen, S., Mende, G., & Wieseke, J. (2018). Are two reasons better than one? The role of appeal type in consumer responses to sustainable products. *Journal of Consumer Psychology, 28*(4), 644–664.

Fuller, D. A. (1999). *Sustainable marketing: Managerial-ecological issues*. Thousand Oaks, CA: SAGE Publications.

Gilsenan, K. (2019). *Lifting the lid on sustainable packaging* [Blog]. Retrievable from: https://blog.globalwebindex.com/chart-of-the-week/lifting-the-lid-on-sustainable-packaging/

Gordon, R., Carrigan, M., & Hastings, G. (2011). A framework for sustainable marketing. *Marketing Theory, 11*(2), 143–163.

Green, T., & Peloza, J. (2014). Finding the right shade of green: The effect of advertising appeal type on environmentally friendly consumption. *Journal of Advertising, 43*(2), 128–141.

Kemper, J. A., & Ballantine, P. W. (2019). What do we mean by sustainability marketing? *Journal of Marketing Management, 35*(3–4), 277–309.

Kotler, P. (2011) Reinventing marketing to manage the environmental imperative. *Journal of Marketing, 75*(4), 132–135.

Kristofferson, K., & White, K. (2015). Interpersonal influences in consumer psychology. In M. Norton, D. Rucker, & C. Lamberton (Eds.), *The Cambridge Handbook of Consumer Psychology* (pp. 419–445). Cambridge: Cambridge University Press. doi: 10.1017/CBO9781107706552.016

Laroche, M., Habibi, M. R., & Richard, M. O. (2013). To be or not to be in social media: How brand loyalty is affected by social media? *International Journal of Information Management, 33*(1), 76–82.

Laroche, M., Habibi, M. R., Richard, M. O., & Sankaranarayanan, R. (2012). The effects of social media based brand communities on brand community markers, value creation practices, brand trust and brand loyalty. *Computers in Human Behavior, 28*(5), 1755–1767.

Lau, D. (2016). *The problem with sustainability marketing: What's in it for me?* [Blog]. Retrievable from: https://blogs.ubc.ca/daniellau/2016/01/21/the-problem-with-sustainability-marketing-whats-in-it-for-me/

Lim, W. M. (2016). A blueprint for sustainability marketing: Defining its conceptual boundaries for progress. *Marketing Theory, 16*(2), 232–249.

Luchs, M. G., Naylor, R. W., Irwin, J. R., & Raghunathan, R. (2010). The sustainability liability: Potential negative effects of ethicality on product preference. *Journal of Marketing, 74*(5), 18–31.

Lyon, T. P., & Maxwell, J. W. (2011). Greenwash: Corporate environmental disclosure under threat of audit. *Journal of Economics & Management Strategy, 20*(1), 3–41.

Mortimer, N. (2015). *Lessons from WWF's #EndangeredEmoji campaign*. The Drum. Retrievable from: https://www.thedrum.com/news/2015/07/28/lessons-wwf-s-endangeredemoji-campaign.

Neilsen. (2018). *Global consumers seek companies that care about environmental issues*. Nielsen.com. Retrievable from: https://www.nielsen.com/eu/en/insights/article/2018/global-consumers-seek-companies-that-care-about-environmental-issues/

Niemtzow, E. (2015). *Want to sell sustainability? Give consumers what they want [Blog]* BSR. Bsr.org. Retrievable from: https://www.bsr.org/en/our-insights/blog-view/want-to-sell-sustainability-give-consumers-what-they-want.

Nyilasy, G., Gangadharbatla, H., & Paladino, A. (2014). Perceived greenwashing: The interactive effects of green advertising and corporate environmental performance on consumer reactions. *Journal of Business Ethics, 125*(4), 693–707.

Patel, D. (2017). *The millennial marketplace and the propagation of the triple bottom line.* Forbes. Retrievable from: https://www.forbes.com/sites/deeppatel/2017/07/28/the-millennial-marketplace-and-the-propagation-of-the-triple-bottom-line/#641d8681d04a

Peattie, K. (2001). Towards sustainability: The third age of green marketing. *The Marketing Review, 2*(2), 129–146.

Peattie, K., & Crane, A. (2005). Green marketing: Legend, myth, farce or prophesy? *Qualitative Market Research: An International Journal, 8*(4), 357–370.

Pichert, D., & Katsikopoulos, K. V. (2008). Green defaults: Information presentation and pro-environmental behaviour. *Journal of Environmental Psychology, 28*(1), 63–73.

Rettie, R., Barnham, C., & Burchell, K. (2011). *Social normalisation and consumer behaviour: Using marketing to make green normal.* Kingston Business School, Kingston University.

Rettie, R., Burchell, K., & Riley, D. (2012). Normalising green behaviours: A new approach to sustainability marketing. *Journal of Marketing Management, 28*(3–4), 420–444.

Rishi, B., & Bandyopadhyay, S. (Eds.). (2017). Contemporary issues in social media marketing. London: Routledge.

Rishi, B., & Kuthuru, N. R. (2021). A review for managerial guidelines for social media integration of IMC in digital era. In *Digital Entertainment* (pp. 187–212). Singapore: Palgrave Macmillan.

Salazar, H. A., Oerlemans, L., & van Stroe-Biezen, S. (2013). Social influence on sustainable consumption: Evidence from a behavioural experiment. *International Journal of Consumer Studies, 37*(2), 172–180.

Savitz, A. W., & Weber, K. (2006). *The triple bottom line: How today's best-run companies are achieving economic, social and environmental success—And how you can too.* New York, NY: John Wiley.

Schwartz, S. (2013). *Sustainability storytelling: Creating a narrative that matters [Blog]* BSR. Bsr.org. Retrievable from: https://www.bsr.org/en/our-insights/blog-view/sustainability-storytelling-creating-a-narrative-that-matters

Stewart, M. C., & Arnold, C. L. (2018). Defining social listening: Recognizing an emerging dimension of listening. *International Journal of Listening, 32*(2), 85–100.

The Guardian. (2015). The problem with sustainability marketing? Not enough me, me, me. Retrievable from: https://www.theguardian.com/sustainable-business/behavioural-insights/2015/mar/09/problem-sustainability-marketing-not-enough-me

Valentine, O. (2019). *Social media's influence on green consumerism – we are social.* We are social. Retrievable from: https://wearesocial.com/blog/2019/11/social-medias-influence-on-green-consumerism

Villarino, J., & Font, X. (2015). Sustainability marketing myopia: The lack of persuasiveness in sustainability communication. *Journal of Vacation Marketing, 21*(4), 326–335.

White, K., Habib, R., & Hardisty, D. J. (2019). How to SHIFT consumer behaviors to be more sustainable: A literature review and guiding framework. *Journal of Marketing, 83*(3), 22–49.

White, K., Hardisty, D., & Habib, R. (2019). *The elusive green consumer*. Harvard Business Review. Retrievable from: https://hbr.org/2019/07/the-elusive-green-consumer

World Commission on Environment and Development. (1987). *Our common future (the Brundtland Report)*. Oxford, UK: Oxford University Press.

Section VII

Advances in Knowledge of Social and Sustainable Marketing: Addressing Global Crises

16

Management of Shocking Global Crises: The Use of Public Marketing 4.0 within a Social Responsibility and Sustainability Approach

Manuel Antonio Fernández-Villacañas Marín
Consultant in Logistics & Management at M&M Planning and Project Management, Spain

Ignacio Fernández-Villacañas Marcos
Consultant in Technology at M&M Planning and Project Management, Spain

CONTENTS

Learning Objectives 436
Themes and Tools Used 437
Public Marketing Concept 437
Conceptual Chapter Framework 438
Public–Private Response to the New Impacting Global Crisis ... 438
Introduction 441
Modernization of Public Management and the Application
   of Public Marketing 442
   The Urgent Modernization of Public Management Is Necessary
      in the New Disruptive Environment 444
   Public Marketing: A Framework for the Modernization
      of Public Management 448
   Key Strategies of Public Management Modernization 452
   Research Methodology 454
Summary of the Main Problems to Be Solved 455

Discussion: Application of Public Marketing 4.0 through Social
    Utility Functions, and Proposal of a Generic Model of Planning
    and Management for Shocking Global Crises 455
Public Marketing 4.0: The Optimal Combination
    of Traditional and Digital Marketing with New
    Disruptive Technologies 455
Applying Public Marketing 4.0: Social Welfare
    Functions 457
A Model of Social Welfare Function 461
Strategic Intelligence Management of Shocking
    Global Crisis: Maximization of the Social Utility
    Index-Function 463
Determining a Planning and Management Generic Model
    of Shocking Global Crisis: Scenario Planning 464
Conclusion 465
Lessons Learned 466
Additional Content 467
    Key Questions 467
    Key Definitions 468
    Key Author Mentions 468
Credit Authors Statement 469
References 470

LEARNING OBJECTIVES

After reading the chapter, the reader should be able to:

- Analyse the situation of social threat posed by the crises generated by the current disruptive environment, the consequent need for the urgent modernization of public management through public marketing and its inclusion within the framework of social responsibility that reinforces its legitimacy both social and democratic guarantees in governance.
- Study the concept and scope of public marketing 4.0, within a socially responsible and sustainable approach, as a framework for the modernization of public management, through key public–private strategies for this, and using and optimizing welfare functions within the framework of the Sustainable Development Goals of the UN 2030 Agenda.

THEMES AND TOOLS USED

Themes	Tools
1. Modernization of public management	New public management
	Post-new public management
2. Public sector marketing application	Social marketing
	Socially responsible marketing
	Global sustainability
3. Development of public marketing 4.0	Public marketing
	Digital marketing
	Lean and agile management
	Digital technologies
4. Management of shocking global crisis	Strategic intelligence
	Data science
5. Optimization of social utility index-function	Social utility functions
	Cost-benefit analysis
	Sustainable development goals
6. Planning model of shocking global crisis	Prospective planning
	Strategic anticipation

PUBLIC MARKETING CONCEPT

Marketing has been occupying a prominent place in the field of business management for decades. However, its application in the public sector is still very limited, and incipient in most Western (Chias, 2011) and Eastern countries, such as China, whose progressive application raises good expectations to guide the challenges it faces today, the most populous country, second economy in the world (Zaheer et al, 2015).

However, various aspects are prompting public organizations to adopt basic marketing approaches in their management, by considering the needs and demands of citizens as the central axis of their approach. All of them are interrelated and linked to the processes of reform and modernization of the public sector, as has been shown previously. In addition to the increasing budgetary restrictions of the governments, the evolution and social development are causing progressively greater demands in democratization and citizen participation. Likewise, the lack of representativeness of the traditional political parties has led to the emergence of populism and the appearance of globalist movements posed by supranational and even world governments. The alternatives in the political debate and the apparent greater competition between the parties at the different levels of government are giving rise to new government offers that are closer to and apparently more interested in citizens' problems.

This new sensitivity of those responsible for public management toward the needs of citizens and their expectations implies the convenience of assuming marketing as a main methodology for the management of the public sector. In this sense, marketing represents for the public administration a philosophy that tries to strengthen its spirit of civic responsibility, in contrast to traditional, insensitive, autocratic or routine-dominated bureaucratic models. The implementation of marketing in the public sector represents a new approach that fosters the spirit of service and that provides it with highly effective tools for the design, management, and control of public policies, fully adapted to the demands of citizens. Its traditional concepts work well in the public sector, and the application of marketing to public management can help governments act with greater efficiency, quality, speed, fairness and comfort (Kotler & Lee, 2007). Therefore, public marketing is considered a very suitable framework to develop the management of the new shocking global crises, whose conceptual framework and research process for its application is summarized in the diagram in Figure 16.1.

The free market has consumers as its core and is based on their voluntary exchange. Therefore, it focuses on individual interest. However, public entities represent, in addition to individuals, communities. Therefore, they follow a common interest, taking on a special importance the concept of public value that is generated as a result of the actions of the public administration, under the direction and authorization of the representative government, and always taking into account the economy, the effectiveness and the efficiency with which public policies should be designed and public services generated (Rainey, 2014).

CONCEPTUAL CHAPTER FRAMEWORK

PUBLIC–PRIVATE RESPONSE TO THE NEW IMPACTING GLOBAL CRISIS

The current COVID-19 pandemic is spreading human suffering to all areas of the world, destabilizing the global economy like never. By affecting all countries in a general way, the profound inequalities between

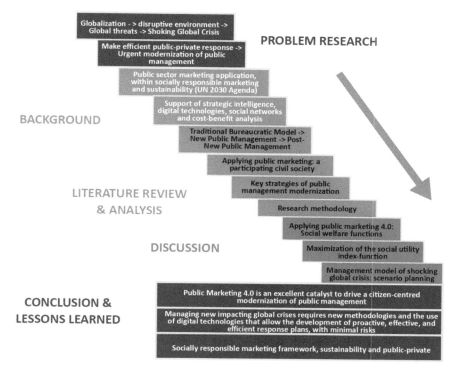

FIGURE 16.1
Diagram of conceptual chapter framework and research process. (From own elaboration).

the actions of different governments are evident, as well as the failures with which the 2030 Agenda for Sustainable Development has been addressed.

This progressive increase in risk situations associated with infectious diseases could be the result of a rapid global change that is altering the relationship between humans and our environment (World Health Organization, 2020). Factors such as the concentration of the population in large urban centres and its progressive aging, the predatory use of resources by humanity that generate profound imbalances in ecosystems, the consequent climate change and increased temperatures, or the increase in population displacements induced by the uncontrolled development of tourism, among others, seem to favour both the contagion and the spread of these deadly diseases. In this way, this crisis is highlighting the great interdependence between the different elements of sustainability, from the integrity of the ecosystem to health and well-being, and the socioeconomic prosperity that follows.

The 2019 UN Global Sustainable Development Report highlighted that we were not on track to achieve the Sustainable Development Goals (SDGs) and called for more substantial and accelerated action to achieve the transformative changes necessary to achieve the 2030 Agenda. But are the actors correctly interpreting this request and implementing public policies and concrete actions to respond to the crisis that allow strengthening the achievement of these goals and the 2030 Agenda? Or is the COVID-19 crisis causing the 2030 Agenda to be neglected and countries to focus solely on the recovery of traditional predatory economic models and to turn a blind eye to the clear interdependence among sustainability factors revealed by the COVID-19 crisis?

The truth is that the pandemic has revealed a high international vulnerability to these types of threats, as well as the inability of many nations to respond effectively to environmental uncertainty (Haarhaus & Liening, 2020). It seems clear that this crisis will imply a profound change in lifestyles and, quite possibly, in traditional economic systems, with a structural change that will lead us to a new global situation. In our opinion, the response of the governments of many countries and of civil society has been lengthy, improvised, ineffective and insufficient.

But these threatening triggers are also present in other equally devastating phenomena: failure of climate action, extreme weather, water crisis, natural disasters, man-made environmental disasters, political corruption and crisis of democratic systems, cyber-attacks, failure of the global governance, loss of biodiversity, involuntary migration, social instability, etc. (World Economic Forum, 2020).

Generalizing to these other shocking global crises, it seems advisable to undertake a restructuring of the methodology to manage them. In addition to reducing the vulnerability of the population through preventive actions and the management of sectoral policies in the face of each threat, it is necessary to develop multidisciplinary and multi-sectoral action strategies for early detection and strategic anticipation, response planning and crisis management, in terms of prospective (Rose, 2017). Coordination between all the organizations involved, the necessary public–private collaboration and the necessary international cooperation must be ensured

(Fernández-Villacañas, 2020a). In these new strategies of disruptive management, the political ideology should be less prominent, and the inclusion of marketing models is essential, identifying and optimizing the corresponding social welfare function. Ultimately, it would try to avoid surprises and improvisations, as well as deviant and insufficient responses, developing digital systems of strategic anticipation and prospective planning that facilitate the most effective and efficient response with minimal risk (Schuhly et al, 2020).

INTRODUCTION

This new situation advises the urgent modernization of public management in most countries, within the permanent transformation of the public sector, through the implementation of public marketing, guaranteeing the democratic values of public governance. This would avoid a series of quite generalized phenomena, totally inadequate to face the crises that are taking place induced by the current pandemic, or other shocking global crises that may arise: highly politicized systems that formulate corporatist and fragmented policies, dominated by influential politicians with interests outside the public interest, and in which the opinions of civil society and experts practically do not count; avoid a hierarchical, individualistic and short-term culture, with little planning and implementation capacity and low regulatory quality; and, in general, avoid opacity and lack of transparency that make it impossible for the public to trust public management.

On the other hand, socially responsible marketing is a voluntary decision of organizations to guide the fulfilment of business economic objectives together with a series of moral, ethical, legal, and environmental aspects, to guide their efforts in the maximization of benefits, in addition to generating the least social impact and the greatest possible social benefits. In its development, social marketing is applied as a concept of seeking the best quality of life for society, balancing the interests of organizations with those raised in society. The key to socially responsible marketing will depend on the ability to achieve a dynamic balance between commercial benefit and social benefit, in solidarity, commitment, support and awareness campaigns. The approach proposed for the implementation of public marketing supposes the consolidation of social responsibility as an innovative trend in the public sphere, which affects the planning, organization and operation of institutions, fully integrating into socially responsible marketing. Further, the assumption of the social responsibility model reinforces the social legitimacy of the public sector through its adaptation to the demands of citizens.

Public marketing, unlike the marketing models that are currently used exclusively for electoral purposes, would allow a more dynamic, exhaustive and real identification of social needs. In addition, it would enable the planning and management of a supply of public services totally focused on the true needs of citizens, as well as a more effective, efficient, and transparent management of public resources, adjusted to the achievement of the Global Sustainability Goals established in the UN 2030 Agenda.

Ultimately, it would be possible to strengthen and improve capacities for more democratic governance. The support of strategic intelligence, the use of social networks, cost-benefit analysis (CBA) and the massive use of disruptive data science technologies, would transform traditional public marketing models into a new public marketing 4.0, to improve management in response to the new and foreseeable shocking global crises that the international community will surely suffer soon.

The research has been developed based on the application of new public-private strategies in response to new shocking global crises in a disruptive environment. This requires the urgent modernization of public management whose main reference framework is considered the application of public marketing 4.0.

The work has been structured in two main parts. First, the review of the literature on the problem investigated and its analysis, regarding the modernization of public management and the application of public marketing. Second, the discussion, relative to the application of public marketing 4.0 through social utility functions, and proposal a generic planning and management model of shocking global crises.

MODERNIZATION OF PUBLIC MANAGEMENT AND THE APPLICATION OF PUBLIC MARKETING

The true economic and social reason for the existence of a private organization must be the satisfaction of the needs, demands and desires of its clients (Crompton & Lamb, 1986). If you change the term private organization to public administration and client to citizen, the result will be socially responsible public administration, the essence of one of the key concepts in terms of public ethics. A bureaucratic public administration is the exact opposite of a socially responsible public administration. The difficulties in connecting with the social needs of citizens in this type of public administration, especially when political corruption and waste of public resources have been installed in many governments by the hand of populism, becomes unaffordable when the social environment turbulent changes rapidly and public organizations remain immovable and unable to capture emerging needs.

In recent years, the concept of socially responsible marketing has emerged that focuses, in addition to consumer satisfaction, on the contribution to society, the community and the environment, creating value for both the company and society. Social responsibility goes beyond philanthropic acts, since companies have focused on environmental problems, and the implementation of socially responsible marketing allows companies and public entities to meet the demands and needs of consumers and citizens, causing an increase in the demand for their products or services, and at the same time generating a sustainable circle with the environment (Barragán et al, 2017).

On the other hand, social marketing appeared as a non-profit extension of marketing that had a very specific general purpose, voluntary social change, and that was implemented through complex, personal and anticipatory exchange relationships that focused on ideas and specific social causes (Kotler & Andreasen, 1991). Subsequently, it was considered that it could generate a positive change in socially rejected behaviours and practices, such as tobacco consumption or environmental pollution, reaching more citizens (McAuley, 2014). In recent years, social marketing has been defined as a set of marketing strategies focused on creating value for customers, in a way that guarantees both their well-being and the community in general, turning citizens into consumers of social goods (Pykett et al, 2014), seeking that society adopts beneficial behaviours and moves away from harmful or harmful ones, with a close relationship with legal, technological, economic and informational factors. Currently, the objective of social marketing is to improve people's quality of life through initiatives and actions that allow achieving a responsible and sustainable social transformation (Saunders et al, 2015; Kennedy & Parsons, 2014).

Even though there is a great connection between the two concepts and that both work collaboratively, social marketing and socially responsible marketing are not the same, they are not equivalent concepts. Thus, a social marketing campaign does not return to a socially responsible organization. There are differences between the two terms.

- First, while social marketing is an instrument that is managed in a department for which it is responsible, social responsibility marketing involves the entire organization, regardless of whether it has its own department.
- Second, the main objective of social marketing campaigns is to reach citizens and consumers, while socially responsible marketing cares about and takes care of all stakeholders or interest groups, as is the case with public marketing.
- Third, social marketing is a tactic within the socially responsible marketing strategy carried out by the different initiatives, which in the public sector would be equivalent to the public marketing strategy.

The Urgent Modernization of Public Management Is Necessary in the New Disruptive Environment

During the last two decades, very rapid and intense changes have been generated in a global environment that is considered very disruptive, whose generating factors act without adjusting to a known historical pattern and in circumstances of total uncertainty (Mufudza, 2018). This is a new characteristic of today's world that has great importance as it fully conditions the establishment of strategic plans in all areas. The possible shocking triggers that have been presented in the introduction to the chapter, act globally and are very much alive today. The result is that the environment in which companies, the public sector and civil society operate is very different from what was usual, and presents new risks and threats, as well as opportunities, to which we must react in advance applying new methodologies (Fernández-Villacañas, 2020b).

Strategic anticipation is the ability of organizations and individuals to explore and monitor the future, allowing decision-makers to seek and seize opportunities, protect their interests and ensure the achievement of their objectives and goals (Balbi, 2015). Deeply related to strategic anticipation is the concept of foresight, which we can define as the participatory, systematic and multidisciplinary process to collect information and intelligence about the future and build visions with the aim of informing the decisions that will be made in the present (Gavigan, 2001). Among the purposes of prospective planning, it is worth highlighting that of generating alternative visions of desired futures for the planner, that of promoting action, that of promoting relevant information under a long-term approach, that of creating explicit alternative circumstances of possible futures, and establishing values and decision rules to achieve the most desired possible future (Berger, 1964).

On the other hand, strategic intelligence is an increasingly demanded methodology, which has been evolving and assuming a polysemic value (Palacios, 2018). It allows the obtaining, integration and analysis of a wide variety of information that offers key conclusions about a problem, in its relationship with other existing problems. It consists of the use of methodologies and techniques that make it possible to reduce the great uncertainty that affects strategic decision-making processes. Strategic intelligence provides great utility to organizations derived from its ability to assign critical and transformational information to those responsible for strategic decisions (Balbi, 2015), from its ability to early detection of risks, threats and opportunities present in a difficult disruptive environment (Gilad, 2004), and in general its ability to improve organizational performance and levels of success (Liebowitz, 2006).

Several expert authors have criticized the strategic planning technique for considering that it should be alternatively continuous rather than at certain intervals (Vecchiato, 2015), and have proposed the convenience of evolving toward prospective planning, making possible a more comprehensive perspective of the organization (Bratianu, 2017). To achieve those capabilities of strategic intelligence related to being able to penetrate the future, it is necessary to assume a change in strategic attitude, abandoning the reactive attitude and developing a proactive, anticipatory and preventive attitude.

This new disruptive situation advises the urgent modernization of public management within the process of permanent transformation of the public sector, to enable the implementation of these new methodologies as well as the changes necessary to apply them. We are going to present the main initiatives of the transformation process that have been proposed, and then analyse public marketing as a framework for action to develop the process of modernization of public management.

In 1991 the so-called New Public Management was proposed as a public sector transformation program that was implemented in virtually all OECD countries and beyond (Hood, 1991; Hood, 1995; Lindberg et al, 2015). This new approach emphasized from the beginning the application of the concepts of economy, efficiency and effectiveness in government organizations, as well as in political instruments and their programs, trying to achieve total quality in the provision of public services, and providing less attention to procedures and requirements dominated by legal and non-optimized management approaches, standards and recommendations (Leeuw, 1996). The objective of this initiative was the creation of an efficient and effective administration, favouring the implementation of competition mechanisms, promoting the development of higher quality public services and favouring citizen participation (García Sánchez, 2007).

The implementation of this approach implies a profound transformation of the patterns of behaviour that govern the relations between the agents of the public sector and between them and the private sector. This dynamic of change tended to generate resistance among the agents that could see their niches of power in the state compromised. By putting the citizen at the centre of the model, a fundamental break with the traditional perspective is introduced: the final objective of the public administration will be to efficiently process the demands of citizens, guaranteeing the provision of quality services adjusted to their needs.

However, the reforms that were proposed in the organizational structures to disaggregate the entities in the public sector led to a greater fragmentation of the functions to be performed and a certain ambiguity in their specification. In addition, there was an excessive level of specialization that resulted in a lack of coordination and cooperation between the different agencies and organizations. On the other hand, decentralization led to new control and accountability systems, as opposed to the intended flexibility that limited the management autonomy of those responsible for public administration activities (Christensen & Laegreid, 2007a; Diefenbach, 2009; Goldfinch & Wallis, 2010). Finally, the focus excessively focused on efficiency has led to the marginalization of the traditional principles of the public sector regarding the provision of public goods and services, such as neutrality, social justice or social welfare (Arellano & Cabrero, 2005).

These difficulties led to the beginning of a review process of the new public management that gave rise to new approaches that are grouped under the so-called post-new public management (Christensen & Laegreid, 2007b; De Vries & Nemec, 2013). This new reform has generated a hybridization of the public sector and the private sector during the last fifteen years. If the previous approach generated greater flexibility and disaggregation in large bureaucratic organizations, this meant the existence of smaller control units, with lower levels of coordination and cooperation between them, which value a reconfiguration that provides greater interconnection (Goldfinch & Wallis, 2010; Christensen & Laegreid, 2008; Pollit, 2009).

For this reason, a revision of the postulates of Weber's bureaucratic theory has been promoted, insisting on the decentralization and reorganization of public administrations, but without losing the validity and growing application of the public–private collaboration models. Figure 16.2 summarizes the

	TRADITIONAL BUREAUCRATIC MODEL	NEW PUBLIC MANAGEMENT	POST-NEW PUBLIC MANAGEMENT
SETTING	Centralization: horizontal coordination	Decentralization: vertical coordination	Re-centralization: horizontal and vertical coordination
CONCEPTION OF THE CITIZEN	Application of legality: Citizen - Managed	Customer Satisfaction: Citizen - Client	Citizen orientation. Responsibility and accountability. Citizen participation..
REGULATION	Administrative law	Flexibility: escape from the administrative law. Contract formalization.	Administrative Law. Openness and transparency.
PROCESSES	Standardization: administrative procedures	Process orientation: techniques and procedures of the private sector.	Management professionalization
STRUCTURE	Hierarchical structure	Control units	Work networks. Cooperation between organizations.
EVALUATION	Authority and control	Control and evaluation of results (output)	Management control and evaluation (outcome)
HUMAN RESOURCES	Bureaucratization of public workers	Professionalization of human resources	Professionalization of human resources

FIGURE 16.2
Traditional bureaucratic model vs new public management vs post-new public management. (Adapted from García Sánchez, 2007).

main characteristics that differentiate the traditional bureaucratic model, that of new public management and that of post-new public management.

The development of post-new public management has been based for some years on digital technologies to achieve a more coordinated and integrated public management and services (Dunleavy et al, 2006). All these changes and the great capacities derived from the application of new disruptive technologies have generated a new approach called *Governance of the Digital Age*. This implies the reintegration of functions in the governmental sphere, the adoption of holistic structures oriented to the needs of citizens and the advancement of the digital transformation of public bodies and administrative processes. This approach offers an ideal opportunity to generate self-sustaining change, with a wide range of very positive cultural, technological, organizational and social effects, all closely related (De Vries & Nemec, 2013).

Public Marketing: A Framework for the Modernization of Public Management

Public marketing is the element that complements new public management and post-new public management as it allows the introduction of market instruments, tools and techniques in public management, as advocated (Antoniadis et al, 2019). This application of marketing allows strengthening the action of the State, under more technical, effective and efficient parameters, as well as results-based management. It also offers greater coherence to the idea of modifying the management structure of public administration, guiding the common welfare and social development with greater precision through the inclusion of all citizens (Escourido, 2017). We can affirm that marketing represents a frame of reference for the modernization of public management, within a socially responsible marketing and sustainability approach.

Thus, the advantages of applying marketing to public sector are multiple by concentrating the effort on the social needs and expectations of citizens: public resources are managed in order to meet and respond to the problems and demands of citizens, the supply of public services permanently improve in objective and subjective quality by increasing the level of satisfaction, the democratic legitimacy of public entities is increased, it becomes possible to serve with equity the population sectors with particular characteristics and in a situation of social disadvantage, etc. However, concepts, methodology and tools cannot be mechanically transferred from private to public marketing, but it is necessary to develop models that consider the particular characteristics of the public environment (Puig et al, 1999; Puig, 2004).

Public marketing has had various applications since its inception, among which the improvement of the management and provision of public services stands out. Its most significant applications are the following (Kotler & Lee, 2007).

- **Market research:** It consists of the design, compilation, analysis, and systematic information on a situation that affects a public entity that allows knowing the demographic and socioeconomic characteristics, values, attitudes, opinions as well as the behaviour patterns of the target social group to which the public service is directed. Likewise, it allows to know the competitive public–private environment in which the offer of said public entity and its perceived image will be inserted. The information generated is essential for making decisions about the management of the supply of public services.
- **Development and improvement of public programs and services:** It involves a systemic process of generation and selection of ideas to create an offer of services aimed at satisfying social needs, development and proof of the concept, the design of the service considering its central and peripheral attributes as well as its derived advantages for the target social group, pilot test of the service, and finally creation of distribution strategies and promotion of the offer of programs and public services designed.
- **Segmentation and differentiation:** These techniques allow introducing modifications, especially in the price and distribution of public services, considering the diversity of social needs, as well as the profiles of the target audience based on sociodemographic criteria. Adapted alternatives for access to services are also being developed.
- **Determination of public rates and prices:** Public rates and public prices are not usually established in relation to the costs incurred to produce them, both direct and indirect. However, marketing provides significant information to create both economic and non-economic incentives that maximize the use of a service, as well as to design disincentives that try to eliminate socially unwanted behaviours.

- **Optimization of distribution channels for the provision of public services:** Aspects such as schedules, conditions of the place and the mode of provision ... are determined with the design and management of distribution channels, conditioning the choice of response of the citizens and with it, affecting their demand and user satisfaction.
- **Effective communication:** It allows the target audience to be informed of the attributes of the public service offered, emphasizing its benefits and motivating potential users to act. The achievement of effective communication is the result of an orderly and systematic process that begins with the study of the target audience, continues with the creation of key messages with an appropriate style for the users of the service, continues with the selection of the message transmitter and the most suitable channels, and finally manages the distortions that occur in the encoding and decoding of messages.
- **Modification of social behaviours:** It consists of persuading and motivating a target social group to voluntarily accept, reject, modify or abandon behaviours that are considered beneficial for society. Normally, the techniques and methodologies of social marketing are usually applied for this purpose.

One of the most relevant conceptual limitations of public marketing is the incorporation of citizen participation in its methodology. For this, it is necessary to enrich and develop psychosocial research techniques on social needs and citizen expectations, beyond surveys. Thus, the necessary concurrence and participation of civil society in public management is one of the main improvements derived from the application of marketing to public management. Citizenship must not only be an end and an objective of its execution but also a means of action in all the activities of the sequential and recurring phases: *Consultation, Agreement, Decision-making*, and *Action* (Figure 16.3). Based on this approach, the entire process should be aimed at maximizing the collective social utility function freely established by civil society, with the limitations of available public resources.

The viability of the proposed improvements has as a necessary condition prior to the application of public marketing models, the implementation of lean and agile management techniques, tools and methodologies. Both methodologies are simultaneously opposing and complementary. Strategies should be combined lean and agile management since it is

FIGURE 16.3
Public sector marketing approach—A participating civil society. (From own elaboration).

necessary to answer to different needs, low costs (lean) and quick responsiveness (agile), both highly valuable for efficient and effective performance (Vazquez-Bustelo & Avella, 2006; Aronsson et al, 2011).

The integrated application of the lean and agile methodology generates a new leadership approach that creates and maintains a dynamic culture of continuous improvement through a synergistic system with the best strategies, techniques and ideas. Therefore, very relevant approaches are considered to implement the change toward organizations oriented toward excellence, whose application to the public sector will lead to the improvement of the effectiveness and efficiency of the organization, the improvement of customer orientation, and a better overall performance of the core of its activities.

The benefits of implementing both approaches in public management positively affect all members of society. Users of public services develop increasingly demanding expectations, faster changes, more innovative designs, fewer tax desires, ease of processing and always perfect quality. Faced with such expectations, faced with strong pressures to reduce costs, and with increasingly reduced reaction time margins, lean and agile management becomes a key factor to streamline and eliminate the waste of operational and administrative processes of the different public bodies (Radnor & Walley, 2008; Radnor & Boaden, 2008).

Key Strategies of Public Management Modernization

A series of key guiding principles for public sector reform have been identified within The Commonwealth countries that can serve as a guide in actions to modernize public management throughout the international community (Commonwealth Secretariat, 2016).

- **A new pragmatic and results-oriented framework:** It is the response of governments to the growing demand for a higher quality of public service offerings by citizens, focused entirely on society, and to the growing scarcity of public resources. Public planning and management have evolved toward results-oriented approaches, whose objectives are to optimize value for money, determine priorities in strategic objectives as well as maximize satisfaction in the provision of services by measuring levels of effectiveness and efficiency accomplished.
- **Clarification of objectives and administrative structures:** The capacities of public entities to face the challenges of a highly disruptive global environment continue to be limited. The development of these capacities represents a priority of government actions, previously evaluating the need and feasibility of transforming the organizational structures of the public sector.
- **Intelligent political strategies and engagement:** It is necessary to guarantee a dynamic balance between politics (politicians, political organizations, ideological programs, and political power), public policies (goals, objectives and action programs) and public management capacities (public administration), that allow the progressive transformation of the public sector and the necessary reform of the supply of public services, improving the provision, performance and citizen satisfaction.
- **Goal-oriented competencies and skills development:** The skills development programs of public servants must be oriented to the satisfaction of specific needs that define the objectives that guide the actions of public entities, in the short, medium and long term. These capacities

must cover both the legal, financial and administrative problems as well as the social skills of any managerial action.
- **Experimentation and innovation:** The capacity development processes that enable the progressive transformation of public management toward greater effectiveness and efficiency should be promoted through innovation, experimentation, and the application of digital technologies, enabling organizational and individual learning. The comparative analysis of the leading countries has shown that public management is more effective and efficient when innovative formulas are applied to solve problems. Despite this reality, implementing innovation is not easy given the inertia of a public sector culture of risk aversion and high regulation. For this reason, the transformation of public management must propose cultural changes that must be supported by political action.
- **Professionalization and improved morale:** The loss of morale and commitment in public conduct is a practically widespread phenomenon in the public administrations of most countries. Although palliative measures have been proposed to reverse this process, such as salary improvements or greater social recognition, the results have not been good enough. It is necessary to apply new measures and incentives to improve the perception of public servants, improve their performance and promote their social inclusion. One of the most negative trends that have occurred has been the politicization of the public administration and the management of appointments and promotions of civil servants with subjective criteria not linked to merits and abilities, placing trust in the leader before professionalism. All this undermines government democracy and trust in a totally discredited political class, and in which corruption is prevalent. Therefore, the best and broadest technical and moral training of public servants is essential, with a professionalization process that guarantees appointment and promotion procedures based on personal merits and skills and makes the modernization of public management feasible.
- **Effective and pragmatic anticorruption strategies:** The implementation of strategies to fight corruption have proliferated in practically all developing countries, but they have not achieved much success, perhaps because they are too ambitious in those countries where institutional changes deep draft is not viable. It is known that the

most degrading and critical aspects of corruption are linked to the most important public policies, and that the implementation of digital technologies in the most critical areas can be used to decisively promote strategies to eliminate corruption.
- **Effective public financial management:** The planning and management of public finances is at the core of the transformation of the public sector and the modernization of public management. Effective and efficient management of public resources is essential for societies to advance and nations to achieve their goals and objectives, including the SDGs. The digital transformation of public procurement and financial management systems will support the elimination of corrupt practices and the inappropriate use of increasingly scarce public sector financial resources. However, the fight against corruption must strengthen and depoliticize financial oversight agencies. There is also a need to improve the effectiveness, efficiency, and transparency of public procurement systems.

Research Methodology

The research has been developed from the application of public marketing as a strategy to modernize public management, in response to the new shocking global crises that are being generated in the current disruptive environment.

New approaches have been generated to make the response to these new crises more effective and efficient, proposing the development and implementation of new strategies for the modernization of public management, making use of public marketing models enriched by digital marketing, in a framework of social responsibility and sustainability.

Specifically, the application of public marketing 4.0 has been proposed through the maximization of social satisfaction, through the optimization of social welfare functions, as a basis for prospective planning and crisis management. This problem has been studied by previously reviewing an extensive specific bibliography, analysing the contents of the most revealing works.

The originality of the study is reflected in the proposed approaches, the criteria used, the new conceptualizations and the findings. Among these, we highlight, on the one hand, the use of the social utility functions, as a public marketing instrument to optimize social welfare within the framework of global sustainability, and the proposal of a specific model of social welfare function and the maximization of the equivalent social utility function-index. On the other hand, the determination of a generic model for the management of shocking global crises through scenario planning.

SUMMARY OF THE MAIN PROBLEMS TO BE SOLVED

- Conceptual proposal of new public marketing 4.0, as the optimal combination of traditional and digital marketing, for the development and implementation of new strategies for modernizing public management, within a framework of social responsibility and sustainability.
- Within new public marketing 4.0, application of social welfare functions, setting the objective of maximizing social satisfaction through optimization of social welfare functions, as a basis for prospective planning and crisis management.
- Specification and proposal of an analytical model of a social utility function-index, equivalent to social welfare functions, which allows the transformation of qualitative into quantitative, and with it, mathematical calculation, and optimization.
- Specification and proposal of a generic model for the planning and management of new shocking global crises, through scenario planning.

DISCUSSION: APPLICATION OF PUBLIC MARKETING 4.0 THROUGH SOCIAL UTILITY FUNCTIONS, AND PROPOSAL OF A GENERIC MODEL OF PLANNING AND MANAGEMENT FOR SHOCKING GLOBAL CRISES

Public Marketing 4.0: The Optimal Combination of Traditional and Digital Marketing with New Disruptive Technologies

Marketing 4.0 can be considered as the current stage in the evolution of the concept, in which traditional marketing approaches are optimally combined with new digital marketing approaches. The goal is to gain the trust and support of customers using online and offline channels.

In the field of a digital environment of exponential change, public and private entities have to actively interact with social networks, resorting to the use of new digital tools, applying data science methodologies and technologies to their management to make it more effective, efficient and agile, but maintaining a good part of the traditional management and marketing models, mixing the old and new forms of the traditional and the digital (Kuazaqui & Lisboa, 2019).

Traditional marketing focuses its attention on the product or service and the sales action, for which it influences the consumer using advertising with the main objective of selling. It uses traditional media such as television, radio or the newspaper, and takes place in an offline environment. Since you are trying to reach a mass audience, you need a larger budget due to the high cost of these media. Communication is one-way since there is no interaction between the parties, although companies can respond to consumers if they consider it necessary. Furthermore, there is great difficulty in being able to measure the results obtained in this approach.

In the case of digital marketing, the so-called digital media (search engines, email, social networks, blogs, ...) are used through online channels, concentrating their focus on the interests, feelings and tastes of consumers. Digital marketing tries to attract people on a voluntary basis, offering them entertainment, education, and valuable content, developing a two-way communication, direct, simple and in real time with the target audience, facilitating feedback. Social networks are the main tool in its performance, eliminating demographic and geographical barriers, facilitating communication efficiency and promoting innovation. In addition, the planning, management, and control of marketing actions are much more agile, allowing a better adaptation of the campaigns, a greater effectiveness of them with reduced costs, and being able to know and measure the results effectively and immediately. Finally, online marketing enables customers to obtain information online, helping organizations better understand them and create more efficient actions (Başyazıcıoğlu & Karamustafa, 2018).

Marketing 4.0 is integrating online channels with offline ones, trying to hybridize the best of both areas, the immediacy and intimacy of online channels with the differentiating force that offline actions represent. Through this so-called omnichannel strategy, it is expected to obtain a transparent and consistent experience. In addition, the effectiveness of these new relationships is complemented by machine-to-machine connections through artificial intelligence, improving marketing productivity and facilitating person-to-person connection to improve transparency, engagement, and consumer loyalty.

The evolution of public marketing toward public marketing 4.0 through the inclusion of digital marketing and the use of new strategies, methodologies and tools, represents an immense potential to catalyse the modernization of public management. The transformation of the public sector requires maximizing the effectiveness and efficiency in meeting the needs of citizens, offering appropriate services and programs for this, which implies knowing very well what the target audience needs and if the desired added value is being generated. The correct implementation of the marketing-mix tools offered by public marketing 4.0 will allow effective and permanent two-way communication, helping to achieve interest, interaction and greater credibility of public entities on the part of citizens. Public management must adapt to the use of digital media, participating in the new online communication channels, but without abandoning the traditional channels that are still significant to connect with an important part of the target audience.

In any case, the maxim that must prevail in the transformation processes must be a "governments at the service of the citizens" and not "citizens at the service of the governments". Public marketing 4.0 must guarantee the democratic strengthening of societies, avoiding becoming political-electoral marketing, and avoiding that societies end up evolving toward Orwellian models with a totalitarian profile, as Orwell describes in his dystopian fiction novel, in which it is manipulated information, and massive surveillance develops along with political and social repression (Orwell, 2012).

Applying Public Marketing 4.0: Social Welfare Functions

Well-being and social satisfaction are the set of aspects that allow satisfying the needs and quality of life of the members of a society (Sen & Naussbaum, 1998). Social well-being is a not directly observable condition, which must be interpreted and measured through behaviours, judgments and opinions that reflect the level of social satisfaction, that allow comparisons from one place to another, and from one moment to another. It has a very significant load of subjectivity of the individuals that make up a community, although it is correlated with certain social, economic, and environmental factors. However, when deciding what is beneficial or harmful for an individual, the ultimate criterion is the autonomy of their own preferences. Given that the preferences of individuals are often irrational and do not follow the principle of maximizing subjective well-being, preferences revealed by behaviour should not be considered true preferences (Harsanyi, 1955).

One of the most significant concepts that should be incorporated in the analysis of welfare and social satisfaction is that of Pareto Optimum. This principle determines that situations are efficient, if when there is a change in that situation, someone benefits without harming anyone (Pareto, 1938). According to this principle, an allocation of resources is more efficient than any other if the parties involved are at least in the same condition as before and at least one of them is in a better situation than they were initially. Pareto adds that if the welfare (utility) of one individual increases without decreasing that of another, the social welfare of the group of individuals increases, if the rest of the aspects that affect it remain constant. This social welfare would not increase when the situation of one and the other, that is, of the group of individuals, evolves in a divergent way to any change in the economy. And finally, Pareto concludes that in an economy in which individuals are endowed with a certain allocation of goods and services, the rationality in the decisions of individuals would necessarily lead them toward equilibrium positions of exchanges that would be optimal states.

Traditionally, three different approaches have been proposed to specify and determine the social welfare functions that allow optimizing the satisfaction of social groups.

- **Welfare economics approach:** Basically, it is based on the identification of welfare with individual wealth using as an implicit reasoning that the more wealth increases, the more happiness increases. Since the level of wealth can be quantified, the approach would involve using this quantification to measure the degree of happiness and, by extension, the level of well-being. In this way, it would be appropriate to use the Gross Domestic Product per capita or more elaborate composite indicators, such as the Human Development Index.

    However, in the definition of the concept of well-being, both objective and subjective elements intervene, many of which are

not very rigorous to quantify them exclusively in monetary terms. Furthermore, not all values measurable in monetary terms affect welfare in the same direction, nor could they be accepted without considering the disruptive effect of price changes. Therefore, this approach can only be used with a very general character in the measurement of social welfare, although it is evident that the economic component has less and less representativeness as the income level of individuals increases. It is necessary to include other aspects related to the social and environmental field.
- **Measurement of social well-being with social indicators:** The approach of social indicators as an instrument for measuring social well-being is because it is a multidimensional concept. It is necessary to decompose it into its component elements, so that its integration covers the entire concept, assigning to each decomposed element a series of representative sub-indicators of its social involvement, which meet certain conditions of overall synergy.

  In 1969, the US Department of Health, Education and Welfare initially defined social indicators as a statistic of direct normative interest, which facilitates comprehensive and balanced judgments about the condition of the main aspects of a society. Subsequently, the UN considered them summary series related to the status and trends of living conditions and the availability and performance of social services. Finally, the OECD determined that a social indicator is a direct and valid measure that reveals levels and changes over time in a fundamental social concern (Pena-Trapero, 2009).
- **Approach to social utility functions:** Social welfare is closely related to the satisfaction of human needs, both individual and collective. It follows that to measure social welfare, the measure of the degree of utility provided by the goods and services made available to individuals and society can be used, that is, to use social utility functions.

  In the framework of the Pareto approach, the determination of alternative resource allocations that represent the Pareto optimum will be in the maximum utility curve that represents the maximum possible social welfare, which will allow to trace the social welfare functions (Salvatore, 1992). In this way, a social welfare function establishes the different combinations that pose the same level of satisfaction or social welfare.

As a starting point, the utilitarian social welfare function would be the sum of the utility functions of all the individuals belonging to that social group:

$$W = U1 + U2 + \cdots + Un$$

According to this initial approach, the relative weight of the utility of everyone would be the same, regardless of the situation of each one, an issue that provides a democratic approach that, however, does not adapt to a social reality in which individual and leadership differences are clear. The conceptual difficulties are many more. Utility is not the only source of well-being, well-being is not always what people seek, and their success and desires cannot be evaluated exclusively in terms of their well-being (Sen, 2000). However, for utilitarianism, well-being is associated with the maximization of a social utility that results from the simple sum of individual utilities, regardless of the way in which they are distributed among the different members of society. The application of utilitarianism has been based on the arbitrary assumption that if two people have the same demand function, they must obtain the same level of utility both from the consumption of the same goods and services and from the same level of income.

Within this approach, the proposal made by Arrow is noteworthy in which the participating agents will express the values they assign to social situations through an ordering of preferences, said values being the relevant information when making social welfare judgments, trying not to include interpersonal comparisons of utility and cardinal measures of utility (Arrow, 1963).

From our point of view, satisfaction and social well-being should be interpreted by means of metrics that admit interpersonal comparisons, which are based on the functioning, capacities and specificities of individuals. This approach reflects the freedom of human behaviour in the processes of choice, as well as the diversity and reality of the phenomenon of social inequalities. When the problem of maximizing social welfare is posed, it refers to the maximum for society. However, societies are made up of multitudes of individuals, and the problem has always been to determine how to maximize satisfaction or social well-being for all members of society (Case et al, 2012). In this regard, Pareto's reflection on these two ideas can shed light.

First, if a situation is not optimal, we will be sure that none of the individuals involved will be interested in remaining in it. Second, an economically optimal situation is not necessarily socially desirable, and on the practical level there is room for the no-choice approach.

From a microeconomic point of view, utility functions have played a very important role in the analysis at the individual level of consumer demand. Based on the axiom of ordered preferences, according to which any set of alternative consumption situations can be ordered following a coherent and unique order of ascending preferences. With this, it is possible to obtain ordinal measures of the utility provided by a set of goods and services, at least at the theoretical level of individual utility. This measurement requires accepting that ordered increases can be obtained that would not be empirically observable, for which it is necessary to implement a stochastic measure of utility in relation to risk acceptance, provided that an individual faces uncertain results and their consumption probabilities are valued.

A Model of Social Welfare Function

In the late 1930s, Bergson proposed a social welfare function like individual utility functions, with an approach later completed by Samuelson. The approach constructed social indifference curves that were like indifference curves for individual consumption, from these individual utility functions. In the development of these functions of social utility, it seems necessary to integrate the approaches of the economy of well-being, the development of social indicators today strongly linked to the paradigm of sustainable development and its objectives, as well as the approaches of utilitarianism.

According to the Bergson-Samuelson approach, a social welfare function is a qualitative mathematical construction that considers welfare derived from a given set of individual preferences, assigning social utility values to possible reasonable alternative states of feasible associations of its economic system, and associating a social preference to each possible conformation of individual preferences (Kakwani & Son, 2016; Sen, 2018). This function represents the possible patterns of collective choice and the alternative social states of allocation, trying to achieve the optimal allocation of resources based on the preferences of the individuals of that society with respect to collective decisions.

Collective decision models, currently linked to interactive models of democracy (Brill, 2019), attempt to obtain valid and robust criteria for the aggregation of individual preferences and transform them into social preferences. Analytically, the utility level function is obtained from the problem of maximizing the social welfare function W_{Social}, which is qualitative, defined by the integration and interaction of individual utilities Ui, i from 1 to the n members of society:

Maximise $W_{Social}\left(U1, U2, \ldots Un\right)$ subject to restrictions minimum individual utility, feasibility of allocations and availability of resources.

In 2015 the UN established the SDGs within 2030 Agenda. Since then, sustainability has become an important aspect of management (Rosati & Diniz, 2019), which requires that the public and private sectors implicitly carry out the critical evaluation of their activities in economic, social and environmental terms (Yun et al, 2019), including responsibility and the ethics of social behaviour as well as the new requirements of citizens in the welfare economy. The SDGs, as a plan for global development, demand the interconnection of all social sectors as key development actors, and an unprecedented level of cooperation and collaboration between civil society, business, government, NGOs, foundations and others for their achievement. Partnership and collaboration between the social sectors have become an essential paradigm of sustainable development (Stibbe et al, 2018), perhaps in response to the limitations of traditional methods of development led by states.

On the other hand, scenario planning is a tool that involves generating a series of scenarios described in depth, each of which talks about a possible different future for society, and considers how each different future could influence decision-making of global crisis management (Sayers, 2011). The focus of the strategy is on decision making and assumes that the future that will follow will be the result of decisions made in the present. It is therefore a sequential linear approach, but it assumes that in complex situations the future is unpredictable, changes rapidly, and that decision-making is strengthened by taking a more open and flexible approach to the future.

The integration into a joint model of strategic intelligence management, within the framework of public marketing 4.0 and that uses the maximized social utility functions as well as the objectives, variables, and indicators of sustainable development, is considered an excellent basis to assume the prospective planning of response scenarios to a possible shocking global crisis, and for its management.

Strategic Intelligence Management of Shocking Global Crisis: Maximization of the Social Utility Index-Function

Before proposing such a model, it is necessary to assume a systemic conception of sustainability and sustainable development (Gallopín, 2003), and the deployment of the 17 SDGs and 169 objectives associated with them in measurable and integral composite indicators (Lorenzo, 2020). Then, the strategic intelligence management model for both prospective planning and management of possible shocking global crises would be based on the maximization of the utility index-functions that would use the quantitative (and not qualitative) variables of the deployment of the SDGs:

$$Max\{W_S(U_1, U_2, \ldots, U_n)\} \approx Max\{U_S\} = Max\left\{\sum_{i=1}^{n} Prob(U_i) * U_i\right\}$$

$$= Max\left\{\sum_{j=1}^{169} p_j * f(SDG)_j\right\}$$

where W_S is a qualitative social welfare function, U_S is the quantitative individual utility function, $Prob(U_i)$ represent the individual probability of acceptance of consumer risk in an environment of uncertainty based on the functioning, capacities and specificities of individuals, p_j is the weighting index for each objective and $f(SDG)$ is a function of the composite indications of the deployment of each objective.

This optimization is subject to the following restrictions:

$$U_i \geq K \text{ (minimal quality of life restriction)}$$

$Max\left\{\sum_{j=1}^{169} p_j * f(SDG)_j\right\}$ involves the resource allocation, which must be feasible and adjusted to the available resources.

To apply this model, it is necessary to know the economic and social impact of each alternative. The CBA methodology (Nas, 2016; Boardman et al, 2018) estimates and adds the equivalent monetary value for the social group studied, of the benefits and nonexclusively monetary costs of different alternative public, private or public–private projects. To reach a conclusion about its suitability, all aspects of the initiative, both positive and negative, must be expressed in terms of a common unity. Therefore, both financial and social costs and profits, headings must transform their units of measurement and be estimated as "opportunity income" and "opportunity costs". The equivalent monetary value is based on the information derived from the real choices made in the markets by consumers and producers, analysing historical data and future estimates of the demand and supply of goods and services affected by the analysed alternative, updated over time using a discount rate.

Determining a Planning and Management Generic Model of Shocking Global Crisis: Scenario Planning

The strategic intelligence management model, within the framework of public marketing 4.0, must permanently monitor the indicators that comprehensively explain the behaviour of the possible triggers that generate vulnerability. As can be seen in Figure 16.4, the potential impact of the

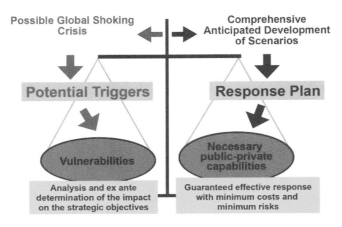

FIGURE 16.4
Scenario planning: generating alternative response plans. (From own elaboration.)

state and the evolution of the different crisis triggers and the potential impact on the established strategic objectives (SDGs) must be analysed and determined ex ante, evaluating direct preferences and behaviours of individuals by applying CBA models.

In all cases in which there is a possible shocking crisis, comprehensively understood but prioritized by the estimated level of probability of occurrence, all possible prospective scenarios would be fully generated, with the help of Big Data, Business Analytics, Machine Learning, Artificial Intelligence, Simulation Systems, etc. Each scenario will imply the determination of a specific strategic response plan that will identify the public–private capacities that are necessary to guarantee an effective response, based on the maximized social utility function, and generating the maximum difference between social benefits and costs, with minimal risk of error.

Thus, action strategies and capacity development for strategic anticipation will depend on the dynamic balance between the comprehensive social benefits of the consequences of each possible impacting global crisis and the comprehensive social cost of generating these capabilities. At the same time, direct action plans and the deployment of these capacities must be developed, which make it possible to convert the most desirable scenarios into those most likely to become reality. It is considered that the proposed model improves the models that are being used, which do not guarantee the maximization of social utility.

CONCLUSION

- The situation generated by the pandemic makes the governance of many countries difficult, whose governments have proposed late, improvised, ineffective and insufficient responses to the pandemic crisis. This high revealed vulnerability shows the existence of a very threatening and disturbing environment, which will foreseeably continue to generate shocking global crises generated by future pandemics or other triggers.
- To face this high social vulnerability, it seems necessary to urgently modernize public management from the digital transformation of their organizations whose main conceptual framework of reference

is considered public marketing, and in which political ideology should be less and less present.
- The advantages of applying public marketing are multiplied by concentrating the management effort on the social needs and expectations of citizens: public resources are managed to meet and respond to the problems and demands of citizens, the supply services public organizations permanently improve their objectives and subjective quality by increasing the level of social satisfaction, public entities increase their democratic legitimacy, population sectors with particular characteristics and in a situation of social disadvantage can be approached with equity, etc.
- In addition to the application of tools such as economic intelligence, strategic anticipation, prospective planning and Data Science, social utility functions have been studied as an instrument of public marketing for the optimization of social welfare. This approach integrates most of the significant variables of the social, economic, and ecological dimensions of global sustainability, transforming the traditional models of public marketing into public marketing 4.0, integrating the culture of lean and agile management, and ensuring a more effective, efficient and democratic public management.
- A model for maximizing social welfare has been proposed, optimizing the functions-index of social utility based on the SDGs of the UN 2030 Agenda, and the use of CBA. A planning model of global crisis scenarios and a general action scheme adjusted to the defined framework have been determined, which contextualizes the importance of the proposed approach.

LESSONS LEARNED

- Managing the shocking global crises produced because of the current disruptive environment requires the implementation of new methodologies, such as strategic anticipation, prospective planning and strategic intelligence. This new approach should guide, and not the other way around, the application of the most appropriate digital technologies that allow the development of proactive, efficient and sustainable response plans, and a more effective crisis management with minimal risks.

- The research has been developed from the application of the new response strategies to the new global crises, and has shown that marketing, international cooperation and public–private collaboration, are fundamental, making necessary a global standardization of actions.
- For the development of this new approach, we have learned the convenience to modernize public management with the use of public marketing 4.0 models, within a framework of social responsibility and sustainability. And within public marketing 4.0, social welfare functions have been considered as an instrument, specifying the proposal as an application of a strategic intelligence management model related to the global crisis studied, which allows maximizing social satisfaction by optimizing welfare functions, as well as a scenario planning and crisis management model.
- The analysed field suggests a clear evolution toward the development of strategic, integrated and digital management structures and models, which raises interesting and leafy lines of interdisciplinary research. Among them, on the one hand, it is necessary to develop the specific framework for the application of public marketing 4.0 in each sector of public administration, with empirical criteria, specifying the strategic, organizational and functional aspects. On the other hand, it is also necessary to develop the strategic intelligence model proposed, which allows orienting decisions in the field of strategic anticipation, prospective planning and crisis management shocking global.

ADDITIONAL CONTENT

Key Questions

Why is the response being raised to the shocking global crises generated in the new disruptive environment (such as the COVID-19 pandemic) being late, improvised, ineffective and insufficient?

Why is public marketing an excellent framework to promote the modernization of public management in a socially responsible and sustainable approach?

What are the most significant aspects raised by Post-New Public Management and the Governance of the Digital Age?

What are the differences between social welfare functions, social utility functions and social utility index-functions?

What is Cost-Benefit Analysis?

How does the proposed model for planning and managing shocking global crises work?

Key Definitions

Public sector marketing is a two-way catalyst that increases and accelerates the exchange processes between users of public services and the agents responsible for their offer, planning and design.
(Larry Coffman, 1986)

Social marketing differing from other areas of marketing only with respect to the objectives of the marketer and his or her organization. Social marketing seeks to influence social behaviours not to benefit the marketer, but to benefit the target audience and the general society.
(Kotler & Andreasen, 1991)

Corporate social responsibility is the continuing commitment by business to behave ethically and contribute to economic development while improving the quality of life of the workforce and their families as well as of the local community and society at large.
(World Business Council for Sustainable Development, 1999)

Crisis management is the process by which an organization deals with a disruptive and unexpected event that threatens to harm the organization or its stakeholders.
(Jonathan Bundy et al, 2017)

Sustainability is the integration of environmental health, social equity, and economic vitality to create thriving, healthy, diverse and resilient communities for this generation and generations to come. The practice of sustainability recognizes how these issues are interconnected and requires a systems approach and an acknowledgement of complexity.
(UCLA Sustainability Committee, 2020)

Key Author Mentions

From the perspective of history, the mighty force of social change is the unproclaimed power of society's citizens to challenge and withdraw legitimacy from any or all of society's institutions.
(Willis Harman, 1976)

Sustainable development, rather than representing a major theoretical breakthrough, is very much subsumed under the dominant economic paradigm. As with development, the meanings, practices, and policies of sustainable development continue to be informed by colonial thought, resulting in the disempowerment of most of the world's populations, especially rural populations in the Third World. Discourses of sustainable development are also based on a unitary system of knowledge and, despite its claims of accepting plurality, there is a danger of marginalizing or co-opting traditional knowledges to the detriment of communities who depend on the land for their survival.

(Bobby Banerjee, 2003)

Marketing is much more than advertising; it is about knowing your customers, partners, and competitors; segmenting targeting and positioning; communicating persuasively; innovation and launching new services and programs; developing effective delivery channels; forming partnerships and strategic alliances; performance management and pricing/cost recovery. Marketing turns out to be the best planning platform for a public agency that wants to meet citizens' needs and deliver real value. In the private sector, marketing's mantra is customer value and satisfaction. In the public sector, it is citizen value and satisfaction.

(Kotler & Lee, 2007)

We need to turn the recovery into a real opportunity to do things right for the future.

(UN Secretary-General António Guterres, 2020)

CREDIT AUTHORS STATEMENT

Manuel Antonio Fernández-Villacañas Marín: Term, Conceptualization, Methodology, Validation, Investigation, Resources, Writing-original draft preparation, Project administration, Writing-review and editing, Supervision.

Ignacio Fernández-Villacañas Marcos: Methodology, Formal analysis, Validation, Investigation, Resources, Writing-review & editing, Visualization.

REFERENCES

Antoniadis, I., Stathopoulou, M. & Trivellas, P. (2019): Public Sector Marketing in a Period of Crisis: Perceptions and Challenges for the Public Sector Managers. In: D. Sakas & D. Nasiopoulos (Eds.): *Strategic Innovative Marketing*. IC-SIM 2017, Proceedings in Business and Economics, Springer.

Arellano, D. & Cabrero, E. (2005): La Nueva Gestión Pública y su teoría de la organización: ¿son argumentos antiliberales? Justicia y equidad en el debate organizacional público. *Gestión y Política Pública*, Volume 14 Issue 3. México DF (México): Centro de Investigaciones y Docencia Económicas.

Aronsson, H., Abrahamsson, M. & Spens, K. (2011): Developing lean and agile health care supply chains. *Supply Chain Management, International Journal*, Volume 16 Issue 3.

Arrow, K.J. (1963): *Social Choice and Individual Values* (2nd Ed.). New York: Wiley.

Balbi, E.R. (2015): *Anticipación estratégica el mayor desafío para la prevención y gestión de riesgos*. http://www.academia.edu

Banerjee, S.B. (2003): Who sustains whose development? Sustainable development and the reinvention of nature. *Organizations Studies*, Volume 24 Issue 1.

Barragán, J., Guerra, P. & Vilalpando, P. (2017): Marketing and corporate social responsibility: Proposal of strategic business model. *Daena: International Journal of Good Conscience*, Volume 12 Issue 1, March.

Başyazıcıoğlu, H. & Karamustafa, K. (2018): Marketing 4.0: Impacts of technological developments on marketing activities. *Kırıkkale University Journal of Social Sciences (KUJSS)*, Volume 8 Issue 2, July.

Berger, G. (1964): *Phénoménologie du temps et prospective*. Paris, France: PUF.

Boardman, A.E., Greenberg, D.H., Vining, A.R. & Weimer, D.L. (2018): *Cost-Benefit Analysis. Concepts and Practice* (5th Ed.). Cambridge University Press.

Bratianu, C. (2017): Strategic Thinking in Turbulent Times. In: C. Bratianu, A.M. Dima & S. Hadad (Eds.): *Proceedings of the 11th IC on Business Excellence*. Bucharest University of Economic Studies.

Brill, M. (2019): Interactive Democracy: New Challenges for Social Choice Theory. In: J.F. Laslier, H. Moulin, M. Sanver & W. Zwicker (Eds.): *The Future of Economic Design*. Studies in Economic Design, Cham, Switzerland: Springer.

Bundy, J., Pfarrer, M.D., Short, C.E. & Coombs, W.T. (2017): Crises and crisis management: Integration, interpretation, and research development. *Journal of Management*, Volume 43 Issue 6.

Case, K., Fair, R. & Oster, S. (2012): *Principios de Microeconomía* (10th Ed.). México: Pearson Educación.

Chias, J. (2011): Marketing público en España y Latinoamérica, algunas consideraciones. In: P. Kotler (Ed.): *Marketing en el sector público*. México: Fondo de Cultura Económica.

Christensen, T. & Laegreid, P. (2007a): Reformas Post Nueva Gestión Pública. Tendencias empíricas y retos académicos. *Gestión y Política Pública*, Volume 16 Issue 2. México DF (México): Centro de Investigaciones y Docencia Económicas.

Christensen, T. & Laegreid, P. (2007b): The whole of government approach to public sector reform. *Public Administration Review*, Volume 67 Issue 6, Nov/Dec.

Christensen, T. & Laegreid, P. (2008): NPM and beyond – structure, culture, and demography. *International Review of Administrative Science*, Volume 74 Issue 1.

Coffman, L.L. (1986): *Public Sector Marketing: A guide for practitioners*. New Jersey: John Wiley & Sons.
Commonwealth Secretariat (2016): *Key Principles of Public Sector Reforms. Case Studies and Frameworks*. UK: The Commonwealth.
Crompton, J.L. & Lamb, Ch.W. Jr. (1986): Marketing Government and Social Services (1st Ed.): *Wiley Series on Marketing Management*. New York: Wiley.
De Vries, M. & Nemec, J. (2013): Public sector reform: An overview of recent literature and research on NPM and alternative paths. *International Journal of Public Sector Management*, Volume 26 Issue 1
Diefenbach, Th. (2009): New Public Management in Public Sector Organizations: The Dark Sides of Managerialistic 'Enlightenment'. *Public administration*, Volume 87 Issue 4, Wiley Online Library.
Dunleavy, P., Margetts, H, Bastow, S. & Tinkler, J. (2006): New Public Management is Dead: Long Live Digital-Era Governance. *Journal of Public Administration Research and Theory*, Volume 16 Issue 3.
Escourido. M. (2017): *El marketing de ciudades como una herramienta de gestión pública local: una aplicación al caso de As Pontes de García Rodríguez (A Coruña)*. Doctoral Thesis, Facultad de Economía y Empresa, Universidad de A Coruña (Spain).
Fernández-Villacañas, M.A. (2020a): Inteligencia estratégica logística frente a crisis globales emergentes. Anticipación estratégica y planificación prospectiva. *VII Simposio Online de Logística y Competitividad*. Medellín (Colombia): High Logistics Group.
Fernández-Villacañas, M.A. (2020b): Strategic intelligence and decision process: Integrated approach in an exponential digital environment. In: K. Sandhu (Ed.): *Leadership, Management, and Adoption Techniques for Digital Service Innovation*. Hershey: IGI-GLOBAL.
Gallopín, G. (2003): *Sostenibilidad y desarrollo sostenible: Un enfoque sistémico*. División de Desarrollo Sostenible y Asentamientos Humanos. Chile: UN CEPAL.
García Sánchez, I.M. (2007): La nueva gestión pública: evolución y tendencias. *Presupuesto y Gasto Público*, Volume 47, pages 37–64. Madrid (Spain).
Gavigan, J.P. (2001): A Practical Guide to Regional Foresight. *FOREN – Foresight for Regional Development Network*. Brussels: European Commission, Strata Programme.
Gilad, B. (2004): *Early Warning: Using Competitive Intelligence to Anticipate Market Shifts, Control Risk, and Create Powerful Strategies*. New York, NY: AMACOM.
Goldfinch, S. & Wallis, J. (2010): Two Myths of Convergence in Public Management Reform. *Public Administration*, Volume 88 Issue 4, December.
Haarhaus, T. & Liening, A. (2020): Building dynamic capabilities to cope with environmental uncertainty: The role of strategic foresight. *Technological Forecasting and Social Change*, Volume 155, June.
Harman, W.W. (1976): *An Incomplete Guide to the Future*. San Francisco, CA: San Francisco Books.
Harsanyi, J. (1955): Cardinal welfare, individualistic ethics, and interpersonal comparisons of utility. *The Journal of Political Economy*, Volume 63 Number 4, August.
Hood, C. (1991): A public management for all seasons? *Public Administration*, Volume 69 Issue 1.
Hood, C. (1995): The new public management in the 1980's: Variations on a theme. *Accounting Organizations and Society*, Volume 20 Number U3. Pergamon (UK).
Kakwani, N. & Son, H.H. (2016): *Social Welfare Functions and Development. Measurement and Policy Applications*. UK: Palgrave Macmillan.

Kennedy, A.M. & Parsons, A. (2014): Social engineering and social marketing: Why is one "good" and the other "bad"? *Journal of Social Marketing*, Volume 4 Issue 3.

Kotler, Ph. & Andreasen, A. (1991): *Strategic Marketing for Nonprofit Organizations* (4th Ed.). Englewood Cliffs, NJ: Prentice-Hall.

Kotler, Ph. & Lee, N. (2007): *Marketing in the Public Sector: A Roadmap for Improved Performance*. Upper Saddle River, NJ: Pearson Education, Inc.

Kuazaqui, E., & Lisboa, T.C. (2019): Marketing: The evolution of digital marketing. *Archives of Business Research*, Volume 7 Issue 9.

Leeuw, F.L. (1996): Performance auditing, new public management and performance improvement: Questions and answers. *Accounting, Auditing & Accountability Journal*, Volume 9 Issue 2.

Liebowitz, J. (2006): *Strategic Intelligence: Business Intelligence, Competitive Intelligence, and Knowledge Management*. Boca Raton, FL: Auerbach Publications.

Lindberg, K, Czarniawska, B. & Solli, R. (2015): After NPM? *Scandinavian Journal of Public Administration*, Volume 19 Issue 2. Gothenburg (Sweden): School of Public Administration.

Lorenzo, C. (2020): *Medición de los Objetivos de Desarrollo Sostenible en la Unión Europea a través de indicadores compuestos*. Documentos de Trabajo nº especial (2ª época). Fundación Carolina, Madrid (Spain).

McAuley, A. (2014): Reflections on a decade in social marketing. *Journal of Social Marketing*, Volume 4 Issue 1.

Mufudza, T. (2018): Dynamic Strategy in a Turbulent Business Environment. In: *Strategic Management. A Dynamic View*. London, UK: IntechOpen

Nas, T.F. (2016): *Cost-Benefit Analysis. Theory and Application* (2nd Ed.): Maryland: Lexington Books.

Orwell, G. (2012): *1984* (1st Ed.). Penguin Classics, Penguin Books.

Palacios, J.M. (2018): The role of strategic intelligence in the post-everything age. *The International Journal of Intelligence, Security, and Public Affairs*, Volume 20 Issue 3.

Pareto, W. (1938): *Manual of Political Economy. A Critical and Variorum Edition*. Translation of original book in English of 1938. In: A. Montesano, A. Zanni, L. Bruni, J.S. Chipman & M. McLure (Eds.) (2014). Oxford, UK: Oxford University Press.

Pena-Trapero, B. (2009): La medición del Bienestar Social: una revisión crítica. *Estudios de Economía Aplicada*, Volume 27 Issue 2.

Pollit, C. (2009): Editorial: public service quality — between everything and nothing? *International Review of Administrative Sciences*, Volume 75 Issue 3.

Puig, T. (2004): *Marketing de servicios para administraciones públicas con los ciudadanos. En red, claves y entusiastas*. Seville, Spain: Junta de Andalucía.

Puig, T., Rubio, L. & Serra, A. (1999): El marketing, el marketing de servicios públicos, y la gestión pública. In: C. Losada i Marrodán (Ed.): *¿De burocrátas a gerentes? Las ciencias de la gestión aplicadas a la administración del Estado*, Washington, DC: Banco Interamericano de Desarrollo.

Pykett, J., Jones, R., Welsh, M. & Whitehead, M. (2014): The art of choosing and the politics of social marketing. *Policy Studies*, Volume 35 Issue 2.

Radnor, Z. & Boaden, R. (2008): Does lean enhance public services? Editorial: Lean in public services – Panacea or paradox? *Public Money & Management*, Volume 28 Issue 1, February.

Radnor, Z. & Walley, P. (2008): Learning to walk before we try to run: Adapting lean for the public sector. *Public Money & Management*, Volume 28 Issue 1, February.

Rainey, H.G. (2014): *Understanding and Managing Public Organizations* (5th Ed.). San Francisco, CA: Jossey-Bass.

Rosati, F. & Diniz, L.G. (2019): Addressing the sustainable development goals in sustainability reports: The relationship with institutional factors. *Journal of Cleaner Production*, Volume 215, 1 April.

Rose, A. (2017): *Defining and Measuring Economic Resilience from a Societal, Environmental and Security Perspective*. Singapore: Springer.

Salvatore, D. (1992): *Microeconomía* (4th Ed.). Mexico DF, Mexico: McGraw-Hill.

Saunders, S., Barrington, D. & Sridharan, S. (2015): Redefining social marketing: Beyond behavioural change. *Journal of Social Marketing*, Volume 5 Issue 2.

Sayers, N. (2011): Maximising the effectiveness of a scenario planning process. *Perspectives*, Volume 15 Issue 1.

Schuhly, A., Becker, F. & Klein, F. (2020): *Real Time Strategy: When Strategic Foresight Meets Artificial Intelligence*. Bingley: Emerald Publishing.

Sen, A. (2018): *Collective Choice and Social Welfare*. An Expanded Edition. Cambridge, MA: Harvard University Press.

Sen, A. (2000): *Social Exclusion: Concept, Application, and Scrutiny*. Social Development Papers No. 1. S. McMurrin (Ed.). Office of Environment and Social Development, Asian Development Bank.

Sen, A. & Naussbaum, M.C. (1998): *La Calidad de Vida*. México: The United Nations University, Fondo de Cultura Económica.

Stibbe, D.T., Reid, S. & Gilbert, J. (2018): *Maximising the Impact of Partnerships for the SDGs. A Practical Guide to Partnership Value Creation*. The Partnering Initiative and United Nations Department of Economic and Social Affairs.

UCLA Sustainability Committee (2020): *Charter for the UCLA Sustainability Committee*. Los Angeles, CA: UCLA.

UN Secretary-General António Guterres (2020): *Climate Change and COVID-19: UN Urges Nations to 'Recover Better'*. UN Department of Global Communications.

Vazquez-Bustelo, D. & Avella, L. (2006): Agile manufacturing: Industrial case studies in Spain. *Technovation*, Volume 26 Issue 10.

Vecchiato R. (2015): Strategic planning and organizational flexibility in turbulent environments. *Foresight*, Volume 17 Issue 3.

World Business Council for Sustainable Development (1999): *Corporate Social Responsibility: Meeting Changing Expectations*. Geneva, Italy: World Business Council for Sustainable Development.

World Economic Forum (2020): *The Global Risks Report 2020*. WEF.

World Health Organization (2020): *Climate Change and Human Health. Risks and Responses*. UN.

Yun, G., Yalcin, M.G., Hales, D.N. & Kwon, H.Y. (2019): Interactions in sustainable supply chain management: A framework review. *International Journal of Logistics Management*, Volume 30 No 1.

Zaheer, A., Wei, S., Chong, R., Abdullah, M. & Khan, K.U. (2015): High aiming in public sector marketing: A way forward to boost China's economy. *Mediterranean Journal of Social Sciences*, Volume 6 Issue 6, November.

Section VIII

Advances in Knowledge of Social and Sustainable Marketing: Understanding the Benefits of Sustainability Reporting Practices by Social Enterprises

17

Assessing Sustainable Outcomes of Reporting Practices by Social Enterprises

Judith M. Herbst
Centre for a Waste-Free World, Queensland University of Technology, Gardens Point Campus, Australia

CONTENTS

Learning Objectives 477
Introduction 478
Literature Review 480
Methodology 484
Research Findings 486
    Internal Reporting 490
    External Reporting 492
    Integrated Reporting 495
Discussion 497
    Theoretical and Practical Contributions 498
Conclusion 499
Lessons Learned 500
Discussion Questions 500
Exercise 500
References 502

LEARNING OBJECTIVES

After reading the chapter, the reader should be able to:

- Analyze different styles of reporting practices by profit-for-purpose organizations, and understand the consequent need for sharing the

'pulse' of a company to show how it is faring across multiple areas, including financial, natural, human, social, and manufactured capital in meeting its purpose
- Study the concept of integrated reporting and scope of this socially responsible and sustainable approach, illustrated by case studies, to utilize as a framework for optimizing the capabilities of companies and to meet societal needs

INTRODUCTION

Social enterprises are classified as hybrid organizations (Doherty, Haugh, & Lyon, 2014; Peattie & Morley, 2008). Positioned on a continuum between for-profits and nonprofits, they are characterized as businesses with "primarily social objectives whose surpluses are principally reinvested for that purpose in the business or in the community, rather than being driven by the need to maximize profit for shareholders and owners" (UK Government, 2011). To maintain commercial viability, a social enterprise may operate within a larger business, affiliate with a for-profit enterprise as a subsidiary, or establish divisions whereby one part functions "for-benefit" and the other part acts "for-profit" (Alter, 2004; Pitta & Kucher, 2009). In the latter instance, each side cooperates for joint survival.

While social enterprises continue to evolve with new legal forms and recent third-party certification schemes (Brewer, 2016), traditional financial metrics are not deemed sufficient to fulfil the array of needs of constituents, spanning donors, business partners, clients, customers, or volunteers in social enterprises (McLeish, 2011). An implication is that there is a relationship between parties, and the organization as the primary actor is supposed to inform the other about its strategies and actions, as well as to provide justification or explanation for its conduct or misconduct which happens through reporting. Additionally, there has been a greater call for measurement and evaluation of performance in social enterprises (Arvidson, Lyon, McKay, & Moro, 2010) to contribute to theory and accounting practice. To be accountable to stakeholders, an organization needs to have a system to capture and convey its performance.

Since the nature of social enterprises involves balancing both social and financial objectives, both of these areas need to be specifically assessed (Tyler III, 1983). By providing this information, it will help constituents to understand how these organizations reconcile their accounts, identify which reporting tools are being utilized, and explicate about returns on investment.

While social enterprises make money as commercial businesses, many of them obtain additional grant funding to conduct social programs, so it is important for them to be accountable to their grantors, donors, and other groups that contribute such resources to demonstrate that investments are producing expected results for those businesses and their intended beneficiaries. Nevertheless, there is no uniform standard in social enterprises for measuring financial and social impact.

As primarily small businesses that are not publicly listed, many are not compelled to disseminate financial reports. Apart from grants that require an acquittal on how funds were spent and associated returns, there are generally no other requirements imposed on social enterprises to track and release data on their results. Few countries subscribe to mandatory reporting despite the link between integrative management of sustainability performance, measurement, and reporting (Schaltegger & Wagner, 2006). Indeed, social enterprises should be able to show they meet the needs and expectations of the multiple entities that they serve (Brown & Forster, 2013; Hasnas, 2013). It is feasible to obtain added value from carrying out this exercise because reporting is a medium to promote how an organization is achieving its cause and can be a way to connect stakeholders with its mission (Jones & Mucha, 2014).

Although there is an increased trend to discuss the plethora of existing tools to capture social and economic performance in the field of social entrepreneurship (Öncer, 2019), or to suggest developing more useful frameworks (Crucke & Decramer, 2016; Luke, 2016), performance reporting in these third sector organizations is still under-researched (Connolly & Kelly, 2011; Ridley-Duff & Bull, 2016). Considering the lack of empirical data, especially evidence grounded in newer types such as integrated reporting (IR) (Berg & Jensen, 2012), the learning objectives of this chapter is to show the results of how multiple cases, social enterprises, disseminate information about their social and financial performance to their stakeholders, and to discover how their practices may affect their sustainable missions.

This chapter first reviews the body of literature on measurement and performance reporting within social enterprises. Next, it focuses on the tools for accountability that are utilized by seven heterogeneous organizations dedicated to the triple bottom line (Elkington, 1997). Then themes are uncovered from analyzing the data, noting similarities, and differences across these organizations. Lastly, theoretical and practical implications are drawn, demonstrating that by engaging in a more comprehensive style of performance report instead of issuing a standalone financial, social, or sustainability account, it can bring greater benefits for the social enterprises and their constituents over other available reporting methods.

LITERATURE REVIEW

Conventional methods for conducting performance in social enterprises are inadequate. First, they do not capture the value of commitments to achieve both social and commercial objectives. Second, it is difficult to capture this information using traditional means because performance can include: "any metric that organizations, or those who evaluate them, use to ascertain progress toward stated goals and objectives, be it behavioral, skill based, or assessment of outcomes" (Lane & Casile, 2011, p. 241). Third, auditing with older, restricted approaches is rife with challenges because there is no concrete method to identify and quantify intangible benefits of company operations (Tyler III, 1983), or to quantitatively measure data that reveal macro-level performance about products and programs. Of the various frameworks that measure financial and nonfinancial returns, some of the most popular types that show complementary microlevel performance at the market and customer level include the practice-based social return on investment (SROI), social accounting and audit (SAA), and the balanced scorecard, a type that integrates social and economic aspects for blended value (Öncer, 2019).

Notwithstanding that multiple methods exist to evaluate overall performance, the most common global frameworks utilized in third sector organizations (Nicholls, 2009), particularly in social enterprises, are SROI and SAA (Ridley-Duff & Bull, 2016). SROI is a top-down and system-based approach to evaluate outputs, outcomes, and impacts. Adapted from the concept of cost-benefit analysis, it attempts to translate significant social impacts such as assessing the value of human resources using proxies to derive quantifiable value to inform interested social investors and public representatives. The SAA framework, alternatively, is a bottom-up and mainly qualitative method to assess an organization's triple bottom line. It entails gathering data from within and outside of an organization's operations by working in consultation with stakeholders to prepare a report and a governance statement that are subject to verification by independent auditors. The SAA is seen as a helpful tool for managerial planning and guiding improvements besides being useful for stakeholder engagement. Yet, SROI and SAA have limitations.

SROI cannot calculate the nonfinancial achievements of an organization because they cannot be quantified (Arvidson et al. 2010), and it omits giving an internal perspective of a company's activities. Although SAA, on the other hand, can express a narrative of organizational performance, it offers insufficient financial data for social investors. These frameworks therefore fail to convey a comprehensive account of value to conform to public (European Commission, 2011) and institutional pressure to measure and account for social impact in social enterprises (Vik, 2017).

Newer reporting practices offer a wider scope of value created by organizations. These tools are being utilized to holistically account for pertinent nonmonetized contributions and financial returns. They factor in human, social, natural, manufactured, and financial sources of capital (Parkin, Johnson, Buckland, & White, 2004) described in Table 17.1, which are imperative to exist. They also recognize measures beyond gross domestic product (GDP), allowing companies to report on wider areas of their business operations in one complete document. It is relevant to social enterprises because these types of organizations have broader goals than solely turning a profit.

IR (International Integrated Reporting Council, 2011) aligns with this inclusive approach to reporting. It is garnering support across small and large corporate entities from various sectors as a way to explain how a company creates value across multiple dimensions through a whole-of-system analysis (CPA Australia, 2016). This method offers the ability to explain how a company is commercially, socially, or environmentally resilient, and it can guide investors in learning about the benefits to society that derive from an organization's operations (Adams & Simnett, 2011) by providing insight into whether a business is meeting its short- and long-term targets. Thus, IR can relay how a company's business model and strategic direction create value for stakeholders over time (KPMG, 2012) by giving a succinct narrative that factors in an organization's resources and relationships while making future projections.

Although researchers contend IR is progressive (Watson, 2011), an evolution in corporate communication (Chartered Institute of Management Accountants, 2015) because it presents value and reduces risk for

TABLE 17.1

Overview of the Five Capitals Model

Capitals	Descriptions	Examples of Stocks of Capitals	Examples of Flows From Healthy Stocks
Human	Motivation, energy, and the ability to form relationships as well as the health and intellect of individuals	Health, spirituality, knowledge, and skills	Creativity, happiness, participation, and innovation in work
Social	Groups that bring added value to individuals	Families, communities, schools, workplaces, and government	Justice and inclusion, security and satisfaction in school and work
Natural	Resources and the services that flow from nature	Vegetation and ecological systems in land, sea, and air	Improved energy, climate, food, water, and waste
Manufactured	Materials and built infrastructure from industry	Tools, power supplies, roads, and buildings	Space to live and work as well as access and distribution of real products
Financial	Money that attaches value to products and services and enables buying and selling of the other sources of capital	Stocks, bonds, and cash	Means to value, own or exchange other stocks of capital

Source: Adapted from Parkin et al. (2004, pp. 12, 39).

decision-making (Stubbs & Higgins, 2014), IR is criticized for lacking standards of quality assurance for the five capitals (Cheng et al. 2014; Frias-Aceituno, Rodríguez-Ariza, & Garcia-Sánchez, 2014). The International Integrated Reporting Council (2011) rationalizes that assurance on IR will evolve as the practice continues to unfold. Others contest integrated reports for different reasons (Flower, 2015). They claim it may reveal partial or one-sided viewpoints (Brown & Dillard, 2014; O'Dwyer & Boomsma, 2015), so they believe it does not contribute to integrative management of sustainability (Stacchezzini, Melloni, & Lai, 2016). Yet, IR has received prominence since its introduction (de Villiers, Rinaldi, & Unerman, 2014). Adopters are required to be transparent about what information managers apply to decision-making to reflect their integrative thinking, sustainable actions, and to avoid siloed approaches to measurement and reporting practices (Giovannoni & Fabietti, 2013).

There are only limited systematic forms of evaluation of performance (Arena, Azzone, & Bengo, 2015). In social enterprises, it is important to implement a sound measurement framework to unify social and/or environmental data with economic accounts (Schaltegger & Wagner, 2006) to arrive at blended value to drive competitive advantage (Lodhia, 2014), and it is important to tell a complete story of a company's performance (Mook, Richmond, & Quarter, 2003). Irrespective of which framework is selected by a social enterprise, however, it is generally agreed what is crucial is for an organization to divulge how it is addressing its social impact.

Theoretical underpinnings on performance evaluation refer to the need to show the adequacy of an organization in measuring and comparing its performance against established benchmarks for satisfying its value-laden motives and agendas (Connolly & Kelly, 2011). For this study on performance of social enterprises, stakeholder theory was the most applicable lens to view how these organizations disclose their performance since these organizations carry out activities on behalf of many constituents. It is necessary for a business to not only elucidate how it is acting ethically in reporting on the interdependent parts of its social, political, and economic interests (Deegan & Unerman, 2011), but it should show how it is meeting the needs of governmental agencies, funding bodies, and other key stakeholders. By prioritizing its salient stakeholders according to attributes of power, legitimacy, and urgency (Mitchell, Agle, & Wood, 1997), then an organization can determine how to portray content to account for the internal and external parts of company operations. Those groups perceived as holding higher degrees of influence in power, for instance, are more likely to receive the greatest attention.

Ideally, a performance report should bring enhanced transparency with key stakeholder groups and benefit company managers or directors for conducting more informed decision-making. By adopting a stakeholder orientation to assemble a performance report, it can become a marketing vehicle to forge deeper connections with investors, intermediaries, policymakers, beneficiaries, and members of the public, to uphold credibility or build a company's reputation to engender lasting relationships (Parrot & Tierney, 2012) to fulfill company objectives and open avenues for growth.

METHODOLOGY

Qualitative research was employed to capture the scope of financial and social measurement and reporting practices in social enterprises. Since there is little scholarly research on these organizations devoted to the triple bottom line, the study adhered to a phenomenological paradigm to guide the process. Inductive research (Blaikie, 2009) was employed through a comparative approach of cases to explore theory and practice of how reporting to various stakeholders is conducted and may bring certain outcomes. Multiple case study was appropriate. This method was utilized in prior social enterprise research, and it presents opportunities to capture measurement of social impact using IR by related nonprofit organizations (Adams & Simnett, 2011). Further, it is a pathway to discover holistic forms of accounting by aggregating market value with nonmonetized inputs that are missing in conventional accounting (Mook et al. 2003). Yin (2009) recommends looking at multiple cases over a single case for accumulating a better understanding of real-life events. Additionally, it is more likely to yield generalizable conclusions from replication of data that facilitates corroboration (Eisenhardt & Graebner, 2007).

A purposeful sampling strategy (Yin, 2009) was used to identify and negotiate access to Australian social enterprises that actively work toward achieving sustainable development (United Nations, 2002). Business database research, ABI/INFORM, and credible publications were consulted to find applicable organizations (Kernot & McNeil, 2011). They all adopted strategies to enhance environmental quality and social well-being which was embedded in their core business activities or missions (Parrish, 2010). Descriptions of their business activities and sustainability practices follow.

First, AAC is the parent company of multiple social enterprises including Ashoil and Ashlinen, which both aim to broaden local employment opportunities across the remote Pilbara region of Western Australia (Ashburton Aboriginal Corporation, 2013). Under agreements with a mining company and its associated suppliers (Minerals Council of Australia, n.d.), Ashoil collects used cooking oil and converts it into biodiesel to resell this cleaner fuel for drill and blast operations (Lee, 2013) and to decrease its own greenhouse gas emissions. Ashlinen, alternatively, collects and recycles used mining uniforms for resale or for donation to charity shops to extend the life cycle of textiles (Ashburton Aboriginal Corporation, 2014).

Second, Abbotsford Convent is a historic compound located beside the Yarra River of Melbourne, Victoria. The organization has received international recognition as a thriving arts and cultural precinct (Abbotsford Convent, 2014b). Eleven historic buildings are gradually being restored within an enclave of 6.5 hectares of gardens. A variety of tenants occupy leases and run activities that cater to an array of professional and personal pursuits.

Third, Hepburn Wind is a community-invested wind farm that harvests renewable energy in rural Victoria. It offers carbon offsets and sells clean power through a partnership with an energy reseller, plus it uses its profits to disburse grants through a community fund, especially to start up more renewable energy projects in nearby townships (Hepburn Community Wind Park Co-operative Limited, 2014).

Fourth, Infoxchange is as an ICT service provider. Originally it was set up to connect databases of government agencies to place homeless people in temporary accommodation. Gradually, their services expanded to link wider public services and help nonprofit organizations become more digitally proficient (Infoxchange, 2014). In the aftermath of damage from earthquakes in Christchurch, New Zealand, the company extended overseas to shift business operations online. Then in 2018, Infoxchange merged with Connecting Up to enlarge its capacity to serve disadvantaged people through technology (Infoxchange, 2018).

Fifth, Perth City Farm was initiated to establish a green zone in an urban setting of Western Australia. Members received permission from the East Perth Redevelopment Authority to turn a toxic strip of land into an organic agricultural area and community hub (Perth City Farm, 2014). The social enterprise offers horticultural education, leases space, and hosts events to supplement their produce sales. It is well known for partnering with locals and neighboring institutions on social projects.

Sixth, Resource Recovery began by managing the rubbish disposal services for a regional council in New South Wales (Great Lakes Community Resources, 2014). In the process, it diverted and repurposed a range of material streams that were destined for landfill, and it became a best practice waste management leader (Resource Recovery, 2015). Eventually, a consulting branch was spun off, RRA, and the company geographically extended their business model in other states with tip collection and industrial ecology services.

Seventh, WorkVentures was a pioneer of profit-for-purpose businesses. It was founded on the principle that social enterprise can translate into an innovative business model (Hetherington, 2008). Today, the company performs IT repairs for major corporate clients, it refurbishes and resells computers to low-income earners, and it provides vocational training programs to integrate youth, migrants, and refugees into jobs (WorkVentures Ltd, 2014).

Although these social enterprises all operate within different industries, they are commonly small- and mid-size enterprises (SMEs), and they constitute respected business entities with at least a five-year track record. Two suitable respondents were identified from each company based on their experience and ability to capture different perspectives and insights into organizational performance reporting practices. In total, 14 founders, senior executives, or managers who occupied paid or voluntary positions participated. The researcher also scrutinized more than 300 pieces of secondary data on the organizations to triangulate the findings, drawing from peer-reviewed literature, government documents, company websites, project and annual reports, and articles from industry associations. Hence, the evidence was gathered from primary and secondary sources, and the interviews were held at the business settings.

A brief survey was administered to learn the profile of the organizations listed in Table 17.2. Then respondents answered semi-structured questions that were designed to obtain consistent and pertinent information for making comparisons among the cases. This method also let the researcher observe firsthand details such as how the organizations engage in water or land conservation, and waste management practices.

The interviews were audio-recorded, transcribed, and analyzed through a two-step process, using manual and NVivo coding to arrive at higher-level themes from the evidence. It involved an iterative approach to extract patterns and build explanations (Welsh, 2002). When interpretations were reached, reports were written and confirmed with participants in 2016 to ensure robust results.

RESEARCH FINDINGS

The organizations pursue sustainability in distinctive ways to reach their missions given in Table 17.3. In the course of doing business, they enact marketing activities to obtain revenue for their operations. Moreover,

TABLE 17.2
Profile of Social Enterprises

Companies and Respondents	Legal Status	Industries	Locations	Paid Staff	Annual Turnover
AAC/AAC1—Ashoil	Subsidiaries of proprietary limited corporation	Biodiesel	Tom Price and Wangara, Western Australia	<5	$700–800 K
AAC2—Ashlinen		Apparel			
Abbotsford Convent 1	Not-for-profit limited by guarantee	Public arts	Abbotsford, Victoria	20	$2.6 M
Abbotsford Convent 2					
Hepburn Wind 1	Co-operative	Renewable energy	Leonards Hill, Victoria	<5	$941 K
Hepburn Wind 2					
Infoxchange 1	Not-for-profit limited by guarantee	ICT	Richmond, Victoria, Brisbane, Queensland, and Christchurch, New Zealand	100	$8 M
Infoxchange 2					
Perth City Farm 1	Branch of incorporated not-for-profit	Agricultural and community hub	East Perth, Western Australia	<10	$400 K
Perth City Farm2					
Resource Recovery 1	Subsidiaries of incorporated not-for-profit	Waste management and recycling	Tuncurry, New South Wales, Queensland	<35	$2 M
Resource Recovery 2/RRA					
WorkVentures 1	Not-for-profit limited by guarantee	IT	Mascot, New South Wales	100	$15 M
WorkVentures 2					

Source: Herbst (2017, 2019).

TABLE 17.3

Cases with their Mission Statements

Cases	Missions
Ashoil and Ashlinen	To benefit members and other Aboriginal people across the Pilbara through culturally appropriate employment, education, training services, and programs (Ashburton Aboriginal Corporation, 2014).
Abbotsford Convent	To restore the infrastructure of a national heritage site for Victorians; to build a safe, accessible public hub for connectivity and well-being; to create a cultural platform of appealing and inspiring content; and to catalyze a creative cluster to harness skills, and share ideas and resources among the community (Abbotsford Convent, 2014b).
Hepburn Wind	To create a cooperatively owned renewable energy project that operates as an exemplary wind farm for the community (Hepburn Community Wind Park Co-operative Limited, 2014).
Infoxchange	To create social equality and opportunity by empowering people through access to information and communications technology (Infoxchange, 2014).
Perth City Farm	CROP – To showcase innovative urban farming and foster partnerships and community participation in food production workshops, interactive tours, and social enterprises that promote sustainable food systems; COOK – To use the farm-based cooking school and community kitchen to bring people together to share and prepare healthy meals using seasonal produce harvested on site; CONNECT – To encourage community engagement through hands-on learning and social education designed to inspire dialogue, debate and awareness around local, national, and global food issues (Perth City Farm, 2014).
Resource Recovery and RRA	To use local knowledge and networks in partnership with government, industry, and community for maximizing community resources and to assist people experiencing disadvantage to develop social and economic livelihoods (Resource Recovery, 2015).
WorkVentures	To connect people with choices so they can improve their own lives, using the benefits of technology (WorkVentures Ltd, 2014).

TABLE 17.4

Social Marketing Interventions

Cases	Social Marketing Programs or Advocacy
Ashoil and Ashlinen	Offering apprenticeships to Aboriginal students to counter low rates of school attendance, and to prepare youth for employment
Abbotsford Convent	Holding health and well-being events to encourage healthier lifestyles
Hepburn Wind	Coordinating grassroots campaigns to overcome political opposition to community renewable energy
Infoxchange	Teaching ICT skills to marginalized people residing in public housing projects to bridge the digital divide
Perth City Farm	Giving permaculture workshops and hosting weekend markets to sell local produce
Resource Recovery and RRA	Setting up a bike repair and resale shop to teach skills to at-risk high schoolers and developing a community garden with bush regeneration to the community
WorkVentures	Running an IT academy to teach skills to youth, and overseeing an afterschool program to assist incarcerated or troubled youth to become integrated into society

Source: Herbst (2017, 2019).

the social enterprises implement social marketing interventions to drive social or environmental change. Table 17.4 highlights social programs or advocacy initiated by each company.

Several themes emerged about the social marketing. It was clear the organizations offer either external programs for social impact or it is an intrinsic part of their daily operations. Abbottsford Convent did both—they cooperated with the state Office of Environment and Heritage and architects to incorporate environmentally sustainable design principles in built structures, and they allied with the Yarra Energy Foundation, corporates and academics to strategize how to install solar panels. The Convent separately negotiated with tenants to maintain ventilation and find ways to shade building interiors for energy efficiency. Hepburn Wind led a political action campaign for renewable energy. They suggested for shareholders to write to their government leaders to allow communities the right to harness small-scale commercially renewable power. By galvanizing an effective movement for change, it made members and outsiders want to be connected with a more responsible organization.

Being viewed in a positive light resulted in generating more volunteers for the organizations' community work. Infoxchange and WorkVentures were able to obtain hardware and software donations to teach digital skills to migrants and refugees. Perth City Farm sourced people who contributed to replanting trees in deforested areas.

By engaging with their communities, the organizations could achieve milestones: to bring different clans together to work in harmony; they could broker deals with businesses for a social license to operate; and they could break cycles of unemployment. Ashoil and Ashlinen subsidized cultural events and offered a training ground to high-schoolers as a springboard for long-term employment. Resource Recovery acted similarly to give training to people from all walks of life and they sponsored workshops or social get-togethers in their outdoor hub.

The cumulative effect of these social marketing efforts translated into higher sales for the organizations. Employees, customers, and the public wanted to be affiliated with good corporate citizens, but it was still important for the social enterprises to produce a good product at a good price and deliver it through the right distribution channels or place.

This research examined the fourth marketing mix element of promotion, to see how the organizations communicated about their sales, social, or environmental endeavors with their target audiences. Evidence showed management would monitor company performance and, in turn, disseminates data to executives or board members for internal purposes; information would be released across multiple formats for external audiences; and at times, it was amalgamated into an integrated report. These accountability categories are exhibited in Figure 17.1. When the social enterprises compiled their reports into an inclusive, singular document rather than fragmented analyses, it seemed to produce greater social, environmental, and economic value to maximize impacts for those organizations.

Internal Reporting

The social enterprises exchanged valuable information about performance among their organizational leadership. Staff members periodically informed their parent company or board members on the status of meeting company objectives with key performance indicators (KPIs) to benchmark their effectiveness in delivering goods and services, such as measuring the amount of income received, the number of staff members

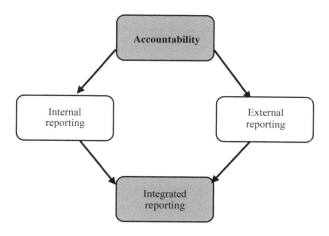

FIGURE 17.1
Types of reporting for accountability.

trained, and the number of visitors received. Then the organizations were able to perform cost-benefit analyses to see what areas they could make modifications to improve their potential gains.

> Last year we restored three new spaces ... So, I guess it's a big thing in the organization about developing that understanding of the need for the commercial side of the business in supporting a lot of the community and arts programming.
>
> (Abbotsford Convent, Respondent 1)

Perth City Farm used the evaluations of its financial accounts to identify which areas of its business brought in higher cash flow, facilitating the decision to focus more efforts to increase venue hire of its premises, and to hire new management for its operations. Some of the other organizations used the data about their internal performance for the purpose of tendering or renegotiating business contracts.

Several social enterprises also used the assessments to determine how to fund their social endeavors. Managers would look at their revenue per project to figure out where they could divert surplus toward funding social programs which reflects literature stating that nonprofits resort to cross-subsidization for such purposes (James, 2017).

We've got some areas doing better than others. We measure each of them independently and then make those decisions as to when and where we want to cross-subsidize.

(Infoxchange, Respondent 2)

Coupled with the quantitative data on market performance, the social enterprises employed Internet tracking to collect qualitative data of consumer opinions to indicate to leaders how target audiences responded to their products and services. This information helped to justify expenditures in marketing and advertising campaigns via online media platforms, and it was useful to capture data for strategic planning.

External Reporting

The social enterprises engaged with their external stakeholders over multiple channels to share company information. Representatives of Ashoil and Ashlinen regularly attended Aboriginal conferences to explain their business model as social enterprises in the mining sector. Similarly, Resource Recovery and RRA attended waste management conferences to discuss how they function, communicating best practices to their associates within the Community Recycling Enterprise Network. The other organizations similarly networked with colleagues about their performance at industry events.

Part of that's by going to conferences—talking to other players in the waste industry and just looking at what we do.

(Resource Recovery, Respondent 1)

While the social enterprises acted openly and transparently to divulge information about their performance to colleagues or the public, they did not have a uniform system in place for gathering data. Table 17.5 reviews different systems and types of disclosure by each social enterprise. When an acquittal was needed to be prepared for a funding body to justify that money was spent as stipulated in a contract, then an organization would often circulate this report for wider reach. WorkVentures and Infoxchange published information to show the financial and social benefits of computer-based training programs that were subsidized via corporate sponsorships and in-kind support (Piccone, 2011; Wheadon, 2010; WorkVentures Ltd, 2011). Positive effects included attaining higher education, employment, and satisfaction from beneficiaries and service providers.

TABLE 17.5

Summary of Systematic Forms of Disclosure

Organizations	Print or Online	Internal on Social or Financial Data	External on Social or Financial Data	Integrated Data	Extent and Types of Publicly Available Performance Reports
Ashoil and Ashlinen	O	S, F	S	X	Concise annual report of social impacts, Indigenous presentations
Abbotsford Convent	P, O	S, F	S, F	√	Annual review and comprehensive financial and social performance report prepared by staff and independent consultants, regular posts, and presentations, e-newsletters, stakeholder meetings and site tours
Hepburn Wind	O	S, F	S, F	X	Annual financial and social report, regular posts and presentations, member e-newsletters, site tours, intellectual property shared on a wiki
Infoxchange	P, O	S, F	S, F	√	Annual financial and social report, comprehensive special financial and social project reports prepared by independent consultants, white papers on social endeavors, regular posts, and presentations
Perth City Farm	O	S, F	S	X	External evaluation report on a collaborative project, regular posts
Resource Recovery, RRA	P, O	S, F	S, F	√	Annual financial and social report, special project evaluation and funding acquittal report prepared by an independent consultant, regular posts and presentations, site tours
WorkVentures	P, O	S, F	S, F	X	Concise annual financial and social report, special project evaluation and funding acquittal reports, regular posts

Source: Herbst (2017).

Additionally, the organizations distinguished between short-term outcomes and long-term impacts. WorkVentures and Perth City Farm measured the number of participants that completed certified training programs. Then the social enterprises continued to monitor their trainees in a longitudinal evaluation process to assess how many participants remained in the workforce. This information was tabled and written up for public dissemination and feedback for the government.

> Well I could give you more long-term data in terms of our group-training organization…we get them trained up and they get their qualifications while they're inside a host employer. So, they start working full-time…and we track these kids for two, three, four, five years because they're on social media. We link with them on LinkedIn, Facebook. They're our alumni…
>
> (WorkVentures, Respondent 1)

Benchmarking results for at least three-to-five years also enabled the organizations to keep in touch and develop long-term relationships with their stakeholders. Hepburn Wind maintained a regular e-newsletter and other online formats to communicate about their progress or challenges. It helped them to foster so much public goodwill that its executives decided to express their gratitude by hosting special on-site events such as sustainability education days. These events heightened awareness and support for their mission to advance an exemplary wind farm.

Spinoffs of delivering products and programs were mainly reported online to stakeholders via e-newsletters or annual reports. Only for a special occasion would the social enterprises distribute a special printed report such as when Infoxchange wanted to commemorate its 10-year anniversary, writing about their achievements. The social enterprises frequently posted videos or operated their own YouTube channels. They preferred video as a platform because they found it is an effective medium to articulate narratives of their journeys, show images, and obtain comments from program recipients.

The social enterprises would determine whether to release a concise financial cost-benefit analysis or social account depending on whom they wanted to address. At a minimum, most of the organizations seemed to primarily respond to the need to divulge financial information to capital investors. For public audiences, on the other hand, they generally released broader information. Of greater interest though, several social enterprises presented a comprehensive picture of their organizational performance using IR.

Integrated Reporting

Abbotsford Convent continuously collected data to assess their performance. They hired quantity surveyors to map the costs and benefits at different stages of their redevelopment—buildings that were restored and they received estimates of those structures that needed to be completed across their site (Essential Economics Pty Ltd, 2011; Essential Economics Pty Ltd, 2014). Additionally, they gathered qualitative and quantitative data from an independent economic impact study undertaken by PricewaterhouseCoopers to assess their returns on investment to transform the Convent into a thriving multipurpose complex. The Convent then merged all of this data with content on the outcomes from receiving a raft of public, private and nonprofit support over their first decade of operation (Abbotsford Convent, 2014a). This major report doubled as a prospectus to seek future funding for ongoing rehabilitation.

Infoxchange similarly provided annual reports that went beyond conventional accounting measures. However, their integrated style of reports focused on ICT and social justice through promoting policies for digital inclusion (Dickins, 2014; Walton, Kop, Spriggs, & Fitzgerald, 2013).

Resource Recovery paralleled this pattern of disseminating integrated reports, but they published less regularly, giving information on specific projects. A commissioned report on "The Green" assessed the value of developing a community center and regenerating the bushland around their industrial park (Hastings, 2013). The remaining social enterprises were investigating what they could do to improve their future reporting practices. Hepburn Wind was even analyzing how to upgrade their future annual reports toward something closer to IR.

> We participate as a kind of benchmarking/monitoring/evaluation system that is being set up at the moment through the Coalition of Community Energy…we're defining what we should be capturing as projects come onto the Australian landscape…
>
> (Hepburn Wind, Respondent 2)

Thus, findings indicated the social enterprises used a range of reporting formats to collect and communicate with their external stakeholders. The continuum pictured in Figure 17.2 reflects this evidence, demonstrating the organizations released a range of incremental amounts of information

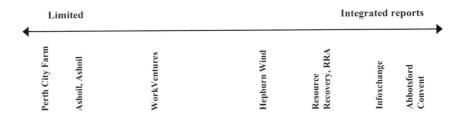

FIGURE 17.2
Levels of reporting by the cases.

that comprised brief financial statements to more integrated types of social and financial accounts.

Although the reporting was inconsistent and often incomplete on the social enterprises' impacts, Table 17.6 reveals a scope of triple bottom line data that was disseminated by three of the social enterprises that reported in line with IR.

TABLE 17.6

Cases Disseminating Integrated Data

Case	Social Indicators	Environmental Indicators	Economic Indicators
Abbotsford Convent (2014b)	Annual growth of events hosted, records of visitor numbers and satisfaction from programs delivered on site	Efficiencies in energy and water consumption post-restoration of buildings, e.g., water tanks and solar energy	Return on investment from venue hire and costs of capital investments
Infoxchange (2014)	Number of participants trained in digital literacy for improved health and well-being	Number of families and businesses linked to cloud computing to save transport emissions	Profit and loss including total sales of refurbished computers
Resource Recovery (2015)	Number of trainees certified, percentage of Indigenous employees and reduction in recidivism of locals who were employed by the company	Percentage of waste recovered from landfill and amount of m^2 of land regenerated	Income and assets from awarded contracts and payments from job service agency referrals

Source: Herbst (2017).

DISCUSSION

This chapter focused on research that explored the use of various accounting tools by seven Australian social enterprises to discharge their accountability to stakeholders. These organizations face special challenges in reporting due to needing to recount information about their dual social and commercial pursuits. Nevertheless, accountability can be a means to reconcile both of these remits, and greater transparency can bring increased recognition of organizational impacts, resulting in increased interest from investors and the public (Wilkinson et al. 2014).

Although performance reporting fluctuated with multidimensional systems of reporting being employed (Hynes, 2009), this finding is in line with a large global study to assess what types of reporting are carried out by social enterprises (Huysentruyt et al. 2016). In general, the cases reported about the number of target markets they served as a benchmark of their performance through the products or services they provided, and their rates of success in helping their clients or beneficiaries to attain social objectives. The results show all of the social enterprises were at least meeting their designated missions.

Those cases that leaned toward IR demonstrated more benefits were realized. Abbotsford Convent reached every goal it set in its initial five-year plan. Infoxchange partnered with new government agencies to provide ICT services across technological platforms in Australia and abroad. And Resource Recovery became nationally renowned and sought after for its waste management practices due to promoting their achievements which prompted company expansion. This evidence confirms showcasing more comprehensive company results can yield enhanced impacts—a win-win for beneficiaries and the social enterprises (Pritchard, Ní Ógáin & Lumley, 2012).

It is important to acknowledge that IR is a worthwhile tool because it overcomes limitations associated with established reporting methods (Kay & McMullan, 2017). It considers what is important to stakeholders rather than the companies electing to simply report on finances or aspects that are limited. Moreover, it was apparent from the comprehensive reports by Abbotsford Convent and Infoxchange that monitoring organizational performance has become a core business practice, and independent auditing has been implemented for quality assurance to bolster trust and credibility of the metrics.

Theoretical and Practical Contributions

In accordance with stakeholder theory (Deegan & Unerman, 2011), giving a broader quantitative and qualitative overview of performance through IR can be a vehicle for companies to connect with multiple audiences. Literature (Soyka, 2013) emphasizes the advantages, including the ability for a company to present more than typical balance sheet figures. An organization can use this mechanism to relate to its buyers, suppliers, customers, and regulators among other constituents. These target publics are more likely to respond to the communication. For this study, audiences seemed to be responsive to those organizations engaged in deeper reporting on the five capitals as company growth was achieved and forecast. Abbotsford Convent managed to acquire donations that they pooled with revenue to pay for ongoing redevelopment. Infoxchange expanded and acquired resources through its merger that was publicized in its annual reports, and they anticipated servicing more clients ahead. The empirical evidence thereby extends stakeholder theory in the third sector.

According to the International Integrated Reporting Council (2013), once a company participates in IR, it is regularly expected to release data on its operations, strategies, and impacts, but there is a lack of enforcement by governments. Unless statutory reporting is imposed as a mandatory requirement, this situation is likely to continue to prevail which is unfortunate due to potential opportunities to be gained.

Taking a holistic approach to accountability is shown to be more meaningful than issuing separate social and financial reports to external audiences. In addition, the evidence confirms that deeper reporting can be constructive for internal purposes since it helps managers or board members to execute strategic planning and decision-making. None of the conventional frameworks are sufficient to achieve such purposes. This practical knowledge can be an incentive for diffusion of IR by other social enterprises. The case studies in this book chapter can be used as exemplars, to demonstrate to marketers and social marketers how to reach socially responsible consumers through effective performance reporting on sustainable practices.

CONCLUSION

This study provides insight about performance reporting, especially the early stages of IR for seven Australian SMEs that promote sustainable development. Notwithstanding the demands of preparing financial, social, or combined accounts that primarily limit this application to large corporations (Carnegie & Burritt, 2012), momentum is building for instituting IR, particularly from reputable national accounting agencies (e.g., CPA Australia) and global accounting firms (e.g., KPMG International). Statutory reporting for it is growing as well in various countries. Legislation passed in South Africa requires all publicly listed companies to issue an integrated report. This country encourages all public, private, or nonprofit organizations to adopt IR (SAICA, 2009). There is also a push toward greater measurement and reporting practices by social enterprises with recent requirements being introduced in Italy and the United Kingdom. Discussions were even held in Finland and Norway to consider introduction of future legislation (European Commission, 2016; Jardine & Whyte, 2013; Nordic Council of Ministers, 2015), signaling further changes there.

Research in this area of the social enterprise field is overwhelmingly conceptual. It mostly proposes dimensions on what types of data would be useful to capture to meet the needs of clusters of stakeholders (Arena et al. 2015). Yet, some case studies reinforce the development of IR (Frostenson, Helin, & Sandström, 2012; Lodhia, 2014; Soyka, 2013; Stubbs & Higgins, 2014). And the International Integrated Reporting Council (2011) recommends its value for factoring in the stocks presented in the five capitals model to move beyond measurements of GDP.

Integrity of the results of this study was maintained by interviewing and confirming findings among respondents, and by cross-checking numerous pieces of data over several years to detect patterns for interpretation. Also, the researcher adopted trustworthy provisions (Shenton, 2004) to prevent bias. The main limitation though is the results might not be representative of the international situation because the study was confined to a relatively small sample in Australia. However, it is reasonable to infer that social enterprises carry out similar reporting practices in other developed countries where the results could be tested in a cross-country study for validation.

Recommendations for future research therefore include undertaking a larger study, notably in those countries that are legislating for greater accountability through IR. Another investigation could attempt to determine how deeper reporting brings competitive advantages, specifically linked to the Sustainable Development Goals.

LESSONS LEARNED

This chapter illustrates that the more a social enterprise is inclusive and responsive to its socially responsible stakeholders when communicating on its performance, the greater the likelihood that it will realize its objectives. While mechanisms to discharge accountability to stakeholders vary, by addressing the five capitals through IR, a company can precipitate a higher contribution to sustainability.

DISCUSSION QUESTIONS

It is important for a company to report about all sources of capital. Traditionally, a company only covers information on its manufactured and financial stocks.

1. How might a company address its human capital in its reporting practices?
2. How might a company address its social capital in its reporting practices?
3. How might a company address its natural capital in its reporting practices?

EXERCISE

Select a social enterprise such as The Big Issue. Now design a framework of an integrated report for them by following these steps recommended by the International Federation of Accountants (Thompson, 2017).

1. **Prepare a statement of intent for stakeholders.** Briefly explain the aims, ambitions, and rationale of the business, reinforcing a commitment to good governance, transparency, and long-term strategies.

Then outline how the organization will implement its performance reporting practices, including a timeframe and specific milestones to reach its mission.
2. **Map the key stakeholders.** Organizations need to understand their target audiences, and to be clear about their expectations of the business for the present and future. Identify the principal stakeholders such as customers, employees, and community members. Determine how your company uses its products and services to satisfy each group.
3. **Consider how the business creates value.** In what ways does the business generate value with its goods or services? Be concise but address all these material matters: the financial base for trading and growth; the infrastructure and assets for production; the intellectual property, processes, skills, expertise, and knowledge of management and employees; the access and proximity to water and energy; and the connections with customers and suppliers. Do not forget to state in what ways the business also invests in its community to cater to social or environmental needs.
4. **Look at the business model.** Does the business model and its associated strategies support value creation? Does it reflect stakeholders' expectations? Should the business make any adjustments to factor in risks and opportunities?
5. **Determine what resources are needed.** Will the company require additional resources to implement any changes to its business model or strategies? Think about whether the business may need to purchase new equipment, hire new staff, upgrade its product design, innovate, or source additional financial capital to accommodate business growth and to help society if demand for services heightens.
6. **Tailor your performance report to appropriate communication channels.** Ensure the messages that you formulate in a report will be effective. Do the company's existing platforms enable messages to be delivered to the organization's internal and external stakeholders? Should different types of social media be utilized to encourage cross-organizational communication and richer engagement with business strategy?

Remember every organization is structured and functions differently. An organization can leverage its available resources to adapt to IR over several business cycles as new systems for reporting are implemented and best practices evolve.

REFERENCES

Abbotsford Convent (2014a). *Let's finish the job*. Melbourne, VIC: Abbotsford Convent.

Abbotsford Convent (2014b). *This is Abbotsford convent*. Melbourne, VIC: Abbotsford Convent.

Adams, S., & Simnett, R. (2011). Integrated reporting: An opportunity for Australia's not-for-profit sector. *Australian Accounting Review*, 21(3), 292–301.

Alter, K. (2004). Social enterprise typology. *Virtue ventures LLC*. Washington, DC.

Arena, M., Azzone, G., & Bengo, I. (2015). Performance measurement for social enterprises. *VOLUNTAS: International Journal of Voluntary and Nonprofit Organizations*, 26(2), 649–672.

Arvidson, M., Lyon, F., McKay, S., & Moro, D. (2010). *The ambitions and challenges of SROI* (Third Sector Research Centre Working Paper No. 49). University of Birmingham.

Ashburton Aboriginal Corporation (2013). *Annual report 2012-2013*. Tom Price, WA: AAC.

Ashburton Aboriginal Corporation (2014). *Corporate profile*. Tom Price, WA: AAC.

Berg, N., & Jensen, J. C. (2012). Determinants of traditional sustainability reporting versus integrated reporting. An Institutionalist Approach. *Business Strategy and the Environment*, 21(5), 299–316.

Blaikie, N. (2009). *Designing social research*. Cambridge and Malden, MA: Polity Press.

Brewer, C. V. (2016). The ongoing evolution in social enterprise legal forms. In D. R. Young, E. A. M. Searing, & C. V. Brewer (Eds.), *The social enterprise zoo: A guide for perplexed scholars, entrepreneurs, philanthropists, leaders, investors, and policymakers* (pp. 33–64). Cheltenham: Edward Elgar Publishing.

Brown, J., & Dillard, J. (2014). Integrated reporting: On the need for broadening out and opening up. *Accounting, Auditing & Accountability Journal*, 27(7), 1042–1067.

Brown, J. A., & Forster, W. R. (2013). CSR and stakeholder theory: A tale of Adam Smith. *Journal of Business Ethics*, 112(2), 301–312.

Carnegie, G. D., & Burritt, R. L. (2012). Environmental performance accountability: Planet, people, profits. *Accounting, Auditing & Accountability Journal*, 25(2), 370–405.

Chartered Institute of Management Accountants (2015, August). *Integrated reporting for SMEs - Helping businesses grow*. CIMA. https://www.cimaglobal.com/Documents/Thought_leadership_docs/reporting/integrated-reporting/Case_study_IR_SMEs.pdf

Cheng, M., Green, W., Conradie, P., Konishi, N., & Romi, A. (2014). The international integrated reporting framework: Key issues and future research opportunities. *Journal of International Financial Management & Accounting*, 25(1), 90–119.

Connolly, C., & Kelly, M. (2011). Understanding accountability in social enterprise organisations: A framework. *Social Enterprise Journal*, 7(3), 224–237.

CPA Australia (2016). *Integrated reporting: Delivering market resilience*. https://www.intheblack.com/

Crucke, S., & Decramer, A. (2016). The development of a measurement instrument for the organizational performance of social enterprises. *Sustainability*, 8(161), 1–30.

de Villiers, C., Rinaldi, L., & Unerman, J. (2014). Integrated reporting: Insights, gaps and an agenda for future research. *Accounting, Auditing & Accountability Journal*, 27(7), 1042–1067.

Deegan, C., & Unerman, J. (2011). *Financial accounting theory* (2nd ed.). Berkshire, England: McGraw-Hill Education.

Dickins, M. (2014). *Digital inclusion: Taking up the challenge*. Hobart, TAS: Social Action and Research Centre, Anglicare Tasmania Inc.

Doherty, B., Haugh, H., & Lyon, F. (2014). Social enterprises as hybrid organizations: A review and research agenda. *International Journal of Management Reviews*, 16(4), 417–436.

Eisenhardt, K. M., & Graebner, M. E. (2007). Theory building from cases: Opportunities and challenges. *Academy of Management Journal*, 50(1), 25–32.

Elkington, J. (1997). *Cannibals with forks: The triple bottom line of the 21st century*. Oxford: Capstone.

Essential Economics Pty Ltd (2011). *Abbotsford Convent development - Economic impact assessment*. Melbourne, VIC: Essential Economics.

Essential Economics Pty Ltd (2014). *Abbotsford Convent restoration project - Employment impacts*. Melbourne, VIC: Essential Economics.

European Commission (2011). *Social business initiative: Creating a favourable climate for social enterprises, key stakeholders in the social economy and innovation*, (COM/2011/0682). Brussels: EC.

European Commission (2016). *Social enterprises and their eco-systems: Developments in Europe*. C. Borzaga and G. Galera (Authors). Brussels: Directorate-General for Employment, Social Affairs and Inclusion, EC.

Flower, J. (2015). The International Integrated Reporting Council: A story of failure. *Critical Perspectives on Accounting*, 27, 1–17.

Frias-Aceituno, J. V., Rodríguez-Ariza, L., & Garcia-Sánchez, I. M. (2014). Explanatory factors of integrated sustainability and financial reporting. *Business Strategy and the Environment*, 23(1), 56–72.

Frostenson, M., Helin, S., & Sandström, J. (2012). Sustainability reporting as negotiated storytelling. In *EBEN Annual Conference 2012* (pp. 1–35). Barcelona, Spain.

Giovannoni, E. & Fabietti, G. (2013). What is sustainability? A review of the concept and its applications. In: C. Busco, M. Frigo, A. Riccaboni & P. Quattrone (Eds.), *Integrated Reporting* (pp. 21–40). Switzerland: Springer.

Great Lakes Community Resources (2014). *Annual report 2014*. Great Lakes, NSW.

Hasnas, J. (2013). Whither stakeholder theory? A guide for the perplexed revisited. *Journal of Business Ethics*, 112(1), 47–57.

Hastings, C. (2013). *Draft evaluation report: 'The green'*. Tuncurry, NSW: Great Lakes Community Resources.

Hepburn Community Wind Park Co-operative Limited (2014). *Annual report for the year ended 30 June 2012*. Daylesford, VIC: Hepburn Wind.

Herbst, J. (2017). *How Australian social enterprises use strategic marketing and social marketing to drive accountability and change for sustainable development* [Doctoral dissertation, QUT Research Repository]. https://eprints.qut.edu.au/103631/

Herbst, J. M. (2019). Harnessing sustainable development from niche marketing and coopetition in social enterprises. *Business Strategy and Development*, 2(3), 152–165.

Hetherington, D. (2008). *Case studies in social innovation: A background paper*. Per Capita. http://percapita.org.au/wp-content/uploads/2014/11/Social-Innovation-Cases.pdf

Huysentruyt, M., Le Coq, C., Mair, J., Rimac, T., & Stephan, U. (2016). *Social entrepreneurship as a force for more inclusive and innovative societies: Cross-country report*. Berlin, Germany: Hertie School of Governance.

Hynes, B. (2009). Growing the social enterprise–Issues and challenges. *Social Enterprise Journal*, 5(2), 114–125.

Infoxchange (2014). *Annual report 2014*. Melbourne, VIC: Infoxchange.

Infoxchange (2018, November 28). *Infoxchange and Connecting Up join forces to deliver greater impact*. Melbourne, VIC: Infoxchange. https://www.infoxchange.org/au/news

International Integrated Reporting Council (2011). *Towards integrated reporting: Communicating value in the 21st century*. ACCA. https://www.accaglobal.com/africa/en/technical-activities/technical-resources-search/2011/december/iirc-towards-integrated-reporting.html

International Integrated Reporting Council (2013). *Consultation draft of the international <IR> framework*. https://integratedreporting.org/

James, E. (2017). Commercialism and the mission of nonprofits. In P. Frumkin & J. Imber (Eds.), *In Search of the Nonprofit Sector* (pp. 87–98). Oxon and New York: Routledge.

Jardine, C., & B. Whyte. (2013). Valuing desistence? A social return on investment case study of a throughcare project for short-term prisoners. *Social and Environmental Accountability Journal*, 33(1), 20–32.

Jones, K. R., & Mucha, L. (2014). Sustainability assessment and reporting for nonprofit organizations: Accountability "for the public good". *VOLUNTAS: International Journal of Voluntary and Nonprofit Organizations*, 25(6), 1465–1482.

Kay, A., & McMullan, L. (2017). Contemporary challenges facing social enterprises and community organisations seeking to understand their social value. *Social and Environmental Accountability Journal*, 37(1), 59–65.

Kernot, C., & McNeil, J. (2011). *Australian stories of social enterprise*. Kensington, Australia: University of New South Wales.

KPMG (2012). *Integrated reporting: Performance insight through better business reporting: Issue 2*. KPMG International Cooperative. https://assets.kpmg/content/dam/kpmg/pdf/2013/03/integrated-reporting-2012.pdf

Lane, M., & Casile, M. (2011). Angels on the head of a pin. *Social Enterprise Journal*, 7(3), 238–258.

Lee, A. (2013). *Energy from waste (EfW): The SEQ journey* [CSIRO EfW Seminar]. Brisbane, QLD: Field Services Group, Brisbane City Council.

Lodhia, S. (2014). Exploring the transition to integrated reporting through a practice lens: An Australian customer owned bank perspective. *Journal of Business Ethics*, 129(3), 585–598.

Luke, B. (2016). Measuring and reporting on social performance: From numbers and narratives to a useful reporting framework for social enterprises. *Social and Environmental Accountability Journal*, 36(2), 103–123.

McLeish, B. (2011). *Successful marketing strategies for nonprofit organizations: Winning in the age of the elusive donor*. New York, NY: Wiley.

Minerals Council of Australia (n.d.). *Getting it right: Indigenous enterprise success in the resource sector*. Forrest, ACT: Australian Government.

Mitchell, R. K., Agle, B. R., & Wood, D. J. (1997). Toward a theory of stakeholder identification and salience: Defining the principle of who and what really counts. *The Academy of Management Review*, 22(4), 853–886.

Mook, L., Richmond, B. J., & Quarter, J. (2003). Integrated social accounting for nonprofits: A case from Canada. *Voluntas: International Journal of Voluntary and Nonprofit Organizations*, 14(3), 283–297.

Nicholls, A. (2009). We do good things, don't we? *Accounting, Organization and Society*, 34, 755–67.
Nordic Council of Ministers (2015). *Social entrepreneurship and social innovation in the Nordic countries* (2015:562). Copenhagen, Denmark: Norden.
O'Dwyer, B., & Boomsma, R. (2015). The co-construction of NGO accountability: Aligning imposed and felt accountability in NGO-funder accountability relationships. *Accounting, Auditing & Accountability Journal*, 28(1), 36–68.
Olson, M., & Attolini, G. (2017, August 3). *Integrated thinking & reporting requires trusted advisors: Guiding your SME clients.* International Federation of Accountants. https://www.ifac.org
Öncer, A. Z. (2019). Performance measurement in social enterprises: Social impact analysis. In N. O. Iyigun (Ed.), *Creating business value and competitive advantage with social entrepreneurship* (pp. 205–231). Hershey, PA: IGI Global.
Parkin, S., Johnson, A., Buckland, H., & White, E. (2004). *Learning and skills for sustainable development: Developing a sustainability literate society.* London: Higher Education Partnership for Sustainability (HEPS).
Parrish, B. D. (2010). Sustainability-driven entrepreneurship: Principles of organization design. *Journal of Business Venturing*, 25(5), 510–523.
Parrot, K. W., & Tierney, B. X. (2012). Integrated reporting, stakeholder engagement, and balanced investing at American Electric Power. *Journal of Applied Corporate Finance*, 24(2), 27–37.
Peattie, K., & Morley, A. (2008). Eight paradoxes of the social enterprise research agenda. *Social Enterprise Journal*, 4(2), 91–107.
Perth City Farm (2014). Home. https://www.perthcityfarm.org.au
Piccone, V. (2011). *iGetIT! project 2010 WV internal evaluation report.* Sydney, NSW: WorkVentures.
Pitta, D. A., & Kucher, J. H. (2009). Social enterprises as consumer products: The case of vehicles for change. *Journal of Product and Brand Management*, 18(2), 154–158.
Pritchard, D., Ní Ógáin, E., & Lumley, T. (2012). *Making an impact: Impact measurement among charities and social enterprises in the UK.* London: New Philanthropy Capital.
Resource Recovery (2015). Home. http://www.resourcerecovery.org.au/
Ridley-Duff, R., & Bull, M. (2016). Measuring social value: Outcomes and impacts. In *Understanding social enterprise: Theory and practice* (2nd ed.). (pp. 131–154). London: SAGE Publications.
SAICA (2009). *Draft code of governance principles for South Africa.* Johannesburg: Institute of Directors in Southern Africa.
Schaltegger, S., & Wagner, M. (2006). Integrative management of sustainability performance, measurement and reporting. *International Journal of Accounting, Auditing and Performance Evaluation*, 3(1), 1–19.
Shenton, A. K. (2004). Strategies for ensuring trustworthiness in qualitative research projects. *Education for Information*, 22(2), 63–75.
Soyka, P. A. (2013). The International Integrated Reporting Council (IIRC) integrated reporting framework: Toward better sustainability reporting and (way) beyond. *Environmental Quality Management*, 23(2), 1–14.
Stacchezzini, R., Melloni, G, & Lai, A. (2016). Sustainability management and reporting: The role of integrated reporting for communicating corporate sustainability management. *Journal of Cleaner Production*, 136, 102–110.

Stubbs, W., & Higgins, C. (2014). Integrated reporting and internal mechanisms of change. *Accounting, Auditing & Accountability Journal*, 27(7), 1068–1089.

Thompson, P. (2017, December 14). *How can SMEs implement integrated reporting? A starter kit*. International Federation of Accountants. https://www.ifac.org

Tyler III, J. E. (1983). How nonprofits grow: A model. *Journal of Policy Analysis and Management*, 2(3), 350–365.

UK Government, Department of Business, Innovation and Skills (2011). *A guide to legal forms for social enterprise*. London: UK Government.

United Nations (2002). Plan of implementation of the World Summit on Sustainable Development and Johannesburg *Declaration on Sustainable Development*. World Summit on Sustainable Development.

Vik, P. (2017). What's so social about social return on investment? A critique of quantitative social accounting approaches drawing on experiences of international microfinance. *Social and Environmental Accountability Journal*, 37(1), 6–17.

Walton, P., Kop, T., Spriggs, D., & Fitzgerald, B. (2013). A digital inclusion. Empowering all Australians. *Australian Journal of Telecommunications and the Digital Economy*, 1(1), 9.1–9.17.

Watson, A. (2011). Financial information in an integrated report: A forward looking approach, *Accountancy SA*, December, 14–17.

Welsh, E. (2002). Dealing with data: Using NVivo in the qualitative data analysis process. *Forum Qualitative Social Research* 3(2), 1–9.

Wheadon, G. (2010). *Evaluation of the PCs into homes initiatives*. Canberra, ACT: Elton Consulting.

Wilkinson, C., Medhurst, J., Henry, N., Wihlborg, M., & Braithwaite, B. W. (2014). *A map of social enterprises and their eco-systems in Europe: Executive Summary*. Brussels: ICF Consulting Services, European Commission.

WorkVentures Ltd (2011). *i.settle.with.it: SIPRY program evaluation report*. Sydney, NSW: WorkVentures.

WorkVentures Ltd (2014). *Social inclusion through technology: 50,000 low-cost computers delivered across Australia to low income households*, Sydney, NSW: WorkVentures.

Yin, R. K. (2009). *Case study research: Design and methods*. Thousand Oaks, CA: SAGE Publications.

Section IX

Advances in Knowledge of Social and Sustainable Marketing: Safeguarding against Unsocial and Irresponsible Customers

18

Unsocial and Irresponsible Behaviour: What Happens When Customers Lie?

M. Mercedes Galan-Ladero
University of Extremadura, Spain

Julie Robson
Bournemouth University, UK

CONTENTS

Learning Objectives	510
Themes and Tools Used	510
Theoretical Background	510
Introduction	511
Theoretical Background	513
Service Failure and Service Recovery	513
Misbehaving Customers	514
Business Ecosystems	516
Social Responsibility and Social Marketing	517
Research Design	518
The Case Study	520
The Origins of the Holiday Sickness Scam	520
Impact of the Scam	522
Resolution of the Scam	523
How the Industry Worked with Members of the Ecosystem to Address and Resolve the Issue	523
Reactions in the United Kingdom: British Government's Reaction and Others	526

Conclusions 527
Lessons Learned 528
Discussion Questions 529
Project/Activity-Based Assignment/Exercise 529
Notes 530
References 530

LEARNING OBJECTIVES

After reading the chapter, the reader should be able to:

- Stimulate discussion on unsocial and irresponsible customer behaviour, by recognizing that not all customers are socially responsible and indeed can be encouraged by others to behave in this way.
- Promote critical thinking on the effectiveness of existing business practices in addressing unsociable and irresponsible behaviour
- Recognize the need to monitor closely the wider business ecosystem, to identify and respond to changes to protect an industry and its stakeholders.
- Learn from real-life case studies, i.e., the "holiday sickness scam" involving British tourists and Spanish hotels that how theory can be used in practice.

THEMES AND TOOLS USED

Case study – secondary data

THEORETICAL BACKGROUND

- Service Failure and Service Recovery
- Misbehaving Customers
- Business Ecosystems
- Social Responsibility and Social Marketing

INTRODUCTION

Service failure is an issue that most organizations will face at some time (Galan-Ladero & Galera-Casquet, 2018), regardless of how good their service is. When this happens, it is important to implement service recovery measures and reinstate customer satisfaction as quickly as possible. Monetary compensation (e.g. a discount on the price or payment of a determined amount of money) and/or non-monetary compensation (such as, an additional free or discounted services or a formal apology letter) are required to repair the damage caused due to failed services. But, what happens when the service has not failed, and misbehaving customers engage in unsocial and irresponsible behaviour, or to put this question more bluntly, what happens when customers lie?

Misbehaving customers include those that lie and break the accepted norms of behaviour in consumption situations (Fullerton & Punj, 1997). While initially research believed that misbehaviour was perpetrated by just a minority, later research demonstrated that it is more widespread (Fisk et al, 2010; Fullerton & Punj, 2004; Greer, 2015) and represents a major issue for most companies regardless of industry (Daunt & Harris, 2011). Indeed, misbehaving customers pose significant financial, psychological, and physical cost to organizations, their employees, and their customers (Daunt & Harris, 2011). Such customers are unsocial and irresponsible as they lack the ethical values that would normally constrain such behaviour and do not consider the consequences of their behaviour on others. Fraudulent claims cost money to settle, damaging the financial (and reputational) standing of those firms involved. The costs are also often passed on to other consumers as the affected firms seek to recover their costs. In the end, everyone suffers as a consequence.

Research has identified that the cause of customer misbehaviour varies (Rummelhagen & Benkenstein, 2017), It can for example be due to: (1) failure by a service employee; (2) customer profile (some customers are more likely to misbehave than others – Daunt & Greer, 2015); and (3) situational factors (e.g., it can even be motivated by the actions of other customers encourage or provoke this behaviour). In the tourism sector, where best practice in service recovery management can be found (Singh & Crisafulli, 2017), many establishments have frequently experienced a broad and varied range of customer misbehaviours. One of these customer misbehaviours has been especially serious in recent years: when, despite the fact that the good service was delivered, misbehaving customers lied and fraudulently claimed monetary compensation. This was what happened in the Spanish holiday sickness scam, whereby claims management companies (CMCs) encouraged English tourists to make fraudulent claims.

This chapter examines this event taking an ecosystem perspective, i.e., considering the wider, business context in which firms operate and the direct and indirect links between different actors within that ecosystem (Moore, 1993; Iansiti & Levien, 2004). Understanding the nature and cause of changes within an ecosystem can provide important insights for organizations that are located within these systems and help them to identify ways to protect themselves, their industry, and their ecosystem. This case therefore serves to stimulate discussion on socially responsible consumers by addressing the converse – misbehaving customers. It identifies the role of the firm in changing consumer behaviour from unsocially responsible to socially responsible by examining the effectiveness of existing business practices and demonstrating the benefits of adopting a business ecosystem approach. The empirical setting of this chapter is the Spanish tourism ecosystem with a particular focus on three key actors: the hotelier, the customer, and the claims management firm. A single exploratory case study is used, based on secondary data collected from online and publicly available data published in Spanish and in English, and included Spanish and English newspapers and tabloids, industry reports, and British and Spanish governmental publications.

The paper is structured as follows: first, we provide a theoretical background that examines relevant literature on service failure and recovery, customer misbehaviour, business ecosystems, and corporate social responsibility (CSR) and social marketing. Next, we explain the research methodology used in this case. Then follows, a discussion of the case analysis which is broken down into three key stages: (1) it traces the origins of the scam back to 2015 when the increasing volume in claims was first identified; (2) it identifies the impact of the scam on the Spanish tourism industry in 2016 at the height of the scam; and (3) it identifies how the industry worked with distant members of the ecosystem in 2017 and 2018 to address and resolve the issue. Finally, we finish this chapter with a conclusion and implications of the research.

THEORETICAL BACKGROUND

In this section, the theoretical background is mapped out under the headings of service failure and service recovery, misbehaving customers, business ecosystems, and social responsibility and social marketing.

Service Failure and Service Recovery

Service failure has been defined as a problem (Bitner et al, 1990), a situation where the customer's basic service needs are not met (Smith et al, 1999) and when something goes wrong (Ayertey, 2018) with the service being delivered. This *problem* concerns the service product, the facility in which the service is being delivered, the behaviour of the employees delivering that service, or the behaviour of other customers (Bitner et al, 1990).

Most companies seek to provide their customers with a good customer service, nevertheless, service failures are common across all organizations as the causes are many and difficult to completely eliminate. In part, this is because of the very nature of services, their characteristics, namely, intangibility, inseparability, variability, perishability, and a lack of ownership (Grande, 2005; Ayertey, 2018). For example, in the case of *inseparability*, it is difficult to uncouple the interaction between a service provider and the customer, and this interaction can result in problems due to the human interaction (Ayertey, 2018).

Both service failure and the way in which that service is recovered have a considerable effect on the relationship between a customer and an organization (Ayertey, 2018). Service failure will result in customer dissatisfaction, leading to customer complaints, negative word of mouth and ultimately customers will switch to another service provider (Albus, 2012). Indeed, the most frequent cause of customers terminating their relationship is service failure (Keaveney, 1995). Service failure, therefore, adversely affects company market share and profits as well as the longer term damage to its brand and reputation (Ayertey, 2018).

Service recovery seeks to rectify a service failure by changing the negative attitudes the customer has following the service failure (Miller et al, 2000). Organizations can choose from a wide range of service recovery measures that are categorized as either financial or non-financial.

Financial includes discounts, coupons, a lower price and refunds; non-financial includes apologizing, a replacement service or correction of the problem that has caused the service failure. In the hospitality industry, the most effective service recovery measures have been found to be an apology to the customer, restoration of the service and compensation (Albus, 2012). However, the effectiveness of these measures varies by context, for example, the gravity of the failure and how important it is perceived by the customer and the type of service.

Misbehaving Customers

Customer misbehaviour, that is, "acts which violate the generally accepted norms of conduct in consumption situations" (Daunt & Greer, 2015, p. 1506), is undesirable and distinct from other behaviours such as making legitimate complaints (Hu et al, 2017). Levels of customer misbehaviour, whether it is intentional (Fullerton & Punj, 1993) or unintentional (Harris & Reynolds, 2004), are increasing worldwide.

The causes of customer misbehaviour vary depending on context (Rummelhagen & Benkenstein, 2017). Daunt and Greer (2015) identified two key drivers: (1) a response to a service failure where the consumer misbehaves to punish the firm and (2) the specific nature of the service environment where ambient dimensions such as temperature, noise, or crowding, stimulate misbehaviour. However, these authors also identify an additional key driver in the extant literature: personal or internal factors (e.g. personality traits, socio-demographics, and emotional states). Here, misbehaving customers are characterized as younger and male with a low income, low-educational attainment, low levels or morality, and preferring to hide behind the anonymity of misbehaviour, for example, in a crowd of unknown people. Customer misbehaviour has also been associated with three different motives (Kashif & Zarkada, 2015): financial gain, ego, and revenge. In the case of *financial gain*, the customer is typically seeking financial recompense for damages that have not been experienced; misbehaviour *motivated by ego*, often involves property damage to gain self-esteem from peers, but can also include verbal abuse and sex crimes; finally, misbehaviour motivated by *revenge*, aims to "punish" a staff member (e.g. because of a bad service). Rummelhagen and Benkenstein (2017) also identify that the actions of other customers can encourage misbehaviour.

The actions of misbehaving customers have an impact on the organization, employees, and other customers. The impact on the *organization* can be considerable because customer misbehaviour increases the costs to organizations (in terms of the resources required to resolve and compensate the customer), and also damages the firm's reputation (Kashif & Zarkada, 2015) which can result in a decrease in revenue. Customer misbehaviour can also negatively affect *employees'* health and work motivation (Salomonson & Fellesson, 2014). Staff may experience stress and anxiety (Grandey, 2003), demotivation, job dissatisfaction, and low morale as they struggle to perform their role (Kashif & Zarkada, 2015), resulting in high absenteeism, burnout, and increased staff turnover (Huang et al, 2010). *Other customers* may also be affected by misbehaving customers as their consumption experience is damaged (Rummelhagen & Benkenstein, 2017), which in turn can impact their perception of the firm and the quality of it as a service provider (Kashif & Zarkada, 2015; Grove et al, 2015).

It is therefore important that strategies for dealing with misbehaving customers are identified and put in place before the event occurs. Some service businesses have therefore adopted a proactive approach by implementing rules and protocols to avoid or reduce customer misconduct (Cai et al, 2018). For instance, some hotels do not allow children to stay as guests, some restaurants do not allow children at specified times and/or exclude them from of the restaurant to maintain a calm, relaxed, noise-free environment. In addition, some airlines, transport companies, and hotels also have "blacklists", and refuse to serve disruptive customers due to previous bad behaviour (Cai et al, 2018).

The actions taken in response to customer misbehaviour are very different to those in a service failure context. These can range from no action being taken to relocating (e.g. separating the misbehaving customer from other customers) or preventing further misbehaviour (e.g. using security and record-based systems to prevent the return of the misbehaving customer) to confronting the deviant customer via punishment or reporting the misbehaviour to the police (Boo et al, 2013). Where the customer misbehaviour involves fraud, as in the case of the Spanish holiday sickness scam, the initial response will be to implement service recovery activities for a service failure as the fraud is as yet unknown. It is only when the scam is identified as a scam by the service provider that a different approach will be considered and the misbehaviour addressed.

Business Ecosystems

The concept of the ecosystem originated in the biological sciences, where it comprises a range of species that are dependent upon each other (Tansley, 1935). In this system, each species contributes to the wellbeing and ultimate survival of the ecosystem as a whole. Ecosystem composition can change overtime, as the strength of one species increases and becomes more dominant and others weaken and may die out.

Ecosystems were first adopted by strategic management in 1993 in the seminal paper by Moore (1993). He used this concept to explain the reciprocity that exists within the business environment in which firms operate. Understanding the ecosystem and the ways in which it operates can provide important insight for businesses as it encourages them to recognize that their success (and indeed ultimate survival) is not dependent solely on their own actions, but the actions other actors within the ecosystem (Iansiti & Levien, 2004). Any changes within the ecosystem will have an impact on all ecosystem members. The business ecosystem, therefore, has the advantage that it can explain change dynamics and the consequences of these changes on ecosystem members (Moore, 1993).

Members, or actors, of a business ecosystem include a wide range of organizations and individuals. Moore defined an ecosystem as an "extended system of mutually supportive organizations; communities of customers, suppliers, lead producers, and other stakeholders, financing, trade associations, standard bodies, labour unions, governmental and quasi-governmental institutions and other interested parties" (Moore, 1998, p. 168). An ecosystem does not, therefore, mirror an industry, but can extend across different industries and actors in one ecosystem can also be a member of other ecosystems (Moore, 1993).

An ecosystem is not static, and it will evolve according to external and internal forces. A key *internal force* is co-evolution as actors affect and are affected by other actors within the ecosystem (Basole, 2009; Moore, 1993). Mäkinen and Dedehayir (2012) describe this process of co-evolution as organizations feeding-off, supporting, and interacting with one another; this usually occurs when they are exchanging knowledge and resources, and manufacturing products and services. Organizations within the ecosystem may therefore have a co-operative relationship, a competitive relationship, or a co-opetition relationship (Basole, 2009; Isckia, 2009). *External factors* are outside of the ecosystem and primarily originate from the environment in which the ecosystem is located. For example, changes in society or the economy will have an impact on the actors within the ecosystem bringing about change (Nehf, 2007).

In this chapter, an ecosystem perspective is adopted when examining the changes that took place in the Spanish tourism market. We identify the actors involved in the initial change that disrupted and destabilized the Spanish tourism ecosystem; the impact of this change which almost wiped out the Spanish hotels and resorts; and finally, we identify the ways in which ecosystem actors worked together to re-stabilize the ecosystem.

Social Responsibility and Social Marketing

Social responsibility has acquired great importance today, especially from a business, or corporate, perspective. Thus, CSR is "a concept whereby companies integrate social and environmental concerns in their business operations and in their interaction with their stakeholders on a voluntary basis" (Commission of the European Communities, 2001, p. 7). CSR can increase customers' positive attitudes and identification with a company, creating and promoting a positive corporate image. But it can also be useful in customer retention, protecting companies from losing customers when a service failure happens, by affecting consumer trust during service recovery evaluations (Albus, 2012): CSR enriches the positive effects of good service recovery; or, conversely, CSR has a halo effect on negative service recovery and mitigate its adverse effects (such as negative word-of-mouth). Thus, CSR has great importance in the hospitality industry since 1990s (Albus, 2012), focusing on community involvement, environmental management, and customer and employee relations. Many hoteliers have considered CSR very seriously, and the adoption of efficient service recovery policies to resolve service failures is an example of their societal approach.

In addition to companies, social responsibility also extends to governments and consumers. On the one hand, governments have focused their social responsibility on protecting consumers, defending their rights and interests through legislation (e.g. avoiding misleading advertising, preventing the marketing of unsafe products, demanding that complete information be provided to the consumers, etc.). On the other hand, social responsibility also concerns consumers. The responsible consumer incorporates social, environmental, and/or ethical considerations in their purchasing behaviour (Galan-Ladero, 2012), in addition to purely economic criteria.

But, what happens when a consumer is not socially responsible? Customer misbehaviour, and specifically fraudulent behaviour, is an example of this whereby customers try to benefit from socially responsible companies and their compensation policies (Kotler & Keller, 2016), without being entitled to them. In this context, social marketing can be a key instrument (Alves & Galan, 2019). Kotler and Zaltman (1971, p. 5) defined social marketing as "the design, implementation and control of programs calculated to influence the acceptance of social ideas, and implies considerations of product planning, price, communication, distribution, and marketing research". Social marketing, therefore, promotes social behaviour, a specific value, or a specific attitude, facilitating a behaviour or trying the acceptance of a certain idea, all considered beneficial for society, or for part of it. Or, on the contrary, it can also try to discourage or prevent behaviour, an attitude or idea that is socially undesirable, because it is considered harmful (Galan-Ladero & Galera-Casquet, 2019).

Social marketing often combines different complementary approaches to achieve its objective (Santesmases, 1999; Penelas et al, 2012), including legal (e.g. laws, regulations, recommendations, etc.), technological (e.g. product innovation to help carry out the desired behaviour), economic (e.g. affecting the price of cost of carrying out, or not, a certain behaviour), and informational (e.g. providing information to the target audience about the benefits or harms of a certain behaviour; and using different styles for messages – e.g. humour, fear, irony, persuasion, storytelling, etc.) – Alves & Galan (2019).

In this chapter, we will consider how social marketing was used to raise awareness of the problem of customer misbehaviour, informing and, in addition, warning about the legal consequences that false claims can have.

RESEARCH DESIGN

A single case study (Easton, 2010; Stake, 2003) was used. According to Yin (2014), case studies place an object in context, and therefore allow detail, depth, and richness of data (Yin, 2014). Our selected case was the

Spanish tourism market centring on the Spanish hotels/resorts, the CMCs and fraudulent tourists. Given our ecosystem perspective, we were also alert to the identification of any other actors who contributed to the issue under study, i.e., the holiday sickness scam. Data on this case was collected using secondary data from publicly available online documents relating to the holiday sickness scam.

Secondary data can be biased depending on the source (Farquhar, 2012) and therefore, a wide range of sources were examined to moderate this. These included articles from the popular press which provided in-depth details of the scam process and interviews with the perpetrators outlining their motivations and what influenced them to behave in this way. In addition, official online publications, such as industry and government reports, were reviewed which provided detailed descriptions of the measures that were adopted by the Spanish resorts as well as policymakers in Spain, the United Kingdom and the European Commission (a full list of the sources accessed can be found in Table 18.1). As the research team comprised both native Spanish and English speakers, publications in Spanish and English were collated and reviewed. This was considered important to ensure a balanced view and to understand the perspective of each nation.

Data collection covered a four-year period from 2015, when the claims volume began to be significant, to 2018, when it was essentially resolved. The search terms used (in Spanish and English) included "*reclamaciones falsas*" and "*demandas falsas*"/"false claims", "*falsas intoxicaciones*"/ "fake sickness claims", "*turistas ingleses*"/"English tourists", "*turistas británicos*"/"British tourists", "*denuncias falsas*"/"false allegations". The

TABLE 18.1

Secondary Data Online Sources

Sources (Online Versions)	Examples
Industry reports	ABTA (UK), Hotelier Associations (Spain).
Government reports	Foreign Office (UK), Spanish Government reports
Main national Spanish newspapers	ABC, El País, El Mundo
Other national online Spanish newspapers	El Confidencial, Última Hora, Diario Información.
Regional/Local Spanish newspapers	Diario de Mallorca, El Nacional.
Spanish tabloids/Magazines	Lecturas
British tabloids	The Sun, The Mail on Sunday

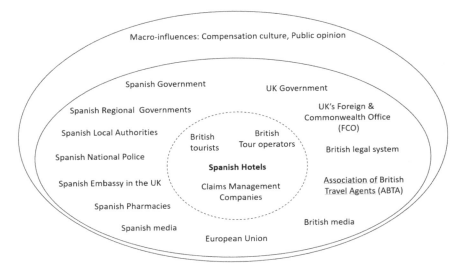

FIGURE 18.1
Spanish holiday sickness scam ecosystem from the perspective of the Spanish hotels.

final database of documents, therefore, provided a very detailed account of the scam and the activities of all relevant actors.

Data analysis was conducted in two phases. In *phase one*, all documents were read several times in order to gain familiarity with data. Following the repeated readings, a timeline was constructed of the events and actions that took place. This timeline was subsequently divided into three key stages: (1) the origins of the scam; (2) the impact of the scam; and (3) the resolution of the scam. In the *second phase* of the analysis, all actors within the case were identified in order to understand the ecosystem. Each actor was classified according to their role within the scam (see Figure 18.1).

THE CASE STUDY

The Origins of the Holiday Sickness Scam

CMCs have long been a part of the Spanish holiday ecosystem, initially playing a relatively small role within the Spanish Tourism ecosystem. CMCs act as intermediaries between claimants (the holidaymaker) and the companies being claimed against (in this case the UK tour operators who then reclaimed any money they lost from the Spanish hotels). CMCs help claimants with the whole claim dealing with the paperwork and administrative details for a fee which is either a percentage of the final payout or a flat lump sum, or both (MoneyExpert, 2020).

In 2015 the position of the CMCs within the ecosystem began to change as a growing number of CMCs started to operate in the main Spanish tourist areas, where they encouraged English holidaymakers to complain and make false claims against the hotels where they were holidaying in order to obtain a full refund plus compensation (Sorroche, 2016; Ruiz, 2016). The process was simple and involved a number of actors within the ecosystem (see Figure 18.1).

The CMCs recruited customers via sales representatives at all-inclusive establishments in tourist resorts during their holiday or after their holiday via the Internet searching social networks for English people to contact who had uploaded a photo of their holiday in Spain. They presented a standard form with details of an alleged illness – usually, gastroenteritis or digestive system disorders – and their potential compensation on a "no win no fee" basis[1] (Pascual, 2017).

The "British tourist holidaymaker" submitted a claim to the "British tour operator" from which he/she had purchased their holiday. In the claim, the holidaymaker declared they had suffered food poisoning from food eaten at the hotel they had stayed at. Claims were mainly made against "all-inclusive hotels" as it was more difficult for these businesses to prove that holidaymakers could fall ill by eating food outside of their facilities. "British judges" considered that these tourists spent all their time in the hotel, and only ate in its restaurant; therefore, if they fell sick, liability automatically fell on these hotels (Pascual, 2017).

A "pharmacy" receipt was sufficient to provide evidence that the customer had been ill and that a drug was needed. The receipt was attached to the claim, which was presented to the "British Justice". Sometimes, medical reports were also attached, but they were usually written several months later. Although most British tourists were aware of this practice, the fraud was usually carried out by younger people (Pascual, 2017; El Nacional, 2017), with scarce academic and economic resources.

Under British Law, the travel agent had to prove that there was no damage. This was the key to the scam. Clients could submit their claim for food poisoning up to five years later. The tour operator had to prove that they had not been the source of the food poisoning.

The "British tour operators" waived litigation, to avoid the high court costs of British Justice (even without clear evidence as litigation could cost up to £50,000 per claim in the United Kingdom – Magro, 2017). They paid the claims (including the legal assistance expenses) and then sought to recover the cost from the Spanish hotel that was compelled to pay the claim as the contract signed with the British tour operators included liability clauses for any damages.

The "Spanish hoteliers" had no previous knowledge of the individual claims and no opportunity to challenge the details or payments to customers. They only discovered the claim when they suffered a direct discount in the periodic payments they received from the British tour operators.

From an ecosystem perspective, the CMCs changed their role within the ecosystem, from a relatively minor, reactive player responding to holidaymaker claims due to sickness to a proactive role seeking out valid claimants and encouraging others to make fraudulent claims. In so doing the CMCs disrupted the ecosystem homeostasis (Morgan & Brown, 2001), i.e., the prevailing stability between the actors, and in particular that relating to the Spanish hotels. This was largely made possible due to a key change in the external environment (Nehf, 2007). In this specific case, there was a changing attitude among the British public as they became more litigious and a compensation culture was created (see macro-influences in Figure 18.1). This change in culture was best described by Young (2010) in his report to Government that stated "Today accident victims are given the impression that they may be entitled to handsome rewards just for making a claim regardless of any personal responsibility" (Young 2010, p. 7). The act of claiming compensation when a loss had been experienced not only became more acceptable, but perceptions of the unacceptability of fraudulent claims were also being challenged.

Impact of the Scam

At the beginning of 2017, the hotels were alerted through the industry's main tourist associations of the problem with misbehaving customers. Fraudulent claims had increased from under 1,000 in 2015, to more than 10,000 in the first few months of 2016, mostly due to digestive disorders and food poisoning claims, despite no food alerts having been reported in those areas (Pascual, 2017). Hoteliers in Benidorm, a popular town with English tourists, estimated that they alone had lost several million Euros due to the false claims in just a few years. By the end of 2016, British tourists' claims year-on-year increased 700% in the province with a cost of 13–15 million Euros (Pascual, 2017). In Majorca, another popular tourist destination, the estimated figure reached 50 million Euros in the first nine months of 2016 (Magro, 2017). Over the coming year the fraudulent behaviour extended to most Spanish resorts, with the main targets being all-inclusive establishments (Magro, 2017).

The CMCs not only claimed for their client's refund (each Briton usually spent an average of between £800 and £1,000 per holiday – Pascual, 2017), they also included a large amount for legal assistance (ABC, 2016): around £3,000 as a minimum per claim (Ellery et al, 2016). As a result, a typical claim per customer was between £5,000 and £6,000 (Pascual, 2017), well in excess of the original cost of the holiday. Some CMCs guaranteed the client up to 18,000 Euros net of compensation and 98% success in the lawsuit, being the distribution of the compensation 60% for the CMC and 40% for the client (Bohorquez, 2018). This situation became unsustainable for the sector and many hoteliers considered cancelling their agreements with the main British tour operators and thereby British market tourists (Dollimore, 2016; Sorroche, 2016).

Within a healthy ecosystem, participants recognize their interdependency, they understand the system-wide impact of their actions (Iansiti & Levien, 2004). In this case, the actions of the CMCs in conjunction with the British tourists and British tour operators resulting in an imbalance in the ecosystem, impacting negatively on other actors, namely the Spanish hotels. The very survival of this ecosystem was threatened. If the ecosystem was to continue, then action needed to be taken to rebalance the ecosystem and this would require the co-operation of different members of the ecosystem in order to resolve the issue (Moore, 1993).

Resolution of the Scam

How the Industry Worked with Members of the Ecosystem to Address and Resolve the Issue

The "modus operandi" was of such importance that the situation was considered an emergency (El Mundo, 2016). "Hoteliers" reacted firmly and called for greater involvement of government authorities as they sought to raise awareness of the behaviour of the British judicial authorities

(Sorroche, 2016). Turespaña, the organization that represents the interests of Spanish tourism around the world, had tried and failed to raise the issue by itself from its London delegation (Magro, 2017). It was clear that only with the co-operation and support of different actors within the ecosystem that stability would be returned, and the survival of the Spanish hotels secured. "Local Authorities", "Regional Governments", "Spanish Government", the "UK Government and the European Union" were all called upon to help resolve the situation.

In response, the "Spanish Secretary of State for Tourism" took charge of the matter personally due to the Government's concern that the fake claims from British holidaymakers could damage Spain's reputation as a tourist destination. The Spanish Embassy in London, through the Minister of Tourism, wrote a report about the crisis of gastric complaints and concluded that "the Spanish hotel sector should seek alternatives to the British client that comes through the tour" (Magro, 2017).

In the United Kingdom, the Government was also taking action (Europa Press, 2018). The Spanish Government consulted with the "UK's Foreign & Commonwealth Office (FCO)" on potential measures to solve the problem, and the FCO itself issued a statement confirming that Spain was a secure destination and warning that to make fake claims was a crime that would result in prosecution (Dollimore 2016).

The battle was also taken to the "European Union" in Brussels, to challenge the British tour operators' clauses that enabled the tour operators to decline all responsibility for damage that holidaymakers could suffer in the hotels, although these travellers paid directly to the tour operators. But both the British consumer protection system and the EU Directive regulating package travel (2015/2302) at that time made the tour operator responsible for everything that happened during the trip, including what happened at the hotel (Magro, 2017). Hoteliers were required to waive Spanish legislation and submit to the "British Justice" in cases of litigation which favoured the English customer (since a 2013 reform and the introduction of "no win, no fee", consumers were required to spend nothing to pursue their claims and only paid a fee if the case was won and they received compensation – Magro, 2017). In summary, the key to resolving this issue was for the British tour operators to give up the clause.

In the meantime, the "hoteliers" implemented their own measures. For example, when holidaymakers had finished their vacations, they were required to sign a statement stating that they had not suffered any illness and they were reminded that false claims were a punishable offense in Spain (Pascual, 2017). Although the document was only valid in Spain, this was done to try to change the judicial mechanism that operated from the United Kingdom (Pascual, 2017). Most all-inclusive Spanish hotels also introduced electronic wristbands for holidaymakers, which were scanned when they ordered any food or drink (Ellery & Harwood, 2017). Other hotels started to offer half-board packages instead of all-inclusive ones to reduce the opportunity for holidaymaker to claim due to food poisoning (Jowaheer, 2017).

The hoteliers also contacted "pharmacies" in the tourist areas and asked them not to sell stomach remedies to British holidaymakers without a prescription (Powell & Ellery, 2017). In addition, the hoteliers enlisted the national and international media (the British media in particular) to help them (El Mundo, 2016). The British newspapers, such as The Mail On Sunday, conducted their own investigations, as they were concerned about how this issue was impacting the perception of English tourists and the possibility that the scam could result in an increase in the price of Spanish package holidays.

"Hoteliers" sought to remove the CMCs representatives from their resorts via legal as they were in effect inciting a crime of fraud (El Mundo, 2016). The representatives recruited potential claimants on the street, without any type of license or professional qualification as lawyers (they are unlicensed businesses). Furthermore, they did not comply with tax obligations and regulations (they were not registered in the Spanish Social Security System and, in consequence, they were not making social security contributions – Dollimore, 2016). The first representative to be arrested for encouraging tourists to make false food poisoning claims was in Majorca, in June 2017. In September 2017, another seven were arrested, and were considered members of a criminal gang (Bohorquez, 2018).

In addition, HOSBEC, a hotelier association, launched a social marketing campaign in the United Kingdom (*#BeHonest*), in collaboration with the Valencian Tourism Agency, in 2017, to raise awareness of the consequences of false claims (Martinez, 2018) using social networks and influencers in this campaign.

Finally, the "British tour operators" cooperated with hoteliers and sought to strengthen their relationship by (1) curbing claims at source, especially when sharing information. One of largest British travel operators hired private investigators to probe ongoing fraud cases (Ellery, 2016). Another tour operator wrote to specialist holiday sickness claim companies to caution them that fraudulent claims were unacceptable (Cain, 2016); (2) renegotiate contracts and limit their liability in case of a new wave of claims; and (3) pressurize the British Government to change the consumer law.

Reactions in the United Kingdom: British Government's Reaction and Others

Following Spanish pressure and findings from a popular UK newspaper (the Mail on Sunday), the "British Ministry of Justice", which regulates the claims industry, launched an investigation. It considered that the claims were serious (British tourists were being given a bad name abroad – Morris, 2017), and promised tough action against firms that broke the rules (Ellery et al, 2016). The scam had spread globally, with similar stories at all-inclusive destinations including in Portugal, Greece, Cyprus, and Turkey, Egypt, and Tunisia, and even the Caribbean. In American hotels, $100,000 per month had been claimed by British clients (Cain, 2016).

In response, the UK Government stopped lawyers paying claims firms money for each case they passed on and considered of introducing an upper limit or "cap" on solicitors' fees (Ellery et al, 2016). Under such proposals, a solicitor would not be able to recover their costs from the losing party if the claim was valued at under £5,000. This would be a disincentive for lawyers to take on such cases. This important change in law was approved in April 2018 in advance of the summer holiday season.

In addition, the Government forewarned Spanish holidaymakers travelling that fraudulent sickness claims could result in a custodial sentence. The Foreign Office added the following wording to its webpage: "There have been reports of a rise in claims companies targeting holidaymakers at resorts in Spain inviting them to make false insurance claims regarding holiday sickness. Making a fraudulent insurance claim is illegal and a criminal offence in Spain and can incur penalties such as heavy fines and imprisonment".

The "Claims Management Regulator" also met with the "Association of British Travel Agents (ABTA)", the leading association of travel agents and tour operators in the United Kingdom, and was working with "tour operators", "Spanish hoteliers", and the "Spanish police" to crack down on the rogues. ABTA had also been calling for the British Government to take further quick and decisive action to close the legal loopholes that we were being exploited by these "cowboy companies". In June 2017, ABTA and travel industry partners launched "Stop Sickness Scams" campaign (ABTA, 2017).

The campaign was supported on social media by sharing materials and using #StopSicknessScams. Thus, for example, ABTA explained the problem and the consequences of the fake claims in its website[2] ("submitting a fraudulent claim is a criminal offence in the United Kingdom"), and offered a variety of videos that had been also uploaded to YouTube. Some videos were cartoons, warning about the scam ("Don't get involved" and "Report the tout"); other videos were about holidaymakers speaking about holiday sickness scams; and, finally, other videos showed industry representatives discussing the holiday sickness scams and its destructive effect on the travel industry. Thus, this social marketing campaign combined informational, legal, and economic approaches (Santesmases, 1999), to persuade and avoid British holidaymakers to be involved in this scam. Although the message was merely informative, it also stimulated feelings of fear (it is "a criminal offence which could result in a criminal record and fine").

By the end of 2017, there were the first tangible signs that the battle with the fraudsters was at last showing signs that a victory was in sight when the first "British Court judgements" for false claims imposed prison sentences and fines to the British tourists involved. These judgements began to dissuade potential claimants, who retracted their claims (Hosteltur, 2018). This led to a drastic reduction in false claims, until they had almost disappeared in 2018.

CONCLUSIONS

Many CSR companies aim to improve the quality of the services they provide to their customers and operate good service recovery policies to recover customer satisfaction after a service failure. Misbehaving customers seek to exploit socially responsible companies when their service has not failed. The Spanish holiday sickness scam is an example of this and highlights that not all customers are socially responsible and can be encouraged by others to behave in this way. By examining misbehaving customers, the opposite of socially responsible customers, this study aimed to stimulate a wider discussion on socially responsible consumers.

Extant literature points to a range of strategies that companies can adopt to recover service failure and address misbehaving customers. In this scam, the false claims were handled as "business as usual", claimants were compensated for their "loss". Although many studies have supported monetary compensation as a means to retain clients (Singh & Crisafulli, 2017), this case identifies the weakness of this strategy. It was only when the volume of claims became untenable, that action was taken to address the problem. Social marketing can be a key tool to educate consumers, to raise awareness about consumer misbehaviour, influencing their attitudes and behaviours so that they act ethically and responsibly, and to warn to the legal consequences that fake claims can have. This case has reviewed how social media was included as part of a wider strategy to return misbehaving consumer to more socially responsible customers.

Application of the business ecosystem provided valuable insights, it clearly identified the need to closely monitor the wider business ecosystem to identify and respond to changes to protect an industry. In addition, it highlighted the strength of adopting a business ecosystem perspective as the Spanish hotels drew on different actors (e.g. governments, police, and business associations) from across their ecosystem to address the issue and return stability within their industry.

LESSONS LEARNED

1. Service failure is unavoidable for all organizations and industries; service recovery is important to reinstate customer satisfaction after a service failure.
2. CSR companies should improve the quality of the services that they provide to their customers, and also consider good service recovery policies to maintain and recover customer satisfaction after a service failure.
3. Misbehaving customers try to exploit socially responsible companies and their service recovery policies and, although the service has not failed, they fraudulently claim compensation.
4. Extant literature points to a range of strategies that companies can adopt to recover service failure and also address misbehaving customers.

5. Although many studies have demonstrated the importance and usefulness of the service recovery strategy based on monetary compensation in the tourism sector to retain clients and build customer loyalty (see, e.g. Singh & Crisafulli, 2017), this case reveals that managers should apply it carefully.
6. The business ecosystem can provide valuable insights to businesses. It clearly identifies the need to monitor closely the wider business ecosystem, to identify and respond to changes to protect an industry and its stakeholders.
7. Social marketing can be a key tool to educate consumers, to raise awareness about consumer misbehaviour, influencing their attitudes and behaviours so that they act ethically and responsibly, and to warn of the legal consequences that fake claims can have.

DISCUSSION QUESTIONS

1. Identify the consequences that unsocial and irresponsible customer behaviour had for each of the different actors in the business ecosystem.
2. Discuss the possible business practices that could be adopted in addressing unsociable and irresponsible behaviour. Estimate the effectiveness that they could have.
3. Identify a similar scam in your own country, or in other countries. Explain what happened and comment on the consequences.

PROJECT/ACTIVITY-BASED ASSIGNMENT/EXERCISE

1. Role-playing
   Divide the class in different groups, each group representing one different actor in this business ecosystem. Each group should defend the assigned actor's interests (first, without considering CSR and ethics; later, adopting socially responsible behaviour). What are the differences and similarities?
2. Exercise
   Search other similar cases of scams in your country, or in other parts of the world. How were those cases resolved?

NOTES

1. In a "no win no fee" contract the fee for services is only payable when the client has a favourable result.
2. More information in: https://www.abta.com/tips-and-advice/staying-safe-on-holiday/stop-sickness-scams-and-what-do-when-you-have-genuine-claim

REFERENCES

ABC (2016, September 15). *Abogados británicos incitan a sus compatriotas a hacer reclamaciones falsas a hoteles de Benidorm*. Retrievable from: http://www.abc.es/espana/comunidad-valenciana/abci-abogados-britanicos-incitan-compatriotas-hacer-reclamaciones-falsas-hoteles-benidorm-201609152058_noticia.html

ABTA (2017). *Stop sickness scams*. Retrievable from: https://abta.com/tips-and-latest/abta-campaigns/holiday-sickness

Albus, H. (2012). *The effects of corporate social responsibility on service recovery evaluations in casual dining restaurants* [Master Thesis]. University of Central Florida, USA.

Alves, H.M., & Galan-Ladero, M.M. (2019). Theoretical background: Introduction to social marketing. In M.M. Galan-Ladero & H.M. Alves, (Eds.), *Case Studies on Social Marketing. A Global Perspective* (pp. 1–10). Switzerland: Springer.

Ayertey, S.B. (2018). *An evaluation of online service failures and recovery strategies in the UK fashion industry* [Doctoral Thesis]. Plymouth Business School. University of Plymouth, UK.

Basole, R.C. (2009). Visualization of interfirm relations in a converging mobile ecosystem. *Journal of Information Technology, 24*, 144–159.

Bitner, M.J., Booms, B.H., & Teatreault, M.S. (1990). The service encounter: Diagnosing favorable and unfavorable incidents, *Journal of Marketing, 54*(January), 71–84.

Bohorquez, L. (2018, January 19). *Así funcionaba la trama de las falsas intoxicaciones de británicos en hotels de Mallorca*. Retrievable from: https://elpais.com/economia/2018/01/18/actualidad/1516291141_447366.html

Boo, H.C., Mattila, A.S., & Tan, C.Y. (2013). Effectiveness of recovery actions on deviant customer behavior – The moderating role of gender. *International Journal of Hospitality Management, 35*, 180–192.

Cai, R., Lu, L., & Gursoy, D. (2018). Effect of disruptive customer behaviors on others' overall service experience: An appraisal theory perspective. *Tourism Management, 69*, 330–344.

Cain, K. (2016, November 16). Bogus Comp Warning. Brits face ban from Spanish resorts after sharp spike in number of fake sickness claims. *The Sun*.

Commission of the European Communities (2001). *Green Paper. Promoting a European framework for Corporate Social Responsibility*. Brussels, 18.7.2001 COM (2001) 366 final.

Daunt, K.L., & Greer, D.A. (2015). Unpacking the perceived opportunity to misbehave: The influence of spatio-temporal and social dimensions on consumer misbehavior. *European Journal of Marketing, 49*(9/10), 1505–1526.

Daunt, K.L., & Harris, L.C. (2011). Customers acting badly: Evidence from the hospitality industry. *Journal of Business Research, 64*, 1034–1042.
Dollimore, L. (2016, November 16). Britons to be banned from holidays in Spain over fake sickness claims. *The Olive Press*. Retrievable from: http://www.theolivepress.es/spain-news/2016/11/16/britons-to-be-banned-from-holidays-in-spain-over-fake-sickness-claims/
Easton, G. (2010). Critical realism in case study research. *Industrial Marketing Management, 39*(1), 118–128.
El Mundo (2016, September 16). Los hoteleros plantan cara a la 'picaresca' británica". Retrievable from: http://www.elmundo.es/comunidad-valenciana/alicante/2016/09/16/57db9b7346163f50518b4570.html
El Nacional (2017, March 31). Las falsas gastroenteritis de los turistas británicos en la Costa Dorada. Retrievable from: http://www.elnacional.cat/es/sociedad/fraude-costa-daurada-hoteles_148212_102.html
Ellery, B. (2016, May 12). Caught on camera, the holiday bug cowboy urging tourists to lie for cash: Travel firm uses 'spies' to rumble rogue company over bogus sick claims as experts warn rise in practice could cost the industry millions. *The Mail on Sunday*.
Ellery, B., & Harwood, A. (2017, February 19). The (fake) sick man of Europe. The Scottish mail on Sunday. Retrievable from: https://www.pressreader.com/uk/the-scottish-mail-on-sunday/20170219/281655369836149
Ellery, B., Manning, S., Gallagher, I., & Wace, C. (2016, September 25). Cowboy firms are systematically coaching British tourists to lie about holiday sickness to win thousands of pounds in compensation. *The Mail on Sunday*.
Europa Press (2018, April 18). Nuevas medidas en Reino Unido para frenar las reclamaciones falsas por enfermedad en vacaciones. Retrievable from: https://www.europapress.es/turismo/mundo/noticia-nuevas-medidas-reino-unido-frenar-reclamaciones-falsas-enfermedad-vacaciones-20180418142341.html
Farquhar, J.D. (2012). *Case study research for business*. London, UK: SAGE.
Fisk, R., Grove, S., Harris, L., Keeffe, D., Daunt, K.L., Russell-Bennett, R., & Wirtz, J. (2010). Customers behaving badly: A state of the art review, research agenda and implications for practitioners. *Journal of Services Marketing, 24*/6, 417–429.
Fullerton, R.A., & Punj, G. (1993). Choosing to misbehave: A structural model of aberrant consumer behavior. In L. McAlister & M.L. Rothschild (Eds.), *NA – Advances in Consumer Research* (Vol. 20; pp. 570–574). Provo, UT: Association for Consumer Research.
Fullerton, R.A., & Punj, G. (1997). What is consumer misbehavior. *Advances in Consumer Research, 24*, 336–339.
Fullerton, R.A., & Punj, G. (2004). Repercussions of promoting an ideology of consumption: Consumer misbehavior. *Journal of Business Research, 57*, 1239–1249.
Galan-Ladero, M.M. (2012). *Variables que influyen en la actitud hacia el marketing con causa y determinantes de la satisfacción y la lealtad en la 'compra solidaria'* [Doctoral Thesis]. Caceres: Servicio de Publicaciones de la Universidad de Extremadura.
Galan-Ladero, M.M., & Alves, H.M. (2019). *Case studies on social marketing. A global perspective*. Switzerland: Springer.
Galan-Ladero, M.M., & Galera-Casquet, C. (2018). What happens when customers lie? The case of Holiday Sickness Scam. *17th International Congress on Public and Nonprofit Marketing*. 6–7 September. Bournemouth, UK.

Galan-Ladero, M.M., & Galera-Casquet, C. (2019). Marketing para el cambio social. In L.M. Cerdá & M. Ramírez (Coord.), *Fundamentos para un Nuevo Marketing* (pp. 251–284). Madrid: Sinderesis.

Grande, I. (2005). *Marketing de los Servicios* (4th ed.). Madrid: ESIC.

Grandey, A.A. (2003). When the show must go on: Surface and deep acting as determinants of emotional exhaustion and peer-rated service delivery. *Academy of Management Journal, 46,* 86–96.

Greer, D.A. (2015). Defective co-creation: Developing a typology of consumer dysfunction in professional services. *European Journal of Marketing, 49*(1/2), 238–261.

Grove, S., Fisk, R., Harris, L., Ogbanna, E., John, J., Carlson, L., & Goolsby J. (2015). Disservice: A framework of sources and solutions. In L. Robinson, Jr. (Ed.), *Marketing Dynamism & Sustainability: Things Change, Things Stay the Same.... Developments in Marketing Science: Proceedings of the Academy of Marketing Science* (pp. 169–172). Heidelberg: Springer.

Harris, L.C., & Reynolds, K.L. (2004). Jaycustomer behavior: an exploration into the types and motives in the hospitality industry. *Journal of Services Marketing, 18*(5), 339–357.

Hosteltur (2018, May 11). Adiós a las reclamaciones falsas: los juicios y la cárcel surten efecto. Retrievable from: https://www.hosteltur.com/128088_adios-reclamaciones-falsas-juicios-carcel-surten-efecto.html

Hu, H., Hu, H., & King, B. (2017). Impacts of misbehaving air passengers on frontline employees: Role stress and emotional labor, *International Journal of Contemporary Hospitality Management, 29*(7), 1793–1813.

Huang, W.H., Lin, Y.C., & Wen, Y.C. (2010). Attributions and outcomes of customer misbehavior. *Journal of Business Psychology, 25,* 151–161.

Iansiti, M., & Levien, R. (2004). Strategy as ecology. *Harvard Business Review, 82*(3), 68–81.

Isckia, T. (2009). Amazon's evolving ecosystem: A cyber-bookstore and application service provider. *Canadian Journal of Administrative Sciences, 26,* 332–343.

Jowaheer, R. (2017, June 10). Brits arrested in Spain 'for encouraging fake illness claims'. Retrievable from: www.aol.co.uk

Kashif, M., & Zarkada, A. (2015). Value co-destruction between customers and frontline employees: A social system perspective. *International Journal of Bank Marketing, 33*(6), 672–691.

Keaveney, S. M. (1995). Customer switching behavior in service industries: An exploratory study. *Journal of Marketing, 59*(2), 71–82.

Kotler, P., & Keller, K.L. (2016). *Dirección de Marketing* (15th ed.). Mexico: Pearson.

Kotler, P., & Zaltman, G. (1971). Social marketing: An approach to planned social change. *Journal of Marketing, 35*(3), 3–12.

Magro, A. (2017, February 26). Los turistas ingleses hurtan a los hoteleros 50 millones al año con demandas falsas. *Diario de Mallorca, Economía.* Retrievable from: http://www.diariodemallorca.es/mallorca/2017/02/26/fraude-masivo-lleva-hoteleros-pensar/1192824.html

Mäkinen, S.J., & Dedehayir, O. (2012). Business ecosystem evolution and strategic considerations: A literature review. In the *18th International ICE Conference on Engineering, Technology and Innovation,* 1–10. IEEE.

Martinez, D. (2018, October 31). Prisión para cuatro turistas británicos por una denuncia falsa contra un hotel de Benidorm. *Alicante Plaza*. Retrievable from: https://alicanteplaza.es/prision-para-cuatro-turistas-britanicos-por-una-denuncia-falsa-contra-un-hotel-de-benidorm

Miller, J.L., Craighead, C.W., & Karwan, K. (2000). Service recovery: A framework and empirical investigation. *Journal of Operations Management*, 18(4), 387–400.

MoneyExpert (2020). Claims management companies. Retrievable from: https://www.moneyexpert.com/credit-cards/claims-management-companies/

Moore, J.F. (1993). Predators and prey: A new ecology of competition. *Harvard Business Review*, 71(3), 75–86.

Moore, J.F. (1998). The rise of a new corporate form. *Washington Quarterly*, 21(1), 167–181.

Morgan, E.S.K., & Brown, J.H. (2001). Homeostasis and compensation: The role of species and resources in ecosystem stability. *Ecology*, 82(8), 2118–2132.

Morris, H. (2017, June 21). Compensation culture is ruining Britain's reputation abroad, tour operators warm. *The Telegraph*. Retrievable from: https://www.telegraph.co.uk/travel/news/false-sickness-claims-hurting-british-reputation-abroad/

Nehf, J.P. (2007). Shopping for privacy on the internet. *The Journal of Consumer Affairs*, 41(2), 351–375.

Pascual, A. (2017, February 19). Los hoteles de Benidorm buscan sangre en las toallas para atajar la estafa británica. *El Confidencial*. Retrievable from: http://www.elconfidencial.com/espana/2017-02-19/benidorm-hoteles-estafa-claim-farms-britanicos-rincon-de-turismo_1334195/

Penelas, A., Galera, C., Galan, M., & Valero, V. (2012). *Marketing Solidario. El marketing en las Organizaciones No Lucrativas*. Madrid: Piramide.

Powell, M., & Ellery, B. (2017, May 28). Hotel sues 'fake' holiday bug pair for. *Pressreader. The Scottish Mail on Sunday*. Retrievable from: https://www.pressreader.com/uk/the-scottish-mail-on-sunday/20170528/281758449240258

Ruiz, J.L. (2016, September 5). Abogados británicos incitan a turistas a reclamar contra hoteles de Mallorca. *Ultima hora*. Retrievable from: https://ultimahora.es/noticias/local/2016/09/05/218414/abogados-britanicos-incitan-turistas-reclamar-contra-hoteles-mallorca.html

Rummelhagen, K., & Benkenstein, M. (2017). Whose fault is it?: an empirical study on the impact of responsibility attribution for customer misbehavior. *European Journal of Marketing*, 51(11/12), 1856–1875.

Salomonson, N., & Fellesson, M. (2014). Tricks and tactics used against troublesome travelers—Frontline staff's experiences from Swedish buses and trains. *Research in Transportation Business & Management*, 10, 53–59.

Santesmases, M. (1999). *Marketing. Conceptos y Estrategias* (4th ed.). Madrid: Piramide.

Singh, J., & Crisafulli, B. (2017). Case Study 4: Managing customer complaints: The case of Imperial Orchid Hotels in Thailand. In S.K. Roy, D.S. Mutum, & N. Bang (Eds.), *Services Marketing Cases in Emerging Markets*. Switzerland: Springer International Publishing.

Smith, A.K., Bolton, R.N., & Wagner, J. (1999). A model of customer satisfaction with service encounters involving failure and recovery. *Journal of Marketing Research*, 36(3), 356–372.

Sorroche, A.S. (2016, September 17). Los hoteles denuncian pérdidas millonarias por las reclamaciones falsas de los turistas ingleses. *Diario Información*. Retrievable from: http://www.diarioinformacion.com/benidorm/2016/09/16/hoteles-denuncian-perdidas-millonarias/1806227.html

Stake, R. (2003). Case studies. In N.K. Denzin & Y.S. Lincoln (Eds.), Strategies of qualitative inquiry (2nd ed.) (pp. 134–164). Thousand Oaks, CA: SAGE.

Tansley, A.G. (1935). The use and abuse of vegetational concepts and terms. *Ecology, 16*(3), 284–307.

Yin, R.K. (2014). *Case study research design and methods* (5th ed.). Thousand Oaks, CA: SAGE.

Young, D.Y.B. (2010). *Common Sense Common Safety: A Report by Lord Young of Graffham to the Prime Minister Following a Whitehall {u2011} wide Review of the Operation of Health and Safety Laws and the Growth of the Compensation Culture*. Cabinet Office.

Section X

Pedagogical Directions and Best Practices: Imparting Social and Sustainability Marketing Competencies

This section discusses pedagogical approaches that can be employed by teachers and trainers in their classrooms or workplaces to effectively impart the social and sustainability marketing competencies.

19

Case Method as an Effective Pedagogical Tool: Some Insights for Better Learning Outcomes for Social and Sustainability Marketing Educators

Chandana R. Hewege
Faculty of Business & Law, Swinburne University of Technology, Melbourne, Australia

CONTENTS

Learning Objectives 537
FIDSAI Framework 541
    Step 1: Familiarise 541
    Step 2: Immerse 541
    Step 3: Dissect 541
    Step 4: Sense-Making 543
    Step 5: Appraise Alternative Solutions 543
    Step 6: Internalise 543
Funding Declaration 544
Acknowledgment 544
Credit Author Statement 544
Reference 544

LEARNING OBJECTIVES

After reading the chapter, the reader should be able to:

- Understand the deep learning aspects of case method as a pedagogical tool

- Demonstrate how FIDSAI tool can be applied for achieving effective learning outcomes
- Gain an overview of the case studies included in the book

Social and Sustainability Marketing: A Casebook for Reaching Your Socially Responsible Consumers through Marketing Science contains valuable resources for "those who want to know" as well as for "those who want to understand" the intricacies of social and sustainability marketing approaches. Acquiring knowledge (knowing) and understanding are two distinct, but related cognitive processes. These two are not mutually exclusive. Knowledge is a prerequisite for understanding, yet a person can know something without understanding. In the modern-day world, teachers and learners are constantly being overwhelmed by colossus amount of information millisecond by millisecond through diverse virtual media. Making a good sense of this information is essential to understand the big picture related to a life event or business decision-making scenario. We, as teachers and trainers, are confronted with the daunting task of guiding and helping our learners to understand and develop wisdom going beyond memorising knowledge, *Knowing*. I believe that proper understanding of a phenomenon by a learner or a trainee would help him or her internalise the multiple perspectives presented by a given situation and thereby would help a learner undergo a positive attitudinal change, resulting in significantly modified behavioural intentions or actions. As such, pedagogical tools that enhance holistic understanding of intricate and multifaceted forces operating simultaneously at a given scenario need to be used by educators with a view to inducing their learners to produce intended outcomes. Case studies used in the case method of teaching could be considered an effective pedagogical tool in this regard.

In the field of medicine, students aspiring to become medical doctors dissect bodies and examine live patients to understand how the human body functions in real-life situations. Likewise, law students work with their seniors and engage in actual "court cases" to understand the complex legal procedures. Students of science perform laboratory experiments to create and simulate real-life conditions to understand how various elements behave. However, business students may not be able to experience simulations as effectively as medical, law and science students would do. The closet approximation to real-world business/marketing situations can be created through case studies. This is why, case method as a pedagogical tool is still considered by teachers and trainers as an effective way of imparting competencies that lead to proper understanding of the whole scenario presented to a learner or trainee. A case study used for pedagogical purposes can simply be described as a written (or multimedia illustrated) account of a problem situation or a scenario describing various complex interactions among actors and things at a given context and time. However, case study research methodology should be distinguished from a case study teaching methodology. Our focus here is on pedagogical aspects of case studies. According to Yin (1994, p. 13), in the context of case study research method, a case is referred to as "an empirical inquiry that investigates a contemporary phenomenon within its real-life context, especially when the boundaries between phenomenon and context are not clearly evident…relies on multiple sources of evidence". From this definition, it is clear that actual, real-life cases are written as contemporary phenomena locating them within a real-life context. In actual complex problem situations, phenomenon or the decision situation faced by key actors cannot easily be distinguished from its context. By reviewing and analysing a case study, a learner or a trainee could understand the multiple causes affecting a situation. Clear identification of multiple causes of a situation would enable a learner to understand the situation and generate appropriate solutions.

Albert Einstein once said "problems cannot be solved at the same level of awareness that created them". To find solutions to complex problems, it requires a higher level of awareness and understanding of the problem context and the causes. Marketing discipline and practice have come of age over the last few decades. Nevertheless, social and sustainability issues tend to pose new challenges for marketing academics and practitioners in the contemporary world requiring them to think of a paradigm shift which demands a new level of awareness and understanding. Expanding boundaries of the current thinking process to develop innovative solutions to wicked problems can be considered the main outcome of a paradigm shift. Thinking, feeling and acting in a way that it preserves the social wellbeing of consumers and the sustainability of Mother Nature while generating a surplus (profit) for a firm needs to be at the centre of marketing strategy making. To inculcate social and sustainability perspectives in the mindset of marketing thinkers and practitioners, it is essential that they are exposed to diverse contexts of social and sustainability issues through real-life case scenarios. We believe that this edited book serves this purpose by presenting short and longer versions of real life case studies.

Prior theory or conceptual frameworks play a significant role in analysing a case study. Making sense of multiple forces operating in a given scenario described in a case study is fundamental to understanding the situation holistically. This is an essential prerequisite for generating valid and effective solutions. In the first section of this edited book, we present some original research papers describing various theoretical approaches and conceptual frameworks to be used as lenses through which the case events should be viewed in order to make sense of the overall situation. Case analysis without any theory or perspectives would be just "knowing" not "understanding". Theories or conceptual frameworks can shed more light on the issues being presented in a given case study. This is an essential part of the case analysis.

Original research papers on social and sustainability marketing that we present in Section II of this book include themes such as understanding sustainability marketing, sustainable consumption and consumer behaviour, understanding sharing economy and marketing, understanding social marketing, harnessing the power on online media, addressing global crises, sustainability reporting practices by social enterprises and safeguarding against unsocial and irresponsible customers. These scholarly work by authors having expertise in diverse aspects of social and sustainability marketing would provide you with some theoretical perspectives. I believe that these works could serve as theoretical lenses for your sense-making process when analysing case studies. Please be aware that this is not an exhaustive list of scholarly works on social and sustainable marketing. You are encouraged to consult other related resources suited to the scenarios described in a case.

Case analysis is an art. This exercise partly offers a learner the same mental satisfaction he or she could derive from solving a complex puzzle. Moreover, case analysis can be intellectually rewarding when possible solutions are found to resolve the issues presented in the case scenario. For a general reader who wants to understand as to how social and sustainability marketing is applied in practice, the cases contained in this book would be of immense value. For educators who want to impart knowledge, attitude and skills related to social and sustainability marketing, the discussion questions and learning activities provided at the end of each case can be very handy.

FIDSAI FRAMEWORK

Many experts of case methods have introduced various steps to be followed when analysing a case. All these analytical approaches have merits on their own account. However, I would like to propose a simple stepwise approach to make the best use of the cases presented in this book. These steps could be followed by self-learners as well as educators in a class room environment. Figure 19.1 illustrates the suggested tool – Familiarise-Immerse-Dissect-Sense Making-Appraise-Internalise (FIDSAI).

The FIDSAI analytical approach suggested above offers a systematic approach to reap significant learning outcomes.

Step 1: Familiarise

In this step, a learner is advised to read the case as if he or she was reading a short story! This will allow the learner to appreciate the scenario from his or her immediate perception about the events and people described in the case. Depending on the length and complexity of the case, a learner may read the case several times to fully familiarise with the context and events.

Step 2: Immerse

Here, the learner will deep dive into the facts, events, roles, emotions, issues and dilemmas presented both at surface and hidden levels. In this stage, it is necessary to link the macro case context with the micro elements which are specific events, roles, actions and outcomes presented in a time line. An in-depth understanding of the situation with the aid of facts presented in the case should be achieved. It may also be necessary to refer to some external sources to update the background knowledge of the context within which the case scenario is located.

Step 3: Dissect

In this step, the learner embarks on the real analysis of the case. It is necessary to identify and classify key events and people. Also, it is essential to develop cause-effect relationships if the case presents a complex problem situation affecting a firm or some actors. Mind mapping and use of other popular tools such as SWOT analysis, fishbone diagrams and flowcharts always come handy. If the case is about a complex problem situation caused

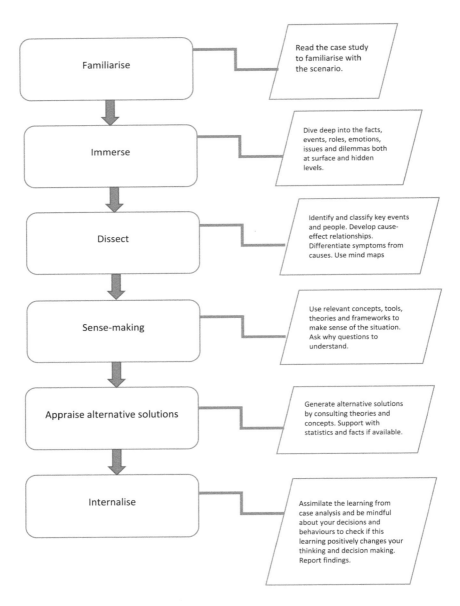

FIGURE 19.1
Suggested methodology for case analysis – FIDSAI framework. (From author's own).

by multifaceted factors, it is ideal if the learner differentiates the symptoms from the causes. This exercise will facilitate generating appropriate solutions.

Step 4: Sense-Making

This step is crucial for understanding the case scenario by asking how and why questions particularly. Prior theory plays a salient role in this stage. A learner is advised to use relevant concepts, tools, theories and frameworks to make sense of the situation. By doing this, the learner moves from the "knowing" stage to the "understanding" stage. Also, the learner can develop a higher awareness level to view the holistic picture. This can be compared to what Albert Einstein once said; "Problems cannot be solved at the same level of awareness that created them". So, we need to rise several levels above the problem scenario to generate effective solutions.

Step 5: Appraise Alternative Solutions

Having dissected the complex problem scenario and understood the possible root causes and why events occurred and actors behaved in a certain manner, the learner is now ready to develop alternative solutions to the problems presented. These alternatives can be ranked by using some criteria which can be developed by combining facts given in the case and found from external sources and theoretical viewpoints.

Step 6: Internalise

This stage is very important for assessing one's learning out of this whole process. A learner may experience a cognitive level change after assimilating the learning from case analysis. Also, the learner can track his or her effectiveness of learning by being mindful or aware of decisions and behaviours to gauge if this learning positively change his or her thinking and decision-making. If required, a proper report of findings of the case analysis needs to be developed.

This book includes 17 cases addressing several key aspects of social and sustainability marketing. These cases are further classified into five categories: short case studies from emerging economies, long and complex case studies from emerging economies, short cases from developed economies, long and complex cases from developed economies, and cases from global context.

From the developing country context, both short and long case studies cover industries such as beauty and skincare, textile, dairy, power meters, medical and personal hygiene. From the developed country context, case studies cover industries such as sport, pet products, food and leisure, education, charity, public service and chemicals. All these cases report real-life social and sustainability marketing scenarios that offer deep insights for learners. Paying attention to the variation of developing and developed country contexts is helpful to understand as to how social and sustainability initiatives and their outcomes vary depending on the country context. The nature of issues and the potential solutions tend to vary according to the industry and the country context.

Educators and learners may select as they think fit the discussion questions and learning activities provided at the end of each case study. These questions and activities are formulated by individual authors of the case studies keeping in mind that the cases would be used by educators and learners for various learning scenarios.

I fervently believe that social and sustainability conscious marketing educators and learners would create intellectually rewarding learning outcomes out of these pedagogical tools.

FUNDING DECLARATION

No funding received.

ACKNOWLEDGMENT

None.

CREDIT AUTHOR STATEMENT

The chapter was written in its entirety by Dr Chandana R. Hewege, PhD (Monash), MBA, BSc.Bus.Ad. (Hons-1st class)(J'pura), Accredited Teacher in Higher Education (SEDA, UK).

REFERENCE

Yin, R.K. (1994). Case study research: Design and methods (Vol. 5). Thousand Oaks, California: Sage.

Section XI

Selected Case Studies to Reflect on Practice and Use as Learning Tools: Case Studies from Emerging Economies

The section is composed of long and short real-life cases with varying degrees of complexity across sectors. These case studies come with discussion questions, teaching materials, real life social and sustainability scenarios, and activities that can be utilised by educators and trainers to develop relevant competencies of learners. These cases introduce learners to diverse social and sustainability marketing contexts and range of issues across several industries and countries.

20

From Skin Whitening to Skin Brightening and, Now, Skin Glowing: How L'Oréal Sustains Its Skincare Line from Colourism and Genderism to Racism and Classism

Huey Fen Cheong and Surinderpal Kaur
Universiti Malaya, Malaysia

CONTENTS

Learning Objectives	548
Themes and Tools Used	548
Tools: Case Study and Role Play	548
Themes	548
Theoretical Background	548
The Case Study	551
Background/Introduction	551
The Social Demands for/against Skin Whitening: From Colourism and Genderism to Racism and Classism	552
L'Oréal Paris' Product Packaging: The Silent Salesman on the Shelf	555
The Problems	558
Discussion	558
Conclusion	559
Lesson to be Learned	559
Discussion Questions	559
Credit Author Statement	560
Note	560
References	560

LEARNING OBJECTIVES

After reading the chapter, the reader should be able to:

- Evaluate and critique marketers' (re)actions to socio-political issues or controversies
- Perform and understand key roles/positionalities in marketing controversies, i.e. marketers, social activists, consumers, legislators, and also, researchers/analysers
- Defend own role/position and listen to others' positions
- Assemble different positionalities and construct a comprehensive, effective action plan

THEMES AND TOOLS USED

Tools: Case Study and Role Play

This *case study* shows a real-life example of how marketers are divided in fulfilling different demands: (1) their corporate social responsibility in addressing current issues (commonly known as brand activism or "woke" marketing; also social demands); (2) their responsibility in meeting consumers' demands (market demands); and (3) the sustainability of their business (business demands), which all may or may not align with each other, but intersect. This provides a platform for role play, in which students position and self-reflect from different roles, i.e. marketers, social activists, consumers, legislators, and also, analysers/researchers. These roles have their respective issues, challenges, interests, agendas, needs, demands, politics, approaches, etc.

Themes

Woke; brand activism; woke-washing; CSR (corporate social responsibility); role play

THEORETICAL BACKGROUND

The case study suggests the *intersectionality theory* on identities, and also on perspectives.

The choice of the intersectionality theory is first inspired by Sobande (2019), who uses the theory to address the same issues, i.e. gender and anti-Black racial issues, in marketing. The concept, intersectionality, basically links to identity construction and the understanding of it. It challenges simple generalisations of identities (plural), which do not take into account their complex intersections among race, ethnicity, class, gender, and sexuality. Such generalisations are typically criticised as uncritical and essentialist. Intersectionality is prevalent in social studies that emphasise on identities, e.g. Butler's (1999, 2006) *Gender Trouble*, which dominates gender studies and problematises (the term "trouble") generalisations of gender identities.

Nevertheless, Cheong (2020) problematises the over-emphasis of identity and anti-generalisation (usually via intersectionality) as products of Western individualism, which values one's individuality (see also her thesis precis, Cheong, 2021). According to the Oxford Advanced Learner's Dictionary,[1] the term "identity" is generally defined as "who or what somebody/something is". However, it can be individualistically defined as "the characteristics, feelings or beliefs that make people different from others" (usually as a countable noun, identities), or collectivistically defined as "the state or feeling of being very similar to and able to understand somebody/something" (usually as an uncountable noun, identity). Nevertheless, according to the same dictionary, the origin of "identity" is from the "late 16th century (in the sense 'quality of being identical'): from late Latin *identitas*, from Latin *idem* 'same'". To sum up, the term "identity" can be individualistic "I-dentity" or collectivistic "identi-ty". For collectivists, the concept of identity, itself, is grouping individuals, which is often deemed as uncritical in Western-dominated (individualistic) academia (Cheong, 2020). Some relevant concepts to consider: social grouping, categorisation, division, and diversity.

The above paragraph shows intersectionality between two cultural perspectives or values, i.e. Western individualism and Eastern collectivism. This leads to intersectionality of different perspectives, which is an important element in the case study that narrates the complexity behind sociopolitical issues, i.e. having different voices from various parties. From the intersectionality of identities to intersectionality of perspectives, we adapt the intersectionality theory to suit the case study.

Unlike the critical approach that often leads to one-sided deconstructions (as shown in many activisms), the concept of intersectionality in perspectives encourages one's flexibility, reflexivity, and even, empathy in thinking across different positionalities. These positionalities refer to the typical roles involved in social marketing, i.e. marketers, social activists, consumers, legislators, and also, analysers/researchers. These roles have their respective issues, challenges, interests, agendas, needs, demands, politics, approaches, etc, which may or not align with other roles. If one could reflect and think from different positionalities, one could construct a comprehensive action plan, which helps to resolve an issue and, hopefully, satisfy all parties (if possible). As such, the critical sense will decrease and the focus will be more on problem-solving, rather than problematising (see Cheong, 2020, about problematising the critical approach, i.e. being critical of critical). Some concepts to consider: brand activism, woke bravery, woke marketing or advertising, and the negative, woke-washing (see Sobande, 2019).

The above invites reflection on Sobande's (2019) critical perspective on woke advertising, i.e. how she critically sees it as "woke-washing" instead. This echoes what Vredenburg et al. (2018) observe: while consumers (preferably, "society") expect brands to involve in activisms for social change, they are sceptical that the brands are authentic and genuine when doing it. Nevertheless, Cheong (2020), being critical of "critical", sees the dominance of the critical approach in academia as a Western cultural product, which may not align with Eastern collectivistic values that favour collective harmony and stability. While Western individualism favours challenging social systems that are often perceived as social oppression on individuals, it may be too aggressive to the Asian liking, whose collectivism prioritises social/collective stability and harmony. This seems to reflect on fewer activisms, including brand activisms, in the Asian contexts, which is often understood as "problems of (individual) power and democracy" (Piper & Uhlin, 2004). Another study, Durkin (2008, p. 45), relates the "aggressive" argumentation of critical thinking with higher masculinity in the West, whose masculine interaction favours "polarised argumentation" instead of the feminine "intuitive reasoning" in the Asian context. Most East Asian students in his study "rejected full academic acculturation into Western norms of argumentation" and created a "Middle Way" that included more empathy and reasoning (Durkin, 2008, p. 38; see also Tan, 2012, in "Critical Thinking: The Chinese Way"). The intention is to maintain harmony by avoiding unnecessary offence and confrontation.

All in all, besides critical (evaluative) argumentation, we aim to promote reflexive (empathetic) reasoning or understanding on the marketers, who not only need to address intersectionality in identities but also intersectionality of perspectives from various parties.

THE CASE STUDY

This section presents the case study, which provides a story plot for students' role-playing and reflection to understand the challenges of marketers in social and sustainable marketing.

BACKGROUND/INTRODUCTION

This case study shows how one of the world's largest beauty producers, L'Oréal, acted swiftly according to current social demands to sustain its skincare line – skin whitening. It narrates how L'Oréal remains socially responsible (meets social demands, usually social justice and activism), while meeting the market/consumer demands (e.g. for skin-whitening products), *corporate/brand demands* (e.g. corporate social responsibility [CSR] and brand image), and *business demands* (e.g. profit), which all demands may or not align with each other, but intersect.

The most recent example is its prompt response to the *Black Lives Matter* movement, which started since late May 2020 and is still ongoing, as of September 2020. L'Oréal announced that it would stop labelling its products with words like "white", "fair", "light" from all its "skin evening products" (reported by Geller, 2020). L'Oréal company owns some famous skincare brands, including L'Oréal Paris, Garnier, Maybelline, Lancôme, and Biotherm. The issue has now extended to the caste system in South Asia (reported in Javed, 2020).

Nevertheless, skin-whitening controversies go beyond white-black racism. There remain market demands, particularly in Asia, for products that protect the skin from the sun and reduce or prevent dark spots.

This case study focuses on L'Oréal's mother brand, L'Oréal Paris. It looks at how L'Oréal Paris sustains its skin-whitening line from female beauty products (L'Oréal Paris' *White Perfect*) to male grooming products (L'Oréal [Paris] Men Expert's *White Activ*) and now, as non-racial-discriminatory symbols (awaiting its new brand name), using naming strategies and packaging designs. It shows how L'Oréal Paris has been sustaining its market and social responsibility across social controversies, from colourism and genderism to racism and classism. The data was taken from Malaysia – a multiracial country in South East Asia, i.e. having people of different skin colours.

THE SOCIAL DEMANDS FOR/AGAINST SKIN WHITENING: FROM COLOURISM AND GENDERISM TO RACISM AND CLASSISM

The long-standing debate on skin whitening has been addressing the social discrimination of people with darker skin. This is often known as colourism, which is a practice of prejudice, favouritism, and/or discrimination of people – within a group (of the same race or class) or between groups (of different races or classes) – due to skin colour/tone. The reasons behind the fair-beauty ideal ranges from socio-economic status (do not need to work under the sun) and patriarchy (women staying at home) to Western cultural imperialism (cf. Cheong & Kaur, 2019). Until recently, there are still public health concerns over the harmful skin-whitening ingredients, which warn against the purchase of fairness products (reported in Sparks, 2019). Nevertheless, it is important to point out that skin tanning in the West also involves colourism and health concerns.

Despite the controversies over colourism and its health concerns, the market demand for skin-whitening products remains high, especially in Asia (Noble, 2019). Whether or not beauty companies use dangerous skin-bleaching agents, they are facing the challenges of fulfilling the market demand (also consumers' demand), while overcoming social controversies.

Another challenge of L'Oréal's skin-whitening line is to address the metrosexual trend, which witnesses the rise of men's interest in their physical appearance, leading to a growing demand for male grooming products. While metrosexuality challenges the norms of beauty practice, which is traditionally feminine, men's skin-whitening practice challenges the gender norms further by claiming the feminine beauty ideal, i.e. breaking the dark-men/fair-women tradition (Frost, 1990, 2005). Numerous studies discovered that men try to avoid associating with femininity (i.e. anti-femininity) in their grooming practice, including purchasing grooming products (cf. Cheong & Kaur, 2015).

A news article, *Male Enlightenment* (Cheong, 2013), shows how L'Oréal, together with other companies or brands, introduced *skin-brightening* products for men. As a collective, they redefined the feminine skin whitening to skin brightening for men. This helps to reduce or avoid association with femininity, which makes men feel more comfortable. Here, we see how the male grooming industry suits the market demand for genderism (gender differentiation) to separate men's products and practices from women.

However, gender issues behind men's skin whitening remains a topic (i.e. not seen as a norm), as shown in recent studies mainly in Asia (Cheong & Kaur, 2019; Lasco & Hardon, 2019; Mukherjee, 2020). Nevertheless, gender-related activists or researchers typically problematise such genderism (usually in gender binaries) in marketing for stereotyping or even reinforcing it. Yet, Cheong (2021) sees the emphasis of anti-stereotyping and (self-)identity as a product of Western individualism, which values individuals' freedom and individuality. Her survey on British and Malaysian literature about metrosexuality shows that British studies see metrosexuality as an identity or rather, identities (plural; anti-stereotyping), while Malaysian studies see it as a social phenomenon, e.g. the rise of male grooming practices or the change of masculinity (singular; in general).

The most recent skin-whitening controversy is caused by the anti-racist movement, Black Lives Matter, which has swept across white-dominant countries in the United States, the United Kingdom, European countries, Australia, and New Zealand. Started since May 26, 2020, the movement was sparked by the death of a black American man, George Floyd, during a police arrest, in which he was pinned on the ground with a white police officer knelt on his neck (reported in Bennett et al., 2020). Protesters marched on the street the next day, shouting "Black lives matter" and Floyd's final words, "I can't breathe!" The protest turned violent with looting and setting buildings on fire, including police stations, all over the United States (see images in Bell et al., 2020). More countries joined in the protest and destroyed the statues of historical white figures such as Christopher Columbus (reported in Grovier, 2020). During this period,

L'Oréal and Unilever were under scrutiny, being the big players in the skin-whitening market.

> The L'Oréal Group has decided to remove the words white/whitening, fair/fairness, light/lightening from all its skin evening products.
>
> (L'Oréal's public statement, cited in The Star, June 29, 2020)

On June 26, 2020, L'Oréal publicly announced to stop labelling its 'skin evening' products with words suggesting "white", "fair", and "light" (reported by Geller, 2020). Other beauty producers such as Unilever did the same; Johnson & Johnson even planned to discontinue selling fairness products in Asia and the Middle East under its Clean & Clear and Neutrogena brands (reported in The Star, 2020). Few days later on June 30, L'Oréal's CEO Jean-Paul Agon mentioned the company's plan in a shareholder meeting to change the labelling of "white" or "fair" to "glow" or "even" "to meet changing expectations and attitudes" (cited in *Reuters*, 2020). The aim is to rethink products that provide skin protection from the sun (UV rays) and dark spots, particularly in Asia. Similarly, Unilever also changed its "Fair & Lovely" brand to "Glow & Lovely", and its men's range from "Fair & Handsome" to "Glow & Handsome" (reported in Business Insider India, 2020).

Nevertheless, critics were still putting pressure on the beauty producers to stop selling skin-whitening products altogether – to "quit the multi-billion-dollar market" that is "popular in Africa, Asia and the Middle East" (reported in McEvoy, 2020). These critics fail to acknowledge the demand-and-supply theory, which explains why the market exists and becomes popular in certain contexts, which have different environment and skin issues. This anti-skin-whitening sentiment had spread to the South Asia to challenge their caste systems (reported in Javed, 2020).

The controversy behind skin whitening is never-ending from colourism and genderism to racism and now, classism, which may occur and gain attention at different times, not one replacing the other. Together with other beauty companies, L'Oréal renamed and reconceptualised its skincare line from *skin whitening* for women to *skin brightening* for men and now, *skin glowing* for all races to meet different social demands, while sustaining their skincare line. This discussion contains the latest updates until the point of writing in September 2020.

L'ORÉAL PARIS' PRODUCT PACKAGING: THE SILENT SALESMAN ON THE SHELF

Besides naming, L'Oréal has been appropriating its packaging designs to meet current social demands. Widely known as the silent salesman on the shelf, packaging represents the brand image and personality, which attract consumers owning or desiring a similar image and personality. This is particularly important to beauty/grooming products, which work on enhancing one's image. This case study focuses on L'Oréal's mother brand, L'Oréal Paris, which are sold in Malaysia, and looks at how its packaging designs change over time.

For women's skin-whitening range, L'Oréal Paris *White Perfect*'s current packaging added a metallic "plate" to the previous packaging, which was purely white (see Figure 20.1).

This changes its fair beauty image to a professional image, using the elements of science and technology. The emphasis of "L'Oréal Paris Laboratories" on the back description of the current packaging further proves its intention to project the company's R&D. This seems to address the health concern behind skin whitening. It sends a message that the company is constantly

FIGURE 20.1

L'Oréal Paris' *White Perfect* cleansers and toners [manufactured around 2012 (left) and around 2015 until now (right)].

improving its products through R&D. Also, the emphasis of Tourmaline Gemstone as its key ingredient, in both packaging, projects an image of an expensive product, which will not use cheap skin-bleaching agents.

For the men's skin-brightening range, L'Oréal (Paris) Men Expert's *White Activ*, the issue is with genderism between female beauty and male grooming practices, and between female skin whitening and male skin brightening. Also in Malaysia, Cheong and Kaur present a detailed analysis on *White Activ*'s product packaging to analyse how it legitimises male grooming (Cheong & Kaur, 2015) and male skin whitening (Cheong & Kaur, 2019) among men (see also Cheong, 2014). This case study looks at the diachronic set of *White Activ*'s Brightening Foam, manufactured from 2010 to present (see Figure 20.2). Brightening Foam is one of the types of cleanser in the *White Activ* range. It was first introduced as the only cleanser, named *Cleansing Foam*, when *White Activ* was first introduced (see the first product in Figure 20.2).

The diachronic set of the packaging began with a soft tone between white and grey, followed by a sudden addition of the darkest colour, black, and bright orange in late 2012. The latest packaging replaces bright orange with black. The soft beginning suggests that L'Oréal started with more feminine men, who were more ready to accept the products. When more men could accept its products, male grooming, and/or skin brightening, it increases the masculine image to suit more men. The use of white,

FIGURE 20.2
Four generations of L'Oréal Men Expert's *White Activ* cleanser, Brightening Foam [manufactured (from left): around 2010, early 2012, late 2012, and 2017 until now].

grey, and black shows how L'Oréal presents skin whitening differently through colour tones (signifying skin tones), not purely or mainly white like L'Oréal Paris' *White Perfect* that signals fair beauty (see Figure 20.1). The sudden changes around 2012 from "Cleansing Foam" to "Brightening Foam", followed by adding bright orange, align with the skin-brightening concept. This concept was reported in Malaysia the next year in *The Star, Male Enlightenment* (published on March 7, Cheong 2013).

As for L'Oréal's skin-glowing products in response to the Black Lives Matter movement, the new product or packaging is yet to be found (updated in August 2020). Nevertheless, L'Oréal Malaysia has stopped showing its White Perfect or White Activ products in its Facebook brand page since the Black Lives Matter Movement at the end of May 2020. To continue catering for its consumers' need for sun protection, it promotes its sunscreen series, UV Perfect, instead (see Figure 20.3).

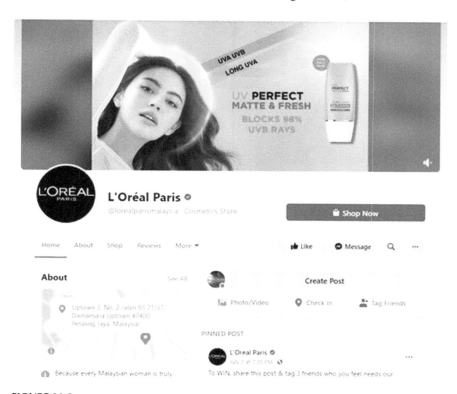

FIGURE 20.3
The cover video of L'Oréal's UV Perfect on L'Oréal Malaysia's Facebook page (uploaded July 2-22, 2020; https://www.facebook.com/lorealparismalaysia/?ref=page_internal).

THE PROBLEMS

1. How could marketers satisfy ALL parties with various issues, challenges, interests, agendas, needs, demands, politics, approaches (ways of doing), etc. intersecting?
2. How could marketers find ONE way and/or solution, which fulfils the changing and intersecting demands across times?

We highlight "ALL" and "ONE" to make students realise the complexities behind marketing. It is (almost) impossible to satisfy everyone and there is no definite way or solution that works for all. This again call for more reasoning over critiques, which are more critical, evaluative, and/or judgemental. Nevertheless, both reasoning and critical approaches can be taught here.

DISCUSSION

The case study outlines the problems and actions taken by L'Oréal to sustain its skin-whitening line, while trying hard to fulfil its CSR (referring to social marketing). One can analyse its actions from a critical perspective and/or reflective reasoning, which leads to deconstructive problematisation and constructive problem-solving, respectively. Both skills are important thinking skills.

However, the critical approach is more favourable in the Western individualistic contexts, whereas the less-confrontational reasoning is more favourable in the Eastern collectivistic contexts, where collective/social harmony and stability is the priority. Nevertheless, such an East/West binary is, itself, perceived as uncritical and problematic in the Western-dominated academia, which is anti-binaries/categories and favours seeing people or things in continua and/or intersections. This reflects Cheong's (2021) observation of the British academic context:

> There is culture in one's thinking;
> That is how knowledge is socially constructed.
> But there is also culture in one's (dis)belief of culture;
> That is the power of culture.

CONCLUSION

The case study on L'Oréal's skin-whitening line shows how it acted solidarity with other companies to meet different social, market, and brand/corporate demands across times, which keep changing and may/not align with each other, but intersect. This creates trends in the market, which a company needs to keep up to sustain or even to survive through various critical voices, which, again, may/not align with each other, but intersect. To sum up, the company, together with other companies, need to find ways to address these intersections of demands and voices.

LESSON TO BE LEARNED

At the end of the lesson, readers learn:

1. To understand the complexities behind social marketing
2. To reflect from different positionalities: marketers/L'Oréal, activists, consumers, legislators, and analysers
3. To critique and evaluate the (in)appropriateness and/or (un)workability of a brand's action, based on the consideration of the complexities involved
4. To be appreciative and understanding towards a brand's expertise and challenges in addressing social complexities

DISCUSSION QUESTIONS

1. List the factors that *intersect* and form one's (personal) identity.
2. List the demands from various parties, i.e. also factors, that *intersect* and influence the marketer's action and decision.
3. Do you think L'Oréal's actions in the case study are appropriate/inappropriate or workable/unworkable in addressing the intersections above?

Note: The focus on appropriate/inappropriate refers to the critical perspective, while workable/unworkable leads to reasoning.

CREDIT AUTHOR STATEMENT

This case study has started since 2014 as part of Cheong's Master's dissertation (Cheong, 2014), which was supervised by Assoc. Prof. Dr. Surinderpal Kaur. From the dissertation, both supervisor and supervisee produced two journal articles (Cheong and Kaur, 2015, 2019). As stated in this case study, the past dissertation and publications only used L'Oréal's packaging within a single period (found within 2014–2015) and focused on metrosexuality, which compared male and female skin-whitening product packaging. The current case study, on the other hand, presents a diachronic analysis of L'Oréal's male and female skin-whitening product packaging with a new discussion focus – Black Lives Matter.

With the new issue, Black Lives Matter, both ex-supervisor and ex-supervisee opened the case study again, in which the ex-supervisee did the research and writing, with the supervision of the former supervisor.

Cheong, H. F. (2014). Mediated discourse analysis: Negotiating metrosexuality within L'Oréal Men Expert's packaging discourse. (Master's dissertation). Master of Linguistics, University of Malaya, Malaysia.

Cheong, H. F., & Kaur, S. (2015). Legitimising male grooming through packaging discourse: a linguistic analysis. *Social Semiotics*, *25*(3), 364–385. doi:10.1080/10350330.2015.1026650

Cheong, H. F., & Kaur, S. (2019). Mirror, mirror on the wall, who's the fairest 'hunk' of them all? Negotiating a masculine notion of skin whitening for Malaysian men. *SEARCH Journal of Media and Communication Research*, *11*(1), 57–76.

NOTE

1 Retrieved from https://www.oxfordlearnersdictionaries.com/

REFERENCES

Bell, B., Ngala, F., Jarrus, S., & Cherry, J. (2020). Capturing the cry for change: Photographers on the BLM protests. *The Guardian*. Retrieved from https://www.theguardian.com/media/2020/jun/12/capturing-the-cry-for-change-photographers-on-the-blm-protests

Bennett, D., Lee, J. S., & Cahlan, S. (2020, May 30). The death of George Floyd: What video and other records show about his final minutes. *The Washington Post*. Retrieved from https://www.washingtonpost.com/nation/2020/05/30/video-timeline-george-floyd-death/?arc404=true

Business Insider India. (2020, July 2). HUL's Fair & Lovely to now be called 'Glow & Lovely'. *Business Insider*. Retrieved from https://www.businessinsider.in/advertising/brands/news/huls-fair-lovely-to-now-be-called-glow-lovely/articleshow/76750953.cms

Butler, J. (1999). *Gender trouble: Feminism and the subversion of identity*. New York; London: Routledge.

Butler, J. (2006). *Gender trouble: Feminism and the subversion of identity*. New York: Taylor & Francis.

Cheong, B. (2013, 7 March). Male enlightenment. *The Star*. Retrieved from https://www.thestar.com.my/Lifestyle/Archive/2013/03/07/Male-enlightenment/

Cheong, H. F. (2014). *Mediated discourse analysis: Negotiating metrosexuality within L'Oréal Men Expert's packaging discourse*. (Master's dissertation). Master of Linguistics, University of Malaya, Malaysia.

Cheong, H. F. (2020). *A problematised critical approach: Constructions of metrosexualities in the UK and Malaysia*. (PhD thesis). Lancaster University, UK.

Cheong, H. F. (2021). Let the culture speak: How gender is socially/culturally constructed by Western individualism and Eastern collectivism. *Gender, Place & Culture*, 28(2), 299–304. doi:10.1080/0966369X.2020.1789072

Cheong, H. F., & Kaur, S. (2015). Legitimising male grooming through packaging discourse: A linguistic analysis. *Social Semiotics*, 25(3), 364–385. doi:10.1080/10350330.2015.1026650

Cheong, H. F., & Kaur, S. (2019). Mirror, mirror on the wall, who's the fairest 'hunk' of them all? Negotiating a masculine notion of skin whitening for Malaysian men. *SEARCH Journal of Media and Communication Research*, 11(1), 57–76.

Durkin, K. (2008). The middle way: East Asian master's students' perceptions of critical argumentation in UK universities. *Journal of Studies in International Education*, 12(1), 38–55.

Frost, P. (1990). Fair women, dark men: The forgotten roots of colour prejudice. *History of European Ideas*, 12(5), 669–679.

Frost, P. (2005). *Fair women, dark men: The forgotten roots of color prejudice*. Christchurch, New Zealand: Cybereditions.

Geller, M. (2020, June 26). L'Oreal to drop words such as 'whitening' from skin products. *Reuters*. Retrieved from https://www.reuters.com/article/us-l-oreal-whitening/loreal-to-drop-words-such-as-whitening-from-skin-products-idUSKBN23X224

Grovier, K. (2020, June 12). Black Lives Matter protests: Why are statues so powerful? *BBC*. Retrieved from https://www.bbc.com/culture/article/20200612-black-lives-matter-protests-why-are-statues-so-powerful

Javed, S. (2020, July 23). To truly overcome Anti-Black racism, we must also challenge South Asia's caste system. *The Independent*. Retrieved from https://www.independent.co.uk/life-style/colourism-racism-anti-black-george-floyd-skin-whitening-unilever-south-asia-a9625076.html

Lasco, G., & Hardon, A. P. (2019). Keeping up with the times: Skin-lightening practices among young men in the Philippines. *Culture, Health & Sexuality*, 22(1), 1–16.

McEvoy, J. (2020, June 26). L'Oreal, Unilever reassess skin lightening products—But won't quit the multi-billion dollar market. *Forbes*. Retrieved from https://www.forbes.com/sites/jemimamcevoy/2020/06/26/loreal-unilever-reassess-skin-lightening-products-but-wont-quit-the-multi-billion-dollar-market/#7a02cdb5223a

Mukherjee, S. (2020). Darker shades of "fairness" in India: Male attractiveness and colorism in commercials. *Open Linguistics*, 6(1), 225–248.

Noble, A. (2019, August 9). Skin lightening is fraught with risk, but it still thrives in the Asian beauty market—Here's why. *Vogue*. Retrieved from https://www.vogue.com/article/skin-lightening-risks-asian-beauty-market

Piper, N., & Uhlin, A. (Eds.). (2004). *Transnational activism in Asia: Problems of power and democracy*. London: Routledge.

Reuters. (2020, June 30). 'Glow' to replace 'whitening' in some L'Oreal skin products. *Reuters*. Retrieved from https://www.reuters.com/article/us-l-oreal-whitening/glow-to-replace-whitening-in-some-loreal-skin-products-idUSKBN241288

Sobande, F. (2019). Woke-washing: "Intersectional" femvertising and branding "woke" bravery. *European Journal of Marketing*, 54(11), 2723–2745.

Sparks, H. (2019, September 30). Officials warn skin-lightening creams 'should be avoided at all costs'. *New York Post*. Retrieved from https://nypost.com/2019/09/30/officials-warn-skin-lightening-creams-should-be-avoided-at-all-costs/

Tan, C. (2012). 16 Critical Thinking: The Chinese Way. In *Learning from Shanghai: Lessons on Achieving Educational Success* (pp. 171–184). Singapore: Springer.

The Star. (2020, June 29). L'Oreal removes words like 'whitening' from its products. *The Star*. Retrieved from https://www.thestar.com.my/lifestyle/family/2020/06/29/l039oreal-removes-words-like-039whitening039-from-its-products

Vredenburg, J., Kapitan, S., Spry, A., & Kemper, J. (2018, December 5). Woke washing: What happens when marketing communications don't match corporate practice. *The Conversation*. Retrieved from https://theconversation.com/woke-washing-what-happens-when-marketing-communications-dont-match-corporate-practice-108035

21

Fashion Accessory Brand Development via Upcycling of Throwaway Clothes: The Case of Chapputz

Selcen Ozturkcan
School of Business and Economics, Linnaeus University, Kalmar, Sweden

CONTENTS

Learning Objectives	563
Themes and Tools Used	564
Theoretical Background	564
Fashion Industry and Its Impact on the Environment	564
Upcycling	566
Fashion Industry in Turkey	568
An Upcycling Fashion Initiative in the Making	568
Chapputz – An Upcycling Brand of Fashion Accessories	570
Contributions to UN's Sustainable Development Goals	572
Credit Author Statement	573
References	573

LEARNING OBJECTIVES

After reading the chapter, the reader should be able to:

- Comprehend the environmental effects of fast fashion
- Assess the challenges fast fashion managers face
- Analyze market and innovation capabilities within the circular economy
- Develop and propose alternative sustainable fashion marketing strategies

THEMES AND TOOLS USED

PESTEL, SWOT, WWHD

THEORETICAL BACKGROUND

Fast fashion, which is known for its introduction of new lines very frequently, impacted the fashion market and production on several dimensions. First, it boosted the adverse environmental impact of the fashion industry both in consuming natural resources and polluting the environment. Second, the rapid introduction of new fashion forms shifted rather recent purchases of the consumer toward the unfashionable form, resulting in dramatic increases in throwaway clothing. This chapter introduces the reader to these elements through accumulated facts and figures. In addition, the approach of upcycling is introduced, particularly with relevant literature background that applies to fashion in terms of circular economy. This is followed by the development of a fashion accessory brand, namely Chapputz, with its core brand elements built on upcycling of throwaway clothing. In addition, the brand offered income generation to rural female artisans involved in the upcycling process, thus extended itself to a social entrepreneurial example. Lastly, Chapputz attributions to the United Nation's sustainable development goals (SDGs) are presented.

This chapter focuses on a young social entrepreneur, Yasin, who introduces a fashion accessory brand that upcycles throwaway clothing to reduce their adverse environmental impacts while also empowering rural females in its production processes. Thus, the case has many linkages to the United Nation's SDGs (UN, 2020).

FASHION INDUSTRY AND ITS IMPACT ON THE ENVIRONMENT

The global apparel market accounted for 4% of the world's Gross Domestic Product (FashionUnited, 2020). Like most other contemporary production, fashion market products are designed with relatively shorter life spans to consume fast and throwaway the goods. Fast fashion's continual increase in industrialization-fueled production-consumption, coupled with the assumption of infinitely available feed of unlimited raw materials, results in a dramatic throwaway mentality. Fueled by the fall in prices and increased speed in fashion delivery, individuals purchase clothes swiftly increase in recent times, resulting in clothing to account for between 2% and 10% of the environmental impact of consumption (Šajn, 2019). Fashion developed such a fast cycle that the concept of fast fashion emerged. Simultaneously, some companies increased their frequency of offering from two in the year 2000 to five in 2011, with some brands such as Zara and H&M offering as much as 24 collections annually (Šajn, 2019). Only about 20% of all textiles are recycled or reused; the rest is directed to landfills (McCarthy, 2018). That is a garbage truck of clothes burnt or landfilled every second, making approximately 2,600 kg. It is enough to fill one and a half empire state buildings every day, and Sydney Harbor every year (Morlet et al. 2017). On a medium scale, 4% of global landfills are filled with clothing and textiles (Roberts, 2015).

The fashion industry also consumes a substantial amount of natural resources in its production and has an enormously detrimental impact on the environment via its carbon emission. About 2.6% of global water is used for growing cotton, which is an important raw material. Producing a single cotton t-shirt consumes 2.7 tons of water, while a pair of jeans consumes 7.5 tons (UNEP, 2018). On the other hand, 2.7 tons of water is enough for one person to drink in two and a half years (Reichart & Drew, 2019). In some countries such as Uzbekistan, cotton farming used up so much water from the available resources that resulted in drying up and drought. Following 50 years of aggressive water use in cotton production, the Aral Sea, one of the world's four largest lakes, dried up, turning into a desert and a few small ponds (McFall-Johnsen, 2019).

Fashion also causes some water-pollution problems. The estimated percentage of industrial wastewater produced due to textile dyeing and treatment is approximately 20% of global wastewater (Drew & Yehounme, 2017; UNEP, 2018), which places the industry second-largest consumer of water worldwide (UNECE, 2018). Manufacturing of textiles from raw materials is a tedious process involving around 8,000 synthetic chemicals. Textile dyeing is regarded as the world's second-largest polluter (Kaye, 2013). Dumping water leftover from the dyeing process into ditches, streams, or rivers is very common worldwide (UNEP, 2018). Each year about 2 million

Olympic-sized swimming pools could have been filled by the water used in the dyeing process alone (Drew & Yehounme, 2017).

In terms of carbon emission, the fashion industry's impact on the environment is more than international flights and maritime shipping combined (UNEP, 2018). Moreover, if the fashion industry's current trajectory continues, estimates indicate that it may be responsible for 26% of global carbon emissions by 2050 (Morlet et al. 2017). On the other hand, approximately 30 kg of clothing per person is discarded every year in the developed world. The value of clothing that goes into landfills is estimated at around USD 180 million (Wrap, 2020). Reports indicate that more than 134 million tons of textiles might be discarded annually as of 2030 (Kerr & Landry, 2017).

There is no doubt about the apparent need to consider sustainability awareness among the apparel industry's producers, marketers, and consumers. Since fast fashion is also a pursuit of modernity, the emergent requirement to preserve the invaluable natural resources involves many facets of social, economic, and environmental issues and presents a global challenge.

> Most fashion retailers now are doing something about sustainability and have some initiatives focused on reducing fashion's negative impact on the environment. However, there is still a fundamental problem with the fast fashion business model where revenues are based on selling more products, and therefore retailers must constantly offer new collections. It would be unrealistic to expect consumers to stop shopping on a large scale, so going forward, I would expect to see more development and wider adoption of more sustainable production methods such as waterless dyeing, using waste as raw material, and the development of innovative solutions to the textile waste problem.
>
> Patsy Perry, senior lecturer in fashion marketing at the University of Manchester (UNEP, 2018)

UPCYCLING

A sound strategy alternative capable of protecting the ecosystem might be offered by introducing upcycling, which uses the existing, potentially useful materials to prevent them from ending up as waste. Upcycling, different than recycling, does not break the consumer materials down, so their base material can be used as raw material to manufacture new consumer products. It instead refrains from breaking down the materials (Habitat,

2020). By contrast to recycling, upcycling adds value to materials through their creation into new modes (James & Kent, 2019). In upcycling, materials are reused without any degrading to their quality or decomposition. Upcycling is defined as an improvement to the value/quality of the product by making a superior product (Dervojeda, Verzijl, Rouwmaat, Probst, & Frideres, 2014) by providing new value to materials that were not in use or would be discarded (Fletcher & Grose, 2012), where lower-value items are repurposed for creating higher-value end-use things (Janigo & Wu, 2015). It is also referred to as "an effort to achieve the better or the same functionality of the product by minimal consumption of raw material and energy" (Paras & Curteza, 2018). In clothing, upcycling also helps prevent pollution in redirecting the garments from landfills to the supply chain (Bhatt, Silverman, & Dickson, 2019; Janigo, Wu, & DeLong, 2017). Moreover, reducing the need to purchase new product results in reduced raw material use and help to conserve precious natural resources (Ali, Khairuddin, & Abidin, 2013). Both environmental and social justice activists have been reported to resort to upcycling in launching sustainable fashion innovations (Odabasi, 2020) in the developed world.

Upcycling is a term referring to "value increase through altering or customizing" products (McDonough & Braungart, 2009; Thomas, 2008), which offers benefits to the circular economy (Figure 21.1), since waste in

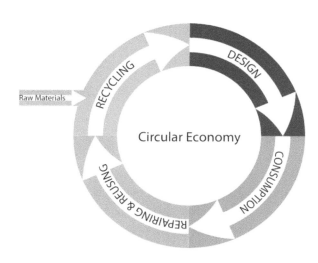

FIGURE 21.1
Operating system of circular economy. (From Atalay Onur, 2020, p. 58).

fashion is unfashionable or out-of-date clothing (Fletcher, 2008). In upcycling, waste transforms into a source for producing new clothing (Atalay Onur, 2020).

FASHION INDUSTRY IN TURKEY

As of the year 2019, Turkey ranked 5th as a global textile exporter, with 3.9% of all global exports (Sabanoglu, 2019). Despite this volume, typical in an emerging country context, the local sectoral reports and statistics concerning the industry are scarce and not available in a structured healthy manner for depicting any long-term macro analysis. Yet, there is the affirmed growth trend with figures ranging to a certain extend. The attributed growth is often explained by fashion gaining importance in daily life, especially in Turkey's urban areas. The emerging "street fashion" in urban areas defines an accessible fashion concept, which is claimed to offset the desire and importance of luxury branded goods amongst Turkish consumers, particularly among youth (Euromonitor, 2014). Vintage and second-hand clothing took advantage of the novel attitude that emerged toward fashion. Some designers respond to this emerging demand by introducing "vintage" clothing lines. However, there was no systematic approach proposed toward the dramatically increasing textile waste (Chapputz, 2015) in the form of an upcycling novelty that would take advantage of the circular economy.

An Upcycling Fashion Initiative in the Making

A young Turkish social entrepreneur named Yasin Sert, who was deeply concerned with the environment, assumed a social mission in creating an upcycling brand that not only benefited the ecosystem but also enriched a long-forgotten old artistic technique starting from the year 2016 (Sert, 2016). As a sophomore undergraduate student in an urban university, Yasin was highly concerned about the fast-fashion trend's environmental consequences in his immediate environment. He sought a way to reduce the waste piling from throwaway clothing since he realized that too many clothing items were becoming unfashionable very fast. Simultaneously, Yasin was interested in tapping into the idle rural female workforce to

Fashion Accessory Brand Development • *569*

FIGURE 21.2
United Nation's sustainable development goals. (From UN, 2020).

improve the widening income gap urban vs. rural, and male vs. female, where the later was often disadvantageous. In this light, he started seeking alternatives to merging multiple aims, which also have appeared in the United Nation's SDGs (Figure 21.2) (UN, 2020).

One day, Yasin discovered the "çaput kilim" weaving, which had its roots in the Turkish Nomad culture as an upcycling method, for producing light carpets form throwaway textiles. This method relied on fabric stripes prepared from thrown away clothing as its raw material by simply cutting them into long stripes. Most rural women used to know how to prepare raw material fabric stripes. They also got the stands for hand-weaving as they all received in-person tutoring from the previous female generations during their upbringing. However, in recent times there were not many that still kept the practice. In his search for an experienced and knowledgeable person on the topic, Yasin met Faize Yoldaş.

Faize Yoldaş, who lived in her village in the rural Turkey in her all life since she borne, kept on with weaving light carpets from the fabric stripes she made out of worn-out textiles. She was in her late 50s and was the only person in the village who kept on with the traditional practice that she learned when she was a teenager. Yet, she has never had the occasion of putting her artisan at work in return for income. Females in the village were all housewives with no hindsight of income-generating work opportunities in the vicinity. When Yasin showed up in her village looking for a "çaput kilim" artisan, villagers had pointed him to her house. She welcomed Yasin as it was a long-standing tradition of hospitality in their village. She showed her equipment and explained Yasin how she produces the light kilims.

Yasin realized that he found the idea that he was looking for. However, the light kilims would not be sufficient to attract urban customers. He opened the idea of transforming the light kilim to fashion accessories. He would collect throwaway clothing from large cities, which was in abundance due to fast fashion making most clothing unfashionable in the blink of an eye. He would bring what he collected to Faize, who would then prepare the fabric stripes from this thrown away clothes and then weave them on her traditional equipment to hand in the woven product to Yasin. Yasin would sew the woven product into fashion accessories by using sustainable materials along the process. He would also setup an online retailing business and develop agreements with other businesses targeting the environmentally concerned customers such as cafes and gyms. The two shook hands agreeing to share the possible earnings, and Yasin left the village heading back to the city. He had an important mission for collecting as much thrown away clothing as he could so that Faize would be able to prepare the raw materials.

Chapputz – An Upcycling Brand of Fashion Accessories

The established business developed the brand Chapputz, which identified itself with its environmentally concerned social mission to re-define a sustainable fashion by (1) reducing the negative impacts of fast fashion on the environment and (2) promoting welfare for the society at large, simultaneously. Chapputz was a social entrepreneurial partnership between a rural elderly female artisan Faize Yoldaş, who was specialized in the long-forgotten weaving of thorn throwaway clothing, and an entrepreneurial environmentalist college-sophomore Yasin Sert, who collected used and throwaway clothing. Chapputz introduced handcrafted backpacks, clutches, iPad cases, messenger bags, and lace pillows weaved from the collected throwaway clothes (Figure 21.3). Despite the many challenges in the market where giants with massive marketing budgets competed, Chapputz's handmade craft designs became popular as it paved its way as

FIGURE 21.3
Chapputz iPad case and iPad clutch case.

an upcycling fashion apparel brand with a unique social mission. Thanks to Chapputz, a vanishing artisanship received a boost in demand for many rural females that started to make a living by taking part in its production processes. The thrown away and worn clothing were upcycled and introduced to the fashion-conscious consumers as an apparel item with exceptional value.

Chapputz aimed to provide a systemic approach to recycling the worn clothes, which did not exist to a similar extent for the well-conditioned ones. It helped to raise awareness that there was an alternative to throwing away the worn cloths. Chapputz transformed the to-be-thrown away to some stylish handmade kilim products via "çaput kilim" weaving, which had its roots in the Turkish Nomad culture as an upcycling method (Chapputz, 2016). In addition to its environmental benefits, weavers were mostly rural village women who were empowered through the production process. Lastly, a new amalgamation of traditional and modernity was made possible through the product variety that involved iPad cases, clutches, and backpacks. Chapputz helped protect the environment, empowered its weaver females, and restored traditional folk culture into fashion accessories (BinYaprak, 2017; KU, 2017).

As a case, Chapputz is an exemplary upcycling apparel brand that refrained from landfilling but fulfilled both an ecological and a social mission. Notably, the distribution channels that Chapputz chose distinguished its products from other fast fashion brands. As a revolutionary social innovation, it initiated an upcycling fashion trend, which resulted in an informed customer group that started to search for an authentic story in every possible step of the value chain. This has benefitted Chapputz in differentiating its products in its niche marketplace, thus serving its sustainability.

CONTRIBUTIONS TO UN'S SUSTAINABLE DEVELOPMENT GOALS

It is "clear that the development of the fashion industry has a significant impact on the achievement of the UN Sustainable Development Goals" (UNECE, 2018). Chapputz contributed to several SDG of the United Nations (UN, 2020). Responsible consumption and production (SDG 12), which is structured in several subfields such as use of natural resources, chemical waste, fossil fuels, and integration of sustainable practices into the production cycles, applied in multidimensions to the Chapputz's deliverables as a fashion accessory brand. Moreover, consumers actions are also considered as possible vectors to make a change in this goal. Customers of Chapputz, who were informed of an alternative way to soothe their fashion needs in a sustainable manner, had also indirectly contributed to the goal. Decent work and economic growth (SDG 8) can be achieved by improving the working conditions of workers in fashion supply chains. Chapputz introduced the possibility to unemployed female artisans to take part in the upcycling process, and thus promoted inclusive and sustainable economic growth in the rural areas. Clean water and sanitation (SDG 6) was directly served by the Chapputz due to its substitution of throwaway clothing with that of virgin material in production processes. Climate action (SDG 13) calls for urgent action to combat the climate change. Similar to its contribution to SDG 6, Chapputz provided direct positive impacts on reducing carbon emission prevalent to textile production from raw materials by simultaneously contributing to SDG 13. The social dimension of the Chapputz as an upcycling fashion accessory brand that utilized rural female artisans was also directly linked to gender equality (SDG 5). The income generated for these otherwise unemployed females also contributed to eradicating poverty (SDG 1), which remains as a problem in fashion industry especially in developing countries. While majority of the SDGs are directly linked to the fashion industry, Chapputz tackled many of them either directly or indirectly by providing a very unique brand establishment.

CREDIT AUTHOR STATEMENT

All relevant due credit of this work belongs to Selcen Ozturkcan as the single author.

REFERENCES

Ali, N. S., Khairuddin, N. F., & Abidin, S. Z. (2013). *Upcycling: Re-use and recreate functional interior space using waste materials*. Paper presented at the International Conference on Engineering and Product Design Education, Dublin, Ireland.

Atalay Onur, D. (2020). Integrating circular economy, collaboration, and craft practice in fashion design education in developing countries: A case from Turkey. *Fashion Practice*, *12*(1), 55–77. doi:10.1080/17569370.2020.1716547

Bhatt, D., Silverman, J., & Dickson, M. A. (2019). Consumer interest in upcycling techniques and purchasing upcycled clothing as an approach to reducing textile waste. *International Journal of Fashion Design, Technology and Education*, *12*(1), 118–128. doi:10.1080/17543266.2018.1534001

BinYaprak (2017). *Kendi kültürüne dön!* Retrieved from https://youtu.be/nK-lEIW9o-s

Chapputz (2015). *Arzu Kaprol'un Chapputz Yorumu*. Retrieved from https://youtu.be/erEKheehpgY

Chapputz (2016). *Chapputz English*. Retrieved from https://youtu.be/49UWqqtbmV8

Dervojeda, K., Verzijl, D., Rouwmaat, E., Probst, L., & Frideres, L. (2014). *Clean technologies: Circular supply chains*. Retrieved from http://ec.europa.eu/DocsRoom/documents/13396/attachments/3/translations

Drew, D., & Yehounme, G. (2017). *The apparel industry's environmental impact in 6 graphics*. Retrieved from https://www.wri.org/blog/2017/07/apparel-industrys-environmental-impact-6-graphics

Euromonitor. (2014). *Consumer Lifestyles in Turkey*. Retrieved from http://www.euromonitor.com/consumer-lifestyles-in-turkey/report

FashionUnited. (2020). *Global fashion industry statistics – International apparel*. Retrieved from https://fashionunited.com/global-fashion-industry-statistics

Fletcher, K. (2008). *Sustainable fashion and textiles: Design journeys*. Malta: Earthscan.

Fletcher, K., & Grose, L. (2012). *Fashion & sustainability: Design for change*. London: Laurence King Publishing.

Habitat. (2020). *What is upcycling?* Retrieved from https://www.habitat.org/stories/what-is-upcycling

James, A. S. J., & Kent, A. (2019). Clothing sustainability and upcycling in Ghana. *Fashion Practice*, *11*(3), 375–396. doi:10.1080/17569370.2019.1661601

Janigo, K. A., & Wu, J. (2015). Collaborative redesign of used clothes as a sustainable fashion solution and potential business opportunity. *Fashion Practice*, *7*(1), 75–98.

Janigo, K. A., Wu, J., & DeLong, M. (2017). Redesigning fashion: An analysis and categorization of women's clothing upcycling behavior. *Fashion Practice*, *9*(2), 254–279. doi:10.1080/17569370.2017.1314114

Kaye, L. (2013). *Clothing to dye for: The textile sector must confront water risks*. Retrieved from https://www.theguardian.com/sustainable-business/dyeing-textile-sector-water-risks-adidas

Kerr, J., & Landry, J. (2017). *Pulse of the fashion industry.* Retrieved from https://globalfashionagenda.com/wp-content/uploads/2017/05/Pulse-of-the-Fashion-Industry_2017.pdf

KU. (2017). *Social entreprise fair.* Retrieved from https://kusif.ku.edu.tr/en/conferences/social-impact-social-finance-conference/sosyal-girisim-fuari/

McCarthy, A. (2018, 22 March). *Are our clothes doomed for the landfill?* Retrieved from https://remake.world/stories/news/are-our-clothes-doomed-for-the-landfill/

McDonough, W., & Braungart, M. (2009). *Cradle to cradle: Remaking the way we make things.* North point press: New York.

McFall-Johnsen, M. (2019). *The fashion industry emits more carbon than international flights and maritime shipping combined. Here are the biggest ways it impacts the planet.* Retrieved from https://www.businessinsider.com/fast-fashion-environmental-impact-pollution-emissions-waste-water-2019-10

Morlet, A., Opsomer, R., Herrmann, D. S., Balmond, L., Gillet, C., & Fuchs, L. (2017). *A new textiles economy: Redesigning fashion's future.* Retrieved from https://www.ellenmacarthurfoundation.org/assets/downloads/publications/A-New-Textiles-Economy_Summary-of-Findings_Updated_1-12-17.pdf

Odabasi, S. (2020). Upcycling as a practice for decolonializing fashion: An interview with Ngozi Okaro. *Fashion Theory, 24*(6), 975–977. doi:10.1080/1362704X.2020.1800998

Paras, M. K., & Curteza, A. (2018). Revisiting upcycling phenomena: A concept in clothing industry. *Research Journal of Textile and Apparel, 22*(1), 46–58. doi:10.1108/RJTA-03-2017-0011

Reichart, E., & Drew, D. (2019). *By the numbers: The economic, social and environmental impacts of "fast fashion".* Retrieved from https://www.wri.org/blog/2019/01/numbers-economic-social-and-environmental-impacts-fast-fashion

Roberts, E. (2015). *Thank you, conscious company magazine.* Retrieved from https://www.indigenous.com/blog/article/conscious-company-magazine-features-indigenous-eco-ethical-fashion

Sabanoglu, T. (2019). *Textiles and clothing industry in Turkey - Statistics & facts.* Retrieved from https://www.statista.com/topics/4844/textiles-and-clothing-industry-in-turkey/

Šajn, N. (2019). *Environmental impact of the textile and clothing industry.* Retrieved from https://www.europarl.europa.eu/RegData/etudes/BRIE/2019/633143/EPRS_BRI(2019)633143_EN.pdf

Sert, Y. (2016). *Sürdürülebilirliğe Çıkan Çaputtan Bir Yol (A shoddy road to sustainability).* Retrieved from https://youtu.be/5OKIjUJxIdI

Thomas, S. (2008). From green blur to ecofashion: Fashioning an eco-lexicon. *Fashion Theory, 12*(4), 525–540.

UN. (2020). *Sustainable Development Goals.* Retrieved from https://sdgs.un.org/goals

UNECE. (2018). *Fashion and the SDGs: What role for the UN?* Retrieved from https://www.unece.org/fileadmin/DAM/RCM_Website/RFSD_2018_Side_event_sustainable_fashion.pdf

UNEP. (2018). *Putting the brakes on fast fashion.* Retrieved from https://www.unenvironment.org/news-and-stories/story/putting-brakes-fast-fashion

Wrap. (2020). *Household waste prevention hub – clothing.* Retrieved from http://www.wrap.org.uk/content/clothing-waste-prevention

22

Sustainable Marketing in China: The Case of Monmilk

Ruizhi Yuan
Nottingham University Business School,
University of Nottingham, Ningbo, China

Yanyan Chen
Toulouse Business School, Toulouse, France

CONTENTS

Learning Objectives	576
Theoretical Background	576
Background	576
The Significance of Sustainability	577
Triple Bottom Line	579
3BL in Practice	581
Sustainability in China	582
The Case—The Monmilk Group and Its Deluxe Milk	585
The Sustainable Marketing of Deluxe Milk	587
High-Quality Milk Sources	588
Green Manufacturing and Harmonious Working Environment	589
Monmilk's 3BL Challenges	590
Conclusion and Lessons Learned	591
Discussion Questions	593
Group Activity-Based Exercise	593
Acknowledgments	593
References	593

576 • *Social and Sustainability Marketing*

LEARNING OBJECTIVES

The learning objective of this case is to understand the "triple bottom line framework" and apply it to analyze firm's sustainable performances. The case is for a Bachelor degree level as well as MSc-level introductory marketing course that focus on "sustainability" marketing issues.

After reading the chapter, the reader should be able to:

- Understand the role of sustainability and establish its importance to company performance according to the triple bottom line framework
- Explore the challenges and current situation associated with the company regarding its sustainability practices
- Critically reflect the triple bottom line framework in the context of sustainability among Chinese companies

THEORETICAL BACKGROUND

The triple bottom line (3BL) framework was developed by Elkington (1998) to evaluate corporate performance from three dimensions: social/ethical, environmental, and financial. That is, an organization's performance and its business value should be measured from a broad perspective which includes its social, environmental, and economic performances. It represents bottom lines for the achievement of a goal of sustainability when enterprises conduct their business to consider the 3Ps of people, planet, and profit. This 3P concept indicates the deep roots of each bottom line and provides guidance for enterprises to balance monetary profit, social equity, and natural capital with a long-term outlook. The 3BL framework is now adopted by many companies and enterprises to direct their sustainable practices and to measure their performances (Norman & MacDonald, 2004; Shnayder, Rijnsoever, & Hekkert, 2015).

BACKGROUND

Sustainability as a concept is not new, and it has existed for hundreds of years. However, since 1987, academics, business leaders, and public policymakers started to pay attention to it. The UN Commission on Economic

Development, in its 1987 Brundtland report titled *"Our Common Future"*, claims sustainability as satisfying the current needs without sacrificing the needs for the future generations (Brundtland Commission, 1987). Munier (2005) adds that sustainability concerns with future based on personal values and apply ethical and moral principles to guide human behaviors. While conventional marketing is often perceived as profit-driven, the objective of sustainable marketing is to advocate sustainable customer behavior and offer sustainable marketing practices. Sustainability development involves finding the dynamic balance between the environment and the well-being from the economic and social perspectives. Researchers argue that sustainable marketing derives from corporate social responsibility (CSR) and includes three dimensions (environmental, economic, and social) (Kotler, 2011; Roy, Verplanken, & Griffin, 2015). These sustainable marketing dimensions are originated from Elkington's (1994) 3BL framework which offers a balanced view on an organization's environmental and social actions associated with its economic performance.

This chapter aims to provide a comprehensive understanding of the scope of the implementation of sustainable marketing practices. The focus is on the identification of the 3BL in companies. The case of Monmilk—the top dairy brand in China—is analyzed in order to understand key questions of how companies employ different sustainable marketing practices and what might be the key challenges. The results emphasize the different elements of 3BL which influence successful pathways for sustainability. However, sustainable strategies face not only issues of balancing the proportion of investment in different dimensions of 3BL but also the difficulty of measuring the return on investment in these sustainable practices.

THE SIGNIFICANCE OF SUSTAINABILITY

Sustainability marketing strives to achieve marketing aims and objectives and, at the same time, bear in mind to leave zero, or minimal, negative footprints on the local, community, and global environment that might eventually result in both financial and economic loss for present and future generations (Cooney, 2009). Previous studies argue that corporate sustainability is parallel to CSR (e.g., Montiel, 2008; Lloret, 2016). Gladwin, Kennelly, and Krause (1995) argue that sustainability can be achieved by people actions in a prudent, equitable, inclusive, and secure manner. Shrivastava (1995) discusses the concept of sustainable development with an environmental emphasis, including the practices of total sustainable competitive strategies, quality environmental management, corporate population impact control, and technology investment. Bansal (2005) applied three principles of corporate sustainable development: economic, social, and environmental integrity. Belz and Peattie (2012) also point out that sustainability marketing refers to "building and maintaining sustainable relationships with customers, the social environment and the natural environment" (p. 29).

For the measures of corporate sustainability development, scholars have not agreed on a universal way to measure, assess, or monitor a firm's sustainability practices. Epstein and Roy (2001) focus on the sustainability performance indicators, such as bribery, community involvement, product safety, environmental impacts, work force diversity, corruption, ethical sourcing, human rights, and usefulness. Based on the stakeholder model, Ameer and Othman (2012) suggest the use of four dimensions to measure sustainability development: community, environment, diversity, and ethical standards. Engida, Rao, Berentsen, and Lansink (2018) further articulate three main dimensions: (1) environmental (i.e., contractors and supply chain, operations, and products); (2) social (i.e., consumers, employees, society contractors and supply chain and community, and philanthropy; (3) governance (i.e., corporate governance, business ethics, and public policy).

As a key to transformation, sustainability marketing practice has the potential to change business and the communities in which they operate. Firms that already adopt sustainability marketing practices are keen to highlight the benefits of the 3P approach (Emery & Trist, 2012). Extant research has demonstrated that sustainable marketing yields significant advantages for firms, as well as for consumers. For the firm, sustainable marketing can improve consumer attitudes toward a company's product or service (Sun, Garrett, Phau, & Zheng, 2018). Today, sustainable marketing is viewed as companies to differentiate with competitors and provide additional values to company brand. Hunt (2011) investigates the tensions in the sustainable development and found that the key focus of public policy should be the economic growth in the relatively poor areas. Most research explains that consumers have more favorable attitudes toward a firm if a sustainability strategy is implemented (e.g., Sun et al., 2016). For instance, brands with eco-label or sustainability labels can create a favorable customer attitudes toward the company (Liu et al., 2014), enhance business-to-customer relationships (Kim, Taylor, Kim, & Lee, 2015), and improve customers and stakeholders' attachment toward the company (Bansal, 2005). Vanhamme and Grobben (2009) further explain that corporate engagement in social efforts could help a company build its brand reputation, and translate this into brand memorability by the customers.

Based on the discussion above, the concept of corporate sustainability is well articulated with Elkington's (1998) 3BL framework because it pays attention to the need for a proper balance among the three parameters—environmental, social, and economic—to facilitate sustainability in firms. The details of 3BL are discussed in the following section.

TRIPLE BOTTOM LINE

Traditionally, business uses "bottom line" to refer to the "profit" or "loss" based on a company's revenue and expenses. However, environmentalists and humanitarians suggest a broader definition of "bottom line" from a societal cost-benefits perspective. In addition to monetary profit, social and environmental concerns should also be considered with a long-term outlook. In the mid-1990s, Elkington (1998) developed the concept of "the triple bottom line" (see Figure 22.1) to measure corporate performance in *Cannibals with Forks: The Triple Bottom Line of 21st Century Business*. This framework considers corporate performance from three dimensions: financial, social/ethical, and environmental (Elkington, 1998). The ideas behind this are that a company's performance should not only be measured by its *profit*, but should include considerations of *people* and *planet*. The term 3BL has, since then, become increasingly popular and spread rapidly in the last few decades (Norman & MacDonald, 2004).

Referring to the dimensions in 3BL, the financial bottom line is well understood. It is seen as the economic value created by corporations. In contrast with the traditional view which limits financial benefits only enjoyed by companies themselves, this *profit* aspect in 3BL also covers the real economic benefits that have an impact on the whole of society. From a financial sustainable perspective, together with corporations' revenues and expenditures, business diversity indicators and business climate

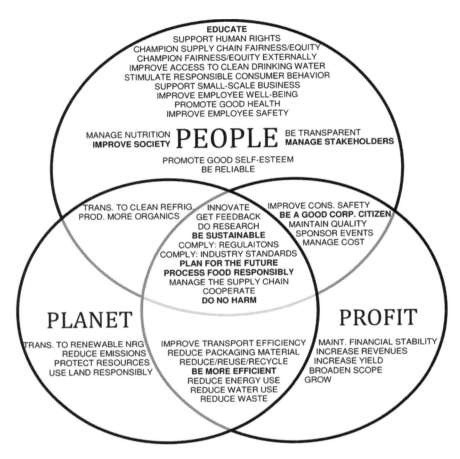

FIGURE 22.1
Triple bottom line model. (From Shnayder et al., 2015).

factors, such as job growth and employment distribution, are included as related variables (Slaper & Hall, 2011).

Social/ethical bottom line refers to the social equity and welfare that corporations produce. It fulfils the obligations to all the shareholders and related stakeholders, including suppliers, employees, customers, and communities. Specifically, the focus of social sustainability lies in both the internal community (i.e., employees) and various external communities (e.g., government, suppliers, minority groups, etc.) (Gimenez, Sierra, & Rodon, 2012). The social/ethical perspective in 3BL emphasizes *people*'s individual universal rights, such as equality, fairness, and justice (Shnayder et al., 2015). For example, a company that values the social/ethical bottom line often takes good care of its employees' safety and aims to improve their well-being. Meanwhile, the company also stimulates responsible consumer behavior and supports local communities.

Environmental bottom line refers to the natural capital of the *planet* that requires environmentally sustainable practices. It is akin to the concept of eco-capitalism. With the dual purpose of benefiting the planet and minimizing harm to the environment, companies are required to reduce the ecological footprint they leave as a result of their operations (Gimenez, Sierra, & Rodon, 2012). Generally, an environmentally sustainable company works on operations which reduce waste and emissions, improve energy efficiency to protect the natural environment, and prevent environmental degradation. More importantly, environmentally sustainable practices help to improve businesses' profitability in the long-term. A reduction in the usage of toxic materials and better governance in manufacturing processes help companies establish a sustainable reputation which, in turn, impacts consumers' attitudes and generates positive word of mouth. Additionally, it also helps companies suffer less from reputational damage when confronted with brand crises or service failures and increases their resilience to critical events.

3BL in Practice

However, despite the distinct classification of inherent responsibilities in each bottom line, sometimes there are overlapping responsibilities in regard to different bottom lines when actual tasks and initiatives are conducted in the real business world (Shnayder et al., 2015). For example, the code of conduct which relates to the environmental bottom line only covers environmentally sustainable practices that help protect the planet. However, with the aim to meet this particular bottom line, a food company may offer CSR initiatives to advance local farmers' planting techniques by offering energy and water-saving planting equipment. This initiative actually covers both environmental and social/ethical bottom lines as it protects the local environment via a subsequent reduction in water waste, and improves both energy efficiency and the competitiveness and welfare of the local farmer community in the market. More details and examples of overlapping cases are illustrated in Figure 22.1 (adopted from Shnayder et al., 2015). Though the overlap in actual practices in different bottom lines in some cases adds difficulties to the 3BL measurement in academic research, the flexibility of combing these bottom line responsibilities in one task allows organizations to apply them in a particular manner that helps to fit specific needs in practice. More specifically, different customers may have their own preferences toward a particular bottom line. The combination of different bottom lines into one initiative enables organizations to maximize its impacts on customers with different preferences.

The 3BL is an important tool in the provision of fundamental and conceptual support for sustainability goals (Slaper & Hall, 2011). Given the rapid uptake of the 3BL concept, corporations, governments, NGOs, and activist groups are inspired to participate into CSR movements. Accordingly, a variety of standards (e.g., ISO standards; SA 8000) and standard-setting bodies (e.g., International Organization for Standardization, Social Accountability International) have changed to identify, measure, and audit sustainable performances, especially from social and environmental aspects (Norman & MacDonald, 2004). These standards have profound impact on businesses' ethical practices among corporations. Companies shift from profit-maximizers to responsible corporations that take into account long-term business success and encourage ethical and sustainable performances. Particularly, many corporations use the language of 3BL and devote huge investment to social- and environmental-related campaigns.

SUSTAINABILITY IN CHINA

In China, sustainable marketing has shifted from ideology to reality, but is marked by immature conceptions and unbalanced patterns (Yan & She, 2011). In 1993, the National People's Congress (NPC) Environmental and Resources Protection Committee was formed, and a proposal named "China Environmental and Resources Protection legal framework" was issued. This proposal is considered the starting point for China's environmental legislation to promote environmental protection (Bai, Sarkis, & Dou, 2015). However, for the first decade, China found it difficult to assimilate the fundamental premises of sustainability in business operations. A 2006 report concluded that, from among 127 emerging-market companies, 19 companies in China were recognized by an "especially low take up" of sustainability practices (Baskin, 2006, p. 31). Since then, sustainability becomes an emerging concept around the country. Recognizing the challenges to the sustainable development of both the economy and society, the Communist Party of China (CPC) announced a "Socialist Harmonious Society" strategic document on 11 October, 2006 (Yan & She, 2011). As a result, due to the advocacy of China's government, social organizations, and scholars, sustainability or CSR practices are booming; in 2010, for the first time, over 600 companies in China issued sustainability reports (Lee, 2011). Despite the government's strategic change, more actions are needed for the whole society. In a 2006 survey of 890 companies based in China, including all private, state-owned, and multinational enterprises, most companies have mistaken ideas about sustainability or CSR (Yan & She, 2011). They perceive sustainability to be a key issue for large organizations rather than the SMEs. Moreover, sustainability is also perceived as a burden or a distraction for many companies as their main focus is on "building their business".

Increasingly, 68% of companies in China identified that the most significant concerns of the organization and its stakeholders are to "create wealth for the society" and to "promote the nation's development" (Lee, 2011). Such development should be placed in the context of the current situation in China. It is a generally held perception that people's attentions to a variety social issues, including poverty, disasters, and lower level of food safety, has been a hot topic in today's China. Furthermore, a 2011 report found that, among those companies engaging in CSR campaigns in China, Chinese domestic companies (including China Mobile, Lenovo, and Haier) continue to have the strongest presence in terms of awareness among consumers (R3 En-Spire, 2011). In a 2019 HSBC Navigator: Trading with China report, which surveyed more than 1,700 decision-makers in 34 countries, it is claimed that Chinese businesses are more likely to view sustainability as vital to their long-term prosperity compared with decision-makers in other countries (HSBC, 2019). China has been strong in recent years in its creation of sustainable and creative growth through structural changes in both manufacturing and service sectors. Following the global sustainability footprint, China has initiated development of 10 demonstration areas across the nation to utilize innovative developments to facilitate sustainability by the end of 2020 (Wang, 2019).

On the other hand, from the consumer's perspective, Chinese consumers are willing to support firms that organize sustainability activities. According to the 2012 Goodpurpose Survey, 50% of Chinese consumers consider that business is required to balance the focus between company profits and societal well-being. Nearly 67% of Chinese consumers prefer to purchase products from a company with a strong philanthropic reputation, and most of them state that they would like to choose certain brands which are supporting sustainability with the same quality products (Goodpurpose, 2012). However, at the same time, the Chinese public are becoming as cynical as Western consumers. In China, food scandals, rumors of charity money embezzlement, and fraudulent financial statements have been exposed in recent years. Another trend in customers' perceptions of sustainability in China is that much younger consumers are more likely to encourage an environmentally friendly shopping behavior. For example, according to the 2018 China Sustainable Consumption Report, the individuals who are younger than 20-year-old usually express most concerns about sustainable practices in China (Smith, 2019).

In summary, due to rapid economic growth, China has come under various pressures to address its environmental footprint. Consumers, marketers, and public policymakers all strive to engage in the promotion of sustainability practices. It is inevitable that China will explore industrial sustainability in a different way than Western nations because of economic, social, and environmental differences. Although the term "sustainability" was not fully understood or implemented in China in earlier years, the idea of sustainable development has evolved with people's minds and behaviors in China because people are increasingly aware of the widespread environmental issues around the world during the last 20 years. To address their sustainability challenges, such as creating public awareness and facilitating industrial revolution, China has employed several measures, including public policies, technological innovation developments, and awareness and engagement campaigns, to further raise the awareness and call for future actions toward the sustainability issues.

One fundamental question to be drawn from this background is how can a firm engage in the adoption of sustainability practices? How does a firm overcome the challenges of sustainability? These issues are discussed in this chapter using the case of Monmilk in China to examine the adoption of sustainability practices by business. More specifically, the case focuses on the external and internal pressure placed on firms to adopt sustainability practices and investigates the motivation and impacts of sustainability adoption.

THE CASE—THE MONMILK GROUP AND ITS DELUXE MILK

In 1999, Niu Gensheng, the founder of Monmilk, started his business legend in a small factory building of tens of square meters. In 2002, its sales exceeded 2.1 billion RMB, ranking it fourth among the national dairy companies. Monmilk's sales doubled again in 2003 as a result of its excellent marketing strategy cooperating with the Shenzhou 5 Manned space mission. With its great potential, Monmilk was then officially listed on the Hong Kong Stock Exchange in 2004. Due to the Sanju Qingan scandal in 2008, as an alternative to bankruptcy, Monmilk became part of the COFCO group in 2009, and by 2016 it ranked 10th in the global dairy industry. In 2018, Monmilk was awarded the "Global Dairy Top 10" by Rabobank and ranked one of the top 20 most valuable Chinese brands.

Over the last 21 years, Monmilk has become one of the largest dairy processing organizations in China and the annual production capacity of dairy products has reached 9.5 million tons with an annual revenue of RMB 80,000 million. By the end of 2019, the Group had set up more than 40 production bases in 19 provinces and autonomous regions. The product line covers milk formula, ice cream, liquid milk, and other dairy products. Lessons learned from the 2008 Chinese milk scandal, which involved milk, infant formula, and other food materials being contaminated with melamine, means Monmilk has been actively devoting itself to greater sustainability. The tainted milk powder mainly produced by Monmilk and Yili Industrial Group—the two largest dairy producers in China in 2008—was blamed for the deaths of six infants and illnesses in 54,000 others. As a direct consequence of the scandal, Monmilk has paid 948.6 million RMB in compensation and, more importantly, it lost most of its brand loyalists at the time. In order to rebuild customer trust and position itself as respectful of human health, Monmilk significantly

586 • Social and Sustainability Marketing

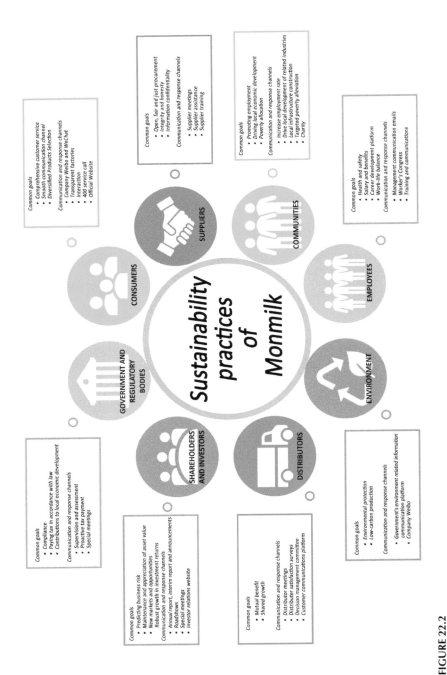

FIGURE 22.2
Sustainability practices of Monmilk. (Adapted from Monmilk official website).

increased its investment in sustainability and claims its own responsibility is to provide nutrition and health to more people (especially youth). The CEO of Monmilk, Liu Minfang, believes that guiding consumers toward a healthier life is a major responsibility for his company. The group has made every effort to promote sustainable solutions in the marketplace (see Figure 22.2). Aligning with the "Healthy China" strategy and the call for targeted poverty alleviation, Monmilk established a partnership with the United Nations World Food Program to roll out the "Inclusive Nutrition Plan". In 2006, Monmilk donated nearly 110 million RMB to help local children who live in poverty-stricken areas in 500 primary schools for one year (Li, Haywood-Sullivan, & Li, 2012). Furthermore, Monmilk has established agreements with world's top aseptic packaging material manufacturing company, and being the first one in the market to use aseptically recyclable packaging materials for its variety of milk products. Since 2005, Monmilk has promoted high-end products with differentiated features; its two premier brands of milk are Milk Supreme and Milk Deluxe.

THE SUSTAINABLE MARKETING OF DELUXE MILK

As a sub-brand of Monmilk, Milk Deluxe focuses on producing high-end and premier quality products. In the 2000s, there was an increasing trend in the consumption of milk in the Chinese market alongside economic development and consumers' increasing power in terms of their consumption. In the meantime, consumers' demands for quality milk were also increasing constantly. Observing this trend, Monmilk established Milk Deluxe in 2005 to meet the rising demands of the evolving milk market. The brand name "Deluxe" originates from Mongolian and means *top grade* or *premier class*, which indicates a high quality and superior identity. Reflecting its brand name, Milk Deluxe provides premier quality milk with better nutrition and higher protein than other varieties. Every 100 mL of Deluxe milk contains 3.6 g natural protein and 120 mg natural calcium, compared to the national standard of 2.9 g/100 mL protein in the Chinese milk industry.

Holding a vision that promises a healthier world, Monmilk follows commitments which are guided by *The UN Sustainable Development Goals* to produce "more nutritious products", to encourage customers toward "a healthier life", and to build "a sustainable planet" (Monmilk, 2018, p. 13) based on the 3BL framework. In order to maintain its superior identity and retain its CSR image, Milk Deluxe has made great efforts throughout its milk producing and processing operations to follow CSR guidelines. Particularly, a three-level sustainable development system is adopted to ensure its sustainable strategic planning, starting from decision-making at management level through to implementation level. First, both medium-term and long-term sustainable plans are discussed, developed, and decided on. Then, following its strategic planning and sustainable principles, effective management and eco-friendly operations are conducted to fulfill its ethical and sustainable development goals. A series of practices are adopted based on its "from grass to glass" management, which aims to ensure the quality and safety of its milk, from production, to manufacture, and finally to customers' tables. These ethical and sustainable practices cover different aspects ranging from its *milk source selection* to its *manufacturing process*.

High-Quality Milk Sources

To improve "national nutrition and health with the care for children's nutrition and health in poor areas" (Monmilk, 2018, p. 82), the very first step in Deluxe's quality control, in accordance with its CSR guidelines, is the strict selection of its milk source. To ensure high-quality milk sources, Deluxe controls every possible aspect that helps improve the quality of milk sources: the ranch, the cow, and the forage grass. The milk source for Deluxe is Northern Organic Ranch, an exclusive organic pasture operated by the Monmilk Group. The ranch is located at 40-degrees north latitude in Hebei Province, which is considered the "golden belt" of milk sources. In 2013, a "Rancher University" program was launched to educate local ranchers and herdsmen in upgrading technology and advanced management. As a result, ranchers managed to transform their ranch into a modern and intensive digital farm. Now, modernized and standardized farming, as well as eco-system management, are carried out in all processes undertaken at the ranch. For example, the "ecotype energy" treatment process is adopted to transfer fecal sewage produced by cows into biogas. Then, the biogas is returned to the field to fertilize the farmland and, in turn, it cultivates forage grass. In doing so, Deluxe achieves its aim of reducing waste and protecting the ecological environment by successfully recycling its resources through its "cow-biogas-grass" agro-ecological cycle.

In addition, apart from providing comprehensive support to ranchers for sustainable ranch management, Deluxe adopts refined cow management to ensure animal welfare and the exclusive quality of its milk sources. Following the belief that "high quality milk comes from happy cows" (p. 27), Deluxe respects the characteristics of the dairy cow and has improved their welfare in a range of ways. That is, from cow breeding (e.g., independent delivery room, refined milking equipment) to safety and epidemic prevention (e.g., confirmed drinking water quality, fly-killing, heat stress prevention). More importantly, different types of forage grass are cultivated with better nutrition (e.g., alfalfa with protein as high as 18–23%). These high-quality organic forage grasses are provided to better satisfy the different tastes of dairy cows.

Green Manufacturing and Harmonious Working Environment

Apart from the full-lifecycle management in the Northern organic ranches, Deluxe also adopts a green manufacturing approach in the whole process of its production management. Specifically, Deluxe implements a series of energy saving projects during its manufacturing process. Clean energy sources are used extensively to fulfill its goal of low carbon production. For example, in its energy-optimizing workmanship and energy saving boilers projects, renewable energy, such as solar, wind, biogas, and biomass are used to reduce waste and greenhouse gas emissions. As a result, over 20,000 tons of fossil fuels (e.g., coal) were saved in 2019 to meet the goal of energy conservation and emission reduction. Meanwhile, employees are assured with a sense of well-being in their working environment. To build a happy and harmonious working environment, employees are encouraged to communicate with the company on its communication platform. Employee representatives listen to and review the comments from general employees to identify any issues regarding work and life and to ensure their interests are known and acted upon in the workspace. Additionally, special attention is paid to employees' rights and needs as well as their work-life balance. Two special funds, an employee care fund and an employee care mutual support fund, have been launched to help those who have difficulties in their family, especially when spouses and/or children are in need.

Furthermore, the concept of environmental protection and lean management is advocated in its industrial chain, from packaging to storage and transportation. For example, green package materials, certified by the FSC or SFI forest system, are used to ensure resource conservation as well as package quality. Meanwhile, Deluxe minimizes the road travel of its milk to reduce the energy consumption and waste emissions that can be caused by transportation. A green storage strategy of *"production and sales in the same place"* is adopted to ensure the milk's freshness and to reduce potential environmental impact from product transportation. More importantly, a digital ecosystem is applied to increase the efficiency of transportation and storage systems. Resource Planning (ERP), Laboratory Information Management System (LIMS), and Intelligent Warehouse Management System (WMS) are used to increase energy efficiency and product traceability, as well as to reduce operating and product management costs. Alongside road travel, raw material providers, downstream distributors, and logistics companies interact with each other efficiently to achieve a win-win cooperation among all stakeholders. Indeed, Deluxe also provides professional financial support to these stakeholders when they face financing difficulties.

MONMILK'S 3BL CHALLENGES

As Deluxe continues to perform sustainably, from production to manufacture, and to improve the welfare of society as well as that of the planet, it has gradually recovered from the 2008 milk scandal and rebuilt its brand image. In May, 2018, Milk Deluxe won the only "gold award" at the 12th Biofach in China. Having regained the trust of the Chinese customer, Deluxe's sales volume exceeded 11 billion RMB in 2016 and 12 billion RMB in 2017. It maintained this high level of growth at 15% in 2018.

However, as a whole group, Monmilk did face some challenges in its performance of sustainable and ethical practices which follow the 3BL framework. First, the 3BL framework suggests an integration of social, environmental, and financial dimensions. This is difficult to achieve, especially after the 2008 milk scandal. Specifically, a critical question for Monmilk to concern that time is how to balance the proportion of their investments in different dimensions. At that time, Monmilk has invested mainly on the social and environmental pillars, and sacrificed the financial pillar to re-build its brand image. Apart from ensuring milk quality via its milk safety program, a main focus of investment to improve social welfare was applied as a strategic move to gain customers' trust and rebuild its reputation. Now, Monmilk has emerged from the 2008 milk scandal and gradually re-built its reputation in the milk industry in China. Consequently, a new question to Monmilk in the current stage is how to adjust its CSR strategy and achieve a new balance within the three dimensions of the 3BL framework.

Another challenge for Monmilk to follow the 3BL principle is that it has difficulties in measuring the ROI (return on investment) of its CSR practices. The profit dimension in 3BL is easy to measure, while this is not the case for the people and planet dimensions. The first difficulty lies in the overlapping dimensions in one practice, i.e., that one CSR initiative may cover duties from both social and environmental dimensions. As a result, the actual investment on a particular dimension cannot be easily calculated. More importantly, the return on these investments are difficult to measure since these ethical and/or sustainable practices often help increase brand awareness and create brand equity. Generally, the value of brand awareness and brand equity cannot be measured on the basis of CSR initiatives. In addition, CSR practices often have delayed impacts, and thus the ROI represents the influence from the last year's, or last few years', accumulated inputs.

CONCLUSION AND LESSONS LEARNED

Due to the increasing discussions concerning food safety, Monmilk finds itself difficult to gain customer confidence and the pressures from government has urged them to move into an environmental-friendly organization. This needs huge investments and efforts. Pursuing sustainable management and employee satisfaction is essential to Monmilk's survival. The implementation of Monmilk's sustainability practices is positively associated with "People" and "Planet" in 3BL.

The first 3BL value proposition represented by Monmilk case is "People". Monmilk provides guaranteed higher quality organic milk, which implies social benefits. Previous studies show that Chinese customers pay more attention to organic products and other forms of eco-labelling now than in the past (e.g., Yu, Yan, & Gao, 2014). They perceive that organic food is more nutritious than conventional farmed food. Monmilk has seized this opportunity and effectively connected itself with "safer and higher quality" and "organic milk". By promoting its Deluxe milk, customers are educated with eco-labelling and learn about the nutrition of the milk. They tend to be more satisfied with, and thus favor, products from firms with sustainable management practices.

In terms of promoting organic milk, there is also a challenge to the supply side of the industry. Environmental problems, such as non-point source of pollution, are areas of organization' focus that needs to be addressed in the sustainability development. Consequently, regarding performance outcomes, Monmilk's sustainability efforts which target suppliers can also contribute to the firm's 3BL. As Monmilk emphasizes, its "Planet" orientation is geared toward improving its sources of clean energy, which are used extensively to fulfill its goal of low carbon production, and green packaging and transportation. As a result, these practices have made improvements to Monmilk's environmental performance and helped the firm to alleviate the sustainability challenges.

In conclusion, in recent years, issues of sustainability have gained increasing attention on the business research agenda. Firms act sustainably when they support the three dimensions of 3BL (i.e., the 3 Ps of people, planet, and profit). Changes in consumption patterns and recent food scandals in China concern both Chinese public policymakers and marketers. Integrated sustainability programs and projects are both crucial and challenging for firms to address the complex issues around environmental remediation, food safety, and future sustainability management. China, therefore, has much to gain by shifting to more sustainable production methods. This chapter has considered a case study of a Chinese dairy processing organization—Monmilk—and explored the trajectories of sustainability practices and the 3BL dimensions (i.e., profit, people, and planet). The benefits of these sustainability practices include improvements in environmental and public health awareness in China. The lessons emerging from the case provide insights for researchers, practitioners, and policymakers into how sustainable development can be better supported, both in China and elsewhere.

DISCUSSION QUESTIONS

1. What is the key challenge faced by Monmilk in the past?
2. How does Monmilk deal with this challenge?
3. Based on the triple bottom line model, what suggestions can be given to Monmilk to sustain its position in the market in the future?

GROUP ACTIVITY-BASED EXERCISE

Take a company of your choice which you consider practices sustainably.

1. Justify your choice by detailing characteristics which you consider they show to practice sustainability.
2. Use the triple bottom line framework to evaluate the objectives of the company's sustainable development.
3. Consider the benefits and the potential issues for the company's implementation of its sustainability practices.
4. What do you think this company should do in the future to enhance its sustainability development?

ACKNOWLEDGMENTS

The authors acknowledge the financial support from National Natural Science Foundation of China (NSFC) no. 71804149.

REFERENCES

Ameer, R., & Othman, R. (2012). Sustainability practices and corporate financial performance: A study based on the top global corporations. *Journal of Business Ethics*, 108(1), 61–79.

Bai, C., Sarkis, J., & Dou, Y. (2015). Corporate sustainability development in China: Review and analysis. *Industrial Management & Data Systems*, 115(1), 5–40.

Bansal, P. (2005). Evolving sustainably: A longitudinal study of corporate sustainable development. *Strategic Management Journal*, 26(3), 197–218.

Baskin, J. (2006). Corporate responsibility in emerging markets. *Journal of Corporate Citizenship*, 24(Winter), 29–47.

Belz, F. M., & Peattie, K. (2012). *Sustainability marketing: A global perspective* (2nd ed). UK: John Wiley & Sons, Ltd.

Brundtland Commission. (1987). *Report of the world commission on environment and development: Our common future* [online]. Available at: https://sustainabledevelopment.un.org/content/documents/5987our-common-future.pdf [accessed 17th May 2020]

Brundtland, G., & Khalid, M. (1987). *UN Brundtland commission report. Our Common Future*. Available at: https://sustainabledevelopment.un.org/content/documents/5987our-common-future.pdf

Cooney, S. (2009). *Build a green small business: Profitable ways to become an ecopreneur*. New York, NY: McGraw-Hill.

Elkington, J. (1994). Towards the sustainable corporation: Win-win-win business strategies for sustainable development. *California Management Review*, 36(2), 90–100.

Elkington, J. (1998). Partnerships from cannibals with forks: The triple bottom line of 21st century business. *Environmental Quality Management*, 6, 37–51.

Emery, F. E., & Trist, E. L. (2012). *Towards a social ecology: Contextual appreciation of the future in the present*. Berlin/Heidelberg: Springer Science & Business Media. [online]. Available at: https://books.google.co.jp/books?hl=zh-TW&lr=&id=aUHUBwAAQBAJ&oi=fnd&pg=PR10&dq=Towards+a+social+ecology:+Contextual+appreciation+of+the+future+in+the+present&ots=vyz9_LDYLr&sig=1vhyhx3crUO2FeOJlKEbxbpkLbc&redir_esc=y#v=onepage&q=Towards%20a%20social%20ecology%3A%20Contextual%20appreciation%20of%20the%20future%20in%20the%20present&f=false [accessed 10th May 2020]

Engida, T. G., Rao, X., Berentsen, P. B., & Lansink, A. G. O. (2018). Measuring corporate sustainability performance–the case of European food and beverage companies. *Journal of Cleaner Production*, 195, 734–743.

Epstein, M. J., & Roy, M. J. (2001). Sustainability in action: Identifying and measuring the key performance drivers. *Long Range Planning*, 34(5), 585–604.

Gimenez, C., Sierra, V., & Rodon, J. (2012). Sustainable operations: Their impact on the triple bottom line. *International Journal of Production Economics*, 140, 149–159.

Gladwin, T. N., Kennelly, J. J., & Krause, T. S. (1995). Shifting paradigms for sustainable development: Implications for management theory and research. *Academy of Management Review*, 20(4), 874–907.

Goodpurpose. (2012). *Goodpurpose: Edelman goodpurpose 2012 global consumer survey* [online]. Available at: http://www.amcham.or.id/auto_/intranett1/Edelmangoodpurpose2012-IndonesiaFindings.pdf [accessed 6 June 2020]

HSBC. (2019). *Sustainability key to Chinese markets* [online]. Available at: https://www.hsbc.com/insight/topics/sustainability-key-to-chinese-markets [accessed 19th June 2020]

Hunt, S. D. (2011). Sustainable marketing, equity, and economic growth: A resource-advantage, economic freedom approach. *Journal of the Academy of Marketing Science*, 39(1), 7–20.

Kim, J., Taylor, C. R., Kim, K. H., & Lee, K. H. (2015). Measures of perceived sustainability. *Journal of Global Scholars of Marketing Science*, 25(2), 182–193.

Kotler, P. (2011). Reinventing marketing to manage the environmental imperative. *Journal of Marketing*, 75(4), 132–135.

Lee, S, Y. S. (2011). *Fortune China CSR Ranking 2011 Report* [online]. Available at: http://www.fortunechina.com/rankings/c/2011-03/15/content_51879.htm [accessed 5th June 2020]

Li, S. F., Haywood-Sullivan, E., & Li, L. (2012). Made in China: The Mengniu Phenomenon. *Management Accounting Quarterly*, 13(3), 9–9.

Liu, M. T., Wong, I. A., Shi, G., Chu, R., & Brock, J. L. (2014). The impact of corporate social responsibility (CSR) performance and perceived brand quality on customer-based brand preference. *Journal of Services Marketing*, 28(3), 181–194.

Lloret, A. (2016). Modeling corporate sustainability strategy. *Journal of Business Research*, 69(2), 418–425.

Monmilk. (2018). *2018 China Mengniu dairy company limited: Sustainability report (ESG report)* [online]. Available at: http://www.mengniuir.com/attachment/20190724184801566635666627_en.pdf [accessed 5th June 2020]

Montiel, I. (2008). Corporate social responsibility and corporate sustainability: Separate pasts, common futures. *Organization & Environment*, 21(3), 245–269.

Munier, N. (2005). *Introduction to sustainability*. Dordrecht, the Netherlands: Springer.

Norman, W., & MacDonald, C. (2004). Getting to the bottom of "triple bottom line". *Business Ethics Quarterly*, 14(2), 243–262.

R3 En-spire. (2011_. *Benchmarking CSR in China. Fall 2011* [online]. Available at: http://www.rthree.com/ [accessed 17th June 2020]

Roy, D., Verplanken, B., & Griffin, C. (2015). Making sense of sustainability: Exploring the subjective meaning of sustainable consumption. *Applied Environmental Education & Communication*, 14(3), 187–195.

Shnayder, L., Rijnsoever, F. V., & Hekkert, M. P. (2015). Putting your money where your mouth is: Why sustainability reporting based on the triple bottom line can be misleading. *PLoS ONE*, 10, 1–24.

Shrivastava, P. (1995). The role of corporations in achieving ecological sustainability. *Academy of Management Review*, 20(4), 936–960.

Slaper, T. & Hall, T. (2011). The triple bottom line: What is it and how does it work? *Indiana Business Review*, 86, 4–8.

Smith, T. (2019). *Do Chinese consumers care about sustainable fashion?* [online]. Available at: https://jingdaily.com/do-chinese-consumers-care-about-sustainable-fashion/ [accessed 13th June 2020]

Sun, Y., Garrett, T. C., & Kim, K. H. (2016). Do Confucian principles enhance sustainable marketing and customer equity? *Journal of Business Research*, 69(9), 3772–3779.

Sun, Y., Garrett, T. C., Phau, I., & Zheng, B. (2018). Case-based models of customer-perceived sustainable marketing and its effect on perceived customer equity. *Journal of Business Research*, 117, 615–622.

Vanhamme, J., & Grobben, B. (2009). "Too good to be true!". The effectiveness of CSR history in countering negative publicity. *Journal of Business Ethics*, 85(2), 273.

Wang, Y. (2019). *Economic Watch: Sustainability pursuit drives China's innovation* [online]. Available at: http://www.xinhuanet.com/english/2019-11/22/c_138574770.htm [accessed 17th June 2020].

Yan, J., & She, Q. L. (2011). Developing a trichotomy model to measure social responsible behavior in China. *International Journal of Market Research*, 53(2), 253–274.

Yu, X., Yan, B., & Gao, Z. (2014). Can willingness-to-pay values be manipulated? Evidence from an organic food experiment in China. *Agricultural Economics*, 45(S1), 119–127.

23

Nurpu: A Dream towards a Sustainable Handloom Weaving Society

Sathyanarayanan Ramachandran
IFMR Graduate School of Business, Krea University,
Sri City, Andhra Pradesh, India

S. A. Senthil Kumar
Department of Management, School of Management,
Pondicherry University, Karaikal Campus, Puducherry, India

CONTENTS

Learning Objectives	598
Themes and Tools	599
Relevant Theory	599
Handloom Industry in Shambles	600
Sivagurunathan's Dream	600
Gandhian Inspiration	601
Attempt at Revival	601
Sustainable Production	602
Customer Perception, Pricing and Business Dilemma	604
Marketing	605
Ecology, Economy and the Enterprise	607
Restart Framework	607
Questions for Discussions	609
Funding Declaration	610
Acknowledgment	610
Credit Author Statement	610
Notes	610

The handloom industry is in a dying condition. I took special care during my wanderings last year to see as many weavers as possible, and my heart ached to find how they had lost, how families had retired from this once-flourishing and honourable occupation.

Mahatma Gandhi in his Speech on Swadeshi
at Missionary Conference, Madras on February 14, 1916

The handloom weaving is in a dying condition. Everyone admits that whatever may be the future of the mill industry, the handlooms ought not to be allowed to perish.

Mahatma Gandhi, in his Letter dated July 3, 1917
to V. S. Srinivasa Sastri

...Nature enlists and ensures the co-operation of all its units, each working for itself and in the process helping other units to get along their own too-the mobile helping the immobile, and the sentient the insentient. Thus, all nature is dovetailed together in a common cause. Nothing exists for itself. When this works out harmoniously and violence does not break the chain, we have an economy of permanence.

Shri J. C. Kumarappa in Economy of Permanence

LEARNING OBJECTIVES

After reading the chapter, the reader should be able to:

- Understand sustainable business in the apparel industry through handlooms
- Apply various sustainable business concepts like the flourishing business canvas, RESTART framework, etc
- Understand the challenges and opportunities of the handloom industry
- Apply the Gandhian thoughts in modern business scenario
- Use digital marketing, cause marketing and public relations
- Enable scope for field study with the case scenario as a starting point
- Enable mini projects at MBA level courses based on the case scenario
- Discuss the dilemma of expansion and growth vis-à-vis sustainable long-term progress

THEMES AND TOOLS

- Flourishing Business Canvas
- Circular Economy
- RESTART Framework
- Economy of Permanence
- Cause-Based Digital Marketing

RELEVANT THEORY

1. Economy of Permanence towards Sustainability
2. Flourishing Business Model Canvas
3. RESTART Framework

India, the second-largest country in terms of population, is one of the major producers as well as consumer of textiles and fashion apparels. It is essential to have innovative sustainable solutions and innovations pertaining to the fashion industry. Indian textile industry has a long history and cultural heritage and has been one of the world's significant producers of textiles for several millennia. There are archaeological pieces of evidence about the presence of the cotton-based textile industry in the Indus valley civilization towns of Harappa and Mohenjo-Daro.[4] In the medieval period, Indian weavers were quite famous for their sophisticated handloom materials, design and artistry. Dacca[5] Muslins were one of the finest and lowest weighing muslins.[6] Till the industrial revolution and the arrival of power looms and sewing machines, Indian handloom sector thrived in terms of business and trade. Many families were traditionally involved in weaving activity and passed on the tacit knowledge within the family and the clan.

Nurpu is an exciting, sustainable fashion brand from the state of Tamil Nadu in India. Nurpu means weaving in Tamil language, one of the oldest surviving classical languages in the world, spoken mainly in the state of Tamil Nadu and by the diaspora in countries like Malaysia, Singapore, Canada and Europe. Tamil Nadu is one of the highly industrialized states in India with globally competitive textile clusters like Coimbatore and Tirupur and the adjoining districts like Karur and Erode. While mechanized production processes and power looms dominate the sector, Nurpu ventured into handlooms, which is one of the vanishing, traditional arts that use sustainable ways to make fabrics. Since the conventional handloom weaving families were finding it difficult to get new businesses amidst the volume-driven, highly scaled-up power loom mills, many next-generation family members stepped out of the family ventures. Some of them got employed in sunrise industries like the software industry.

HANDLOOM INDUSTRY IN SHAMBLES

Mr. Sivagurunathan is one such member of the family of traditional weavers who took up IT as a profession after his MBA graduation and his family also quit handloom weaving and ventured into trading business for better revenue and livelihood.[7] Many such weaving families went in search of greener pastures and became workers in power loom industries and other sectors. It also means a slow death of traditional art and a vanishing act of one significant cultural aspect of a community. This was also a demotion of sorts of the master craftsmen and weavers into mere labourers working with the machines, thus losing the pride and self-respect of being a creator.[8] One entire colony of weavers called 1010 Colony in Chennimalai of Erode district, which used to inhabit 1010 busy weavers' families, became silent without any sound of the active shuttle that continually moves in the loom.

SIVAGURUNATHAN'S DREAM

Some of the next generation members like Mr. Sivagurunathan were relatively well-off due to a better paying Information Technology industry. Sivagurunathan worked in the IT industry for 11 years in Bangalore, the Infotech powerhouse of India.[9] Yet, he learnt about the traditional nuances of handlooms from his father, whenever he visited home during weekends and holidays. It was in 2016, Sivagurunathan decided to quit his high-paying job to start NURPU to revive the handloom industry and to help the traditional weavers. He was inspired by Mr. Nammalwar,[10] Mr. Theodore Bhaskaran,[11] Mr. Sivaraj of cuckoo movement for children[12] and by the sustainable living ideals of Mahatma Gandhi and the Gandhian economist and freedom fighter[13] Shri J. C. Kumarappa.[14]

GANDHIAN INSPIRATION

Like many of his classmates, Sivagurunathan also initially had the regular career dreams. Being an avid book reader, some of the books on virtual water and novels centred around the theme of "Values" made him to inquire about purpose of life. A meeting with the founders of Cuckoo forest school gave a different perspective to life and provided the opportunity to learn more about the Gandhian ideals. These interactions culminated in Sivagurunathan teaching weaving to the children in the forest school. Another meeting with Mr. Nammalwar strengthened the resolve to follow Gandhian ideals towards sustainable living. Later, Sivagurunathan got introduced to the Gandhian Mr. Surendra Koulagi, who was doing handloom weaving in Melkote, Karnataka, and was striving to promote Khadi through Janapada Seva Trust. Well-respected Mr. Koulagi was an octogenarian who has directly worked with Mahatma Gandhi and his associates like Acharya Vinobha Bhave. During Sivagurunathan's interaction with Mr. Koulagi, he has advised that handloom is not a sector for profiteering and one needs lot of patience to grow slowly.

When he heard the stories of backbreaking work done by the traditional weavers from his neighbourhood in the spinning mills and the power looms, he decided to help them out to revive the handloom business in an eco-friendly way. Both friends and family were surprised by this decision, but Sivagurunathan persisted and travelled a lot to learn about weaving and colouring. After a lot of trials and tribulations, brought six to ten weavers on board to make them believe in his ideas of reviving the dying art. Nurpu came in to existence as a registered Micro, Small and Medium Enterprises (MSME) on October 2, 2016, which was again, Mahatma Gandhi's birthday. Investments came from his own savings.

ATTEMPT AT REVIVAL

With a dream to revive the 1010 Weavers Colony, Sivagurunathan first attempted talking to couple of weavers to convince them to return to weaving. His initial overtures weren't taken seriously by the weavers as they already had bad experiences with many such revivers who started with

602 • *Social and Sustainability Marketing*

| Handloom and Weaver | The Shuttle, Warp and the Weft |

FIGURE 23.1
The handloom.[24]

fanfare and left without a trace soon. But, Sivagurunathan kept on visiting Chennimalai for several months to meet some of the weavers and was able to build confidence. Many of the remaining master weavers are above 50 years of age and the next generation is not in to weaving at all. So, there is a huge threat of the art and tradition of handloom weaving unique to Erode disappearing with the passing away of current generation. There is also a social pressure for the current generation to get educated in schools and colleges and go for a salaried job and weaving as a profession has been looked down upon. So, it is a monumental challenge to convince many weavers and to bring them in to a single society of weavers. This is the grand dream of Sivagurunathan in the long term, i.e., to form a cooperative society and bring in more weavers. Towards this, he started Nurpu as almost a single-man show and got in to sustainable handloom production. Please refer to Figure 23.1 for an illustration on handloom.

SUSTAINABLE PRODUCTION

A business model for sustainability should have a balance between Economic, Social and Natural values while creating the economic value (see Figure 23.2). Nurpu uses locally produced cotton yarn, and they avoid using chemicals. There are no synthetic threads. The dyes are organically

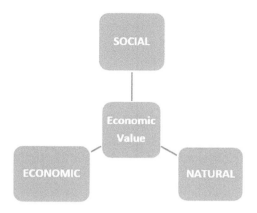

FIGURE 23.2
Business model for sustainability.

made on the traditional way using plant leaves, flowers, fruits and seeds or using azo-free low-impact dyes. He is also planning to set up a handloom weaving centre to teach the art and techniques of handloom to the next generation and preserve the tradition.[15] Location wise this sustainability action is even more significant because Erode district had continuous issues of river water and groundwater pollution due to the illegal discharge from the consented bleaching, dyeing and tannery units.[16]

The chemical dye is a severe threat to the environment as it resists biodegradation due to high thermal photostability.[17] Please refer to Figure 23.3 for the traditional product portfolio of Nurpu.

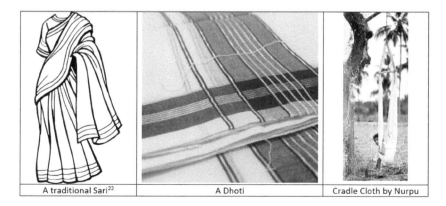

FIGURE 23.3
Traditional product portfolio of Nurpu.

CUSTOMER PERCEPTION, PRICING AND BUSINESS DILEMMA

- Sivagurunathan wanted to launch only unbleached shirt as it is more environment-friendly. When he initially launched the same, the customer response was lukewarm as they preferred a brighter shirt that underwent a bleaching process. The first dilemma was whether to go for coloured cloth as unbleached cloth did not find favour with many customers. Some of the mentors were advising him not to be too idealistic in producing unbleached cloth as we cannot change customer's perception and habit system overnight. The larger issue was to survive in the short term and work for organic changes in the long term.
- In colour there were three choices:
    - **Azo-dye** – Synthetic dye made of chemicals. Cheaper in cost and offers brighter colours and will not run during washing. But, harmful for health and environment. Banned in several countries.[18]
    - **Azo-free dye** – These are low-impact dyes made of chemicals. But, without the azo-group which are more harmful. Azo-free dyes are relatively better than the azo-dyes and cheaper than the natural dye. Quite common in export garments.
    - **Natural dye** – Skin and environment friendly as it is made from plants (mix of iron rust and mustard powder – Terminalia Chebula, Clitoria ternatea, Eclipta prostrata, leaf of ebony tree, Butea monosperma, etc.). In this process, dying alone will cost Rs. 200 per meter. Overall, the shirt will be three times costlier than when it is normal dyed.

The dying process is outsourced from a local expert. A shirt could be sold for Rs. 890 with some profit, if it is dyed with azo-free dye, whereas usage of natural dye would make the price close to Rs. 2400. With thin profit margin and astute management, the price could be brought up to Rs. 1600. Nurpu decided to do natural-dye based garments only based on advance orders. Costing is very tight in this field as weavers will not take up a job below a threshold and pricing of yarn is also tightly controlled by the suppliers. Only areas where some cost cutting could be done is by the choice of dying process and reducing the marketing spend.

MARKETING

Nurpu sell through its own online shopping page and through stalls in exhibitions. Nurpu has built a Facebook-based social media community of environmentally conscious people who value sustainable fashion. Nurpu also has a twitter presence, but not as active as on Facebook. In terms of product portfolio, Nurpu produces traditional attires like sari[19] (price ranging from Rs. 1500 to 2990), dhoti[20] (price ranging from Rs. 380 to Rs. 1280), shirts (priced around Rs. 800–900) and towels (Rs. 170–350). Another exciting product which Nurpu sells is the unbleached, handloom cradle cloth for newborn babies. Catering to another traditional method of making the babies to sleep using a long cloth or a soft sari hung from the ceiling. Nurpu also sells hooded towels for the babies. Since it is organic, smooth and unbleached, it is harmless on the tender skin of the babies. He is also informally supported by the family – his wife, mother and sister on some of the works like packing, accounting, etc. Please refer to Figures 23.4 and 23.5 for screenshots of Nurpu's ecommerce shopping pages and online community page.

Nurpu also does cause marketing with a concept called "Gift a cradle scheme" through crowd-funding. According to Nurpu, many pregnant women from the lower-income strata who use the rural government

FIGURE 23.4
Nurpu's shopping page – www.nurpu.in.

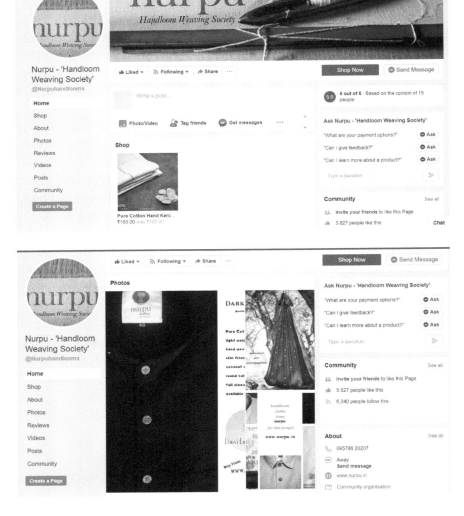

FIGURE 23.5
Nurpu's Facebook community page.

hospitals do not have the purchasing power to buy a cradle for the newborn baby. Through this donation scheme, Nurpu is supplying their cradle cloth as a sponsored gift to the needy. A win-win for the weavers and the beneficiaries. This idea came due to the exigencies forced by the COVID-19 pandemic, which squeezed the business and the new avenue created by the introduction of cradle cloth gave the working opportunity to the weaver.[21]

Sivagurunathan also visited several high-end boutiques and apparel retail outlets in Chennai to understand the prevailing price ranges of such organic apparels. As many of these organic apparels were sold at a higher premium, Sivagurunathan had another dilemma based on ideology. One of the major aims of Gandhian like J. C. Kumarappa was to make such organic products affordable and available for the masses. However, the current offerings were costly and were catering to the higher income groups.

ECOLOGY, ECONOMY AND THE ENTERPRISE

There are several stakeholders including weavers, customers and the civil society, regulators, Governments, pollution control boards, channel partners, next-generation children, farmers and suppliers in this value-network. As a single-owner MSME, Nurpu operates at a small scale, mitigating the input costs by balancing it with human efforts in its grander objective of making eco-friendly fabrics by reviving the handloom weaving tradition with a badge of honour. The tacit knowledge of weaving methods from the traditional handloom weavers is a key resource in this revival process.

Looking from the business model perspective, Nurpu is a classic case of integrating ecological and social issues into the core business and value creation. This could be bucketed and analyzed with the flourishing business canvas.[22] It is a robust tool that helps organizations to "design economic, social and environmental aspects of the business model". The canvas template could be accessed from http://www.FlourishingBusiness.org.

RESTART FRAMEWORK

Nurpu has inherently absorbed the ways and means to make the business model more sustainable, and some aspects of circularity are still open-ended, which could be explained by the RESTART framework.[23]

- **REDESIGN** – Nurpu has redesigned the business model by bringing in the elements of environmental and the societal needs into the business model canvas.

- **EXPERIMENTATION** – The protagonist had done extensive travel, training and comparative study of business processes in the experimentation phase before implementing the redesigned approach.
- **SERVICE-LOGIC** – This is a new vista for Nurpu to promote resource-efficient consumption which is not yet explored.
- **THE CIRCULAR** – Though the sustainable processes are in place, the overall approach is still linear in Nurpu's case. One factor that is advantageous to Nurpu in this context is the fact that there is a focus on end-product being bio-degradable. Still, bringing in circularity in the activities would be the next orbital challenge for Nurpu.
- **ALLIANCES** – Nurpu is a genuine start towards alliances. The business is right now a single-member entity with bootstrapped support and aims to build an alliance in the future. Nurpu has also allied with a crowd-funding platform for business enablement.
- **RESULTS** – Nurpu has shown that a nimble operation is possible within the framework of sustainability. Social media has enabled a no-cost marketing model to create both awareness as well as preference within the small scale of operations.
- **THREE-DIMENSIONALITY** – In terms of revival of traditional handlooms, less environmental pollution and the dream of providing livelihood opportunity with dignity, Nurpu has worked on the three-dimensional model of social, environmental and financial performance.

Nurpu is one of the best examples of a genuine attempt by a business player to integrate social and environmental canvas. In this passionate pursuit, Nurpu has indirectly and positively impacted the society on many of the UN's sustainability development goals (SDGs; refer to Figure 23.6), viz., no poverty, zero hunger, clean water and sanitation, decent work and economic growth, responsible consumption and production, climate action, life below water and life on land. While the production level sustainability is achieved, the RESTART framework further guides the business towards achieving circularity which could become a part of the next level business design for Nurpu. By working on the circularity and service-logic aspects, phase two operations of Nurpu shall be more polished towards sustainable fashion in its real sense.

Note: This case study is written based on the depth interview with the proprietor of Nurpu Handlooms Mr. C. Sivagurunathan and with the

FIGURE 23.6
United Nation's sustainable development goals.[25]

publicly available information about the enterprise and does not illustrate either effective or ineffective handling of a business situation.

QUESTIONS FOR DISCUSSIONS

1. How do you look at the two dilemmas of Nurpu and Sivagurunathan in terms of business model? What will be your suggestion to him?
2. Comment on the current sustainability marketing strategies. What would be your marketing approach?
3. As a marketer, how would you change the negative consumer behaviour towards unbleached cloth?
4. What is the way forward for Nurpu? Is there a scope for circularity in its business?
5. Discuss the impact and validity of Gandhian thoughts in the context of current day sustainability movements like SDGs and Circular Economy.

According to United Nations, "The 2030 Agenda for Sustainable Development, adopted by all United Nations Member States in 2015, provides a shared blueprint for peace and prosperity for people and the planet, now and into the future. At its heart are the 17 Sustainable Development Goals (SDGs), which are an urgent call for action by all countries - developed and developing - in a global partnership. They recognize that ending poverty and other deprivations must go hand-in-hand with strategies that improve health and education, reduce inequality, and spur economic growth – all while tackling climate change and working to preserve our oceans and forests."[26]

FUNDING DECLARATION

Not Applicable.

ACKNOWLEDGMENT

The authors would like to thank Mr. Sivagurunathan, Proprietor of Nurpu Handlooms, for generously sharing information about his venture during writing this case-study.

CREDIT AUTHOR STATEMENT

Sathyanarayanan Ramachandran: Conceptualization, framework, depth interview with protagonist and drafting.

S. A. Senthil Kumar: Inputs in conceptualization, reviewing, perspectives and ideas.

NOTES

1. The collected works of Mahatma Gandhi, Volume 15, P162, https://www.gandhiheritageportal.org/the-collected-works-of-mahatma-gandhi, Accessed on September 3, 2020.
2. Ibid, P459.
3. Kumarappa JC, Economy of permanence. 1945. P11, https://www.mkgandhi.org/ebks/economy-of-permanence.pdf, Accessed on September 3, 2020.
4. From Wilson, Kax. *History of Textiles*. (1979). http://char.txa.cornell.edu/IndianTex.htm, Accessed on August 1, 2020.

5. The City was a part of undivided India before independence and currently the capital of Bangladesh.
6. From Wilson, Kax. *History of Textiles*. (1979). http://char.txa.cornell.edu/IndianTex.htm, Accessed on August 1, 2020 (Bhagavatula, 2010).
7. TN: A former IT professional is reviving the lost art of handloom weaving in the organic way by Sayantani Nath, https://effortsforgood.org/social-good-businesses/nurpu-handlooms/, Accessed on August 1, 2020.
8. Ibid.
9. Threads of tradition, revival and humility bring handloom biz back by SP Kirthana, February 23, 2019. https://www.newindianexpress.com/cities/chennai/2019/feb/23/threads-of-tradition-revival-and-humility-bring-handloom-biz-back-1942694.html, Accessed on August 1, 2020.
10. Dr. G. Nammalvar was an Indian green crusader, agricultural scientist and an environmental activist.
11. Mr. Sundararaj Theodore Baskaran is an Indian film historian and wildlife conservationist.
12. Mr. Sivaraj, a naturalist, started this movement to provide natural learning opportunities to underprivileged children.
13. *Threads of Life* by Akila Kannadasan, March 23, 2017, https://www.thehindu.com/society/threads-of-life/article17609999.ece, Accessed on August 1, 2020.
14. Shri J. C. Kumarappa was also the author of the book *"An Economy of Permanence"* in the 1940s, which is one of the classic texts with early ideas of sustainable living and circular economy.
15. Threads of tradition, revival and humility bring handloom biz back by S. P. Kirthana, February 23, 2019. Accessed on August 1, 2020.
16. Tamil Nadu Pollution Control Board Action Plan on Rejuvenation of River Cauvery Mettur to Mayiladuthurai Stretch (Priority-I), P61, 2019.
17. Periyasamy A. P., Militky J. (2020). Sustainability in textile dyeing: recent developments. In: Muthu S., Gardetti M. (eds.) Sustainability in the Textile and Apparel Industries. Sustainable Textiles: Production, Processing, Manufacturing & Chemistry. Springer, Cham. https://doi.org/10.1007/978-3-030-38545-3_2
18. Schroecker, J. (2018). Why we need to get rid of azo dyes, https://www.trustedclothes.com/blog/2018/05/08/why-we-need-to-get-rid-of-azo-dyes/, Accessed on September 3, 2020.
19. Long piece of cloth with colourful designs, draped around the body by women.
20. Long white cloth worn by men by draping it around the waist.
21. Weavers of Erode's 1010 Colony now weave cradle clothes for newborns by Akila Kannadasan, July 15, 2020. https://www.thehindu.com/society/weavers-of-erodes-1010-colony-now-weave-cradle-cloths-for-newborns/article32090736.ece, Accessed on August 1, 2020.
22. The Flourishing Enterprise Innovation Toolkit. http://www.flourishingbusiness.org/the-toolkit-flourishing-business-canvas/, Accessed on August 1, 2020.
23. Jørgensen S., Pedersen L. J. T. (2018). Roadmap to a RESTART. In: RESTART Sustainable Business Model Innovation. Palgrave Studies in Sustainable Business in Association with Future Earth. Palgrave Macmillan, Cham. https://doi.org/10.1007/978-3-319-91971-3_4
24. Illustrative images from www.pixabay.com, Accessed on August 1, 2020.
25. The 17 Goals. https://sdgs.un.org/goals, Accessed on August 1, 2020.
26. Ibid.

24

Social and Sustainability Marketing: Secure Meters: The Dharohar Case

Kirti Mishra
Indian Institute of Management, Udaipur, India

Shivani Singhal
Dharohar, Udaipur, India

CONTENTS

References 617

In an increasingly impact conscious world, both employees and customers are sensitive to the ethical credentials of a business, voting with their feet (Edelman Trust Barometer, 2020). There has been a significant increase in the level of social consciousness of consumers, making it imperative for marketeers to pay close attention to socially responsible consumer behavior (Han & Stoel, 2017) and to develop novel ways to tap the potential of this growing consumer segment. In this case study, we present employee volunteering as a strategy that allows companies to reach and understand socially responsible consumers through community engagement. Community engagement has a positive impact on internal brand awareness and also enables employees to satisfy community needs and interests, which has been reported to help reinforce the credibility of organizational social responsibility efforts and improve brand equity (see Torres et al. 2012).

Secure Meters Ltd is a family owned, multinational solutions provider for revenue management, power quality and energy efficiency, working across 50 countries, with over 6,500 employees (Semsites). They have led the field in energy measurement, management and control since 1987.

Group CEO, Suket Singhal, said a key factor in their success is having the best of a family business; "lots of close ties, very low employee turnover, people with extreme trust with each other" alongside "all the processes of multinational organization. It's that kind of balance that we have between the two, which makes Secure relatively unique." This is supported by their long-term focus, with their vision to endure over generations driving daily decisions.

Mandatory community engagement reinforces these cultural values, strengthening the close ties within and outside the organization. All employees spend three working days volunteering in the community each year. When asked why community engagement for the team matters, the CEO said "human beings generally don't tend to be selfish on their own but when they get into organization they tend to become very selfish for themselves or their organization and by providing them this outlet you get them to realize actually its more important to be selfless, it's more important to be collaborative, it's more important to think about more about yourself as a company and that's a benefit which many other things will not be able to provide."

This idea of volunteering reinforcing core business beliefs was echoed by Ananya Singhal, Chief Services, People, Communications and Infrastructure, as he said "It's an inherent part of our existence, the same way we mandate financial good practice and high-quality manufacturing, design and everything else. It has the part of the same arc and therefore mandating working in society is incredibly useful ... and therefore that becomes something which must be done. Same way saying you need to do quality work in the office."

Most Semsites choose to volunteer through Dharohar, a nonprofit that runs structured programs for Secure. After four years of mandated volunteer engagement (optional programs ran since 2012), we can see its impact on how Semsites feel about and take ownership for Secure, its values and its success. It has reinforced trust and collaboration in team members as well as deepening a sense of belonging. According to an organization wide survey (12% response rate), 89% of respondents noted that volunteering had increased their commitment to Secure's values and culture. A total of 85% reported that through volunteering they acquired skills, which they use at work.

According to 88% of Semsites, volunteering has reinforced their understanding and awareness of Secure's values, and 94% reported that it reinforced the commitment to and pride in working at Secure. Volunteering has also increased employees' organizational identification (OI) and ownership; which have been shown to reduce attrition and bolster strong citizenship behavior. (Punjaisri et al. 2009).

These conclusions were also borne out in interviews. A senior Semsite said that volunteering helped sustain Secure's values and culture; "what happens is people like me with a long association with the company automatically form a strong bonding with the new team members and ... then you don't have to teach values.... Automatically they get into life. This is something amazing." The same volunteer spoke both of the joy he felt through volunteering and the development of his communication skills.

Even new team members become brand ambassadors through community engagement. A recent addition to the team shared that on community projects, "people tend to open themselves up ... they tend to put down those barriers ... so you try to get connect with that core part and that connection that builds. You will not believe that that actually repays when you come back to your office ... when I had to work with them (co-volunteers) afterwards, it was so smooth the tuning set up, the understanding was made and the major reason for that we were actually not talking official. We were not talking something work when we were actually doing community services, we were really being [ourselves]." He felt that "donating your time is the most important thing where you actually get into it and you take the benefit out of it for yourself as well. Not just for the company."

Semsites clearly see the value in the community engagement program; 90% reported it was a strong differentiator between Secure and its competitors. Volunteering has built deep social awareness and knowledge, facilitated good corporate citizenship behavior, embedded sustainability and strong HR practices within the business and fostered a deep community connect. These factors help build ownership of the brand, engagement with the community and a sense of belonging. At Secure, employee volunteering evokes organizational identity, by building employees' perceived self-brand fit, brand knowledge, and belief in the brand. By converting brand values into individual actions, volunteering turns employees into brand champions and potentially drives other desired employee behaviors (e.g. employee performance, citizenship behaviors and retention) that benefit the organization beyond brand building. The lower than industry average turnover rate at Secure is a testament to its strong culture and engaged employees.

This is in addition to the direct market benefits of deep community engagement. In an increasingly impact conscious world, Secure is able to report its deep community engagement when pitching for public and private projects both in India and overseas. Often this information is asked for formally in the tender process, whilst in other cases, there is an implicit desire to work with businesses that are well connected with the community. Especially in service provision, this connection is not only good for development but it also allows work to be more effective, and therefore is seen as a competitive advantage by current and potential customers.

Summarizing the benefits of employee volunteering, Ananya Singhal, Chief Services, People, Communications and Infrastructure noted,

> ... (employee volunteering) it does create a connect of the team with community and given that our products and services are used in the community I'm sure it will also be giving insights to our people in terms of other viewpoints ... people were not exactly like them but how they are working with in the communities so the business sees an immense benefit not just in terms of knowledge about community but a sense of wellbeing which is important and that then again feeds in to our culture, a sense of change in possible. So all of these things are incredibly positive drivers for us.

Community engagement at Secure creates pathways for reaching to socially responsible consumers by (a) building a deep community connect, (b) fulfilling the needs and interests of community and (c) improving stakeholder perception of Secure's corporate social responsibility (CSR) efforts. Insights from Marketing Science reveal that deep level of community engagement affects consumer's judgment and feelings toward companies, and these influence buying behavior, buying decisions, willingness to pay premium prices and increase purchases (Pivato et al. 2008).

Highlighting the synergy between understanding customers and understanding community the CEO said, "If I come back down to our vision, our vision is very clear we want to be a business that endures. You can't endure unless you think more holistically about your impact and influence in the world. How are you influencing and impacting your customers. So, we have to look at what is that influence and impact with the communities that we are working with, with our suppliers, service providers.. the environment that we are working in and our culture, again coming back to the question of engagement, it's not just engagement with the business, it's not just our team engaging the business, it's about the business engaging with the community."

Through engagement with communities as part of its various volunteering programs, Secure is able to effectively meet the needs of local communities which in turn increases brand value and brand equity. Meaningful interactions with community help Semsites understand their existing and potential customers. Working with community engenders empathy which is crucial for future growth and development.

Community engagement through employee volunteering encapsulates an inside-out and outside-in approach to social and sustainable marketing. It enables organizations to build a strong internal brand and also serves the needs of socially responsible consumers. This case study demonstrates that building an internal brand through socially responsible activities like employee volunteering not only promotes citizenship behaviors but also helps marketers to reach socially responsible consumers and develop customer relationships.

REFERENCES

Edelman Trust Barometer (2020). *Edelman Trust Barometer Special Report: Brands and the Coronavirus*. Retrievable from: https://www.edelman.com/sites/g/files/aatuss191/files/2020-06/2020%20Edelman%20Trust%20Barometer%20Specl%20Rept%20Brand%20Trust%20in%202020.pdf

Han, T. I., & Stoel, L. (2017). Explaining socially responsible consumer behavior: A meta-analytic review of theory of planned behavior. *Journal of International Consumer Marketing*, 29(2), 91–103.

Pivato, S., Misani, N., & Tencati, A. (2008). The impact of corporate social responsibility on consumer trust: The case of organic food. *Business Ethics: A European Review*, 17(1), 3–12.

Punjaisri, K., Wilson, A., & Evanschitzky, H. (2009). Internal branding to influence employees' brand promise delivery: A case study in Thailand. *Journal of Service Management*, 20 (5), 561–579.

Torres, A., Bijmolt, T. H., Tribó, J. A., & Verhoef, P. (2012). Generating global brand equity through corporate social responsibility to key stakeholders. *International Journal of Research in Marketing*, 29(1), 13–24.

Section XII

Selected Case Studies to Reflect on Practice and Use as Learning Tools: Case Studies from Emerging Economies (Complex and/or Long)

25

Saheli: The Zero-Side-Effect Pill—Marketing of Oral Contraceptives in the Context of Sexual Education to Create Socially Responsible Consumers

Neharika Binani, Anshika Singh and Palakh Jain
Bennett University, India

CONTENTS

Learning Objectives	622
Tools Used	622
Chapter Summary	623
Background/Introduction	623
Purpose of the Study	623
Scenario	623
Context of Research	624
Issues/Findings	624
Assumptions	624
Body	625
Importance of Sexual Education for Creating Socially Responsible Consumers	625
Methodology Used	627
How the Pill Became a Lifestyle Drug	630
Side-Effects, I-Pill and the Morning after Pill Controversy	631
Addressing the Gap	634
About Saheli and Analyzing Opportunities for Growth	635

Conclusion, Recommendations, and Marketing
    Interventions 638
Future Research: Alternative Solutions for Male
    and Female Contraception 642
Acknowledgment 644
Credit Author Statement 644
References 644

LEARNING OBJECTIVES

This chapter aims to define oral contraceptives in the context of sexual education and create awareness about safer alternatives through social and sustainable marketing.

After reading the chapter, the reader should be able to:

- Define oral contraceptives and categorize them as a necessity or luxury good
- Know about the history of marketing contraceptives—how they became a lifestyle drug—what is a lifestyle drug
- Know about marketing of sexual education to reach socially responsible consumers, and its importance
- Know about Saheli—the world's first nonsteroidal contraceptive
- Know about influence of social media in creating awareness of a product

TOOLS USED

Qualitative methods:

- Surveys
- Interviews

Quantitative tools:

- Bar Graphs
- Pie Charts

CHAPTER SUMMARY

Through an online survey conducted with more than 100 respondents, this chapter aims at evaluating the consumer decision-making process for oral contraceptives and examines factors that drive their choices. We will also look at how awareness for safer alternatives like Saheli, the world's first nonsteroidal contraceptive pill, can be created through social and sustainable marketing interventions.

BACKGROUND/INTRODUCTION

Purpose of the Study

This chapter aims to influence and support the use of safe oral contraceptive options for women by equipping them with the right knowledge and creating a socially responsible community.

SCENARIO

Oral contraceptive pills are known to have multiple side effects, which can be mild or extreme. This chapter takes a closer look at the following:

1. What are the steps that consumers take in consuming contraceptive pills, and how is their experience?
   To understand touchpoints in the consumer decision-making journey and finding areas to implement new marketing strategies
2. Why has Saheli, a "safer" nonhormonal contraceptive pill, taken almost 30 years to be recognized as an alternative?
   Here the focus would be on the lack of awareness due to lack of marketing, how a product like a contraceptive pill's consumption is directly related to its marketing, and how social media is relevant to bridge the gap between the two.

CONTEXT OF RESEARCH

Lack of awareness of Saheli: World's first nonsteroidal contraceptive pill developed in India. How can a social and sustainable marketing approach help Saheli reach socially conscious consumers and create growth opportunities?

ISSUES/FINDINGS

1. Is the pill still regarded as a lifestyle drug?
2. Are women fully aware of all existing alternatives to contraception?
3. If the purchase of contraceptives is influenced by mediums other than personal research, is there a scope for marketing interventions?
4. Do women and men feel equally responsible while taking decisions relating to contraceptives?
5. How does sexual education impact decisions regarding contraception and safe sex?

ASSUMPTIONS

1. Women in the age group of 18–45 often consume birth control pills in the form of emergency contraception as an impulsive measure taken to terminate conception.
2. Women indulge in this practice assuming that oral contraceptive pills can treat acne, PMS-induced cramps, and extensive blood clotting, being completely unaware of their side effects.
3. Lack of accessibility and awareness among women about alternative, safe contraception options have led to low demand for Saheli's oral contraceptive pill.

BODY

Importance of Sexual Education for Creating Socially Responsible Consumers

India is quite prudish in discussing matters related to sexuality. Two out of three girls are unaware of changes their body goes through at puberty in case of menstruation and even less know about the risks of sexually transmitted diseases (STDs) such as HIV. (1) There is a lack of sexual health awareness due to the absence of formal sex education imparted by families, schools, or other institutions. So, it is not surprising that only 11% of women and 20% of men in India know about the existing oral contraceptive pill market, (2) where most of the information they get is not from healthcare officials, but rather click-bait advertisements.

Therefore, it is crucial to understand the need for proper sexual education through curriculums or informed marketing of products by brands for this book's context and reaching or creating socially responsible consumers.

One must understand that the importance of proper sexual education is not limited to the act; it is equally important to promote health, well-being, dignity, and respect for human rights, gender equality, and, most importantly, teaching and empowering the young to lead a safe, healthy, and productive life. In 2018, UNESCO updated its International guide to sex education to address the relevance of issues like well-being, personality development, and creating respectful social relationships.

The Comprehensive Sexual Education (CSE) program is designed to learn about cognitive, emotional, physical, and social aspects of being sexually active or inactive. This curriculum aims to ensure that the youth has right knowledge and the right awareness about the impact of their decisions and equips them to become socially responsible consumers of tomorrow.

Generally, sex education programs touch upon these topics if introduced properly, but when our system fails to do so, young people tend to get confused and conflicted with information regarding sexual activity or relationships. This leads to bad decision-making.

Eleven percent of the world's teen pregnancies happen in India, which translates to approximately 16 million women between the ages of 15–19 who go through childbirth each year. According to the National Family Health Survey, even now, 27% of the girls marry before the age of 18. The reason is the lack of use of contraception. (2)

The larger population of our research consisted of young adults unaware of these decisions and choices; due to the lack of such an education system, there is no governance over how a consumer may approach purchasing a contraceptive product.

We can attribute this to the strict governance of such advertising and its subsequent withdrawal in the late 1990s. It was resulting in a lack of information, usage, and acceptance of the drug even now. In 2005, when the Ministry of Health and Family Welfare finally realized that banning advertising for such crucial medication had led to a complete lack of awareness, they relaxed their regulations with a strong emphasis on the need to vigorously promote oral contraceptives as a part of their family planning initiatives. Pharmaceuticals could finally sell their oral contraceptives over the counter (OTC), and a shift in their perspective on marketing contraceptives came about. (3)

This shift introduced a variety of definitions and meanings to oral contraceptives which further confused the consumer.

According to a research paper on National Centre of Biotechnology Information (NCBI), contraception is defined as intentional prevention of conception through the use of various devices, sexual practices, chemicals, drugs, or surgical procedures and any device or act whose purpose is to prevent women from becoming pregnant can be defined as a contraceptive. Few known contraception methods are withdrawal, male condom, female condom, oral contraceptive pills, injectables, emergency contraceptive pills, IUDs, vasectomy, and tubectomy. An oral contraceptive is defined as a pill that helps in preventing pregnancy. It consists of hormones that prevent the release of eggs and thickens the cervical lining. Most oral contraceptives include estrogen and progestin. The oral contraceptive is also commonly known as a birth control pill. (3) Contradicting this, the US department of health and human services defined a birth control pill or a contraceptive pill as merely a medicine used by women at a reproductive stage in life to prevent pregnancy or conception of a fetus and said that there are two types of birth control pills: (1) Regular oral contraceptives (ROCs) for daily consumption with two hormones, estrogen and progestin, also known as combined pills and (2) Emergency oral contraceptives (ECs) that consisted of a single hormone and were consumed within a time frame of 72 hours post-coitus to prevent ovulation. (4) There is no clarity given on the actual function and use. According to WHO, the birth control pill is categorized as an "essential medication" since it concerns "the priority needs of the population" and has a high "public health relevance." However, in most countries, to obtain a contraceptive pill, one must show a prescription. (5)

The irregularities in these definitions have created confusion among women worldwide, making the consumer's decision-making journey to purchase an oral contraceptive a confused one with multiple challenges and questions but no clear answers and giving marketers a motive to intervene to create their own communication.

METHODOLOGY USED

Therefore, to clearly understand the consumer journey of purchasing and using a contraceptive pill we have used a mixed research methodology, examining both qualitative and quantitative aspects of the process. Mixed research method is a type of research where the researchers use a combination of qualitative and quantitative components to gain a broad and deep understanding. The design requires at least one qualitative and one quantitative component.

Our use of a mixed research method has strengthened our study's conclusion, not limiting it to only the users of oral contraceptives but also including viewpoints and interferences of their partners, assuming that the topic of contraception is a sensitive one involving both the parties. We have used surveys, interviews, and merged the data sets from both for the purpose of our analysis.

The purpose of using this model is that both the methods help us discover and confirm our hypothesis and analysis. The data that we got from our surveys allowed us to generate our conclusion, to explore the initial hypothesis and the data that we got from our interviews of gynecologists and consumers helped us test it. A mixed approach let us enhance and clarify the results from one method with the results from the other method and also extended the breadth and range of inquiries by adding the interview responses and understanding of medical terms. We witnessed a diversity of views which uncovered the relationships between variables through quantitative research while also revealing meanings among research participants through qualitative research. The method not only helped in expansion but also helped in development, as results from one method helped develop or inform the sampling and implementation of the other method.

While designing a mixed method study, we were advised to start with one research question and consider what the purpose is for using the model.

Besides purposes, this method also has an overall theoretical drive. We chose qualitative dominant mixed research method as our overall drive. It is a type where one relies on a qualitative, post positivist view of the research process and recognizes the addition of quantitative data to benefit the paper.

For analysis and conclusion, we integrated the components. Deciding where the point of integration will be and how the integration of results will be done are important decisions. So, we used the primary way to do that. We merged the two data sets, connecting from the analysis of one set of data (Surveys) to the collection of the second set of data (Interviews) and finally used a theoretical framework to bind together the data sets (Table 25.1).

In conclusion, to create this research paper our methodology was:

1. Defining our topic and our research objective
2. Conducting a preliminary research of 15–20 articles to create a solid base for our literature review and to check if there is any matter to support our observation/question
3. A full-fledged literature review on topics relating to oral contraceptives and their marketing
4. Formation of hypothesis and assumptions along with future research capabilities
5. Creating a survey—for oral contraceptive users and their partners along with interview questions for gynecologists who have been in the field for 10+ years
6. Data collection and compilation
7. Analysis of data as well as cross-examination of results according to surveys and interview responses
8. Combining our primary data with our secondary data of literature review and previous research work
9. Proving or disproving hypothesis
10. Forming conclusions and implications

However, our research is limited to only contraceptive pills, due to COVID-19, it is also limited to video interactions, online surveys, and communication with respondents as well as doctors. It focuses primarily

TABLE 25.1

Methods Used

	What	Why	How
Qualitative Research Method [26]	Focuses on words and digs up the opinions, feelings, and experiences of the respondents	In depth understanding of the respondent's experience. Have the opportunity to ask follow-up questions. Have a real valuable conversation around the subject of the research.	Using common techniques like focus groups, interviews where discussions happen around the subject of the research allowing respondents to give their opinions and get insights
Quantitative Research Method	Focuses on numerical data to quantify the problem which can be transformed into statistics	Quantifies attitudes, opinions, behaviors, and other variables. Generates results from a large sample of population. It uses numerical data to formulate facts and patterns.	Using techniques like surveys, regression, and polls which are structured with close ended questions. It is circulated within the large sample population and their answers are collected and analyzed in form of numbers, percentages, etc. It shows the behavior of the respondents which helps in understanding the pattern.
Mixed Research Method	Combination of qualitative and quantitative methods so that the conclusion of the study is strengthened and expanded. (30)	Clarification and enhancement of results. Expansion of range of enquiry. Use of results from one method to develop the other method. Diversity of views. Confirm and discover by using a quantitative method to create a hypothesis and testing it by using a qualitative method.	Merging the data sets and connecting the analysis from both the sets. Using a framework (theoretical or program) to bind the data.

on urban women and women aging between 18 and 45 with a strong focus on the use of oral contraception only. For our future research, we have dived into male contraceptives and believe that it is the next revolution for the contraceptive industry to create socially responsible consumers.

We feel that using this model of mixed research methods was most important due to the large amount of information present online as well as the lack of information and existing biases in consumers' minds. In the following paras, we have explained how our collected data ties with our research to give the reader a complete picture of the current scenario of the contraceptive pill market and the advertising done for the same.

HOW THE PILL BECAME A LIFESTYLE DRUG

With the opportunity in place and the rise in advertising and promotion of oral contraceptives post-2005, the pill strayed away from "the essentials category," grabbed much attention in the media, and became a gateway for female empowerment; providing women with a sense of liberation by allowing them to decide between the path of family planning or focusing more on their career, ambitions, and goals all while enjoying the pleasures of a healthy and active lifestyle. This led to them believing that consumption of oral contraceptives had a direct relation to improved life quality. (6)

Large pharmaceuticals started taking advantage of the situation and started using marketing tactics rather than scientific explanations as a measure to create a positioning strategy for their product by referring to oral contraceptives as "a lifestyle drug." (7) More than ever, pharmaceuticals started promoting the perceptual benefits of consuming contraceptive pills rather than mentioning their core functional benefit of reducing unwanted pregnancy chances. Bringing in streams of revenue for the pharmaceuticals as, in the minds of women, contraceptive pills did reduce not only the chances of pregnancies but also solved multiple problems relating to acne, cramps, their menstrual cycle, and a new but commonly found hormonal problem known as the polycystic ovary disease (PCOD) in women. The positioning and the perception it created, limited pharmaceuticals spending on the innovation behind these pills, and the market flooded with contraceptive pills without low or no variation.

When we asked our respondents to tell us about their consumption pattern of oral contraceptives based on their intake frequency, we found that about 72% of women consume these pills at the time of emergency and about 66% of their consumption was in the form of a morning-after pill. Adding to that, when we interviewed gynecologists and asked them to describe their most common patients, they told us that most of them were young adults and recently married women—telling us that even now, consumption of contraceptives is casual and frequent. Women were

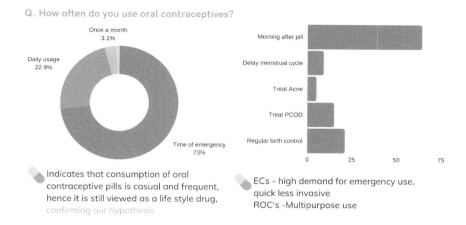

FIGURE 25.1
Time and way of using an oral contraceptive as answered by our respondents.

still viewing these pills as a lifestyle drug to indulge in unplanned sexual encounters and for other reasons, such as delaying their menstrual cycles for important occasions (See Figure 25.1).

SIDE EFFECTS, I-PILL AND THE MORNING AFTER PILL CONTROVERSY

Quoting Marian Wright Edelman, "In every seed of good, there is always a piece of bad" the same as in the case of side effects that come with the convenience of consuming an oral contraceptive. These side effects include weight gain, increased acne, hormonal imbalance, nausea, dizziness, cramps, irregularities in the menstrual cycle; in extreme cases it can also cause depression, heart disease and cancer. Contraceptive pills like YAZ have also been responsible for causing up to 15 deaths worldwide. (8, 9)

Overlooking this information resulted in the pill being advertised as a lifestyle drug, to help "improve the quality of life"—but in actuality, it could very well diminish it.

Today, about 81.3% of our respondents said that they were conducting prior research before purchasing oral contraceptives, and 84.4% claimed to know all about the possible side effects; this told us that our respondents were making a conscious decision; they were aware, yet they were still purchasing hormonal oral contraceptives (Figure 25.2). About 47%

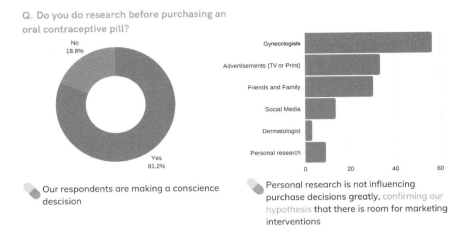

FIGURE 25.2
Research done and points of influence during consumer decision-making journey of oral contraceptives.

experienced mild symptoms like headaches and nausea, 40% felt terrible with depression and mood swings, and only 10% said that they felt excellent post-consumption. This led us to question – why were they using oral contraceptives even though the side effects were so severe, did they search for more options?

Companies like Cipla continue to spend close to $2 Million (USD) each year for creating visually appealing ad campaigns for promoting their emergency contraceptive, the I-Pill. These ads are targeted to women between the age groups of 16 and 26 years and focus heavily on the perceptual benefits of consuming ECs. These ads use celebrity figures to enhance their credibility, phrases like "tension-free," "most convenient," and "the best option for your contraceptive needs" while neglecting the possible mention of side effects or that consumption of these pills are strictly for emergency use. (3) This miscommunication, along with the pills being available OTC, often confuses women on the safety and dosage, which results in them consuming the EC 2–3 times a month (which is extremely unsafe, as per our conversations with experts in the field!) See Figure 25.3.

Having several implications; in our research, we found that about 60% respondents have seen an ad for the I-pill, making it the most known brand in our sample space. About 75% who are aware of the I-pill,

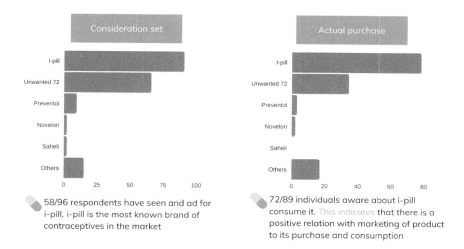

FIGURE 25.3
A positive relationship between the promotion and purchase of a product.

regardless of seeing an ad, consume it, telling us that the marketing of a product has a positive relationship with its purchase and consumption. To address this, the intervention made by ECs in the oral contraceptives market, Business Standard says that promotional activities have "made contraceptives of the last resort, the first choice." (10) Circling back to the dosage, the majority of our respondents consist of young women aging between 18 and 22. We learn that with early and frequent ECs, women often develop problems in their regular cycle due to multiple listed side effects causing hormonal imbalances and problems relating to the same, leading to the identification and acceptance of side effects of the lifestyle drug mentality. (11)

Over the years, with the diminishing use of condoms and vasectomies, falling by 52% and 73%, birth control pills as a measure of emergency contraception have become one of the most advised, first choices for population control and family planning in India. (6) In addition to increased marketing of ECs, the inability of government for promoting conventional methods of contraception like IUDs, vasectomies, or give value to their own found nonsteroidal treatment has also strengthened consumers belief of relying on the ease of consuming an emergency contraceptive pill, giving low importance to the effects on a woman's physical and mental health.

ADDRESSING THE GAP

So far, we have addressed the fact that the oral contraceptives market lacks innovation and initiative. This leads us to believe that as women are conducting preliminary research and are aware of side effects, they are also searching for safer alternatives but cannot access the right solutions or are unaware of the same. There is a lack of government intervention, and oligopolistic pharmaceuticals have taken over the contraceptive market, practicing push marketing and eating up the entire market share. (6) This clearly conveys us that there is a need for a consumer to be made aware of the existing alternatives through focusing on promoting benefits for women at large to lead a safe and healthy life.

It is essential to create a planned approach focusing on the consumers' needs, rather than sales. We can do this by reviewing the qualitative aspects of the collected data and evaluating it using the past and current findings. Our research study used a mixed research method, including qualitative and quantitative aspects. (12) This was done by thoroughly examining existing research, using a questionnaire survey (112 sample size), and conducting personal interviews to understand the consumer's decision-making journey. We used the marketing funnel; the AIDA model, and multi-dimensional scaling to create broad conclusions. Doing the above is difficult but not impossible. In such cases, marketers need to take a selfless approach (13, 14) focusing on influencing society; by advocating for ideas, consumer beliefs, and triangulating data to form long-term solutions and recommendations for optimizing the marketing of any product.

One such product is India's "Saheli" contraceptive pill. When we asked our respondents if they were aware of nonsteroidal oral contraceptive solutions, about 74% responded with a no (Figure 25.4); when asked if they knew about Saheli, 45% said yes, and 30% did not know, but showed interest to know more—leading us to believe in the existence of a definite gap in the marketing initiatives of alternative contraceptive pills among urban women. This informed us that even though our sample space claimed to know about Saheli, it was not a part of their consideration set of oral contraceptives, and they were not aware of its benefits as a nonsteroidal pill.

Q. Are you aware of any non-steroidal oral contraceptive pill for women?

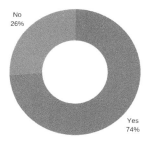

 Individuals nonsteroidal pills which concluded that people did research about the side effects of the pill, but never tried to find a safe alternative.

 Lack of awareness about nonsteroidal pills due to low advertisements and marketing campaigns

FIGURE 25.4
Respondents lack knowledge of nonsteroidal contraceptive pills.

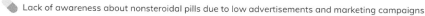

ABOUT SAHELI AND ANALYZING OPPORTUNITIES FOR GROWTH

Saheli is the world's one and the only nonsteroidal contraceptive pill. It is a regular intake birth control pill created by Dr. Nitya Anand. Launched in 1991, it was developed by the Central Drug Research Institute (CDRI), Lucknow, and is currently produced by Hindustan Latex Ltd. It contains Ormeloxifene also known as Centchroman, a Selective Estrogen Receptor Modulator (SERM) as the main ingredient and only requires one dose per week after a biweekly dosage for the first three months of usage. (15) Women above the age of 45 also use Centchroman to combat menorrhagia, leading to excessive or heavy bleeding during the menstrual cycle. The difference is that Saheli works unusually than any other steroidal birth control pill since Saheli does not use hormones in the actual pill, it prevents estrogen from building the uterine lining, halting the process of fertilization without causing any hormonal imbalance, because of which; Saheli claims to have low to no side effects on consumption.

Saheli is also a government of India initiative (16) and is included in the Indian family welfare and planning program. On top of that, it is also WHO approved and sold as Novex-DS in certain places. (16) It is the go-to option for women in smaller cities and rural areas as it is easily accessible through government initiatives in those areas. The awareness about this pill is more in rural areas than the urban areas. Even though Saheli is one of the safest contraception options with a unique proposition, it still lacks awareness at a commercial level. Since most of our respondents are from tier 1 cities like Delhi, Chandigarh, Kolkata, and Mumbai, we can say that they lack this information due to not being a part of Saheli's target audience.

Contraceptive pills are one of the most used methods of contraception across the globe. Keeping in mind that our research assumed that our sample space consisting of urban women aged between 18 and 45, used oral contraceptive as one of the first two choices of contraception methods, when asked what other contraceptive methods they preferred to use, a staggering 93% people answered using a condom (Figure 25.5). According to WHO and AIIMS, a condom is a barrier method of contraception. (17, 18) When used correctly, it protects consumers against the transmission of STDs with no mention of safeguarding one against the possibility of conception. Hence, using a condom can be attributed to the increasing awareness in the prevention of STDs rather than preventing an unwanted pregnancy.

When asked about their use of condoms, our respondents included statements like "condoms are the best way for birth control," "Condoms are best for intercourse, they prevent STDs. Even if you have a trusted partner, my preference would be the same because of its easy availability, being economical, no side effects, greater efficiency" and "If it works and if it is safe, then I can be sure that my partner will not get pregnant. You do not have

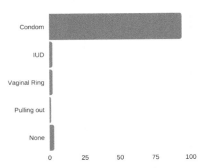

Q. What are other methods of contraception used by you?

Condom is the most preferred choice after contraceptive pills as a contraception method because they are readily available and can be purchased over the counter.
Other reason was that it protects women from STDs.

FIGURE 25.5
Showcasing other methods preferred by sample space as contraceptive solutions.

to worry about whether they are taking the pill on time. It will allow me to have more control rather than relying on my partner for birth control."

When we weighted the effectiveness of using the condom versus using the pill, we found that condoms are only 85% effective compared to a 99.7% effective contraceptive pill to prevent pregnancy. Pointing to cognitive dissonance, as most people claimed that they felt condoms were more reliable and safer as protection. Adding to our analysis, even though consumers were conducting research, they were only aware of the options they had researched. Linked to their social knowledge, other factors regarding their choice of contraception included; the convenience of purchase, use, availability, price, and self-protection as they did not want to be dependent on others. We had to find out if our respondents were doing anything about this lack of safer options. When asked what factors influenced their purchase of a contraceptive, about 56% consulted gynecologists, 44% relied on advertisements and social media platforms, and others included advice from family and friends and personal research modes. This confirmed our hypothesis that since personal research was not a significant factor influencing consumer behavior, there was room for marketing interventions.

To improve our research, we conducted telephone interviews with gynecologists and both male and female respondents; the doctors told us about their knowledge of pills like Saheli and why it has not reached its potential in the urban market. They pointed out that this was due to a lack of marketing initiatives and distribution in only rural areas. With the lack of a proper marketing campaign addressing the benefits of Saheli (Centchroman), there was no way to create attention or interest for urban women to desire or act on purchasing the product. The gynecologists could recommend the Saheli pill, but women would be comfortable using a product they had some awareness level.

The lack of growth of the Saheli (Centchroman) pill in the Indian market can be attributed to inabilities of the government and pharmaceuticals for marketing the product using the right kind of techniques and a clear depiction of Saheli's benefits. When searched, the only marketing initiative taken to promote this product boils down to two different advertisements, including Sridevi in 2005 and the other that includes Raveena Tandon. (23, 24) Both ads address a fight against the side effects of consuming hormonal birth control pills, unlike any other ads that mention no side effects at all and end by providing the consumers with a toll-free number to call on for procuring the pill. The ad mentions Saheli as a no side effect pill, since contraceptive pills are OTC to ensure consumers are taking completely conscious decisions, mentioning such crucial after-effects of consumption of a product shows the value of brands' social responsibility like Saheli. Compared to an I-pill ad, which focuses on establishing a fear factor regarding unwanted pregnancy and its effects on social life, (25) the Saheli ad carries an uplifting tone and conveys a message of self-awareness and empowered women. Many pharmaceuticals even sell the Saheli pill under different names such as Centron, Chhaya, and Novax-DS, which further confuses consumers. Eighty-three percent of our respondents said it is easy to purchase and consume oral contraceptives. Since we have established the fact that these pills are OTC, using a toll free number to procure a pill without any mention of it being available in regular medical stores increases the inconvenience adding to the judgment as well as lack of discretion the respondents felt during their purchase.

With the lack of awareness on Saheli and the Indian oral contraceptive market dominated by big pharmaceutical companies like Cipla, did Saheli have any scope for growth?

Since the beginning of 2020, after almost 30 years, the Saheli contraceptive pill has gained media attention worldwide. Recently, Saheli made its appearance on a Netflix documentary called, "Sex, explained!" produced by Vox Media. The documentary sheds light upon the various forms of oral contraceptives, their side effects, and highlighted Saheli as the world's foremost nonsteroidal birth control pill, as a result creating interest in the consumers' minds, wanting to know all about the product, and purchase it. Not only that, but Ms. Sophie Dowding also filed a petition on change.org to Mr. Boris Johnson (UK Prime Minister) for making the nonsteroidal pill available in the United Kingdom. (As on May 23, 2020, 1300+ have signed in support). (26) Ormeloxifene, a molecule in Saheli, has been discovered to have cancer-curing properties against breast, ovarian, and neck cancer and improve Osteoporosis. Therefore, the meaning of Saheli—being a friend, is indeed no coincidence. (27)

CONCLUSION, RECOMMENDATIONS, AND MARKETING INTERVENTIONS

With a reading of this chapter, the user's understanding of the contraceptive industry increases, they are more aware about marketing of oral contraceptive pills, and the user's lack of information decreases. The user now understands the real reason behind consumers' emotions about the different types of contraceptives and the involvement of both parties in making a decision.

We have also discussed why there is a need to market sexual health education at an early stage in life and how creating socially responsible consumers from the initial years of puberty for sexual health products is essential. This chapter gives an insight into the journey of a woman while making a purchase decision on procuring a contraceptive option and promotes Saheli as a comparatively safer method for oral contraception. Through the study, the user also understands the importance of social media marketing with the example of, "Sex, Explained", a Netflix documentary portraying how women worldwide learn about products like Saheli. We also learn how influential social media can be when we take the example of the petition created to bring Saheli to the United Kingdom. The following are some critical conclusions found while conducting our research that explains how a product's marketing is related to its consumption.

1. Individuals are researching online, but their research is incomplete because they are unaware of nonhormonal contraceptive pills.
2. Even if individuals claim to know about Saheli, it is not at the top of their mind or consideration set. When asked which contraceptive pills are in their consideration set, which pills they use more or are aware of, Saheli belongs at the bottom of the pile, and this can be attributed to the marketing strategy on behalf of the government for Saheli in Urban areas.
3. A large population of females is interested in a male contraceptive pill as they believe that it is an equal decision for both parties, so both parties should be responsible (refer to future research).
4. About 56% of people are influenced by their gynecologists while deciding on which contraceptive methods to use, 33% take action after seeing an advertisement, 29% take suggestions from family and friends and a few about 1% conduct personal research—showing us that women do not rely on their research and find it essential to get an expert's opinion or go by reviews and feedback of their known circle.
5. More and more women still rely on the consumption of emergency contraceptives (ECs) or other alternatives like condoms.
6. With the increase in the frequency of marketing a product, to ensure retention, chances of usage increase. It was found that 86 individuals

said they are aware of I-Pill, 79 said they use I-Pill, 72 said they had seen ads related to contraceptive pills. When asked to name which contraceptive pill they have seen most ads for, out of 58 responses, 54 said that the brand they have seen ads for is I-Pill.
7. Naturally, consumers remember those brands whose advertisements they have seen and can recall. During the start of this reading, we reflected upon how the I-Pill has spent large sums of money marketing throughout its release. Compared to this, Saheli, according to our interviews with Dr. Deepa and Dr. Manju, states that Saheli has not marketed itself or built much reach in the whole country. Only targeted rural areas have been the saving grace for Saheli, where all consume it.
8. There is an increase in demand for Saheli post releasing of a Netflix documentary by Vox Media named "Sex, Explained." This indicates that once educated, women are keen to find safer options.
9. Women still identify the pill as a lifesaving drug; in our survey, many respondents still use it to delay their menstrual cycle or treat acne. However, the number is decreasing as more and more women are realizing the negative effects of using the pill as an emergency contraceptive.

In conclusion, we can confirm that Saheli as a nonsteroidal contraceptive pill has much scope for growth in the urban market due to its benefits of easy consumption, availability, and lack of multiple side effects.

We can also say that through our analysis of consumers' journey of purchasing oral contraception, specifically in the case of Saheli, to reach its potential in the rural market, there is a lot of room for marketing intervention which can be done by increasing the presence of Saheli across traditional and new-age social media, through internet marketing. The existing inclusion of Saheli in the family welfare product gives the government more than enough reason to increase the traction of Saheli through a relevant social media strategy, digital marketing through targeted google ads, presence on e-commerce medical websites, print media and TV.

In this day and age, with society becoming more fluid, concepts of sexual health, knowledge, and well-being are important as individuals choose to talk about them more freely in all settings. With a rise in casual relationships, women's orientation toward their own goals and achievements in life also makes Saheli a good option for those leading a busy, fast-paced life as the Saheli pill's consumption is convenient. Telling us that it can be promoted as a lifestyle drug for all the good reasons. Other than that, the most significant factor for Saheli's growth is the fact that Saheli has cancer-curing properties, useful for women reaching the stage of menopause, giving relief from cramps and other symptoms of the menstrual cycle. As an added benefit, women above 45 can also use Saheli to combat menorrhagia, excessive or heavy bleeding during menstrual cycles.

The only place where consumers cannot use Saheli is like an emergency contraceptive or a cure for PCOD as it does not have the hormones necessary to balance the condition of a patient suffering from PCOD. Even though Saheli is a pill to be used once a week, it cannot be compared to an EC as it does not have similar side effects and is not used as an EC because the full dosage of Saheli as regular birth control is consuming one pill on a once-a-week basis, thus giving women more flexibility in indulging in sexual activities, planned or unplanned, without worrying about contraception daily or otherwise. However, it is advised to use a condom to prevent any STDs, as oral contraceptives do not prevent those.

All our research points toward increasing frequency, awareness and giving a strong focus on the right marketing initiatives, to create opportunities for nonsteroidal oral contraceptives across the globe. Through our research, we take a deep dive and look at factors such as awareness, accessibility, and the desire to find alternatives to safer oral contraceptives that can ensure the purchase of a high involvement product like a birth control pill—affirming a positive relation between the marketing of a product and its consumption in this case. We also understand that qualitative research, using mixed marketing methods such as creating touch points toward online communication base for customers, target marketing through blogs, social media, google adverts, reviews, and video content, is crucial for catering to the needs of consumers.

Through our research, we can firmly say that to reach customers efficiently, conducting online surveys and personal interviews to gain insights into their actual consumption is essential along with educating people involved in the decision-making process to be conscious and well aware of their partners and patients in the case of doctors. Allow researchers and users to focus on defining the research question, in this case, on the opportunities for nonsteroidal contraceptives. (14) The best way to do so is by acquiring primary and secondary data, forming assumptions followed by hypothesis, and analyzing data to prove or disprove the hypothesis. In our case, we firmly proved our hypothesis about marketing and the presence of demand in nonsteroidal pills. However, our research, to an extent, was limited to only knowledge of urban men and women, along with sole focus on oral contraceptives. (12)

Recommendations formed by us included the use of marketing campaigns and more vital government initiatives focusing not only on safe sex but also on safe birth control options, new sustainable packaging for the Saheli pill to appeal to the modern consumer along with an emphasis on its cancer-curing and no side-effect benefits. We also believe that spreading positive word of mouth with the help of women NGOs like Pratisandhi—focusing on sexual health and well-being—can prove to be a game-changer for the nonsteroidal pill. In today's day and age, brands need to find socially responsible, loyal customers who are willing to advocate for their products and bring a paradigm shift in other consumers' minds, which is essential for products like Saheli that are clean and focus on improving life as a whole. Using modern methods of marketing like google ads to target the youth (women aging between 18 and 22) to spread awareness about pills like Saheli as a better option for regular contraception with a once a week use compared to regular birth controls with daily usage is vital to reduce hormonal problems in the future.

FUTURE RESEARCH: ALTERNATIVE SOLUTIONS FOR MALE AND FEMALE CONTRACEPTION

Now that we know that contraceptive pills have been dominating the contraceptive market for 60 years, it is not only the pills that can be nonhormonal; we found that gels can also act on this. A birth control gel, which is non-hormonal, is set to hit the stores soon. The gel Phexxi by Evofem Biosciences Inc. got FDA approval on May 22, 2020 is a vaginal gel formulated to create an acidic environment that is inhospitable to sperms by maintaining the pH levels of the vagina. The application involves using a single dosage before sex. This gel is said to be 86% effective. (28)

About 1400 women were clinically studied, and then Evofem declared that it prevents pregnancy 93.3% if used correctly and 86% of the time overall. The gel is to be made available in September. (28) Till then, we can just hope that it will have a positive impact on the contraceptive market. Currently, the company is planning to take advantage of social media promotions and online influencers. Learning from Saheli, we can say that marketers of the gel need to ensure that they create the right kind of awareness among users and address all possible questions.

Talking about alternatives for men, we came across a thought of why only females practice methods of avoiding pregnancy at the cost of their health? Is there a solution in the market that allows men to share this responsibility?

Commonly, most contraception methods for men are limited to the use of condoms, withdrawal, or in rare cases, vasectomies. Due to this, women feel burdened and often neglect to discuss birth control with their partners, resulting in women's decisions to bear the trouble of prevention and, in some cases, the extreme case of going through an abortion. When asked women if they felt men should also have a more active part in sharing this responsibility, 92% said yes, statements included:

- "I 100% agree with the fact that men should also be given an option like this. Often times, during sex, the pressure of taking care of not getting pregnant is on the woman, and if not all, some men do still complain about using a condom."
- Definitely, contraception is a responsibility that belongs to both.
- Definitely! It's always the women's headache, especially in the case of oral contraceptive pills. And as they have a negative impact, it's not the best practice if it doesn't suit you. Have often wondered and been angry about the fact that there is nothing synonymous for men in the market.

To find out what men thought about this, we surveyed 16 men to understand how much they know about contraceptive pills if they know their partner's consumption, what is their involvement in these decisions, and would they ever consider consuming a similar alternative if made available. As it turns out, they have basic knowledge about how contraceptive works. Some even knew about the side effects. However, out of 16, only 6 were aware that their partner is/was taking any birth control pills. When asked if they ever bought contraceptives for their partner, only seven said yes, showing us that men are not a part of the discussion related to contraceptives.

However, with a rise in men wanting a fairer conversation regarding contraception and more birth control options, various research institutes have started working on alternatives. One such alternative, currently in the making, is a gel known as Nestorone—Testosterone, the first of its kind hormone birth control for men that consists of hormones such as testosterone and progestin. Once applied to the skin, it seeps into the skin and starts using its properties to reduce and stop the production of sperm. However, it remains 10 years away from getting released in the market for consumption. (28) When we asked them about how open they would be to use it, 70% said they would be open to it if there were no side effects like oral contraceptives for women, and about 30% said that they were happy with existing options.

The use of improved male contraception will not only lead to equal rights over the offspring and a fairer conversation, but it will also lead to decreasing the number of unwanted births and conception due to unintended encounters and sexual practices. An international reproductive health journal on contraception states that the number of births will decrease by almost 30% in certain countries. It also talks about the acceptance of newer male contraception methods, along with its acceptability and success rate. (29) Our future research would be related to this topic. So far in our research, we have come across limited research that sheds light upon the need and effects of a better, newer method of male contraception.

ACKNOWLEDGMENT

We would like to dedicate this chapter to our parents and our wonderful sisters.

CREDIT AUTHOR STATEMENT

The authors are responsible for conceptualization, hypothesis, data collection, analysis, and final implications. They are responsible for the draft and design, as well as for the revision of the final case study.

All authors have approved the final manuscript.

REFERENCES

1. UNESCO. (February 2, 2018). Why comprehensive sexuality education is essential. *UNESCO*.
2. Tatter, Grace. (November 28, 2018). Sex education that goes beyond sex. Harvard Graduate School of Education.

3. Jain, Rakhi, Muralidhar, Sumathi. (2012). Contraceptive methods: Needs, options, and utilization. National Center for Biotechnology Information.
4. Chaudhuri, Angela. (September 27, 2018). Why India has 16 million teenage pregnancies. FIT the Quint.
5. National Family Health Survey 3 (2005–2006). Key indicators for India from NFHS-3. http://www.rchiips.org/nfhs/pdf/India.pdf
6. Sheron, Nayantara. "Global Media, Culture, and Identity." Edited by Rohit Chopra and Radhika Gaijala, Google Books, Google, Taylor & Francis Group, books.google.co.in/books?id=xD6BV0n3Fk8C&pg=PA84&lpg=PA84&dq=Chapter%2B6%3A%2BReading%2Bthe%2BIPill%2BAdvertisement%2C&source=bl&ots=hXjXJNkqKY&sig=ACfU3U2K6_kJ5NVoAfGttLDNaTu_HstB6Q&hl=en&sa=X&ved=2ahUKEwjrsJei3YjqAhWBbX0KHaleDN0Q6AEwAXoECAkQAQ#v=onepage&q=Chapter%206%3A%20Reading%20the%20I-Pill%20Advertisement%2C&f=false
7. Butler Tobah YS (expert opinion) (6 November, 2019). Mayo Clinic.
8. Emergency contraception. World Health Organization, Essential Medicines. https://www.who.int/news-room/fact-sheets/detail/emergency-contraceptionorld
9. Raina Paul (23 February, 2017). Business Standard "Thanks to Indian men's reluctance to use contraceptives, population surges." http://www.rchiips.org/nfhs/pdf/India.pdf,
10. Am J Public Health. (August, 2012). https://www.ncbi.nlm.nih.gov/pmc/articles/PMC3464843/
11. Mridu Khullar Relph. (26 May, 2010). In India, Banking on the "morning after" pill, time. http://content.time.com/time/world/article/0,8599,1991879,00.html
12. Jagriti Gangopadhyay. (18 January, 2018). The emergency contraceptive pill is not debated enough. *DownToEarth*.
13. (16 November, 1960). Enovid advertisement obstetrics and gynaecology. https://ajph.aphapublications.org/doi/full/10.2105/AJPH.2012.300706
14. Ellen Scott. (01 July, 2016). Everything you need to know about taking a pill to delay your period for special occasions and holidays. *Metro*.
15. Petrescu, M., & Lauer, B. (2017). Qualitative Marketing Research: The State of Journal Publications. The Qualitative Report, 22(9), 2248–2287. https://nsuworks.nova.edu/tqr/vol22/iss9/1
16. Bloomberg. (May 23, 2020). Birth Control Gel Approved As No-Hormone Option In Decades. *Fortune*. https://fortune.com/2020/05/22/birth-control-gel-no-hormone-fda-approval/
17. Banu, Kulter, & Özdemir, Sami. (2014). Social Marketing Concept & Application of Social Marketing on Organizations. Management Journal, 3, 1–7.
18. Minton, Elizabeth, Lee, Christopher, Orth, Ulrich, Kim, Chung Hyun, & Kahle, Lynn. (2013). Sustainable Marketing and Social Media. Journal of Advertising, 41, 69–84. 10.1080/00913367.2012.10672458.
19. Saheli. (August 27, 2016). Saheli – safe oral contraceptive pill. *YouTube*. https://www.youtube.com/watch?v=ObfH_KVV8eQ
20. Ipillwomen. (September 13, 2007). Ipill Films. *YouTube*. https://www.youtube.com/watch?v=R37VoYw6g8M
21. Dowding, Sophie. (February, 2020). Make the only non-hormonal birth control pill Centchroman (Saheli) available in the UK. *Change.org*. https://www.change.org/p/jonathan-sheffield-make-centchroman-saheli-available-in-the-uk

22. Balusubramanian, D. (June 24, 2017). On conception and contraception: The story of Saheli. *The Hindu*. https://www.thehindu.com/sci-tech/science/on-conception-and-contraception-the-story-of-saheli/article19140909.ece
23. Srilakshmi, B., Tasneem, S. A., & Sundari, T. (November, 2014). Ormeloxifene in DUB. *ResearchGate*. https://www.researchgate.net/publication/270330244_Ormeloxifene_in_DUB
24. Pardes, Arielle. (March 27, 2019). Male birth control could actually happen, but do men want it? *Wired*. https://www.wired.com/story/how-we-reproduce-male-contraceptive/
25. Schoonenboom, Judith, Johnson, R. (July 5, 2017). How to construct a mixed method research design. *ResearchGate*. https://www.researchgate.net/publication/318228078_How_to_Construct_a_Mixed_Methods_Research_Design
26. DeFranzo, S. (September 16, 2011). What's the difference between qualitative and quantitative research? *SnapSurveys*. https://www.snapsurveys.com/blog/qualitative-vs-quantitative research/#:~:text=Quantitative%20Research%20is%20used%20to,from%20a%20larger%20sample%20population

26

Much Needed 'Pad Man' for Indian Females to Be Dignified: A Case Study on Period Poverty

Sneha Rajput and Pooja Jain
Prestige Institute of Management, Gwalior, India

CONTENTS

Learning Objectives 649
Themes and Tool Used 649
    Theme 649
    Tools Used 649
Introduction 650
MHM Programs in India 651
    MHM Journey in India 652
    MHM Programs in India: A Lacuna 653
    Menstrual Hygiene and India's Shrinking Sex Ratio 654
Understanding 'Menstruation' 655
    Mythological Religion Front of Menstruation 656
    Restrictions and Myths: Physical, Social, and Religious Boundaries 656
    Menstrual Product Market Players 657
    Menstrual Product Glitches 657
    Safe Disposal: A Major Challenge for Government 658
    No GST: GOI Move in a Positive Direction 659
Government and Menstrual Hygiene Management 660
    Highlights: Steps Taken at Policy Level 660
        Ministry of Women and Child Development (MWCD) 660
        Ministry of Human Resource Development (MHRD) 661
        Ministry of Drinking Water and Sanitation (MDWS) 661
        Ministry of Health and Family Welfare (MHFW) 661
        Tribal Development Department (TDD) 662

- Rural Development and Panchayat Raj Department (RDD) 662
- State Women Development Corporation (SWDC) 662
- Major Awareness Interventions in India: Efforts by Ministries, Government of India 663
  - ASHA (Accredited Social Health Activist) 663
  - Swachh Bharat Mission-Gramin (SBM-G) 663
  - The Adolescent Reproductive and Sexual Health (ARSH) and the Adolescence Education Programme (AEP) 664
  - Rashtriya Kishore Swasthya Karyakaram (RKSK) 664
  - Rajiv Gandhi Scheme for Empowerment of Adolescent Girls (RGSEAG) SABLA 664
  - Sarva Siksha Abhiyan (SSA, 2000–2001) and Rashtriya Madhyamik Shiksha Abhiyan (RMSA, 2009) 665
  - Aajeeveka-National Rural Livelihood Mission (NRLM) 665
  - Stree Swabhiman Scheme by CSC 665
- Major Schemes for Increasing Availability of Menstrual Products 665
  - Janaushadhi Suvidha 665
  - UDITA 666
  - Asmita Yojana 666
  - Khushi Scheme 667
- Sanitation: A Major Barrier in MHM 667
  - Sanitation at Home 667
  - Sanitation in Societies 667
  - Sanitation at School 668
- Significant Setbacks for Poor MHM 668
- Stakeholders in MHM 670
- Hurdles and Omissions 672
- National Family Health Survey (2015–2016): A Real Face 673
  - State's Position: NFHS-4 2015–2016 674
    - Menstrual Product Usage: Age-Wise 677
    - Menstrual Product Usage: Rural vs. Urban 677
    - Menstrual Product Usage: Illiterate vs. Educated Females 677
  - MHM Reality: GOI-NFHS-4 Survey 684
- NGOs and Corporates: A Boon on GOI 687
- MHM Problems in Rural India: Least Researched 704
- Actions to Overcome MHM Challenges 704
- Cognitive Dissonance: A Dilemma 709
- Implications 710
- Conclusion 711

Questions..712
Definition 713
Terminologies That Are New to You...713
Authorship Statement 713
Authorship Contributions..713
    Category 1..713
    Category 2..713
    Category 3..714
Acknowledgment...714
Conflict of Interest 714
References..714
Further Readings 716

LEARNING OBJECTIVES

After reading the chapter, the reader should be able to:

- Identify the steps taken at various ministerial levels in India.
- Evaluate the National Family Health Survey (NFHS) 2015–2016 exclusively related to MHM, to understand the overall outcomes of the policies and programs of various ministries.
- Evaluate the contribution of NGOs, SHGs, and corporates as a stakeholder in the MHM program.
- Study exclusive schemes related to MHM.
- Identify the gaps present at policy-level action.

THEMES AND TOOL USED

Theme

Menstrual Hygiene Management, National Family Health Survey, Policy-level Actions, Campaigns, Stakeholders.

Tools Used

Policy analysis framework (Collins, 2005)

    Girja, a 21 years old married female, mother of 3 kids, was standing outside the temple. When someone asked her the reason, she stereotypically replied, "I don't have to." The answer is well understood to many but seems illogical to none. She is also restricted from entering the kitchen and at many more religious events. She works in small appetizers making factory from where she gets an enforced off due to the same as she cannot touch pickle and other food items. All this is catastrophic but true and cannot be altered as far as Indian culture is concerned.

Yogita, a 14-year-old girl working as a housemaid, did not go to school after her first menses. Her family considered this dangerous, not due to health issues but due to so-called "bad spirits" chasing for such girls. Her father mentioned that "where is the money to buy menstrual products," And the story goes on…

INTRODUCTION

India, commonly known as 'Bharatvarsh', is a holy land of villages, diverse religion, and practices. The faiths are well accepted and recognized under the Indian constitution, and are deeply rooted in the hearts of Indians. It is not just more than double the Indian population resides in villages, it is 'soul of India lives in its villages' as said by Mahatma Gandhi. The Indian constitution has laid several fundamental human rights, including the right to freedom of religion, equality, right against exploitation, cultural and educational rights, and constitutional remedies. The right to sanitation stands above all fundamental rights but still needs to be substantiated. Every fundamental right is well justified in India, be it the right to freedom of religion or education. For instance, being the land of diverse religion, Hinduism, Jainism, Buddhism, Sikhism, it constitutes the population in varying ratios (Figure 26.1). The literacy rate in states is ranging from 75.5% to 82.34%, which is above the average literacy rate of the country, i.e., 75%. Despite decent literacy statistics, the country has numerous social taboos existing in the country like live-in relationships, divorces, sex, LGBTQ, love marriages, feminism, career-oriented women, to name a few. The said list is incomplete without 'menstrual' taboo dominating all with no reason and logic behind.

Menstrual cycles are understood as taboos first and then as a biological blessing toward healthy womanhood and motherhood. Irrespective of countries, menstruation is a natural process related to female natal.

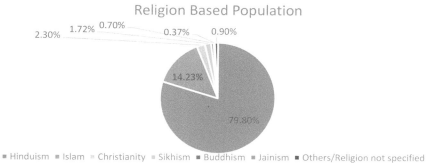

FIGURE 26.1
Population (region wise). (From CensusInfo India, 2011).

Unfortunately, in India, it is considered burdensome, awful, and 'to be just whispered' element at many levels of society irrespective of their financial and educational background. It would be wrong to say its existence as taboo is only below the poverty class; instead, many up class society perceived it as a taboo equally. The country is full of stories like 'Girja' and 'Yogita'. While keeping the taboos aside, poor Menstrual Hygiene Management (MHM) makes extremely difficult for girls residing in rural and urban slums. The policies and programs run by the Government of India (GOI) need to be reframed, as females are the pillar of a healthier society, city, state, and country. The government must receive support from every stakeholder present who is responsible as well as a beneficiary from the success of mission 'MHM'.

MHM PROGRAMS IN INDIA

'Menstrual Hygiene Day' was initiated by WASH United Initiated in 2013 and is celebrated on May 28 to raise awareness of the challenges faced by the female fraternity. However, the journey of menstrual hygiene in India started in 2000. Officially, Menstrual Hygiene Day came into existence in 2014 with the launch of Swachh Bharat Mission (SBM-G) in 2014. 'Menstrual Hygiene Day' was celebrated for the first time on May 28, 2014.

The objective was to address and resolve the problems faced by females during menstruation and to make females more dignified, secure, and safe. Usually, 'menstruation' is considered taboo, and project MHM focuses on breaking this silence to empower females. Another focus area of MHM in India is to enhance WASH infrastructure and to make menstrual hygiene products available to girls so that they continue their journey toward education, empowerment, and enrichment.

MHM Journey in India

UNICEF and WaterAid (2018) have described the entire journey of MHM in India.

- **2000** – Introduction of WASH facility throughout the country.
- **2009** – Safe drinking water and separate toilets for male and females under RTE Act.
- **2011** – Ministry of Health and Family Welfare (MHFW) launched a scheme for promoting and generating awareness on menstrual hygiene management.
- **2014** – Milestone in the journey of MHM in India, with the launch of SBM-G, the funds allocated for MHM.
- **2015** – Ministry of Drinking Water and Sanitation (MDWS) launches the first National Guidelines for MHM. A Communication Framework: Menstrual Management, published by UNICEF. In the beginning of NFHS survey 2015–2016 in India (state-wise), the vital elements included in the survey were menstrual product usage and hygiene level of females.
- **2016** – Ministry of Human Resource Development (MHRD) takes charge of MHM guidelines at the state level and also launched the guidelines in Maharashtra and Uttar Pradesh.
- Funding was done by the Department of Foreign Affairs, Trade and Development (DFAD), Government of Canada through a grant to UNICEF for research related to MHM.
- NFHS survey 2015–2016 continued.
- **2017** – NFHS survey 2015–2016 continued.
- MDWS considers menstrual waste to be a part of solid-liquid resource management.

2018 – Release of NFHS survey 2015–2016. Tax exemption for sanitary napkins.

2019 – Construction of 90 million toilets to support hygiene and better sanitation in India under SMB (G).

MHM Programs in India: A Lacuna

MHM programs are not only popular in India; instead, they are very popular across the globe. In India, menstruation is considered a taboo, making adolescent girls face stigma and feel socially excluded from society in those days. Poverty adds more trouble to the girls as it is difficult for them to choose between other essentials and sanitary napkins. It is to be noted that they do not choose between the types of sanitary napkin; instead, they have to select between bread or a sanitary napkin. The harassment does not end here, Government has made provision for free education, but the right is snatched from them once they hit puberty. Due to the lack of infrastructure present in school, they decide to dropout (Ben Amor et al., 2020). Dropping out of school is another upsetting setback of inadequate MHM facilities. UNICEF/WHO highlighted

> menstrual blood absorbing product, a washroom for changing the absorbent if needed and water and soap for washing the body and, a place to dispose of pads of safely as essential of better MHM.
>
> (Ballys, 2017)

The apprehension is – do girls in rural India have access to elements mentioned by UNICEF/WHO for MHM? There is a vast dilemma present to evaluate the success of such programs, campaigns, and policies. 'What is the real status of such programs? What is the fundamental objective of campaigns for corporates to serve or gain popularity?' are specific unexplored questions.

In the current chapter, the objective is to identify and focus on the problems and stigma faced by females living in rural and urban slum populations related to menstrual hygiene. Still there is a dilemma present toward probable and significant causes that remain unexplored exclusively for rural and urban slums. Although the government and other responsible groups are incredible, unfortunately, the ground reality does not match with what they claim it to be as per numerous researches. The case attempts to sync the relevant studies and campaigns done in India exclusively on rural and urban slum females, connecting the theory and the ground reality. Therefore, the following are the main objective of this study:

- To identify the steps taken by various ministerial levels
- To analyze the statistics of the National Family Health Survey (NFHS) 2015–2016 exclusively related to MHM to understand the overall outcomes survey
- To evaluate the contribution of corporates, NGOs, and SHGs as a stakeholder in the MHM program
- Although the MHM program in India contributes from various responsible groups, there is still a need to study the ground reality from the researchers' point of view. Therefore, the chapter also focuses on the outcomes of the research done on rural and urban slum females in India
- To identify the gaps, present policy-level action and alternates to bridge these gaps

Menstrual Hygiene and India's Shrinking Sex Ratio

Insufficient access to menstrual hygiene and lack of sanitation is one reason for female deaths worldwide (India Today, 2019). World Bank mentioned that 'One out of ten deaths in country like India is linked to poor sanitation'. It is essential to understand how MHM is related to female deaths in India. In India, proper sanitation infrastructure available for girls can reduce this mortality by retaining adolescent girls at school, who otherwise take a drop. This dropout from school results in early marriages followed by early pregnancy. Early pregnancy leads to both maternal as well as newborn deaths (Joy & Bhagat, 2016, p. 62). CensusInfo India's (2011) survey has shown a worrying fact of declining sex ratio across the country, be it state-wise or religion-wise. In India, there is a decline seen in 2015–2017 from 2014 to 2016 in females per 1000 males (Figures 26.2 and 26.3). Decline from 900 (2013–2015) to 896 (2015–2017) depicts the ground reality for generating more concern toward the health and hygiene of females in the country. Out of 29 states, Haryana, Uttarakhand, Delhi, Gujarat, Rajasthan, Uttar Pradesh, Punjab, and Maharashtra have shown the most disappointing results. Haryana is 'worst among' survey states

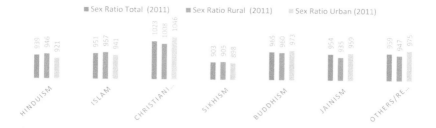

FIGURE 26.2
Sex ratio (religion wise-rural and urban). (From CensusInfo India, 2011).

FIGURE 26.3
Sex ratio (state-wise). (From CensusInfo India, 2011).

with a ratio of 833 females per 1000 males. Religion-wise, other than the Christian community, the sex ratio shows a range of 903 to 956 females per 1000 males in total. In urban areas, it is 898 to 975 females per 1000 males, and in rural, 905 to 957 per 1000 males.

UNDERSTANDING 'MENSTRUATION'

> Menstruation (men-STRAY-shuhn) is a woman's monthly bleeding. When you menstruate, your body sheds the lining of the uterus (womb). Menstrual blood flows from the uterus through the small opening in the cervix and passes out of the body through the vagina-most menstrual periods last from 3 to 5 days.
>
> (Nordqvist, 2016, p. 1)

Human life is critically revolving around primary activities like birth, death, reproduction, food, clothing, and shelter. Activities other than these are merely supporting the primary activities. The 'menstruation' is the most crucial element present on this entire planet causing human life to come alive in the form of pregnancy, be it any country, culture, religion, or caste. It is a misrepresentation that 'menstruation' is essential to just females; nevertheless, it is crucial for a family, a couple, a parent, and for the entire society too. Despite its importance to every individual, the burden is laden only and only on females.

Mythological Religion Front of Menstruation

Indian mythological statement exhibits that since Vedic times it is associated with Indra's guilty of killing Vritra, a Brahmana. Droplets of his blood have taken place in the body of a female as 'menstrual'. This guilt has made female dirty and impure during flow days. Quran mentions, 'go apart from women during the monthly course, do not approach them until they are clean'. Bible projects that a female is unclean, 'whoever touches... shall be unclean and shall wash his clothes and bathe in water and be unclean until evening' (Druet, 2017).

Restrictions and Myths: Physical, Social, and Religious Boundaries

In addition to the mythological stories, 'menstruation' as a taboo has many myths in India. Hence, the MHM is a burning issue in India (Gera, 2019). There are certain restrictions and myths associated as enlisted by the Garg and Anand (2015), i.e., a menstruating girl is not allowed to participate in any holy event, females are not allowed to enter into the kitchen, and cooking is another in the list. Touching sacred books or seeing the sour food items like 'pickle' will turn them bad or awful. During menstruation, a female is not allowed to roam out in the market, especially at night, due to the fear of 'bad' or 'evil' spirit that might attract her. Her used pads and clothes are kept secret so that nobody uses it for black magic or for hypnotizing the female. The female during menstruation day is recommended not to wash her hair frequently and restricted from doing exercise, which will cause her more trouble. There is a myth of skipping food items like pickles and curd as they provide balminess to the body. Apart from restrictions, people across societies feel uncomfortable talking about the topic. It is essential to ensure the necessary support and facilities for adolescent girls and women. It is also crucial that society, communities, and family challenge the status quo and break the silence around menstruation. Hence, it is the duty of those with authority. Out of 336 million, 4% of girls have enough knowledge about menstruation before menarche. More than 60% of girls skip school due to this and most of them face issues like low confidence, embarrassment, and humiliation because of these restrictions (Lufadeju, 2018).

Menstrual Product Market Players

Out of 336 million menstruating women-only, 36% of females use napkins produced either at the commercial level or at the local level in India. Among all menstrual products, the sanitary napkin is holding a significant share and generates the most substantial revenue. However, in the rural market, there is low consumption due to awareness and high cost. Therefore, it would be correct to say that 'sanitary napkin' generates its primary revenue from urban areas because of awareness level, financial independence, more of working females, and freedom of product choices. The rural female is still struggling for primary household products like groceries, vegetables, etc. as their basic needs.

As reported by IMARC (2020), the Indian market for a sanitary napkin in 2019 has already reached US$499.8 million. The market is concentrated on a few manufacturers, competing in prices, quality, and innovation. Leading players operating in the market are P&G Hygiene and Health Care Ltd, Johnson & Johnson Pvt. Ltd, Unicharm India Pvt. Ltd, Emami, Ltd, Mankind, Kimberly-Clark Pvt. Ltd, and Edgewell Personal Care. In biodegradable players, Carmesi, Saathi, Heyday, Anandi, Aisha Ultra, and Sakhi are leading ones. However, as per research, only less than 18% of Indian women use sanitary pads (Ramchandran, 2019). More than 77% of menstruating girls and women in India use old cloth, ashes, newspapers, dried leaves, and husk sand during periods. The Indian government is not far behind in the race by launching Janaushadhi Suvidha oxo-biodegradable sanitary napkin, only for Rs. 1/pad, to make it accessible, affordable, and available.

Menstrual Product Glitches

Kaur et al. (2018) reviewed various menstrual products available in the market like tampons, reusable tampons, menstrual cups, bamboo fire pads, banana fiber pads, and water hyacinth pads. In the Asian market, sanitary pads are still typical. The cost of the same varies from brand to brand and from the raw material used in manufacturing. Unfortunately, besides these products, females living in rural areas also use sacks, cow dung, leaves, and mud during their menses. Government, as well as NGOs, is on their toes to creating awareness on the benefits of using hygiene products such as sanitary napkins among women through various modes like social media, etc. Each of these menstrual products has its advantages and disadvantages and is grouped as hygienic and unhygienic products. For instance, absorbents like newspapers, plastic bags, mud, cow dung, and leaves might be available for free but are highly infectious and negatively affect health. Sometimes, they are even difficult to use due to less absorbing capacity.

Similarly, cloth from domestic waste is, although reusable and readily available, suffer from the disadvantage as they need water and soap for washing. This absorbent also requires a single spade for drying up for the next use. The local or hand-made napkin is on the list with less cost and benefit of reusing for the subsequent few cycles but again with a high need for proper washing and drying up. Commercial reusable or disposable napkins are considered as the most hygienic but are beyond the affordability of rural females. The major challenge is the safe disposal of these napkins. Therefore, these napkins may be hygienic but are not absolutely environment friendly.

Safe Disposal: A Major Challenge for Government

In India, a female is using eight non-degradable pads every month. The menstrual waste generated in India is estimated to be 113,000 tons annually. One of the most crucial elements in the entire MHM is the safe and eco-friendly method for disposal of menstrual waste approved under legal norms. The MHM also includes training on the techniques and ensuring that this is imparted well among females and supporting staff for further sharing with the society. In India, there is a general practice of disposing of such waste, which is entirely unsafe and unacceptable. In cities, however, this waste is handed over to garbage collecting vans to much extent, but on the other hand, throwing away in open areas is commonly adopted practice. Unsafe disposal, i.e., throwing such waste into ponds, rivers, or in the open fields, leads to other severe and series of infections.

The proper and safe disposal ensures the process of destruction of the absorbents and contaminated material with no human contact, keeping in mind the lowest loss for the environment. The municipal corporation of each area can play a significant role in reducing such waste through waste collection systems. Unfortunately, in India, there are many challenges present.

Although there is the Public Health Engineering Department (PHED), which continuously guides on various methods for safe disposal options, this is more challenging in rural areas where the bins and vans do not reach. The waste bin is a good and affordable option, subject to their placement at every nook and corner. It would probably reduce the problems related to the seepage, odor, or waste before mass disposal. In a rural setting, making pit is a common practice. Burning in an open heap should be avoided or must be done in a bottomless hole. There are conventional as well as modern methods for disposing of the waste. In traditional methods like making a pit pot, burying it deep, burning in a pit is commonly recommended, whereas, in modern ways, an incinerator is a good option. However, in rural settings, where females are more using cloth and cotton, conventional methods can work wherein 'pit burning' is the least recommended method in all kinds of material used. The incinerator is suitable in urban settings and one of the best options among all others in plastic liner napkins.

Looking at the actual scenario, majorly absorbent is wrapped in newspaper and is thrown in open fields and rivers. There are instances where it is thrown without any wrapping openly in empty grounds nearby, leading to sever health issues for anybody in contact. The saddest part of such practice is the impact on the health of waste management workers. It is waste management workers who manually segregate the waste.

No GST: GOI Move in a Positive Direction

Indian government scrapped its 12% tax on all menstrual products. More than 400,000 activists signed the petition to remove the tax and named it 'blood tax'. Then finance minister Piyush Goyal announced the exemption of 100% tax on sanitary napkins. Since 2018, the GOI is monitoring the cost of sanitary pads. The objective was to ensure the reduction in the prices of the menstrual product after the tax exemption of goods and services tax (GST). Although the government has clarified that 'nil' tax means that the input will not receive any tax credit benefit, it is not sure that there will be a significant reduction in the prices. Anti-profiteering tax authorities are keeping a check on big manufacturers to ensure that exemptions leading to benefits accruing to companies are passed on to consumers. However, the manufactures will still be burdened with the supply chain tax issues. Other than supply chain taxes, the tax on input like adhesive can be up to 18%. Therefore, still, the tax issue needs to be pondered again.

GOVERNMENT AND MENSTRUAL HYGIENE MANAGEMENT

Menstruation hygiene management is the responsibility of various ministerial levels. In 2005, MHM became the critical area of operations. The launch of the National Rural Health Mission (NRHM) and ASHA (Accredited Social Health Activist) took charge to promote the same at various levels. Various departments working under the direct control of government are working effectively to upgrading the status of women across the country. Central departments holding the baton for this upgradation are Ministry of Women and Child Development (MWCD), MHRD, MDWS, MHFW, Tribal Development Department (TDD), Rural Development & Panchayat Raj Department (RDD), and State Women Development Corporation (SWDC).

Highlights: Steps Taken at Policy Level

Ministry of Women and Child Development (MWCD)

Significant activities by MWCD include the formulation of schemes for adolescent girls. Some exclusive activities undertaken are listed below:

- Scheme for Adolescent Girls (SAG), SABLA, replaced the Nutrition Program for Adolescent Girls (NPAG) and Kishori Shakti Yojana (KSY) in many cities. MWCD has successfully implemented the SABLA scheme launched in 2011, i.e., Rajiv Gandhi Scheme for Empowerment of Adolescent Girls (RGSEAG).

- They are providing training to Anganwadi workers (AWW). Along with this, it also monitors and supervises the performance of AWW.
- MWCD is actively involved in promoting the production, promotion, and sales of sanitary napkins at the village level and ensures the availability of sanitary napkins in interior areas.
- It also promotes MHM in rural areas and the NARI (National Repository of Information for Women) scheme at the village level.

Ministry of Human Resource Development (MHRD)

MHRD has taken steps ahead to deal with poor MHM practices across the country. It works on the following fronts:

- Appointing nodal officers for training purposes to adolescent boys and girls on puberty issues
- Promoting MHM at school levels
- Counselling parents toward MHM is another task in the list

Ministry of Drinking Water and Sanitation (MDWS)

MDWS focuses mainly on the availability of water and sanitation facilities in the deprived area. It is actively involved in:

- Creating awareness relating to MHM
- Ensuring the availability of hygiene-related facilities
- Safe disposal of menstrual waste
- Granting funds related to menstrual hygiene to various campaigns and activities
- Promoting MHM awareness schemes under 'Gramalaya'

Ministry of Health and Family Welfare (MHFW)

MHFW has contributed enormously by putting its efforts in the following manner:

- Running effectively 'Rashtriya Kishor Swasthya Karyakram (RKSK)'
- Counseling on aspects like reproduction and sexual health with adolescent girls

- Giving advice to teenage girls on puberty and MHM
- Creating awareness by delivering lectures at the school level related to MHM through audio and video content
- Working on the nutritional status of teenage girls
- Conducting the event of the 'ASHA' program virtually
- Ensuring the availability and distribution of sanitary napkins 'freedays' at a special price
- Working on safe disposal of menstrual absorbent waste

Tribal Development Department (TDD)

The role of TDD is limited to tribal development schemes. It had played a significant role by providing training to teachers and residential staff in 'ashram schools' and 'madrassas.' It has actively undertaken promotional activities and ensured a regular supply of sanitary napkins.

Rural Development and Panchayat Raj Department (RDD)

RDD ensures the availability of absorbents through SHGs under the fold of the National Rural Livelihoods Mission (NRLM). It is also involved in manufacturing the absorbents at the local level. Generating awareness among females for MHM and ensuring its basics for like water and sanitation are regular activities undertaken by RDD.

State Women Development Corporation (SWDC)

SWDC is another pillar of strength in MHM; it undertakes the following measures to ensure better MHM.

- They are promoting and supporting MHM awareness activities.
- They are ensuring the regular supply of sanitary napkins.
- The establishment of disposal mechanisms and facilities is another concern area.
- They are providing training to staff.
- It is ensuring availability of water, sanitation, and hygiene-related (WASH) infrastructure.

Major Awareness Interventions in India: Efforts by Ministries, Government of India

ASHA (Accredited Social Health Activist)

The introduction of ASHA in 2006 is one of the great actions taken by NRHM, as it is the one which is the closest to the rural population. Intensive training is given to ASHA from using to disposing of sanitary napkins. ASHA conducts monthly meetings, home visits, ensuring the availability of sanitary napkins and their stock keeping. They coordinate with Village Health and Sanitation Committee (VHSC) and AWW

Swachh Bharat Mission-Gramin (SBM-G)

MDWS/Ministry of Urban Development (MoUD) collaboratively launched the SBM. World Bank has approved loan of US$1.5 billion toward SBM under rural component, known as SBM – Gramin (SBM-G); SBM ensures that benefits delivery to the 60% of India's rural population. MHM is a critical area of operation in Swachh Bharat Mission Guidelines established by the MDWS to support all adolescent girls and women in 2014. SBM-G has clearly stated the role of state governments, district administrations, engineers and technical experts' inline departments, and school head teachers and teachers separately. State governments have more to do with designing and budgeting part of MHM for districts for areas like Information, Education, and Communication (IEC) content, functional toilets for girls, teachers training, and disposal of menstrual waste. Districts have a significant role in SBM-G. They are responsible for disseminating awareness on MHM at school levels and making them familiar in terms of misconceptions and myths using IEC content developed at the state level. Districts have more intervention in sensitizing parents, stakeholders, infrastructure, disposal, and training at the school level. MDWS has also provided guidelines for infrastructure associated with MHM, which highlights 'access to a clean and well-maintained toilet' i.e., there should be one separate toilet over 40 girls. Therefore, toilets must be in the ratio of 1:40 with privacy and space, and facilities like shelf, mirror, and hooks. In 2017, the Ministry also launched Guidelines for gender issues in sanitation.

The Adolescent Reproductive and Sexual Health (ARSH) and the Adolescence Education Programme (AEP)

NRHM launched ARSH and AEP programs in 2011 for rural girls in the age group of 10–19, intending to enhance awareness on menstrual hygiene and to empower girls. Further, the plan is for increasing the use of good quality sanitary napkins and their safe and environment-friendly disposal. The scheme is for 150 districts working through ASHA, SHGs, and Kishori Mandals (KM), ensuring the supply of reasonably priced and good quality sanitary napkins. As per guidelines, sourcing of sanitary napkins will be done through a bidding process exclusively in Central, Northern, and North-Eastern states, where SHGs may not have matured yet or are non-existent.

Rashtriya Kishore Swasthya Karyakaram (RKSK)

MFHW launched RKSK in 2014. It is more into information (IEC & IPC) dissemination on MHM issues. MHM is not the core area of the RKSK program. However, it works for counseling on aspects like nutritional status, disorders related to menstruation, personal & menstrual hygiene, and the use of sanitary napkins. RKSK is also handling the curative services like treatment of the menstrual disorder. Adolescent Friendly Health Clinics (AFHCs) under the ages of RKSK have three major operational zones, i.e., District Hospitals, Community Health Centers, and Primary Health Centre, wherein menstrual hygiene is one of the activities. To summarize, RKSK works on 7c's, i.e., coverage, content, communities, clinics (health facilities), counseling, communication, and convergence.

Rajiv Gandhi Scheme for Empowerment of Adolescent Girls (RGSEAG) SABLA

RGSEAG-SABLA program was initiated in 2011 by the MWCD. It is an integrated service to improve health, nutrition, and empowerment for girls. It provides awareness about MHM to adolescent girls through Anganwadi Centers. Significant areas of SABLA promote awareness about health, hygiene, nutrition, adolescent reproductive, sexual health (ARSH), and family and child care. The scheme also disseminates information/guidance about existing public services such as Primary Health Centers (PHC), Community Health Centers (CHC), District Hospital (DH), etc.

Sarva Siksha Abhiyan (SSA, 2000–2001) and Rashtriya Madhyamik Shiksha Abhiyan (RMSA, 2009)

SSA in 2000–2001 and RMSA in 2009 schemes were initiated by the MHRD. The program aims to provide elementary education for all and enhance access to secondary education, respectively, prioritize sanitation infrastructure in schools to improve school retention. Moreover, the disposal mechanism, like the incinerator establishment in schools, is among the critical tasks of the scheme.

Aajeeveka-National Rural Livelihood Mission (NRLM)

NRLM came into existence in 2011 and as an initiative by the Ministry of Rural Development (MRD). NRLM focuses more on awareness generation and supports SHGs and small manufacturers to produce sanitary pads for more access to sanitary napkins.

Stree Swabhiman Scheme by CSC

Although CSC is not a government organization, it works under the flagship of the Ministry of Electronics & Information Technology and also uses its domain. CSC supports more than 35,000 women entrepreneurs providing various G2C and B2C. CSC provides digital inclusiveness and support for digital literacy, financial inclusion, skill development, etc. CSCs have launched the social initiative 'Stree Swabhiman', which is into the manufacturing of sanitary napkins and to make females more aware of health and hygiene. It is encouraging women entrepreneurs to manufacture and educating females at centers. CSC is more focused on Bio-degradable, eco-friendly sanitary pads. CSC has a guideline for providing Rs. 500 per girls per year to their village-level entrepreneurs.

Major Schemes for Increasing Availability of Menstrual Products

Janaushadhi Suvidha

Under the flagship of Pradhan Mantri Bhartiya Janaushadhi Pariyojana, the GOI has launched 'Janaushadhi Suvidha – Oxo-biodegradable sanitary napkins' on World Environment Day, June 4, 2018. The

'Janaushadhi Suvidha' napkins aim to provide affordable sanitary napkins. More than 3600 Janaushadhi Kendras were made to avail of the facility and work in 33 States/UTs across India as per GOI. Suvidha sanitary napkins initially were priced at Rs. 2.5/pad, and most importantly, the napkins are biodegradable. The scheme revises the price of sanitary napkins to Rs. 1/pad. Therefore, the Janaushadhi Suvidha scheme seems promising in terms of providing affordable and environment-friendly sanitary napkin. More than 6600 such Janaushadhi Suvidha centers are operational and the sale was approximately 3.43 crore pads till June 2020.

UDITA

The Government of Madhya Pradesh (GMP) launched UDITA in 2015. UDITA is working on awareness, accessibility, and affordability of sanitary napkins and is operational in 97,135 Aganwadis. It operates through SHGs and NGOs and also ensures the participation of local sanitary pad manufacturing companies. Other activities like the disposal mechanism of hygienic napkins are significant areas of operation. UDITA works under the guidelines of SABLA and Kishori Shakti Yojana and the ARSH clinic for counseling and consultation.

Asmita Yojana

The Government of Maharashtra launches Asmita Yojana under flagship of the Minister of Rural Development and Women and Children Development, announced by Smt Pankaja Munde. Under the scheme 'ASMITA Yojana,' the government has decided to distribute subsidized napkins to the rural women in the state. This scheme is for rural women and adolescent girls of the age group 11–19 years. For teenage girls, a pack of 8 sanitary napkins is available at a subsidized rate of Rs. 5/each. For every packet, there is a subsidy of Rs. 15.20. Anyone can become the sponsor for the project by donating money toward the scheme by selecting the number of girls and the number of months to be sponsored by the individual on the website https://mahaasmita.mahaonline.gov.in/. There are currently 636 sponsors with total girls sponsored 11897 till date. As per the scheme, SHGs can buy sanitary napkins at subsidized rates through ASMITA mobile app.

Khushi Scheme

'Khushi Scheme' is an initiative launched by the Government of Odisha under the leadership of Chief Minister Odisha Mr. Naveen Patnaik. The scheme aims to promote female health and hygiene and distribute free sanitary napkins to school girls. Amount of Rs. 70 Crore is allocated for the project by the Government of Odisha. More than 1.7 million girls in grades 6th to 12th and girls in government-aided schools will benefit from the scheme. The core focus area of the project is to retain more and more girls in school.

Sanitation: A Major Barrier in MHM

Sanitation in India is a prominent black spot in the entire Swachh Bharat Mission. The country either don't have proper toilets or is not maintained correctly. Using a public toilet in India is considered as the most unhygienic thing. People often prefer using roadside or railway tracks and view them more hygienic than public restrooms. It is undoubtedly the most infectious facility, suffering from a lack of maintenance. Expecting public toilets to be cleaned and sanitized is a big dream for Indians.

Sanitation at Home

Movies like 'Toilet' have highlighted the issues of not having toilets at home. Females feel awkward and embarrass for using public toilets outside the home. Especially at night, the fear of anti-social activities increases. The other options like going to dense plantation areas, fields, railway tracks are also not safe. Many cases of rapes and violence while using toilets at night are present in India. Till 2019, the Modi government has declared the construction of 9.2 crore toilets in the homes. This program covered more than 5.5 lakh villages in 28 states and union territories and has made them open defecation-free. The rank of India on the sanitation index is still low despite the programs like Swachh Bharat (Clean India) Mission.

Sanitation in Societies

Open defecation remains a problem across the country. However, India has organizations like 'Sulabh Complex.' 'Sulabh Complex' is one of the most prominent international and largest nonprofit organizations in India, founded in 1970. It has complex constructed across the country with the facilities of bathing, laundry and urinals. After the Swachh Bharat Mission launch in 2014, over 90 million toilets have been built across rural India so far. Even the country has the Community Sanitary Complex in market places, public places, places like bus stand, etc. where there is usually a high footfall of the public. Unfortunately, India, the world's second-largest country by population, has 732 million people without access to toilets. Therefore, the issue remains unsolved, even at the societal level.

Sanitation at School

The awareness generated at the school level can help girls to overcome stigma and shame associated with menstruation. Proper sanitation facility at schools is required to be worked upon so that adolescent girl can become more confident. Poor sanitation is the root cause of girls dropping out of school. Along with sanitation, schools must provide training sessions for better MHM practices to students irrespective of their gender. WASH facilities can help girls to continue their education even after they hit puberty. The availability of sanitary napkins and their safe and private disposal remains a missing element in the sanitation issue. Moreover, schools lack ideal toilets (as described under SBM-G), which must be designed separately for staff and students, also specifically for males and females

Significant Setbacks for Poor MHM

Geertz et al. (2016) reviewed the significant areas where the females are facing challenges during their menstruation. As mentioned in the previous paragraphs, poor MHM not only drags the confidence level of females down but also makes them feel less dignified as compared to males. In India, be it rural areas or urban areas, do not have any restrictions on moving out at night and joining the event. On the other hand, females, even on her regular days, are not allowed to move out of home in the late evenings. This discrimination between genders hits the psychology of a female so profoundly that she starts feeling deprived of self-esteem and confidence. However, this lack of mobility is more common in villages as compared to urban areas.

As mentioned in 'MHM Programs in India: A Lacuna', education is another area that gets a severe hit by the poor MHM. In rural areas, girls are not allowed to go out, making even going school during these days out of the question. Many girls leave their education after they start getting their menses. India's school infrastructure is still far away to be considered safe and comfortable for girls during menstruation. Lack of water, sanitation is still not considered as necessary as a classroom in any rural setting school. Leaving the school at an early age, in return, severely affects many dimensions associated with females like early marriages, early pregnancy, and ultimately becoming a reason for increased female death ratio in the country.

Females not only degrade in their physical health but also in their mental health due to the unavailability of proper menstrual hygiene. Lack of awareness and unavailability of appropriate absorbents at an affordable price depreciates the nutritional status of rural females. As mentioned in 'Menstrual Hygiene and India's Shrinking Sex Ratio', practices like early marriages are already present as a burden on Indian rural culture, which imposes early motherhood, miscarriages, which results in problems like dysmenorrhea, hormonal disturbances among females. Several issues of physical health are directly associated with mental health too. As poor MHM results in poor physical health and badly affect the mindset of a female. Hormonal disturbances, at times, severely lead to stress, depression, and anxiety among females, followed by mood swings and mental disorders.

The disadvantages of poor menstrual hygiene do not end here. Further, it also affects the financial stability and dependence of a female. As several times, due to poor health, the females are absent from the workplace. Further, there are regions where females get forced-leave during her menses, which results in a reduction in wages, which is a setback to her financial stability also. Another dimension is that many workplaces do not have proper washrooms and restrooms, which makes it utmost difficult for any female to go to work on her periods.

Females not only feel uninvolved in the society but also in her own family due to various restrictions present like not entering into the kitchen or 'pooja room.' Not touching the sour food items or looking at them is restricted for a female. Moreover, this kind of standoffish restriction makes a female more falter in her life.

Stakeholders in MHM

Superficially it seems like only females of the society are the real stakeholder to be benefitted from better MHM practices. Undoubtedly females are the primary stakeholders. The section unfolds the reality showcasing a series of stakeholders from various perspectives like policy formations, decision-makers, etc. As poor MHM practices increase the workload of the stakeholders, and on the contrary, better MHM practices can help every section of society. The section enlists the following significant stakeholders (Figure 26.4).

- Adolescent girls/females: Every setback associated with poor MHM is faced directly by females. Poor MHM affects females not only physically but also mentally, emotionally, and financially. The majority of the females living in rural/urban slums dropout of school after their first menses, which leads to early marriages, early motherhood, and various hormonal issues thereon.

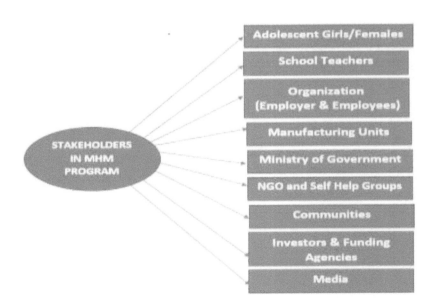

FIGURE 26.4
Stakeholders in successful MHM program. (Synthesized by the authors).

- School Teachers: The teachers fail to disseminate awareness on various dimensions to such girls. Better MHM can help the girls to continue with their schools and gain essential knowledge and understanding. Therefore, the contribution of school teachers can be significant with such improved practices.
- Organizations: The female employees will have low absenteeism and can get better pay for their work. On the other hand, the employer can get the benefit of enhanced performance, which can be profitable for the company or venture.
- Manufacturing units: MHM can lead to increased awareness of the usage of absorbent instead of other unhygienic products. It increases the demand for sanitary napkins and other menstrual products. Thus, manufacturing units get directly influenced by better MHM practices.
- Government: Improved MHM practices can lower down the burden of government and can help to reformulate the related policies focusing on the gaps present. The government gets benefitted on two dimensions one is the policymaking, and the other is decision making. The Ministry of Health and Ministry of Education are prime stakeholders among all different ministerial levels.
- NGOs and SHGs: NGOs and SHGs have played a significant role across the country by increasing awareness, conducting donation drives, and manufacturing low-cost absorbent. It has worked and supported all government programs and policies of MHM. To a greater extent, the status of females has gone up with the efforts of NGOs and SHGs.
- Communities: Better MHM practices affect communities and their surrounding in varieties. The improved sanitation, waste disposing facilities, availability of affordable and eco-friendly sanitary pads can help the communities to remain waste-free and healthy. It, in turn, improves the health of the citizens of a nation.
- Investors and funding agencies: Investors and funding agencies are keen to invest in nations with high health index. Better MHM, as mentioned in earlier sections, can reduce the infections, diseases, and mortality rates, which helps any country to move up in the health index. In 2020, India ranks 42nd and has a 66.21 score in health care index, making it a promising place for various health and other related projects.

- Media: Media is more into benefitting MHM programs positively. It plays a responsible role in creating awareness among the general public (Rahman et al., 2018). Conventional modes of media help generate awareness in rural and urban slums.
- Celebrities: In recent trends, stars are coming up as brand and campaign ambassadors to support MHM practices in India. Manushi Chillar, Akshay Kumar, Vidya Balan have successfully endorsed various campaigns. Being a public figure, the citizens worldwide admire their efforts and even follow the messages spread by them.

Hurdles and Omissions

Although the time has come for more actions in the country. Despite all ministries working on MHM, but few seem to reach to the actual problem persisting. Significantly few programs are focusing on the availability of affordable and eco-friendly sanitary napkins. For few schemes like Janaushadhi Suvidha, which is under the flagship of Pradhan Mantri Bhartiya Janaushadhi Pariyojana, the GOI can focus on the production and distribution of low-cost sanitary napkin at Rs. 1/pad. Similarly, RKSK is also working on the core issue of affordable sanitary napkins. Other ministries like the MDWS have targeted sanitation-related matters in the country. Therefore, they have come up with the satisfying statistics of building 90 million toilets across the country.

The country demands more action than planning. Major schemes and projects are still in the phase of drawing guidelines rather than their implementation. The policies designed by various ministries level are not apparent in the process part, like how the objective mentioned will be achieved and the resources allocated for each purpose. Similar kinds of objectives are present in all the guidelines presented by different ministries, creating a role conflict. The awareness creation is looking like over-talked issues in all the ministries, and the other two issues of accessibility and affordability remain faded. The guidelines do not provide a unique way to achieve the objective, and the entire burden is kept repeatedly on the shoulder of ASHA, ANM, and AWW. This resource sharing reflects a lack of resources in the MHM as a whole program. Tax on the inputs remains another loophole in the entire MHM guideline. The ministries must relieve the manufacturers, especially SHGs, small and medium entrepreneurs, with the liability of such input tax. Merely exempting the final product has not solved the issue. Even the input which bears the significant tax burden must be made tax free.

No data represents the success of these programs as if the ministries have not worked beyond guidelines. The NFHS statistics mention the current status of rural females and their product usage ratios. The findings exclusively related to MHM programs are not present under any head. Although infrastructural data do exist, that does not prove the improvement in MHM practices in the country.

NATIONAL FAMILY HEALTH SURVEY (2015–2016): A REAL FACE

National Family Health Survey (NFHS) is one of the biggest national-level surveys conducted at the state level organized by the GOI under the stewardship of the MHFW in nodal collaboration with IIPS, Mumbai. In NFHS-4 2015–2016 survey report, 29 states statistics on menstruation and the hygiene levels of females are present under section methods of 'menstrual protection.' The survey in total included urban 78,417 (32%) and rural 166,100 (68%) females (Table 26.1). Females of age 24–49 areas unweighted cases throughout the survey. 68% of rural and 32% of urban respondents participated in the survey. Majorly cloth, locally prepared napkins, sanitary napkins, and tampons are considered as menstrual protection options. Out of this cloth is considered unhygienic, and the rest all are hygienic methods. The data reflected that females in urban areas prefer using sanitary napkins (59.2%) as compared to cloth (42.6%). At the same time, this ratio was the opposite in the case of rural females. In rural females prefer using cloth (71.4%) compared to sanitary napkins (33.6%). Tampons are not so common as other menstrual products but are more common among urban females (3.4%) than rural (1.95) females. Locally made napkins are preferred almost equally by urban (19.5%) and rural (14.8%) females.

The survey mentions the role of education (Table 26.2), where females with no schooling used more of cloth (88.8%) as compared to sanitary napkins (13%) and locally prepared napkins (6.8%). In contrast, females

TABLE 26.1

State-Wise Total Sample Size

S.No	States	Age 15–19 yr	Age 20–24 yr	Urban	Rural
1	Andhra Pradesh	1329	1742	953	2118
2	Arunachal Pradesh	2369	2280	1364	3283
3	Assam	4693	4961	1245	8409
4	Bihar	10059	7778	2343	15494
5	Chhattisgarh	4694	4542	2130	7107
6	Goa	237	227	284	181
7	Gujarat	3708	3825	3094	4439
8	Haryana	3382	4299	2872	4809
9	Himachal Pradesh	1318	1402	259	2461
10	Jammu & Kashmir	3972	4222	2110	6084
11	Jharkhand	5589	5320	2941	7968
12	Karnataka	3716	4348	3393	4671
13	Kerala	1504	1519	1418	1605
14	Madhya Pradesh	11624	11642	6691	16576
15	Maharashtra	4604	5195	4794	5005
16	Manipur	2009	2061	1568	2502
17	Meghalaya	1767	1626	800	2593
18	Mizoram	2038	1924	2471	1491
19	Nagaland	1742	1708	1410	2040
20	Odisha	5572	5774	1911	9435
21	Punjab	2617	3212	2088	3741
22	Rajasthan	8136	8230	3875	12491
23	Sikkim	837	845	545	1137
24	Tamil Nadu	3901	4450	4036	4314
25	Telangana	997	1355	1096	1256
26	Tripura	727	756	406	1077
27	Uttarakhand	3250	3189	2247	4193
28	Uttar Pradesh	22015	18542	10175	30382
29	West Bengal	2933	3196	1771	4357

Source: NFHS Survey 2015–2016 (Compiled by the Authors).

with school education had a significant difference in using sanitary napkins (60.7%) as compared to cloth (43.1%) (Figure 26.5).

State's Position: NFHS-4 2015–2016

Out of the four methods mentioned above, the last one, i.e., tampons, are holding merely a 1.8% average share in total usage in India. The use of

TABLE 26.2
India's Overall Position in Menstrual Product Usage

Menstrual Products Used	15–19 yr	20–24 yr	Urban	Rural	No Schooling	>5 yr	05–07 yr	08–09 yr	10–11 yr	<12 year
Cloth	61.9	62.4	42.6	71.4	88.8	83.8	77.7	69.7	54.2	43.1
Locally Prepared Napkins	16.4	16.1	19.5	14.8	6.8	8.8	12.3	14.8	18.9	21.5
Sanitary Napkins	41.8	41.8	59.2	33.6	13	18	25.1	35.1	51	60.7
Tampons	2.4	2.4	3.4	1.9	1	1.6	1.4	2	2.6	3.5
Other	0.1	0.1	0.1	0.1	0.1	0.1	0.1	0.1	0.1	0
Nothing	0.4	0.5	0.3	0.6	1.1	1	0.6	0.4	0.3	0.3
Total Females	121552	122966	78417	166100	25324	8728	30573	56890	51587	71415

Source: NFHS Survey 2015–2016 (Synthesized by the Authors).

676 • Social and Sustainability Marketing

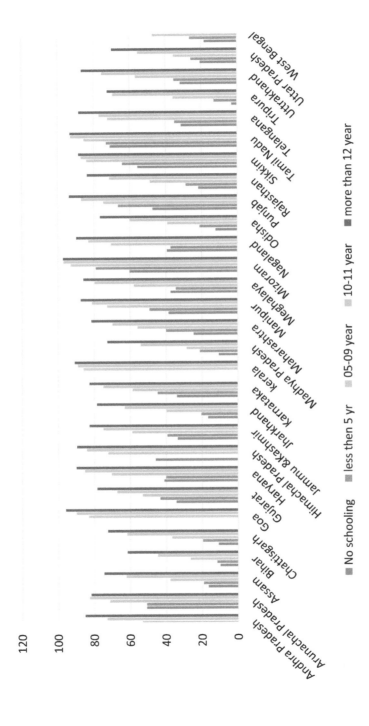

FIGURE 26.5
Hygiene level as per schooling. (From CensusInfo India, 2011).

cloth still has a significant percentage with 55.05% and followed by sanitary napkins (50%) and locally prepared napkins (14%) on average. Many states are leading in the usage of sanitary napkins as compared to cloth. Andhra Pradesh, Arunachal Pradesh, Goa, Haryana, Kerala, Mizoram, Nagaland, Punjab, Sikkim, Tamil Nadu, Telangana, and other states are still having usage of cloth as a menstrual product. Locally prepared napkins are uncommon in many states yet.

Menstrual Product Usage: Age-Wise

Females aged between 15 and 19 years (Figure 26.6) in Andhra Pradesh, Arunachal Pradesh, Goa, Haryana, Himachal Pradesh, Kerala, Maharashtra, Manipur, Mizoram, Nagaland, Punjab, Sikkim, and Telangana were high in the use of sanitary napkins. Females aged 20–24 years (Figure 26.7) in Andhra Pradesh, Goa, Haryana, Himachal Pradesh, Kerala, Mizoram, Nagaland, Punjab, Sikkim, Tamil Nadu, Telangana are high in the usage of sanitary napkins as compared to cloth. The level of hygiene (Figure 26.8) is more in females aged between 15 and 19 years in all the states of India except Haryana, Madhya Pradesh, Punjab, Rajasthan, Uttar Pradesh.

Menstrual Product Usage: Rural vs. Urban

States where even the rural area females (Figure 26.9) using sanitary napkins more than cloth are Andhra Pradesh, Goa, Haryana, Himachal Pradesh, Kerala, Punjab, Tamil Nadu. Out of these, the ratio is significantly high in Mizoram as the usage of the sanitary napkin is more than three times that of cloth. In Gujarat, the locally prepared napkins are more than other sanitary napkins not in other states. In the case of urban living females, Assam, Bihar, Chhattisgarh, Madhya Pradesh, Tripura, Uttar Pradesh showed that despite living in urban areas, the usage of cloth was high as compared to sanitary napkins and other options available (Figure 26.10). The level of hygiene among rural females has shown a severe setback as compared to urban females in all the states (Figure 26.11).

Menstrual Product Usage: Illiterate vs. Educated Females

The NFHS-4 2015–2016 has shown the real problems related to the MHM awareness programs. The role of schooling plays a significant role in

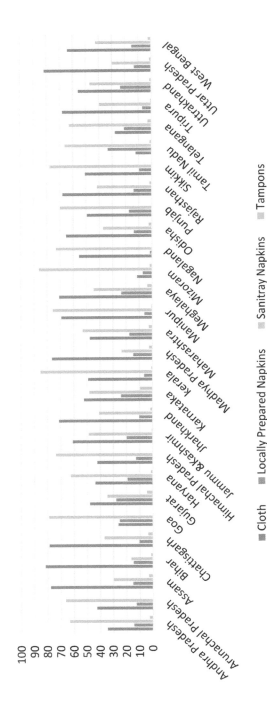

FIGURE 26.6
Menstrual products used among 15- to 19-year-old females. (From NFHS 2015–2016 (Synthesized by the Authors)).

Much Needed 'Pad Man' for Indian Females • 679

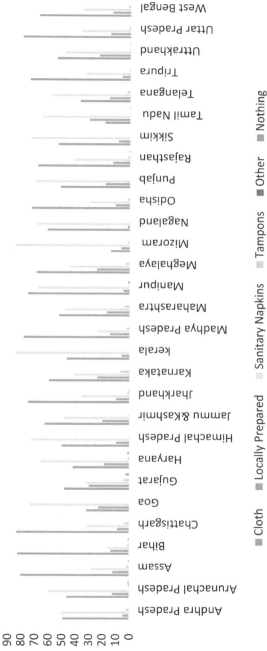

FIGURE 26.7
Menstrual products used among 20- to 24-year-old females. (From NFHS 2015–2016 (Synthesized by the Authors)).

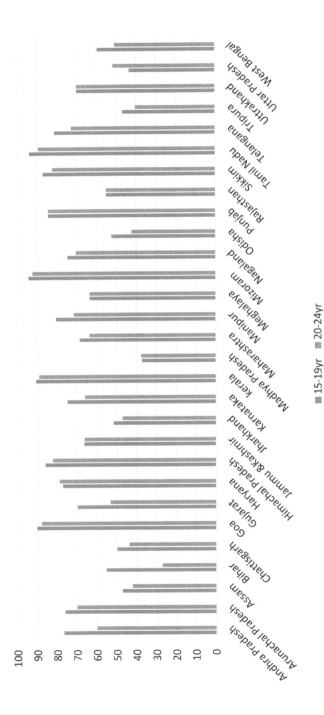

FIGURE 26.8
Level of hygiene-age wise. (From NFHS 2015–2016 (Synthesized by the Authors)).

Much Needed 'Pad Man' for Indian Females • 681

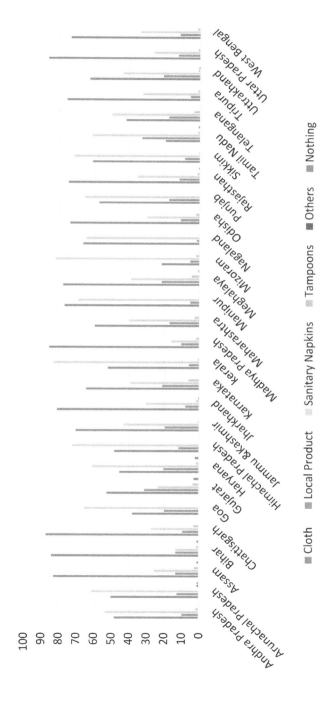

FIGURE 26.9
Menstrual product used-rural females. (From NFHS 2015–2016 (Synthesized by the Authors)).

682 • Social and Sustainability Marketing

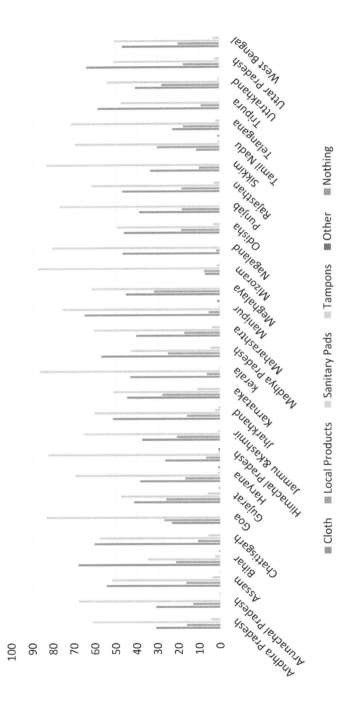

FIGURE 26.10
Menstrual product used – urban females. (From NFHS 2015–2016 (Synthesized by the Authors)).

Much Needed 'Pad Man' for Indian Females • 683

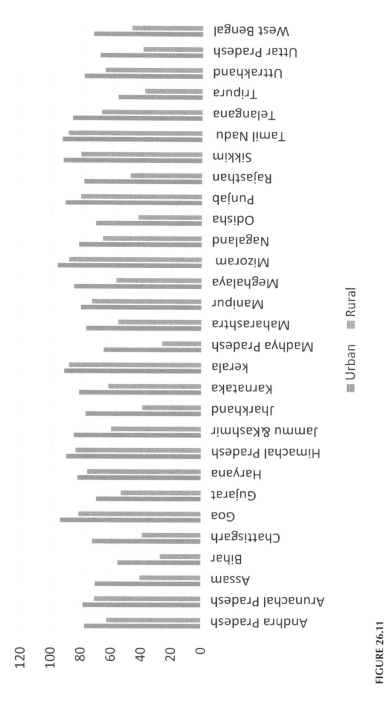

FIGURE 26.11
Level of hygiene-rural vs. urban. (From NFHS 2015–2016 (Synthesized by the Authors)).

improving the hygiene status of females across the country, be it a rural or urban living female. The facts and figures showed that only two states Goa and Kerala, were high in education status as there was no respondent from the 'no schooling' group. In other states, the females with no schooling education used cloth more than sanitary napkins (Figure 26.12). Females with schooling education of more than 12 years have shown a significant difference in using sanitary napkins compared to cloth in all the states (Figure 26.13). Females use even the locally prepared napkins, which are considered hygienic with 'school education of more than 12 years' in all the states except Arunachal Pradesh, Mizoram, and Nagaland. In these states, females from 'no schooling' and 'schooling education of more than 12 years' use locally prepared napkins equally. In the case of the level of hygiene of females, there is a constant increase seen with the rise in education level from no schooling to <5 years, 05–09 years, 09–11 years, and more than 12 years in all the states of India.

MHM Reality: GOI-NFHS-4 Survey

NFHS surveys familiarize the reality existing in the country at the ground level. Despite various campaigns and ministerial level schemes and efforts, the government is severely lagging in MHM practices. India, where the rural population is the backbone of the country, needs upliftment in many aspects like education, hygiene, social stigmas, and taboos. Menstrual practices are below average in terms of hygiene in rural areas as compared to urban settings. Sixty-eight percent of the rural region population are still living deplorable life. More than 71.4% of females are still using unhygienic methods like cloth during menstruation. Other products are still uncommon and unaffordable for them. Andhra Pradesh, Goa, Haryana, Himachal Pradesh, Kerala, Punjab, Tamil Nadu were excellent in terms of menstrual product usage among rural females. In these states, rural area living females used sanitary napkins more as compared to the cloth, which is the reflection of the effectiveness of awareness and training efforts by various groups like public, private and social. Other states need serious threadbare analysis regarding menstrual product usage differences.

NFHS-4 has declared education as the essential element in improving hygiene and changing preference for hygienic menstrual products among rural living females. The facts highlighted that the level of hygiene among

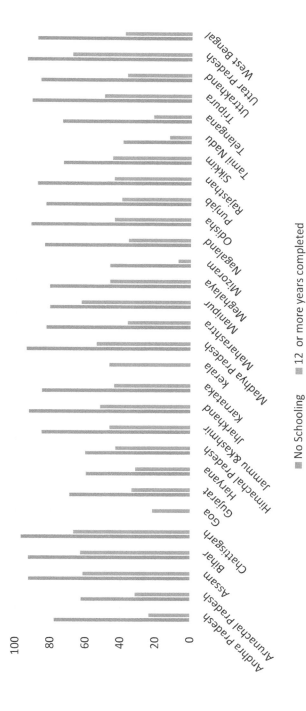

FIGURE 26.12
Menstrual product usage (cloth). (From NFHS 2015–2016 (Synthesized by the Authors)).

686 • *Social and Sustainability Marketing*

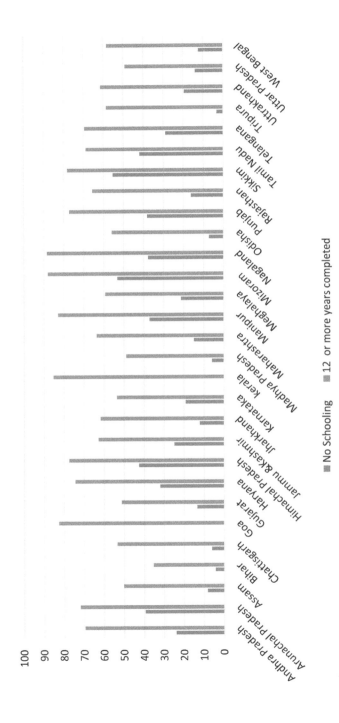

FIGURE 26.13
Menstrual product usage (sanitary napkins). (From NFHS 2015–16 (Synthesized by the Authors)).

females increases with the rise in education level from no schooling to < 5 years, 05–09 years, 09–11 years, and more than 12 years in all the states of India. In every state of India, females with some primary education have preferred using sanitary napkin over a cloth. Therefore, it is education and awareness that has helped the females to improve their menstrual hygiene standards and not their age. Thus, the government must establish a system to interconnect the education with menstrual awareness, which will result in enriching the female status in rural areas. The MHM schemes and policies are calling for a significant amendment and demands for keeping primary education as the foremost factor to improve the mental, physical, and social tallness of females and society both.

Other than Andhra Pradesh, Goa, Haryana, Himachal Pradesh, Kerala, Punjab, Tamil Nadu, in other states, the cloth is a primary menstrual absorbent leading to a low level of hygiene and making females prone to infections. A low level of hygiene is common among rural females. However, with a rise in education, the type of menstrual product usage and hygiene level goes up (Sommer, 2009). There are various ministerial departments where the issues of MHM addressed, but unfortunately, the ground realities demonstrated in the NFHS survey is saddened.

NGOS AND CORPORATES: A BOON ON GOI

The mission has gained momentum among international organizations, government, small and medium-sized enterprises and NGOs to address problems related to menstrual health. Several MHM campaigns by corporates and NGOs are working for better MHM practices in different manners. These campaigns have supported and enhanced the efforts put in by the GOI. For instance, the real 'Pad Man' of India, Arunachalam Muruganantham, founder and owner at Jayashree Industries, has contributed by installing mini-machines, which can manufacture sanitary pads for less than a third of the cost of commercial pads, in 23 of the 36 states and Union Territories of India. The organization not only works for producing low-cost sanitary napkins but is also employing many females as they earn more than Rs. 2000 every month. Organizations also promoted MHM through short movies and some Oscar-winning short documentaries. With the establishment cost of Rs. 5 lacs, the unit has the objective of making low-cost sanitary pads and producing. Similarly, the country is blessed to have responsible corporates and NGOs. They are actively involved in responsible marketing of such products and working rigorously on improving female status in India (Tables 26.3 and 26.4).

More than 45 prominent MHM campaigns have played a significant role in bringing up the status of females in rural and urban slums. Objectives on which campaigns work are serving different aspects and different regions of the country. Corporates have undertaken menstrual-related programs as a part of CSR since the last decade. Organizations like Gandhigram (est. 1947), Gramalaya (est. 1987), CWS (est. 1992), Vatsalya (est. 1995), and Goonj (est. 1998) have journeyed MHM since independence. There are campaigns of the new world, adopted the latest practices and methods to promote MHM across and outside the country too. NGO's like Aakar, Aaina, Amari foundations have taken MHM to new heights by getting into E-learning, launching mobile apps, presenting MHM awareness through videos, announcing 'period leaves' for females, etc. Campaigns in India are working on 12 different parameters: (a) understanding, (b) donation drives of menstrual kit, (c) manufacturing low-cost sanitary pads, (d) environmental concerns, (e) safe disposals, (f) E-learnings, (g) training, (h) taboos and silence, (i) WASH infrastructure development, (j) financial assistance (fair pay, period leaves, lowering taxes on menstrual products, developing entrepreneurship skills, help to leverage govt. schemes), (k) supporting government departments and SHGs, and (l) school absenteeism during menstruation.

Corporates associated campaigns majorly worked on advocating breaking taboos and silence (52%) and awareness (47%). Other parameters followed were conducting donation drives, manufacturing low-cost sanitary products, and WASH infrastructural development. Other significant parameters like vending machines and incinerators, environmental concern, E-learning (10%) have received the least attention. NGOs consider awareness as a significant task (76%), followed by conducting workshops and training programs (42%), advocating breaking taboos and silence (34%), supporting government departments, and SHGs (34%). Although NGOs have paid more attention to manufacturing low-cost sanitary pads (26%), environmental concerns (26%), and financial assistance (26%) as

TABLE 26.3

Corporate Associated Campaigns

Organization	Key Person/s	Headquarters and Website	Location/ Area Served	Establishment Type	Campaign	Objectives and Achievement of Campaign
Ace Group	Vijay Agarwal	Faridabad, India; www.ace-cranes.com	Worldwide	Private Company; 1995		• Spread awareness on MHM • Donating sanitary napkins across schools in Noida and Greater Noida
Assocham (one of the apex trade associations of India)	Niranjan Hiranandani (President)	New Delhi (India); www.assocham.org	Ahmedabad, Bengaluru, Kolkata	Non-governmental trade association; 1920	'Need to Break Silence and Build Awareness' (May 2019)	• In association with sanitary napkin brand 'Whisper', it pledged to reach o • In 5 Crore girls by the end of 2022 under the flagship 'Mother-Daughter Menstrual Health & Hygiene Program.' • It has reached to more than 40 thousand schools in Pan India.
Balmer Lawrie & Co. Ltd. (A Government Of India Enterprise)	Scotsmen: George Stephen Balmer and Alexander Lawrie.	Kolkata; www.balmerlawrie.com	Kolkata	Government of India; 1867	'Menstrual Hygiene Management' (under Nirman Foundation)	• To install of sanitary napkin vending machine & incinerator in four schools of Kolkata • To promote through educative materials like the comic. • To deliver training programs in schools.

(*Continued*)

TABLE 26.3 (Continued)

Corporate Associated Campaigns

Organization	Key Person/s	Headquarters and Website	Location/ Area Served	Establishment Type	Campaign	Objectives and Achievement of Campaign
Chimp&z Inc (A Digital Marketing Agency)	Angad Singh Manchanda and Lavinn Rajpal (Co-Founders)	Mumbai; https:// www. chimpandzinc. com/	Mumbai, New Delhi and Toronto	Private Company; 2013	'We Don't Whisper. We Stay Free.'	• To create & advocate open conversations about the period to end the 'shame' associated. • Under the campaign, it Launched a video titled 'We Don't Whisper, We Stay Free';
Dharma Life (a social enterprise): in collaboration with Spain-based women's sportswear brand Believe Athletics	Gaurav Mehta (founder and CEO	Delhi; www. dharmalife.in	Delhi	Social Enterprise; 2009	'Wind Beneath Her Wings'	• To spread awareness about menstrual health. • To ensure every sale on the Website of 'Believe Athletics' educates one young girl about menstrual hygiene and provides her with a sanitary kit. • To date, it has educated over 2000 adolescent girls in various parts of rural India were provided with sanitary kits as part of a new campaign.

(*Continued*)

TABLE 26.3 (Continued)

Corporate Associated Campaigns

Organization	Key Person/s	Headquarters and Website	Location/Area Served	Establishment Type	Campaign	Objectives and Achievement of Campaign
Essar Global Fund Limited ('EGFL')	Shashi Ruia (Chairman) Ravi Ruia (Vice-Chairman) Prashant Ruia (Group CEO)	Mumbai; www.essar.com	Mumbai, Maharashtra	Private Company; 1969	'Essar Foundation' (under Kaustubh Sonalkar -CEO)	• To distribute free sanitary pads to women in Mumbai slums & Mumbai police. • To support 'Aanganwadi' toward educating girls • Using video and game-based learning to promote MHM. • It launched the Sahej app to promotes best menstrual hygiene practices under the 'Kavach A Movement'.
Unicharm India and Spark Minda Group in collaboration with Global Foundation hunt	Takahisa Takahara (President and CEO) (Foundermembers- Mr. Praveen Karn/Ms. Radhika Ralhan)	Mita, Minato, Tokyo, Japan; http://www.unicharm.co.jp/english/index.html	Worldwide	Public (K.K); 1961	'Bringing Empowerment to Women'	• To have open dialogs on MHM Awareness, especially among the semi-urban women of Delhi NCR.
Himalaya Drug Company, in collaboration with the Rotary Club of Bombay	Shailendra Malhothra (Global CEO, Rest of World), [1] Philipe Haydon (CEO, India)	Bengaluru, Karnataka, India; www.himalayawellness.com	Worldwide	Private Company; 1930	'Menstrual Hygiene Campaign'	• To make sanitary pads and make them available to girls at an affordable price of Rs 2. • To install sanitary pad dispensers in schools and colleges across Mumbai.

(Continued)

TABLE 26.3 (Continued)

Corporate Associated Campaigns

Organization	Key Person/s	Headquarters and Website	Location/Area Served	Establishment Type	Campaign	Objectives and Achievement of Campaign
Kent RO	Mahesh Gupta (Founder)	Noida, India; www.kent.co.in	Worldwide	Private Company; 1999	'Kadam Sapno Ki Aur' (March 2017 under the leadership of Sunita Gupta)	• To Promote MHM awareness among male and female both. • To support the manufacturing of reusable and washable and hygienic hand-stitched sanitary pads; distribution of these pads to 'Yuvati' foundation for weaker sections of society.
Navigant' (The software company)	Julie M. Howard (CEO)	Chicago, Illinois, US; www.navigant.com	Worldwide	Subsidiary; 1983	'Padaid' (February 2018)	• To run the sanitary napkin donation drive and hand over the collection to 'Sakhi', an NGO working for women.
NDTV in collaboration with Dettol	Prannoy Roy (Co-Chairperson) Radhika Roy (Co-Chairperson)	New Delhi, India; www.ndtv.com	Mass media India	Public; 1988	'Banega Swasth India' (October 2019; Brand ambassadorship-Mr. Amitabh Bachchan)	• To help in building smart toilets in rural areas. • To conducts clean-up drives and create awareness on WASH across India.
Shudh Plus Hygiene Products, Niine Sanitary Napkins	Amar Tulsiyan (Founder & Director), Sharat Khemka (Co-Founder), Gaurav Bathwal (Co-Founder)	Gurgaon, Haryana; http://www.niine.com	Pan India	Partnership Firm; 2016	'Niine Movement-Menstrual Awareness campaign & Run4Niine' (2018)	• To educate on taboos and stigma attached • To enhance WASH facilities • To ensure availability of sanitary napkins during emergencies like lockdown or strikeouts

(Continued)

TABLE 26.3 (Continued)

Corporate Associated Campaigns

Organization	Key Person/s	Headquarters and Website	Location/ Area Served	Establishment Type	Campaign	Objectives and Achievement of Campaign
Nobel Hygiene	Kartik Johari (vice-president), KK Johari (MD)	Mumbai, Maharashtra; https://www.nobelhygiene.com/	Nashik Mumbai	Private Company; 1999	'Nobel Hygiene Campaign.'	• To tackle the taboos on periods and menstrual hygiene. • To create awareness on pressing health issues.
Paree Sanitary Napkins (Soothe Healthcare Pvt. Ltd.)	Sahil Dharia (Founder)	New Delhi; http://www.soothehealthcare.com	Pan India	Private Company; 2012	'#SheFirst #PadsAreEssential campaign'	• To Create Awareness for MHM. • To promote and create an identity by being the first-ever human resource landmark announces 'Period Leave' for all its female employees.
Procter & Gamble (Whisper in collaboration with Leo Burnett	David S. Taylor (CEO) Jon R. Moeller (COO) Chetna Soni, Category Leader, P&G Indian sub-continent)	Cincinnati, Ohio, United States; www.in.pg.com	Worldwide (except Cuba and North Korea)	Public; 1837	'#TouchThePickle' (2014)	• To spread awareness on taboos on mensuration. • To create awareness for ensuring that the girls do not drop the school. • It has launched a film that creates awareness 'girls across India dropping out of school' once they hit puberty • It pledged to reach out to 5 crore girls by 2022 for educating on MHM.
					KeepGirlsInSchool Mobileshaala	• To promote phone-based learning on menstrual hygiene education

(*Continued*)

TABLE 26.3 (Continued)

Corporate Associated Campaigns

Organization	Key Person/s	Headquarters and Website	Location/ Area Served	Establishment Type	Campaign	Objectives and Achievement of Campaign
Shomota Women Care Private Limited 'Partners with Luke'	Madhuri Shaha, Deepak Pal, Meghan Norean (Directors)	Kolkata; info@shomota.com	Pan India	Privately Held; 2014	'Menstrual Hygiene Campaign'	• To fight for taboos, silence, period shaming, and filling the earth with trash. • To create awareness on ecofriendly MHM. • To ensure fair pay to the women manufacturing menstrual products.
SINORA (the makers of PeeBuddy)	Deep Bajaj (Founder);	New Delhi, Delhi NCR; http://thesirona.com/	Delhi NCR	Privately Held; 2015	'fundraiser campaign for menstrual hygiene.'	• To promote sustainable environmental practices by developing and distributing biodegradable pads. • To cater to financial assistance from groups/individuals/ organizations to help as many girls and women as possible.

(Continued)

TABLE 26.3 (Continued)

Corporate Associated Campaigns

Organization	Key Person/s	Headquarters and Website	Location/ Area Served	Establishment Type	Campaign	Objectives and Achievement of Campaign
Tata Groups	Natarajan Chandrasekaran (chairman)	Bombay House, Mumbai, Maharashtra, India; www.tata.com	Worldwide	Private; 1868	Tata Water Mission initiative (June 2014)	• To promote safe and effective menstrual hygiene management in around 900 villages. • To build a conducive socio-cultural atmosphere for females to manage menstruation • To ensure women have clean absorbents and a basket of related products.
Versatile Enterprises (P) Ltd	Late Shri O.P Seth and Late Shri D.V Seth (Founder)	Ludhiana, Punjab: www.versatilegroup.in	Germany, Denmark, U.K., and southeast Asia	Privately Held; 1974	'Happy Girl Amodini'	• To create awareness for breaking the menstrual taboos. • To spread knowledge of myths and beliefs. • To conduct workshops by a panel of gynecologists.

Source: Compiled by Authors.

TABLE 26.4
NGO's Associated Campaigns

NGO (Campaign If Any)	Key Person/s	Headquarters/Website	Location/Area Served	Establishment Type and Year	Objectives/Achievements
Aaina (In collaboration with Water Aid)	Sneha Mishra	Odisha; www.aaina.org.in	Bhubaneshwar (Odisha)	Non-profit; 1998 (MHM activities 2013)	It currently covers more than seventy villages in Odisha. • To focus on spreading awareness about acceptable MHM practices. • To networks with family members, schools, ASHA/ANM/Anganwadi employees, teachers • To make a platform for delivering lectures, IEC materials, demonstrating facts and videos • It works with 130 Kishori manuals.
Aakar Innovations Pvt. Ltd	Somebody Ghosh	New Delhi; http://www.aakarinnovations.com	Pan India	Social Enterprise; 2011	• To manufacture and sell eco-friendly sanitary napkins at a low cost using SHGs at various local points. • To generate awareness in Bihar, Gujarat, Maharashtra, Karnataka. (mainly Pan India). • To date, it has enabled 25,000 females to get benefited from such kind of products each month.
Amari Foundation	Pallavi Arya and Amandeep Gautam tax	Faridabad(India); https://www.amarifoundation.in/	Across India	Non-Profit; 2018	• To educate young girls and their families about acceptable MHM practices. • To break the taboo and remove myths surrounding. • To donate menstrual kits of one year with booklet – 'Sakhi.'

(Continued)

TABLE 26.4 (Continued)

NGO's Associated Campaigns

NGO (Campaign If Any)	Key Person/s	Headquarters/Website	Location/Area Served	Establishment Type and Year	Objectives/Achievements
Centre for World Solidarity (CWS)	Dr. G. Prakasam	Secunderabad; http://www.cwsy.org	Pan India	Non-Profit; 1992	• To provide training on manufacturing, funding, and capacity building support for better MHM. • To conduct donation drives.
EcoFemme (Pad for a Pad)	Kathy Walking, Anbu Sironmani	Auroville; www.ecofemme.org	Tamil Nadu	Non-Profit; 2010	• To promote the use of eco-friendly pads by 'kudumbashree' groups. • To provide pads at discounted rates for rural women. • To conduct sessions in various and versatile organizations like offices, orphanages, colleges, yoga centers across Chennai
Epic Humanitarian World Tour	Venkaiah Naidu/ Manushi Chillar	New Delhi	Pan India	2018	• To promote the importance of hygiene among females.
Gandhigram Trust	M.R. Rajagopalan	Gandhigram; www.gandhigram.org	Tamil Nadu	Non-Profit; 1947	• To promote and conduct water and sanitation programs. • To conduct rural development activities and awareness sessions. • It has reached thousands of schoolgirls and women with the help of SHGs in Gandhigram and also worked along MNREGA to improve village sanitation infrastructure.
Goonj	Anshu Gupta	Delhi; http://www.goonj.org	Pan India	Non-Profit; 1998	• To manufacture and distribute low-cost menstrual products named 'My Pad' in remote areas. • It distributed more than two million such reusable and low-cost sanitary pads. It has dispatched 260,000 napkins. It also processed over one thousand tons of old recyclable material.

(Continued)

TABLE 26.4 (Continued)

NGO's Associated Campaigns

NGO (Campaign If Any)	Key Person/s	Headquarters/Website	Location/Area Served	Establishment Type and Year	Objectives/Achievements
Gramalaya	Sait Damodaran (The Toilet Man) and Sneha Shergill	Musri Town (Tiruchirappalli City)	Tamil Nadu	1987 (Under Indian Trust Act of 1882)	• To ensure females using 100% reusable sanitary pads in rural and urban slums. • To promote AWASH facilities and smart toilets. • It has educated more than six lac girls for MHM, established 40,000 incinerators across areas of Tamil Nadu.
Greentree (Pinkathon)	Shilpi Sahu and Sindhu Naik	Bangalore; https://www.greenthered.in/	Bangalore	Non-Profit; 2016	• To educate on sustainable MHM and menstrual products like menstrual cups and cloth pads. • It has delivered more than seventy workshops and lectures at various levels and places.
HappytoBleed	Nikita Azad,	Patiala, Punjab; NA	Pan India	Self Help Group; 2015	• To end the silence and taboos associated. • To fight for the 'Sabrimala Temple' movement (temple has a machine detecting menstruating female before entering the temple)
Healing Fields Foundation	Mukti Bosco	Hyderabad; http://www.healing-fields.org	Pan India (Community Health Facilitator (CHFs) program)	Non-Profit; 2000	• To provide training on health issues and government schemes on health financing. • To conduct activities for spreading MHM awareness. • It is actively engaged in sanitary napkin production and distribution across Bihar and Andhra Pradesh and has enabled over two lac women to get sanitary napkins at an affordable cost.

(Continued)

TABLE 26.4 (Continued)

NGO's Associated Campaigns

NGO (Campaign If Any)	Key Person/s	Headquarters/Website	Location/Area Served	Establishment Type and Year	Objectives/Achievements
Indian Development Foundation (IDF) (Mission one million – Dignity project)	Dr. A. R. K. Pillai, Founder President	Mumbai; http://www.idf.org.in/contact-us.html	Multiple States	Charitable Trust under the Societies Registration Act of 1860 and Bombay Public Trusts Act of 1950; 2012	• To mentoring females • To raise awareness on stigma, taboos, and misunderstanding. • To ensure the availability of low-cost sanitary pads. • To promote safe disposal of menstrual waste. • It conducts donation drives and converts money into 'The Dignity Kit.'
Jayashree Industries	Arunachalam Muruganantham (a Padma Shri awardee)	Coimbatore, India; http://newinventions.in/	Pan India	Social Enterprise; 2009	• To make machines for producing sanitary napkin machines and thus help in making low-cost sanitary pads. • It creates change agents for spreading awareness and reach poor communities through SHG. • It helps to leverage government schemes across 14 states in India. • To work for lowering the taxes on sanitary napkins so the usage and production both can go up.
Kasturba Gandhi National Memorial Trust	Padmavathi Pamarthy W	Hyderabad: http://www.kgnmthyd.org	Andhra Pradesh	Non-Profit; 2011	• To generate awareness for using sanitary pads by working with 'Anganwadi' workers and CDPOs so to reach rural areas. • To manufacture low-cost sanitary pads. • It has reached hundreds of schoolgirls and girls living in urban slums and has also worked for Integrated Child Development Scheme (ICDS).

(Continued)

TABLE 26.4 (Continued)

NGO's Associated Campaigns

NGO (Campaign If Any)	Key Person/s	Headquarters/Website	Location/Area Served	Establishment Type and Year	Objectives/Achievements
Muse (A Period of Sharing – Maasika Mahotsav)	Amritha Mohan and Nishant Bangera	Mumbai; https://musefoundation.in/about-us/	Mumbai, Ahmedabad, Uttarakhand,	Non-Profit; 2012	• To focus on environmental aspects of menstruation and safe disposal • 'Maasika Mahotsav' is celebrated under the flagship of NGOs by keeping various activities. • To bring the discussion of menstruation to the forefront. It has also conducted events like 'the happy girls' and 'a period of sharing' at Dahanu.
Myna Mahila Foundation, in association with Goodera (Meghan Markle)	Suhani Jalota	Mumbai; https://mynamahila.com/	slum redeveloped areas in Mumbai	2015	• To break and discuss aloud mensuration taboos like the chatty Myna bird. • To conduct workshops on manufacturing cost-effective sanitary pads. • To ensure girls do not miss schools due to mensuration. • To cater donations for providing free kits to the girls and mentor more on MHM.
Nirman foundation	Ashwini Mishra (Director), Syamali Mishra and Pradip Mishra (Board Of Trustee),	Kolkata; http://nirmanfoundation.org/	West Bengal	Non-Profit; 2009	• To increase awareness. • It works with NGOs and government bodies such as the PHED, PHRD, PNB Metlife India, etc. • To create MHM and acceptable practices among Girls in schools in Purulia district. • It has launched a short film on MHM Break the silence – A Step toward https://www.youtube.com/watch?v=b9FdgQCP334&t=22s

(Continued)

TABLE 26.4 (Continued)

NGO's Associated Campaigns

NGO (Campaign If Any)	Key Person/s	Headquarters/Website	Location/Area Served	Establishment Type and Year	Objectives/Achievements
Orikalankini	Dr. Sneha Rooh,	https://www.facebook.com/orikalankini/?ref=page_internal	Across India	Non-Profit (Social Media); 2017	• To promote awareness of MHM through social media campaigns.
Sahyog Foundation (Oxygen – Women Hygiene Drive)	Meher Sarid	New Delhi; http://www.sahyogfoundation.in/women-hygine-awareness	Rajasthan, Haryana, Uttar Pradesh, Odisha, and Delhi NCR slums.	2003	• To facilitate the programs on health and hygiene awareness associated with poor menstrual hygiene. • To work on the reduction of absenteeism of females from the workplace and schools. • Collaboratively with SHGs, NGO – Urjaw Energy, to set up low cost sanitary napkin manufacturing plants, which will also help in employing women employment, especially in Gorakhpur. • To date, it has conducted more than 318 camps and has benefitted more than 58,865 girls across.
Swayam Shikshan Prayog (SSP)	Prema Gopalan	Mumbai; http://www.sspindia.org	Multiple States	Non-Profit; 1993	• To establish the concept of 'Arogya Sakhi', i.e., a mentor for providing health counseling and menstrual product information on sanitary napkins. • It has developed a good network of more than 300 Arogya Sakhis across many villages in Maharashtra. • It also sensitized girls on nutritional status and MHM across 100 villages under a project funded by World Bank. • It also works on enhancing female entrepreneurship skills in a menstrual product like sanitary pads.

(Continued)

TABLE 26.4 (Continued)
NGO's Associated Campaigns

NGO (Campaign If Any)	Key Person/s	Headquarters/Website	Location/Area Served	Establishment Type and Year	Objectives/Achievements
The red cycle	Arjun Unnikrishnan	https://theredcycle.in/	NA	Volunteer-Based Organization; 2015	• To promote open discussion about mensuration and menstrual hygiene management.
UGER 'new beginning' by Jatan Sansthan	Lakshmi Murthy – Udaipur, Rajasthan	NA	Na	Na	• It ensures to empower women from lower socioeconomic groups by giving them training for doing business related to sanitary napkins to make them financially independent as well as making low-cost sanitary napkins. • To end the silence and taboos by delivering lectures across universities and colleges in rural and urban community groups.
Vasudha Vikas Sansthan (VVS)	Gayatri Parihar	Dhar (MP); www.vasudhavikassansthan.org	Madhya Pradesh	Non-Profit; 2000	• To generate awareness on better MHM practices, myths, and taboos on menstruation. • To ensure females get sanitary napkins, as it works on generating entrepreneurship skills in females for making pads using a small, low-cost machine developed by VVS. • It works on promoting the importance of WASH facilities; to date, it has ensured that 78 panchayats are thoroughly sanitized and open defecation-free and more than 50 self-help groups marketing pads that serve as a livelihood option for the women.

(Continued)

TABLE 26.4 (Continued)
NGO's Associated Campaigns

NGO (Campaign If Any)	Key Person/s	Headquarters/Website	Location/Area Served	Establishment Type and Year	Objectives/Achievements
Vatsalya (the 'Breaking Silence' program)	Nee lam Singh	Lucknow (UP); www.vatsalya.org.in	Uttar Pradesh	Non-Profit; 1995	• It also works with a network of Kishori Mandal and to effectively reach out to more than 8,500 girls and 5,000 women through MHM workshops and has enabled the construction of 2,700 toilets and 1,100 incinerators. • It works for 66 villages in Lucknow (Uttar Pradesh) on the 'Breaking Silence' program to promote MHM, WASH, construction of toilets, and incinerators. • It supports the community in leveraging government schemes such as Nirmal Bharat Abhiyan. • It has also worked to promote the programs like Increasing Awareness and Safe Sanitary Practices among Adolescent Girls (IAAAG) program implemented in partnership with P&G.
YesIBleed	Maneka Gandhi (Union Minister for Women and Child Development)	New Delhi	Pan India	2018	• To promote MHM on social media platforms. • To ensure the supply of inexpensive and eco-friendly sanitary pads. • To create awareness on the safe disposal of used products.

Source: Compiled by Authors.

compared to corporates. Donations drives (23%), E-learning methods (15%), WASH infrastructure development (15%), vending machines, and incinerators (11.5%) are still lagging. A severe issue of 'school absenteeism during menstruation' has received the least importance from both corporates and NGOs (5% and 7%, respectively).

MHM PROBLEMS IN RURAL INDIA: LEAST RESEARCHED

In India, there is a lacuna present when it comes to the studies related to menstrual practices or menstruation problems, specifically on rural or urban slum females. Till 2010 not much attention has been paid to topics like menstrual hygiene practices. However, in other countries like the UK, this ratio is slightly higher. Although after 2011, the subject has gained much popularity. It is a matter of concern, a country like India where three-fourth of the population lives in villages or urban slums. Still, the subject and community remain unexplored on such sensitive issues. States like Assam, Haryana, Maharashtra, Odisha, Uttarakhand, Puducherry, and Rajasthan had a fair amount of research work. Gujarat and Punjab with moderate and Goa, J&K, Jharkhand, Karnataka, MP, UP, Tamil Nadu, Telangana had few studies. In other states, periodic reviews were found on MHM exclusively in rural and urban slums (Table 26.5).

The present research works majorly focused on various combinations of 6 different parameters: a) menstrual product usage, b) awareness level, c) hygiene level, d) health level, e) restrictions, and f) open defecation/in-house sanitation and rare work on the open discussion. There is occasional empirical work present on policy-level and is another shocking fact (Sharma et al., 2020). No review has ever evaluated the impact of such policies on MHM.

ACTIONS TO OVERCOME MHM CHALLENGES

The objective of better MHM is possible through two aspects: the facilities, i.e., accessibility and affordability, and the second is familiarity, i.e., awareness on the subject. Facilities like access to the low cost and

TABLE 26.5

Characteristics of the Studies Done (*n* = 38)

S. No	Characteristics	Number %
1	**Year of Publication**	(*N* = 38)
	2011–2015	8 (21.05)
	2016–2020	30 (78.9)
2	**Region of Study**	
	Rural/Rural Slum	21 (55.26)
	Urban Slum	5 (13.1)
	Semi-Urban	4 (13.15)
	Combination (Rural/Urban/Tribal)	6 (15.78)
	Tribal	1 (2.6)
	General	1 (2.6)
3	**Region**	
	North India	19 (50)
	South India	6 (15.7)
	East India	4 (10.5)
	West India	5 (13.15)
	Central India	1 (2.6)
	India	4 (10.5)
4	**Method of Data Collection**	
	Interview	8 (21.05)
	Questionnaire (Semi structured/structured)	23 (60.5)
	Mix Method	5 (13.15)
	Case/review based	2 (5.2)
5	**Research Domains***	
	Product Usage	26 (68.4)
	Awareness	18 (47.3)
	Hygiene Level	6 (15.7)
	Health Issues	8 (21)
	Restrictions	8 (21)
	Open Defecation/In House Sanitation	5 (13)
	*List Available	
	**as one study covers more than one aspect/method of data collection/regions therefore the total can be greater than 100%	

Source: Compiled by Authors.

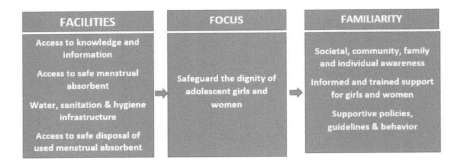

FIGURE 26.14
Effective menstrual hygiene management. (From Vikaspedia, 2019).

ecofriendly menstrual products, especially sanitary napkins, as they are the most preferred among females in India irrespective of their residential status (Vikaspedia, 2019). In the absence of such availability, knowledge, and awareness remain unused. Another most important aspect i.e., familiarity, majorly includes the availability of sufficient readable material or any mode which can be demonstrated to the females to enhance their knowledge (Figure 26.14). Therefore, the following action plan is needed to achieve the core objectives

- **Awareness:** Major research works revealed a lack of awareness in rural and urban slum populations of the country (Garg & Anand, 2015; Deshpande et al., 2018; Gupta, 2019). Due to a lack of awareness, the hygiene levels go down and raise much concern on the topic. Females are not familiar with the in-depth knowledge of menstruation. The reappearance of MHM contents through different ways will act as a catalyst and will help to develop more awareness. Adding MHM in school-level education and advancement through integrated methods will work in MHM. Along with school staff, SHG's, NGOs, AWW, and ASHA, karyakartas can bridge the gap between menstrual knowledge and adolescents. Both genders should know menstrual hygiene looking at the long-term benefits from the same. It is time to involve males in the entire information dissemination to have more sensitization on the topic. It is a must to note that IEC material needs to be in local languages for higher impact.
- **Accessibility and affordability:** NFHS results state failure of MHM practices in India, i.e. the lack of accessibility to the better and

affordable menstrual product. There are various glitches present in the use of different menstrual products (see 'Menstrual Product Glitches'). Majorly the policies and schemes are focusing more on awareness rather than accessibility. Other than this, the schemes are also depending on the sponsorships for raising funds. The government must make it mandatory for big corporates to make MHM as part of their CSR. The government must monitor NGOs for exerting a high level of effort toward ensuring the accessibility of menstrual products to adolescent girls. Corporates and NGOs focus on breaking the taboos and creating awareness, respectively, which is already being done by several schemes but unfortunately, it is all overlapping. India needs many more initiatives by responsible groups. However, many commercial brands are existing but are beyond the affordability of rural and slum population.

- **Tax exemption:** Sanitary napkins may not be essential in the list of the commercially sold product, but it is an utmost vital product for a female. Therefore, instead of burdening it with taxes and duties, the government must lower taxes on sanitary napkins. To date, the government has worked superficially on removing the tax burden. However, the central part of the tax still exists on inputs and other logistics part. The government must count sanitary napkins as life-saving instead of essential

- **Safe disposal infrastructure:** Many NGOs and corporates have taken footsteps, ensuring to increase awareness, but environmental concerns related to the safe disposal of menstrual waste is a big gap existing in India. However, in many schemes, safe disposal is a primary concern, but the concern needs to grow from policy to its implementation. 'Menstrual waste bins,' segregation of menstrual waste (at home) for contactless disposal., pits near toilets is need of the hour. Moreover, Indian toilets need incinerators devices which is one of the safest methods of managing waste. Responsible groups like corporates, NGOs and SHG's must focus on arranging for such devices. It is essential to understand that the availability of infrastructure will lead to awareness and vice versa. Lack of sanitation infrastructure and disposal facility at the village level leads to other infections and open defecation (Sommer, 2010) as elaborated by Vikaspedia that water, sanitation, and handwashing facilities, which contribute a lot to MHM. The country is still lagging in providing this infrastructure to the females.

- **More funded research on the policy-level:** The impact of various schemes on females residing in rural and urban slums is still unknown. The current status of core issues is still under the veil and is visible only through ground level works. The research will reveal the actual gaps and lacunas and, therefore, will act as the path creator for policymakers. It is a must for the government to promote more research in MHM exclusively. The GI must announce various funds and grants solely for research on the MHM domain.
- **Rewards and incentives:** The rewards at the state level and district level for encouraging better performance can make differences. Individuals can be motivated by particular incentives for working as volunteers in MHM programs at village levels. Rewards and incentives will help the villages to compete and perform better. Simultaneously, the reward system will resolve the issue of a lack of human resources faced by GOI.
- **Projects in higher education:** Research-based social projects/internships mandatory at the college level for students irrespective of their domain will help the GOI to get the research team and to develop socially responsible citizens of the country.
- **Inclusion in National Education Policy (NEP) 2020**: MHM has always been part of NEP (Sommer et al., 2017). Therefore, MHM inclusion in the latest NEP 2020 is essential but on a broader level. NEP must make it mandatory for schools, institutes, and universities to provide WASH infrastructure with great sensitivity. As mentioned by UNICEF/WHO MHM friendly toilets are essential to be included vigilantly in NEP 2020.
- **Encouragement to entrepreneurs:** The GOI must focus on attracting more entrepreneurs for menstrual products. Adding the product to the list of special economic zones will help the entrepreneur gain more advantages subject to no import of inputs. As per the finance ministry, the firm not importing inputs has no obligation for exports. Therefore, the benefits of SEZ can do wonders for entrepreneurs.
- **Government partnering with corporates, NGOs, and SHG's:** Like other developmental programs, MHM also needs the collaborative efforts but by way of a partnership between government, corporates, NGOs, and SHG's to have a synergy effect and utilize the available resources properly

- **Convergence at various ministerial level:** MHM is not associated with health-related ministries; therefore, all the ministries must come together and a single framework to avoid clashes and redundancy. Moreover, monitoring will become easier. Departments like the Ministry of Environment and Forest and the Ministry of Urban Development must take part in MHM, former for waste management and later for urban slums.

COGNITIVE DISSONANCE: A DILEMMA

The GOI has a vast area of operation wherein MHM is just a sub-function and not an exclusive area to work. Therefore, at the time of fund allocation, no special privilege is given to MHM. From the allocated fund, the ratio between various activities is not precise. The ministries are into ethical dilemmas associated with MHM practices. For instance, up to what degree the use of sanitary napkins be promoted so that the environment is on the safer side when it comes to the disposing of the menstrual waste.

On the other hand, awareness and usage are correlated to each other but more depending on the education level of females and the income of the family. Inline, manufacturing low-cost sanitary napkins may not be much attractive for big giant corporates, as it can harm their other superior products. India, the second-largest country population-wise, where SHG's, NGOs, small and micro enterprises neither can fulfill the demand nor can achieve economies of scale by giving a handful of production. Government with vast operational areas with limited schemes and programs is not able to unravel the situation. Other stakeholders (see 'Stakeholders in MHM') are in an ethical dissonance to pick which dimension of MHM practices for serving. Therefore, they end up with the initial step of generating awareness. Stakeholders need to comprehend that understanding in isolation cannot work. It needs to be assisted by the availability of economic, sanitary napkins. The vicious circle of dilemma goes on.

While listing the taboos associated with menstruation, some taboos are religious-oriented, like not entering the temple or attending religious events. It, at times, hurt the emotions of citizens hence could be rejected. The situation seems to be quite puzzling, as the time demands a serious redo but the road map to this redo is a blur. Here the 'value approach theory' needs to be adopted to choose the alternate which offers more good or less wrong.

IMPLICATIONS

The case exhibits the various efforts taken by the government for improving the MHM practices across the country. It is not only the government, but various NGO's and SHG's also have to come forward for supporting the objective of the government. The action plan suggests different alternates for the government to be kept in mind while designing the policy framework for MHM.

Managerial implications: As mentioned in 'Stakeholders in MHM', multiple stakeholders gain from such programs. Menstrual product marketers, especially those in sanitary napkin manufacturing units, can be highly satisfied by the lacuna present in the MHM programs. The case has highlighted the unavailability of affordable sanitary napkins and poor sanitation as a robust understanding of the poor MHM practices in India. In rural India, the lack of cheap and economic sanitary napkins forces females to adopt unhygienic practices (Kaur et al., 2018). The marketers can develop a product at a bare minimum cost to reach the below poverty segment. There are possibilities of a new segment for the marketers to serve, which helps to improve their market share and help to develop a favorable brand image (Haque & Kumar, 2018). Biodegradable pads, incinerators can be another new product, which is the need of the hour for a country like India. This need provides marketers with a golden untapped segment and gains popularity as socially responsible marketers. The gap present in the MHM practices provides marketers with an opportunity to have a broader scope of CSR activities like collaborating with NGOs, SHGs, donation drives, sponsoring events, and conducting free workshops & sessions for rural and urban slum females. It is a must to note here that the MHM needs more movements related to donation than generating awareness. Therefore, corporate is the position to suffice the need will win the game. CSR not only helps in enhancing brand image but will also help to position the upcoming products in customers' minds. Looking at the promotional aspects, just by recreating the promotional content on media, the marketers can help the society to come out of logjams (mentioned in 'MHM Programs in India: A Lacuna') and to rethink menstruation in a new and more scientific way. Although overcoming such a stereotyped mentality would be extremely challenging, but in turn, such marketers can become real heroes in society. To date, no sanitary napkin has ever focused on taboos associated other than physical restrictions like playing sports, etc. in their ads. Besides, Indian marketers can collaborate with celebrities.

Research implications: This case identifies the gaps at the policy level, and thus, there is a scope for research organizations, students, and scholars to thinking toward research on steps taken by the government and its effect on improvements in MHM practices. Along with the policy-level analysis, the contribution of various campaigns is a subject to study. Identifying the orientation of movements is still a gap for investigation. It is clear from the NFHS-4 survey that efforts by corporates, NGOs, and SHGs are not enough. Therefore, the content can be utilized further for extending the research work in the states where there is no or less research work is done. Other menstrual product usage remained unexplored, which is a significant area to be studied in the future. Along with this many developed cities can be analyzed to identify differences existing.

CONCLUSION

In this study, stakeholders are streamlining the efforts to establish better MHM practices in the country. However, the policies and programs run by the government do not suffice to the problems existing. The support from corporates, NGOs, and SHGs is limited to a few geographical regions with rare exceptions. There are evidence and research which has proved the declining status of females in rural and urban slum due to economically underprivileged population. The case emphasizes identifying appropriate and feasible approaches so that it can benefit the economically disadvantaged community. India needs many more 'Padman' like Arunachalam Muruganantham to help the nation overcome serious and often under-addressed issues like poor MHM. Nevertheless, the government must sustenance the efforts of such 'Padmans' by redesigning the policies and entrepreneurship framework keeping dual aspects of the growth of such firms for employment generation and health improvement of rural and urban slum females in India.

QUESTIONS

1. Suggest capacity building measures for policy actors.
2. 'Putting policies or programs on the charge is like looking at the one side of a coin and not reading between the lines.' Justify.
3. What do you think is the root cause of the issues presented in the case? How can it be solved keeping the 'Value approach' theory in mind?

EXERCISE: TEAM TIME

Use the power of social media promotion

> Let' come together, join hands, and be a responsible citizen of a country.
> Let's design a video message that can change social perception.
> Lets' call it 'Mensuration is normal'.
> Let this go viral on social media

Remember: Together we can bring this change.

DID YOU KNOW

- Just like we celebrate other days, we also have 'World Menstrual Hygiene Day' celebrated globally on May 28
- A female spend nearly ten years of her life on her period
- Mary Beatrice Davidson Kenner, a female African-American inventor, patented the sanitary belt in 1956, the first product featuring an adhesive to keep the pad in place.
- 23 million girls in India dropout of school annually because of the lack of menstrual hygiene management facilities,
- Only 36% of females use sanitary napkins in India

DEFINITION

House et al. (2012) *defined menstruation as 'an integral and normal part of human life, indeed of human existence, and menstrual hygiene is fundamental to the dignity and wellbeing of women and girls and an important part of the basic hygiene, sanitation, and reproductive health services to which every women and girl has a right.' (p. 8)*

TERMINOLOGIES THAT ARE NEW TO YOU

Metrorrhagia (irregular menstruation)
Oligomenorrhea (light menstruation)
Polymenorrhea (cycles with intervals of 21 days)

AUTHORSHIP STATEMENT

We, at this moment, undertake that all authors have participated sufficiently in work and has taken responsibility for the content and revision of the manuscript. Furthermore, we declare that this material or similar material is not published. This work will not be submitted to or published in any other publication before its appearance in the book entitled 'Social and Sustainability Marketing: A Casebook for Reaching Your Socially Responsible Consumers through Marketing Science.'

AUTHORSHIP CONTRIBUTIONS

Category 1
Conception and design of the study: Dr. Sneha Rajput
Acquisition of data: Dr. Sneha Rajput, Ms. Pooja Jain
Analysis and interpretation of data: Ms. Pooja Jain

Category 2
Drafting the manuscript: Dr. Sneha Rajput, Ms. Pooja Jain

Revising the manuscript critically for important intellectual content: Dr. Sneha Rajput

Category 3
The names of all authors
Dr. Sneha Rajput, Ms. Pooja Jain

ACKNOWLEDGMENT

We are thankful to Mr. Jishnu Bhattacharya for his continuous guidance and motivation to complete this work. We sincerely acknowledge the mentorship of Dr. S.S. Bhakar (Director Prestige Institute of Management, Gwalior) and Prof. Manoj Das for teaching the core concept of the case at various platforms. I am also thankful to Prof. Sathyaprakash Balaji Makam for giving guidance with brilliant suggestions to revise the topic. We are grateful to Dr. Neelam Rajput Verma (Sr. Gynecologist, Associate Professor, GRMC, Gwalior, MP) for guiding us on the subject and Dr. G.S Rajput (Ex-Director State Institute of Health Management and Communication, Gwalior) for his valuable guidance. We extend our sincere gratitude to Prof. Rachit Jain (Assistant Professor, ITM University, Gwalior) for his tremendous contribution to manuscript editing and formatting.

CONFLICT OF INTEREST

Authors of the paper had no conflict, neither financially nor academically.

REFERENCES

Ballys, E. (2017). Menstrual hygiene management: Policy brief. *Share*. 1–7. www.shareresearch.org

Ben Amor, Y., Dowden, J., Borh, K. J., Castro, E., & Goel, N. (2020). The chronic absenteeism assessment project: Using biometrics to evaluate the magnitude of and reasons for student chronic absenteeism in rural India. *International Journal of Educational Development*, 72, 102140. https://doi.org/10.1016/j.ijedudev.2019.102140

CensusInfo India. (2011). *Demography*. Census of India 2011 (Final Population Totals). https://censusindia.gov.in/2011census/censusinfodashboard/index.html

Collins, T. (2005). Health policy analysis: A simple tool for policy makers. *Public Health*, *119*(3), 192–196. https://doi.org/10.1016/j.puhe.2004.03.006

Deshpande, T.N., Patil, S.S., Gharai, S.B., Patil, S.R., & Durgawale, P.M. (2018). Menstrual hygiene among adolescent girls – A study from urban slum area. *Journal of Family Medicine and Primary Care*, *7*, 1439–1445.

Druet, A. (2017, September 8). *How did menstruation become taboo?* HelloClue. https://helloclue.com/articles/culture/how-did-menstruation-become-taboo

Garg, S., & Anand, T. (2015). Menstruation related myths in India: Strategies for combating it. *Journal of Family Medicine And Primary Care*, *4*(2), 184. https://doi.org/10.4103/2249-4863.154627

Geertz, A., Iyer, L., Kasen, P., Mazzola, F., & Peterson, K. (2016). Menstrual health in India: Country landscape analysis. *Reproductive Health Matters*, *51*(4), 1–25. https://doi.org/10.1016/S0968-8080(13)41712-3

Gera, N. (2019, September 27). *Menstrual hygiene: A challenging development issue*. DownToEarth. https://www.downtoearth.org.in/blog/health/menstrual-hygiene-a-challenging-development-issue-66973

Gupta, S. (2019). An empirical study of managing menstrual hygiene in schools (A special reference to Government Upper Primary Schools in District Sambhal (Uttar Pradesh). *Integrated Journal of Social Sciences*, *6*(2), 39–43. http://pubs.iscience.in/journal/index.php/ijss/article/view/887

Haque, A. U., & Kumar, A. (2018). Communication strategies for raising awareness about menstrual hygiene at the bottom of the pyramid. *International Journal of Advanced Multidisciplinary Research*, *5*(6), 45–55. https://doi.org/10.22192/ijamr

House, S., Mahon, T., & Cavill, S. (2012). Menstrual hygiene matters: A resource for improving menstrual hygiene around the world. *Reproductive Health Matters*, *21*(41), 257–259. https://doi.org/10.1016/S0968-8080(13)41712-3

IMARC. (2020). *Indian sanitary napkin market research report 2020–2025*. IMARC Group. https://www.imarcgroup.com/indian-sanitary-napkin-market

India Today. (2019, June 20). No access to menstrual hygiene is the fifth biggest killer of women in the world. *India Today*. https://www.indiatoday.in/education-today/gk-current-affairs/story/no-access-to-menstrual-hygiene-fifth-biggest-killer-of-women-in-the-world-1552450-2019-06-20

Joy, K. J., & Bhagat, S. (Eds.). (2016, December). *Right to sanitation in India: Nature, scope and voices from the margins*. Forum for Policy Dialogue on Water Conflicts in India, Pune, https://www.soppecom.org/pdf/Right-to-sanitation-in-India-Nature-scope-and-voices-fro-the-margins.pdf

Kaur, R., Kaur, K., & Kaur, R. (2018). Menstrual hygiene, management, and waste disposal: Practices and challenges faced by girls/women of developing countries. *Journal of Environmental and Public Health*, *2018*, 1–9. https://doi.org/10.1155/2018/1730964

Lufadeju, Y. (2018, May 25). *FAST FACTS: Nine things you didn't know about menstruation*. Unicef.Org. https://www.unicef.org/press-releases/fast-facts-nine-things-you-didnt-know-about-menstruation

Nordqvist, C. (2016). Menstruation and the menstrual cycle. *Medical News Today*. https://doi.org/10.1109/ATEE.2017.7905023

Rahman, S., Islam, H., Rodrick, S. S., & Nusrat, K. (2018). The role of media in creating social awareness about the female hygiene practices during menstruation cycle in Bangladesh. *IOSR Journal of Business and Management*, *20*(5), 4–15. https://doi.org/10.9790/487X-2005030415

Ramchandran, R. (2019, September 25). *Why do government efforts to ensure menstrual hygiene focus on sanitary napkins?* The Wire. https://thewire.in/women/why-do-government-efforts-to-ensure-menstrual-hygiene-focus-on-sanitary-napkins

Sharma, S., Mehra, D., Brusselaers, N., & Mehra, S. (2020). Menstrual hygiene preparedness among schools in India: A systematic review and meta-analysis of system-and policy-level actions. *International Journal of Environmental Research and Public Health, 17*(2), 647. https://doi.org/10.3390/ijerph17020647

Sommer, M. (2009). Where the education system and women's bodies collide: The social and health impact of girls' experiences of menstruation and schooling in Tanzania. *Journal of Adolescence, 33*(4), 521–529.

Sommer, M. (2010). Putting 'menstrual hygiene management' into the school water and sanitation agenda. *Waterlines, 29*(4), 268–278. http://dx.doi.org/10.3362/1756-3488.2010.030

Sommer, M., Figueroa, C., Kwauk, C., Jones, M., & Fyles, N. (2017). Attention to menstrual hygiene management in schools: An analysis of education policy documents in low- and middle-income countries. *International Journal of Educational Development, 57*, 73–82. https://doi.org/10.1016/j.ijedudev.2017.09.008

Vikaspedia, (2020). *Menstrual hygiene management*. Ministry of Electronics and Information Technology. https://vikaspedia.in/health/women-health/adolescent-health-1/menstrual-hygiene-management

FURTHER READINGS

Gramalaya. (2020). *Goodwill ambassador*. Gramalaya India. https://gramalayaindia.org/goodwill-ambassador/

Ministry of Drinking Water and Sanitation. (2012). *Handbook on scaling up solid and liquid waste management in rural areas*. Ministry of Drinking Water and Sanitation. https://swachhbharatmission.gov.in/sbmcms/writereaddata/images/pdf/technical-notes-manuals/Scaling-up-SLWM-in-Rural-areas.pdf

Ministry of Drinking Water and Sanitation Government of India. (2015, December). *Menstrual hygiene management*. http://www.ccras.nic.in/sites/default/files/Notices/16042018_Menstrual_Hygiene_Management.pdf

Ministry of Health and Family Welfare (MOHFW), Government of India. (2016). *NFHS-4 fact sheets for key indicators based on final data*. International Institute for Population Sciences (IIPS). http://rchiips.org/NFHS/index.shtml

Ministry of Women & Child Development, GOI. (2017, September). *Scheme for adolescent girls (SAG)*. https://wcd.nic.in/schemes/scheme-adolescent-girls-sag

NARI (National Repository of Information for Women). (2017, March). *Schemes for women*. National Informatics Centre (NIC). http://www.nari.nic.in/

National Health Mission, Ministry of Health and Family Welfare, GOI. (2020, May). *Menstrual hygiene scheme (MHS)*. National Health Mission Department of Health & Family Welfare, Ministry of Health & Family Welfare, Government of India. https://nhm.gov.in/index1.php?lang=1&level=3&sublinkid=1021&lid=391

National Health Portal. (2018, June 8). *Menstrual hygiene day*. https://www.nhp.gov.in/menstrual-hygiene-day_pg

NRMC. (2018). *2018 India: End line evaluation of GARIMA project in Uttar Pradesh (2013–2016)*. https://www.unicef.org/evaldatabase/index_103312.html

Rashtriya Kishor Swasthya Karyakram. (2014, January). *Operational framework: Translating strategy into programmes*. Adolescent Health Division Ministry of Health and Family Welfare Government of India. https://nhm.gujarat.gov.in/images/pdf/RKSK_Operational_Framework.pdf

Sharma, K. (2018). *Assessing the menstrual hygiene management practices in urban and rural areas of Madhya Pradesh*. Atal Bihari Vajpayee Institute of Good Governance and Policy Analysis. http://bhabha-coe.mapit.gov.in/aiggpa/uploads/project/Assessing_the_Menstrual_Hygiene_Management_Practices_in_Urban_and_Rural_areas_of_Madhya_Pradesh_(1).pdf

WaterAid. (2018). *Menstrual hygiene management in schools in South Asia*. UNICEF. https://washmatters.wateraid.org/sites/g/files/jkxoof256/files/WA_MHM_SNAPSHOT_INDIA.pdf

World Bank. (2015, December 15). *World Bank approves US$1.5 billion to support India's universal sanitation initiatives*. https://www.worldbank.org/en/news/press-release/2015/12/15/world-bank-approves-usd-1point5-billion-support-india-universal-sanitation-initiatives#:%7E:text=WASHINGTON%2C%20December%2015%2C%202015%20%2D,latrine%20with%20a%20focus%20on

Section XIII

Selected Case Studies to Reflect on Practice and Use as Learning Tools: Sustainability Marketing in the NFL

27

Sustainability Marketing in the National Football League (NFL): The Case of the Philadelphia Eagles

Jairo León-Quismondo
Universidad Europea de Madrid, Faculty of Sport Sciences, Madrid, Spain

CONTENTS

Learning Objectives	722
Themes and Tools Used	722
Theoretical Background	722
How Do Sports Influence Society?	723
Introduction	723
NFL	724
Context: History and Evolution	724
NFL by the Numbers	725
Marketing Strategy within the NFL	725
NFL Teams: The Case of the Philadelphia Eagles	729
The Beginning of a "Green Era"	730
"Go Green" Program	730
Summary of the Major Problem	732
Discussion	733
Conclusion	733
Lessons Learned	734
Discussion Questions	734
Activity-Based Assignment	734
Additional Content	735
Additional Readings	735
Credit Author Statement	736
References	736

LEARNING OBJECTIVES

After reading the chapter, the reader should be able to:

- Introduce the five main initiatives of the NFL regarding sustainability: solid waste management, material reuse, food recovery, sports equipment, and book donations, and greenhouse gas reduction
- Explore how NFL events embrace sustainable environmental principles and best practices, from the largest (Super Bowl) to smaller ones (regular season)
- Understand how professional sport teams can reach socially responsible consumers. For that purpose, the specific case of Go Green program by Philadelphia Eagles is analyzed

THEMES AND TOOLS USED

- Marketing mix (4ps strategy)
- SWOT

THEORETICAL BACKGROUND

Currently, consumers are willing to consume green products. In services, the scientific literature is not as extended as in products. However, an increase in the offer of green services and green events is evident.

An event that promotes environmentally friendly habits not only benefits from the reduction of the carbon footprint but also affects society. Through this theory, green events are considered influential for future behaviors of consumers.

Additionally, the sport industry is one of the most dominant in the world. Fans feel linked to their teams as they are part of them. Although sporting events are not yet totally embracing sustainable practices, some programs have started to run. The power of sporting events together with sustainable marketing can be a successful connection since fans from all over the world are potential socially and environmentally responsible consumers.

HOW DO SPORTS INFLUENCE SOCIETY?

Can a sport event influence the future behavior of fans? How important is the input they receive from the league or the team? This chapter explores the case of Go Green Program by Philadelphia Eagles. Their innovative venue management strategy together with the marketing strategies adopted is a reference in terms of sustainability marketing.

INTRODUCTION

The sports industry is one of the most influential businesses in the world. The great visibility that sporting events have acquired in the last two decades, makes this sector one of the most powerful in terms of marketing. Specifically, in the United States, the National Football League (NFL) is the professional league with the highest average attendance, with more than 66,000 spectators per game during the 2019 regular season.

The most important sport competitions in the world are the Big Four American leagues. The NFL, Major League Baseball (MLB), National Basketball Association (NBA), and National Hockey League (NHL) are the top professional leagues in the world by revenue, only matched by the English soccer league, the Premier League (Deloitte, 2020).

Several studies have explored the social outcomes of environmental sustainability initiatives (Trail & McCullough, 2020). Positive impacts have been demonstrated on the beliefs and behaviors of their fans both in their everyday actions and when attending sport events (Greenhalgh & Drayer, 2020).

Thus, the purpose of this chapter is to introduce the initiatives of the NFL regarding sustainability, to explore how NFL events embrace sustainable environmental principles and best practices, and to understand how professional sport teams can reach socially responsible consumers through sustainable marketing. This vision will help marketers to reach socially responsible consumers.

NFL

Context: History and Evolution

The NFL is the most important and profitable American professional league. This league was created in 1920, then known as American Professional Football Association (APFA), and later in 1922 renamed as NFL.

From its first years, the NFL games have been highly influential among the American culture. For the first decades, American football was mainly developed in colleges. Such is the case that the league was dominated by the Big Three American Universities, Harvard, Princeton, and Yale, located in the east of the United States (Beldon et al., 2020). Today the rivalry between those colleges remains.

The first season started with 14 teams of 6 different states. Despite the low popularity of this league in comparison with others, a second season was organized, with 21 teams (Crepeau, 2014). Later in 1950, the NFL far exceeded the MLB demand and reputation. This built interest of investors in buying franchises or creating new ones. The growth of this competition is exhibited in Figure 27.1.

FIGURE 27.1
Evolution of the number of NFL franchises per year. (Compiled from www.nfl.com, accessed July 2020).

NFL by the Numbers

The wide audience reached that the NFL reaches makes it one of the most influential entertainment show in the world:

- The first professional American football game broadcasted on television was in 1939 on NBC, with 2 cameras and 8 staff. In 2020, each game needs up to 20 cameras and about 150–200 staff (NFL Football Operations, 2020).
- The XLIV Super Bowl, in 2015, was the most-watched show on television in the United States History, with a total of 114,4 million spectators.
- The NFL is the richest league in the world. The most valuable franchise is the Dallas Cowboys, with $5,5 million in 2019 season (Forbes, 2019) (Figure 27.2).
- The total average in the 2019 NFL season was 16,894,586 people, with an average attendance of 66,151 spectators. Despite the good numbers, this is the lowest record in the last 10 years (ESPN, 2019) (Figure 27.3).

Marketing Strategy within the NFL

The NFL marketing strategy has a stable long-term life. Currently, the NFL brand is one of the most powerful in the United States. What are the basics of its success? Figure 27.4 exhibits the Marketing Mix Strategy of this league. Currently, the target markets of NFL are players, teams, and fans. The NFL fan profile is male with an average age of 50–60 years (people in their 20s are losing interest in the game).

Several specific strategies have been implemented since its origin. Social media and the openness to international markets are two of the most important ones. The development of social media as a marketing tool has increased the pace of change in the sport industry (Filo et al., 2015). Several studies have been completed regarding the use of social media in sport organizations, including the four North American professional sport leagues (Abeza et al., 2019; Hambrick & Kang, 2015). The main marketing strategies in sport are related to relationship marketing (Abeza et al., 2017, 2019), sponsorship (Weimar et al., 2020), or for displaying information (Reed & Hansen, 2013).

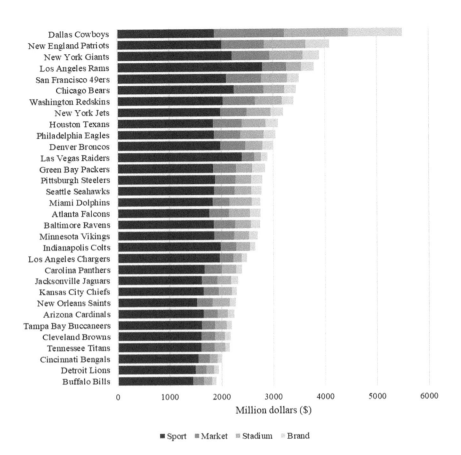

FIGURE 27.2
Franchise value in the 2019 NFL season. (Compiled from www.forbes.com/nfl-valuations/list, accessed July 2020).

Regarding the global expansion to international markets, the NFL has a rich history, especially during the last three decades. However, the inexistence of a wide culture of American football outside the United States, as well as the different time zones makes it difficult for the sport to flourish. The international marketing strategy followed by the NFL includes the American Bowl (1986–2005), World League of American Football (WLAF) (1997–1998), NFL Europe (1998–2007), and NFL International Series (2007–today) (Schwarz, 2018).

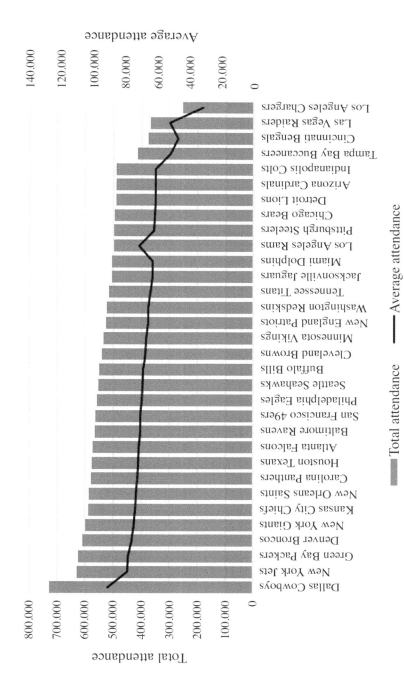

FIGURE 27.3
Total attendance and average attendance in the 2019 NLF season. (Compiled from www.espn.com/nfl/attendance, accessed July 2020).

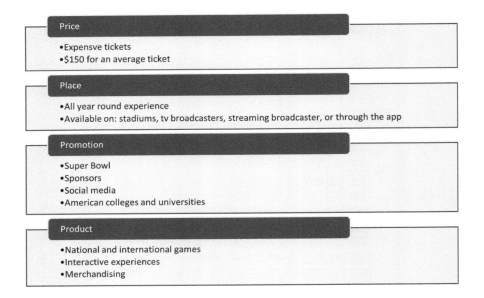

FIGURE 27.4
Marketing Mix Strategy of the NFL.

During last decades, sustainability is becoming increasingly important in today's society. In this line, the sports industry is progressively embracing sustainable goals (Kellison & Kim, 2014; Trendafilova et al., 2013). The rising importance and popularity of sports across the globe provide great opportunities for influencing society.

In addition to the efforts of the NFL for ensuring the long-term sustainability of the brand in national and international markets, the competition has also introduced several sustainability marketing strategies in the last decades. Environmental principles are present in NFL facilities and events, such as the Super Bowl, including solid waste management, material reuse, food recovery, sports equipment, and book donations, and greenhouse gas reduction (NFL, 2011):

- Waste management: It is controlled in NFL stadiums, media centers, and accommodations, substantially limiting the production of waste.
- Material reuse: most of the reusable items in the events are donated to local organizations. These include office items and building materials.

- Food recovery: residual food is donated to local organizations, who distribute the food to needy people.
- Sports equipment and book donations: this is part of a larger project, called Super Kids-Super Sharing. It consists of a call for usable items in schools of the Super Bowl area. The items are collected the weeks before the event. Finally, donations are transferred to other children in need.
- Greenhouse gas reduction: it consists of an extensive tree plantation in for the benefit of the Super Bowl area inhabitants. The NFL continuously tracks the benefits performed by the trees.

The aforementioned strategies turn the most important league in the world even more aligned with sustainability goals, influencing consumers' behavior and consumption patterns.

Thus, sustainability marketing can be a positive tool for influencing behavioral interventions and change for social good, especially for youth (Montez de Oca et al., 2016). This includes positive attitudes towards product disposal and waste management, green practices, responsible consumption, and measuring sustainability practices.

NFL TEAMS: THE CASE OF THE PHILADELPHIA EAGLES

Once the context has been settled, you are invited to reflect about some questions: How can NFL teams embrace sustainable marketing strategies? Would you be able to define a long-term sustainability marketing strategy for your team? Would this strategy affect the daily behaviors of fans?

The complexity of the case is evident. NFL clubs are starting to embrace sustainable marketing strategies (Gregory Greenhalgh et al., 2015). Among the 32 franchises, the Philadelphia Eagles are the pioneers of this movement, as an attempt to influence their customers to become greener (Blankenbuehler & Kunz, 2014). The Eagles' goal is to contribute to the community wellbeing through programs that improve the quality of life of society. Subsequently, the society gets influenced by sustainability marketing and, subsequently, they adopt green practices in their daily routine. The marketing strategy of the Philadelphia Eagles is highly centered on the power of green.

The Beginning of a "Green Era"

The Philadelphia Eagles is an NFL team of the National Football Conference, East division. They are headquartered in Philadelphia since 1933 and they wear a midnight green uniform. Throughout their history, they have had six home fields: Baker Bowl (1933–1935), Philadelphia Municipal Stadium (1936–1939), Connie Mack Stadium (1940–1957), Franklin Field (1958–1970), Veterans Stadium (1971–2002), and Lincoln Financial Field (2013–today).

The last stadium has led to an unprecedented change in the marketing strategy of the team. A sustainability marketing strategy is now established, creating value for the benefit of society.

As aforementioned, in 2003 the Philadelphia Eagles play their first game in the Lincoln Financial Field. At that time, the project started with several recycling bins. However, in 2020, this movement is part of one of the most important and influencing sustainability programs of any professional sport club.

"Go Green" Program

Nowadays, the Philadelphia Eagles are the reference in terms of sustainability. The Lincoln Financial Field can produce about 33% of its entire yearly energy usage. But what are the basics of the Go Green program? The complete list of the practices is exhibited in Table 27.1. They include installation of solar panels, optimal energy usage, field goal forest program, bike share program, food waste management, compostable straws, bottle caps collection, waste diverted from landfills, water-saving, and cleaning and hygiene (Philadelphia Eagles, 2019).

How have the Philadelphia Eagles reached this point? This is mainly a marketing effort to increase their sustainability strategy. All this began in 2003. Starting as an under-the-desk recycling program, the Go Green program has grown into the reference sustainability program in any worldwide sport league. This culture has made an impression not only in fans but also in the whole city of Philadelphia. Consequently, the level of awareness of the importance of maintaining the competition and the planet sustainable.

For example, the bike sharing program encourages people to use alternative means of transportation when traveling to and from the Lincoln Financial Field. However, this habit is adopted by the consumers, who are now more willing to use more environmentally friendly means.

TABLE 27.1

Practices of Go Green Program, Philadelphia Eagles

Practice	Details
Solar panels	10,456 solar panels around the stadium
Optimal energy usage	Low-wattage light bulbs
Field goal forest program	In collaboration with Philadelphia Gas Works (PGW), 10 trees are planted for every field goal. 29 goals in 2019 translate into 290 trees
Bike share program	Convenient transport for employees from the practice field to the stadium
Food waste management	24 tons of food waste decomposed
Compostable straws	Plastic alternative made from 100% renewable resources
Bottle caps collection	Turned into new material for use in the stadium
Wasted diverted from landfills	More than 99% of the waste is diverted from landfills
Water-saving	Water filtration fountains saved than 174,000 water bottles since 2015
Cleaning and hygiene	Sustainable solutions for seating areas cleaning

Another example is the field goal forest program. In 2019, the Eagles, in collaboration with Philadelphia Gas Works (PGW), planted 290 trees. The team is committed to planting 10 trees for every field goal in the Lincoln Financial Field. This situation increases the awareness of the community about the benefits of taking care of the environment. The important point is: Would this awareness be the same without the marketing efforts made by the Eagles? Would their consumption behavior be the same without the Go Green program?

Lastly, practices such as the installation of solar panels, energy-saving, recycling and so, have also penetrated the consumption habits of the city (mostly) and the country (to a lesser extent).

The aforementioned practices have led the Philadelphia Eagles to reach important standards, including Leed Gold Status, for implementing sustainable, measurable, and practical strategies, or ISO 20121, allowing them to be the first professional sport to be recognized for integrating sustainability into management practices (Philadelphia Eagles, 2019). These two certifications have boosted their marketing campaign, reaching significant partners in their Go Green campaign. This is also been used as a marketing tool to expand the awareness of people in terms of sustainable practices.

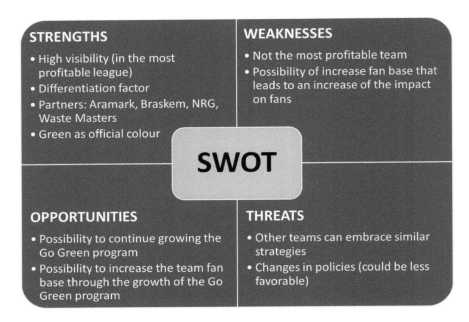

FIGURE 27.5
SWOT analysis of the Go Green program by Philadelphia Eagles.

Figure 27.5 shows a SWOT analysis of the Go Green program by the Philadelphia Eagles.

SUMMARY OF THE MAJOR PROBLEM

In this chapter, the case of the Go Green program by the Philadelphia Eagles is presented. As part of the NFL, the Eagles have embraced sustainability marketing, leading to more responsible consumers.

The main problem is: How can NFL teams embrace sustainable marketing strategies? Would this strategy affect the daily behaviors of fans? Specific strategies and practices are presented, as part of the Go Green program. Starting in 2003, the Philadelphia Eagles have revolutionized the understanding of sustainability marketing in sport venues. The Lincoln Financial Field began with recycling bins in every desktop. Nowadays, the stadium is completely green. This fact makes it a reference to sport venues in the world and influences the behaviors of fans not only in the stadiums but also in their daily routine.

DISCUSSION

The sustainability marketing strategy followed by the Philadelphia Eagles has important implications:

First, the sustainability impact over the environment is clear. Would it be possible a future in the NFL with a complete list of teams following the Lincoln Financial Center example? Although major sporting events such as the Super Bowl are also making big efforts regarding sustainability, there is still a way to go. The attendance of big masses of fans has an important effect on the environment.

Second, social outcomes are extremely important. The Go Green program pursuit is not limited to reduce the carbon footprint. What is more important, it has the power to change the intentions and future behaviors of consumers. The sports industry is an excellent way of changing society. People feel identified with teams, players, and colors. Why not use this power for changing the world for the better?

CONCLUSION

This chapter introduces a real-life case of sustainable marketing of a professional sport league, the NFL. While is true that the league is making big efforts for contributing to the social good, the most representative case is the Go Green program, by Philadelphia Eagles.

Since 2003, Philadelphia Eagles are running a program called Go Green. Throw this initiative, the team is changing the future behaviors of society. The benefits have an effect on the community themselves in two directions: first, direct benefits for the reduction of the carbon footprint, and second, a progressive change in the society, reaching socially responsible consumers.

This scenario provides an illustrative marketing real case that engages fans on sustainability while maintaining the excitement of the show in the stadium. Thus, this case is unquestionably a great example of successful sustainability marketing in the scope of the sports.

LESSONS LEARNED

- The five main sustainability initiatives of the NFL are aligned with teams.
- Leagues and, subsequently, teams, should embrace sustainable strategies and practices. They should be committed to achieve a better world.
- The sports industry is an excellent way to influence society and reach socially responsible consumers.
- Sport fans perceive the efforts made by the teams in terms of sustainability.
- Go Green program by Philadelphia Eagles is an excellent real case that allows to explore the possibilities of social and sustainability marketing in the sport.

DISCUSSION QUESTIONS

- What is the role of the NFL in the Go Green program by Philadelphia Eagles?
- What are the stakeholders of Go Green program by Philadelphia Eagles?
- What is the target market of Go Green program?
- To what extent can Go Green program extend the NFL market to younger people?
- What do you think is the future of social and sustainability marketing in sports?

ACTIVITY-BASED ASSIGNMENT

- Complete a Go Green program for a local sport team. Compare this with the Go Green program by Philadelphia Eagles. Consider how big the differences between the two programs are.
- As a manager in a company of other industry, how could you increase your consumers' identification with your brand? Make

a list, considering the specific features of sport leagues, such as the NFL.

ADDITIONAL CONTENT

Did you know?

- Marketing communications are highly effective when the message is adopted by a leading or influential person or group. Sports organizations meet this condition. Subsequently, messages on sustainability topics wield extraordinary within the audience. In this regard, the NFL has tested the power of new media technologies to engage children in pedagogical projects. For that reason, it seems obvious that sports such as the NFL have a substantial potential of influencing the behavioral intention of customers.

Interesting terms

- Go Green: It is a program launched by the Philadelphia Eagles in 2003. It includes a series of sustainability strategies directed to improve the environment and the society.

Abbreviations

- APFA – American Professional Football Association
- NBA – National Basketball Association
- NFL – National Football League
- NHL – National Hockey League
- MLB – Major League Baseball
- SWOT – Strengths, Weaknesses, Opportunities, and Threats

ADDITIONAL READINGS

McCullough, B., & Kellison, T. (2018). *Routledge Handbook of Sport and the Environment*. London, UK: Routledge; Taylor & Francis Group.

Millington, R., & Darnell, S. C. (2019). *Sport, Development and Environmental Sustainability*. London, UK: Routledge.

CREDIT AUTHOR STATEMENT

Conceptualization, J.L.-Q.; Resources, J.L.-Q.; Writing – Original Draft, J.L.-Q.; Writing – Review & Editing, J.L.-Q.; Visualization, J.L.-Q.

REFERENCES

Abeza, G., O'Reilly, N., & Seguin, B. (2019). Social Media in Relationship Marketing: The Perspective of Professional Sport Managers in the MLB, NBA, NFL, and NHL. *Communication & Sport*, 7(1), 80–109. https://doi.org/10.1177/2167479517740343

Abeza, G., O'Reilly, N., Seguin, B., & Nzindukiyimana, O. (2017). Social Media as a Relationship Marketing Tool in Professional Sport: A Netnographical Exploration. *International Journal of Sport Communication*, 10(3), 325–358. https://doi.org/10.1123/ijsc.2017-0041

Beldon, Z., Weiller-Abels, K., & Nauright, J. (2020). American Football. In J. Nauright & S. Zipp (Eds.), *Routledge Handbook of Global Sport* (pp. 7–17). Abingdon, England: Routledge.

Blankenbuehler, M., & Kunz, M. B. (2014). Professional Sports Compete to Go Green. *American Journal of Management*, 14(4), 75–81.

Crepeau, R. C. (2014). *NFL football: A history of America's new national pastime*. University of Illinois Press. https://doi.org/10.1080/17460263.2017.1315024

Deloitte. (2020). *Football money league*. Manchester, United Kingdom: Deloitte.

ESPN. (2019). *NFL Attendance 2019*. http://www.espn.com/nfl/attendance

Filo, K., Lock, D., & Karg, A. (2015). Sport and Social Media Research: A Review. *Sport Management Review*, 18(2), 166–181. https://doi.org/10.1016/j.smr.2014.11.001

Forbes. (2019). *Sports Money: 2019 NFL Valuations*. https://www.forbes.com/nfl-valuations/list/

Greenhalgh, G., & Drayer, J. (2020). An Assessment of Fans' Willingness to Pay for Team's Environmental Sustainability Initiatives. *Sport Marketing Quarterly*, 29(2). https://doi.org/10.32731/SMQ.292.062020.04

Greenhalgh, G., LeCrom, C., & Dwyer, B. (2015). Going Green? The Behavioral Impact of a Sport and the Environment Course. *Journal of Contemporary Athletics*, 9(1), 49–59.

Hambrick, M. E., & Kang, S. J. (2015). Pin It. *Communication & Sport*, 3(4), 434–457. https://doi.org/10.1177/2167479513518044

Kellison, T. B., & Kim, Y. K. (2014). Marketing Pro-Environmental Venues in Professional Sport: Planting Seeds of Change Among Existing and Prospective Consumers. *Journal of Sport Management*, 28(1), 34–48. https://doi.org/10.1123/jsm.2011-0127

Montez de Oca, J., Meyer, B., & Scholes, J. (2016). Reaching the Kids: NFL Youth Marketing and Media. *Popular Communication*, 14(1), 3–11. https://doi.org/10.1080/15405702.2015.1084623

NFL. (2011). *NFL Green*. https://www.nfl.com/news/nfl-green-09000d5d8205a0e7

NFL Football Operations. (2020). *Impact of Television*. https://operations.nfl.com/the-game/impact-of-television/

Philadelphia Eagles. (2019). *Philadelphia Eagles Go Green*. https://www.philadelphiaeagles.com/community/go-green

Reed, S., & Hansen, K. A. (2013). Social Media's Influence on American Sport Journalists' Perception of Gatekeeping. *International Journal of Sport Communication*, 6(4), 373–383. https://doi.org/10.1123/ijsc.6.4.373

Schwarz, E. C. (2018). Marketing implications of playing regular season games in international markets. In M. Dodds, K. Heisey, & A. Ahonen (Eds.), *Routledge Handbook of International Sport Business* (pp. 79–86). New York, NY, USA: Routledge.

Trail, G. T., & McCullough, B. P. (2020). Marketing Sustainability through Sport: Testing the Sport Sustainability Campaign Evaluation Model. *European Sport Management Quarterly*, 20(2), 109–129. https://doi.org/10.1080/16184742.2019.1580301

Trendafilova, S., Babiak, K., & Heinze, K. (2013). Corporate Social Responsibility and Environmental Sustainability: Why Professional Sport Is Greening the Playing Field. *Sport Management Review*, 16(3), 298–313. https://doi.org/10.1016/j.smr.2012.12.006

Weimar, D., Holthoff, L. C., & Biscaia, R. (2020). When Sponsorship Causes Anger: Understanding Negative Fan Reactions to Postings on Sports Clubs' Online Social Media Channels. *European Sport Management Quarterly*, 1–23. https://doi.org/10.1080/16184742.2020.1786593

28

Leave It: Upskilling a Dog Owning Community

*Jessica A. Harris, Sharyn Rundle-Thiele,
Bo Pang, Patricia David and Tori Seydel*
Griffith University, Nathan, Queensland, Australia

CONTENTS

Learning Objectives	740
Themes and Tools Used	740
Theoretical Background	740
Introduction	742
Method	745
Findings	746
Discussion	750
Conclusion	751
Lessons Learned	752
Discussion Questions	752
Practical Activity	753
Acknowledgments	753
Financial Support	753
Conflict of Interest	753
Ethical Standards Disclosure	753
Authorship	753
Case Study References	754

LEARNING OBJECTIVES

After reading the chapter, the reader should be able to:

- Understand how social marketing principles can be applied to conserve wildlife
- Develop an understanding of the Co-create, Build and Engage social marketing framework (C-B-E), and how to apply C-B-E across iterative project cycles
- Use positive marketing to change behaviour

THEMES AND TOOLS USED

- C-B-E (co-create, build, engage)
- RE-AIM framework

THEORETICAL BACKGROUND

Based on the understanding that engagement is enhanced when programs are designed with people and not for them, marketing seeks to place people at the heart of program design. The Co-create, Build and Engage (C-B-E) process was developed by social marketers to guide marketers through a process that can be applied to design, plan, implement and evaluate a program. C-B-E is a human centred approach that ensures the co-creation and implementation of programs are centred around understanding what people need and want in order to successfully engage them to change a behaviour (Roemer et al. 2020). The C-B-E framework is different to early social marketing frameworks, which outlined a range of activities marketers typically apply without demonstrating which order they should be conducted in for initial program design. Co-creation is the first step in the process. Specifically, co-design seeks to understand key stakeholders and it seeks to gain insights to understand how to move and motivate socially responsible people, and other harder to reach people, to understand how to change a particular behaviour (Roemer et al. 2020). A range of research techniques are carried out in this stage to ensure a deep understanding is gained. For this project, stakeholder interviews, literature reviews of past programmes and surveys were used to gain insight into what had or hadn't worked in the past and identified programs were used in co-design to sensitise the target audience (e.g. dog owners) who were asked in small groups to design a behaviour change campaign for themselves and dog owners like them (see Rundle-Thiele et al. 2019 for full detail).

The next phase in the CBE framework is the Build phase. This involves converting insights identified on completion of the co-creation phase into a program that is designed to meet people's needs. Key insights are used to guide strategy ensuring program elements are aligned to program user needs. The build stage involves strategy, program build ensuring the program is offered at a time and place that is convenient for people. To successfully implement a set program activity that need to be scheduled, budgeted and detailed plans and processes are outlined delivering clear understanding of how the programme will run. On completion of the build step, the program is market ready (Roemer et al. 2020).

The third and final phase of the three-step CBE framework is Engage. In this stage programs are delivered. In the engagement phase, the focus is on ensuring the market knows about the programme through awareness campaigns, and ensuring the programme is fun and engaging. The programme is carefully monitored in this phase and evaluation of the programme success is the final step to the framework indicating recommencement of the CBE cycle as evaluations are used to measure program success and failure. As the CBE cycle recommences (Figure 28.1) insights,

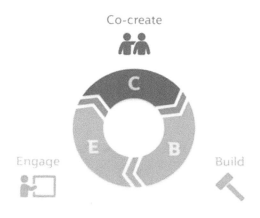

FIGURE 28.1
C-B-E framework.

data and evaluation are used to increase programme effectiveness going forward (Roemer et al. 2020).

Throughout the *Leave It* program, the CBE framework was utilised across all 3 stages informing *Leave It* pilot and city-wide roll-out program design, building on the evaluation and insights from the previous stage. This helped target the audience, build awareness and reach for the programme and ensure all stakeholders were involved over time to extend programme effectiveness. An outline of how the framework for used for each phase is given in Table 28.1.

The second framework that was utilised in *Leave It* programme monitoring and evaluation was the RE-AIM framework (Glasgow, Vogt, & Boles, 1999). This framework is a method that systematically uses five dimensions of reach, effectiveness, adoption, implementation and maintenance. This dimension uses project aims at the start of each project to focus program team efforts on realising targets. RE-AIM focuses efforts to ensure the programme will reach certain milestones and outcomes provided by stakeholders, investors and program management facilitators. Table 28.1 is an example of the RE-AIM framework used by stage 2 in the city-wide rollout programme. As shown in Table 28.2, the project aims are clearly outlined aiming to ensure programme effectiveness and outcomes can be reported to measure program success and failure.

INTRODUCTION

Koala populations are declining all over Australia (Rhodes, Hood, Melzer, & Mucci, 2017), with recent bushfires decimating many habitats (Dickman, 2020). Koalas move on the ground, instead of through the canopy, when looking for mates or new habitats which leaves them at risk of attack from domestic pets (dogs). Next to disease and roads, dogs are the third highest risk factor for koalas cohabiting in urban areas (Rhodes et al. 2017). Programs aiming to decrease dog and koala interactions were needed and *Leave It* was established in 2017 following co-creation with the Redland City Council community (Rundle-Thiele et al. 2019). *Leave It* is a social marketing project that has been conducted over a 4-year period across 3 stages. Supported by Redland City Council, Social Marketing @ Griffith applied the 3 step Co-create – Build – Engage (C-B-E) process (see Roemer

TABLE 28.1

CBE Framework

Stages	Co-Create	Build	Engage
Stage 1 (David et al. 2019)	• Systematic literature review on the interaction between wildlife and domestic pets • Expert interviews (n = 14) • Co-design with dog owners (n = 41) • Online questionnaire (n = 635) Reported in Rundle-Thiele et al. (2019)	• DogFest – a 1-day community event featuring fun, interactive dog-related activities • *Leave It* website • Communication materials – flyers, posters, etc.	• Promotion of *Leave It* – mailing list, media release, social media • Public seminars • Outcome evaluation – dog abilities • Process evaluation – event attendance, website traffic, stakeholder satisfaction, etc.
Stage 2	• Learnings from Stage 1 • Social advertising message testing (n = 538)	Same as Stage 1 plus the following: • Blogs • Located dog training businesses servicing the city area • Attraction of local dog training businesses to join *Leave It* • Bonus benefits negotiated for dog owners – discounted training fees, discounted dog registration, etc.	• City wide roll-out • *Leave It* Accreditation – Train the trainer • Public seminars • Outcome evaluation – dog abilities • Process evaluation – event attendance, website traffic, stakeholder satisfaction, etc.
Stage 3	• Learnings from Stages 1 and 2 • Co-design with dog trainers (n = 6)	Same as Stage 2 plus the following: • *Leave It* online dog training videos • Continuation of attracting local dog trainers for *Leave It* accreditation • Newsletters	• Continuation of city-wide roll-out • Promotion of *Leave It* – mailing list, radio interview, social media advertising • Facebook live dog training webinars • Outcome evaluation – dog abilities • Process evaluation – attendance to events, website traffic, etc.

TABLE 28.2

Example of Leave It Citywide Rollout RE-AIM Framework (Stage 2 Aims and Outcomes)

RE-AIM Dimension	Project Aims	Outcomes
Reach	10% increase in unique visits on the *Leave It* website 10% increase in people reached on Facebook 10% increase in likes and comments on *Leave It* Instagram account	2,037 unique visits on the *program* website (29% increase from pilot) Over 350 flyers distributed 48,694 people reached on Facebook (31% increase from pilot) 199 likes, 45 shares, and 96 comments for organic workshop and seminar Facebook posts 198,000 people reached on radio advertisement and bus shelters.
Effectiveness	Dog obedience abilities have significantly improved through participation in the *Leave It* program (Sit, stay, heel, aversion/not chase things, come back when called, stay quiet on command, crate use)	While increases in come back when called, aversion and stay quiet on command are higher in 2018 and 2019 when compared to pre *Leave It* levels, the results of the citywide roll out indicate that the addition of the *Leave It* branded dog training delivered superior outcomes when compared to city-wide roll out where trainers may (or may not) embed koala aversion into training offerings. The low level of free 6-month registration uptake indicates koala aversion may not have been embedded to a full extent.
Adoption	Achieve 85% uptake for local dog trainers and dog breeders in train the trainer sessions over the 18-month period (minimum of 12 of the 14 dog training companies in Redland City Council undertake Leave It - Train the Trainer sessions) A minimum of 4,500 dog owners participate in individual or group training sessions, talks or workshops that incorporate koala aversion training	46% of dog training companies from the Redland City Council area became *Leave It* accredited by undertaking dog training with either Steve Austin or Ryan Tate. A precise number of dog owners trained in koala aversion skills cannot be directly confirmed. 11 dog training businesses undertook *Leave It* Train the Trainer sessions, and 169 dog owners RSVP'd to free public seminars.

(Continued)

TABLE 28.2 (Continued)

Example of Leave It Citywide Rollout RE-AIM Framework (Stage 2 Aims and Outcomes)

RE-AIM Dimension	Project Aims	Outcomes
Implementation	Satisfaction for *Leave It* program participants to remain high Stakeholder participation benefits to remain high >6.0 out of 7	Over 87% participants in the train the trainer workshops reported they were satisfied, and over 93% reported they were likely to attend another workshop. Over 84% public seminar attendees reported high levels of satisfaction. 12% of the surveyed community reported recalling *Leave It*.
Maintenance	85% of participants express their intention to attend *Leave It* seminars again	83% respondents attending Steve Austin's public seminar reported high likelihood they would attend the seminar again. 96% respondents attending Ryan Tate's public seminar reported high likelihood they would attend the seminar again.

et al. 2020) to co-create, implement, evaluate and iterate a program aiming to decrease dog and koala interactions.

METHOD

The C-B-E process was designed to provide a clear step wise framework for social marketers to design, implement and evaluate social marketing campaigns ensuring a strong marketing orientation. Co-creation will ensure program developers work with stakeholders and program end users to understand what they want and need, and to collect insights that can be used for campaign build. The Build stage is where insights are converted into market ready strategies and actions which can be implemented by project teams to deliver intended outcomes. The last step – Engage – consists of delivering the program in community that has been tested and built. Programs are monitored and evaluated to examine effects and outcomes. The C-B-E process was applied for each stage of the *Leave It* project. Please see Table 28.3 for more details.

TABLE 28.3

C-B-E Framework for *Leave It*

	Co-Create	Build	Engage
Stage 1.0	Systematic Literature Review, Survey, Expert Interviews, Co-Design	DogFest, *Leave It* 4-week dog training program and Train the Trainer Workshop	2,500 owners attended DogFest, 19 dogs completed the *Leave It* pilot dog training program
Stage 2.0	Post-survey, dog trainer observation notes	Public Seminars, Train the trainer workshops	• Over 350 flyers distributed • 48,694 people reached on Facebook 198,000 people reached on radio advertisement and bus shelters. 169 dog owners RSVP to the seminars and 11 dog training companies were trained.
Stage 3.0	Post-survey, Co-design	Public Seminars, Train the trainer workshops, Online public workshops, website, video tutorials	Promotion of *Leave It* – mailing list, radio interview, social media advertising reached a total of 27,840 impressions, 1,167 unique clicks 21,000 views on paid media. Train the trainer workshops had a further 6 companies have *Leave It* training. Total of 71% have now been trained. Facebook live dog training webinars had over 6,000 unique views

FINDINGS

Co-creation indicated that dog owners sought a fun, dog-centred program that showed them what they could do to protect wildlife. The Co-design discussions demonstrated the need to keep a broad mindset and to ensure that the program implemented avoided a koala focus. Dog owners indicted that dogs chased (and killed) other forms of wildlife and they sought a program that was not 100% koala focussed. Dog owners said:

> And, a lot of people will just look at it and be, 'They're just wanting us to not have dogs', and all this kind of thing. Or, they're just going to be

> thinking straight away that anybody who's pro looking after koalas is negative towards animals we keep as pets.
>
> I just wonder if the campaign should be about positive things rather than negative things, that's all. On some of the stuff I think, 'You must lock your dog up. You must do this. You must do that'. Whereas, at least with number 2 they've got tree planting. We destroyed the entire local council area, chopping down every tree so that there's no koalas, but yet we're talking about dogs.
>
> (Rundle-Thiele et al. 2019)

For this reason, *Leave It* did not picture koalas or focus promotion effort on the wildlife aversion ability. In line with dog owner designed elements the *Leave It* program promotional materials featured dogs and a festival celebrating man's best friend was conceived. The pilot program (Stage one) ran in 2017 with local dog trainers receiving professional training from experts on wildlife aversion. *Leave It* dog trainers learned how to embed koala aversion skills training into their current dog training programs. The *Leave It* pilot training programs were run over a 4-week period. Within this period, a number of dogs learnt basic commands of sit, stay and come. Additionally, dogs completing *Leave It* training also learnt aversion skills. Over the course of 4 weeks, 19 dogs participated in the pilot training program. Trainers recorded the sessions and performance of each dog and owner. For example, trainer notes indicated:

> took about 20 mins to settle in. Uses Hand signal however lure was needed to get focus at certain points when distracted by dogs next to her.
>
> (Cashew, Week 1)

There were also successful aversion skills noted at the end of the 4-week course, for example:

> Able to leave moving koala toy.
>
> (Momo, Week 4)

> Can leave koala toy dropped beside her.
>
> (Roxie, Week 3)

After completion of training dog trainers were certified as *Leave It* dog trainers. The promotional program that was implemented in community

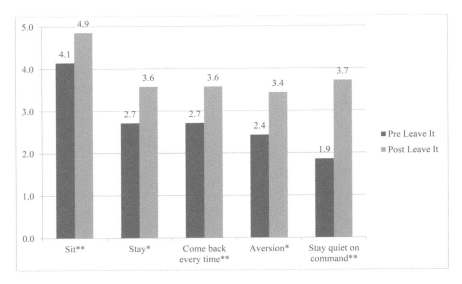

FIGURE 28.2
Stage 1 outcome results.

to raise awareness for *Leave It* included an event *DogFest* supported by a communication program featuring PR, flyers distributed through various local business and community areas, website, mailing list and social media. The pilot study demonstrated the ability of the *Leave It* program to increase koala aversion skills in dogs (the ability to be recalled away from wildlife on command from an owner). In Figure 28.2, the evaluation of dog abilities is provided. All dog abilities were greater post *Leave It* training than pre *Leave It* training.

Over 85% of participants were satisfied indicating they had a positive experience with the program and two-thirds of the participants would do the program again in the future. For full results of the program, please see (David et al. 2019). C-B-E was again applied to iterate and extend on pilot program success in Stages 2 and 3.

Stage 2 involved a citywide roll out. This stage aimed to extend koala aversion training skills for trainers servicing the local government area and to extend communication efforts to raise demand for dog training with the dog owning community. A key focus in this stage was to locate, attract and engage dog trainers to equip them with koala aversion skills. In this 18-month project, 9 out of the 21 dog training companies were engaged and trained using a 'train the trainers' approach.

This method was utilised to upskill the dog training community adding koala aversion as a dog ability. The project aimed to empower dog trainers to improve dog abilities (training, wildlife aversion and denning) within the council area. Free 1-day training sessions were made available to dog trainers in the area. The *Leave It* team hired 2 renowned professional wildlife aversion dog trainers to hold Train the Trainer workshops. Once the dog trainers were trained in the above-mentioned protocols, they were *Leave It* certified. The aim was for these 9 dog training companies to then embed this particular training into their programs throughout the Redland City Council area to ensure this was normal practice for dog training going forward.

Stage two used the RE-AIM framework (Glasgow, Vogt, & Boles, 1999), a multimethod process and evaluation to note the effectiveness of the city wide rollout. The overall aim of the city-wide roll-out was to ensure more dogs in the Redland City Council were trained in aversion skills. Base line surveys at 3 time points throughout the program were utilised to track dog abilities over time. Aversion and come back when called (two key *Leave It* training abilities) were higher post *Leave It* city-wide rollout. A total of 2,013 dog owners were surveyed throughout the city-wide roll-out across 3 time points. A survey also assessed program outcomes with a total of 31 dog trainers and 49 dog owners. Program reach was assessed with a satisfaction rate of 100% for the Train the Trainers and 86% for public seminars. Overall likelihood to re-attend and recommendation to tell others were both very high with an average of 96%. The effectiveness of communication channels utilised during the city-wide roll out were evaluated. Overall the organic social media reach totalled 521 people and our paid advertising (social media, bus shelters and radio) reached over 243,000 people. The full evaluation for Stage 2 is reported in Harris, Rundle-Thiele, David and Pang (2021).

Stage 3 focussed on extending the citywide rollout effort and commenced with co-creation to understand why some dog trainers did not engage with the program in Stage 2. Deep insights were uncovered through these co-design sessions from key stakeholders indicating changes that needed to be made. The main aim was to understand from dog trainers and other related businesses who didn't participate in previous years. Timings (from weekend to mid-week) of workshops were changed to allocate time for businesses whose main earnings were weekend based. Inclusion of the Animal Management team at Redland City Council ensured materials were distributed to all dog and cat owners and *Leave It* information was included in packs for dogs that were adopted from the Animal Management shelter. *Leave It* accredited trainers offered up their time to train the animal management team and dogs in care on the *Leave It* principles of aversion to ensure adopted dogs were trained to avoid koalas. The *Leave It* program also extended work into puppy training and pre-school. Education is one of the biggest dog owner needs. *Leave It* is working on a number of partnerships with local vets and puppy pre-schools to educate staff and subsequently dog owners about the importance of having well trained dogs around urban areas. Communication promoting the benefits of dog training to reach dog owners in the local community was extended. Effort continued to ensure that the remaining dog trainers and breeders within Redlands City Council were contacted and trained in wildlife aversion. A further 6 dog training companies received *Leave It* training during Stage 3. By the end of Stage 3 the *Leave It* program had trained 71% of the dog training companies servicing the Redland City Council area in two wildlife protocols (aversion and denning).

DISCUSSION

Over a 4-year period, *Leave It* used online and offline promotional activities to engage the public. Press releases reached 480,000 people while 175,000 and 60,000 people were reached through radio and bus shelter advertising, respectively. More than 3,800 flyers were delivered to vets, dog retail shops, parks and other retailers changing community norms and attitudes to dog training. With more than 165 attendees in 9 face-to face workshops, 6,300 views for two online webinars and over 1,500 attendees at DogFest, *Leave It* has demonstrated an ability to engage a broad cross section of dog owners and trainers. More than 7,900 people have actively participated in *Leave It* events across the life of the 4-year project.

The *Leave It* program changed community attitudes towards dog training and increased dog abilities. Over the course of the 4-year project dog owners engaged with online and offline training events and support materials and increases in dog abilities over time was observed. To date, *Leave It* has engaged more than 71% of dog trainers servicing the community, demonstrating how capacity can be built ensuring new abilities are

TABLE 28.4
Overall Evaluation Results (Leave It Stage 1–3)

Leave It	Outcome Results
Reach	- 18,500 website page views - 100,000 social media views - 235,000 people reached on radio advertisements and bus shelters - 480,000 views of press release - 3,800 flyers - 1,500 DogFest attendees
Effectiveness	- There were increases of 40% in come back when called, 15% aversion and stay quiet on command increased by 4.5% overall in (2017–2019).
Adoption	- 71% of dog trainers are *Leave It* accredited. They can teach wildlife aversion. - 177 people attended a free dog training seminar and more than 6,300 viewed online webinars
Implementation	- Over 87% of participants were satisfied with the workshops and over 93% would attend another. - 84% of attendees to the public seminars reported high satisfaction
Maintenance	- 96% of attendees expressed their intention to attend the program again.

embedded within a local government area. Dogs receiving training from a *Leave It* accredited trainer report improved koala aversion skills post the training program. There has been a 15% increase in wildlife aversion and a 40% increase in dogs coming back when called. Redlands Afterhours Wildlife Ambulance data (July 2013–March 2020) reported a decrease in dog related koala mortalities from an average 5 koala deaths from dogs per year before inception of the program to 3 koala deaths per year following *Leave It* implementation. Using the RE-AIM framework, the overall achievement of the *Leave It* project can be found in Table 28.4.

CONCLUSION

Leave It can be applied within local government areas to increase dog abilities, change community attitudes towards wildlife and importantly reduce koala deaths in the local area. *Leave It* demonstrates the effectiveness of applying a social marketing approach (CBE) to co-create programs with people ensuring buy in and program adoption. C-B-E serves as a clear stepwise process that professionals can apply to initially design, implement, and evaluate programs that deliver the desired behavioural changes and outcomes. Moreover, C-B-E can be applied over time to iterate and improve on program success.

LESSONS LEARNED

Throughout the 4-year program, community attitudes and conversations changed as a result of *Leave It* implementation. By initiating dog focussed promotional events celebrating man's best friend and embedding wildlife aversion as one ability within dog training programs delivered across the community *Leave It* was able to deliver a 40% reduction in koala deaths from dog attacks. While considerable progress has been made with 70% of dog trainers receiving initial training on wildlife aversion more work is needed into embed the principles of *Leave It*, into dog training, vets, puppy schools and breeding programs so all dog owners have dogs who are trained to leave wildlife alone ensuring species survival. Threatened species, such as koalas, many of whom coexist in urban and semi urban areas face many threats, many of which are preventable. Application of social marketing can reach consumers (e.g. dog owners) directly (e.g. online Facebook live sessions and training videos) and indirectly through the dog training community who can teach koala aversion. By applying social marketing principles of co-creation, programs can be built to meet the needs of the multitude of different stakeholders, ensuring program engagement is maximised over time.

DISCUSSION QUESTIONS

- Who is *Leave It's* target audience?
- What is the core product *Leave It* provided to the target audience in the pilot program?
- What promotional activities did *Leave It* implement in Stage 2? What do you think would have worked better?
- What social marketing principles and theories were used to develop *Leave It*?

PRACTICAL ACTIVITY

Form groups of 4 and work with your team members to design your version of *Leave It* using four Ps frameworks:

- What product/service will you offer to dog owners to reduce dog/koala interactions?
- What cost will your *Leave It* campaign incur and how much? How will you reduce the cost?
- How will you distribute your product/service and make it easier to get to the dog owners?
- What promotional platform will you use to promote *Leave It*? Why?

ACKNOWLEDGMENTS

The authors wish to thank dog owners, dog trainers and the business community who supported research and delivery of the *Leave It* program.

FINANCIAL SUPPORT

Redland City Council and Griffith University funded the research program that led to implementation of *Leave It* in the Redland City Council area.

CONFLICT OF INTEREST

There is no conflict of interest within this paper.

ETHICAL STANDARDS DISCLOSURE

Not Applicable.

AUTHORSHIP

All authors have read and agreed to the published version of the manuscript.

CASE STUDY REFERENCES

David, P., Rundle-Thiele, S., Pang, B., Knox, K., Parkinson, J., & Hussenoeder, F. (2019). Engaging the dog owner community in the design of an effective Koala Aversion program. *Social Marketing Quarterly*, 25(1), 55–68.

Dickman, C. (2020). More than one billion animals killed in Australian bushfires. Retrieved from https://www.sydney.edu.au/news-opinion/news/2020/01/08/australian-bushfires-more-than-one-billion-animals-impacted.html#

Glasgow, R. E., Vogt, T. M., & Boles, S. M. (1999). Evaluating the public health impact of health promotion interventions: the RE-AIM framework. *American Journal of Public Health*, 89(9), 1322–1327.

Harris, J., Rundle-Thiele, S., David, P., & Pang, B. (2021). Engaging dog trainers in a city-wide rollout of koala aversion skill enhancement: A social marketing program. *Australasian Journal of Environmental Management*.

Rhodes, J. R., Hood, A., Melzer, A., & Mucci, A. (2017). Queensland Koala Expert Panel: A new direction for the conservation of Koalas in Queensland. *Queensland Koala Expert Panel*.

Roemer, C., Rundle-Thiele, S., Pang. B., David, P., Kim, J., Durl, J., ... Carins, J. (2020). Re-wiring the STEM pipeline: Applying the C-B-E framework to female retention. *Journal of Social Marketing*, 10(4), 427–446. https://doi.org/10.1108/JSOCM-10-2019-015.

Rundle-Thiele, S., Pang, B., Knox, K., David, P., Parkinson, J., & Hussenoeder, F. (2019). Generating new directions for managing dog and koala interactions: A social marketing formative research study. *Australasian Journal of Environmental Management*, 26(2), 173–187.

29

Coexisting: The Role of Communications in Improving Attitudes towards Wildlife

Bo Pang, Patricia David, Tori Seydel and Sharyn Rundle-Thiele
Griffith University, Australia

Cathryn Dexter
Redland City Council, Australia

CONTENTS

Learning Objectives 755
Themes and Tools Used 756
Introduction 756
Community Baseline Survey 757
Campaign Implementation 758
Campaign Evaluation 760
Lessons Learned 761
Discussion Questions and Activities 762
Funding Declaration 762
References 763

LEARNING OBJECTIVES

After reading the chapter, the reader should be able to:

- Use social marketing to gain insights to inform campaign design
- Implement engaging social marketing campaigns in the community
- Effectively and accurately evaluate the outcomes of the campaign

756 • *Social and Sustainability Marketing*

THEMES AND TOOLS USED

Formative research, social advertising

INTRODUCTION

The koala (*Phascolarctos cinereus*) is an Australian wildlife icon listed in Queensland, New South Wales and the Australian Capital Territory as vulnerable to extinction since 2012. The increased urbanisation of the koalas' natural habitat has resulted in numerous threatening processes that have contributed to their ongoing decline. These include habitat destruction, domestic dog attacks, disease, bushfires and vehicle strike (Rhodes et al., 2015; Australian Koala Foundation, 2020). See Figure 29.1 for a photo of a koala.

Koalas are especially vulnerable to injury or mortality from vehicle strike and dog attack during dispersal and breeding seasons when they travel to seek out a new home range or mate (Queensland Department of

FIGURE 29.1
Koala (Photo credit: Haruka Fujihira).

Environment and Science, 2020). Hence, it is important to develop strategies to raise awareness during peak movement periods. Understanding how residents who coexist with koalas think and feel and what they know and believe is crucial to garner community support to undertake actions needed to help protect koalas.

In 2018, Redland City Council implemented a city-wide communication campaign to activate community driven koala conservation actions. The aim of the campaign was to increase local residents' awareness and improve community information on how to successfully coexist and protect koalas. An integrated communication approach was implemented and advertising channels including billboards, social media, bus backs, bus shelters, cinema, print media, flyers and posters were used to disseminate messages throughout the city area.

COMMUNITY BASELINE SURVEY

A baseline survey was conducted in July 2018 to gain insights into the community's level of awareness of and attitudes towards koalas to inform the planning of the 'Koala Awareness Campaign'. Face-to-face intercept surveys were conducted in public parks, markets, train stations, bus interchanges and ferry terminals in the Redland City Council area. Utilisation of face-to-face intercept surveys ensured access to members of the target audience who are less likely to engage in other forms of surveys (e.g. online), therefore delivering less-biased results (Bush & Hair, 1985). See Figure 29.2 for the data collection locations.

Baseline questionnaires included questions capturing community awareness regarding knowledge of and attitudes towards koalas (Pang et al., 2018), as well as respondent demographics. A total of 502 surveys were collected. Data was entered into SPSS Ver. 25 and data was cleaned prior to analysis. Coding was undertaken to identify themes for open-ended questions. Quantitative data analysis was performed including descriptive statistics, chi-square, t-tests and ANOVAs to identify any possible group differences regarding gender and age.

The baseline survey results indicated that koalas are top of mind for a large proportion (84.2%) of respondents and 85.3% of respondents agreed that Redlands Coast was home to a significant koala population, and this was slightly stronger among females. Although 90.1% of respondents

FIGURE 29.2
Data collection locations (pre and post).

believe koalas are worth the time and effort required for conservation, knowledge of koala movement is relatively low with over half the population having no understanding of movement associated with dispersal and breeding season. Moreover, 18–24 years old scored lowest in considering koalas had a place in the urban landscape and this age group showed low confidence in knowing how to protect koalas; though this age group said they would be more vigilant if they had more information. Most importantly, more than a third (37.2%) of respondents did not know how to protect koalas despite indicating they would like to take action. Specifically, they indicated they wanted general information about koala habitat and threats as well as easy and practical ways to help koalas. The insights guided campaign message design and message placement.

CAMPAIGN IMPLEMENTATION

The koala awareness media campaign was run from September to December 2018. Based on the insights generated by the baseline survey, core campaign messages were designed to promote the actions that

TABLE 29.1

Campaign Messages

Theme	Message and Tagline
Koala Dispersal and Breeding (Bachelor/Bachelorette in Paradise concept)	Hi handsome. Meet our Bachelorette in paradise – Look out for her on roads. Let's care for koalas – its breeding time
Driver Awareness (I'm off to school too)	Koalas are on the move – Let's keep them safe
Den your dog – secured pets	Where's Buster, Let's care for koalas – especially around pets
Koala breeding Bachelor/Bachelorette cinema ads	Guess who's having an affair at Redlands Coast – Let's care for koalas – its breeding time
Targeting men	It's a Bromance (check list) – Let's care for koalas – you know it's right
Koala movement	Like you, I move house. Let's care for koalas – they move about
General (Have you seen my web page)	Drive to the web page for more info

residents could take to help koalas. The campaign was targeted at all Redlands Coast residents, and targeted drivers, young people, pet owners and residents in areas with the highest number of koalas. Social media was specifically geo-targeted to postcode areas where koala sightings are highest. The following messages (Table 29.1) were disseminated throughout the community.

Multiple channels were used including print media, cinema, billboards, social media as well as buses and bus shelters. See Figure 29.3 for examples.

FIGURE 29.3
Billboard and bus ads.

CAMPAIGN EVALUATION

Campaign evaluation was undertaken with a follow-up survey collected in December 2018 after the peak koala movement season. The follow-up survey was conducted to examine whether community attitudes changed as a result of the koala awareness campaign. The follow-up questionnaire included repeated measures of awareness and attitudes to examine whether the community changed their attitudes towards koalas after seeing the campaign. Added questions included measures to examine campaign recall and self-reported behaviour changes. A total of 596 surveys were collected. Campaign effectiveness was evaluated using independent sample t-tests and descriptive statistics.

The Koala Awareness Campaign delivered positive outcomes, increasing residents' awareness about koalas' changing attitudes and behaviours. Approximately one-third (30%) of all respondents reported having seen the campaign. To understand which media channels delivered community awareness, respondents were asked to report in which communication channels they had seen the ads. Results showed that the most successful medium was Billboards (28.5%), followed closely by social media (27%). Over 20% of respondents recalled seeing the campaign in print channels and/or on bus shelters. Approximately, 42.8% of the sample reported seeing one or more messages through various channels. Communication message content was recalled accurately with respondents stating they saw messages about breeding seasons, bachelor koalas as well as koala habitat.

The data on the reach of the campaign indicated wide reach. The billboards reached 122,900 people, which is approximately 80% of all residents living in the council area. Statistics indicate that each of them was exposed to a message 5.9 times on average. The cinema ads were viewed 22,363 times over the period of the campaign, and social media generated a total of 31,211 impressions.

Outcome evaluation results indicated that attitudes towards koala conservation improved after the campaign. Attitudinal questions administered on a 7-point scale where 1 was strongly disagree and 7 was strongly agree, focussed on understanding survey respondent attitudes on the sense of belongingness of the koala in the community, koala conservation, and knowing how to help protect koalas. Results of the data analyses

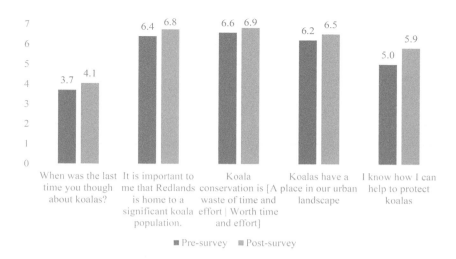

FIGURE 29.4
Attitudes pre- and post-campaign.

indicate positive changes from the baseline intercept surveys to the post campaign intercept surveys. See Figure 29.4.

The communication campaign has prompted residents to behave differently with driving slower and being more alert on the road reported by residents after the 'Koala Awareness Campaign'. For example, 43% of the respondents reported that they have done something differently in terms of road and speed-related actions while 39% reported that they have increased their awareness/alertness after seeing the campaign message.

LESSONS LEARNED

The results from this study confirm the effectiveness of implementing an advertising campaign during peak koala movement periods to remind residents that koalas are around them and to adjust their behaviour, such as slowing down when driving. Insights from this study can be used to design campaigns to improve coexistence between residents and wildlife. For example, given the positive attitudes towards koala conservation in the community and community attitudes that koala protection is a shared responsibility, additional emphasis on actions that community members can take to protect koalas was indicated as a need in future communication campaigns. Additionally, 37.2% of respondents didn't know how to protect koalas. Survey results indicated that future campaigns need to communicate what the Council is doing to help protect koalas and campaigns are needed to clearly communicate how measures implemented have had positive effects on koala populations and mortality reduction.

Based on the early success experienced in this communication campaign a follow up campaign was undertaken in the 2019/2020 breeding season. Serving as a pilot program of a multi-year campaign, this study confirms the necessity of re-occurrence of the koala awareness program to sustain long term impact on local residents' positive attitudes towards koala conservation. The evaluation results shaped who the campaign should target next and how the messages can be customised in the next iteration. Recommendations made by this work were accepted by the City Council and the program has been again implemented in the community during the 2019/2020 koala breeding season.

DISCUSSION QUESTIONS AND ACTIVITIES

- What are the main causes of koala fatality?
- What media channels were used in the campaign?
- What psychographic factors have been improved by the campaign?
- What is the most needed information for the residents to help protect koalas?
- Based on the insights provided in the chapter, what recommendations can you make to improve the effectiveness of the social advertising campaign?

FUNDING DECLARATION

Redland City Council funded and supported this research. The funder played no role in study design, collection, analysis and interpretation of data or the decision to submit the case study for publication.

REFERENCES

Australian Koala Foundation (2020) The koalas – endangered or not. Retrieved from: https://www.savethekoala.com/about-koalas/koala-endangered-or-not#:~:text=Koalas%20are%20in%20serious%20decline,we%20determined%20those%20figures%20here.

Bush, A. J., & Hair Jr, J. F. (1985). An assessment of the mall intercept as a data collection method. *Journal of Marketing Research*, *22*(2), 158–167.

Pang, B., Rundle-Thiele, S., & Kubacki, K. (2018). Can the theory of planned behaviour explain walking to and from school among Australian children? A social marketing formative research study. *International Journal of Nonprofit and Voluntary Sector Marketing*, *23*(2), e1599.

Queensland Department of Environment and Science. (2020). Koala Facts. Retrieved from: https://environment.des.qld.gov.au/wildlife/animals/living-with/koalas/facts

Rhodes, J. R., Beyer, H. L., Preece, H., & McAlpine, C. (2015). South East Queensland Koala Population Modelling Study. Technical Report, UniQuest, Brisbane, Australia.

30

Closing the Confidence Gap in STEM: A Social Marketing Approach to Increase Female Retention

Carina Roemer, Bo Pang, Patricia David, Jeawon Kim, James Durl and Sharyn Rundle-Thiele
Social Marketing @ Griffith, Griffith University, Nathan, Australia

CONTENTS

Learning Objectives	765
Themes and Tools Used	766
Introduction	766
Method	767
Pre-Survey	768
Social Advertisement Campaign Development	768
Self-Efficacy Messaging	768
Injunctive Norm Messaging	769
Post-Survey	770
Results	772
Conclusion	773
References	774

LEARNING OBJECTIVES

After reading the chapter, the reader should be able to:

- Understand how social marketing can contribute to gender equity
- Design and evaluate similar social advertising experiments to facilitate behavioural changes have a better understanding of how to use behavioural change theories to design communication messages

THEMES AND TOOLS USED

Tools for analysing the business environment (PESTEL) are given below:

- The theory of planned behaviour was used to establish the baseline and inform the communication campaign.
- Message framing (positive-negative) was used in a two-by-two experiment.
- Evaluation of the social advertisement campaign.

INTRODUCTION

Females remain underrepresented across the fields of science, technology, engineering and mathematics (STEM). According to the National Science Foundation, computer sciences and engineering have some of the lowest shares of female representation in the field. In 2016, females who attained a bachelor's degree in computer sciences and engineering were 18.7% and 20.9%, respectively (National Center for Science and Engineering Statistics, 2019). Despite substantial efforts into reversing this trend (Carver et al., 2017; Doerschuk et al., 2016; Drew et al., 2015), females are outnumbered by their male counterparts calling for greater advancements in the field. Numerous barriers have been identified impeding females to pursue STEM degrees including, negative academic climate (Settles et al., 2016) and sexism and negative biases (Blackburn, 2017). Low self-efficacy and underestimation of female STEM capability are among the key barriers preventing females from excelling in STEM (Fuesting & Diekman, 2016; Herrmann et al., 2016). Social marketing is capable of addressing complex social issues (Kubacki et al., 2015); however, there is a limited application in STEM fields (Roemer et al., 2020).

The importance of theory is noted in social marketing; however, recent reviews demonstrate that theory application is vague and/or not reported in sufficient details (Truong & Dang, 2017; Willmott et al., 2019). Of the studies reporting on theory (Willmott et al., 2019), the theory of planned behaviour (TPB) was heavily resorted to. The TPB (Ajzen et al., 1980) derives from value expectancy theory (Rosenstock et al., 1988) and assumes that behavioural intention is the central determinant of a specific behaviour being carried out. Intention is, in turn, determined by attitudes, subjective norms and perceived behavioural control (PBC) (Ajzen et al., 1980). Two types of norms are distinguished in subjective norms, namely injunctive norms and descriptive norms (Cialdini & Goldstein, 2004). PBC can also be distinguished by two components, namely self-efficacy and controllability (Trafimow et al., 2002). While the TPB has received criticism (French & Hankins, 2003; Sniehotta, 2009), other scholars highlight the applicability to a range of behaviours (McEachan et al., 2011; Pang et al., 2018). By understanding which factors explain behaviour focus can be directed to the constructs which have the highest prospects of resulting in behavioural change (David & Rundle-Thiele, 2018).

Effective communication serves as a mean of behavioural change. Effective communication is audience-centred, targeting different segments to prompt behaviour change (Albrecht, 1996) and essential to convey messages (Daellenbach et al., 2018). Message framing can either emphasise the advantages of taking certain action or focus on the negative consequences of not engaging in the proposed action (Gallagher & Updegraff, 2012). Both approaches (loss versus gain framing) have been adopted successfully in various campaigns. However, evidence about the effectiveness of message framing is mixed (Homer & Yoon, 1992) and more work is needed to understand whether sustained behaviour changed is enabled as a result of the framing of messages received (Snyder et al., 2004).

This study is one component of a larger research agenda with the overall aim of increasing female retention in STEM university degrees. Adopting a social marketing approach, the aims of this study are threefold. First, this study seeks to increase female retention in STEM in a higher education sector; second to advance the use of theory in social marketing and third, to determine the effectiveness of message framing.

METHOD

This study consisted of a quantitative method conducting a pre-survey to establish the baseline and identify constructs to focus the social advertisement on and a post-survey to determine the effectiveness. Methodological details are provided in turn.

Pre-Survey

A self-report online survey was used to establish the baseline for this study and sent to 9,438 STEM students. Informed by the TPB (Ajzen et al., 1980), the survey features psychological constructs including attitudes, PBC, self-efficacy, perceived social norms including descriptive and injunctive norms, intention to continue studying in their chosen field, sense of belongingness and perceived support. Survey items ($n = 64$) were measured using 7-point Likers scales, retrieved from validated scales. The survey was distributed to STEM university students at one Australian University between November 2018 and January 2019. The baseline sample included 526 STEM students with 291 females, 216 males and 19 unspecified. Almost 70% were aged 18–25 years old. Data analysis was undertaken in SPSS to determine which constructs could potentially predict intentions to stay enrolled in STEM. Self-efficacy (beta = 0.392, $p < 0.05$) and injunctive norm (0.375, $p < 0.05$) were found to be the strongest predictors of intention to stay enrolled.

Social Advertisement Campaign Development

Based upon the theoretical constructs of *self-efficacy* and *injunctive norms*, four social advertisements were developed to communicate existing support services to STEM university students. A two-by-two experiment was designed featuring the two constructs paired with message framing. The social advertisements were distributed via email in a controlled experiment at two time-points over a four-week period. STEM students were divided into five groups, four groups receiving one treatment and a control group with no exposure. Each group received the same email twice with a two-week interval.

Self-Efficacy Messaging

Self-efficacy involves the degree at which an individual is confident in their abilities to perform a behaviour or undertake a task through to completion. Context-specific, *self-efficacy* is related to the student's self-reported ability to continue in their STEM degree. Positive and negative *self-efficacy* messages were developed to improve females' *self-efficacy* while expose them to existing university support services.

Self-efficacy, positive: *I want to do STEM and I know I can!*
Self-efficacy, negative: *Studying STEM is hard am I up to the task?*

Injunctive Norm Messaging

Social norms relate to the influences of others on an individual's behaviour and how these are perceived by the individual to act (e.g. my peers' study hard) and how others believe an individual should act (e.g. my parents believe studying hard at university is important for my future). Injunctive norms informing the social advertisement messages.

Injunctive norm, positive: *The people that matter want to celebrate my achievements in STEM*
Injunctive norm, negative: *The people that matter want you to succeed in STEM – don't let them down*

The social advertisement campaign targeted both genders, as previously formative research had uncovered female-only campaigns were patronising and even sexist at times (Roemer et al., 2020). In addition, females associated their STEM profession with *knowledge, desire to innovate* and *identified as scientists*, which in turn led to a visual representation of a human brain to be featured on all advertisements. Without explicitly targeting females, certain aesthetic choices were made to appeal to females. These include using burgundy red as the main colour of the posters. Light pink was chosen for the brain, appropriate to how brains are generally portrayed. Informed by the constructs, a two-by-two experiment, featuring four messages, was developed with positively or negatively framed messaging. See Figure 30.1 for the two-by-two experiment. Design and layout were consistent across the four posters to ensure testing of framed messages and to minimise any confounding effects.

A website was created to combine all the existing university support services in one place to help students navigate the various offerings. Specifically, the website linked to *social clubs, scholarships, peer-assistant study session, work integrated learning (WiL) program, calendar with upcoming events, mentoring, bridging short courses* and *international student support*. Clicking on the social advertisement would take students to the website. See Figure 30.2 for the frontpage of the website.

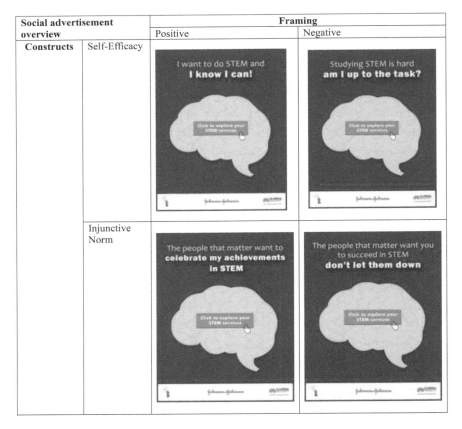

FIGURE 30.1
Social advertisement overview.

Post-Survey

A post-survey was developed to test the effectiveness of the social advertisements featuring recall questions in relation to the social advertisement. All other questions remained the same with the pre-survey to permit repeated measures. The survey was active from February to March 2019 with a total of 434 participants completing the post survey. The majority of the sample was females (60%) with 68% belonging to the 18- to 25-year-old age group. A total of 204 participants were matched from the pre- and the post-surveys. To test the effectiveness of positive and negative framing, the psychological items were analysed using t-test for each social advertisement. Pre- and post-completed respondents were evenly distributed across the five groups with approximately 33 respondents per group (see Figure 30.3 for an overview).

FIGURE 30.2
Website frontpage.

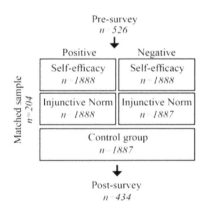

FIGURE 30.3
Overview of experiment.

RESULTS

Self-efficacy (confidence in the ability to perform a behaviour) and *injunctive norms* (perception of whether a behaviour will be approved by important others) were found to be significant predictors for retention. Further analysis determined that various levels of support from peer groups, family units, friends and from the institution at large will contribute to participants self-reported levels of self-efficacy and social norms.

Over the course of the social advertising campaign, 6,008 unique opens (i.e. at least one open per email address – some participants returned to the email later) by STEM2D undergraduates were observed. Compared to the original email list of 9,438 contacts, this shows more than 50% of the total sample saw a poster content and message. Each email was sent at 9:00 am on a Monday morning, with the majority of unique opens occurring between 9:00 am and 10:00 pm. This characterises the participant base as one that actively checks university information sent to them by email, as open rates were very low at other times of the day

Overall, findings suggest positively framed self-efficacy ($p = 0.07$, mean increase = 0.25) and positively framed injunctive norms ($p = 0.07$, mean increase = 0.41) significantly increased PBC ($p = 0.00$) between pre- and post-surveys. No significant results were observed for the negatively framed social advertisements.

Evaluation of the social advertisements indicated that a total of 6,008 unique email opens (i.e. at least one open per email address – some participants returned to the email later on) were obtained during the pilot program. Over 60% of the sample were females and 68% of survey participants belonged to the 18- to 25-year-old age group.

Analysis of the website traffic for all social advertisements, *Social Clubs* ($n = 605$) was the most frequent opened link, followed the *Project Website* ($n = 443$), *Scholarships* ($n = 193$), the *Work Integrated Learning* (WiL) program ($n = 162$) and the *Bridging Short Courses* ($n = 122$). Many participants returned to emails for multiple opens, with some participants clicking on available links up to 13 times.

CONCLUSION

This study reports a successful social marketing campaign that aimed to firstly encourage female students to remain enrolled in their STEM degrees in a higher education sector. Limited to no application of social marketing is evident in the context of female underrepresentation in STEM in higher education sectors (Roemer et al., 2020). This study extends social marketing's capability to tackle underrepresentation of females in STEM.

Second, this study responded to the call to advance theory use in social marketing by adopting the TPB. While the importance of theory has been highlighted, which in turn may offer an important role in the delivery of effective social marketing programmes (David & Rundle-Thiele, 2018), limited application is evident (Truong & Dang, 2017; Willmott et al., 2019). A priori approach advanced theory application in the social marketing domain by identifying the strongest predictors of female STEM retention to focus the social advertisement on.

Third, this study determined message framing effectiveness. Positively framed messaging successfully increased female PBC indicating positively framed messages could enhance communication effectiveness. Behaviour change communication is essential to promote uptake of behaviours (Briscoe & Aboud, 2012). Social marketers should focus attention toward positively framed messages to influence female retention in STEM.

Internal drivers, such as *self-efficacy* and *injunctive norms* are constructed on an individual level. External factors including need to be considered (Blackburn, 2017). Previously, perceptions of institutional climate have found to impact performance ability (Settles et al., 2016). Correlations have been found to influence university climate and sense of belongingness for under-represented groups (Johnson, 2012).

While positives were observed, it is recommended to upscale the social advertisement in combination with other social marketing strategies to support implementation. Single use of a marketing strategy limits program effectiveness (Lahtinen et al., 2020). In addition, involving a broader range of stakeholders is recommended considering the complexities of the issue. A limitation to this pilot is the involvement of students already enrolled in university disregarding other external factors. A holistic and long-term approach is required to increase females' self-efficacy and addressing the gender gap in STEM.

REFERENCES

Ajzen, I., Fishbein, M., & Heilbroner, R. L. (1980). *Understanding attitudes and predicting social behavior* (Vol. 278). Englewood Cliffs, NJ: Prentice-Hall.

Albrecht, T. L. (1996). Advances in segmentation modeling for health communication and social marketing campaigns. *Journal of Health Communication, 1*(1), 65–80.

Blackburn, H. (2017). The status of women in STEM in higher education: A review of the literature 2007–2017. *Science & Technology Libraries, 36*(3), 235–273. doi:10.1080/0194262X.2017.1371658.

Briscoe, C., & Aboud, F. (2012). Behaviour change communication targeting four health behaviours in developing countries: A review of change techniques. *Social Science & Medicine, 75*(4), 612–621.

Carver, S. D., Van Sickle, J., Holcomb, J. P., Jackson, D. K., Resnick, A., Duffy, S. F., … Quinn, C. M. (2017). Operation STEM: Increasing success and improving retention among mathematically underprepared students in STEM. *Journal of STEM Education: Innovations and Research, 18*(3), 20.

Cialdini, R. B., & Goldstein, N. J. (2004). Social influence: Compliance and conformity. *Annual Review of Psychology, 55*, 591–621.

Daellenbach, K., Parkinson, J., & Krisjanous, J. (2018). Just how prepared are you? An application of marketing segmentation and theory of planned behavior for disaster preparation. *Journal of Nonprofit & Public Sector Marketing, 30*(4), 413–443.

David, P., & Rundle-Thiele, S. (2018). Social marketing theory measurement precision: A theory of planned behaviour illustration. *Journal of Social Marketing, 8*(2), 182–201, doi: 10.1108/JSOCM-12-2016-0087.

Doerschuk, P., Bahrim, C., Daniel, J., Kruger, J., Mann, J., & Martin, C. (2016). Closing the gaps and filling the STEM pipeline: A multidisciplinary approach. *Journal of Science Education and Technology, 25*(4), 682–695.

Drew, J. C., Oli, M. W., Rice, K. C., Ardissone, A. N., Galindo-Gonzalez, S., Sacasa, P. R., … Triplett, E. W. (2015). Development of a distance education program by a land-grant university augments the 2-year to 4-year STEM pipeline and increases diversity in STEM. *PLoS ONE, 10*(4), e0119548.

French, D. P., & Hankins, M. (2003). The expectancy-value muddle in the theory of planned behaviour—and some proposed solutions. *British Journal of Health Psychology, 8*(1), 37–55.

Fuesting, M. A., & Diekman, A. B. (2016). Not by success alone: Role models provide pathways to communal opportunities in STEM. *Personality and Social Psychology Bulletin, 43*(2), 163–176. doi:10.1177/0146167216678857.

Gallagher, K. M., & Updegraff, J. A. (2012). Health message framing effects on attitudes, intentions, and behavior: A meta-analytic review. *Annals of Behavioral Medicine, 43*(1), 101–116.

Herrmann, S. D., Adelman, R. M., Bodford, J. E., Graudejus, O., Okun, M. A., & Kwan, V. S. Y. (2016). The effects of a female role model on academic performance and persistence of women in STEM courses. *Basic and Applied Social Psychology, 38*(5), 258–268. doi:10.1080/01973533.2016.1209757.

Homer, P. M., & Yoon, S.-G. (1992). Message framing and the interrelationships among ad-based feelings, affect, and cognition. *Journal of Advertising, 21*(1), 19–33.

Johnson, D. R. (2012). Campus racial climate perceptions and overall sense of belonging among racially diverse women in STEM majors. *Journal of College Student Development*, *53*(2), 336–346.

Kubacki, K., Rundle-Thiele, S., Lahtinen, V., & Parkinson, J. (2015). A systematic review assessing the extent of social marketing principle use in interventions targeting children (2000-2014). *Young Consumers*, *16*(2), 141–158. doi:10.1108/YC-08-2014-00466.

Lahtinen, V., Dietrich, T., & Rundle-Thiele, S. (2020). Long live the marketing mix. Testing the effectiveness of the commercial marketing mix in a social marketing context. *Journal of Social Marketing*, *10*(3), 357–375.

McEachan, R. R. C., Conner, M., Taylor, N. J., & Lawton, R. J. (2011). Prospective prediction of health-related behaviours with the theory of planned behaviour: A meta-analysis. *Health Psychology Review*, *5*(2), 97–144.

National Center for Science and Engineering Statistics. (2019). *Women, Minorities, and Persons with Disabilities in Science and Engineering (WMPD)*. Retrieved from https://ncses.nsf.gov/pubs/nsf19304/digest/field-of-degree-women#engineering

Pang, B., Rundle-Thiele, S., & Kubacki, K. (2018). Can the theory of planned behaviour explain walking to and from school among Australian children? A social marketing formative research study. *International Journal of Nonprofit and Voluntary Sector Marketing*, *23*(2), e1599.

Roemer, C., Rundle-Thiele, S., Pang, B., David, P., Kim, J., Durl, J., …, Carins, J. (2020). Rewiring the STEM pipeline – a C-B-E framework to female retention. *Journal of Social Marketing*, *10*(4), 427–446, doi:10.1108/JSOCM-10-2019-0152.

Rosenstock, I. M., Strecher, V. J., & Becker, M. H. (1988). Social learning theory and the health belief model. *Health Education Quarterly*, *15*(2), 175–183.

Settles, I. H., O'Connor, R. C., & Yap, S. C. Y. (2016). Climate perceptions and identity interference among undergraduate women in STEM: The protective role of gender identity. *Psychology of Women Quarterly*, *40*(4), 488–503. doi:10.1177/0361684316655806.

Sniehotta, F. F. (2009). Towards a theory of intentional behaviour change: Plans, planning, and self-regulation. *British Journal of Health Psychology*, *14*. doi:10.1348/135910708X389042.

Snyder, L. B., Hamilton, M. A., Mitchell, E. W., Kiwanuka-Tondo, J., Fleming-Milici, F., & Proctor, D. (2004). A meta-analysis of the effect of mediated health communication campaigns on behavior change in the United States. *Journal of Health Communication*, *9*(S1), 71–96.

Trafimow, D., Sheeran, P., Conner, M., & Finlay, K. A. (2002). Evidence that perceived behavioural control is a multidimensional construct: Perceived control and perceived difficulty. *British Journal of Social Psychology*, *41*(1), 101–121.

Truong, V. D., & Dang, N. V. (2017). Reviewing research evidence for social marketing: Systematic literature reviews. In K. R.-T. Kubacki & S. Rundle-Thiele (Ed.), *Formative Research in Social Marketing* (pp. 183–250). Singapore: Springer.

Willmott, T., Pang, B., Rundle-Thiele, S., & Badejo, A. (2019). Reported theory use in electronic health weight management interventions targeting young adults: A systematic review. *Health Psychology Review*, *13*(3), 295–317.

31

GlobalGiving and Performance Metrics

Sivakumar Alur
VIT Business School, VIT Vellore, Tamil Nadu, India

CONTENTS

Learning Objectives 777
Themes and Tools Used 777
Theoretical Background 778
Further Reading 784

LEARNING OBJECTIVES

After reading the chapter, the reader should be able to:

- Understand about performance metrics in non-profit's evaluation
- Compare and contrast types of non-profits and differences in evaluation
- Analyse the various metrics used and their significance in performance evaluation

THEMES AND TOOLS USED

Not Applicable

THEORETICAL BACKGROUND

Crowdfunding operates through four major models namely Crowdlending, Equity crowdfunding, Reward crowdfunding and donation crowdfunding. The first two models namely lending and equity models are financial rewards based. For-Profit businesses use these two models to fund their initiation or expansion. Reward crowd funding is a unique form where for-profit businesses could access funding with just a product or service as possible reward. In the case of donation crowdfunding, individuals and organisations fund social and non-profit projects expecting no rewards in return. Online donation crowdfunding platforms act as a marketing resource for fundraisers. Non-profits' marketing campaigns using donation crowdfunding portals have become common with increased digital marketing activities.

Socially responsible consumers, who are digital natives, consider convenience in contribution as an important factor in supporting worthy causes. Crowdfunding portals act as an intermediary between fundraisers and funders of social projects that work for a cause. These platforms help socially responsible consumers to search and evaluate various social projects requiring financial contribution. In addition, they help in easing payment through their links with payment gateways. Thus, effective online marketing generates funding for social marketing projects. Several social projects vie for such funding through various socially inclined donation crowdfunding platforms. The nature of competition decides whether the social project obtains funding or not.

A donation crowdfunding platform could also be a non-profit. In such a situation, a key question is if the platform can be evaluated the same way as other non-profits that work for a cause in the field. A for-profit platform in this case does not pose a dilemma. However, when a non-profit is involved, the parameters for judging the "facilitating non-profit" vs "an active non-profit" (like ones that use the platform for their campaign) cannot be the same. Impact metrics for different non-profits cannot be the same as the causes they work for affect different populations in varied numbers. The distinction between accountability and evaluation is also an issue. Accountability relates to control and assurance whereas evaluation is used for further learning. In many instances, evaluation tends more towards accountability rather than the broader interest of learning. Assessment of learning with a program involves both formative/process evaluation and summative/end of the program evaluation. The typical performance metrics used for evaluation are objective and financially biased like Gifts Secured, Donor & Donation Growth, Donor Retention Rate, Fundraising ROI, etc. Outcome and impact evaluation with respect to the mission of the non-profit is a long term and time consuming and financially draining project, which is not attempted in a large scale. In addition, as the outcome and impact differs across the different non-profits comparison between them is fraught with danger.

In 2002, Mari Kuraishi and Dennis Whittle founded GlobalGiving as a crowd-funded platform. This platform helped source global donors for various local non-profit projects across the world. These local non-profits operated in several sectors like refugee rehabilitation, healthcare, schools, sports, etc. In July 2020, GlobalGiving an US-based non-profit in its 18 years of existence had cumulatively helped in raising $502M dollars from 1,144,316 donors (including 323 companies) for 26,929 projects in 170 countries. In 2019 alone, it had raised more than $62M from 181,864 donors who funded 7,652 projects in 162 countries.

Non-profits that became GlobalGiving partners were selected through a rigorous process. Once the non-profits applied for partnership, GlobalGiving verified their credentials with an internal team through a site visit. In addition, it sought opinions from other local non-profit and philanthropy community members from which the applicant organisation hailed. Moreover, GlobalGiving also verified self-reported application details of the potential partner's activities. After these primary, secondary and self-report evaluations, potential partners were posed a real-life challenge in fund raising. They were required to raise $5,000 from 40 donors in 3 weeks. In case, small non-profits found the challenge difficult, GlobalGiving provided training on fund raising.

On selection as GlobalGiving's partners, new non-profit partners got access to its corporate donor database and online reach of individual donors globally. They also got an opportunity to interact with non-profits globally and access to organisational development resources including training, advice and support in fund raising. GlobalGiving Accelerator a virtual training program helped partners in innovative crowdfunding.

In addition, Webinars and workshops helped in knowledge transfer. Network mapping, storytelling and use of social media were taught to potential partners. Donations to GlobalGiving were deductible in the United States for US income tax payers. The non-profit was also tax-exempt under in the United States. GlobalGiving operated through its US and UK offices. However, it helped source and raise funds in local currencies of various countries. Thus, credibility and visibility of GlobalGiving helped online fundraising for local non-profit projects.

Non-profits typically needed several funding sources. Traditional funding sources were aid (including official governmental aid, grants and donations) and different forms of philanthropy like corporate giving. Online fundraising and crowdfunding helped to diversify funding sources reducing dependence on a single source. GlobalGiving facilitated small local non-profits to reach beyond country borders for funding for their projects. These projects could also receive matching funds, grant incentives and bonus prizes (Table 31.1) depending on fund availability.

Non-profits like GlobalGiving faced scrutiny on several parameters like transparency of fund use and project impact measurement. Several charity watchdogs like Charity Navigator used performance metrics (Tables 31.2 and 31.3) to rate non-profits. These acted as filters for prospective or

TABLE 31.1

An Example of a Drive for Funding Non-Profits Working for Refugees

Project	Non-Profit	Funding Raised	Donors	Grant Incentive	Bonus	Total
COVID-19: Education and Food for Palestinian Children	Tomorrow's Youth Organisation	$21,110	196	$750	$5,000 $10,000	$36,860
Empower Refugee Youth Through Dance	MindLeaps	$15,675	18	$750	$8,000	$24,425
Transformative Education for Refugees in Zimbabwe	Education Matters	$9,028	151	$750	$3,000 $5,000	$17,778
Action for Education: Back to School	Action for Education	$7,035	174	$750	$4,000 $3,000	$14,785
Feed 100 IDP Families In Cameroon	NDES FOUNDATION	$5,597	60	$750	$2,000	$8,347

Source: https://www.globalgiving.org

TABLE 31.2

Comparison of Non-Profits-1

Charity Name	Thousand Currents	Friendship Bridge	Global Fund for Children	GlobalGiving	Nuru International
Location	Oakland, CA	Lakewood, CO	Washington, DC	Washington, DC	Costa Mesa, CA
Overall Rating	92.89	90.07	86.51	98.23	86.58
Financial Rating	90.78	86.54	80.93	97.50	82.80
A&T Rating	96.00	96.00	100.00	(00.00	92.00
Financial Metrics					
Program Expenses	79.0%	86.8%	75.0%	97.4%	86.7%
Admin Expenses	7.1%	5.8%	6.6%	2.2%	10.0%
Fund Expenses	13.8%	7.3%	18.2%	0.2%	3.2%
Efficiency of Fund use	$0.10	$0.32	$0.14	$0.00	$0.03
Growth in Programs	53.0%	8.0%	0.3%	14.6%	0.3%
Working Capital	1.30	1.15	1.63	0.92	0.30
Liabilities to Assets	7.0%	51.9%	10.4%	3.6%	11.5%
Total Revenue	$6,791,650	$5,541,786	$8,361,619	$60,847,693	$4,833,785
Total Expenses	$5,625,416	$4,962,894	$6,168,839	$53,794,024	$6,243,949
Excess/Deficit	$1,166,234	$578,892	$2,192,780	$7,053,669	$-1,410,164
Net Assets	$5,495,932	$5,510,502	$10,292,736	$43,967,919	$1,860,407
CEO	$135,833	$154,254	$34,688	$58,333	$220,718
Compensation % of Expenses	2.41%	3.10%	0.56%	0.10%	3.53%
Transparency and Accountability (Form 990)					
Independent Members in Board	Yes	Yes	Yes	Yes	Yes
No assets diversion	Yes	Yes	Yes	Yes	Yes
Independent accountant	Yes	Yes	Yes	Yes	Yes
No related party loans	No	Yes	Yes	Yes	Yes
Meeting Minutes available	Yes	Yes	Yes	Yes	Yes
Form 990 copy shared	Yes	Yes	Yes	Yes	Yes
Policy on Conflict of Interest available	Yes	Yes	Yes	Yes	Yes

(Continued)

TABLE 31.2 (Continued)

Comparison of Non-Profits-1

Charity Name	Thousand Currents	Friendship Bridge	Global Fund for Children	GlobalGiving	Nuru International
Policy on whistleblower available	Yes	Yes	Yes	Yes	Yes
Policy on Records Retention exists	Yes	Yes	Yes	Yes	Yes
CEO listed with salary	Yes	Yes	Yes	Yes	Yes
CEO compensation process	Yes	Yes	Yes	Yes	Yes
No Board Member compensation	Yes	Yes	Yes	Yes	Yes
Website Information					
Privacy Policy for Donors available	Yes	Yes	Yes	Yes	No
List of Board Members available	Yes	Yes	Yes	Yes	yes
Financial results are Audited	Yes	Yes	Yes	Yes	No
Form 990 details are available	Yes	Yes	Yes	Yes	
List of key staff	Yes	Yes	Yes	Yes	Yes

Source: https://www.charitynavigator.org data generated using their comparison feature.

TABLE 31.3

Comparison of Non-Profits-2

Charity Name	Root Capital	The Resource Foundation	GlobalGiving	Village Enterprise	Growing Hope Globally, Formerly Foods Resource Bank
Location	Cambridge, MA	New York, NY	Washington, DC	San Carlos, CA	Western Springs, IL
Overall Rating	85.57	95.47	98.23	93.88	89.39
Financial Rating	80.00	95.00	97.50	91.35	85.00
A&T Rating	96.00	96.00	100.00	100.00	100.00

(Continued)

TABLE 31.3 (Continued)
Comparison of Non-Profits-2

Charity Name	Root Capital	The Resource Foundation	GlobalGiving	Village Enterprise	Growing Hope Globally, Formerly Foods Resource Bank
Financial Metrics					
Program Expenses	85.9%	92.6%	97.4%	81.0%	85.9%
Admin Expenses	10.2%	3.7%	2.2%	5.5%	7.6%
Fund Expenses	3.7%	3.5%	0.2%	13.3%	6.3%
Fund Efficiency	$0.08	$0.03	$0.00	$0.10	$0.06
Program Growth	−0.7%	12.8%	14.6%	21.6%	−1.1%
Working Capital	0.63	0.34	0.92	1.02	0.96
Liabilities to Assets	82.6%	0.7%	3.6%	5.6%	8.2%
Total Revenue	$22,275,720	$9,305,314	$60,847,693	$4,427,154	$2,723,780
Total Expenses	$23,083,477	$13,188,781	$53,794,024	$3,172,716	$3,255,435
Excess/Deficit	$-807,757	$-3,883,467	$7,053,669	$1,254,438	$-531,655
Net Assets	$15,718,807	$3,497,803	$43,967,919	$2,747,712	$2,985,776
CEO Compensation	$273,194	$167,000	$58,333	$126,671	$50,022
% of Expenses	1.18%	1.26%	0.10%	3.99%	1.53%
Accountability and Transparency Metrics Information Provided on the Form 990					
Independent Members in Board	Yes	Yes	Yes	Yes	Yes
No assets diversion	Yes	Yes	Yes	Yes	Yes
Independent accountant	Yes	Yes	Yes	Yes	Yes
No related party loans	Yes	Yes	Yes	Yes	Yes
Meeting Minutes available	Yes	Yes	Yes	Yes	Yes
Form 990 copy shared	Yes	Yes	Yes	Yes	Yes
Policy on Conflict of Interest available	Yes	Yes	Yes	Yes	Yes
Policy on whistleblower available	Yes	Yes	Yes	Yes	Yes
Policy on Records Retention exists	Yes	Yes	Yes	Yes	Yes
CEO listed with salary	Yes	Yes	Yes	Yes	Yes
CEO compensation process	Yes	Yes	Yes	Yes	Yes

(Continued)

TABLE 31.3 (Continued)

Comparison of Non-Profits-2

Charity Name	Root Capital	The Resource Foundation	GlobalGiving	Village Enterprise	Growing Hope Globally, Formerly Foods Resource Bank
No Board Member compensation	Yes	Yes	Yes	Yes	Yes
Does the charity's website include readily accessible information about the following					
Privacy Policy for Donors available	Yes	Yes	Yes	Yes	Yes
List of Board Members available	Yes	Yes	Yes	Yes	Yes
Financial results are Audited	Yes	No	Yes	Yes	Yes
Form 990 details are available	Yes	Yes	Yes	Yes	Yes
List of key staff	Yes	Yes	Yes	Yes	Yes

Source: https://www.charitynavigator.org data generated using their comparison feature.

potential donors to choose non-profits. While some metrics made sense, the same would not be applicable for all non-profits. GlobalGiving was wondering if the performance metrics could be applied on their partners and their projects.

FURTHER READING

1. https://www.fastcompany.com/1706758/globalgiving-25-your-thoughts
2. https://www.brandsouthafrica.com/investments-immigration/npos-maximise-online-fundraising
3. https://disrupt-africa.com/2018/08/kenyan-startup-m-changa-partners-global-crowdfunding-platform-globalgiving/
4. https://www.prnewswire.com/news-releases/learnzillion-co-founder-and-president-to-become-ceo-of-globalgiving-300714960.html
5. https://economictimes.indiatimes.com/small-biz/money/crowd-funding-platform-globalgiving-entering-indian-market-via-partnership-with-impact-guru/articleshow/58451076.cms
6. https://dotesports.com/business/news/activision-blizzard-donates-2-million-to-support-military-veterans-finding-jobs

32

Co-Creating and Marketing Sustainable Cities: Urban Travel Mode Choice and Quality of Living in the Case of Vienna

Tim Breitbarth
Swinburne University of Technology, Melbourne, Australia

David M. Herold
WU (Vienna University of Economics and Business), Vienna, Austria

Andrea Insch
University of Otago, Dunedin, New Zealand

CONTENTS

Learning Objectives	786
Compact Case Study	786
Internal Stakeholders and Place Marketing	786
Environmentally Friendly Transport as an Emerging Driver for Progressive and Attractive Cities	787
Changing Mobility Consumption Patterns: Vienna's Bold 365 Euro Annual Public Transport Pass for Citizens	787
Using Price as a Marketing Tactic to Lift Public Transport in Individuals' Mode Choice Set	788
Beneficiaries and Co-Creators of Vienna's Sustainability Policy, Practice and Reputation	789
References	790

786 • *Social and Sustainability Marketing*

LEARNING OBJECTIVES

After reading the chapter, the reader should be able to:

- Understand how sustainable city strategies and affordable public transport are interlinked with quality of living and successful place marketing
- Use the approaches and indicators provided alongside the case in order to analyse other cities and similar settings

COMPACT CASE STUDY

Vienna – the capital city of Austria with a population of nearly 2.5 million in its greater metropolitan area – ranks first in both the 2019 Global Liveability Ranking (The Economist Intelligence Unit, 2019) and the 2019 Mercer Quality of Living Ranking (Mercer, 2019), to name just two of the most renowned evaluations. Such rankings were established in recent decades as the trend of urbanization continues to shape life in the twenty-first century (Insch, forthcoming). Being recognized as a progressive and attractive place can help to promote the city as favorable for tourism, business and investments to external audiences. Likewise, it provides impetus for internal marketing as it can foster community involvement and pride.

Internal Stakeholders and Place Marketing

While external marketing has traditionally been the driver for city marketing and place branding in academia and practice, more recently the role of citizens in cultivating urban vibrancy and sustainability has become more prominent: "Cities, by their very nature, depend on their residents for economic, social, cultural and environmental vibrancy … Residents' quality of life and their satisfaction with their city of residence should be the ultimate aim of place management." (Insch & Florek, 2008, p. 138). Furthermore, "cities are living organisms that draw their shape and energy from the people who live and work in them. They defy top-down planning" (Bernstein, 2013, p. 138).

Hence, the participatory notion of people has gained traction. When the relationship between the city government and citizens was that of "administrator" and "resident" in "Cities 1.0," it is "collaborator" and "co-creator" in today's "Cities 4.0" (Foth, 2018). Generally, support from internal and external stakeholders lies at the core of a city's reputation according to the Reputation Institute's global 2018 City RepTrak analysis, in which Vienna ranks 4th. Internal stakeholders – especially residents – drive decisions made in regard to tourists' image formulation and consumption behavior (Stylidis et al., 2015).

Environmentally Friendly Transport as an Emerging Driver for Progressive and Attractive Cities

Environmental policies and respective actions and outcomes are becoming marketing assets for a city. The prominent Mercer ranking draws on "natural environment (climate, record of natural disasters)" and "public services and transportation (electricity, water, public transportation, traffic congestion, etc.)" as 2 of the 10 categories of quality of living. That these factors increasingly make a difference is illustrated, on the one hand, by Mercer's explicit acknowledgment that Australian cities like Sydney and Melbourne recently dropped out of the Top 10 because of increasing traffic congestion; and on the other hand, by a deeper analysis of Vienna's improved ranking as its car mode share has fallen significantly compared to peer cities (Buehler et al., 2017).

City rankings focused on environmental sustainability supporting Vienna's city marketing. Those rankings may not be produced every year, but Vienna came in 5th in the latest Arcadis Sustainable Cities Index (2018) (Arcadis, 2018) and 4th in the European Green Cities Index (The Economist Intelligence Unit, 2012). Progress of European cities in(to) the top 10 of the IESE Cities in Motion Index (2019) (IESE, 2019) is attributed to environmental and transport efforts: Amsterdam (#3; local government directed at phasing out car transport in the city; 90% of Amsterdam's households use bicycles as a means of transportation), Copenhagen (#8; environmental protection, e.g., city's pledge of becoming carbon neutral by 2025) and Vienna (#10; great steps forward in transportation).

Changing Mobility Consumption Patterns: Vienna's Bold 365 Euro Annual Public Transport Pass for Citizens

Given the state of the planet and climate change projections, distinguished marketing scholar Roland T. Rust proposes changing geographic consumption patterns as a feature of "the future of marketing" (Rust, 2020). More than half of the world's population now living in urban areas and already a decade ago such areas accounted for 80% of humanity's greenhouse gas emissions, with combustion engine driven vehicles being a major contributor, also to poor air quality (European Green City Index, 2012) (The Economist Intelligence Unit, 2012). Hence, in an urban setting travel mode choice of citizens is critical, which is where a number of European cities are setting themselves ambitious targets – and where Vienna has made a distinctive mark. The City of Vienna is committed to prioritizing public transport, walking and cycling as the most environmentally friendly mobility modes (Pamer, 2019).

In the 2010 council elections a key aspect of the Austrian Green Party's campaign was the introduction of an individual public transport annual pass for all citizens for as low as 100 Euro (119 USD). After their election success and negotiations with their coalition partner (Social Democratic Party of Austria), in 2012 the City of Vienna introduced a 365 Euro annual pass for all public transport – still clearly making it the most affordable pass in any comparable city, arguably even worldwide.

This "1-Euro per day" campaign was accompanied by supporting policies like encouraging parking management and further improving walkability/cyclability (96% of the city's residents live within walking distance to a public transport station), and targeted measures like large-scale advertising. Most of the city's internal stakeholders greatly support public transport and support for the transport policy was expressed in the outcome of 2015 and 2020 council elections. Today, there are fewer cars registered in Vienna than there are annual pass holders. This is important as individual and situational mobility decisions are driven by the set of prevalent options perceived to be available.

Using Price as a Marketing Tactic to Lift Public Transport in Individuals' Mode Choice Set

Research shows that when the aim is to achieve lasting modal switching (e.g., from car to public transport), it is important to include the desired mode into consumers' mode choice set. Thereby, the likelihood of the desired situational choice-making and for such to manifest as behavioral routine increases. To achieve lasting change, financial incentives have shown to be successful, for example, the reimbursement of commuting costs by employers (Ton et al., 2020). Certainly, a very affordable pass at 1 Euro (1.19 USD) a day allowing access to an extensive mobility service is a price-based marketing tactic designed to change residents' travel mode choice. The percentage of individual trips by mode has shifted from 40% (1993) to 29% (2018) for cars; 3% to 7% for bikes and 29% to 38% for public transport. Comparable cities in German-speaking countries show only 27% (Berlin) and 18% (Hamburg) use of public transport among residents (Vogt, 2018).

Despite this success, there have been some challenges facing the city administrators. For example, revenue from public transport has gone up due to the high number of riders, but operating costs increased, too, due to added servicing and maintenance required.

However, multiple citizen surveys and advisory public referenda have been used to allow for citizens' input and voice, but also to track support for key decisions (Buehler et al., 2017). Hence, the participatory/collaborative mindset and marketing philosophy embraced as evident for Cities 4.0 aligns well with ambitious policy and action changes to become an internationally renowned place for a high quality of living. Clearly, Vienna demonstrates innovation and leadership with respect to the "1-Euro per day" public transport pass strategy, which few other and usually smaller cities in continental Europe have adopted or trialed since.

Also, criticism aimed at stressing the high operating costs can be countered by the argument that pro-climate policies save costs arising from the carbon offsetting requirements. Also, there is less need to mitigate negative local environmental and health impacts caused by emissions from combustion engines and tire abrasion, for example.

Beneficiaries and Co-Creators of Vienna's Sustainability Policy, Practice and Reputation

With infrastructure and transport being prominent features of cities, the paradigm, policy and perceptional shifts can create favorable halo effects. For example, for Vienna it has created the holistic "Sustainable Vienna" external marketing campaign. Also, it allows individual organizations in the city to become more ambitious around their environmental sustainability targets, strategies and branding. Rapid Vienna, the largest and most successful Austrian football club, is an example.

The largest share of carbon emissions by far at major sports events is caused by fan mobility (Herold et al., 2019; Musgrave et al., 2019). It is estimated that 60% to 80% of all carbon emissions stem from fan mobility. The carbon output of spectator mobility, however, is heavily influenced by modal choice. In the case of Rapid Vienna, 56% of season ticket holders use public transport (39% rely on the car; alone/shared), compared to 40% in England for example. The widespread adoption of Vienna's affordable travel pass and the promotion of public transport availability by the club among its season ticket holders and its home venue Allianz Stadium support this progress.

Due to the location of the stadium in a fringe suburb of the city – something not unusual for European football stadia built over the last two decades – walking as the sole mode of transport is unrealistic. Usually, the fringe location and respective planning and design of the sports venue caters for and attracts a greater share of car usage than in Rapid Vienna's case. Still, Rapid's projected CO2e for one home match due to spectator mobility is 103 tons. It remains both an ambition and challenge that, in general, the reduction potential – and thereby, contribution, to achieve global climate change targets – is very high in the transport sector which takes place in an urban context (UNEP/IPCC, 2018).

REFERENCES

Arcadis (2018). Sustainable Cities Index 2018. Available from: www.arcadis.com/media/1/D/5/%7B1D5AE7E2-A348-4B6E-B1D7-6D94FA7D7567%7DSustainable_Cities_Index_2018_Arcadis.pdf

Bernstein, A. (2013). Cities as ideas. *Harvard Business Review*, April, 138–139.

Buehler, R., Pucher, J., & Altshuler, A. (2017). Vienna's path to sustainable transport. *International Journal of Sustainable Transportation*, 11(4), 257–271.

Foth, M. (2018). Participatory urban informatics: Towards citizen-ability. *Smart and Sustainable Built Environment*, 7(1), 4–19.

Herold, D., Breitbarth, T., Schulenkorf, N., & Kummer, S. (2019). Sport logistics research: Reviewing and line marking of a new field. *The International Journal of Logistics Management*, 31(2), 357–379.

IESE (2019). Cities in Motion Index. Available from: https://media.iese.edu/research/pdfs/ST-0509-E.pdf

Insch, A. (forthcoming). Peoplescapes and Placemaking in a Multicultural World: Residents' Identity, Attachment, and Belonging, In N. Papadopoulos & M. Cleveland (eds.), *Image, Marketing, and Branding of Places and Place-Based Brands: The State of the Art*, Edward Elgar, Cheltenham.

Insch, A. & Florek, M. (2008). A great place to live, work and play: Conceptualising place satisfaction in the case of a city's residents. *Journal of Place Management & Development*, 1(2), 138–149.

Mercer (2019). 2019 Mercer Quality of Living Ranking. Available from: www.mercer.com/newsroom/2019-quality-of-living-survey.html

Musgrave, J., Jamson, S., & Jopson, A. (2019). Travelling to a sport event: Profiling sport fans against the transtheoretical model of change. *Journal of Hospitality and Tourism Research*. https://doi.org/10.1177/1096348020915255

Pamer, V. (2019). Urban planning in the most liveable city: Vienna. *Urban Research & Practice*, 12(3), 285–295.

Reputation Institute (2018). 2018 City RepTrak. Available from: www.reptrak.com/city-reptrak

Rust, R. (2020). The future of marketing. *International Journal of Research in Marketing*, 37, 15–26.

Stylidis, D., Belhassen, Y., & Shani, A. (2015). Three tales of a city: Stakeholders' images of Eliat as a tourist destination. *Journal of Travel Research*, 54(6), 702–716.

The Economist Intelligence Unit (2012). European Green Cities Index. Available from: https://apps.espon.eu/etms/rankings/2012_European_Green_City_Index_sum_report.pdf

The Economist Intelligence Unit (2019). 2019 Global Liveability Ranking. Available from: www.eiu.com/topic/liveability

Ton, D., Bekhor, S., Cats, O., Duives, D., Hoogendoorn-Lanserc, S., & Hoogendoorna, S. (2020). The experienced mode choice set and its determinants: Commuting trips in the Netherlands. *Transportation Research Part A: Policy and Practice*, 132, 744–758.

UNEP/IPCC (2018). Emissions Gap Report 2018. Available from: www.ipcc.ch/site/assets/uploads/2018/12/UNEP-1.pdf

Vogt, J. (2018). ÖPNV in Wien - Im Land der Öffis. In Zeit Online, 18 March, www.zeit.de/mobilitaet/2018-03/oepnv-wien-oesterreich-erfolg-preise

Section XIV

Selected Case Studies to Reflect on Practice and Use as Learning Tools: Case Studies from Developed Economies (Complex and/or Long)

33

Positioning a Company in the Chemical Industry as a Sustainability Driver

Filipa Lanhas Oliveira and João F. Proença
Faculty of Economics, University of Porto, Portugal

CONTENTS

Learning Objectives	796
Themes and Tools Used	796
Theoretical Background	796
Sustainability	797
Circular Economy	798
B2B Markets: Chemical Industry	798
The Impact of Sustainable Approaches in Companies	799
The Roles of B2B Marketing in Sustainability	801
The Paradigm of the Chemical Industry	802
Introduction	802
Sustainability in the Chemical Industry	804
Narrowing the Gap between B2B and B2C Communications	805
Communicating to Organizational Buyers	807
Social Listening	811
Purchasing and Investment Decisions: Organizational Factors	812
Purchasing and Investment Decisions: Personal Factors	812
The Focus on New Generations	813
The Customers	814
The Future of Sustainability Communication	815
Communicating Sustainability	818
Consistent Communication	819
Promoting More Environmentally Sustainable Solutions	821
Sustainability Reporting	824
The Importance of Persuasive Communication for Sustainability	824
The Role of Internal Brand Communication and Education	825

Major Problems 828
Discussion 830
Conclusion 832
Discussion Questions 833
Limitations 833
Credit Author Statement 833
References 834

LEARNING OBJECTIVES

This case study illustrates and discusses a global positioning for Henkel Adhesives Technologies Packaging SBU. The case explores several topics significant to this brand's positioning strategy and discusses the application a conceptual map to support the positioning strategies of other companies under similar circumstances and goals as this SBU. Briefly, the case discusses:

- Sustainability in the chemical industry
- Narrowing the gap between B2B and B2C communications
- Communicating sustainability
- Sustainability reporting
- The future of sustainability communication
- The importance of persuasive communication for sustainability
- The role of internal brand communication and education

THEMES AND TOOLS USED

- Interviews and Corporate report
- Sustainability Positioning Conceptual Map
- Exhaustive PowerPoint case presentation

THEORETICAL BACKGROUND

In an industry that is known for its pollution and in an era where consumers are more concerned about sustainability issues (Zou & Chan, 2019), positioning a company as a Circular Economy expert, is no easy task. The aim of the following literature review is to capture the necessary background to understand how to better achieve this corporate goal from a marketing and communication perspective.

Sustainability

Literature indicates that the concept of sustainability has first received augmented attention during the 1980s with the publicity of the International Union for the Conservation of Nature and Natural Resources or the UNESCO Biosphere (Brown, Hanson, Liverman, & Merideth, 1987). The broader definition is derived from maintaining a certain balance in meeting the needs of the society nowadays without putting at risk the aptitude to satisfy the needs of the future generations (Brundtland et al. 1987; Stirling, 2009).

Summing up the various definitions that exist, it becomes evident that sustainability comprises three important dimensions to ensure its completion – the Triple Bottom Line – changes in social, economic and environmental equity (Murat, 2009; Hunt, 2011; Newport, Chesnes, & Lindner, 2003; Savitz, 2013). Dresner (2012) breaks down the relationship between sustainability and equity even further by explaining the dilemma between the advancement of technology to enable the replacement of natural capital by human capital on the cost of current natural capital in order to keep equity of natural resources. Economists tend to downplay the risk of exploiting present natural capital by not considering worst-case scenarios whereas environmentalists tend to highlight these cases (Menon & Menon, 1997).

For last, literature suggested that the key players towards sustainability are not only the corporations but also consumers and governments (Kotler & Armstrong, 1996). Consumers must be prepared and accept the fact that they have to decrease their consumption levels and governments' policies must take into account ecologically sustainable matters (Shrivastava, 1995). And regarding governments, Sheth and Parvatiyar (1995) noted that they can regulate both consumption and companies by either imposing taxes on pollution, tradable pollution licences, performance bonds, deposit-refund systems and make use of differential prices approaches. Besides this, they can also encourage investments in cleaner technology and options, and offer subsidisations for adopting of environmentally friendly solutions.

Circular Economy

Circular Economy is an approach that aims to include all the abovementioned concepts into a change of the economic system. As suggested by its name, Circular Economy intends to change economic activities by moving them away from the linear system (take, make, dispose) towards a closed loop – which relies on redesigning products and services so that the linear economy waste is replaced by a system that incorporate durability, repair, re-use, restoration and recycling terms or by simply the adoption of product service systems (PSS). This circular approach has been considered the best alternative between these two (Farooque et al. 2019).

The topic has been receiving increasing attention since the European Union report "Closing the loop" in 2015 (Act, 2012) was published. Besides this, the topic was ambitiously integrated and considered as "the origin to boost competitiveness, create jobs and generate sustainable growth" (Commission, 2015). This philosophy creates a business model in which the life circle of materials is prolonged, and the usage of new resources becomes less necessary since recycling, reuse or repair become crucial steps within the business rationale (Den Hollander & Bakker, 2016; Linder & Williander, 2017). This process requires companies to reutilise used products in the production chain (Muranko, Andrews, Chaer, & Newton, 2019). A circular business model (CBM) is when a company looks for solutions to close material loops (Nußholz, 2017, p. 12). To achieve such drastic developments in the economic and societal systems, it will be important to alter consumer behaviour towards a pro-circular behaviour as well (Geissdoerfer, Savaget, Bocken, & Hultink, 2017). Pro-circular behaviour is about establishing the priority to purchase products or services that reduce damage to the environment, economy and society – resources-efficient or remanufactured products (Muranko et al. 2019).

B2B Markets: Chemical Industry

> The reputation of pesticide producer's hovers somewhere between that of al-Qaida and serial killers.
> EU parliamentary employee
>
> Johnson (2012, p. 2)

Although the al-Qaida comparison is exaggerated, it translates the hard battle that the chemical industries' companies have been facing from the public opinion. As a result, businesses within the sectors of chemicals, energy, mining and oil are constantly ranked by its low public trust levels; these brands mostly rely on corporate branding rather than product branding to promote their environmentally sustainable initiatives (Sheth & Sinha, 2015).

Business-to-business (B2B) markets represent, on average, more than half of the GDP of all developed and many developing countries (Wiersema, 2013; Kleinaltenkamp, 2018). However, some authors defend that B2B and business-to-customer (B2C) do not only represent two different marketplaces. Instead, they simultaneously operate in business-to-business-to-consumer (B2B2C) marketing, considering all the parties as customers (Iankova et al. 2019; Mingione & Leoni, 2020).

The Impact of Sustainable Approaches in Companies

Sustainability is a significant trend in the chemical industry (Johnson, 2012). Since the begging of the century, about 80% of the world's 29 largest companies have adopted some kind of sustainable initiative or committed organization (Johnson, 2012). The same also applies to smaller ones and major industry association (Johnson, 2012). However, the best definition of sustainability for the general public seems to be merely the act of showing concern. The next phase is "trying to do the right thing" (p. 44) (Johnson, 2012). Whether or not the action is, in fact, environmentally beneficial and is secondary (Johnson, 2012; Grimmer & Woolley, 2014; Matthes, Wonneberger, & Schmuck, 2014; Yang, Lu, Zhu, & Su, 2015). Literature also states that "probably the largest area of debate surrounds the notion of sacrifice. Must we sacrifice economic growth or personal consumption to protect the environment?" (p. 44).

Despite of some indulgences about the fact that environmental objectives can't be combined with the economic goals of a business and that these could be a direct threat to environmental conservation, Sharma, Iyer, Mehrotra, and Krishnan (2010) views them as "two sides of the same coin" (p. 1). In fact, the motivations associated to this investment can be easily acknowledge while revising some literature. Table 33.1 aims to not only sum up the benefits that these researchers found but also strengthen the credibility on each of the benefits by enumerating the consensus points that various authors that have data to support it.

TABLE 33.1

Benefits of Sustainability Initiatives

Benefits	Authors
Strength competitive advantage	Brown & Dacin (1997); MacArthur (2013); Bilgin (2009); Sharma et al. (2010); Wagner (2005); Rodriguez, Ricart, & Sanchez (2002); Martin & Schouten (2011); Menon & Menon (1997); Shrivastava (1995); Mudambi, (2002); Mariadoss, Tansuhaj, & Mouri (2011); Shahbazpour & Seidel (2006)
Proximity to customers and customer loyalty	Martin & Schouten (2011); Mudambi (2002); Luo & Bhattacharya (2006); Mariadoss et al. (2011); MacArthur (2013); Geissdoerfer et al. (2017)
Boost in economic performance	Papadas, Avlonitis, Carrigan, & Piha (2019); Sharma et al. (2010); Mariadoss et al. (2011); Menon & Menon (1997); Shrivastava (1995)
Improve reputation	Papadas et al. (2019); Sharma et al. (2010); Geissdoerfer et al. (2017); Menon & Menon (1997); Brown & Dacin (1997); Shrivastava (1995); Mariadoss et al. (2011); Moktadir et al. (2018); Rehman, Seth, & Shrivastava (2016)
Cost savings and increased efficiency	Rehman et al. (2016); Popa & Popa (2016); Stahel (1994); Martin & Schouten (2011); Allwood, Ashby, Gutowski, & Worrell (2011); Kumar & Christodoulopoulou (2014)
Innovation and new business opportunities	Papadas et al. (2019); Sharma et al. (2010); Dubey, Gunasekaran, & Ali, (2015); Kolk (2000); Martin & Schouten (2011); Rehman et al. (2016)
Attracts and retains talented employees	Bhattacharya & Korschun (2008); Martin & Schouten (2011)
Market share increase	Savitz (2013); Geissdoerfer et al. (2017); Xie, Huo, & Zou (2019); Papadas et al. (2019); Sharma et al. (2010)
Rising living standards in emerging markets	Hsu, Tan, & Mohamad Zailani, (2016)
Best customer feedback	Mariadoss et al. (2011)
Reduces risks	Martin & Schouten (2011

Source: Own authorship.

According to Kumar and Christodoulopoulou (2014), when a firm includes sustainability practices into their business plan, they translate into direct and indirect benefits. The direct benefits concern the optimization of operations and lower costs. The indirect benefits concern the enhancement of brand value when communicating with the company's stakeholders, which usually require some time to show their results.

The Roles of B2B Marketing in Sustainability

Literature suggests that marketing B2B "will be the glue that will drive successful firms" (Sharma et al. 2010, p. 338) and it is actually a "key driver to sustainable development" (Mariadoss et al. 2011, p. 1306). However, comparing to a B2C context, B2B firms are facing a harder challenge in what comes to promote environmental sustainability: although business customers are more rational, they continue to misperceive circular offers (Houston, Briguglio, Casazza, & Spiteri, 2019) as offers of lower quality (Muranko et al. 2019) and considerably more expensive options (Mariadoss et al. 2011).

Controversially, more radical perspectives condemn the role of marketing in sustainability as it is a vigorous creation of needs that doesn't match sustainable development principles (Rex & Baumann, 2007) – not having into account demarketing and counter-marketing techniques (Achrol & Kotler, 2012). Additionally, they often see marketing as an anti-hero for sustainability – as it can stimulate unmanageable levels of demand and consumption (Peattie, 2001).

Nevertheless, greenwashing is another topic that is easily brought to mind when speaking about green marketing. As environmental problems have crashed the industry for years and its perception of dishonesty, insincerity or uncaring by the industry, greenwashing became a hot topic (Johnson, 2012). It is the act of having poor environmental performances but communicating otherwise by lying, acclaiming compliance, green products, processes, and endorsements, spinning words or science (Chamberlin & Boks, 2018).

In other words, marketing was part of the disease. Now, it is a part of the cure. It can be a way of changing these negative perceptions and consumption patterns that have caused unsustainable lifestyles, towards a more environmentally conscious consumption (Farooque et al. 2019). Thus, it is essential to educate consumers to change behaviours (Smol, Avdiushchenko, Kulczycka, & Nowaczek, 2018) and involve them in this chain in order to keep up the movement of raw materials and energy and to repair and maintain the utility of a product – so that the product can later be remanufactured or reused by another company, creating a closed and long-lasting product and consumption loops (Geissdoerfer et al. 2017).

In green marketing, it is usual to assume that listing functional advantages is enough to persuade consumers to buy the product or service; however, research shows that emotional appeals, or those using both functional and emotional elements, actually make a bigger and impactful effect (Grimmer & Woolley, 2014; Matthes et al. 2014). Green marketers need to emphasize both tangible and intangible value and align it with environmental benefits with consumer self-interest to successfully increase sales and such kind of consumption (Grimmer & Woolley, 2014; Yang et al. 2015). Additionally, businesses are now attempting to establish the "normalization" and integration of sustainability (Martin & Schouten, 2014) by introducing longer-term effects and benefits by addressing business models such as localization or PSS (Dangelico & Vocalelli, 2017).

THE PARADIGM OF THE CHEMICAL INDUSTRY

Adhesives are used across all sectors – from aerospace to food packaging – and they guaranty safety and lightness when bonding different surfaces. However, despite the utility they have in our lives, their manufacture is characterized by the large use of chemicals and petroleum. During the last 15 years, final consumers have also shown an exponential increase in their concerns and awareness about the consequences of an imminent climate catastrophe. For instance, businesses within this industry are suffering from the consequent negative reputation, are recognizing the effects of their externalities and reconsidering their overall strategies. However, regardless of these industries' efforts and their adoption of more sustainable systems, they are continuously observing consequent generalized negative reputation. Although the chemical industry is omnipresent across the whole value chain and it is among the most polluting industries, it translates its great potential and even greater responsibility to make the difference for the environment. Accordingly, Henkel Adhesives Technologies, as a B2B Company in this industry, aims to occupy a pioneering position within its industry as a Circular Economy expert and a sustainability driver.

INTRODUCTION

Henkel is a German chemical and consumer goods multinational company. It extends to the consumer and industrial sector and is divided in three business units (Adhesive Technologies – global leader, Laundry &

Home Care, Beauty Care). Its most renowned brands are Loctite, Persil, Syoss, Schwarzkopf and Fa amongst others. It was founded in 1876 by Fritz Henkel and now is one of the ten biggest familiar companies in Germany. The mission of Henkel Adhesives Technologies is to: deliver solutions to the global challenges includes, to raise awareness about the importance of Circular Economy options and to empower the respective industry to tackle the environmental crisis. This company strives for value creation not only at an organizational level but also for their customers and aims to help lowering and moderating the environmental footprint at the same time – "achieving more with less" (Henkel's Three Factor Strategy). Although this is not the classic "eco-friendly", "green" or "sustainable" company, it is continuously contributing towards a Circular Economy system by offering its customers solutions (in this case, adhesives) that allow the extant materials included in the product to be kept in the loop. By taking responsibility for its products and solutions all along the value chain, Henkel seeks to create transparency and enable sustainable processes, services and consumer goods through many channels. This company aims to become the most sustainable possible while enabling its customers to become more sustainable as well. After all, to maintain the comfort we have on our daily lives, the products that result from this industry will never be 100% indispensable. For this reason, it is up to consumers, Henkel and other similar companies to create smarter alternatives or make use of these products in the most efficient way. And more importantly, to design its production to maintain these products in the loop for as long as it is sustainably possible. However, Henkel's dilemma lays into changing its negative corporate reputation. They are implementing communication strategies that position this SBU as sustainability driver and an expert in Circular Economy.

This is a rich case study that shows the efforts and results of a global company in a delicate position within a highly polluting industry. Besides this, it was noticed that literature has widely explored Circular Economy business models, but there are only a few that relate it with a marketing and communications perspective. The purpose of this case study is to smoothen the comprehension of how the communication of sustainable approaches and visions are being done within a B2B corporation, to understand how it can contribute to a sustainable performance of a company in a similar context and finally to discuss and define what are the best practices. Additionally, this case study investigates and discusses the chemical industry operations concerning environmental topics and, ultimately, investigates how a company, that is making environmental efforts, can mitigate this generalization through stronger positioning and communication strategies. In order to assist similar positioning strategies of other companies under similar circumstances, it was also proposed a conceptual map with key elements to assist such positioning.

SUSTAINABILITY IN THE CHEMICAL INDUSTRY

The companies' short-term profit goals are affecting the welfare of the neighbourhoods they operate in (Porter & Kramer, 2006) and consequently, these self-centred businesses within the chemical industry are continuously facing frustration from these communities and their own stakeholders (Sheth & Sinha, 2015). However, literature suggests that one way to empower their corporate brands is by attributing more importance to sustainability communication approaches, instead of treating sustainability marketing as a liability (Sheth & Sinha, 2015).

Henkel's ultimate goal is to change its corporate reputation as a polluter to a sustainability driver in the chemical industry and implement communication strategies that position this SBU as an expert in Circular Economy.

> It's one thing is to have outsiders, who maybe aren't our industrial customers, looking at Henkel as a chemical company and other thing is the perspective of what is that company looking for sustainability. But for our customers I think this bigger focus on sustainability and Circular Economy with actionable solutions, has been very welcomed by our customers as the competitive landscape and consumer's behaviour is changing.
> Interview to Head of Global MCA for Henkel Adhesives Technologies Packaging SBU.

This new Henkel Adhesive's Packaging positioning strategy consists of proving to be a "sustainability driver that offers its customers sustainable solutions". Although it is not the classic "eco-friendly", "green" or "sustainable" company, it is continuously contributing towards a Circular Economy system by offering its customers solutions (in this case, adhesives) that allow the extant materials of the product to be kept in the loop. It is a company that aims to become the most sustainable possible while enabling its customers to become more sustainable too. After all, to maintain the comfort we have on our daily lives, the products that result from this industry will never be 100% indispensable. For this reason, it is up to consumers, to Henkel and to other similar companies to create smarter alternatives or make use of these products in the most efficient way. And more importantly, to design its production to maintain these products in the loop for as long as it is sustainably possible.

Accordingly, Henkel Adhesives Technologies has shown, in a market analysis, that 91% of its competitors and key accounts were concerned about communicating sustainability approaches and only 42% of them were concerned about communicating Circular Economy approaches (Henkel Adhesives Technologies, 2020). Besides this, it was noted that only the companies communicating about Circular Economy were implicitly communicating their concerns about sustainability. These concerns could be verified by the investments they have made in proper campaigns to communicate their activities, their changes on corporate infrastructure, their adoption of more ethical frameworks projects and their results and impacts. Therefore, these companies were not greenwashing. Lastly, although all the key accounts that were investing in communicating their concerns for sustainability, only half of the competitors were doing so (op. cit). Consequently, based on this report, this Henkel SBU new positioning strategy appears to be a pioneer in its market – which can result in even higher chances to stand out and succeed.

NARROWING THE GAP BETWEEN B2B AND B2C COMMUNICATIONS

Although Henkel Adhesives Technologies is communicating in the B2B marketplace, it agrees that their form of communication is tending towards a B2B(2C) approach. Examples of these can be verified in the campaigns "[be more]" (Figure 33.1), "Where to find Henkel Adhesives" (Figure 33.2), "Sustainability Posts" (Figure 33.3), "Episodes of Circular Economy" (Figure 33.4), etc.

Henkel Adhesive Technologies Packaging SBU considers to have three target interlocutors (Figure 33.5): consumers in general, customers that are also consumers (employees who, outside of their workplace, present

FIGURE 33.1
"[be more]" campaigns. (From Henkel Adhesives Technologies LinkedIn Posts). (*Continued*)

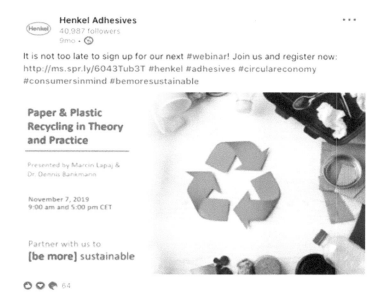

FIGURE 33.1 (Continued)
"[be more]" campaigns. (From Henkel Adhesives Technologies LinkedIn Posts).

a regular consumer behaviour) and organizational buyers (business customers and key decision-makers). This communication approach is supported by three key components: the opportunity for (a) enabling a rapid spread of the brand's values and also the creation of corporate brand co-value, (b) target their customers at all levels and in all of their circumstances, c) leave the "blight of sameness", (d) create a sense of empathy with these business men and e) benefit from the influence that consumers have on such retailers.

Communicating to Organizational Buyers

When considering these three types of different interlocutors, it empowers a rapid spread of the brand's values and also the creation of corporate brand co-value as brands are joining forces between both their stakeholders and final consumers. By focusing on the brand's key decision-makers' customers, employees, investors, suppliers and communities interests, Henkel aims to be truly market-focused (Hult, 2011).

FIGURE 33.2
"Where to find Henkel Adhesives" campaign. (From Henkel Adhesives Technologies LinkedIn Posts).

Henkel Adhesives
40,988 followers
8mo

Businesses have responded to the consumers desire for elevated #convenience by providing various and improved on-the-go options. Amplified #packaging needs should go hand in hand with inherent re-design and #innovation, otherwise 30% of plastic packaging will never be reused or recycled. http://ms.spr.ly/6040Tp6QG #bemoresustainable #itstartswithus #sustainablepackaging #sustainability

93 · 2 Comments

Henkel Adhesives
40,986 followers
7mo

#millenials trust in packaging is diminishing for safety, efficacy and protection. Brands' sustainable practices are the way to nurturing their trust. Connect with us to ensure that the key measures are met, while also enabling the #sustainablepackaging for your products. http://ms.spr.ly/6047TpQO3 #bemoresustainable #itstartswithus #consumersinmind

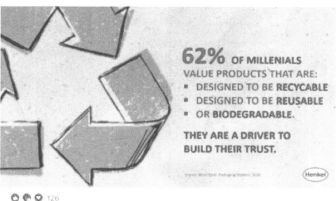

126

FIGURE 33.3
LinkedIn "Sustainability Posts". (From Henkel Adhesives Technologies LinkedIn Posts).

FIGURE 33.4
"Episodes of Circular Economy" campaign. (From Henkel Adhesives Technologies LinkedIn Posts).

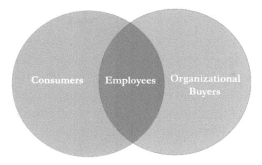

FIGURE 33.5
Three communication interlocutors. (From own authorship).

Social Listening

To efficiently do so, this SBU explores the benefits of social listening to search, list and address the issues that are specifically relevant to all stakeholder groups. By getting closer to related communities and participating in LinkedIn Groups and social discussions, managers can form and establish a strategy that outlines the principles around which their sustainability practices should be centred. As salespeople are becoming significantly more participative on online platforms and are getting increasingly more involved in the social selling concept, Henkel Adhesives Technologies marketers found this an opportunity to easily identify and reach the Key Decision Makers via social media platforms (such as LinkedIn and Twitter) and on business events and exhibitions. In a B2B context, these contacts tend to be less extensive when comparing to a B2C environment but, in this case, they represent the biggest part of Henkel Adhesives Technologies sales. Based on this, managers can form and establish a strategy that outlines the principles around which their sustainability practices should be centred. For a communication plan, it is first based on the analysis of the customers and organizations purchasing process (Lynch & De Chernatony, 2004). The extant literature recommends to analyse the organization's buyer behaviour must be analysed in three stages: by its structure (identifying the key decision-makers, their roles and their influence power), by its process (starting by the problem identification and goes until post-purchase evaluation) and content – that influences the buying decision and is generally evaluated using economic and non-economic criteria (Brassington & Pettitt, 2006).

Purchasing and Investment Decisions: Organizational Factors

In a B2B context, there is a long-held assumption that buyers are generally more well-informed and rational about what they purchase than final consumers (Dibb, Simkin, Pride, & Ferrell, 2012). Nevertheless, even regarding sustainability matters, for organizational buyers individuals in purchasing or investing situations, buying decisions are often made in a group with multiple buying influences (also known as buying center) (Lynch & De Chernatony, 2004). Decision factors such as product type, price, delivery, quality consistency, supplier reliability, customer service, availability and purchasing situation are included in the economic criteria.

Purchasing and Investment Decisions: Personal Factors

Regarding the non-economic criteria, the extant literature includes trust, prestige (the need for status), career, ease of doing business, pride, sentiment, piece of mind, security, friendship and social needs, such as ethics and environmental sustainability (Lynch & De Chernatony, 2004). Regarding the non-economic criteria related to environmental sustainability, the extant literature includes natural disasters, non-sudden pollution and damage, and politically incorrect behaviour and products (Johnson, 2012). It should be kept in mind that these are sensitive subjects and not necessarily caused by this industry (Johnson, 2012).

Additionally, managers also make decisions under an "emotional vacuum" (p. 405) condition (Elsbach & Barr, 1999), also known as lack knowledge, motivation or interest. These managers are moved by emotional factors (Desai & Mahajan, 1998). In what comes to lack of knowledge, it includes situations where there is no market-leading brand or there is a high similarity between brands (MacKenzie & Spreng, 1992; Miniard, Sirdeshmukh, & Innis, 1992), therefore these business managers behave as a normal final consumer. This "emotional vacuum" constitutes a possible opportunity for incorporating the power of emotion and non-economic criteria into the brands' internal and external communications. By keeping in mind that organizational buyers have other soft spots and other purchasing criteria besides functionality, investing in good branding practices and communicating the brand's emotional values, marketers will gain the attention from the organizational buyers under the "emotional vacuum" condition, that posteriorly will recognize of other functional brand values (Lynch & De Chernatony, 2004). Besides this, a good branding practice will also increase the brand's value and loyalty, gives the opportunity to charge premium prices and finally, strengthen a sustainable differential advantage (Lynch & De Chernatony, 2004). However, one must keep in mind that the purpose of good branding practices that involve emotional values is not to stir emotions and lead decision managers into making irrational decisions – it is rather to deliver the stakeholders' most important value proposition (op. cit.). For instance, Henkel's goal is to provide more sustainable solutions with the same product quality that is associated to the adhesives market leader. Furthermore, the usage of the three interlocutors approach allows to tackle these customers at all levels and in all of the circumstances – either inside and outside of their working context or under "emotional vacuum" or under total awareness about the market – and open more opportunities to deliver the value prepositions.

Besides the "emotional vacuum" condition, Hutton (1997) explains that organizational buyers often make decisions with a large number of expenditure. This "fear factor" can be related to the fear of organizational or financial failure or personal risk as their job is at stake. However, this insecurity can give brands an opportunity to finally leave the "blight of sameness" (Hutton, 1997) among other business offers. Communicating by using these three interlocutors approach enables business brands, like Henkel Adhesives Technologies, to create a sense of empathy with these business men that are tirelessly looking for ways to assure they can trust a new business partner and thrive. This technique allows businesses to stress their offers beyond the typical product functionalities and highlight the purchase's unique features such as the need of objective advice and support from a market leader, well-established, highly reputable and environmentally conscious company (Mudambi, 2002).

The Focus on New Generations

This Henkel Adhesives' SBU communication strategy leans towards winning the heart of the future consumers – The Millennials and Pandemic Generation. These two generations are the result of an increasing commitment from their elder generations. They care and understand the relevance of environmental sustainability and are now shifting their preferences towards brands that share the same ideals. They are most concerned about environmental issues and prefer brands that are sharing the same worries (Johnson, 2012). As 53% of the consumers say they have switched to lesser-known brands and organizations whose offers are perceived to be more sustainable (Jacobs et al. 2020), it is important to focus on these communities as soon as possible. According to Sashi (2012), these communities are a superior source of pre-sales information and also a promising source to create a meaningful post-purchase dialogue with other users. As a result, these dialogues and advocacy parties will facilitate a better connection with new prospects (Sashi, 2012).

> They are not going to stay kids forever, they are really growing up during climate change, and they see first-hand the effects. They are very much more willing to pay more for our sustainability aspects of a product when they buy it, then maybe somebody of my generation.
> Interview to the Head of Global MCA for Henkel Adhesives Technologies Packaging SBU

The Customers

As retailers are directly in contact with this generation of final consumers, they are asking their suppliers to help them with their sustainability goals (Iannuzzi, 2012). As a result of the influence that consumers have on such retailers, these retailers are now starting to create their own scorecards and assessments for sustainability (Johnson, 2012). As "'meeting your customers' needs is the inspiration behind business-to-business (B2B) marketing" (Iannuzzi, 2012, p. 149), the B2B companies that are following this trend, will be able to thrive at least for the greener product market (Iannuzzi, 2012) and communicating to them can be a smart strategy to reinforce a B2B brand.

> There is a big push that is coming from consumers [to focus on sustainability and CE] and also brand owners who know they have to react to that. It is also about having to be able to react to new regulations and new legislations that are coming up with stricter guidelines for recyclability.
> Interview to the Head of Global MCA for Henkel Adhesives Technologies Packaging SBU

Henkel Adhesives Technologies has various key accounts that are global retailers, especially in the B2B marketplace. However, although some of Henkel Adhesives Technologies customers are brand owners, they do not represent all of the Henkel Adhesives Technologies customers.

> Most of our customers are one step before us – the converters.
> Interview to the Head of Global MCA for Henkel Adhesives Technologies Packaging SBU

This means that it is on these converters that the community also depends on to make sure that the products delivered to the brand owners can meet those requirements. As the majority of customers represents a big percentage in sales, Henkel Adhesives Packaging focuses in communicating to both business (which represents the vast majority of their customers) and consumers as both are demanding and supporting the creation of more environmentally conscientious solutions and initiatives to one another.

THE FUTURE OF SUSTAINABILITY COMMUNICATION

To confirm the continuous increase of importance that the public is attributing to sustainability, in 2018, the global sustainable investment was $30.7 trillion in the five major markets, such as Canada, United States, Japan, Australia/New Zealand and Europe – which means a 34% increase in two years, with a high tendency to continue to increase (Alliance, 2018). This significant force in financial markets of sustainable investing has grown for most of these markets (except for Europe, where it is believed that the decrease is due to the "stricter standards and definitions for sustainable investing") Figure 33.6 (Alliance, 2018). It translates the steadily growing importance and concern about environmental help for both customers and stakeholders. Additionally, it proves that investing in more sustainable initiatives is not a momentum but rather an increasing and powerful trend that will benefit businesses. Taking that into account, the future of sustainability communication becomes even clearer. Not only consumers but also investors and business customers are interested in hearing about this topic and about the progress and solutions found towards environmental sustainability. Naturally, adopting aspects of environmental

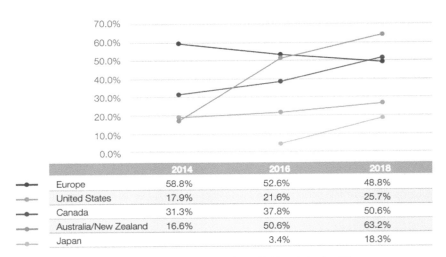

FIGURE 33.6
Percentage of sustainable investing in the total managed assets 2014–2018. (From Alliance, 2018).

sustainability is one of the most strongly supported approaches for companies in the chemical industry. The Sustainability Consortium and Ellen MacArthur foundation support and embrace sustainability and Circular Economy initiatives and they can provide useful information to business and marketing managers. As a result, Henkel focuses on uncovering what will be the next expectations from the chemical industry and networking the within market by setting top-to-top meetings and get to know partner's organizations (Achrol & Kotler, 2012).

As for Henkel Adhesives Technologies, LinkedIn is one of the most important communication channels, we analysed the impact that sustainability topics on LinkedIn Posts had comparatively to any other topics. The data analysed was concerned to all the LinkedIn Posts that were published within August 2019 and December 2019 by this SBU. By categorizing all of the posts between sustainability-related and non-sustainability-related post, we confirmed our initial prognosis. The average values for reach, engagement rate, shares, clicks and reactions were all notably higher for sustainability-related posts than for non-sustainability related posts. Figures 33.7 and 33.8 present the outcomes of this analytics. According to this analysis, we can verify that sustainability is,

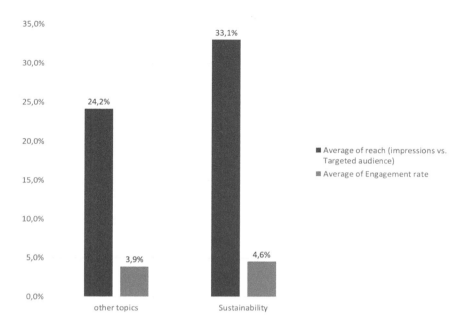

FIGURE 33.7
Reach and Engagement Rate. (From Own authorship).

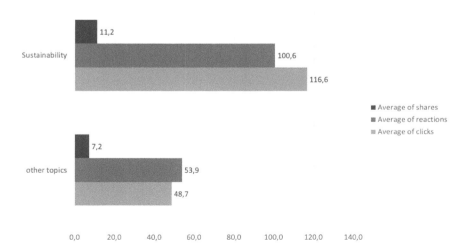

FIGURE 33.8
Shares, Clicks and Reactions. (From Own authorship).

in fact, a topic of general interest to the audience. If a brand is looking for capturing the public's attention, communicating its environmental concerns and initiatives can be a rewarding approach. In general, according to this analysis, communicating about sustainability topics, considerably increases the number of shares, clicks and reactions. This means that people are more reactive and responsive to sustainability topics than to any other topics. Consequently, these reactions will translate in a higher number of impressions (number of people that have seen the post on their feed but were not targeted by the brand's marketers, i.e. someone that saw a post on their feed because their friend liked that exact post). However, for Henkel, there is still some work to be done in what comes to increase even more the engagement rate. Although in this analysis, the engagement rate presents higher values for sustainability topics comparatively to other topics, the challenge for this SBU new positioning goal is to make this difference even more significant. In order to position as a sustainability driver, this SBU must further study how to communicate more impactfully these sustainability initiatives and ultimately, increase even more the engagement rate.

COMMUNICATING SUSTAINABILITY

> We also have to think a little bit more specifically about target groups that we want to reach. And then we can create content that hits that sweet spot – what are the challenges on their desks right now.
> Interview to the Head of Global MCA for Henkel Adhesives Technologies Packaging SBU.

In order to spot the right content that should be created to target the relevant groups, it is important to find the right channel. In an interview with the Head of Global Market and Customer Activation for Henkel Adhesives Technologies Packaging SBU, she proposed to ask the following questions in order to understand these key elements to effectively communicate this positioning. "Who is our target group?", "Where are you reaching these people?", "What kind of content should we be creating?". However, it is worth to note that, this can lead into creating micro-content, which is something very difficult to keep going on long-term basis.

The art in here would be to create content for one topic that would satisfy a broad audience but then to see how you can cut that down into bite size pieces and adapt them for different target groups.
Interview to the Head of Global MCA for Henkel Adhesives Technologies Packaging SBU

Consistent Communication

Besides this intrinsic acknowledgements, Henkel Adhesives Technologies Packaging SBU guarantees consistent communication through all channels, including factsheets, presentations, storytelling, presence at events and fairs, social media platforms, web platforms, trade journals, etc.

We are also pushing this kind of content out in a regular basis which keeps us on top of mind – that we are Circular Economy experts – So when people have questions about it their first thought is to turn to Henkel about it.
Interview to the Head of Global MCA for Henkel Adhesives Technologies Packaging SBU

In this case, this SBU focuses particularly on storytelling as an effective way to strongly communicate the values of a brand. This marketing technique brings cohesion, logic and consistency to the communications. It justifies the introduction of new topics so that every campaign makes sense to the audience and explains its reason for being.

I notice that some of our competitors are trying to follow us but the difference is that they don't seem to have a branding that is specific to Circular Economy or sustainability – there is no roof over it. I can't recognise any certain kind of cohesion that belongs to a bigger strategy. So I think that is also an advantage that we do have a communications framework for Circular Economy and it is very clear that all these pieces belong together even if they are in different channels.
Interview to the Head of Global MCA for Henkel Adhesives Technologies Packaging SBU

"In B2B, we are moving away from brands and towards people – customers are more attached to the industry influencers than the brands that employ them" (Backaler, 2018, p. 74). Henkel Adhesives Technologies chose Nicola Tagliaferro as one of its business influencers despite the fact that B2B companies typically work with journalists and industry analysts. As the "business influencer" concept is relatively new to even the world's most-respected B2B brands (op. cit.) the adoption of one of the B2C marketing techniques translated into an advantage for Henkel Adhesives as it has proved to create more effectively business opportunities.

Henkel frequently reaches its customers by messaging them with the delivery or service details (which results in increased satisfaction). It also uses cookies and re-marketing techniques to be on top of the customers mind (which results in increased retention and return rate) and by creating and promoting content online (social platforms, website, showcase page, blogs), Henkel aims to remind customers of their relationship and connection with the firm. Besides this, Henkel marketers created an online platform, full of content, where the sales people can access and share on their social media platforms those, already carefully prepared, materials to guarantee the correct approach to the customers and potential customers and provide them with relevant content. Additionally, literature actively encourages the creation channels for customers to create meaningful interactions with each other (to increase advocacy), and other channels where customers can directly engage with the firm's members (Sashi, 2012). All of these approaches reinforce the importance and the opportunity that Henkel is creating by providing means to develop existing customers relationships.

Besides this, the fact that this SBU shares very frequently, on its online and social platforms, its webinars, business partnerships, peer testimonials and reviews, also allows the name of the brand to keep on top of mind of its customers and potential customers. Henkel has been associated with Global Initiatives and new influencer collaborations such as Ellen MacArthur Foundation, CEflex, AtEPW – and has been globally recognized by the Global SDG Awards, Global Good Awards, Fortune's Rankings. In a study of the top 29 companies in the chemical industry, sharing these recognitions proved to be positive initiative as nearly two-thirds of the companies reported sustainability or environmental awards, either in a global level or in a regional level (Johnson, 2012). According to Sheth and Sinha (2015) and (Johnson, 2012), these initiatives "can serve several functions: signal to stakeholders that the company is committed to sustainability; provide to the company information and guidance, through joint research and discussion; advertise as a third-party the company's efforts or programs; organize some engagement with stakeholders"

(p. 61). These authors suggest that such initiatives must be communicated through joint press releases, articles, communication and events.

Besides these awards testimonials, Henkel also makes use of customer's testimonials and reviews. This enhances the product/service benefits from the customer perspective. To all the product claims, there must be a positive result from at least one externally recognized test or certificate; hence all the technical requirements to make Circular Economy claims must be assured, in advance, to be fulfilled before published.

Promoting More Environmentally Sustainable Solutions

Although the new generation of consumers is proven to be more aware and concerned about environmental issues, many consumers and business customers are still perceiving circular products negatively (Muranko et al. 2019). The low endorsement of Circular Economy focused products is related to the fact that these products are stereotypically perceived as poorer quality comparing to brand new and traditional products (Muranko et al. 2019), considerably more expensive (Mariadoss et al. 2011) and the benefits of the product are not immediately evident to consumers (for example, general waste reduction) (Marc Atherton FRSA, 2015).

Henkel's three business units present multiple actions towards a more sustainable future. However, selling a product or a service highlighting that benefit may be misperceived as an offer that, in some way, has poorer performance than the traditional one. Professionals in this SBU noticed that various competitors and key accounts were making the same approach – very few communicated in their social media platforms the positive environmental impacts they were practising. Surprisingly, some of them were referring to these environmental changes but in a very understated way – only very few posts were about the subject; the *environmental impact* section in the brand's websites was not disposed in a user-friendly manner and this type of information was slightly harder to find compared to other topics; when mentioning *sustainability*, the focus was in the corporate level of the brand instead of promoting, to their customers, the brand's most sustainable options of their products.

To increase the importance of choosing more sustainable solutions and encourage customers to rethink their assumptions about Circular Economy options, some of the techniques consist of rephrasing, renaming or framing old assumptions. When comparing the new Circular Economy options to the linear ones, choosing the former ones must not be a burden to their users at any point – everything remains "normal". It is important to evoke familiar connotations as the traditional and usual communication approach and some terms must be kept in order to imply "normality". However, some authors suggested that "Sustainability" is a poor word to be used in a Chemical Industry context (Johnson, 2012). The credibility gap issue might be created by emphasizing too much the word *sustainability* (and sometimes *responsibility*) as this wording can appear deliberately misleading for public relations or stakeholder relations (Johnson, 2012). Instead, it is suggested to look for specific terms that describe more suitably the role and approach of the brands. However, despite Henkel Adhesives Packaging SBU marketer's efforts on focusing on the "Circular Economy" and "[be more]" approaches, the word sustainability continues to pop up very often on their communication.

Furthermore, in order to maintain these familiar connotations, this SBU has opted by creating a new naming consistency for new more sustainable offers – by keeping the usual naming of their upgraded products and adding a recyclability prefix to it – and also opted to rephrase its motto: "develop new solutions for sustainable development while continuing to shape their business responsibly and increase their economic success"; become the first-choice partner in all relevant packaging and consumer adhesives markets that sets sustainability benchmarks across the packaging and consumer goods market; fulfil of the latest consumer's and market's trend – the concern for environmental protection and demand for more sustainable solutions.

> However it can't just be communication. It is also necessary to have solutions to back up what we are. Communications really have to be backed up by solid prove points from the business. It has to be a part of the business strategy in order to make this communication effective. Because otherwise it's green washing (...) Communication, in the end, is the reflection of what the strategy is.
> Interview to the Head of Global MCA for Henkel Adhesives Technologies Packaging SBU

Communicating the company's sustainability and environmental concerns allows brands to be seen as, at least, modest and, if the case, as *best-in-class* but issue-focused (Johnson, 2012). Henkel aims to maintain its *best-in-class* positioning as the "Global leader in the adhesives market – across all industry segments worldwide" as mentioned on Henkel's webpage. However, this may imply to communicate in a low-key manner to avoid sounding as if they are over apologizing for the environmental impact, or to avoid being seen as an "iron fist in a velvet glove" (p. 106) (forgiving on some issues and rigid on others) (Johnson, 2012). It is of high importance not to fall into the arrogance of much sustainability reporting.

> Arrogance is not illegal, but in the event it invites disbelief, satire and perhaps ridicule – and it is hardly conducive to building trust.
>
> (Johnson, 2012, p. 112)

Thus, instead of sharing explicitly the brand's benefits, this SBU has to be focused on sharing its rewards to its customers and on answering the question – *What is the reward for the customer or for the potential customer? What would this person benefit be if they were working with Henkel Adhesives Technologies or with its product?* In fact, this SBU recently noticed that the communications were being done very broadly. Although its storytelling and communications concerning the harmful side-effects of wasteful consumption were successful, it wasn't being told in the right perspective. It should be demonstrating how its customers could help prevent such damage (Achrol & Kotler, 2012). This can be achieved through the communication of the specific results of sustainability activities while explaining the impact of those events and actions for helping the planet and mankind. Boasting or "showing off" is not an option but rather leading customers, implicitly and clearly, to understand those benefits and then emotionally engage with the brand.

"Talk is cheap. Money is real." (Johnson, 2012, p. 42) Contrarily to communicating to active green consumers, in B2B context only saying to save the planet rarely works. However, the impact of showing how it saves money often does (Johnson, 2012). Thus, evoking the costs, the financial incentives and the added value to encourage the value-added is especially relevant to the customer cost factor. Nevertheless, a company must show that it wants to help customers to get more opportunities to add value.

> We are not just communicating about sustainability or Circular Economy. We actually have the solutions that pay into that. And it is a very positive thing that we are a company they can work with and help our customers meet their sustainability goals. It is a very positive development.

Interview to the Head of Global MCA for Henkel Adhesives Technologies Packaging SBU

Sustainability Reporting

Moreover, despite the traditional view defending that "to the financial community, sustainability primarily means preventing scandal" (Johnson, 2012, p. 46), sustainability reporting has proven to be is as much important as financial reporting (op. cit.). Sustainability ratings guide investors on identifying environmentally sustainable and responsible companies to invest on. As these ratings are based on the information contained on the companies' reports, brands must guarantee that they not only effectively communicate their related brand value on all their communication channels but also focus on doing so in their sustainability reports. Sustainability reporting can not only be a source for advertising but also as a source of corporate communications, for both internal and external stakeholders, i.e. "opinion leaders who then disseminate the content further" (Johnson, 2012, p. 63). With regards to corporate expenditure, Henkel invests in publishing annual environmental sustainability reports separated from the financial reports and both with a stakeholder focus.

THE IMPORTANCE OF PERSUASIVE COMMUNICATION FOR SUSTAINABILITY

To influence behavioural attitudes and product perceptions, educating business customers, consumers and all the extant stakeholders is the first step (Muranko et al. 2019; Smol et al. 2018; Suárez-Eiroa, Fernández, Méndez-Martínez, & Soto-Oñate, 2019). In accordance with Muranko et al. (2019), marketers and sales professionals at Henkel follow three key components for fruitful persuasive messages (I) advocate a position for a particular problem or recommend specific behaviour, (II) provide arguments reasoning to adopt of the advocated position (III) introduce factual evidence to reinforce the reasoning and argument.

Besides this, this SBU is focused on accepting stakeholder outrage. It becomes important to allow the stakeholders to have a long meeting where they define its end. Literature supports that this shows a serious commitment to address entirely all their concerns and by persuading the subject (not the meeting) to be boring, will make the stakeholders understand that the issue is being properly-managed and doesn't need additional attentiveness from the audience (Johnson, 2012). By allowing them speak first, listening carefully and saying little will show that the business representatives are first interested in what they have to say and consequently, will increase the chance to gain their attention to what the business representatives have to say. When first speaking, literature suggests to rewind their ideas in a questioning way (op. cit.) in order to demonstrate that there was a clear understanding between both parties and by responding only to the criticisms business representatives agree with and giving credit to the critics will make the outraged stakeholders to feel better.

The Role of Internal Brand Communication and Education

"To tap the potential of B2B brands, business marketers must understand and effectively communicate the value of their brands" (Mudambi, 2002, p. 527). Henkel Adhesives Technologies aim to do so by starting to educate and bring awareness to both business customer and the final consumer, about the importance of Circular Economy systems and about how its solutions help, in fact, decreasing the environmental footprint. The final milestone to conquer is to make consumers and business customers understand Henkel's expertise in Circular Economy and by evoking Henkel's name associating it as a trustworthy brand that they can turn to when it comes to increasing their environmental positive impact.

However, against the traditional views, a successful external brand communication is highly related to the employees understanding and their commitment to brand values (Lynch & De Chernatony, 2004). These values must be well represented on trainings, on internal communication media and mainly on the brand's culture (op. cit.).

> There was also a lot of education that had to take place internally and externally about what is Circular Economy, what does it mean for Henkel or the Packaging SBU.
> Interview to the Head of Global MCA for Henkel Adhesives Technologies Packaging SBU

Henkel Adhesives SBU education work happens mainly through webinars but also includes other different vehicles such as the usage of symbols, heroes, activities, rituals, ceremonies and stories. Trainings also include, on-the-job pieces of trainings, learning with access to internal and external courses about the products, company, market and sales negotiation techniques. At Henkel, internal education and communication are seen as important pillars for establishing this new positioning strategy. However, it is believed that internal miscommunication or lack of internal education are slowing down the results on its positioning strategy – challenging obstacles such as a detected small fraction of their employees that remain unrecognizing, unaware or unfamiliar with its accomplishments have to be overcome. Accordingly, Henkel is establishing new educational initiatives to formulate this positioning message and assure it is passed across various management levels. It has been tackling this issue by education from within through an environmental sustainability/Circular Economy awareness programs and creation of internal and external educational webinar series, online content hub, comprehensive global communication package, white papers, expert interviews, articles, blog posts, etc.

> We are realising that, at least externally, the questions are getting much more specific. By laying this base work, it was possible to start to satisfy the need for information about Circular Economy, particularly when it comes to adhesives.
> Interview to the Head of Global MCA for Henkel Adhesives Technologies Packaging SBU

Besides this, to reinforce and maintain the organization on track on the sustainable and Circular Economy path, Henkel Adhesives Technologies Packaging SBU has created and empowered the position of a Chief Sustainability Officer or Senior Sustainability or Circular Economy managers per business units – Ulla Hueppe, Dennis Bankmann and Marcin Lapaj. Sheth and Sinha (2015) also advocated these positions as part of the list of sustainability actions that are worth.

> We have a true expert in our team – Dennis Bankmann – and he is really a respected expert in the industry on CE. And that is gold for me as a communicator. He is really knowledgeable and he is really a respected expert in the industry of Circular Economy and that is the key element that I would say that makes us unique.
> Interview to the Head of Global MCA for Henkel Adhesives Technologies Packaging SBU

Besides, these professionals will also benefit from the freedom to create their own content. This guarantees that customers will receive extremely informative, reliable and concise content, and it will enable a successful positioning, for both the company and the experts themselves, as experts in Circular Economy.

Nevertheless, to implement these changes the CSO certainly needs to maintain constant interactions with functional areas of the company: HR, IT, Finance, Operations, Marketing, etc. The creation of these positions will give the organization better-prepared experts to answer frequently asked questions (FAQ). Not only they give additional incites about complex topics to clients but also diffuses a trusted word throughout the company about the circularity concepts.

Another cornerstone that this SBU is that it has been using solidify its internal and external communications is by being involved in environmental activities. These can include awareness programs, internal and external webinars, present in stands, trade fairs and tradeshows, exhibitions, radio, technical and sensorial presentations. As more and more tradeshows, such as LabelExpo and K-show, are focusing on Circular Economy, one of the successful examples of Henkel is the Henkel Family Info Kreis internal event. In this event in 2019, they finally presented one solution to the long-known-problem of impossibility to recycle black plastic. Its carbon black pigments do not enable the package to be sorted using Near Infra-Red (NIR) technology that is widely used in plastics recycling. These new black plastics are now not disposed in a landfill or incinerated (Wrap.org.uk., 2020). Instead, they are identifiable in the recycling centres by its unique blue colour when exposed directly to the sun (Figure 33.9). Another successful external example was the Henkel Sensory Room experience that occurred in one of the Henkel's exhibitions (Figure 33.10).

Classic Aristotle's senses classical hierarchy implies that *visus* (sight) and *auditus* (hearing) are the two superior senses (commonly expected in media and communications) followed by *odoratus* (smell), *gustus* (taste), and tactus (touch). As the last three are easily forgotten, in order to remarkably promote new products and services, it is worth investing in interactive and sensory customers experiences like these and leaving an unforgettable impression on the clients' mind, (Achrol & Kotler, 2012) (i.e. visits to production facilities, during exhibitions organizing sensorial presentations with effects on chemicals, show the differences between sustainable and traditional products, etc.).

FIGURE 33.9
In collaboration with Ampacet, Henkel is developing an innovative solution for black plastic packaging that is fully recyclable - REC-NIR-BLACK (From Henkel Adhesives Technologies image stock).

Finally, to determine the success of the activities in terms of profitability and contribution and impact to the environment and to society, it is important to establish solid KPIs. These key performance indicators can be: number of new products or projects, number of customer contacts regarding Circular Economy and sustainability, number of new visitors in the new home page, number of environmentally sustainable initiatives per year, reach of the campaigns on social media, number of double opt-in approvals, water used, waste created, CO_2 emissions and number of accidents.

MAJOR PROBLEMS

Considering Henkel Adhesives Packaging SBU's case study, the challenge is in evaluating the sustainability route for brands and corporate reputation building. Firms in similar contexts are in lack of solid KPIs and budget to measure the success of their activities, not only in terms of profitability but also to determine how they are contributing to the environment and to the society. Additionally, low budgets also limit the investments to target internal education and to powerfully impact and sensitize customers and final consumers. Besides this, the coordination

FIGURE 33.10
A communication from Henkel Adhesives Technologies (From Henkel Adhesives Technologies image stock).

of all parties with the responsible marketing team can be quite challenging – i.e. sustainability experts, product developers, suppliers and partners involved. Team collaboration is vital to the success of this project. In what comes to enable a truly Circular Economy system, it is only possible with a strong network of the industry experts and partners – it is truly a task for the entire value chain. Furthermore, there might be some difficulties in analysing the customers' needs for B2B companies, as they are more discrete about their needs and aspirations, and the way they communicate it. Even though this SBU is continuously doing good communications, it hasn't been specific enough in what comes to communicate objectively the reward for the customer or for the potential customer.

DISCUSSION

In order to reach social responsible consumers, throughout this case study we understand that, this Henkel SBU marketers opted by establishing a new positing strategy. While each company can and should define their own strategies according to its own corporate needs, budget, mission, values and goals, Henkel's Global Market and Customer Activation strategy covered mainly three focus areas (Figure 33.11): Internal Focus, Stakeholder Focus and External Focus (External Communications, External Aspects and Key Communication Elements). These three areas aimed to reach both socially responsible consumers and business customers. The strategic solutions that supported Henkel's positioning (and possibly guide other companies under similar circumstances) were (Figure 33.11): adopt and report environmental sustainability aspects, get involved in environmental activities, create new related partnerships, develop existing customers relationships and communicate and benefit from their positive feedbacks, using social listening tools, identify Key Decision Makers and the main topics of concern of the community, focus in addressing the relevant issues for all stakeholder groups, publish annual environmental sustainability reports and awards, analyse customers and key accounts purchasing process, accept stakeholder outrage, create and empower the position of a Chief Sustainability Officer, Senior Sustainability or Circular Economy managers per business units, prepare experts to answer FAQ, establish solid internal and external educational initiatives, establish solid KPIs, rename the sustainable products/services, rephrase the brands motto, evoke familiar connotations, highlight the unique nature of purchase, evoke the simplicity and convenience, share business partnerships, peer testimonials and reviews, storytelling, evoke meaning and stimulate the brands empathy, "Talk is cheap. Money is real." in what comes to convince business customers, "Sustainability" is a poor word which can appear deliberately misleading for public, communicate clearly "What is the reward?" to the customer or stakeholder and guarantee frequent and consistent communications through all the online platforms and events. To sum up, such positionings lead to a very delicate dilemma of converting a negative environmental reputation of an industry

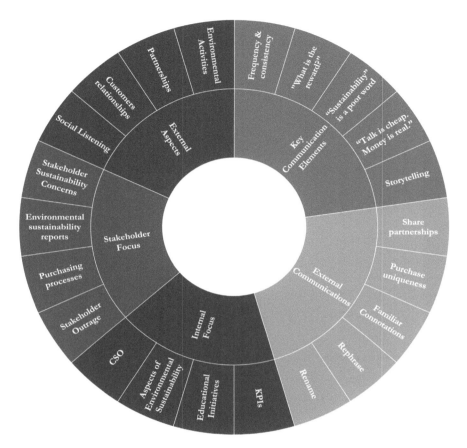

FIGURE 33.11
Conceptual map with key elements to assist sustainability positioning strategies. (From own authorship).

into a positive, ambitious and inspiring one. Thus, brands' strategies and communications must be sensitive to the topics they are going to approach, must be consistent with their standing position and for last but not least, brands must have solutions to back up what is announced.

In the future, it would not only be interesting to deeper analyse the role of SCE within a company under similar circumstances but also to analyse the role of this position within companies in other sectors, industries and geographical areas. Also, it noteworthy to develop further research on companies in different geographical areas in order to compare the different strategies and its outcomes, to understand the impact that these sustainability topics have in each geographical area and to collect more quantitative data about the environmental sustainability communication strategies and their impact inside and outside of the chemical industry. Likewise, it would be interesting to evaluate whether brands in the chemical industry are, in general, afraid of advertising and communicating their sustainable solutions because of the fear of being perceived as a lower performance brand. Furthermore, as in this case it is believed that internal miscommunication or lack of internal education is slowing down the results on its positioning strategy, future research can rely on this occurrence to further study the impact of poor corporate internal communications on the perception of the communicated positioning to the community. It would be interesting to investigate and ascertain how much does internal assumption can influence external perceptions. This information can be very precious when it comes to determining what are the main obstacles for a corporate positioning in sustainability and what should be the managers' main concerns to focus when tackling flaws on their strategies more effectively.

CONCLUSION

This case study has reinforced the importance and great return of corporate investment in environmental activities. The environmental activities and milestones have shown to be a relevant topic of interest to the general public, B2B customers and final consumers. It has also shown significant positive impacts on the corporate economic sustainable development brands whenever their environmental initiatives were well communicated. As a result, the importance of correct brand communications arises and we contribute with key elements to support such brands' positioning, and outline some of the difficulties that may appear along the way.

Even though positioning a chemical industry company, as a sustainability driver can sound confusing, the fact that these industries make huge environmental impacts, also shows their potential to positively affect it. As a matter of fact, to maintain the same levels of comfort and convince on our daily lives, their products will never be 100% indispensable. Thus, Henkel Adhesives technologies aim to provide solutions to guarantee that these products are kept in the loop for as long as it is sustainably possible. The gradual results and outcomes of this new SBU positioning and the issues and improvements are still undergoing. Henkel presents continuous important accomplishments towards its positioning as a sustainability driver and a Circular Economy expert. This SBU shows to be continuously searching for issues and systems to improve and solve them.

DISCUSSION QUESTIONS

1. How can such companies and a global company like Henkel and its SBUs induce and change its polluter reputation towards a sustainability driver and Circular Economy Expert?
2. What are the weaknesses of these programs?
3. Are there ways of moving the self-interested values to the recognition that the Circular Economy approaches can bring advantages in both financial, reputational and environmental terms to customers and society?
4. Are there opportunities to improve the chemical industries companies' reputation towards stakeholders and customers?
5. Are the promotional mechanics adequate and enough to create awareness in the target audience?
6. Are the rewards of more sustainable options communicated adequately to customers?
7. Are there other target incentives that should be approached?
8. Do you have suggestions for improving the positioning efforts?
9. Will marketing campaigns be powerful enough to communicate these expertise and translate into remarkable financial results?

LIMITATIONS

This report is limited by a seven months' experience within the company thus, further conclusions could have been made in a longer period.

CREDIT AUTHOR STATEMENT

The authors would like to express their sincere thanks to the editor and the anonymous reviewers for their constructive and detailed comments on earlier draft of this chapter.

João F. Proença gratefully acknowledges financial support from FCT-Fundação para a Ciencia e Tecnologia (Portugal), national funding through research grant UIDB/04521/2020.

REFERENCES

Achrol, R. S., & Kotler, P. (2012). Frontiers of the marketing paradigm in the third millennium. *Journal of the Academy of Marketing Science*, 40(1), 35–52.
Act, S. M. (2012). Communication from the commission to the European parliament, the council, the economic and social committee and the committee of the regions.
Alliance, T. (2018). 2018 Global sustainable investment review. *Global Sustainable Investment Alliance*, 4, 9.
Allwood, J. M., Ashby, M. F., Gutowski, T. G., & Worrell, E. (2011). Material efficiency: A white paper. *Resources, Conservation and Recycling*, 55(3), 362–381.
Backaler, J. (2018). Business to Business (B2B) Influencer Marketing Landscape. In *Digital Influence* (pp. 69–85). Springer.
Bhattacharya, C. B., & Korschun, D. (2008). Stakeholder marketing: Beyond the four Ps and the customer. *Journal of Public Policy & Marketing*, 27(1), 113–116.
Bilgin, M. (2009). The PEARL Model: Gaining competitive advantage through sustainable development. *Journal of Business Ethics*, 85(3), 545.
Brassington, F., & Pettitt, S. (2006). *Principles of Marketing*: Pearson Education.
Brown, B. J., Hanson, M. E., Liverman, D. M., & Merideth, R. W. (1987). Global sustainability: Toward definition. *Environmental Management*, 11(6), 713–719.
Brown, T. J., & Dacin, P. A. (1997). The company and the product: Corporate associations and consumer product responses. *Journal of Marketing*, 61(1), 68–84.
Brundtland, G. H., Khalid, M., Agnelli, S., Al-Athel, S., & Chidzero, B. (1987). Our common future. *New York*, 8–9.
Chamberlin, L., & Boks, C. (2018). Marketing approaches for a circular economy: Using design frameworks to interpret online communications. *Sustainability*, 10(6), 2070.
Dangelico, R. M., & Vocalelli, D. (2017). "Green Marketing": An analysis of definitions, strategy steps, and tools through a systematic review of the literature. *Journal of Cleaner Production*, 165, 1263–1279.
Den Hollander, M., & Bakker, C. (2016). Mind the gap exploiter: Circular business models for product lifetime extension. *Proceedings of the Electronics Goes Green, Berlin, Germany*, 6–9.
Desai, K. K., & Mahajan, V. (1998). Strategic role of affect-based attitudes in the acquisition, development, and retention of customers. *Journal of Business Research*, 42(3), 309–324.
Dibb, S., Simkin, L., Pride, W. M., & Ferrell, O. C. (2012). *Marketing: Concepts and Strategies*. Cengage.
Dresner, S. (2012). *The Principles of Sustainability*. Routledge.
Dubey, R., Gunasekaran, A., & Ali, S. S. (2015). Exploring the relationship between leadership, operational practices, institutional pressures and environmental performance: A framework for green supply chain. *International Journal of Production Economics*, 160, 120–132.

Elsbach, K. D., & Barr, P. S. (1999). The effects of mood on individuals' use of structured decision protocols. *Organization Science, 10*(2), 181–198.

European Environment Agency (2015). Closing the loop-An EU action plan for the Circular Economy. *Communication from the Commission to the European Parliament, the Council, the European Economic and Social Committee and the Committee of the Regions COM, 614*(2), 2015.

Farooque, M., Zhang, A., Thürer, M., Qu, T., & Huisingh, D. (2019). Circular supply chain management: A definition and structured literature review. *Journal of Cleaner Production, 228*, 882–900. doi:10.1016/j.jclepro.2019.04.303

Geissdoerfer, M., Savaget, P., Bocken, N. M. P., & Hultink, E. J. (2017). The Circular Economy – A new sustainability paradigm? *Journal of Cleaner Production, 143*, 757–768. doi:10.1016/j.jclepro.2016.12.048

Grimmer, M., & Woolley, M. (2014). Green marketing messages and consumers' purchase intentions: Promoting personal versus environmental benefits. *Journal of Marketing Communications, 20*(4), 231–250.

Henkel Adhesives Technologies. (2020). Competitor and key accounts CE comunication. *Henkel Adhesives Technologies Internal Report* (pp. 1–30).

Houston, J., Briguglio, M., Casazza, E., & Spiteri, J. (2019). Stakeholder views report: Enablers and barriers to a circular economy.

Hsu, C.-C., Tan, K.-C., & Mohamad Zailani, S. H. (2016). Strategic orientations, sustainable supply chain initiatives, and reverse logistics. *International Journal of Operations & Production Management, 36*(1), 86–110. doi:10.1108/ijopm-06-2014-0252

Hult, G. T. M. (2011). Market-focused sustainability: Market orientation plus! *Journal of the Academy of Marketing Science, 39*, 1–6. doi: 10.1007/s11747-010-0223-4

Hunt, S. D. (2011). Sustainable marketing, equity, and economic growth: a resource-advantage, economic freedom approach. *Journal of the Academy of marketing Science, 39*(1), 7–20.

Hutton, J. G. (1997). A study of brand equity in an organizational buying context. *Journal of Product & Brand Management, 6*(6), 428–439. doi:10.1108/10610429710190478

Iankova, S., Davies, I., Archer-Brown, C., Marder, B., & Yau, A. (2019). A comparison of social media marketing between B2B, B2C and mixed business models. *Industrial Marketing Management, 81*, 169–179.

Iannuzzi, A. (2012). *Greener Products the Making and Marketing of Sustainable Brands* (Vol. 13). NW, Suite 300: CRC Press, Tailor & Francis Group.

Jacobs, K. R., Robey, J., van Beaumont, K., Lago, C., Rietra, M., Hewett, S., Buvat, J., Manchanda, N., Cherian, S., B Abirami. (2020). *Consumer Products and Retail – How sustainability is fundamentally changing consumer preferences.* Retrieved from https://www.capgemini.com/wp-content/uploads/2020/07/20-06_9880_Sustainability-in-CPR_Final_Web-1.pdf

Johnson, E. (2012). *Sustainability in the Chemical Industry.* The Netherlands: Springer Science & Business Media.

Kleinaltenkamp, M. (2018). Peter LaPlaca–The best marketer of industrial and B2B marketing research. *Industrial Marketing Management, 69*, 125–126.

Kolk, A. (2000). *Green reporting.* Retrieved from hbr.org

Kotler, P., & Armstrong, G. (1996). Livro: Marketing. *Editora Compacta.* São Paulo: Atlas.

Kumar, V., & Christodoulopoulou, A. (2014). Sustainability and branding: An integrated perspective. *Industrial Marketing Management, 43*(1), 6–15.

Linder, M., & Williander, M. (2017). Circular business model innovation: inherent uncertainties. *Business Strategy and the Environment*, 26(2), 182–196.

Luo, X., & Bhattacharya, C. B. (2006). Corporate social responsibility, customer satisfaction, and market value. *Journal of Marketing*, 70(4), 1–18.

Lynch, J., & De Chernatony, L. (2004). The power of emotion: Brand communication in business-to-business markets. *Journal of Brand Management*, 11(5), 403–419.

MacArthur, E. (2013). Towards the circular economy. *Journal of Industrial Ecology*, 2, 23–44.

MacKenzie, S. B., & Spreng, R. A. (1992). How does motivation moderate the impact of central and peripheral processing on brand attitudes and intentions? *Journal of Consumer Research*, 18(4), 519–529.

Marc Atherton FRSA. (2015). Blog: Behaviour Change Towards a Circular Economy – Part 1. Retrieved from: https://www.thersa.org/discover/publications-and-articles/rsa-blogs/2015/10/blog-behaviour-change-for-ce-part-1

Mariadoss, B. J., Tansuhaj, P. S., & Mouri, N. (2011). Marketing capabilities and innovation-based strategies for environmental sustainability: An exploratory investigation of B2B firms. *Industrial Marketing Management*, 40(8), 1305–1318. doi:10.1016/j.indmarman.2011.10.006

Martin, D. M., & Schouten, J. (2011). *Sustainable Marketing*. Pearson Prentice Hall.

Martin, D. M., & Schouten, J. W. (2014). The answer is sustainable marketing, when the question is: What can we do? *Recherche et Applications en Marketing (English Edition)*, 29(3), 107–109.

Matthes, J., Wonneberger, A., & Schmuck, D. (2014). Consumers' green involvement and the persuasive effects of emotional versus functional ads. *Journal of Business Research*, 67(9), 1885–1893.

Menon, A., & Menon, A. (1997). Enviropreneurial marketing strategy: the emergence of corporate environmentalism as market strategy. *Journal of marketing*, 61(1), 51–67.

Mingione, M., & Leoni, L. (2020). Blurring B2C and B2B boundaries: corporate brand value co-creation in B2B2C markets. *Journal of Marketing Management*, 36(1-2), 72–99.

Miniard, P. W., Sirdeshmukh, D., & Innis, D. E. (1992). Peripheral persuasion and brand choice. *Journal of Consumer Research*, 19(2), 226–239.

Moktadir, M. A., Rahman, T., Rahman, M. H., Ali, S. M., & Paul, S. K. (2018). Drivers to sustainable manufacturing practices and circular economy: A perspective of leather industries in Bangladesh. *Journal of Cleaner Production*, 174, 1366–1380.

Mudambi, S. (2002). Branding importance in business-to-business markets: Three buyer clusters. *Industrial Marketing Management*, 31(6), 525–533.

Muranko, Z., Andrews, D., Chaer, I., & Newton, E. J. (2019). Circular economy and behaviour change: Using persuasive communication to encourage pro-circular behaviours towards the purchase of remanufactured refrigeration equipment. *Journal of Cleaner Production*, 222, 499–510. doi:10.1016/j.jclepro.2019.02.219

Murat, A. (2009). An introduction to sustainable development by Peter P. Rogers, Kazi F. Jalal and John A. Boyd. *Development and Change*, 40. 10.1111/j.1467-7660.2009.01519_8.x

Newport, D., Chesnes, T., & Lindner, A. (2003). The "environmental sustainability" problem. *International Journal of Sustainability in Higher Education*, 4(4), 357–363.

Nußholz, J. (2017). Circular business models: Defining a concept and framing an emerging research field. *Sustainability*, 9(10). doi:10.3390/su9101810

Papadas, K.-K., Avlonitis, G. J., Carrigan, M., & Piha, L. (2019). The interplay of strategic and internal green marketing orientation on competitive advantage. *Journal of Business Research*, 104, 632–643.

Peattie, K. (2001). Golden goose or wild goose? The hunt for the green consumer. *Business Strategy and the Environment, 10*(4), 187–199.

Popa, V. N., & Popa, L. I. (2016). *Green Acquisitions and Lifecycle Management of Industrial Products in the Circular Economy*. Paper presented at the IOP Conference Series: Materials Science and Engineering.

Porter, M. E., & Kramer, M. R. (2006). Strategy & society: The link between competitive advantage and corporate social responsibility. *Harvard Business Review, 84*(12), 78–92. Retrieved from https://www.scopus.com/inward/record.uri?eid=2-s2.0-33845336816&partnerID=40&md5=4ca287e3daefa9741ad3513cc56cf1f6

Rehman, M. A., Seth, D., & Shrivastava, R. (2016). Impact of green manufacturing practices on organisational performance in Indian context: an empirical study. *Journal of Cleaner Production, 137*, 427–448.

Rex, E., & Baumann, H. (2007). Beyond ecolabels: What green marketing can learn from conventional marketing. *Journal of Cleaner Production, 15*(6), 567–576. doi:10.1016/j.jclepro.2006.05.013

Rodriguez, M. A., Ricart, J. E., & Sanchez, P. (2002). Sustainable development and the sustainability of competitive advantage: A dynamic and sustainable view of the firm. *Creativity and innovation management, 11*(3), 135–146.

Sashi, C. (2012). Customer engagement, buyer-seller relationships, and social media. *Management decision, 50*(2), 253–272. doi:10.1108/00251741211203551

Savitz, A. (2013). *The Triple Bottom Line: How Today's Best-run Companies Are Achieving Economic, Social and Environmental Success-and How You Can Too*. San Francisco, CA: John Wiley & Sons.

Shahbazpour, M., & Seidel, R. H. (2006). *Using Sustainability for Competitive Advantage*. Paper presented at the 13th CIRP International Conference on Life Cycle Engineering.

Sharma, A., Iyer, G. R., Mehrotra, A., & Krishnan, R. (2010). Sustainability and business-to-business marketing: A framework and implications. *Industrial Marketing Management, 39*(2), 330–341. doi:10.1016/j.indmarman.2008.11.005

Sheth, J., & Parvatiyar, A. (1995). Ecological imperatives and the role of marketing. *Environmental Marketing: Strategies, Practice, Theory and Research*, 3–20.

Sheth, J. N., & Sinha, M. (2015). B2B branding in emerging markets: A sustainability perspective. *Industrial Marketing Management, 51*, 79–88.

Shrivastava, P. (1995). The role of corporations in achieving ecological sustainability. *Academy of Management Review, 20*(4), 936–960.

Smol, M., Avdiushchenko, A., Kulczycka, J., & Nowaczek, A. (2018). Public awareness of circular economy in southern Poland: Case of the Malopolska region. *Journal of Cleaner Production, 197*, 1035–1045. doi:https://doi.org/10.1016/j.jclepro.2018.06.100

Stahel, W. (1994). The utilization-focused service economy: Resource efficiency and product-life extension. *The Greening of Industrial Ecosystems*, 178–190.

Stirling, A. (2009). Direction, distribution and diversity! Pluralising progress in innovation, sustainability and development.

Suárez-Eiroa, B., Fernández, E., Méndez-Martínez, G., & Soto-Oñate, D. (2019). Operational principles of circular economy for sustainable development: Linking theory and practice. *Journal of Cleaner Production, 214*, 952–961. doi:https://doi.org/10.1016/j.jclepro.2018.12.271

Wagner, M. (2005). Sustainability and competitive advantage: Empirical evidence on the influence of strategic choices between environmental management approaches. *Environmental Quality Management, 14*(3), 31–48.

Wiersema, F. (2013). The B2B agenda: The current state of B2B marketing and a look ahead. *Industrial Marketing Management, 4*(42), 470–488.

Wrap.org.uk. (2020, 15 July 2020). Recyclability of Black Plastic Packaging | WRAP UK. Retrieved from http://www.wrap.org.uk/content/recyclability-black-plastic-packaging-2

Xie, X., Huo, J., & Zou, H. (2019). Green process innovation, green product innovation, and corporate financial performance: A content analysis method. *Journal of Business Research, 101*, 697–706.

Yang, D., Lu, Y., Zhu, W., & Su, C. (2015). Going green: How different advertising appeals impact green consumption behavior. *Journal of Business Research, 68*(12), 2663–2675.

Zou, L. W., & Chan, R. Y. (2019). Why and when do consumers perform green behaviors? An examination of regulatory focus and ethical ideology. *Journal of Business Research, 94*, 113–127.

34

Social and Sustainability Marketing and the Sharing Economy in the Coffee Shop Culture

Madhavi Venkatesan
Department of Economics, Northeastern University, Boston, Massachusetts, USA

CONTENTS

Learning Objectives	839
Themes and Tools	840
Introduction: Social and Sustainability Marketing and Sharing Economy	840
Background: Life Cycle of Coffee Consumption	844
The Case of Usefull	848
How USEFULL Works	849
What Determines Sustainability?	852
Life Cycle Impact of Coffee Cups: Disposable vs. Reusable	852
But Is the Sustainability of the Coffee Container Enough?	856
Implications	857
References	859

LEARNING OBJECTIVES

This chapter provides an assessment of sustainability through an evaluation of a business model that facilitates passive sustainability aligned behavior. The discussion details the operations of a service provision that enables the substitution of a reusable rented container in lieu of a disposable single-use coffee cup and compares sustainability of these two products using publicly available data. Readers are familiarized with externalities, life cycle assessment and ecosystem impact and are asked to contemplate the breadth of sustainability assessment to include the beverage consumed as well.

THEMES AND TOOLS

Product life cycle assessment

INTRODUCTION: SOCIAL AND SUSTAINABILITY MARKETING AND SHARING ECONOMY

Marketing has become synonymous with the selling of goods and services; however, in the case of social marketing, the focus is not just the product it is also human behavior. By promoting the perception of an alignment between individual values and business objectives, social marketing campaigns encourage positive behaviors, like caring for the environment. In 2016, Patagonia, a company based in California (USA) announced that it would donate 100% of sales revenue from Black Friday, the unofficial starting day of the Christmas retail season, to grassroots environmental groups (Patagonia, 2020). At the time of the promotion, the revenue projection for Black Friday was approximately 2 million dollars. However, after social media dissemination, increased consumer awareness resulted in one-day sales revenue of 10 million dollars (Marcario, 2016). The campaign also contributed to further solidifying the brand image of the company as environmentally conscious. The latter arguably fortified Patagonia's presence in its sector, as consumer awareness and knowledge of both sustainability and the company's focus on environmental activism increased. Today, Patagonia "holds the No. 1 spot in the $12 billion [dollar] outdoor apparel market" and the company's position has been credited to "sustainability sells" (Reints, 2019; Overfelt, 2020). The latter was made evident in 2020. Patagonia's primary competitor, The North Face, adopted an active and vocal stance on Climate Change as a competitive strategy (The North Face, n.d.). Though these companies represent one sector, there is increasingly more evidence that corporations are valued by consumers and investors for promoting pro-environmental behavior through their supply chain and across their product's life cycle (Handley, 2020).

Related to social marketing is sustainable marketing, which centers solely on the promotion of environmental and socially responsible products, practices, and brand values. In 2017, Patagonia released its first television commercial, a significant communication given the company's 44-year history. The commercial focused not on the company's products but on the company's activism. Featuring a call to action to maintain federal protection of 770 million acres of public lands (Nace, 2017), the commercial explicitly highlighted Patagonia's brand value, it's mission: sustainability. Similar advertising campaigns have launched including companies such as, 84 Lumber, Airbnb, Budweiser and Nike, as well as others, all focused on coupling a brand's recognition with sustainability, specific to at least one of its pillars: social justice, environmental justice, and economic equity (St. Louis, n.d.). However, perhaps the most visible promotion for sustainability appeared in the New York Times in 2019, "more than 30 American business leaders, including the heads of outdoor clothing brand Patagonia, The Body Shop owner Natura, Ben & Jerry's (part of Unilever) and Danone's US business" joined together for a full-page ad in the Sunday edition of the New York Times to champion a more ethical way of doing business. The advertisement, targeted to the Business Roundtable (BRT), a lobbying organization comprised of corporate giants including Apple and Amazon, focused on a plea for BRT's members to adopt the planet over profit stance of B Corporations. The B Corporation Declaration of Interdependence is provided in Table 34.1.

TABLE 34.1

The B Corporation Declaration of Interdependence

We envision a global economy that uses business as a force for good.
This economy is comprised of a new type of corporation – the B Corporation –
Which is purpose-driven and creates benefit for all stakeholders, not just shareholders.
As B Corporations and leaders of this emerging economy, we believe:

- That we must be the change we seek in the world.
- That all business ought to be conducted as if people and place mattered.
- That, through their products, practices, and profits, businesses should aspire to do no harm and benefit all.
- To do so requires that we act with the understanding that we are each dependent upon another and thus responsible for each other and future generations.

Source: The B Corp Declaration of Interdependence (2020).

It is important to highlight that for non-B Corporations, corporate interest in social and sustainability marketing does not necessarily originate with business but rather mimics consumer awareness and knowledge—as noted in the recent actions of The North Face. As these factors and subsequent demand align with sustainability objectives, the brand premium of sustainability increases. The corporation from this perspective has the potential to evolve to a B Corporation when its stakeholders provide the financial incentive for B Corporation adoption. Therefore, the variation in business adoption of B Corporation values may be indicative of the target demographic of a specific company's consumer base. From this perspective consumers drive business orientation and innovation; therefore, their circumstances and perception of value affect the demand for services. This latter aspect directly ties to the emergence of sustainability facilitating businesses, also referenced as collaborative consumption or the sharing economy.

The sharing economy has no common definition, instead it can best be understood as an umbrella concept that includes different types of sharing, anything from collaboration to reuse to rent.

> The term 'Sharing Economy' emerged from the global crisis of 2008-09 and the need to do more with less. Its birth demonstrated that necessity is often the best mother of invention. Fueled by technology that was able to match people who had spare or idle resources with those that wanted or needed them, the term became associated with new types of 'peer-to-peer' or person-to-person online marketplaces like Airbnb. In reality, the Sharing Economy is much more than a collection of new types of Silicon Valley backed ventures; it is wide-reaching and changing society as we know it. It is at once an economic system built around the sharing of human and physical resources and a mind-set .... The Sharing Economy is a system to live by, where we care for people and planet and share available resources however, we can.
>
> (Matofska and Sheinwald, 2019a)

A sharing economy is less resource intensive compared with the present accumulation-focused economy as products are not created for individual ownership but rather are available for common use. This means that the same product can be used by multiple people without any one of them having the full financial responsibility for maintenance and replacement. Instead all of them contribute, based on use, a prorated share of expense and depreciation. Belk et al. (2010) contend that sharing is the natural state of consumption, which has only recently, in historical time, been undermined by individual-targeted consumerism. "Not only is sharing critical to the most recent of consumption phenomena like the Internet, it is also likely the oldest type of consumption" (p. 730).

A sharing economy differs from a B Corporation, because the intention is not to sell more and do good but rather to own less, have more for all, and promote common good. "In a [s]haring world, our cultural norms, traditions and creative activities root themselves in collaboration, cooperation and community. This is an ecosphere far away from inequality, poverty and hyper consumption. For a sharing culture to be mainstreamed, a significant shift is required. It's about changing our value system, from me to we, from consumer to user and from owner to co-operator … sustainability supersedes selfishness and inclusivity is vital" (Matofska and Sheinwald, 2019b). The cultural shift inherent in a sharing economy aligns with sustainability and culture has been acknowledged by the United Nations as a significant component to attaining global sustainability (UNESCO, 2000). Further, with respect to the broader economic framework, the inputs and outputs of economic systems are dependent on the culturally determined value structures of a society. Economic outcomes in essence mimic the values of the participants in an economic system, as these evolve, so does the outcome of the economy.

Arguably from this discussion there are variations in how marketing facilitates sustainability, and these are related to both the type of marketing and the target consumer. Essentially, though, it is the consumer that determines the persistence of the business marketing message and ultimately any resulting transformation of the economic framework. Further, consumer education and awareness campaigns can be a catalyst to brand alignment with sustainability but are not sufficient without consumer demand for sustainability. Since context (i.e., socioeconomic, demographic) affects consumer perception of value, not all consumers are appropriate targets for the same sustainability focused enterprises: sharing economy transportation services may thrive in urban areas, but may have no market and as a consequence be cost prohibitive in rural communities (Davidson and Infranca, 2016). The life cycle discussion and case study that follows provides development and operational insight to a context-driven sharing economy enterprise. Highlighting the product's social and sustainability marketing strategy, the discussion also surfaces the conundrum of sustainability by assessing the resource footprint of consumption itself. The discussion concludes with a commentary on the responsibility of marketing to promote holistic sustainability and asserts the significance of consumer education in shifting the perception of the right to consume to the responsibility in consumption.

BACKGROUND: LIFE CYCLE OF COFFEE CONSUMPTION

Coffee is the most consumed and popular beverage in the world after water (Coltria et al., 2019; Shabendeh, 2019). It is also resource intensive. According to the US Geological Survey (2016), globally nearly 120 billion cubic meters are used to produce coffee; this approximately 2% of the global water use for crop production. Though coffee is consumed globally, the majority is consumed in developed markets (see Table 34.2), while coffee is produced in the developing world, primarily in South and Central America, the Caribbean, Africa and Asia (ISIC, 2020; Shabendeh, 2019). These are the same areas where water is not always easily accessible and where Climate Change has facilitated scarcity (WaterAid, n.d.); related, estimates are that coffee production may be significantly reduced in the next 30 years (Vo, 2020; Social Impact, 2019). Given the resource intensity of coffee production, coffee represents a trade-off between present survival for some people and environmental stewardship. Without the latter, the former, survival, too may be compromised. In addition to environmental degradation and water use, coffee production also includes child labor, slave labor, non-human animal cruelty and maintenance of a vicious cycle of poverty in the Global South (Food Empowerment Project, 2020). Further, plantation cultivation of coffee, which is the primary method of production, eliminates non-human habitat, contributes to soil erosion and has a short-lived production cycle, plantations can produce for 12–15 years before they require replanting (Coltria et al., 2019; The Conversation, 2018).

In addition to growth and harvesting, there are many other stages to coffee production, each with a life cycle inclusive of production to waste. These include growth and harvesting to the final waste of the coffee cup used in coffee consumption. While the production of beans has a supply chain impact that affects global poverty and ecosystem preservation, the consumption stage has a significant production and waste footprint, which both taken collectively exacerbate greenhouse gas emissions and affect environmental resource use and quality.

TABLE 34.2

Top 25 Coffee Consuming Countries

Rank	Country	Coffee Consumption (lbs Per Person Per Year)
1	Finland	26.45
2	Norway	21.82
3	Iceland	19.84
4	Denmark	19.18
5	Netherlands	18.52
6	Sweden	18
7	Switzerland	17.42
8	Belgium	15
9	Luxembourg	14.33
10	Canada	14.33
11	Bosnia and Herzegovina	13.67
12	Austria	13.45
13	Italy	13
14	Brazil	12.79
15	Slovenia	12.79
16	Germany	12.13
17	Greece	11.9
18	France	11.9
19	Croatia	11.24
20	Cyprus	10.8
21	Lebanon	10.58
22	Estonia	9.92
23	Spain	9.92
24	Portugal	9.48
25	United States	9.26

Source: Bernard (2020).

On a global scale, the environmental footprint from coffee production and consumption can be proxied by annual production levels, which from 2012/13 to 2017/18 has steadily increased as depicted in Figure 34.1. The corresponding price of coffee during the same period, as depicted in Figure 34.2, has declined. Though these values depict a macro-level perspective of coffee that does not allow for differentiation in quantity and prices based on cultivation differences (i.e., shade, plantation) or wage differences (i.e., fair trade, regulatory protections), the graphics do highlight that production of coffee has increased and prices in aggregate have not

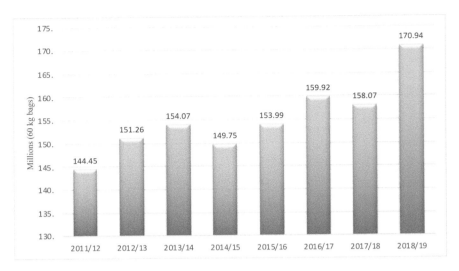

FIGURE 34.1
Annual global coffee production. (From Statista, 2020a).

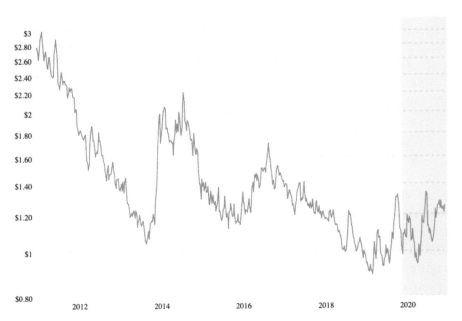

FIGURE 34.2
Global coffee prices. (From Macrotrends, 2020).

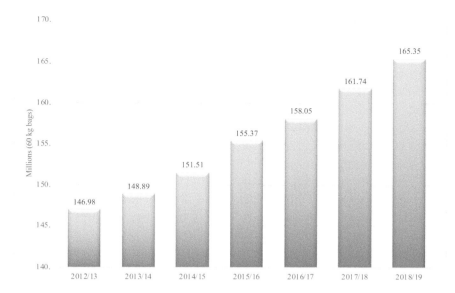

FIGURE 34.3
Annual global coffee consumption. (From Statista, 2020b).

faced upward pressure. From an economic perspective this relationship would be supported if demand for coffee was indeed less than its supply, which is the case. Figure 34.3 depicts the increase annual coffee consumption over the same period. Supply levels of coffee are a attributed to a "record seasonal surplus of beans" in late 2019; however, given climate change and the downward pressure on coffee prices resulting from year-end production, farmers are expected to reduce production going forward; decrease in consumption is at present not anticipated (The Global Warming Policy Forum, 2019).

Prices for coffee are arguably not reflective of the product's resource intensity, ecosystem degradation and contribution to climate change, especially considering the potential compounding of impacts resulting in the geographic spread of cultivation. The supply and demand dynamic for coffee production may offer a significant rationale for the pricing inefficiency. It can be argued that the asymmetry in market power between developing and developed markets has prohibited prices of coffee from fully reflecting their holistic cost. As a result, consumption in developed counties has been facilitated by prices that do not account for the exploitation related to environmental degradation and living wages on the part of the harvester (Bacon, 2010; Breman, 2015; Smith and Loker, 2012); what Jaffee (2014) refers to being an outcome of eco-colonialism. Though there is price differentiation based on specialty coffee shops and types of coffee consumed, the financial returns primarily accrue to the end distributor, in the US market the focus of the present discussion, this includes the branded or specialty coffee shop. Fridell (2007) reflects, "there is a structural dimension to exploitation in the coffee industry, which is driven not primarily by ethical values but by the capitalist imperatives of competition, accumulation, profit maximization, and increasing [labor] productivity. The emergence of ever-larger, monopolistic coffee [Transnational Corporations] TNCs have been due primarily to fierce competitive pressures as corporations have beat out or bought up one another in order to survive. The exaggerated boom and bust cycles of the coffee industry have emerged out of the decisions of millions of large and small coffee farmers, rational on an individual level, to respond to temporary upswings in coffee prices by increasing production. The result has been repetitive crises in overproduction, leading to fierce competition for market shares. Responding to these competitive pressures, wealthy landlords have struggled to force down wages and step up exploitative mechanisms to extract as much wealth from workers as possible, and small farmers have been forced into a variety of survival strategies to cope with near-starvation coffee incomes."

THE CASE OF USEFULL

World Coffee Portal estimates that in 2019 the "47.5 billion dollar US branded coffee shop market grew 3.3% to reach 37,274 outlets. Since then, the United States had been hard hit by coronavirus, which has piled extraordinary pressure on a coffee shop segment already facing intense competition and soaring consumer expectations. Nevertheless, market leaders Starbucks and Dunkin', as well as large boutique operators such as Bluestone Lane, have ramped up digitally integrated ordering, drive-thru and delivery cater for new customer routines" (World Coffee Portal, 2020). Unfortunately, with the adjustment to the pandemic, restrictions were placed on consumer reusable coffee cups. "Major brands like Starbucks and Dunkin' [have] banned the use of personal to-go containers" in March based on coronavirus fears" (Vann, 2020). The elimination of consumer reusables has increased use of disposable coffee cups but has also provided a new entry point to an emerging sharing economy product, a third-party operated, health code-certified, circular economy mechanism, the reusable rental in the form of a coffee cup.

Initially branded as Coffee Cup Collective, USEFULL is one of a few sharing economy firms specializing in the niche reusable coffee cup market. The company, which is based in the Boston metropolitan market, provides reusable stainless steel to-go cups to coffee shop clients for a competitive per unit price compared with the alternative, the disposable single-use coffee cup. Consumers are able to have their coffee in the reusable cup and leave the shop with the expectation that the cup will be returned to a designated location. Several coffee shops and drop-off locations are part of the convenience feature of this reuse facilitating intermediary.

How USEFULL Works

USEFULL is both a direct to consumer and corporate partnering service. It's direct to consumer operation is an app-based service. The first step is in downloading the app to a smartphone or equivalent device. After download the user is prompted to sign-up, which does require a credit card. However, sign-up can be completely free depending on the user's hold time of the reusable coffee cup. There is no charge for a two-day rental but after two days holding the cup does start to accrue charges with a lost fee assessed if the cup does not re-enter the reuse pool. This Freemium membership comes with unlimited use of the product without charge as long as the two-day rental is not exceeded. Recognizing that not everyone may be diligent with respect to timely drop off, USEFULL also has a Premium member ship at the cost of 9.99 dollars a month. This service allows the user a five-day rental period and offers the added incentive of a sustainability marketing aligned donation: "Be an environmental steward by supporting USEFULL as we donate 1% of your subscription to a charity focused on eliminating ocean plastic." The app provides names and locations of participating cafes and drop off locations and the opportunity of online ordering requesting a USEFULL cup.

Charges related to cups are assessed by QR code. All cups have a QR code and a pet name; the latter serves as a novelty marketing component. These are used to track an individual cup from rental to return through

> Our cafe partners share our mission to reduce Greater Boston's waste. Will you join us?

HOW WE WORK TOGETHER

We provide cups, training + materials, and basic marketing.

Offer the USEFULL cups as your primary to-go cup. If a customer requests a single use cup, charge them $0.50 for the use of that cup.

Choose to:
a) manage the cup dishwashing onsite OR
b) have our team manage the dishwashing and logistics for you
(additional fees apply)

We ensure proper inventory of cups across the network.

FIGURE 34.4
USEFULL corporate value proposition. (From USEFULL, 2020).

cleaning and back to inventory and are accessible across the distribution chain by USEFULL. Customers are able to see their specific product use chain by pet name from initial rental to return. The reusable cup inventory are USEFULL's primary asset and represent an initial fixed cost but given the cup in made of stainless steel, this cost is only incurred for replacement and expansion.

USEFULL also offers direct to corporate services. These services are determined on an individual contract basis and no pricing information is available. As depicted in Figure 34.4, these services focus on brand image. Reaffirming that "sustainability sells" (Choi & Ng, 2011; Eccles et al., 2014), USEFULL has positioned its value proposition as being additive to the sustainability image of a corporate client. Corporate partners are teased with the comments:

- 92% of consumers are more likely to trust a company that supports an environmental cause.
- Organizations that focus on sustainability are more likely to attract and retain employees, especially millennials.

In 2019, CGS, a US-based business service provider, released its 2019 Retail and Sustainability Survey, which assessed how sustainable products and business practices affected consumer purchase decisions. The findings supported that in spite of price being an important factor in purchase decisions, consumers were very interested in promoting sustainability and "are also increasingly focused on shopping with brands whose mission they care about" (CGS, 2019; Cho, 2015). A key finding was that "Buyers want sustainable products; and over one-third will pay 25% more for them" (CGS, 2019).

Specific to employee recruitment and retention, a 2017 Chicago Tribune survey highlighted that millennial employees not only sought meaning in their employment but also appeared to be more focused on meaning as conveyed through a company's social impact (Chicago Tribune Staff, 2017). An additional component of meaningfulness can be related to the company's perceived sustainability. A 2015 Federal Reserve Board survey of Young Workers reinforced that money was not the primary focus of young adults. Steady employment was more significant than higher pay at a relative percentage of 62–36. Of those exhibiting a preference for steady employment, 80% reinforced their preference for steady employment noting a preference for one steady job over a stream of steady jobs for the next five years (Federal Reserve Board, 2017).

In both the direct to consumer and corporate market the USEFULL service is based on social and sustainability marketing and leverages the reuse element of the sharing economy. From this perspective it offers the coffee shop the opportunity to brand its coffee product as promoting sustainability and enables the consumer of the coffee to perceive their consumption of coffee as having a lower environmental footprint compared with disposable cups, both aligned with social marketing. The sustainability marketing is the service itself. USEFULL's brand value is in the sustainability of its service offering. For the corporate client, USEFULL's service promotes the corporate brand and operation as being aligned with sustainability, enhancing the favorability rating of both consumers and employees based on present research and aligned from the corporate perspective to sustainability marketing. The value proposition of USEFULL is then tied and sensitive to the perception and reality of their sustainability attributes of their reusable container. Essentially, their reusable cup has to have a lower footprint relative to a disposable cup or arguably even the ceramic cup that the coffee shop may use for in-café dining.

From a revenue perspective, USEFULL generates returns through three channels: coffee shop, per cup use charge; the customer, membership fee; and the corporate client, contract. All channels are directly related to the perception of their reuse offering as being pro-environmental compared with alternatives. The upfront expense of the cups is paid for by a revenue stream that requires repeated reuse of the USEFULL product and service offering. The company's ultimate success is determined by the perceived sustainability of its product and service offering. Ultimately, it is the demand for pro-environmental products on the part of the end user that is the driver of the USEFULL product, its brand and financial success. Increasing demand affects the deployment and return cycle, increasing the reuse life of any given USEFULL cup; and this increase reduces the environmental footprint of the cup, enhancing its overall relative environmental footprint.

What Determines Sustainability?

The definition of sustainability has not been formalized. In discussions related to the attribute, the definition that most commonly is stated is taken from The Brundtland Commission (WCED, 1987) and reflects that consumption and resource use in general needs to be consistent with future as well as present needs. In other words, sustainability can be attributed as a holistic and intertemporal evaluation that includes market values and respective externalized costs; the impact of time with respect to the costs of ecosystem regeneration; and the unknown and complex interconnections inherent in our ecosystem, which can be characterized as risk tolerance. From this perspective evaluation is more than science, it is an art that relies on both quantitative and qualitative parameters.

Sustainability as it relates to USEFULL's product offering relies on a holistic assessment of the life cycle of the stainless cup and plastic lid, the components of the reusable cup, compared with the disposable cup, lid and sleeve, as well as the ceramic cup. The evaluation of the life cycle of these three containers includes the extraction to the production to the consumption and ultimately the waste stage of the product. For the evaluation to be worthwhile it needs to account for more than just the market value of resources used in production but also the ecosystem impacts, and habitat impacts to non-human life, specifically addressing the risk of what is unknown as a constraint on the use of resources.

Life Cycle Impact of Coffee Cups: Disposable vs. Reusable

Nearly 80 years ago, Beulah France (1942) promoted the use of single-use paper cups in health care and even addressed the use of the product as patriotic. Though she addressed post-use waste, her evaluation of the overall cost and benefit of paper cups use was narrow. She focused solely on time spent cleaning ceramic alternatives and the potential spread of communicable disease from ceramic use in a hospital; her life cycle view of paper cup use and disposal did not include an acknowledgement of the cost to the environment, non-human life and habitat loss. Arguably, an anthropocentric view has continued to limit decision-making. At present, specific to disposable cups, "[t]he U.S. accounts for about 120 billion paper, plastic and foam coffee cups each year, or about one-fifth of global total. Almost every last one of them – 99.75% – ends up as trash, where even paper cups can take more than 20 years to decompose" (Chasen and Parmar, 2019). For dominant retailers like Starbucks, which goes through about 6 billion cups a year, and Dunkin', which attributes nearly 70% of its revenue from coffee drink products, the waste associated with their products does pose a brand image issue (Chasen and Parmar, 2019; Simon, 2009; Starbucks, 2020). The same waste footprint is observable on a smaller scale but with the same environmental footprint at private and specialty coffee shops, the target market of USEFULL.

The most common type of disposable coffee cup has a plastic lining; the lining acts as a barrier that prevents the paper cup from becoming saturated and falling apart. However, the lining is co-mingled with paper, which presents a problem with respect to recycling; separation is both difficult and expensive. In fact, most recycling centers cannot process them; as a result, less than 1 percent is presently recycled. In general, though the qualitative, nonmarket environmental costs of single-use coffee cup disposal includes the following:

- Transportation: Greenhouse gas impact
- Landfill disposal: Groundwater contamination (from chemical leaching from plastic)
- Littering: Health and mortality risk of non-human animals
- Incineration disposal: Greenhouse gas impact and air pollution

Given the volume of single-use coffee cups used in an annual basis, the impact of disposable use is significant.

CIRAIG, a research group and center of expertise on sustainability and life cycle thinking (CIRAIG, 2020a), published a study which compared the potential environmental impacts of a 16-ounce, single-use coffee cup made of a mix of cardboard and polyethylene (with a lid made of polystyrene) with a 16-ounce, reusable ceramic cup and also a 16-ounce travelers' mugs made of stainless steel, polypropylene (PP), and polycarbonate. "Over a one-year span (using one cup a day), the reusable cups scored well in the climate change arena—that is, they were associated with fewer greenhouse gas (GHG) emissions than their single-use counterparts. Likewise, they scored better in the human-health category for things such as toxic emissions, smog, and ozone depletion. They also tended to use fewer minerals and fossil fuels than disposable cups did" (CIRAIG, 2020b). However, after including the environmental impact related to cleaning reusables, the sustainability differences were not as clear. The findings as provided below found that ceramic cups fared best in terms of a lighter environmental footprint after multiple uses; the same was true for the stainless steel travel mugs if they were handwashed with cold water.

- Overall, ceramic mugs have fewer environmental impacts than travel mugs or paper cups.
- Ceramic mugs have a smaller potential environmental impact than paper cups with lids when used at least 200 to 300 times.
- Travel mugs generally become preferable to paper cups after a reasonable number of uses, except for the Water Consumption and Quality of Ecosystems categories (for which we cannot identify a winner).
- Most of the potential impacts attributable to travel mugs stem from the fact that they must be hand-washed. A quick rinse in cold water (without soap) would bring travel mugs almost on par with ceramic mugs in terms of their impacts.
- Among travel mugs, stainless steel models with PP lid and handle perform better from an environmental perspective than travel mugs made of PP or polycarbonate.
- Serving coffee in two stacked paper cups (to protect hands from heat) has slightly more potential impacts than using a double-walled cup.
- The use of cardboard sleeves to insulate single-wall paper cups only slightly increases the potential impacts and is preferable to using double-walled cups (CIRAIC, 2020b).

An aspect of the plastic-lined disposable cup that was not considered in the CIRAIC assessment was the relationship between the plastic liner and the petroleum industry, specifically, that the plastic promotes the use of a non-renewable fuel. Further, related and significant is the impact on environmental and human health in the production of plastic from the extraction of petroleum to the creation of plastic pellets that are molded into liners.

However, continuing with the human health perspective, plastic has the potential to leach the chemicals that comprise it into the food and beverages it holds. The most well-known of these chemicals is bisphenol-A, which was first used as a synthetic estrogen in the 1930s. During the plastic manufacturing process, not all BPA gets locked into chemical bonds; as a result, non-bonded, residual BPA can work itself free, especially when the plastic is heated, whether it's a baby bottle in the dishwasher, a food container in the microwave, or a test tube being sterilized in an autoclave. Bao et al. (2020) found a relationship between higher BPA ingestion and higher risk for death. In recent years dozens of scientists around the globe have linked BPA to myriad health effects in rodents: mammary and prostate cancer, genital defects in males, early onset of puberty in females, obesity and even behavior problems such as attention-deficit hyperactivity disorder" (Hinterthuer, 2008). Additional research is being conducted with respect to the connection between synthetic estrogens found in plastic and their impact on the risks of heart attack, obesity and changes in the cardiovascular system (Giuliani et al., 2020; Yang et al., 2011).

Other chemicals in plastic, phthalates are often used as softeners for PVC plastic, to make plastic more flexible. But phthalates have been found to be harmful to human health. Bis(2-ethylhexyl) phthalate (DEHP), Benzyl butyl phthalate (BBP), Dibutyl phthalate (DBP) and Diisobutyl phthalate (DIBP) are classified as endocrine disruptors that are toxic to reproduction, which means that they may damage fertility or the unborn child (Braun et al., 2013). Both BPAs and phthalates are found in plastic containers available on the market today and can be present in the plastic liner of a disposable coffee cup.

Addressing the holistic costs related to the disposal of single-use plastic cups, Chandrasekaran et al. (2011) addressed the impact of disposable cups on the declining population of bees. They observed how bees drawn to the sugar residue left within a cup tossed into an open-air public trash can attract bees, who then subsequently drowned in the coffee residue. This cost is an example of context-driven holistic costs that can emerge in specific locations but are often not included in life cycle assessments; their omission provides an incomplete assessment and provides credibility for a value for risk of what is unknown.

Evans (2019) provided not only a production, consumption to waste footprint in his assessment of reusables versus disposable coffee cups, he also evaluated the environmental impact of waste. With this consideration, according to his estimates, reusable cups have a longer lifespan, which reduces their waste impact relative to disposable single-use cups. Further, "[m]ost reusable cups can be recycled. Glass and ceramic are less of a threat to the natural environment because they will break down over time and do not contain synthetic chemicals, unlike styrofoam or the plastic lining of paper cups which do not biodegrade. Disposal of single use cups poses other threats such as the cost of waste collection and the accumulation of these products in our oceans, when they are not taken to the landfill or incinerated" (Evans, 2019).

However, even with the lower environmental footprint of the reusable container, challenges arise specific to the "on the go" norm. "Coffee to go is frequently a spontaneous purchase rather than a pre-meditated one," says the Paper Cup Recovery and Recycling Group (PCRRG), "some consumers also prefer not to carry a reusable cup, either because they don't carry a bag with them, or they are worried about contamination from a dirty cup." This aspect of coffee consumption ties directly to the value proposition of USEFULL.

The USEFULL model can provide reusable, sanitized products that are efficiently cleaned, and which can be used an indefinite number of times until their physical appearance may impact circulation, at which time they can be recycled. The product caters to the individual who wants gratification with a pro-environmental bias without active engagement (i.e., purchasing a reusable, carrying a reusable) to promote sustainability. Given that the cup can remain in circulation indefinitely, based on calculations from CIRAIG (2020b), if it is able to at minimum be used 150 times, the greenhouse gas footprint will be lower than a disposable cup. Further with high efficiency dish washing, its water footprint can be lowered to enhance its overall sustainability advantage.

But Is the Sustainability of the Coffee Container Enough?

Rosie Frost (2020) notes, "Packaging generally only represents 4 percent of [coffee's] total carbon footprint according to Finnish packaging experts, Huhtamäki". For this reason, if the contents of the coffee cup should be assessed for sustainability, this should include the coffee itself. Given its resource intensity and sensitivity to the speed of climate change as well as its association with climate change, ecosystem impact, and habitat loss, it may be that the container is only a short-sighted salve for a more significant issue, the overconsumption based on underpricing of coffee itself.

DesJardins (2005) defines sustainable production as inclusive of the two endpoints of the production process: the wastes and pollution that comes out of the process. In segmenting the cup from the coffee, the full impact of coffee consumption is not being addressed. Unfortunately, there are only a few studies that have evaluated the life cycle of the production of coffee and these have neglected to holistically integrate ecosystem loss and living wages/standard of living of producers in the analysis. Instead, some studies have highlighted the ecosystem and biodiversity loss (Vogt, 2019), while others have focused on labor (Kitti et al, 2009).

The Sustainable Business Toolkit noted in 2013, "Farmers have been positively encouraged to replace their traditional and supposedly inefficient farming methods with the higher yielding technique of sun cultivation, which has resulted in over 2.5 million acres of forest being cleared in Central America alone to make way for coffee farming in this way. Deforestation trends are serious throughout the coffee producing lands of Latin America and remarkable biodiversity values are at stake. Latin America's tropical forests are critical ecologically for purposes of protection of atmospheric dynamics, water quality, wildlife species, as well as economically … [I]rrespective of how coffee is grown, discharges from coffee processing plants represent a major source of river pollution. Ecological impacts result from the discharge of organic pollutants from the processing plants to rivers and waterways, triggering eutrophication of water systems and robbing aquatic plants and wildlife of essential oxygen." (Moore, 2013). Given the life cycle of coffee as a product, there is a limitation, arguably a significant one, to segmenting the evaluation of sustainability to only one stage of the consumption cycle, production of the product clearly has an impact on sustainability.

IMPLICATIONS

Consumption choices in general are based on the dynamics of demand and supply of a good and are identified with satisfying a need or a want. The impact of consumption decisions can be significant when there is asymmetry of information; fundamentally, there is a relationship between economic and environmental outcomes and consumption choices. Purchases affect labor and environmental resource use. However, most purchase decisions are made through a market mechanism, where the consumer is not aware of the entire production process and waste is not a factor in the consumption decision. This limitation in information transparency often creates a disconnection between the social and environmental justice sensitivities of a consumer, and the realities of their consumption choice in enabling and maintaining the values that they espouse. It is for this reason that marketing has a significant role in an economy. Subject to any prevailing regulatory constraints, in a consumption-led, GDP focused economy, marketing, through establishing demand via marketed wants, promotes purchasing behavior.

Consumers rely on product marketing as a source of information on a product and often default to certifications and brand image as a decision-making tool (Kahn, 2012; Overbeek, 2019). In order for social good and sustainability to be achieved or enhanced through marketing, marketing itself needs to move beyond just the selling phase to the responsibility stage. Marketing needs to consider science, including environmental uncertainties, and take a holistic stance in its relationship with products especially when the product image is sustainability (Whiteman, 1999). Sustainable marketing requires that marketing practitioners "spend an equal, if not greater, amount of time and energy on what happens after the individual or business 'consumes' the product ... [M]arketing's responsibilities and opportunities do not end with the sale (Murphy, 2005). Noting these parameters for the discipline, there is a need to enlarge the discission of sustainability to include a product life cycle and essentially assess the sustainability of a product holistically across the multiple supply chains that are incorporated within it.

In the present global economy, there is an obvious and urgent need to incorporate sustainability into all activities but there is an even more important goal to align all sustainability objectives to the common goal of reducing the human planetary footprint through behavioral change. This change will require a new relationship with the planet that will need awareness, knowledge and action. Marketing, if implemented responsibly, can be instrumental in facilitating all three.

With respect to the discussion of USEFULL, sustainability marketing has an opportunity in educating the user or target consumer and client about the product life cycle of the coffee container and coffee, itself. Education in turn could affect the social marketing aspect of the whole product, providing greater understanding of the footprint of coffee consumption and the vessel of consumption. In turn, this could inspire the adoption of reuse and also responsible consumption (i.e., reduction, selection of coffee based on sustainability criteria). There is a pressing need for marketing to adopt criteria for its own evaluation of success that goes beyond the sale of a product to how the product's deployment promotes social and environmental welfare.

REFERENCES

Bacon, C. (2010). A Spot of Coffee in Crisis: Nicaraguan Smallholder Cooperatives, Fair Trade Networks, and Gendered Empowerment. *Latin American Perspectives*, *37*(2), 50–71.

Bao, W., Liu, B., Rong, S., Dai, S. Y., Trasande, L., & Lehmler, H. (2020). Association between Bisphenol A Exposure and Risk of All-Cause and Cause-Specific Mortality in US Adults. *JAMA Netw Open*, 3(8):e2011620.

Belk, R. (2010). Sharing. *Journal of Consumer Research*, *36*(5), 715–734.

Bernard, K. (2020 August 6). The Top Coffee-Consuming Countries. *World Atlas*. Retrieved from https://www.worldatlas.com/articles/top-10-coffee-consuming-nations.html

Braun, J. M., Sathyanarayana, S., & Hauser, R. (2013). Phthalate exposure and children's health. *Current Opinion in Pediatrics*, *25*(2), 247–254.

Breman, J. (2015). The Coffee Regime under the Cultivation System. In *Mobilizing Labour for the Global Coffee Market: Profits from an Unfree Work Regime in Colonial Java* (pp. 211–254). Amsterdam: Amsterdam University Press.

CGS. (2019 January 10). CGS Survey Reveals Sustainability Is Driving Demand and Customer Loyalty. Retrieved from https://www.globenewswire.com/news-release/2019/01/10/1686144/0/en/CGS-Survey-Reveals-Sustainability-Is-Driving-Demand-and-Customer-Loyalty.html

Chandrasekaran, S., Nagendran, N., Krishnankutty, N., Pandiaraja, D., Saravanan, S., Kamaladhasan, N., & Kamalakannan, B. (2011). Disposed Paper Cups and Declining Bees. *Current Science*, *101*(10), 1262–1262.

Chasen, E. & Parmar, H. (2019 April 28). Berkeley, Others Target a New Environmental Scourge: The To-go Coffee Cup. *The Seattle Times*. Retrieved from https://www.seattletimes.com/business/first-they-came-for-plastic-bags-coffee-cups-are-next/

Chicago Tribune Staff. (2017 November 10). A Survey: What Local Workers Think about Their Companies. Retrieved from http://www.chicagotribune.com/business/careers/topworkplaces/ct-biz-top-workplaces-2017-survey-details-20170929-story.html

Cho, Y. (2015). Different Shades of Green Consciousness: The Interplay of Sustainability Labeling and Environmental Impact on Product Evaluations. *Journal of Business Ethics*, *128*(1), 73–82.

Choi, S., & Ng, A. (2011). Environmental and Economic Dimensions of Sustainability and Price Effects on Consumer Responses. *Journal of Business Ethics*, *104*(2), 269–282.

CIRAIG. (2020a). *Shaping and implementing* Metrics for Sustainability. Retrieved from http://ciraig.org

CIRAIG. (2020b). Life Cycle Assessment (LCA) of Reusable and Single-use Coffee Cups. Retrieved from https://www.recyc-quebec.gouv.qc.ca/sites/default/files/documents/acv-tasses-cafe-resume-english.pdf

Coltri, P., Pinto, H. S., Ribeiro do Valle Gonçalves, R., Zullo, J., & Dubreuil, V. (2019). Low Levels of Shade and Climate Change Adaptation of Arabica Coffee in Southeastern Brazil. *Heliyon*, *5*(2).

Davidson, N., & Infranca, J. (2016). The Sharing Economy as an Urban Phenomenon. *Yale Law & Policy Review*, *34*(2), 215–279.

DesJardins, J. (2005). Business and Environmental Sustainability. *Business & Professional Ethics Journal*, *24*(1/2), 35–59.

Eccles, R., Ioannou, I., & Serafeim, G. (2014). The Impact of Corporate Sustainability on Organizational Processes and Performance. *Management Science*, *60*(11), 2835–2857.

Evans, D. (2019 December 26). Cups: Single Use (Disposable) vs. Reusable – An Honest Comparison. *Plastic Education*. Retrieved from https://plastic.education/cups-single-use-disposable-vs-reusable-an-honest-comparison/

Federal Reserve Board. (2017 February 2). Experiences and Perspectives of Young Workers. Retrieved from https://www.federalreserve.gov/econresdata/2016-survey-young-workers-preface.htm

Food Empowerment Project. (2020). Bitter Brew the Stirring Reality of Coffee. Retrieved from https://foodispower.org/our-food-choices/coffee/

France, B. (1942). Uses for Paper Cups and Containers. *The American Journal of Nursing*, *42*(2), 154–156.

Fridell, G. (2007). Coffee and the Capitalist Market. In *Fair Trade Coffee: The Prospects and Pitfalls of Market-Driven Social Justice* (pp. 101–134). Toronto: University of Toronto Press.

Frost, R. (2020 February 12). Is Your Reusable Coffee Cup Really Making a Difference? *Euronews*. Retrieved from https://www.euronews.com/living/2020/02/12/is-your-reusable-coffee-cup-really-making-a-difference

Giuliani, A., Zuccarini, M., Cichelli, A., Khan, H., & Reale, M. (2020). Critical Review on the Presence of Phthalates in Food and Evidence of Their Biological Impact. *International Journal of Environmental Research and Public Health*, *17*(16): 5655.

Handley, L. (2020 August 10). 'We Can't Run a Business in a Dead Planet': CEOs Plan to Prioritize Green Issues Post-Coronavirus. *Our New Future*. Retrieved from https://www.cnbc.com/2020/08/10/after-coronavirus-some-ceos-plan-to-prioritize-sustainability.html

Hinterthuer, A. (2008). Safety Dance over Plastic. *Scientific American*, *299*(3), 108–111.

ISIC. (2020). Coffee Production Today. *Coffee & Health*. Retrieved from https://www.coffeeandhealth.org/all-about-coffee/coffee-production-today/

Jaffee, D. (2014). A Sustainable Cup?: Fair Trade, Shade-Grown Coffee, and Organic Production. In *Brewing Justice: Fair Trade Coffee, Sustainability, and Survival* (pp. 133–164). Berkeley and Los Angeles, CA: University of California Press.

Kahn, R. (2012). Fair Trade Activists in the United States. In Linton A., *Fair Trade from the Ground Up: New Markets for Social Justice* (pp. 101–119).

Kitti, M., Heikkila, J., & Huhtala, A. (2009). 'Fair' Policies for the Coffee Trade – Protecting People or Biodiversity? *Environment and Development Economics, 14*(6), 739–758.

Macrotrends. (2020). Coffee Prices – 45 Year Historical Chart. Retrieved from https://www.macrotrends.net/2535/coffee-prices-historical-chart-data

Marcario, R. (2016). Record-Breaking Black Friday Sales to Benefit the Planet. *Patagonia.* Retrieved from https://www.patagonia.com/stories/record-breaking-black-friday-sales-to-benefit-the-planet/story-31140.html

Matofska, B., & Sheinwald, S. (2019a). What is the Sharing Economy? In *Generation Share: The Change-Makers Building the Sharing Economy* (pp. 16–24). Bristol, UK: Policy Press

Matofska, B., & Sheinwald, S. (2019b). Is Sharing Cultural? In *Generation Share: The Change-Makers Building the Sharing Economy* (pp. 211–242). Bristol, UK: Policy Press.

Moore, V. (2013). The Environmental Impact of Coffee Production: What's Your Coffee Costing The Planet? *Sustainable Business Toolkit.* Retrieved from https://www.sustainablebusinesstoolkit.com/environmental-impact-coffee-trade/

Murphy, P. (2005). Sustainable Marketing. *Business & Professional Ethics Journal, 24*(1/2), 171–198.

Nace, T. (2017 August 24). After 44 Years Patagonia Released Its First Commercial & It's Not About Clothing. *Forbes.* Retrieved from https://www.forbes.com/sites/trevornace/2017/08/24/44-years-patagonia-released-first-commercial-clothing/#1d5aa9ee3c80

Overbeek, A. (2019). Examining the Efficacy of Fair Trade and Alternative Consumption on Environmental Sustainability and Human Rights in Developing Countries. *Consilience* (21), 158–171.

Overfelt, M. (2020 August 14). As the North Face Battles Patagonia in Outdoors Market, It Bets Tackling Climate Change Will Pay Off. *CNBC.* Retrieved from https://www.cnbc.com/2020/08/14/as-north-face-battles-patagonia-it-bets-climate-change-will-pay-off.html

Patagonia. (2020). We're in Business to Save Our Home Planet. Retrieved from https://www.patagonia.com/activism/

Reints, R. (2019 November 5). Consumers Say They Want More Sustainable Products. Now They Have the Receipts to Prove It. *Fortune.* Retrieved from https://fortune.com/2019/11/05/sustainability-marketing-consumer-spending/

Shabendeh, M. (2019 February 14). Largest coffee producing Countries 2018. *Statista.* Retrieved from https://www.statista.com/statistics/277137/world-coffee-production-by-leading-countries/

Simon, B. (2009). Not-So-Green Cups. In *Everything but the Coffee: Learning about America from Starbucks* (pp. 173–200). Berkeley; Los Angeles; London: University of California Press.

Smith, E., & Loker, W. (2012). "We Know Our Worth": Lessons from a Fair Trade Coffee Cooperative in Honduras. *Human Organization, 71*(1), 87–98.

Social Impact. (2019 February 14). How Climate Change is Killing Coffee. *Knowledge at Wharton.* Retrieved from https://knowledge.wharton.upenn.edu/article/coffee-climate-change/

St. Louis, M. (n.d.). 6 Socially Charged Ads That Caused a Stir. *AdWeek.* Retrieved from https://www.adweek.com/brand-marketing/5-socially-charged-ads-that-caused-a-stir/

Starbucks. (2020). 2019 Global Social Impact Report. Retrieved from https://stories.starbucks.com/uploads/2020/06/2019-Starbucks-Global-Social-Impact-Report.pdf

Statista. (2020a). Coffee Production Worldwide from 2003/04 to 2018/19 (in Million 60 kilogram Bags)*. Retrieved from https://www.statista.com/statistics/263311/worldwide-production-of-coffee/

Statista. (2020b). Coffee Consumption Worldwide from 2012/13 to 2018/19. Retrieved from https://www.statista.com/statistics/292595/global-coffee-consumption/

The B Corp Declaration of Interdependence. (2020). Certified B Corporation. Retrieved from https://bcorporation.net/about-b-corps

The Conversation. (2018 August 22). Coffee Farmers Struggle to Adapt to Columbia' Changing Climate. Retrieved from https://theconversation.com/coffee-farmers-struggle-to-adapt-to-colombias-changing-climate-97916

The Global Warming Policy Forum. (2019 October 7). New World Record: Oversupply of Coffee Beans Sends Global Prices Tumbling. Retrieved from https://www.thegwpf.com/new-world-record-oversupply-of-coffee-beans-sends-global-prices-tumbling/

The North Face. (n.d). Climate Change. Retrieved from https://www.thenorthface.com/en_ca/about-us/responsibility/operations/climate-change.html

UNESCO (Culture Sector). (2000). *World Culture Report*. Paris: United Nations.

US Geological Survey. (2016). The Water Content of Things. *USGS*. Retrieved from https://water.usgs.gov/edu/activity-watercontent.php

USEFULL. (2020). Corporate, Let's Create Impact Together. Retrieved from https://www.usefull.us/corporate

Vann, K. (2020 March 25). COVID-19 Puts BYO Coffee Cups on Hold, but Sanitized Reusable Systems Could Fill the Void. *WasteDive*. Retrieved from https://www.wastedive.com/news/byo-coffe-cup-reusables-coronavirus-covid-19-/574817/

Vo, K. (2020 January 24). Climate Change is Coming for Your Coffee. *Perspectives*. Retrieved from http://www.saisperspectives.com/2020-issue/2020/1/24/climate-change-is-coming-for-your-coffee

Vogt, M. (2019). Coffee: Whose Sustainability? In *Variance in Approach Toward A 'Sustainable' Coffee Industry In Costa Rica: Perspectives from Within; Lessons and Insights* (pp. 197–202), London: Ubiquity Press.

WaterAid. (n.d.). Facts and Statistics. Retrieved from https://www.wateraid.org/facts-and-statistics

Whiteman, G. (1999). Sustainability for the Planet: A Marketing Perspective. *Conservation Ecology*, 3(1).

World Coffee Portal. (2020 August 13). Project Café 2021: World Coffee Portal Announces New Global Coffee Shop Research. Retrieved from https://www.worldcoffeeportal.com/Latest/News/2020/August/World-Coffee-Portal-announces-new-global-coffee-sh

World Commission on Environment and Development (WCED). (1987). *Our Common Future*. Oxford, UK: Oxford University Press.

Yang, C. Z., Yaniger, S. I., Jordan, V. C., Klein, D. J., & Bittner, G. D. (2011). Most Plastic Products Release Estrogenic Chemicals: A Potential Health Problem That Can be Solved. *Environmental Health Perspectives*, 119(7), 989–996.

Section XV

Selected Case Studies to Reflect on Practice and Use as Learning Tools: Global Case Studies

35
Hilton Faces Greenwashing Challenge

Jinsuh Tark
William F. Harrah College of Hospitality, University of Nevada, Las Vegas, USA

Won-Yong Oh
Lee Business School, University of Nevada, Las Vegas, USA

CONTENTS

References 867

Hilton, one of the fastest-growing companies in the world's hotel industry, has its headquarters located in McLean, Virginia. As of 2020, the hotel owns 18 brands and has more than 6,100 hotels in 118 countries.[1] Hilton released a report, "Travel with Purpose 2030 Goals," to mark its 100th anniversary in 2018.[2] The report emphasizes Hilton's goal of reducing the environmental footprint over the next decade and more than doubling investment in environmental activities. In the report, Hilton points out that they would reduce the total amount of carbon emissions in Hilton's properties by 61 percent. Hilton's eco-friendly policies have been in steady progress since 2009. Since then, Hilton has set standards for tracking water, energy, and waste by utilizing LightStay, the software to improve environmental responsibility. Through LightStay, Hilton has reduced carbon emissions, water usage, and waste consumption, each by 34 percent, 20 percent, and 41 percent, respectively, over the past decade.

However, not all consumers are looking at Hilton's effort to preserve the environment positively. Some consumers argue that Hilton's environmental practices are not for the environment, but for its profits, often referred to as greenwashing.[3] There have been few cases showing consumer's distrust of Hilton's environmental activities. In 2006, Tourism Concern, an organization advocating ethical tourism, accused Hilton of its greenwashing behaviors.[4] Tourism Concern argued that Hilton, who claimed to be engaged in eco-friendly environmental protection activities, uprooted numerous palm trees and mangroves to build the Hilton Complex Resort on Mandhoo Island. Although Ian Carter, CEO of the resort, said "false testimony and defamation" against the Tourism Concern's claims, Hilton's doubt grew even more.

Another anecdotal example is when Irene Lane, the founder of Greenloons, a company that supports eco-friendly tourism, stayed at Westin Hilton Head Island Resort & Spa for a conference in 2018.[5] The "Make a Green Choice" campaign was in place at the hotel. If customers decide to participate in the "Make a Green Choice," they are compensated with a $5 food voucher, but there would be no towel or bedsheet replacements. The hotel's placard indicates that the campaign's ultimate goal is to reduce electricity and water consumption. During Irene's stay, the hotel did not replace her towels and bedsheets as agreed. However, Irene felt skeptical about the hotel's campaign when the housekeeping did not even empty her trash can. Irene thought that the more customers join the "Make a Green Choice" campaign, the fewer housekeeping employees are needed. That led her to doubt. By pretending environmentally responsible, Hilton could reduce a considerable amount of labor costs, which will be the hotel's profits.

Computer Generated Solutions (CGS), a company that helps with applications, learning, and outsourcing in business activities, has conducted a survey of 1,000 U.S. individuals between 18 and 65.[6] The survey aimed to find out how sustainable business practices and products drive their purchase decisions. Nearly 70% of the people who participated in the survey responded that sustainability is "somewhat important," and 47% stated that they would pay more for sustainable products. The trend in the hospitality industry shows a similar consumer trend. According to a recent survey, about 79 percent of travelers prefer to choose accommodations that engage in eco-friendly activities.[7] However, most travelers said that programs leading in the wrong direction make them skeptical and unlikely to participate in hotel's initiatives. Consumers' doubts have grown, and they have even come to think that the eco-friendly activities implemented by hotels are greenwashing for the hotel's benefit. Hospitality industry, including Hilton, increasingly engage in eco-friendly activities driven by the growing interest in sustainability, but their intentions are often questioned. Is the hospitality industry implementing eco-friendly programs for protecting the environment or for greenwashing as a means for its profit? How can Hilton convince customers of its green marketing campaigns?

REFERENCES

1. Hilton. (n.d.). *Welcome To Hilton.* https://www.hilton.com/en/corporate/
2. Hospitalitynet. (2019, September 19). *Hilton named 2019 global industry leader for sustainability by DJSI.* Hospitalitynet. https://www.hospitalitynet.org/news/4095113.html
3. Kao, J. (2020, May 05). *Everything You Need to Know About Greenwashing.* Greencitizen. https://greencitizen.com/greenwashing/
4. Simpson, D. (2006, August 09). *War of words on Hilton' greenwash' claims bring spotlight on tourism ethics.* CABI. https://www.cabi.org/leisuretourism/news/15943
5. Lane, I. (2018, September 11). *Latest example of greenwashing found on Hilton head island.* https://greenloons.com/2011/09/27/latest-example-of-greenwashing-found-on-hilton-head-island/
6. Sweeney, S., & Connors, K. (2019, January 10). *CGS survey reveals sustainability is driving demand and customer loyalty.* CGS. https://www.cgsinc.com/en/news-events/CGS-Survey-Reveals-Sustainability-Is-Driving-Demand-and-Customer-Loyalty
7. Rahman, I., Park, J., & Chi, C. G. (2015). Consequences of "greenwashing" *International Journal of Contemporary Hospitality Management*, *27*(6), 1054–1081. https://doi.org/10.1108/IJCHM-04-2014-0202

36

MGM's Dilemma in Responsible Gaming Program

Jinsuh Tark
William F. Harrah College of Hospitality, University of Nevada, Las Vegas, USA

Won-Yong Oh
Lee Business School, University of Nevada, Las Vegas, USA

CONTENTS

References 871

MGM Resorts International (MGM) owns about 30 brands worldwide,[1] mainly in the United States and China, and operates all of the company's businesses while playing a holding company.[2] As a global hospitality and entertainment company, MGM manages several business areas, including gaming, hotels, conventions, dining, and retail. The company owns multiple nightclubs, holds various shows, and events, and has significant ownership of T-Mobile Arena, a multipurpose indoor arena on the Las Vegas Strip.

In Las Vegas, where MGM's headquarter is located at, MGM manages many casinos and hotels (see Figure 36.1) such as Bellagio, ARIA, Vdara, MGM Grand, The Signature at MGM Grand, Mandalay Bay, Delano Las Vegas, The Mirage, New York New York, Luxor, and Excalibur.[3] MGM is not only one of the leading companies in the Las Vegas casino and hotel industry but also makes a significant effort in creating the company's casino business responsible. In 2017, MGM signed a gaming strategy initiative named GameSense,[4] which is adopted by the Massachusetts Gaming Commission to create a 'responsible gambling program.'[5] The program's

FIGURE 36.1
MGM grand hotel in Las Vegas.

main purpose is to inform customers of their positive and transparent responsibility for gambling, so help customers enjoy gaming while carrying responsibilities at the same time. The responsible gaming program provides customers with an opportunity to learn meaningful information in gaming, such as how gambling works. It also includes informing regarding skill-based and chance-based games, gaming tips of GameSense, and setting money and time limits.[6] The 'responsible gambling program' has been successful, with more than 1.25 million customers participating as of 2020.[7] Therefore, MGM expects that responsible gaming enhances its image and reputation by showing its commitment to ethical operation.

Despite MGM's success, the company realized that the program could be a double-edged sword in terms of profitability, given the trend of declining gambling revenues. According to MGM's annual report, casino revenues have declined (see Table 36.1).[8] As casino revenue is an important revenue source, loss of revenue from gaming is detrimental to the MGM's profit. The declining trend is even more worrisome with increasing competition from both domestic and international markets. Since 1978, other states in the United States have legalized gambling, which leads to intensified rivalry to the Las Vegas casino and hotel industry. Jeff Daniels, a reporter

TABLE 36.1

MGM: Financial Information (Las Vegas Strip Resorts)

	Year Ended December 31		
	2019	2018	2017
	(In thousands)		
Table games win	$ 789,330	$ 949,055	$ 931,508
Slots win	1,193,607	1,140,269	1,106,192
Other	64,834	62,249	67,150
Less: incentives	(751,601)	(743,840)	(668,020)
Casino revenue	**1,296,170**	**1,407,733**	**1,436,830**
Rooms	1,863,521	1,776,029	1,778,869
Food and beverage	1,517,745	1,402,378	1,410,496
Entertainment, retain, and other	1,153,615	1,130,532	1,119,928
Non-casino revenue	**4,534,881**	**4,308,939**	**4,309,293**
Total revenue	**5,831,051**	**5,716,672**	**5,746,123**

of the global T.V. business news channel Consumer News and Business Channel (CNBC), also pointed out that many high rollers have moved to Asian markets like Macau,[9] which also leads to a threat to casino revenues in Las Vegas. As a response, Las Vegas Hotels begin to diversify its revenue stream, such as hosting conventions, increasing room rates, and charging fees for parking. For example, due to reduced casino revenue, MGM has started charging customers for parking for the first time in 2016.[10]

As a result, MGM faces a dilemma between profitability and responsibility. MGM should attract as many customers as possible to gamble for the company's profit. At the same time, MGM likes to have an ethical corporate image by promoting the responsible gambling program. MGM is at a crossroads between profitability and responsibility. Should MGM choose to profit by attracting more customers into their casinos? Or, should MGM need to promote a responsible gambling program to be perceived as an ethical company? Where is the optimal balance?

REFERENCES

1. MGM Resorts International. (n.d.). [About MGM Resorts International]. https://www.mgmresorts.com/en.html
2. CNN Business. (n.d.). *MGM Resorts International.* https://money.cnn.com/quote/profile/profile.html?symb=MGM

3. MGM Resorts International. (n.d.). *Resort Reopening Information.* https://www.mgmresorts.com/en/open.html
4. GameSense. (n.d.). *About GameSense.* https://gamesensema.com/about/
5. Massachusetts Gaming Commission. (2018, May 30). *Mission & Values.* https://massgaming.com/the-commission/mission-values/
6. MGM Resorts International. (n.d.). *GameSense.* https://www.mgmresorts.com/en/gamesense.html
7. MGM Resorts International. (2020, March 02). *MGM Resorts' Responsible Gaming Program.* https://investors.mgmresorts.com/investors/news-releases/press-release-details/2020/MGM-Resorts-Responsible-Gambling-Program-Surpasses-125-Million-Customer-Interactions-Nationwide/default.aspx
8. MGM Resorts International. (2020). *2019 Annual Report.* https://s22.q4cdn.com/513010314/files/doc_financials/annual/2019/2019-MGM-Annual-Report.pdf
9. Daniels, J. (2018, January 31). *Las Vegas Strip Casinos See Gambling Revenues Drop for Three Months Straight.* CNBC. https://www.cnbc.com/2018/01/31/vegas-strip-casinos-see-gambling-revenues-dip-for-third-month-straight.html
10. Ho, S. (2016, January 16). *MGM Resorts to Charge for Parking on Las Vegas Strip, Bucking Tradition.* Orange County Register. https://www.ocregister.com/2016/01/16/mgm-resorts-to-charge-for-parking-on-las-vegas-strip-bucking-tradition/

Index

Note: Locators in *italics* represent figures and **bold** indicate tables in the text.

A

Aaina, 688, **696**
Aakar Foundation, 688, **696**
Abbotsford Convent, 485, **487–488**, 489, **489**, 495
ABTA, *see* Association of British Travel Agents
Ace Group, **689**
Active learning, 40–41
Actors/active people, 251
    barriers and resources, 265
    as consumer groups, 259
    and intenders comparison, **267**, 267–270
    intention and action, 263–264
    maintenance self-efficacy, 265–266, 269
    planning, 264–265
    recovery self-efficacy, 266, **266**, 270
    task self-efficacy, 265, 269
Adolescence Education Programme (AEP), 664
Adolescent Friendly Health Clinics (AFHCs), 664
Adolescent Reproductive and Sexual Health (ARSH), 664
Advertising, 90, 121, 133, 456, 639–640, 832, 841
    and bus shelters, 744, 746
    commercial, 365
    eco-friendliness and pro-environmental, 21
    media and marketing, 365
    misleading, 517
    paid, 749
    Patagonia's, campaigns, 841
    and promotion, 630
    restriction, 625
    role of, 301
    social, 314
    social advertisement campaign, 766–773, *770*
    social campaigns tool, 337
    woke, 548, 550
Affordable and clean energy, 26
2030 Agenda for Sustainable Development, 440, 442, 462, 466, 609–610
AH&RA, *see* American Hotel and Restaurant Association
AIDA model, *see* Attention Interest Desire Action
AMA, *see* The American Marketing Association
Amari Foundation, 688, **696**
American Bowl, 726
American Hotel and Restaurant Association (AH&RA), 205
American Marketing Association (AMA), 400
American Professional Football Association (APFA), *see* National Football League (NFL)
Animal welfare, 363, 375, 589
Anticorruption strategies, 453–454
Aristotle, 827
ASHA (Accredited Social Health Activist), 660, 663
Ashburton Aboriginal Corporation (AAC), 484, **487–489**
Asmita Yojana, 666
ASSOCHAM, **689**
Association of British Travel Agents (ABTA), **519**, 526–527
Associative technique, qualitative research, 151–152, *153*
ATLAS.ti Software application, 102
Attention Interest Desire Action (AIDA) model, 313–314, 634

873

Attitude-behavior gap, 132
Autonomy, 124–125, 127, 136
AVE, *see* Average variance extracted
Average variance extracted (AVE), 237, **238**, 239, **239**

B

Balmer Lawrie & Co. Ltd., **689**
Bamboo fiber pads, 657–658
Banana fiber pads, 658
Barrier factors, reducing meat consumption
  economic considerations, 365–366
  media and marketing, 365
  personal attitudes, 364
  social relationships, 364–365
Batellier, P., 72
B Corporation Declaration of Interdependence, 841, **841**
Before the flood, 369, 380
Behavioral beliefs, 215
Behavioral problems, 18
Behaviour influencing factors, reducing meat consumption
  animal welfare factors, 363
  cultural factors, 361
  environmental factors, 362–363
  health-related factors, 361–362
  physiological factors, 362
Belz, Frank-Martin, 23–24, 45, 64
Benzyl butyl phthalate (BBP), 855
Bernatchez, J. V., 59
Beyoncé, 126
Bhusattva, 130
Big Four American leagues, 723
Biodiversity of animals, 26
Birth control pill; *see also* Oral contraceptives
  definition, 626
  types, 626
Bis(2-ethylhexyl) phthalate (DEHP), 855
Black Friday sales, 840
Black Lives Matter movement, 551, 553, 557, 560
3BL framework, *see* Triple bottom line
Blockchain technology, 76
Blood tax, 659

Book donations, 728, 779
Boralex, 60
Brand activism, 548, 550
Brand coolness and Indian sustainable fashion brands, 115–117
  autonomy, 124
  characteristics, **122**, 122–123, *124*
  component characteristics, 123
  culture influence, 125
  defined, 123
  fashion clothing, 118
  grey to green, shift, *129*
  marketing, 116, 134–136
  mass cool brands, 126
  mass production of clothing, 118
  niche cool brands, 126
  subculture, 124, 126
  sustainability in fashion, 117–120, 128–133
Branding for sustainable fashion, 117
Brand management, 121
Brand personality, 128
Brundtland Commission, 19 23, 147, 200, 226, 577, 852
Brundtland Report, 4, 54, 130, 204, **230**, 282, 400, 416, 577
Businesses, 16 17, 18, 20–21
  definition, 516
  ecosystem members, 516
  external factors, 516–517
  internal force, 516
  Spanish tourism ecosystem, *see* Holiday sickness scam
Business marketing, 94
Business Roundtable (BRT), 841
Business-to-business (B2B) markets, 798–799
Business-to-business-to-consumer (B2B2C) marketing, 799
Business-to-customer (B2C), 799

C

CAB model, *see* Cognitive-affective-behavior
Caisse de dépôt et placement du Québec (CDPQ), 60
Cannibals with Forks: The Triple Bottom Line of 21st Century Business, 579

Capital of culture, 88
Carbon emission
    Chapputz, positive impacts, 572
    fashion industry, 565–566
Carbon footprint, 722
Cascades, 56
    Cascades Containerboard
        Packaging, 57
    Cascades Conversion, 60
    Cascades Énergie, 60
    Cascades GIE Inc., 72
    Cascades Groupe Carton Plat Europe, 57
    Cascades Industries Inc., 60
    Cascades Plastics, 60
    Cascades Specialty Products Group, 57
    Cascades Tissue Group, 57, 61
    compensation at, 64–65
    development toward sustainability, **58**
    digital strategy, 76
    DUX Grand Prix, 61
    energy management practices, 73
    environmental practices, 74
    evolution of, 59
    *2020 Excellence in Energy Efficiency
        Award,* 61
    family culture and values, 64–68
    hierarchical levels, 64
    "Hydro-Québec Ecolectric" network, 73
    ISO 26000 standard, 69
    materiality analysis, 69, *71*
    2020 Novae Awards, 61
    open management principles, 77
    overview, 56–58, 74–79
    performance in sustainability, 62
    philosophy, 66
    profit-sharing program with the
        employees, 58
    raw material, 75
    resource optimization, 73–74
    reuse philosophy, 72–73
    shared management, 67
    shared power, 66
    short-circuiting of hierarchy, 64
    stakeholders of, 68–72, *70*
    sustainable development, 57, 59–78, *63*
    Sustainable Development Plan of, 69–71
    will to sustainability, 68–72

Case method teaching, 28–29
    case use and analysis, 34–40
    Grupo Familia mini-case, 29–34
Cause marketing, 605–606
CB, *see* Consumer behavior
CDRI, *see* Central Drug Research
    Institute
CE, *see* Circular economy
CensusInfo India, 654
Central Drug Research Institute
    (CDRI), 635
Centre for World Solidarity (CWS), **697**
CEOs
    on consumer and customer demand, 20
    performance of products and services, 20
    on sustainability, 20, 46
Ceramic mugs, 854
CGS, 850–851
Chapdelaine, Marie-Ève, 75
Chapputz fashion accessory brand
    "çaput kilim" weaving, 569, 571
    contributions, UN SDGs, 569, *569*, 572
    income generation, rural female
        artisans, 564, 569–570
    iPad case and iPad clutch case, *571*
    social mission, environmentally
        concerned, 568, 570
    sustainability, 570–571
    throwaway clothing, 568, 570
Charity Navigator, 780
Chemical industry
    B2B and B2C communications, 805–807
    B2B marketing in sustainability,
        801–802
    communication for sustainability,
        824–825
        internal brand communication and
            education, 825–828
        problems, 828–830
    customers, 814–815
    focus on new generations, 813–814
    organizational buyers, communicating
        to, 807–811
    paradigm of, 802–804
    purchasing and investment decisions
        organizational factors, 812
        personal factors, 812–813

Chemical industry (*Continued*)
  social listening, 811
  sustainability communication, 815–819
    consistent communication, 819–821
    environmentally sustainable solutions, 821–824
    sustainability reporting, 824
  sustainability in, 804–805
  sustainability initiatives, benefits, 815–819
  sustainable approaches in, 799–800
Chernobyl nuclear reactor accident, 21
Chief Marketing Officer (CMO), 20–21
Chimp&z Inc, **690**
China Environmental and Resources Protection legal framework, 582
Choice and ordering techniques, qualitative research, 152–153, *154*
Center for Life Cycle of Products, Services and Systems (CIRAIG), 853–854
Circular business model (CBM), 798
Circular Economy Action Plan for cleaner and more competitive Europe, 317
Circular economy (CE), 73–74, 76, 798; *see also* "Roadmap for the circular economy"
  definition, 315–316
  environmental pollution, 317
  Henkel Adhesives Technologies Packaging SBU on, 803
  operating system, *567*
  social campaign, *see* "Wrocław does not waste" campaign
  upcycling, *see* Upcycling
  waste management, 318
  Wrocław transition, 323–325
  zero waste, 316–317, *317*
  ZWIA, 316
City of Culture, 88
City Strategic Action Plan and Initiative for 2013–2020, 104
Claims management companies (CMCs)
  fraudulent claims, 512, 522–523
  and fraudulent tourists, *see* Holiday sickness scam
Clean water and sanitation (SDG 6), 572
Climate action (SDG 13), 572
Climate change, 26, 844

Clothing; *see also* Fashion industry
  environmental damage, 252, 564–566
  handlooms, *see* Handloom weaving society
  reusing and reselling, *see* Second-hand clothing
CMCs, *see* Claims management companies
Co-create, Build and Engage (C-B-E) process, 740–742, *741*
  framework, **743, 746**
  Leave It program, 742
  RE-AIM Framework, **744–745**
Coffee, 844
  annual consumption, *847*
  annual global production, *846*
  ceramic mugs, 854
  Coffee Cup Collective, USEFULL, 849–852
  coffee cups, disposable *vs.* reusable, 852–856
  container, 856–857
  disposable coffee cups, 849
  drink products, 853
  Dunkin, 848
  environmental footprint, 845
  environmental impact of waste, 856
  global prices, *846*
  polycarbonate mugs, 854
  polypropylene (PP) mugs, 854
  prices of, 847
  single use paper cups, 852
  stainless steel mugs, 854
  Starbucks, 848
  sustainability of coffee container, 856–857
  top 25 countries, **845**
  travel mugs, 854
  World Coffee Portal, 848
Cognitive-affective-behavior (CAB) model, 214–215
Collective decision models, 462
Colourism and genderism
  anti-skin-whitening sentiment, 554
  Black Lives Matter movement, 553
  fairness products, 554; *see also* L'Oréal's skin-whitening line
  market demand, skin-whitening products, 552

men's skin-brightening products, 553
metrosexuality, 553
Commercial marketing, 22
Commercial transactions, 16
Commission Staff Working Document, 98
Communication for sustainability, 824–825
    internal brand communication and education, 825–828
    problems, 828–830
Community engagement, 39, **488**
    brand awareness, 613
    cultural values, 614
    employee volunteering, 614–617
    socially responsible consumers, 616
Community Health Centers (CHC), 664
Competitive products and/services, 17
Completion technique, qualitative research, 151, *152*
Composite reliability (CR) index, 237, **238**, 239
Composter sharing programme, 327
Composting, 325
    free composters, 331
    local, 316
    Wrocław, 320, 331–332
Comprehensive Sexual Education (CSE) program, 625
Computer Generated Solutions (CGS), 866
Construction techniques, qualitative research, 154, *155*, 156, *156–157*, 158
Consumer behavior (CB), 44–46; *see also* Responsible consumer behavior
    CAB model, 214–215
    green consumers, 201, 213, 217–218, 220
    social listening, 420
Consumer-citizens, 188–189
Consumer–firm relationship, 34
Consumerism, 4, 125
Consumers
    active, 259; *see also* Actors/active people
    activism, 4
    attitudes toward sustainability, 21
    awareness, 840
    education and awareness campaigns, 843
    inactive, 259; *see also* Intenders/inactive people
    memory, 128
    online reviews, *see* Online consumer reviews
    orientation, 16
    sustainability expectations, 389
Consumer satisfaction, retailing
    co-creation, 227
    literature review and hypotheses
        brand awareness, 233
        brand image and associations, 234
        consumer awareness, 232
        environmental, social and economic sustainability, 229, 231–232
        findings, sustainability, 229, **230–231**
        positive associations, consumer minds, 232
        proposed model, *235*
        store awareness, 233–234
        store image, 232–234
    quantitative study, *see* Quantitative study
    store awareness and image, 227–228, 241
    sustainable image creation, 227
Consumption, 4
    challenges, modern marketplace, 176–177
    culturally mediated needs, 180
    definition, 176, 190
    dominant social paradigm, 178–179
    live information systems, 179
    materialistic, 179
    post-consumers, 177; *see also* Prosumer identity
    and self-identity, 180–181
    subcultures, 182
    sustainability, 179, 181
    as unfavourable phenomenon, 178–179
Consumption-related emotions, 116–117
Content evaluation (analysis) technique, 87, 106
Contracts, 296–297
Control beliefs, 215–216
Cool branding strategies, *see* Brand coolness and Indian sustainable fashion brands
Coolness, *see* Brand coolness and Indian sustainable fashion brands
Cool trends, 136

Cooperative, 33
Corporate branding, 20
Corporate performance, 54
Corporate social responsibility (CSR), 22, 204, 402–403
  definition, 468
  hospitality industry, 517
  responsible consumer, 517–518
  and social marketing, 517–518
Corporate sustainability
  3BL framework, *see* Triple bottom line
  long-term success, 145
  marketers, 169–170; *see also* Professional marketers
  principles, 578
Cosmopolitanism, 125
Cossette, C., 59
COVID-19 pandemic, 5, 176–177, 273, 341, 440, 606, 628
Cowspiracy: The sustainability secret, 369, 380
Creative prosumers, 186, 189, 191
CR index, *see* Composite reliability
Crisis management
  definition, 468
  public management, *see* Modernization, public management
  public marketing, 437–438; *see also* Public marketing 4.0
  public–private response, 438–441
  scenario planning, 462, *464*, 464–465, 467
  social utility index-function, 463–464
  strategic intelligence, 463–464
Critical thinking, 41–43
Cronbach's alpha, 237–238, **238**
Crowdfunding
  crowdlending, 778
  donation, 778
  equity, 778
  funding non-profits working for refugees, **780**
  non-profit organization, comparison, **781–784**
  platform, 779
  reward, 778
CSE, *see* Comprehensive Sexual Education
CSR, *see* Corporate social responsibility

Cuckoo movement for children, 600
Cultural festivals, 98
  and events, 86
Cultural Manifesto Programs and Initiatives, 86
Cultural tourism, 91
Culture; *see also* European Capital of Culture (ECoC)
  defined, 89
  family and values, 64–68
  focused development, 90
  influence, brand coolness and Indian sustainable fashion brands, 125
  roles and responsibilities, 92
  sustainability and, 62
  sustainable development, 62
Customer adoption intention, 280–281, *281*
Customers, 17; *see also* Misbehaving customers
  chemical industry, 814–815
  empowered, 406

D

Decent work and economic growth (SDG 8), 572
Deluxe milk
  green manufacturing and harmonious working environment, 589–590
  milk source, 588–589
  strategic planning, 588
  UN SDGs, 587–588
DEXi software program, 103, 107
DEX technique, 102–103, 106
Dharma Life, **690**
Dibutyl phthalate (DBP), 855
Digital marketing, 91
Digital transformation, 75
Diisobutyl phthalate (DIBP), 855
District Hospital (DH), 664
DogFest, 748, 750
Dominant social paradigm, 178–179
Domtar, 61
Donation crowdfunding, 778
Doodlage, 130
Dopaco, Inc., 61
Double-leaf Family Towel, 35
Downshifting, 42

Do You Speak Green, 130
Drucker, Peter, 15
DS Smith, 74
Dumpster diving, 187

E

Eastern collectivism, 549–550
EAT–Lancet Commission, 348, 352
Eco-certification
    hotel booking, 206
    hotels, 207
    international and regional eco-labels, 207, **208–210**
Eco-fashion, 132
EcoFemm, **697**
Ecological marketing, 21
Ecological sustainability, 92
Ecology, 101
E-commerce, 297–298
Economic sustainability, 54, 92
    life story, 165–166, **166**
    and personal practices, *167*
Economic well-being, 4
Economy of permanence, 598–599; *see also* Nurpu, sustainable fashion brand
ECO-Schooleducational programme, 327
Ecotype energy treatment process, 588
ECs, *see* Emergency oral contraceptives
Education, environmental protection
    planned measures, "Roadmap for the circular economy," 323
    Poland, 322–323
Electronic word of mouth (eWOM)
    internet forums and online discussions, 394
    online reviews, *see* Online consumer reviews
    and WOM comparison, 395
11.11/eleven eleven, 130
Ellen MacArthur Foundation, 72, 130, 319
Emergency oral contraceptives (ECs), 626, 632–633, 639
Empirical inquiry, 539
Employee volunteering, 614–616
Empowered consumers, 389, 401, 404, 406
#EndangeredEmoji campaign, 422

The end of meat, 369
Environmental damage, 21
Environmental protection, 21
Environmental sustainability, 55
    and customer satisfaction, 227
    reactive to proactive approach, 229
Environmental sustainability initiatives, 723
Epic Humanitarian World Tour, **697**
Equity crowdfunding, 778
Eradicating poverty (SDG 1), 572
ESG (environment, society, governance) standards, 62
Essar Global Fund Limited (EGFL), **691**
Ethical consumption, 253
Ethical fashion, 129
Eudemonic well-being, 4
European Capital of Culture (ECoC), 86–92, 94–96
    analysis
        evaluation of progress and marketing requirements, 97–99
        progressive evaluation and sustainability requirements, 96–97
    applications selections, 99–100
    classification of culture, 100
    content evaluation (analysis) technique, 87, 106
    cultural festivals and events, 86
    economic and social benefits, 86
    European Union action, 98
    evaluation, dimensions for, **101**
    initiative, 94–96
    Kaunas, 99–100, **102**, 104, **105**, 107
    Klaipeda, 99–100, **102**, 104, **105**, 107
    Lithuania city, 92
    Lithuanian candidate cities, *101*
    marketing cohesion of sustainability and culture, 92–94
    marketing requirements, 97–99
    procedures and methods, 99–100
    progressive evaluation and sustainability requirements, 96–97
    research methodology, 100–103
    results, 104–108
    strategy of marketing, 90
    sustainability marketing, 91
    triple bottom line (TBL) technique, 87, 100
    Vilnius, 99

European Free Trade Association (EFTA), 89
"1-Euro per day" campaign, 788–789
EWOM, *see* Electronic word of mouth
Expressive techniques, qualitative research, 153–154, *155*
Extended valence framework, 281, 293, 300
External reporting, social enterprises
    aboriginal conferences, 492
    Ashoil and Ashlinen, 492
    computer-based training programs, 492
    e-newsletter, 494
    short-term and long-term impacts, 494
    systematic forms, disclosure, 492, **493**
    waste management conferences, 492
Exxon Valdez oil spill, 21

F

Familia®, 30, **30**
Familia Institucional®, **30**
Familiarise-immerse-dissect-sense making-appraise-internalise (FIDSAI)
    appraise alternative solutions, 543
    dissect, 541, 543
    familiarise, 541
    framework, 541, *542*
    immerse, 541
    internalise, 543–544
    sense-making, 543
Farmer suicides, India, 283
Fashion industry; *see also* Brand coolness and Indian sustainable fashion brands
    Chapputz, Turkey, *see* Chapputz fashion accessory brand
    consumers, 120
    cotton, production and export, 118
    fast fashion, 564–566, 568, 570–571
    impact on environment, 564–566
    raw materials and garment, producing and export of, 118
    upcycling, *see* Upcycling
Fast fashion
    adverse environmental impact, 564
    environmental consequences, 565, 568
    throwaway clothing, 564–565, 570

Felipe, Andres, 39
Fibers, 119
FIDSAI, *see* Familiarise-immerse-dissect-sense making-appraise-internalise
Flexitarianism, 347, 351, 370, 380
Food choices, 369, 380
Food recovery, 728–729
Food waste
    households, 321
    planned measures, "Roadmap for the circular economy," 322
    reasons, 321
    selective collection, 321
Forest Stewardship Council (FSC) logo, 31
Forma-Pak, **58**, 59–60
For-profit businesses, 778
4P marketing mix, 347
4S marketing mix, 347, 350, 377–378, 380
Francken, M., 22
Frugality, 42
Funding non-profits working for refugees, **780**

G

Game changers, The, 369, 380
Gandhigram Trust, **697**
Gender equality (SDG 5), 572
Gender equity and social marketing, 765–766; *see also* Gender inequality
    injunctive norm messaging, 769–770
    method, 767
    perceived behavioural control (PBC), 767
    post-survey, 770–771
    pre-survey, 768
    self-efficacy messaging, 768–769
    social advertisement campaign development, 768
    STEM, underrpresentation in field of, 766–767
    theory of planned behaviour (TPB), 766
Gender inequality, 25–26, 668
Gender Trouble, 549
"Gift a cradle scheme," 605–606

Global crises management
  modernization, public management, *see* Modernization, public management
  public marketing, 437–438; *see also* Public marketing 4.0
  public–private response, 438–441
  scenario planning, *464*, 464–465
  social utility index-function, 463–464
  strategic intelligence, 463–464
Global crisis of 2008-09, 842
GlobalGiving, 779–780
  donations to, 780
  online fundraising, 780
Global Sustainable Tourism Council (GSTC), 207
Global Web Index survey, 403
Gómez-Salazar, Andres Felipe, 31, 39
Gonzalez, Rodrigo de Jesús, 33
Goodpurpose Survey, 584
Google, 127
Goonj, **697**
Governance of the Digital Age, 448
Government of South Africa Department of Environmental Affairs and Tourism, 204
Gramalaya, **698**
Grand Theft Auto, 126
Grassroot, 130
Green consumers, 21–22, 201, 213, 217–218, 220
Green consumption
  attitude and behavior, consumers, 216
  definition, 253
  HAPA model, 270
  hotels, 211
Green Electricity of Controlled Origin (EVOC), 60
Green environmental marketing, 94
Green gap, *see* Intention-behaviour
Green hotel
  definition, 211
  greenwashing, 212
  practices implemented, **212**
  WTP premium, consumers, 217–218
Greenhouse gas (GHG) emissions, 854
Greenhouse gas reduction, 728–729
Green marketing, 21–22, 201, 203, 417–418, 801
Green products, 722
Greentree, **698**
Greenwashing, 22, 418, 425, 427, 801
  firm-level, 22
  product-level, 22
Grocery retailing, Spain, *see* Quantitative study
Grupo Familia, 29–34
  community empowerment, 34
  Familia®, 30, **30**
  Familia Institucional®, **30**
  Heroes of the Planet, 37
  Innovation Model, 31
  low-income consumers, 33–34, 38
  marketing mix, 35
  Misión + Hogar program, 37–38
  Nosotras®, 30, **30**, 32
  Pequenín®, 30, **30**, 32
  Petys®, 30, **30**, 32
  Pomys®, 30, **30**, 32
  product innovations, 32
  recycling, 32–33, 36
  TENA, 30, **30**
  Triple Bottom Line Metrics, **32**
GSTC, *see* Global Sustainable Tourism Council

H

Handloom weaving society; *see also* Nurpu, sustainable fashion brand
  attempt at revival, 601–602, *602*
  customer perception, pricing and business dilemma, 604
  Gandhian inspiration, 601
  history and cultural heritage, 599
  Sivagurunathan's dream, 600
  sustainable production, 602–603, *603*
  traditional weavers, 600
HAPA model, *see* Health action process approach
HappytoBleed, **698**
Harley Davidson, 127
Harmon's single-factor test, 236
Harvard Business Review, 424
Harvesting, 18–19

Healing Fields Foundation, **698**
Health action process approach (HAPA) model
  actions and action control, 256
  barriers and resources, 256–257
  maintenance self-efficacy, 258
  planning, 255–256
Henkel Adhesives Technologies Packaging SBU, 802–803; *see also* Chemical industry
  B2B and B2C communications, 805–807
  business-to-business (B2B) marketing, 814
  campaigns, *806–808*, *810*, *829*
  on Circular Economy, 803–805, 825–826
  communicating sustainability, 818–819
  communication for sustainability, 824–828
  consistent communication, 819–821
  emotional vacuum, 812–813
  on environmental crisis, 803
  environmentally sustainable solutions, 821–824
  environmental sustainability, 812
  Global Market and Customer Activation strategy, 831
  LinkedIn Sustainability Posts, *809*
  new generations, focus on, 813–815
  organizational buyers, communicating to, 807–811
  purchasing and investment decisions, 812–813
  social listening, 811
  sustainability communication, 815–818
  sustainability in, 804–805
  sustainability positioning strategies, *830*
  sustainability reporting, 824
Hepburn Wind, 485, **487–488**, 489, **489**, 494
"Heroes of the Planet," 35
Heuristic-systematic model (HSM), 298
Hilton, 865–867
  eco-friendly environmental protection, 866
  greenwashing, 866–867
  investment in environmental activities, 865–866
  Make a Green Choice campaign, 866
  Travel with Purpose 2030 Goals, 865

Himalaya Drug Company, **691**
Hippie culture, 125
Hoek, J., 26
Holiday sickness scam
  alleged illness, 521
  amount, legal assistance, 523
  compensation culture, 522
  data analysis, 520
  data collection, 519
  ecosystem, *520*
  false claims, 521, 523
  food poisoning claims, 521–522
  imbalance, ecosystem, 523
  pharmacy receipt and medical reports, 521
  reactions, UK, 526–527
  resolution, 523–526
  secondary data, 519, **519**
H.O.P.E.: What you eat matters, 369, 380
Hotel industry
  consequences on environment, 210–211
  consumer attitude and WTP, green practices, 217–218
  eco-certification, 207
  eco-labels, international and regional, 207, **208–210**
  green hotel, 211–212, 214, 217, 219
  green practices, 212, **212**
  greenwashing, 212
"Housing Estates do not waste" festival, 331–332
HSM, *see* Heuristic-systematic model
Human behavior, 840
Human rights, 18, 229, 578, 625, 650

I

IB gap, *see* Intention–behaviour
IHA, *see* The International Hotel Association
IHEI, *see* International Hotel Environment Initiative
Imperfect (ugly) food, 44
"Inclusive Nutrition Plan," 587
Indian Brand Equity Foundation, 119
Indian Development Foundation (IDF), **699**

Indian textile industry, 119
  cultural heritage, 119
  disposal issues, 120
  maintenance issues, 119
  production issues, 119
  textile mill, Calcutta, 119
Individual-targeted consumerism, 843
Industry; *see also* Chemical industry; Fashion industry; Hotel industry; Textile industry; Tourism and hospitality industry
  cascades, 60
  defined, 17
Industry 4.0, 73, 76
Information, Education, and Communication (IEC), 663
Infoxchange, 485, **487–489**, 490, 492, 495
Injunctive norm messaging, 769–770, 772–773
INSIDE, 126–127
Instagram, 126
Integrated reporting, social enterprises
  annual reports, 495
  cases, integrated data, **496**
  economic impact study, 495
  ICT, 495
  reporting formats, 495, *496*
Intenders/inactive people, 251
  and actors comparison, 267–270
  barriers and resources, 261–262
  as consumer groups, 259
  intention and (in)action, 260–261
  maintenance self-efficacy, 262, 269
  planning, 261
  recovery self-efficacy, 263, **263**, 269–270
  task self-efficacy, 262, 268–269
Intentions
  actors, 263–264
  antecedents, 255
  and behaviours, 255
  intenders, 260–261
  marketing, 254
Intention–behaviour (IB) gap bridging
  actions and action control, 256
  actors, *see* Actors
  barriers and resources, 256–257

comparison, intenders and actors, 267–270
constructs, 254, *255*
HAPA model, *see* Health action process approach
intenders, *see* Intenders
intentions, 254–255
models, 254
planning, 255–256
qualitative research methodology, 259–260, **260**
self-efficacy, 257–258
TPB, 254
Internal reporting, social enterprises
  cost-benefit analyses, 491
  funding, social programs, 491
  information, performance, 490–491
  Internet tracking, 492
  Perth City Farm, 491
The International Hotel Association (IHA), 205
International Hotel Environment Initiative (IHEI), 204–205, 217
International Integrated Reporting Council, 481–482, 498–499
International Paper, 75
International Tourism Partnership (ITP), 204–205
Intersectionality theory
  cultural perspectives, 549
  generalisations of identities, 549
  identity, definition of, 549
  positionalities, 550
Irwin, Susan, 32
ITP, *see* International Tourism Partnership

J

Jack, Lang, 88
Jackson, T., 44
Janaushadhi Suvidha, 665–666
  oxo-biodegradable sanitary napkin, 657
Jayashree Industries, 687–688, **699**; *see also* 'Pad Man' of India, Arunachalam Muruganantham
John, Elkington, 91

K

Ka-Sha India, 130
Kasturba Gandhi National Memorial Trust, **699**
Kaunas, 99–100, **102**, 104, **105**, 107
Kent RO, **692**
Khadi fabric, 130
Khushi Scheme, 667
Kilbourne, W., 44
Kingsey Falls Paper, 60
Kingsey Falls Public Market, 61
Kishori Mandals (KM), 664
Klaipeda, 99–100, **102**, 104, **105**, 107
Koala Awareness Campaign, 757–761
Koalas *(Phascolarctos cinereus)*, 752, *756*, *756*–*757*
    campaign evaluation, 760–761
    campaign implementation, 758–759
    campaign messages, **759**
    community baseline survey, 757–758
    data collection locations, *758*
    Leave It, 742
    populations declining, 742
    Redland City Council community, 742, 757
Kotler, Philip, 17, 23
Krishi Mela (fair for farmers), 292
Kumar, N., 21

L

Leave It program, 742, 748, 750
Lemaire, Antonio, 56
Lemaire, Bernard, 56, 64
Lemaire, Laurent, 56
Levitt, T., 17–18
Levy, Sydney J., 17
Lithuania city, 92
Live information systems, 179
L'Oréal Paris product packaging
    back description, 555–556
    cover video, UV Perfect, 557, *557*
    diachronic analysis, 556, *556*
    men's skin-brightening, 556–557
    White Activ cleanser, 556, *556*
    White Perfect cleansers and toners, 555, *555*, 557

L'Oréal's skin-whitening line
    product packaging, *see* L'Oréal Paris product packaging
    response, Black Lives Matter movement, 551
    social demands, 552–554
Low-income consumers, 33–34

M

Macromarketing, 23, 27, 41, 43
    responsible consumption and production (SDG 12), 43–44
"Made in Wrocław 2019" trade fair, 331
Maintenance self-efficacy
    actors, 265–266, 269
    intenders, 262, 269
    planning and action, 258
Major League Baseball (MLB), 723
Make a Green Choice campaign, 866
Male Enlightenment, 553, 557
Marketers, 6
Marketing, 15–16, 101
    age of sustainability, 400–401
    cohesion of sustainability and culture, 92–94
    decisions, 25
    definition, 23
    of ECoC, 87
    educators, 29
    empowered consumers, 389, 404, 406
    evolution, 388–389
    faculty, 44
    L'Oréal, *see* L'Oréal's skin-whitening line
    online reviews, *see* Online consumer reviews
    pedagogy, 16
    programs, 15
    science, 17
    sustainability, 94
    sustainability initiatives, 149
    WOM communication, 389
Marketing professionals
    economic sustainability, 165–166, **166**, *167*
    initiatives, social sustainability, 159, **160**
    practical implications, 169–170
    social sustainability practices, *see* Social sustainability

"Marketing Week," 146
Market opportunity, 24
Market-oriented organizations, 18
Market-responsive organizations, 18, 46
Mass consumerism, 178–179
Materialism, 178–179
Material reuse, 728
McDonagh, P., 44
McKinsey Center for Business and Environment, 72
Meat the truth, 369, 380
Mega Trend, 116
Melina, Mercouri, 88
Menstrual cups, 657
Menstrual Hygiene Day, 651
Menstrual Hygiene Management (MHM) in India, 649–653
   accessibility and affordability, 706–707
   awareness, 706
   campaigns, 688
   challenges, actions to overcome, 704–709, **705**, *706*
   cognitive dissonance, 709–710
   corporates, NGOs, and SHG's, 708
   encouragement to entrepreneurs, 708
   funded research on the policy-level, 708
   GOI-NFHS-4 Survey, 684–687
   government, role of, 660
      Adolescence Education Programme (AEP), 664
      Adolescent Reproductive and Sexual Health (ARSH), 664
      ASHA (Accredited Social Health Activist), 663
      Asmita Yojana, 666
      Janaushadhi Suvidha, 665–666
      Khushi Scheme, 667
      Ministry of Drinking Water and Sanitation (MDWS), 661
      Ministry of Health and Family Welfare (MHFW), 661–662
      Ministry of Human Resource Development (MHRD), 661
      Ministry of Women and Child Development (MWCD), 660–661
      National Rural Livelihood Mission (NRLM), Aajeeveka, 665
      Rajiv Gandhi Scheme for Empowerment of Adolescent Girls (RGSEAG) SABLA, 664
      Rashtriya Kishore Swasthya Karyakaram (RKSK), 664
      Rashtriya Madhyamik Shiksha Abhiyan (RMSA), 665
      Rural Development and Panchayat Raj Department (RDD), 662
      Sarva Siksha Abhiyan (SSA), 665
      State Women Development Corporation (SWDC), 662
      Stree Swabhiman Scheme by CSC, 665
      Swachh Bharat Mission-Gramin (SBM-G), 663
      Tribal Development Department (TDD), 662
      UDITA, 666
   hurdles and omissions, 672–673
   implications, 710–711
   menstruation, 655–656
      menstrual product glitches, 657–658
      menstrual product market players, 657
      mythological religion front, 656
      no GST on menstrual products, 659–660
      restrictions and myths, 656–657
      safe disposal of pads, 658–659
   ministerial level, 709
   National Education Policy (NEP) 2020, 708
   National Family Health Survey (NFHS), 673, **674**
      menstrual product usage, 677–684, *678–683*
      State's Position, 674–677, **675**, *676*
   NGOS and corporates, 687–688, **689–703**, 704
   as per schooling, *676*
   programs, 651–653
   projects in higher education, 708
   reality, 684–687
   rewards and incentives, 708
   in rural india, 704
   safe disposal infrastructure, 707
   sanitation, 667

Menstrual Hygiene (*Continued*)
   at home, 667
   at school, 668
   at societies, 667–668
  setbacks for poor MHM, 668–669
  sex ratio, 654–655, *655*
  stakeholders in, 670–672
   adolescent girls/females, 670
   celebrities, 672
   communities, 671
   government, 671
   investors and funding agencies, 671
   manufacturing units, 671
   media, 672
   NGOs, 671
   organizations, 671
   school teachers, 671
  tax exemption, 707
Menstrual products; *see also* Sanitary napkins/pads
  glitches, 657–658
  locally prepared napkins, 673, 677
  market players, 657
  menstrual cups, 657
  no GST on, 659–660
  safe disposal of pads, 658–659
  tampons, 657, 674
  usage, **675**, 677, *678–683*
Menstrual protection, 673
Menstruation, 655–656
  defined, 713
  described, 655–656
  menstrual cycles, 650
  menstrual product glitches, 657–658
  menstrual product market players, 657
  'menstrual' taboo, 650–652
  mythological religion front, 656
  no GST on menstrual products, 659–660
  restrictions and myths, 656–657
  safe disposal of pads, 658–659
Metrorrhagia (irregular menstruation), 713
Metrosexuality, 552–553, 560
Micro, small and medium enterprises (MSME), 601, 607
Microentrepreneurs, 36
Millennium Development Goals (MDGs), 2000–2015, 5
Mindful consumption, 253

Mind mapping, 541
Ministry of Drinking Water and Sanitation (MDWS), 660–661
The Ministry of Health and Family Welfare, 626
Ministry of Health and Family Welfare (MHFW), 660–662
Ministry of Human Resource Development (MHRD), 660–661
Ministry of Women and Child Development (MWCD), 660–661
Misbehaving customers
  actions, in response, 515
  causes, 511
  drivers, 514
  employees' health and work, 515
  impact, organization, 515
  monetary compensation, 512; *see also* Holiday sickness scam
  motives, 514
  proactive approach, 515
Misión + Hogar program, 37–38
Mitsky, 126–127
Mixed research method, 627–628, **629**
Modernization, public management
  disruptive environment, 444–448
  public marketing applications, 448–452
  research methodology, 454–455
  socially responsible marketing, 444
  socially responsible public administration, 442–443
  social marketing, 443–444
  social responsibility, 443
  strategies
   anticorruption strategies, 453–454
   experimentation and innovation, 453
   goal-oriented competencies and skills development, 452–453
   intelligent political strategies and engagement, 452
   objectives and administrative structures, 452
   pragmatic and results-oriented framework, 452
   professionalization and improved morale, 453
   public financial management, 454

Monmilk Group, China
    annual production capacity, 585
    3BL challenges, 590–591
    Deluxe milk, 587–590
    donations, 587
    marketing strategy, 585
    Sanju Qingan scandal, 585
    sustainability practices, *586*, 587
    tainted milk powder, 585
Motivational self-efficacy, *see* Task self-efficacy
MSME, *see* Micro, small and medium enterprises
Municipal waste
    landfilling, 320
    planned measures, "Roadmap for the circular economy," 320–321
    thermal transformation, 320
    waste per capita, Poland, 319–320
Muse, **700**
Myna Mahila Foundation, **700**

N

Narayan, R., 75
Narrative inquiry, 150
National Agricultural Market, 284
National Basketball Association (NBA), 723
National Centre of Biotechnology Information (NCBI), 626
National Family Health Survey (NFHS), 625, 654, 673, **674**
    menstrual product usage, 677–684, *678–683*
    State's Position, 674–677, **675**, *676*
National Football League (NFL), 723
    attendance and average attendance in 2019 season, *727*
    fan profile, 725
    franchise value in 2019 NFL season, *726*
    history and evolution, 724
    marketing mix strategy, *728*
    marketing strategy within, 725–729
    number of franchises, *724*
    by numbers, 725
    Philadelphia Eagles, 729–730
        Go Green Program, 730–732
        Green Era, 730
    Super Bowl, 728

National Hockey League (NHL), 723
National People's Congress (NPC) Environmental and Resources Protection Committee, 582
National Rural Health Mission (NRHM), 660
National Rural Livelihood Mission (NRLM), Aajeeveka, 665
Navigant, **692**
NCBI, *see* National Centre of Biotechnology Information
NDTV in collaboration with Dettol, **692**
Newsprint, 74
NFL Europe, 726
NFL International Series, 726
Nike, 126–127
Nirman foundation, **700**
Nobel Hygiene, **693**
No Nasties, 130
Non-profit organizations, 15, 17–18
Nonprofit/social marketing, 27
Norampac, 61, 67
Normative beliefs, 215–216
North Face, 840
Nosotras®, 30
Nurpu, sustainable fashion brand, 599
    business model, sustainability, 602, *603*
    crowd-funding, 605
    dying process, 604
    ecology, economy and enterprise, 607
    "gift a cradle scheme," 605–606
    online community page, 605, *606*
    organic apparels, 607
    RESTART framework, 607–609
    shopping page, *605*
    social media marketing, 605
    traditional product portfolio, 603, *603*

O

Oligomenorrhea (light menstruation), 713
Omnichannel strategy, 456
Omnivore's Dilemma: A Natural History of Four Meals, The, 369
Online consumer reviews
    advantages, 396
    age group, *397*, 398
    companies and brands, 397
    decision-making process, 399

Online consumer reviews (*Continued*)
  empowered customers, 406
  eWOM sources, 399
  first time shopping, 397
  forcing brands, sustainability, 402–404
  motivations to write, 399
  posting of, 398, *398*
  power relations, consumers and brands, 407
  product reviews, 395–396, 398
  purchase decisions, 389–390, 396–397
  and sustainability, 404–406
  tourism sector, 399
  travel reviews, 398
Online fundraising, 780; *see also* Crowdfunding
Oral contraceptives
  birth control gel, 642
  condoms, 636–637
  condom *vs.* pill, 637
  definition, 626
  ECs, 626, 632–633, 639
  factors influencing purchase, 637
  I-pill, 631–632, 638, 640
  as lifestyle drug, 630–631, *631*
  male contraception, 643–644
  marketing interventions
    campaigns, 642
    and consumption, 639–640
    PCOD patients, 641
    positive word of mouth, 642
    rural market, 640
  mixed research methodology, consumer decision-making, 627–630
  Nestorone—Testosterone gel, 643–644
  nonsteroidal, 634, *635*
  ROCs, 626
  Saheli, *see* Saheli, nonsteroidal contraceptive pill
  sharing responsibility, men and women, 643
  side effects, 631–633
Organizational culture, 18
Organizational leadership, 18
Orikalankini, **701**
Our Common Future, Brundtland Report, 4, 204, **230**, 282, 400, 416, 577

Overconsumption, 4, 131, 181, 184, 210–211, 251, 317, 348, 366, 857

P

'Pad Man' of India, Arunachalam Muruganantham, 687, 711
Papel Planeta, 31, 35
Paper Cup Recovery and Recycling Group (PCRRG), 856
Paper Recycling-Coalition, 33
Papier Cascades Inc., 56, **58**, 59
Paree Sanitary Napkins, **693**
Pareto Optimum principle, 458–459
Parsimony, 78
Partial least squares (PLS) regression, 236
Passive responders, 189
Patagonia, 127, 840
  advertising campaigns, 841
  sustainability sells, 840
PBC, *see* Perceived behavioral control
Peattie, Ken, 23–24, 45, 62
Pedagogical tools
  case method, 538–539
  empirical inquiry, 539
  FIDSAI, *see* Familiarise-immerse-dissect-sense making-appraise-internalise
  prior theory/conceptual frameworks, 540
Pequenín®, 30, **30**
Perceived behavioral control (PBC)
  control beliefs, 215–216
  perceived power, 216
Perceived value
  and adoption intention, 280–281
  antecedents, 293
  contractual governance, 296–297
  e-commerce, 297
  information and platform quality, 293, 295
  shared farm equipment, *see* Shared farm equipment services
  users' return on investment, 296
Perfect food, 44
Perkins Papers, 61

Personal context factors, Sydney
longitudinal study
characteristics, study sample, 358, **359**
educational level, participants, 360–361
food preparation, 360
participants' age, 358
socio-economic and income factors, 360
Personal sustainability, marketing
professionals, *see* Marketing
professionals
Perth City Farm, 485, **487–489**, 490–491, 494
PEST, 18
PESTLE, 18
Petys®, 30, **30**
Pew Research Center, 397
Philadelphia Eagles, 729–730
Baker Bowl, 730
Connie Mack Stadium, 730
field goal forest program, 731
Franklin Field, 730
Go Green Program, 730–732
practices of, **731**
SWOT analysis, *732*
Green Era, 730
Lincoln Financial Field, 730
Philadelphia Municipal Stadium, 730
sustainability marketing, 732
Veterans Stadium, 730
Philosophy management, 66
Planetary health, 346, 351–352
Plourde, Mario, 67, 72, 74
PLS regression, *see* Partial least squares
Plüm Énergie, 60
Political reality, sustainable development
capitalist economies, 148–149
economic development, 148
large companies, 147
Pollution, 55
Polycarbonate mugs, 854
Polymenorrhea (cycles with intervals of 21 days), 713
Polypropylene (PP) mugs, 854
Pomys®, 30, **30**
Potential for Substituting Manpower for Energy, 315
Pot Plant, 130
Poverty, 34
Power shift, companies to consumers
bidirectionality, 390
consumer power, 392
crowd-based power, 393
demand-based power, 392
influence, definition of, 393
information-based power, 392–393
social media, 391–392
Word Wide Web, 391, 393
Primary Health Centers (PHC), 664
Private sector marketing, 17
Procter & Gamble, **693**
Product development, 20
Productive capacity, 19
Product service systems (PSS), 798
Professionalization and moral training, 453
Professional marketers
corporate sustainability initiatives, 146
economic sustainability, 165–166, *167*
social sustainability, 160–165
Profitability, 25
Projective techniques, qualitative research
associative technique, 151–152, *153*
choice and ordering techniques, 152–153, *154*
completion technique, 151, *152*
construction techniques, 154, *155*, 156, *156–157*, 158
definition, 150
expressive technique, 153–154, *155*
stimuli, 151
Project Shakti, Hindustan Unilever, 298
Prosumer identity; *see also* Prosumption
creative prosumers, 186, 189, 191
definition, 181
exploitation, 183
groups, 190
sustainability practices, *see*
Sustainability conscious prosumer practices
sustainable identities, 184–185
Prosumption; *see also* Sustainable prosumption
collaborative consumption, 183
coproduction, 183
creative prosumption, 189

Prosumption (*Continued*)
  "double exploitation," 183
  prosumers, definition of, 181–182
  Seyfang's subculture classification, 182
  sustainable prosumer identities, 184–185
Provincial Papers, 61
PSZOK, *see* Selective Municipal Waste Collection Points
Public Health Engineering Department (PHED), 659
Public management
  modernization, *see* Modernization, public management
  new, 446
  post-new, 447–448
  traditional *vs.* new public *vs.* post-new, *447*, 447–448
Public marketing, 437–438; *see also* Public marketing 4.0
  conceptual framework and research process, 438, *439*
  distribution channels optimization, 450
  effective communication, 450
  lean and agile management techniques, 450–451
  limitation, 450
  market research, 449
  needs and demands, citizens, 437–438
  participating civil society, 450, *451*
  public programs and services, development and improvement, 449
  public rates and prices determination, 449
  segmentation and differentiation, 449
  social behaviours, modification of, 450
Public marketing 4.0
  crisis management, 464–465
  digital marketing, 456–457
  marketing-mix tools, 457
  omnichannel strategy, 456–457
  online marketing, 456
  social networks, 456
  social welfare, *see* Social welfare functions
  strategic intelligence management, 463–464
  traditional and digital marketing combination, 455–457

Public–private response, global crisis
  2030 Agenda, 440
  infectious diseases, 439–440
  international vulnerability, 440
  marketing models, 441
  multidisciplinary and multi-sectoral action strategies, 440
  risk situations, 439
Public sector marketing
  definition, 468
  participating civil society, 450, *451*
Pusa Krishi mobile app, 292

Q

Qualitative research
  associative techniques, 151–152, *153*
  choice and ordering techniques, 152–153, *154*
  completion techniques, 151, *152*
  construction techniques, 154, *155*, 156, *156–157*, 158
  cross-sectional study design, 150
  expressive techniques, 153–154, *155*
  narrative inquiry methodology, 150
  participants experience, marketing and sustainability, 158, *158*
  participants profile, 158, **159**
  problems, 159
  projective technique, *see* Projective technique
Quantitative study, retailing
  discriminating validity, 239, **239**
  Harmon's single-factor test, 236
  Likert scale, 236
  measuring instrument, structural model, 237, **238**
  non-probabilistic quota sampling, 236
  PLS regression technique, 236
  questionnaires, 235–236
  reliability, scales, 238–239
  sample distribution, sociodemographic variables, 236, **237**
  structural equation model estimation, 239, **240**
  results, 239, *240*

R

Rajiv Gandhi Scheme for Empowerment of Adolescent Girls (RGSEAG) SABLA, 660, 664
Rashtriya Kishor Swasthya Karyakram (RKSK), 661, 664
Rashtriya Madhyamik Shiksha Abhiyan (RMSA), 665
Recovery self-efficacy, 258
    actors, 266, **266**, 270
    intenders, 263, **263**, 269–270
Recycled fibers, 74
Recyclers, 33
Red cycle, The, **702**
Redland City Council, 750
Redlands Afterhours Wildlife Ambulance data, 751
Regular oral contraceptives (ROCs), 626
Renewable resources, 55
Reporting practices, profit-for-purpose organizations
    accountability categories, 490, *491*
    balanced scorecard, 480
    cases, mission statements, **488**
    external reporting, 492–494
    five capitals model, **482**
    inductive research, 484
    integrated reporting, 495–496, **496**, *496*
    internal reporting, 490–492
    IR, 481–482
    multiple case study, 484
    qualitative research methodology, 484–486
    SAA, 480–481
    sampling strategy, 484
    social enterprises profile, **487**
    social marketing programs/advocacy, **489**
    SROI, 480–481
    stakeholder theory, 483
    theoretical and practical contributions, 498
Report of Palmer, 97–98
Resource pooling, 291–292
Resource Recovery, 485, **487–489**, 490, 492
Resources, village, 291, **291**

Responsible consumer behavior
    CAB model, 214–215
    consumer attitude and WTP, green practices, 217–218
    green consumers, 213
    TPB, *see* Theory of planned behaviour
Responsible consumption
    green marketing, 203
    hospitality industry, 202
    hotel industry, 202
    sustainable practices, 202–203
Responsible consumption and production (SDG 12), 572
Responsible tourism (RT)
    definition, 204
    eco-certification, 206–207, **208–210**, 210
    sustainability, tourism and hospitality industry, 204–206
    sustainable practices, hotel industry, 210–212, **212**
RESTART framework
    alliances, 608
    circular, 608
    circularity, 608
    experimentation, 608
    Nurpu, 608
    redesign, 607
    results, 608
    service-logic, 608
    three-dimensionality, 608
Restrepo, John Gómez, 30–31
Retailing
    co-creation, 227
    consumer satisfaction, *see* Consumer satisfaction, retailing
    and media, 121
    quantitative approach, *see* Quantitative study
    store awareness, 226–227, 233–234, 241
    sustainability development, 226
Reusable tampons, 657
Reward crowdfunding, 778
Right to sanitation, 650
"Roadmap for the circular economy," 318–319
    cycles, 319

892 • *Index*

Roadmap for the circular (*Continued*)
   education, 322–323
   elements, zero waste philosophy, 319
   food waste, 321–322
   municipal waste, 319–321
   Poland's priorities, CE, 318
ROCs, *see* Regular oral contraceptives
Rolland and Paperboard Industries Corporation, 61
RT, *see* Responsible tourism
Rural Development and Panchayat Raj Department (RDD), 660, 662

S

SAA, *see* Social accounting and audit
Saheli, nonsteroidal contraceptive pill
   advertisements, 637–638
   awareness, 636
   dosage, 635
   lack of growth, 637
   Ormeloxifene, 638
   PCOD patients, 641
   rural market, 635–636, 640
Sahyog Foundation, **701**
Samsung Electronics, 127
Sanitary napkins/pads, 657, 673
   biodegradable players, 657
   disposal of menstrual waste, 658–659
   Indian market of, 657
   Janaushadhi Suvidha oxo-biodegradable sanitary napkin, 657
   manufacturers, 657
   reusable, 658
   unsafe disposal, 658–659
Sanitation, 667; *see also* Menstrual Hygiene Management (MHM) in India
   at home, 667
   at school, 668
   at societies, 667–668
Sarva Siksha Abhiyan (SSA), 665
Scenario planning, 462, *464*, 464–465, 467
Scheme for Adolescent Girls (SAG), SABLA, 660

School dropout, adolescent girls, 654
Science, technology, engineering and mathematics (STEM), females in, 766–767
   injunctive norm messaging, 769–770, 772–773
   post-survey, 770–771
   pre-survey, 768
   self-efficacy messaging, 768–769, 772–773
   social advertisement campaign, 769
   social advertisement campaign development, 768
   university degrees, 767
SDGs, *see* Sustainable development goals
Second-hand clothing
   action planning, 256
   actors, 259, 263–264, 267–268, 271
   IB gap, *see* Intention–behaviour (IB) gap bridging
   intenders, 259–262, 267–268, 271–272
   maintenance self-efficacy, 258, 262, 265–266
   recovery self-efficacy, 258, 263, 266, 269
   task self-efficacy, 257, 265
Secure Meters, Dharohar case
   and community engagement, *see* Community engagement
   community projects, 615
   consumer judgment and feelings, 616
   employee volunteering, 614–616
   organizational identification, employees, 615
   success factor, 614
Selective Municipal Waste Collection Points (PSZOK), 331
Self-efficacy
   maintenance, 258, 262, 265–266
   recovery, 258, 263, **263**, 266, **266**
   task, 257, 262, 265
Self-efficacy messaging, 768–769
Self-identity
   definition, 176, 180–181
   extended self, 180
   sustainable consumption, 180–181
Semi-prosumer practices, 186, 190

Service failure
  definition, 513
  relationship, customer and organization, 513
  and service recovery, 513–514
Service recovery, 513–514
"Sex, explained!", Netflix documentary, 638–639
Sexual education
  birth control pills, 626–627
  contraception, 626; *see also* Oral contraceptives
  CSE program, 625
  STDs, 625
  teen pregnancies, 625
  UNESCO, 625
Sexually transmitted diseases (STDs), 625, 636
Shared exchange systems, 188
Shared farm equipment services
  agri mobile applications, 284
  contracts, 296–297
  customer participation, technology-based commerce, 293
  digital/mobile applications, 293–295, 299
  farmer suicides, India, 283
  financial and psychological costs, renting, 295–296
  implications and future research, 300–301
  in-depth interviews, 287–290
  inflated prices, agricultural inputs, 283–284
  information and platform quality, 294–295
  moderators, effect of, 297–299
  perceived value, 281, *281*, 287, 290
  qualitative study coding scheme, 288, **288–289**
  rent machinery, 284
  resource pooling, 291–292
  resources, village, 291, **291**
  seminal studies, 285, **286**
  shared-services platform, 285
  sharing, customer's perspective, 285–286
  TCE, 281–282, *282*
  trust-based relationships, 292
  value-based adoption model, 280–281, *281*
  value drivers and barriers, 281
Sharing economy, 842
  definition, 283
  farm equipment, *see* Shared farm equipment services
Shomota Women Care Private Limited, **694**
Shudh Plus Hygiene Products, **692**
SINORA, **694**
Skills development programs, public servants, 452–453
Skin-whitening controversies; *see also* L'Oréal's skin-whitening line
  female beauty products, 551–552
  product packaging, L'Oréal Paris, 555–557
  social demands, 552–554
SmartCity Wrocław programme, 324
Smurfit Kappa Group, 74
Social accounting and audit (SAA), 480–481
Social and sustainability marketing
  community engagement, 614–616
  community projects, 615
  consumer judgment and feelings, 616
  employee volunteering, 614–616
  organizational identification, employees, 615
  success factor, 614
Social and Sustainability Marketing: A Casebook for Reaching Your Socially Responsible Consumers through Marketing Science, 538
Social and sustainability marketing and sharing economy, 840–844
  case of usefull, 848–852
  coffee cups, life cycle impact of, 852–856
  sustainability, 852, 856–858
  coffee consumption, life cycle of, 844–848
  implications, 857–859
Social belongingness, 189
Social campaign, 313
  sustainable consumption, 323
  zero waste, *see* "Wrocław does not waste" campaign

Social capital, 287, **289**, 298–299
Social consciousness, 133
Social demands, for/against skin whitening
  anti-racist movement, 553
  Black Lives Matter movement, 551, 553
  changes, labelling products, 554
  genderism, 553
  L'Oréal and Unilever, 554
  male grooming, 553
  metrosexual trend, 552–553
  product packaging, L'Oréal Paris, 555–557
Social enterprises, 8
  accountability, 478–479
  performance reporting, *see* Reporting practices
  social objectives, 478
Social entrepreneurship, 27
Social influence
  guidelines, reaching socially responsible consumers, 425, *426*
  peer groups, 425
  sustainable consumption, 424–425
Social Influence Theory, 424–425
Socialist Harmonious Society, 583
Social listening
  consumer and industry trends, 420
  definition, 420
Socially conscious consumption, 253
Socially responsible consumers
  #EndangeredEmoji campaign, 422
  guidelines to reach
    communication message, 423–424
    leveraging, social influence, 424–425, *426*
    normalizing green behavior, 424
    understanding the consumer, 421–422
    value creation, consumer, 422–423
  leveraging social media, *see* Social media, marketing
  safe oral contraceptive options, *see* Oral contraceptives
  sexual education, 625–627
Socially responsible marketing, 441–442, 444

Social marketing, 6, 15, 17–18, 93, 840
  AIDA, 313–314
  behaviour change, recipients, 312–314
  classic 4P model, 315
  CRM, 314–315
  definition, 443, 468, 518
  dietary change, longitudinal case study, *see* Sydney longitudinal study
  goals, 312–313, 336
  L'Oréal, *see* L'Oréal's skin-whitening line
  objectives, 443–444, 518
  4S marketing mix, 347, 350, 377–378, 380
  social behaviour, 518
  social campaigns, implementation of, 313
  socially responsible marketing strategy, 444
  TMC, 313–314, 325, 333
Social media
  company-customer relationship development, 392
  empowered consumers, 404
  engagement, 29
  marketing
    communication message, 423–424
    engaging dialogue, consumer, 420–421
    greenwashing, 418, 423
    leveraging, social influence, 424–425, *426*
    normalizing green behavior, 424
    sense of community, 421
    social listening, 420
    target audience, 420
    understanding the consumer, 421–422
    value creation, consumer, 422–423
  sharing online reviews, 404
  for social change, 428
Social responsibility
  CSR, *see* Corporate social responsibility
  public marketing, 441–442
  and social marketing, 517–518
Social return on investment (SROI)
  description, 480
  limitations, 481
Social service agencies, 18

Social support, definition, 256–257
Social sustainability, 54–55, 92
    engagement, stakeholders and social issues, 229
    experiences, childhood, *162*
    life story, 160–162
    practices, upbringing, **160**, 162, **163–164**
    principal processes, 229
Social taboos, 650
Social utility functions
    individual level, consumer demand, 461
    satisfaction and social well-being, 460
    social utility index-function, 463–464
    strategic intelligence management, 463–464
    utilitarianism, 460
    utilitarian social welfare function, 460
Social welfare functions
    model, 461–463
    Pareto Optimum principle, 458
    social utility functions, 459–461
    social well-being with social indicators, 459
    welfare economics approach, 458–459
Society-business relationship, 22
Solid waste management, 728
Solow, R. M, 19
Sports equipment, 728–729
SROI, *see* Social return on investment
SSMM, *see* Sustainability social marketing model
Stainless steel mugs, 854
Stakeholders, 29
Stakeholder theory, 14–15, 22, 28
State Women Development Corporation (SWDC), 660, 662
Steady Hands, 126–127
"Stop sickness scams" campaign, 527
Strategic anticipation, 444–445, 465–466
Strategic intelligence management
    CBA methodology, 464
    critical and transformational information, 445
    equivalent monetary value, 464
    strategic decision-making, 445
    utility index-functions, 463
Strategic sustainability, 21
Stree Swabhiman Scheme by CSC, 665

Sufficient consumption, 42
Sustainability, 4–5, 18–19, 54, 101, 102; *see also* Economic sustainability; Social sustainability; Sustainability marketing
    agenda, 5
    aims and pillars, 206, *206*
    in China
        CSR practices, 583
        customers' perceptions, 584
        environmental protection, 582
        Monmilk case, *see* Monmilk Group, China
        nation's development, 583
    consumer satisfaction, retailing, *see* Retailing
    corporate sector, 145
    data-driven impact, 427–428
    defined, 19, 128, 797
    definition, 228, 468
    dimensions, 55, **101**, 205
    economic, 5
    environment, 5
    and equity, 797
    in fashion and textile industry, 117
    findings, 229, **230–231**
    implementation in organization, 145–146
    literature review and hypotheses, 228–234, *235*
    marketers, 146–147
    marketing, 6–7
    marketing function, 146
    as marketing strategy, 417–418
    normalization and integration, 802
    and online reviews, 404–406
    in organizations, 145–146
    professional marketers, *see* Professional marketers
    resonate with consumers, 427
    retailing, *see* Retailing
    shaping sustainability behavior, 427
    social well-being, 5
    sustainable development, 19
    three Es, 416
    tourism and hospitality industry, 204–206
    and TPB, *see* Theory of planned behaviour
    Triple Bottom Line, 797
    triple bottom line approach, 228–229, 231

896 • Index

Sustainability conscious prosumer practices
  categories, 188
  consumer-citizens, 188–189
  consumption behaviours, 189
  creative prosumers, 186, 189
  description, 185
  dumpster diving, 187
  empowered consumption, 184
  as evolutionary process, 186
  lifestyle, 184–185
  passive responders, 189
  procurement practices, 185–186
  segments and suggested marketing approach, **192**
  semi-prosumer practices, 186
  shared exchange systems, 188
  swapping activities, 187
  voluntary behaviour, product disposal, 187
Sustainability development
  dimensions, 578
  performance indicators, 578
Sustainability innovations, 24
Sustainability marketing, 14, 16, 23, 29, 91, 93–94, 100, **101**, 149
  applied to Grupo Familia mini-case, 37
  CEOs and corporate strategy formulation, 20–21
  characterization, 24
  in classroom, 26–28
    active learning, 40–41
    case method teaching, 28–40
    consumer behavior, 44–46
    critical thinking, 41–43
    macromarketing implications of SDG 12, 43–44
  consumer attitudes and values, 21–22
  defined, 23–26
  definition, 401, 416–417
  ECoC, 87
  framework, *27*, 46
  fundamentals, 16–18
  green marketing, 418
  greenwashing, 418
  issues raised by Bottled Water in relation to selected SDGs, **42**
  Klaipeda's and Kaunas application, **107**
  marketing decisions, 24
  model and technique of, *106*
  3P approach, 578
  SDGs, 16
  social marketing, 16
  social media, *see* Social media, marketing
  social media strategy, *see* Social media, marketing
  society-business relationship, 22–23
  sustainability, 18–19
  sustainable development, 25
Sustainability Marketing (Peattie and Belz), 62
Sustainability social marketing; *see also* Sydney longitudinal study
  catchphrases, 353, 367–369, 379
  components, 367, *368*
  the EAT Lancet Commission, 348
  environmental element, 368–369
  excessive meat consumption, 347–348
  food preparation part, 369
  health-related module, 369–370
  4P marketing mix, 347
  4S marketing mix, 347, 350, 377–378, 380
  SSMM, *see* Sustainability social marketing model
Sustainability social marketing model (SSMM)
  behaviour change, 348
  meat reduction, 348, *349*
  4S marketing mix, 350
  stages, 348
Sustainable Business Toolkit, 857
Sustainable capitalism, 137
Sustainable consumption, 41–42
Sustainable development, 14, 92
  Brundtland Commission's formulation, 23
  defined, 4–5, 19
  definition, 147, 200–201, 400, 416
  interconnected ring model, 147, *148*
  marketing, role of, 149
  pillars of, 204
  political reality, 147–149
Sustainable Development Goals (SDGs), 5, 16, *25*, 25–28, 39, 41
  business opportunities, 26
  fashion industry, 572
  partnership and collaboration, social sectors, 462

Sustainable fashion, 128–129, 135
    Nurpu, see Nurpu, sustainable fashion brand
    secondary markets, 253
    theoretical contributions, 270–271
Sustainable Fashion Consumption Decisions, *131*
Sustainable food-related behaviour changes
    deterioration, natural environment, 352
    flexitarian options, 351
    individual change, 353
    livestock sector, 352
    longitudinal qualitative study, see Sydney longitudinal study
    meatless days, 353
    socially responsible plant-rich diet, 351
    social marketing, see Sustainability social marketing
    SSMM, see Sustainability social marketing model
    universal planetary health diet, 352
Sustainable prosumption
    collective prosumers, 191
    creative prosumer practices, 191
    ecofriendly commodities, 190–191
    mainstream market mechanisms, 190
    semi-prosumer practices, 190
Sustainable tourism; see also Green hotel; Responsible tourism
    aims of, *207*
    practices, hotel industry, 210–212, **212**
Svenska Cellulosa Aktiebolaget (SCA), 30
Swachh Bharat Mission-Gramin (SBM-G), 663
Swachh Bharat Mission (SBM-G), 651
Swayam Shikshan, **701**
SWOT, 18, 340–341, 541
Sydney longitudinal study, meat reduction
    barrier factors, 363–366
    behaviour influencing factors, 361–363
    challenges, 377
    components, intervention, 354
    data collection and analysis, 356–357
    dietary outcomes, **371–372**
    empirical data collection, 355–356
    external influences and dependencies, 376–377
    game-changing factors, 366–367
    human research ethics permit, 355
    individual responses, 376
    interest and commitment, participants, 376
    men's meat consumption, *373*
    participant conditions, 355
    perceptions and attitudes, meat consumption, 370, 374–376
    personal context factors, 358–361
    questionnaire, 356
    4S marketing mix, 377–378
    social marketing intervention, 353–354
    women's meat consumption, *373*, 376
Symbolic meanings, consumption, 179–180, 190

T

Tampons, 657, 674
Task self-efficacy
    actors, 265, 269
    high-task, 257
    intenders, 262, 268–269
    low-task, 257
Tata Groups, **695**
TCE, see Transaction cost economics
TENA, 30, **30**
Textile industry
    cotton, production and export, 118
    and environment, 119–120
    India, 119
    raw materials and garment, producing and export of, 118
Theory of planned behaviour (TPB), *216*
    behavioral beliefs, 215
    behavioral intention, 215
    control beliefs, 216
    IB gap, 254
    normative beliefs, 215–216
    PBC, 216
    TRA, 215–216
Theory of reasoned action (TRA), 215–216, *216*, 281
3Ps (profit, planet and people), 100, **101**, 102
Tidström, A., 75
TMC, see Transtheoretical model of change

Tourism and hospitality industry
  aims and pillars, sustainability, 206, *206*
  CSR, 204
  IHEI, 204–205
  ITP, 204–205
  sustainable tourism, 205, *205*, *207*
TPB, *see* Theory of planned behaviour
TRA, *see* Theory of reasoned action
Tragedy of the commons, 352, 381
'Train the trainers' approach, 748
Transaction cost economics (TCE), 293, 300
  managing risks, 281, *282*
  online buying behavior, 281, *282*
Transtheoretical model of change (TMC), 313–314, 325, 333
Travel mugs, 854
Tribal Development Department (TDD), 660, 662
Triple bottom line (profit-planet-people) (TBL) framework, 14–15, 23–24, 27–29, 54, 87, 100, *580*
  Cascades company, 57–58
  corporate performance measurement, 579
  environmental bottom line, 581
  financial bottom line, 579–580
  Monmilk, 590–591
  overlapping responsibilities, 581–582
  profit aspect, 579
  social/ethical bottom line, 580–581
  sustainability framework, 55–56
  sustainability goals, 582
  sustainability marketing, 100, **101**
Triple Bottom Line (TBL) of the Twenty First Century Business: Cannibals with Forks, The (John), 91

U

UDITA, 666
UGER, **702**
Unemployment, 34
UNEP, *see* United Nations Environment Program
UNEP DTIE, *see* United Nations environment programs division of technology, industry, and economics
UNESCO, *see* United Nations Educational, Scientific and Cultural Organization
UN Global Sustainable Development Report, 440, 569, *569*
UNICEF, 652
Unicharm India and Spark Minda Group in collaboration with Global Foundation hunt, **691**
Unique selling propositions (USPs), 190–191
United Nations Conference on Environment and Development, 217
United Nations Educational, Scientific and Cultural Organization (UNESCO), 202
United Nations Environment and Development Commission, 400
United Nations environment programs division of technology, industry, and economics (UNEP DTIE), 205
United Nations Environment Program (UNEP), 202, *205–207*
United Nations World Food Program, 587
United Nations World Tourism Organization (UNWTO), 202
Unsocial and irresponsible customer behaviour
  business ecosystems, 516–517
  case study, *see* Holiday sickness scam
  CSR and social marketing, 517–518
  misbehaving customers, 511, 514–515
  service failure and service recovery, 511, 513–514
UN sustainable development goals, 608, *609*
UNWTO, *see* United Nations World Tourism Organization
Upcycling
  benefits, CE, 567, *567*
  clothing, 567–568
  definition, 567
  fashion initiative, 568–570; *see also* Chapputz fashion accessory brand

Uribe, Mario, 30
USEFULL, 849–852
    brand value, 851
    corporate value proposition, *850*
    social and sustainability marketing, 851
    sustainability sells, 850, 852
USPs, *see* Unique selling propositions
Utilitarianism, 460

V

Valence framework theory, 281
Value-based adoption model, 280–281, *281*
"Value co-creators," 181–182
Van Tilburg, R., 22
Van Tulder, R., 22
Vasudha Vikas Sansthan (VVS), **702**
Vatsalya, **703**
Versatile Enterprises (P) Ltd, **695**
Vienna, sustainable city strategies, 786
    Arcadis Sustainable Cities Index, 787
    environmentally friendly transport, 787
    environmental policies, 787
    environmental sustainability, 787
    European Green Cities Index, 787–788
    "1-Euro per day" campaign, 788–789
    Global Liveability Ranking, 786
    internal stakeholders and place marketing, 786–787
    Mercer Quality of Living Ranking, 786
    mobility consumption patterns, 787–788
    price as a marketing tactic, 788–789
    public transport in individuals' mode choice set, 788–789
    sustainability policy, practice and reputation, 789–790
Vilnius, 99
Volitional self-efficacy, 258
Voluntary simplicity, 42

W

Warren's theory of brand coolness, 125
WASH United, 651–652
Waste production, 55
WaterAid, 652
Water hyacinth pads, 658
Water-pollution problems, fashion industry, 565–566
Weber's bureaucratic theory, 447
Weed, Kenneth, 21
Western individualism, 549–550, 553
WestRock, 74
What the health, 369, 380
Willingness to pay (WTP)
    green products, 217–218
Woke marketing/advertising, 548, 550
Woke-washing, 550
WOM communication, *see* Word of mouth
Word associative technique, 152
Word of mouth (WOM)
    communication, 389
    to eWOM, 394–395
Workplace discrimination, 26
WorkVentures, 485–486, **487–489**, 490, 492, 494
World Economic Forum, 5
World League of American Football (WLAF), 726
World Tourism Organization (WTO), 205
The World Wide Fund for Nature (WWF), 404–405, 422
Written assignment, 35
"Wrocław does not waste" campaign
    behavioural objectives, 336–337
    belief objective, 336–337
    communication channels and tools, 329–330
    composter sharing programme, 327
    composting, 331–332
    educational activities, 333
    "giveboxes," 332
    interactive map, 327, *328*, 328–329, 336
    leftover cooking festival, 330
    logo, 326, *327*
    measures, public relations, 330–331
    the Ordinance of the President of Wrocław, 332
    website, 327, 336
    zero waste concept implementation, 326

Wrocław transition, CE;
    see also "Wrocław does not waste" campaign
  municipal waste, 323–324
  residents, 323–324
  urban layout, 324
WTO, *see* World Tourism Organization
WTP, *see* Willingness to pay

Y

YesIBleed, **703**

Z

Zaltman, Gerald, 17
Zero waste; *see also* "Wrocław does not waste" campaign
  city transition, 316, *317*
  definition, 316
  principles, 316
  3R concept, 316
Zero Waste International Alliance (ZWIA), 316
ZWIA, *see* Zero Waste International Alliance